PLEASE STA'

ATE DUE

Tectonic and Stratigraphic Evolution of Zagros and Makran during the Mesozoic–Cenozoic

Geological Society books refereeing procedures

The Society makes every effort to ensure that the scientific and production quality of its books matches that of its journals. Since 1997, all book proposals have been refereed by specialist reviewers as well as by the Society's Books Editorial Committee. If the referees identify weaknesses in the proposal, these must be addressed before the proposal is accepted.

Once the book is accepted, the Society Book Editors ensure that the volume editors follow strict guidelines on refereeing and quality control. We insist that individual papers can only be accepted after satisfactory review by two independent referees. The questions on the review forms are similar to those for *Journal of the Geological Society*. The referees' forms and comments must be available to the Society's Book Editors on request.

Although many of the books result from meetings, the editors are expected to commission papers that were not presented at the meeting to ensure that the book provides a balanced coverage of the subject. Being accepted for presentation at the meeting does not guarantee inclusion in the book.

More information about submitting a proposal and producing a book for the Society can be found on its web site: www.geolsoc.org.uk.

It is recommended that reference to all or part of this book should be made in one of the following ways:

LETURMY, P. & ROBIN, C. (eds) 2010. *Tectonic and Stratigraphic Evolution of Zagros and Makran during the Mesozoic–Cenozoic*. Geological Society, London, Special Publications, **330**.

NAVABPOUR, P., ANGELIER, J. & BARRIER, E. 2010. Mesozoic extensional brittle tectonics of the Arabian passive margin, inverted in the Zagros collision (Iran, interior Fars). *In*: LETURMY, P. & ROBIN, C. (eds) *Tectonic and Stratigraphic Evolution of Zagros and Makran during the Mesozoic–Cenozoic*. Geological Society, London, Special Publications, **330**, 65–96.

GEOLOGICAL SOCIETY SPECIAL PUBLICATION NO. 330

Tectonic and Stratigraphic Evolution of Zagros and Makran during the Mesozoic–Cenozoic

EDITED BY

P. LETURMY
Université de Cergy-Pontoise, France

and

C. ROBIN
Université de Rennes 1, France

2010
Published by
The Geological Society
London

THE GEOLOGICAL SOCIETY

The Geological Society of London (GSL) was founded in 1807. It is the oldest national geological society in the world and the largest in Europe. It was incorporated under Royal Charter in 1825 and is Registered Charity 210161.

The Society is the UK national learned and professional society for geology with a worldwide Fellowship (FGS) of over 9000. The Society has the power to confer Chartered status on suitably qualified Fellows, and about 2000 of the Fellowship carry the title (CGeol). Chartered Geologists may also obtain the equivalent European title, European Geologist (EurGeol). One fifth of the Society's fellowship resides outside the UK. To find out more about the Society, log on to www.geolsoc.org.uk.

The Geological Society Publishing House (Bath, UK) produces the Society's international journals and books, and acts as European distributor for selected publications of the American Association of Petroleum Geologists (AAPG), the Indonesian Petroleum Association (IPA), the Geological Society of America (GSA), the Society for Sedimentary Geology (SEPM) and the Geologists' Association (GA). Joint marketing agreements ensure that GSL Fellows may purchase these societies' publications at a discount. The Society's online bookshop (accessible from www.geolsoc.org.uk) offers secure book purchasing with your credit or debit card.

To find out about joining the Society and benefiting from substantial discounts on publications of GSL and other societies worldwide, consult www.geolsoc.org.uk, or contact the Fellowship Department at: The Geological Society, Burlington House, Piccadilly, London W1J 0BG: Tel. +44 (0)20 7434 9944; Fax +44 (0)20 7439 8975; E-mail: enquiries@geolsoc.org.uk.

For information about the Society's meetings, consult *Events* on www.geolsoc.org.uk. To find out more about the Society's Corporate Affiliates Scheme, write to enquiries@geolsoc.org.uk.

Published by The Geological Society from:
The Geological Society Publishing House, Unit 7, Brassmill Enterprise Centre, Brassmill Lane, Bath BA1 3JN, UK

(*Orders*:   Tel. +44 (0)1225 445046, Fax +44 (0)1225 442836)
Online bookshop: www.geolsoc.org.uk/bookshop

The publishers make no representation, express or implied, with regard to the accuracy of the information contained in this book and cannot accept any legal responsibility for any errors or omissions that may be made.

British Library Cataloguing in Publication Data

A catalogue record for this book is available from the British Library.
ISBN 978-1-86239-293-9

Typeset by Techset Composition Ltd, Salisbury, UK
Printed by MPG Books Ltd, Bodmin, UK

Distributors

North America
For trade and institutional orders:
The Geological Society, c/o AIDC, 82 Winter Sport Lane, Williston, VT 05495, USA
Orders: Tel. +1 800-972-9892
             Fax +1 802-864-7626
             E-mail: gsl.orders@aidcvt.com

For individual and corporate orders:
AAPG Bookstore, PO Box 979, Tulsa, OK 74101-0979, USA
Orders: Tel. +1 918-584-2555
             Fax +1 918-560-2652
             E-mail: bookstore@aapg.org
             Website: http://bookstore.aapg.org

India
Affiliated East-West Press Private Ltd, Marketing Division, G-1/16 Ansari Road, Darya Ganj, New Delhi 110 002, India
Orders: Tel. +91 11 2327-9113/2326-4180
             Fax +91 11 2326-0538
             E-mail: affiliat@vsnl.com

Contents

Acknowledgements

The volume editors would like to acknowledge the following colleagues who kindly helped with reviewing the papers submitted for this volume:

M. Allen	C. Hollis	R. Swennen
M. Bakalovicz	J. Jackson	P. Van der Beek
A. Bally	E. Jaillard	J. Vergne
O. Bellier	E. A. Keller	B. Vincent
R. Bendick	M. Mattei	R. Walker
J. P. Callot	S. Nader	L. Wallace
J. Cosgrove	M. Pagel	A. Zanchi
D. Frizon de la Motte	J. L. Rudkievicz	
K. Furlong	C. Sue	

The Geological Survey of Iran is warmly thanked for its support for the fieldwork performed in Iran.

Geological Survey Of Iran

The following companies are thanked for their participation as sponsors to the MEBE (Middle East Basins Evolution) Programme and for their contributions towards colour printing costs:

Preface

The Middle East Basins Evolution (MEBE) Programme was a 4 year consortium (2003–2006) funded by the major oil companies (BP, ENI, PETRONAS, SHELL and TOTAL) and by the French research organizations (INSU-CNRS and UPMC). This multidisciplinary study of the Middle East, spanning the Arabian–Peri-Arabian and Caucasian–Caspian areas, was led by E. Barrier (CNRS–Université Pierre et Marie Curie, Paris, France) and M. Gaetani (University of Milan, Italy). Its focus was the geodynamic evolution of the area, particularly since the Late Palaeozoic, and emphasized dating. This provided considerable details on the regional kinematics, plate-tectonic models and geodynamic evolutions of the area. The MEBE Programme brought together about 300 scientists from 28 countries, representing 100 universities and research organizations.

To prepare regional syntheses, the MEBE Programme established eight working groups comprising MEBE participants and external regional specialists. The MEBE working groups were focused regionally (Zagros, South Caspian Basin–central Iran, Caucasus, Black Sea, Levantine and East Arabian margins) and on products (stratigraphic comparisons, lithospheric cross-sections). From 2003 to 2005, 26 scientific projects were funded in 14 countries of the Middle East, including the Black Sea, Caucasus, northern Iran, the Zagros, the Arabian margins and the Levant domains.

About half of the MEBE projects were located in Iran because of the pivotal importance of several regions of this country to both scientific and hydrocarbon interests, as well as the continuous support by the Geological Survey of Iran (GSI).

Workshops were a key element in the MEBE Programme. The first one was held in Kiev in February 2006 by the Black Sea Working Group; the Caucasus Working Group met in Ankara in September 2006. A programme-wide workshop in Milan in December 2006 gathered the Zagros, East Arabian margin, South Caspian Basin–central Iran and Stratigraphic Comparisons MEBE Working Groups. The last workshop was held in Paris in December 2006 by the Levant Working Group. An important documentation of the MEBE activities was given during the 2007 EGU General Assembly in Vienna, where a special MEBE session was held.

The results of the programme, as well as the MEBE GIS database, were presented by way of 70 communications during the final MEBE meeting held at the Université Pierre et Marie Curie (Paris) in December 2007. The principal MEBE products are an atlas of 14 palaeotectonic maps, published by the Commission for the Geological Map of the World (CGMW), showing the geodynamic and tectonic evolution of the Middle East between the Late Triassic and the present, and a series of four Geological Society of London Special Publications presenting the results of the regional MEBE working groups.

The four volumes cover the Black Sea–Caucasus, the South Caspian–central Iran, the Zagros–East Arabian margin and the Levant. The present volume presents new data and results on sedimentology, stratigraphy, biostratigraphy, tectonics and kinematics in the Zagros fold belt and the adjacent Makran accretionary prism. The teams involved were sponsored either by the MEBE Programme or by other programmes.

The volume covering the geodynamic evolution of the region from the South Caspian Basin to northern and central Iran, entitled *South Caspian to Central Iran Basins*, is edited by M. F. Brunet (CNRS, Université Pierre et Marie Curie, Paris, France), M. Wilmsen (Friedrich-Alexander-Universität, Erlangen–Nürnberg, Germany) and J. W. Granath (Granath & Associates Consulting Geology, USA). The volume presents detailed results of new fieldwork on the South Caspian Basin, which is among the deepest sedimentary basins in the world, and more precisely on the margins of the South Caspian Basin, where inversion during the Cenozoic stages of the Arabia–Eurasia collision has exposed the rock record in the Alborz, Koppeh Dagh and Binalud mountains in northern Iran, as well as the eastern extent of the Greater Caucasus.

The volume covering the Black Sea to the Caucasus, entitled *Sedimentary Basin Tectonics from the Black Sea and Caucasus to the Arabian Platform*, is being edited by R. Stephenson (University of Aberdeen, UK), N. Kaymakci (Middle East Technical University, Ankara, Turkey), M. Sosson (CNRS, University of Nice, France), V. Starostenko (Institute of Geophysics, National Academy of Sciences of Ukraine, Kiev, Ukraine) and F. Bergerat (CNRS–Université Pierre et Marie Curie, Paris, France). The volume presents detailed results of new fieldwork, as well as syntheses of the tectonic evolution of the Black Sea–Caucasus area (Greater Caucasus, Lesser Caucasus, South and East Anatolia, margins of the Black Sea) constrained by an integration of the newly obtained and previously published data.

The fourth volume, *Evolution of the Levant Margin and Western Arabia Platform since the Mesozoic*, edited by C. Homberg (Université Pierre et Marie Curie, Paris, France) and M. Bachmann (University of Bremen, Bremen, Germany), will cover improvements in our knowledge of the tectonic, stratigraphic and environmental evolution of the Levant Basin and its margins since the Mesozoic.

In summary, the MEBE Programme provides a significant contribution of high-quality geological data at inter-regional scales and a considerable advance in the knowledge of the regional geology. These results will strongly contribute to new interpretations of the geodynamic evolution of the whole Middle East.

PASCALE LETURMY
CÉCILE ROBIN

Tectonic and stratigraphic evolution of Zagros and Makran during the Mesozoic–Cenozoic: introduction

PASCALE LETURMY[1] & CÉCILE ROBIN[2]*

[1]*Université de Cergy-Pontoise, Département des Sciences de la Terre et de l'Environnement, 5 mail Gay Lussac, Neuville/Oise, 95031 Cergy-Pontoise Cedex, France*

[2]*Geosciences Rennes, Equipe des Bassins Sédimentaires, Université de Rennes 1, Campus de Beaulieu, 35042 Rennes Cedex, France*

*Corresponding author (e-mail: Cecile.Robin@univ-rennes1.fr)

The Zagros fold–thrust belt (ZFTB) extends for c. 2000 km from Turkey in the NW to the Hormuz Strait in the SE. This belt results from the collision of the Arabian and Eurasian plates during Cenozoic times and constitutes a morphological barrier (with some peaks exceeding 4000 m) separating the Arabian platform from the large plateaux of central Iran. To the east a pronounced syntaxis marks the transition between the Zagros collision belt and the Makran accretionary wedge. In the ZFTB, the Proterozoic to Recent stratigraphic succession pile of the southern Tethys margin is involved in huge folds detached from the Pan-African basement and offers the opportunity to study the stratigraphic and tectonic evolution of the Palaeo-Tethyan margin over large time periods. Few recent data are widely available on the southern Tethys margin as preserved in the Zagros Mountains. Since the classical works of James & Wynd (1965) and Murris (1980), the most recent synthesis is the palaeogeographical reconstruction of the Arabian platform published by Ziegler (2001). Many petroleum data have been acquired during the last 10 years, but few of these have been published. The Middle East Basins Evolution (MEBE) Programme, coordinated by P. Barrier and M. F. Brunet, in close relationship with colleagues of the Geological Survey of Iran, was an excellent opportunity to go back to the field and to collect new data to better constrain the evolution of this margin. In this volume, the structure of the Zagros Mountains is explored through different scales and using different approaches.

General evolution of the area

In the Zagros Mountains, the geodynamic evolution of the area is mainly linked to opening and closure of the Neo-Tethys Ocean. Prior to the formation of the southern Tethys margin, the area was in an intracratonic setting. During Late Proterozoic–Early Cambrian times, strike-slip and extensional faulting affected the basement and established a structural framework with north–south structures controlling the geometry of the basin and lateral facies changes, and controlling deformation processes through time. The Hormuz salt basin and equivalent series developed in this context and constitute the major décollement level between the Pan-African basement and the overlying sediments. The evaporitic episode was followed by dominantly clastic sedimentation (Cambrian–Ordovician) and Silurian shales linked to a post-glacial sea-level rise. The Hercynian unconformity is visible in the ZFTB and registers the distal effects of the Hercynian orogeny.

The epicontinental sedimentation ceased in the Late Permian, when a rifting episode led to the opening of the Neo-Tethyan Ocean. The geodynamic change was accompanied by a change in sedimentation from dominantly Palaeozoic clastic sediments to marine carbonate sediments during Permian, Mesozoic and Tertiary times. After the rifting period, the passive margin steadily subsided and the carbonate sediments show lateral facies changes with predominantly platform carbonate facies in the SE and basinal facies toward the NW. During this period sedimentation and hiatuses in sedimentation are mainly related to sea-level changes and distant geodynamic events; lateral facies changes may also be related to reactivation of Pan-African structures. Crustal extension is also a significant factor of control on the sedimentation during Mesozoic times.

Around 95 Ma, obduction was initiated along thousands of kilometres on the southern Tethyan passive margin. Docking of ophiolites onto Arabia was diachronous along strike and occurred in the Santonian in the south (Oman) and during the Maastrichtian in the NW Zagros. This major event caused a regional uplift of the margin with emergence of the platform and a change in sedimentation in front of the ophiolite–radiolarite nappes. It was followed by deeper-water conditions and the deposition of deep-water marls and shales from Campanian to Eocene times associated with shallow carbonates deposited along ridges and linked with a bulge in front of the subducting Neo-Tethys. Just before

From: LETURMY, P. & ROBIN, C. (eds) *Tectonic and Stratigraphic Evolution of Zagros and Makran during the Mesozoic–Cenozoic*. Geological Society, London, Special Publications, **330**, 1–4.
DOI: 10.1144/SP330.1   0305-8719/10/$15.00 © The Geological Society of London 2010.

the closure of the Neo-Tethys, shallow conditions were general in the basin, where platform carbonates and evaporitic sediments were deposited. The continent–continent collision probably began in the Early Miocene and was accompanied by a first episode of folding during the deposition of the Gascharan evaporites (Sherkati *et al.* 2005). This first phase of folding was followed by a period of tectonic quiescence during the deposition of the Mishan marls and lower Agha Jari sandstones (of Middle to Late Miocene age). Growth strata in the Mio-Pliocene upper Agha Jari sandstones date the main episode of folding. Along-strike variations in the folding modes and folding geometry are mainly linked to lateral stratigraphic changes and to the presence of deep-seated faults. The last tectonic event was the general involvement of reverse basement faults during Pliocene and Quaternary times and the building of the topography of the Zagros Mountains.

Crustal and lithospheric structure of the fold belt

Paul *et al.* focus their paper on the crustal and lithospheric structure of the ZFTB based on geophysical data. Using receiver function analysis they show that the crust is thickened north of the main Zagros reverse fault (MZRF) under a wide zone including the Sanandaj–Sirjean zone, the Urumieh–Dokhtar magmatic assemblage and the central domain. This fault is well imaged by a low-velocity layer and appears to be a major low-angle thrust crossing the entire crust and allowing thrusting of central Iran onto the Arabian margin. Seismic velocities show lateral contrasts in the upper mantle, with faster velocities in the Arabian margin than in the central Iranian micro-blocks. This contrast allows precise location of the suture between the two upper mantles. The absence of a high-velocity anomaly in the central Iranian upper mantle suggests that the Neo-Tethyan slab is detached from the Arabian margin.

In their paper **Hatzfeld *et al.*** integrate global positioning system (GPS) velocities and work on earthquakes, tectonics and Quaternary geology in support of a compelling reconstruction of Zagros kinematics. At the scale of the Fars arc, the authors find good correlations between surface data (GPS measurements, topography and geomorphology) and seismicity, which reflect deformation in the basement. They confirm that the basement deforms with the cover in the Zagros folded belt and they propose that deformation in the basement is propagating southward. In the second part of the paper, they explore the Kazerun fault system, which distributes the strike-slip motion

along the main Recent fault onto frontal basement faults. Close to the Kazerun fault, several segments branch en échelon. From these data, the authors propose that the oblique convergence in the Zagros folded belt is partitioned between strike-slip movement and shortening perpendicular to the belt.

Regard *et al.* investigate the structure and dynamics of the transition from collision adjacent to the Zagros ranges to subduction at the Makran trench. They present a synthesis of geological and geophysical data acquired during the last 10 years in the area. GPS data and modelling, as well as geomorphological studies, reveal a right-lateral motion at the Makran–Zagros transition. This motion is accommodated in two sharp fault systems. The Zendan–Minab–Palami fault system accommodates 15 mm a^{-1} of movement parallel to the fault (i.e. N160°E) and 6 mm a^{-1} perpendicular to it. Along the Jiroft–Sabzevarn fault system, the displacement rate is estimated to be 3.1 ± 2.5 mm a^{-1}. Seismological studies reveal a crustal plane dipping toward the NE, suggesting thrusting of the Makran above the Zagros. The NE–SW-trending σ_1 axis that has affected the region since the Late Pliocene is contemporaneous with major regional tectonic reorganization proposed by Allen *et al.* (2004) and with the change from thin-skinned to thick-skinned tectonics in the Zagros observed by Molinaro *et al.* (2005), suggesting that these events are closely related.

Structure and palaeostress

Navabpour *et al.* document the tectonic evolution of the southern Neo-Tethyan margin prior to Arabian–Eurasian collision in the High Zagros belt of the Fars province. They present syndepositional normal fault slip data and reconstruct time-constrained palaeostress tensors. The data clearly show two sets of direction of extension. The first, oriented north–south, is oblique to the margin and occurred between 240 and 190 Ma. The second, oriented NE–SW, is perpendicular to the margin and occurred between 180 and 160 Ma. In the last part of the paper, the authors compare their results with other tectonic settings and analogue experiments, and find that the Aden Gulf is a good recent analogue. Finally, the authors propose that during Permian–Triassic times, the continental break-up was oblique in type and initiated NW–SE-oriented normal faults and east–west transform faults in the basement. This early episode was followed by margin-perpendicular extension. All these sets of normal faults were inverted during the Cenozoic compression.

Aubourg *et al.* provide a synthesis of the Tertiary tectonics in the Western Fars arc in Iran based on the comparison of magnetic fabrics with palaeostress results (fault or stria inversion, calcite twins) and active tectonic indicators (focal mechanisms and GPS data). These data are used to define the structural evolution of the chain during the Palaeogene to Quaternary. Magnetic studies in the Western Fars arc show that layer-parallel shortening had a trend of N47 $\pm$ 13° during the Palaeocene before the onset of folding and N38 $\pm$ 32° during the Mio-Pliocene folding phase. Calcite twinning records an average σ_1 trending N25 $\pm$ 15° characterizing the late stage of folding, in agreement with fault slip data. All the methods used indicate a two-phase tectonic evolution of the Western Fars arc and a counter-clockwise rotation of shortening direction from NE to N20° between the two phases.

Folding modes and their relationships with faulting

Leturmy et al. combine a morphological and a structural analysis to characterize the geometry of basement faults involved in the Zagros folded belt during Cenozoic shortening and to propose a relative chronology between folding and basement faulting. In the first part of their paper, the authors focus on three structures of the Eastern Fars arc. They explain the three different geometries by a two-step evolution model with a late basement fault cutting obliquely through already existing detachment folds. In the second part, the authors use the river network and river incision to extend the study on basement faulting to the whole Fars arc and Central Zagros. The authors show that a shift of the actual river bed elevation is observed in the hanging wall of basement thrusts. The authors suppose that part of this tectonic uplift is compensated by river incision and that the value of the compensation is proportional to the age of the activation of basement faults. Because the compensation increases toward the north, the authors conclude that deformation in the basement is propagating toward the south.

The paper by **Burberry et al.** presents a conceptual model to discriminate between different folding mechanisms and they apply it to the Fars arc in the Zagros folded belt. This model is based on the geometric characters of the folds (hinge length, aspect ratio and fold symmetry), folding development (uplift rate, lateral propagation rate) and on the morphology associated with them (geometry of the river network in the vicinity of the fold and distribution of wind and water gaps). With this method, they distinguish three classes in the Zagros folded belt: (1) symmetric and short-length detachment folds; (2) fault-bend folds characterized by long crests with multiple cross-cutting wind gaps; (3)

asymmetric structures with short hinge length that may be either detachment folds or fault-propagation folds. The authors find that fault-bend folds are generally associated with major thrust faults and formed sequentially by footwall collapse as the deformation front migrated towards the foreland basin.

Emami et al. provide a documented analysis of the Anaran anticline, a structure associated with the mountain front flexure in NW Zagros (front of the Pusht-e Kuh arc). In this study, field data are interpreted with the help of analogue experiments to reconstruct the kinematics of the frontal structure. The Anaran anticline is characterized by an S-shaped geometry in map view, a subvertical forelimb, and rotated conjugate normal faults parallel to the strike of the fold, and is associated with a low-angle blind thrust crossing the basement and uplifting the entire Pusht-e Kuh arc. In the second part of the paper, the authors test several tectonic evolutions with analogue models. They conclude their paper with a sequential restoration of the section through time. The evolution of the Anaran anticline started at 7.65 Ma with folding of the cover above the basement and normal faulting at the crest of the structure. This phase was followed by a general uplift of the entire Pusht-e kuh arc around 5.5 Ma produced by the activation of a basement low-angle thrust.

Stratigraphy and geodynamics of the Arabian margin

Robin et al. through a biostratigraphic, sedimentological and sequence stratigraphic study, analyse the Mesozoic deep-water carbonate deposits from the southern Tethyan passive margin in the Pichakun nappes (Neyriz area). They propose a new lithostratigraphic framework, dated by radiolarians from the Middle Jurassic (Aalenian–early Bajocian) to the Late Cretaceous (Turonian–Coniacian). Most of the sediments are deep-sea gravity lobe deposits. Twenty-seven facies, grouped into eight facies associations, are defined. Based on a sequence stratigraphic study (i.e. the stacking pattern), five second-order cycles (of 10–30 Ma duration), defined between two successive distal facies time intervals, are characterized. The most important tectonic event recorded occurred at the Aptian–Albian boundary (the deposition of olistoliths, from a few metres to 100 m thick, in debris flows, related to Austrian deformations). The Arabian-scale late Toarcian and early Tithonian deformations have been recorded as unconformities. The authors consider that another tectonic event occurred during the late Hauterivian.

Piryaei et al. based on the analysis of the sedimentary record (e.g. facies changes, migration

of depocentres) and new chronostratigraphic data
(biostratigraphy, isotope stratigraphy), study the
tectonic evolution of the northeastern Arabian
plate margin (Fars Province, SW Iran) during Late
Cretaceous times, and its evolution from a passive
to an active margin. Three tectonosedimentary
phases, related to obduction processes, are recognized. Phase I (Late Albian to Cenomanian, before
obduction) is characterized by shallow-water platform carbonates and intrashelf basins. Eustatic sea-
level variations and local differential subsidence
controlled the sedimentation during this phase.
Phase II (Turonian to Late Campanian, obduction
phase) consists of pelagic and platform carbonates
in the south, and a foreland basin with obducted
radiolarites, ophiolitic and olistoliths or thrust
slices in the north. Phase III (Late Campanian to
Maastrichtian, after obduction) shows the development of rudist-dominated carbonates associated
with a possible uplift to the NE.

Diagenetic evolution

Hajikazemi *et al.* through a diagenetic study of the
Sarvak Formation (Cenomanian–Turonian) main
oil reservoir in southern Iran, show the importance
of subaerial exposure and meteoric diagenesis,
associated with the regional Turonian unconformity, for the highly variable porosity and permeability of those rocks. Dissolution affected the
entire upper part of the Sarvak Formation; however,
other processes related to subaerial exposure are
brecciation, development of palaeosol and formation of bauxite deposits. Negative $\delta^{18}O$ values (i.e.
$-6.65‰$ to $-1.77‰$) suggest a significant meteoric component. More $\delta^{18}O$-depleted values (e.g.
$-12.31‰$), obtained from late calcite cements,
indicate precipitation from warm fluids. Positive
$\delta^{13}C$ values (i.e. $0.00–3.47‰$), in the various carbonate phases reflect values of seawater coeval
with the Cenomanian–Turonian Oceanic Anoxic
Event (OAE), later modified by meteoric waters
and/or by the extensive erosion and removal of
the uppermost isotopically negative carbonate
layers of the Sarvak Formation.

Hosseini-Barzi analyses the diagenetic evolution, in both time and space, of the Plio-Pleistocene
syntectonic sediment of the Makran subduction
zone of Iran. The two main challenges were to
take into account (1) the high facies diversity, and
then the different porosity pattern and (2) the
masking effect of the pervasive meteoric diagenesis.
The correlation of diagenetic features between
different rock units in space, inferred by using two
distinct dolomite cementation phases, suggests
pulsatory changes in the fluid chemistry related to
eight episodes of seaward and landward migration

of fluids. This could be due to sea-level fluctuation
(eustasy) or uplift, or both. Eastward increase in
the number of phases of cementation and dissolution
suggests that pulsation in fluid migration has developed towards the eastern part of each fault-limited
block. The end of the diagenetic evolution is characterized by reverse fault activities that enhance
methane seepage, as geochemistry of dolomite
cements ($\delta^{13}CO$: -16.15 to $20.16‰$) indicates.

The petroleum systems

Bordenave reviews the five petroleum systems
defined in the Zagros and proposes a scenario for
oil and gas expulsion–migration in a geodynamic
and tectonic scenario that explains oil and gas
field emplacement. The functioning of each petroleum system (i.e. the Palaeozoic, Middle Jurassic,
Early Cretaceous, and Middle Cretaceous to Early
Miocene systems) is described through time.
Modelling of oil and gas maturity through time is
constrained by field work, wells data and geochemical data on source rocks. These data show that two
styles of petroleum systems charged the Iranian
Zagros. In four of them (the older ones) expulsion
occurred before the Cenozoic folding and oil and/
or gas migrated long horizontal distances toward
regional highs and salt-related structures, and some
oil eventually moved into anticlines during Cenozoic
shortening. In contrast, in the younger petroleum
systems, oil expulsion occurred after the beginning
of folding and oil migrated vertically using the fracture system to reach the closest anticlines.

References

ALLEN, M., JACKSON, J. & WALKER, R. 2004. Late
    Cenozoic reorganization of the Arabia–Eurasia
    collision and the comparison of short-term and long-
    term deformation rates. *Tectonics*, **23**, doi: 10.1029/
    2003TC001530.
JAMES, G. A. & WYND, J. G. 1965. Middle East: Strati-
    graphic nomenclature of Iranian oil consortium agree-
    ment area. *AAPG Bulletin*, **49**, 2182–2245.
MOLINARO, M., LETURMY, P., GUEZOU, J. C. & FRIZON
    DE LAMOTTE, D. 2005. The structure and kinematics
    of the south-eastern Zagros fold–thrust belt, Iran:
    from thin-skinned to thick-skinned tectonics.
    Tectonics, **24**, TC3007, doi: 10.1029/2004TC001633.
MURRIS, R. J. 1980. Middle East: Stratigraphic evolution
    and oil habitat. *AAPG Bulletin*, **64**, 597–618.
SHERKATI, S., MOLINARO, M., FRIZON DE LAMOTTE,
    D. & LETOUZEY, J. 2005. Detachment folding in the
    Central and Eastern Zagros fold-belt (Iran): salt mobi-
    lity, multiple detachments and late basement control.
    Journal of Structural Geology, **27**, 1680–1696.
ZIEGLER, M. 2001. Late Permian to Holocene paleofacies
    evolution of the Arabian Plate and its hydrocarbon
    occurrences. *GeoArabia*, **6**, 445–504.

Seismic imaging of the lithospheric structure of the Zagros mountain belt (Iran)

ANNE PAUL[1]*, DENIS HATZFELD[1], AYOUB KAVIANI[2,3], MOHAMMAD TATAR[2] & CATHERINE PÉQUEGNAT[1]

[1]*Laboratoire de Géophysique Interne et Tectonophysique, CNRS—Université Joseph Fourier, BP 53, 38041 Grenoble Cedex, France*

[2]*International Institute of Earthquake Engineering and Seismology, Tehran, Iran*

[3]*Institute for Advanced Studies in Basic Sciences (IASBS), PO Box 45195-1159, Zanjan, Iran*

Corresponding author (e-mail: Anne.Paul@obs.ujf-grenoble.fr)

Abstract: We present a synthesis and a comparison of the results of two temporary passive seismic experiments installed for a few months across the Central and Northern Zagros. The receiver function analysis of teleseismic earthquake records gives a high-resolution image of the Moho beneath the seismic transects. On both cross-sections, the crust has an average thickness of 42 ± 2 km beneath the Zagros fold-and-thrust belt and the Central domain. The crust is thicker beneath the hanging wall of the Main Zagros Reverse Fault (MZRF), with a greater maximum Moho depth in the Central (69 ± 2 km) than in the Northern Zagros (56 ± 2 km). The thickening affects a narrower region (170 km) beneath the Sanandaj–Sirjan zone of the Central Zagros and a wider region (320 km) in the Northern Zagros. We propose that this thickening is related to overthrusting of the crust of the Arabian margin by the crust of Central Iran along the MZRF, which is considered as a major thrust fault cross-cutting the whole crust. The fault is imaged as a low-velocity layer in the receiver function data of the Northern Zagros profile. Moreover, the crustal-scale thrust model reconciles the imaged seismic Moho with the Bouguer anomaly data measured on the Central Zagros transect. At upper mantle depth, P-wave tomography confirms the previously observed strong contrast between the faster velocities of the Arabian margin and the lower velocities of the Iranian micro-blocks. Our higher-resolution tomography combined with surface-wave analysis locates the suture in the shallow mantle of the Sanandaj–Sirjan zone beneath the Central Zagros. The Arabian upper mantle has shield-like shear-wave velocities, whereas the lower velocities of the Iranian upper mantle are probably due to higher temperature. However, these velocities are not low enough and the low-velocity layer not thick enough to conclude that delamination of the lithospheric mantle lid has occurred beneath Iran. The lack of a high-velocity anomaly in the mantle beneath Central Iran suggests that the Neotethyan oceanic lithosphere is probably detached from the Arabian margin.

The Zagros mountain belt results from the collision of Arabia and the micro-plates of Central Iran after the closure of the Neotethyan Ocean. Although no consensus has been reached on the age of the initial collision along the Zagros suture, which ranges from Late Cretaceous to Pliocene (Berberian & King 1981; Allen *et al.* 2004), the Zagros has been considered as an example of a young continent–continent collision belt; for example, in the pioneering thermomechanical models of continental collision by Bird *et al.* (1975) and Bird (1978). The rather small shortening ratios measured in the sediment cover of the Zagros fold-and-thrust belt (between 45 and 85 km, 15–27%; Blanc *et al.* 2003; McQuarrie 2004; Sherkati & Letouzey 2004; Molinaro *et al.* 2005a) suggest that the deep structure of the Zagros is less complicated than that of more evolved belts such as the Alps or the Himalayas. Imaging this structure should give clues to the early phases of mountain building, including the part played by oceanic slab detachment, or the role of continental subduction in the transition from oceanic subduction to continental collision. Indeed, the Zagros is located between the oceanic subduction of Makran and Eastern Anatolia, where a number of seismic analyses suggest that the lithospheric mantle is either thinned or completely removed (Şengör *et al.* 2003).

As reliable data on the lithospheric structure of Iran were scarce, one of the objectives of the Iranian–French co-operative programme in Earth Sciences was to investigate the deep structure of the Zagros, including Moho depth variations. With that aim, we installed temporary seismic stations along two transects of the Zagros for a few months each. The first one (Zagros01 in Fig. 1) was located

From: LETURMY, P. & ROBIN, C. (eds) *Tectonic and Stratigraphic Evolution of Zagros and Makran during the Mesozoic–Cenozoic*. Geological Society, London, Special Publications, **330**, 5–18.
DOI: 10.1144/SP330.2   0305-8719/10/$15.00 © The Geological Society of London 2010.

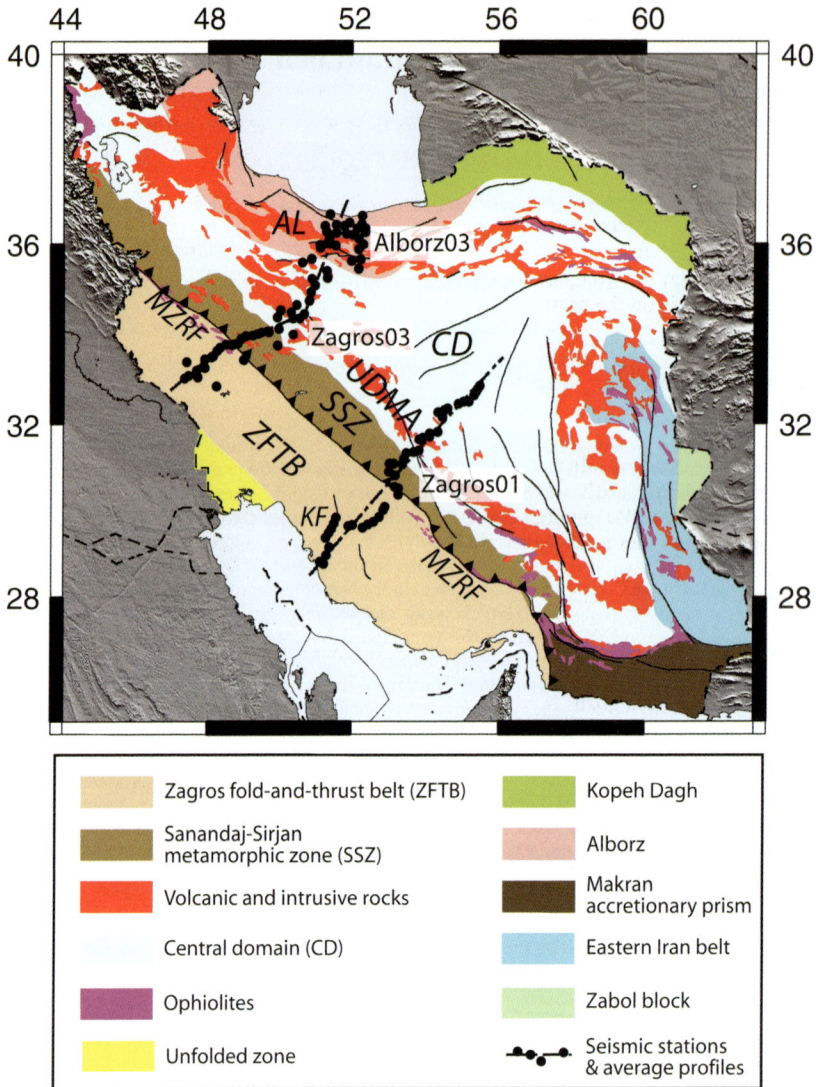

Fig. 1. Location map of the seismic experiments. Temporary seismic stations are shown as black filled circles. The dash-and-dot black lines are the average profiles used to compute the depth cross-sections of Figures 2–6. Geological map modified from the structural map of NGDIR (National Geoscience Database of Iran, http://www.ngdir.ir). MZRF, Main Zagros Reverse Fault; KF, Kazerun Fault; ZFTB, Zagros fold-and-thrust belt; SSZ, Sanandaj–Sirjan Zone; UDMA, Urumieh–Dokhtar Magmatic Assemblage; CD, Central Domain; AL, Alborz.

in the Central Zagros, from the coast of the Persian Gulf (Busher), to the Central domain (Posht-e-Badam). The second one (Zagros03–Alborz03 in Fig. 1) crossed the whole Iranian collision zone from the Northern Zagros (Khorramabad) to the Caspian Sea across the Alborz. The two transects are located on either side of the north-trending Kazerun strike-slip fault system (KF in Fig. 1), which divides the Zagros belt into two domains with different deformation styles. Structural

(Tchalenko & Braud 1974), seismotectonic (Talebian & Jackson 2004) and geodetic (Vernant *et al.* 2004; Walpersdorf *et al.* 2006) observations show that the Northwestern Zagros accommodates the oblique convergence of Arabia and Eurasia by slip partitioning between shortening within the fold-and-thrust belt and dextral strike-slip on the Main Recent Fault, which is parallel and very close to the Main Zagros Reverse Fault (MZRF in Fig. 1). In contrast, convergence in the Central

Zagros is accommodated by pure shortening perpendicular to the MZRF. A comparison of the results of the two seismic experiments could give clues to possible relationships between the deformation style observed at the surface and the lithospheric structure.

The results of the Central Zagros transect have been published elsewhere (Paul *et al.* 2006; Kaviani *et al.* 2007), whereas the results of the Northern Zagros transect are still unpublished. This paper focuses on the comparison of the two profiles, with emphasis on the crustal structure.

Previous data on Zagros lithospheric structure

Crustal structure studies published prior to our seismic transects mainly dealt with gravity data. Dehghani & Makris (1984) produced a Bouguer anomaly map of Iran from a dense set of stations and computed a crustal thickness map from the gravity variations. The main feature of the Bouguer anomaly map is a strong negative anomaly trending NW–SE centred on the MZRF. Its absolute minimum of −230 mgal is located between Shiraz and Esfahan. According to Dehghani & Makris (1984), this negative gravity anomaly gives a maximum crustal thickness of 50–55 km located beneath the MZRF, and a normal thickness of *c.* 40 km beneath the Persian Gulf coast and Central Iran. The more detailed Bouguer anomaly modelling of Snyder & Barazangi (1986) was focused on the Zagros region. They found a maximum crustal thickness of 55–60 km beneath the MZRF, but they argued that the Zagros topographic load is insufficient to explain the observations in terms of simple lithospheric flexure.

All published controlled-source seismic profiles recorded in the Zagros for oil exploration purposes are limited to the sediments above the thick Hormuz Salt layer, which acts as a barrier for seismic waves. Giese *et al.* (1984) reported on refraction profiles recorded from quarry blast sources in southeastern Zagros, but their conclusions are not supported by reverse profiles and the data quality is poor. Hatzfeld *et al.* (2003) used receiver function analysis from earthquake records to infer a crustal thickness of *c.* 46 km beneath a single station in the Central Zagros.

Most results published on the structure of the lithospheric mantle come from seismology. The only exception is the combined modelling of topography, Bouguer anomaly and geoid data by Molinaro *et al.* (2005*b*), who suggested that the lithospheric mantle could be much thinner beneath the Zagros and the Iranian margin than beneath the Persian Gulf. From a regional analysis of

surface-wave records, Maggi & Priestley (2005) showed that the upper mantle (at 100–150 km depth) of the Turkish–Iranian plateau has lower S-wave velocities than surrounding regions such as the Persian Gulf, Northern Arabia or the Southern Caspian. They argued, on the basis of the spatial correlation of this low-velocity anomaly with a positive long-wavelength free-air gravity anomaly indicative of a less dense mantle, and with recent volcanism with a lower mantle signature, to propose a possible delamination of the lithospheric mantle beneath Central Iran. The resolution of their tomography is too low (a few hundred kilometres) to delineate the low-velocity anomaly and relate it to geological features such as the MZRF. This upper mantle low-velocity anomaly has been confirmed for both P and S waves in a recent inversion of body-wave arrival times by Alinaghi *et al.* (2007). Hearn & Ni (1994) and Al-Lazki *et al.* (2004) studied the Pn phase, the conic P wave that propagates in the shallowest upper mantle for regional earthquake–station distances. They showed that P-wave velocities are higher (8.1–8.4 km s^{-1}) beneath the Zagros than beneath most of the Iranian plateau (7.9–8.1 km s^{-1}). The boundary between lower and higher velocities coincides roughly with the MZRF. Al-Lazki *et al.* (2004) interpreted this observation as an indication for absent or limited underthrusting of the higher-velocity upper mantle of Arabia beneath the lower-velocity upper mantle of Central Iran.

Alinaghi *et al.* (2007) showed depth cross-sections of their velocity models to address the question of the Neotethyan slab break-off. Their images show northward-dipping high-velocity anomalies in the mantle beneath the low-velocity anomaly of Central Iran, which can be interpreted as remnants of the oceanic lithosphere. However, the cross-sections are complex, some of them showing a clear continuity of the high-velocity anomaly from shallow (Arabian platform) to great mantle depths, whereas the anomaly is discontinuous in other cross-sections located a few tens of kilometres apart.

Data acquisition and processing

Figure 1 shows a location map of temporary seismic stations installed by French–Iranian teams across the Zagros. The 67 stations of the Central Zagros experiment, named Zagros01, were operated along a 620 km transect from November 2000 to April 2001. From SW to NE, it crosses the Zagros fold-and-thrust belt (ZFTB), the Sanandaj–Sirjan metamorphic zone (SSZ) and the Urumieh–Dokhtar magmatic assemblage (UDMA), and reaches the southern tip of the Central domain (CD). Regarding the logistics, this location was the easiest transect across the most 2D part of the range. However, the southeasternmost part of the profile follows the Kazerun Fault, which is now considered as part of an orogen-scale fault system along which dextral slip on the Main Recent Fault is transferred (Authemayou *et al.* 2006). This major tectonic structure probably has a structural signature that could make the Zagros01 transect less representative of the 2D Central Zagros belt. The second experiment, named Zagros03, crossed the Northern Zagros and the plateau from May to November 2003 (42 stations, 470 km). It was complemented in the Central Alborz by a 2D array of 25 stations from the Seis-UK and Cambridge University pools operated in the same time period (array name Alborz03). Both transects were mostly equipped with sensors with natural period above 5 s, including small subnetworks of broadband sensors suitable for surface-wave studies (natural periods 90 s and 120 s).

The results summarized in this paper were obtained from records of teleseismic earthquakes with magnitude larger than 5.5. The number of events used (*c.* 100 for each experiment) is large enough for a fair resolution in spite of a limited azimuthal coverage (most events have back-azimuths in the range $20-110°$).

Mapping Moho depth is one of the major goals of this type of experiment. This information is obtained by receiver function analysis, an efficient tool to image sharp velocity and density discontinuities at depth beneath seismic stations (Vinnik 1977; Langston 1979). Receiver function (RF) analysis is based on a deconvolution of the first few seconds of signal after the arrival of the P wave in the radial component of teleseismic records. It aims at measuring the time delay between the primary P and the Ps phase, which is a P wave converted to S at velocity discontinuities beneath the station. The Ps–P time difference depends on the depth to the discontinuity, the P-wave velocity (V_p) model and V_p/V_s ratio between the discontinuity and the surface. Thus reasonable assumptions on the values of V_p and V_p/V_s lead to depth estimates from Ps–P delay time measurements. We computed receiver functions using the time domain iterative deconvolution method of Ligorria & Ammon (1999), and migrated depth sections using the common conversion point (CCP) method of Zhu & Kanamori (2000).

The tomography of the upper mantle was achieved by inversion of teleseismic traveltime residuals for lateral variations of P-wave velocity beneath the seismic array. P-wave arrival times are picked at each station for earthquakes at teleseismic distance. Then, traveltime residuals are computed by subtracting theoretical traveltimes computed in a standard spherical Earth model from observed traveltimes, and the average residual for each

event is subtracted from the relevant absolute residuals to give relative residuals. The relative residuals are sensitive to lateral velocity variations beneath the array. Then, the set of residuals is inverted for a 3D model of P-wave velocity relative variations ($\Delta V_p/V_p$) beneath the array. We used the 'tomo3d' implementation (Judenherc 2000) of the widely used and robust ACH inversion method of teleseismic residuals by Aki *et al.* (1977) improved by the 'shift-and-average' procedure of Evans & Achauer (1993) to attenuate the influence of the horizontal sampling in discrete blocks. One of the drawbacks of ACH is that it does not give absolute velocity. Therefore, we also estimated horizontally averaged absolute S-wave velocity–depth models from the inversion of surface-wave dispersion curves for the Zagros01 experiment (Kaviani *et al.* 2007). As absolute S-wave velocities can be interpreted in terms of compositional or thermal changes with depth, they are a good complement to ACH tomography, which detects lateral velocity variations with a fair horizontal resolution of a few station spacings (*c.* 30 km).

Crustal thickness variations from receiver function analysis

Figures 2 and 3 show receiver function depth cross-sections for the two transects. They were computed by projection of the receiver function data on the average profiles shown as dash-and-dot lines in Figure 1. Figures 2b and 3b result from a CCP time-to-depth migration assuming the crustal velocity models of Table 1, which takes the 11 km thick sediment layer of the ZFTB into account for stations located SW of the MZRF according to the model computed by Hatzfeld *et al.* (2003). For stations NE of the MZRF where no information is available, we assumed a two-layer model. The V_p/V_s ratio is a critical parameter in the time-to-depth migration of receiver functions, which can be estimated using the multiples of the Moho conversion (e.g. Zhu & Kanamori 2000). As a result of a poor signal-to-noise ratio, we could not use the multiples in the Zagros01 profile, and we assumed the values given by Hatzfeld *et al.* (2003). The multiples are clearer on the better-quality records of the Zagros03 transect, and we found a value of 1.8 for the crustal average V_p/V_s, without significant variations along the profile. We used this value for the Zagros03 and Alborz03 transects (Table 1). The effect of using different V_p/V_s ratios in the migrations of the two Zagros profiles will be discussed below.

Figures 2c and 3c display stacked radial RF records with the time axis converted to a depth axis using the same crustal velocity models as

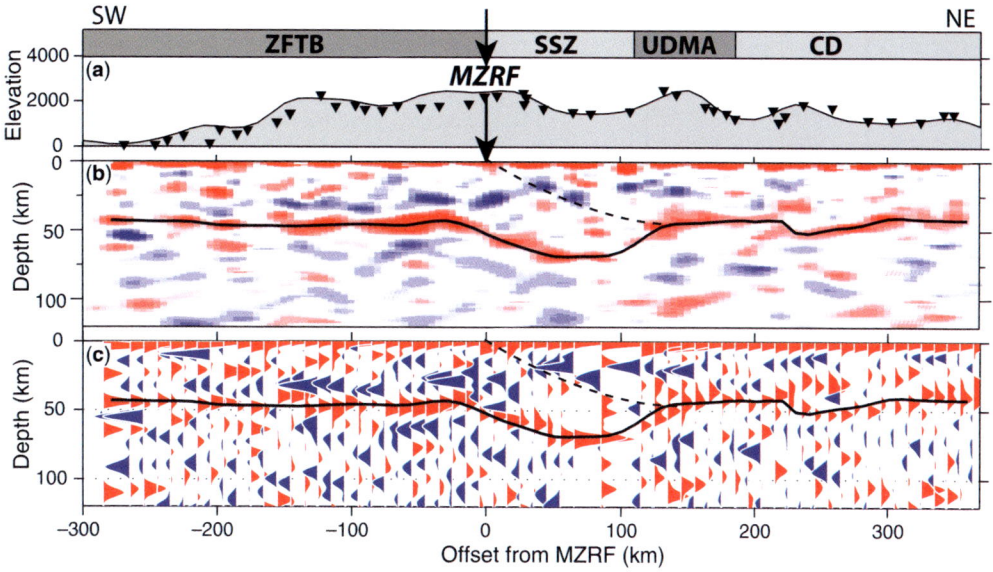

Fig. 2. Depth sections of radial receiver function records of the Central Zagros profile (Zagros01) plotted along the average profile shown in Figure 1. Distance is measured with respect to the intersection of the profile with the surface exposure of the MZRF. (**a**) Topography profile and locations of seismic stations (black triangles). (**b**) Common-conversion point depth migrated cross-section (from Paul *et al.* 2006). (**c**) Radial receiver function records stacked for a common location of the piercing point of the ray at the Moho. Time-to-depth conversion uses the velocity models of Table 1. The continuous black line is the Moho depth profile picked from (b). The dashed line is the crustal-scale thrust hypothesized by Paul *et al.* (2006) to fit RF and gravity data. Same abbreviations as in Figure 1.

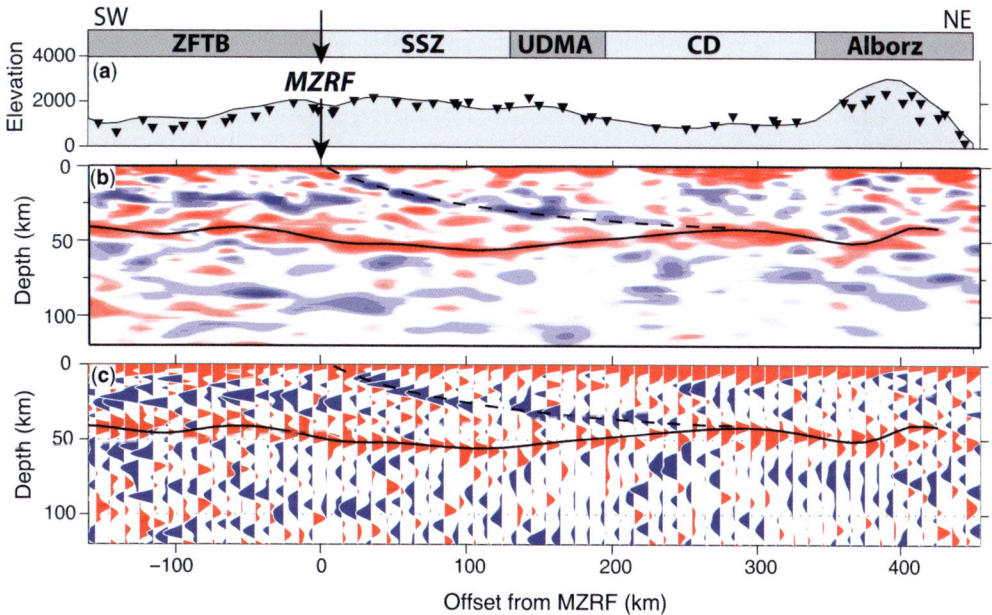

Fig. 3. Same as Figure 2 for the Northern Zagros profile (Zagros03). The dashed line underlines the crustal interface with negative velocity contrast (velocity decreasing with increasing depth) interpreted as the trace of the MZRF.

Table 1. *Crustal velocity models used in the depth migration of the receiver functions*

| | ZFTB ($x \leq 0$ km) | | | | | | SSZ–UDMA–CD–Alborz ($x > 0$ km) | | | | | |
| | Zagros01 | | | Zagros03–Alborz03 | | | Zagros01 | | | Zagros03–Alborz03 | | |
H (km)	V_p (km s^{-1})	V_p/V_s	H (km)	V_p (km s^{-1})	V_p/V_s	H (km)	V_p (km s^{-1})	V_p/V_s	H (km)	V_p (km s^{-1})	V_p/V_s
11	4.7	1.77	11	4.7	1.8	20	5.8	1.73	20	5.8	1.8
9	5.8	1.73	9	5.8	1.8	25	6.5	1.73	25	6.5	1.8
25	6.5	1.73	25	6.5	1.8						

H is the layer thickness.

above. We stacked all RFs with a common conversion point at the Moho ($\pm$5 km measured along the average profile), and plotted them at the abscissa of the conversion point. An average of 14 radial RF records have been stacked to obtain the 62 traces plotted in Figure 2c, and 29 of these 62 traces (47%) result from the stacking of <10 receiver functions. Because of the longer duration of the experiment and better-quality records, the cross-sections of Figure 3 have a better resolution than those in Figure 2, with an average of 37 stacked RFs, and only nine of the 66 traces (14%) resulting from the stacking of <10 RF. The two plots (migrated depth section and section of stacked RFs) are different displays of the same information. Whereas the section of stacked RFs shows waveforms with little processing, the CCP migrated section is smoother, making it easier to pick interfaces. Wavelets of positive amplitude (red) are generated by a velocity increase with depth whereas wavelets of negative amplitude (blue) correspond to a velocity decrease. The laterally continuous positive wavelet at *c.* 50 km depth outlined with a bold black line is the conversion on the Moho. We estimate to $\pm$2 km the uncertainty on the picks of the Moho from the migrated RF depth sections. The uncertainty on absolute depths is larger as we have no precise estimate of the crustal V_p and V_p/V_s models. For example, changing the V_p/V_s model of the Zagros01 transect to a constant value of 1.8 (as for the Zagros03 transect) reduces the estimated Moho depth by 3–6 km.

Assuming a constant V_p/V_s ratio of 1.8, we find an average crustal thickness of 42 $\pm$ 2 km beneath the Zagros fold belt and the Central Domain (CD in Figs 2 & 3) on both transects. In the Central Zagros (Fig. 2), the Moho depth increases strongly from 30 km SW of the MZRF to 140 km NE of it. The maximum depth reached is 69 $\pm$ 2 km (or 63 $\pm$ 2 km assuming $V_p/V_s = 1.8$) between 50 and 90 km NE of the surface trace of the MZRF. No such spectacular bend is visible on the Moho depth profile of the northwestern transect (Fig. 3), where the crust is thickened on a broader length (*c.* 320 km between abscissas −50 and 270 km, instead of 170 km, between abscissas −30 and 140 km, in Zagros01) and the maximum depth reached (56 $\pm$ 2 km at km 130) is smaller. The reliability of the very large crustal thickness found beneath the SSZ on the Zagros01 transect has been thoroughly discussed by Paul *et al.* (2006). Explaining the 20–25 km difference in Moho depth by an artificial pull-down in the migrated depth section would require an average crustal P-wave velocity of 5.2 km s^{-1} combined with a V_p/V_s ratio of 2.0 for the crust of the SSZ, which is unreliable.

Figure 4 shows a comparison of elevation (Fig. 4a), Bouguer anomaly (Fig. 4b) and picked

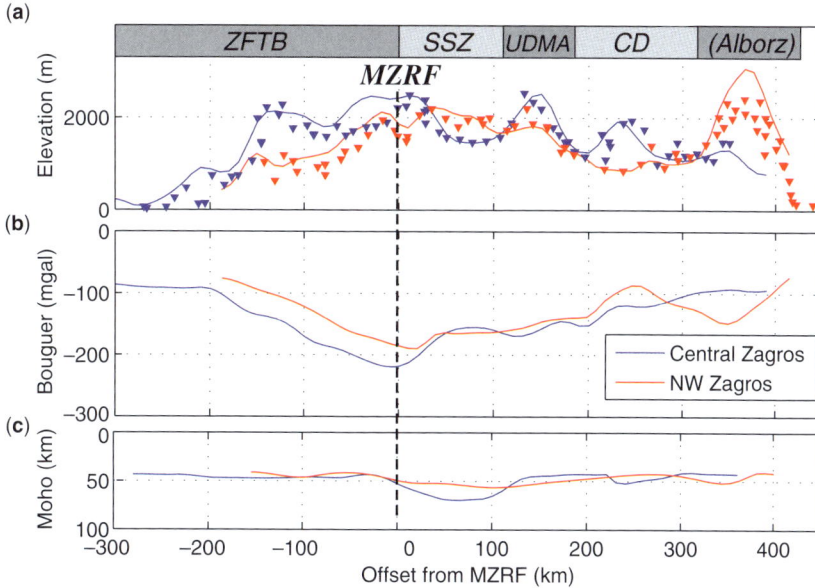

Fig. 4. Comparison of (**a**) the topography profiles, (**b**) the Bouguer anomaly data, and (**c**) the Moho depth profiles picked from receiver functions for the Central Zagros (blue) and Northern Zagros (red) experiments. The distance scale of the Northern Zagros profile was multiplied by 0.94 to correct for the slightly wider SSZ along this transect. The vertical dashed line shows the location of the MZRF.

Moho depth profiles (Fig. 4c) along the two transects. The Moho depth profile of the Zagros01 transect has been corrected to a constant V_p/V_s of 1.8 to allow comparison with the Zagros03 data. The distance scale of the Northern Zagros profile is multiplied by 0.94 to correct for the difference in the width of the SSZ between the two transects. Even after this scale reduction, Figure 4c shows that the region of thickened crust is much wider beneath the Northern than beneath the Central Zagros, but the maximum thickness is smaller. In both cross-sections, crustal thickening starts beneath the northeasternmost part of the ZFTB. The zone of thick crust covers the MZRF region and the SSZ in the Zagros01 transect, whereas it also includes the UDMA and the southern part of the CD in Zagros03. The maximum crustal thickness is shifted toward the NW in Zagros03 with respect to Zagros01. The location of the maximum Moho depth does not coincide with the minimum Bouguer anomaly, which is located in both profiles at the surface outcrop of the MZRF (Fig. 4b). To reconcile the gravity data with the Moho depth profile, Paul *et al.* (2006, fig. 6b) proposed that the crust of Central Iran overthrusts the crust of the Zagros on the MZRF interpreted as a crustal-scale structure rooted at Moho depth (dashed line in Fig. 2b, c). This model explains at least part of the crustal thickening by the superimposition of the two crusts. The almost

double thickness of the dense lower crust beneath the SSZ compensates for the negative gravity anomaly induced by the thick crust and shifts the Bouguer anomaly minimum toward the SE, in agreement with observations. Only the few RFs recorded at stations in the southwestern half of the SSZ display a negative wavelet at depths close to the trace of the MZRF in the model tested by gravity modelling (Fig. 2c). There is no evidence for a major thrust rooted at Moho depth beneath the northeastern half of the SSZ. However, this might be due to a much lower station coverage and poor data quality. Only two stacked traces could be shown in Figure 2c between offsets 65 and 105 km, and they were computed from only one (at offset 65 km) and two (at offset 85 km) RFs. The much better quality sections of Zagros03 (Fig. 3) display a strong amplitude converted phase with negative polarity and northeastward dip, crossing the crust from the surface trace of the MZRF to Moho depth at km 250–270. Figure 3c shows that the wavelet of negative polarity (blue) is followed by a positive wavelet of smaller amplitude (red) in all stacked RFs between km 20 and 110. Waveform modelling shows that these signals can be generated by a low-velocity layer possibly related to the underthrusting of the sediments of the Arabian margin dragged to depth by the subduction of the Neotethyan Ocean (Paul *et al.* 2008).

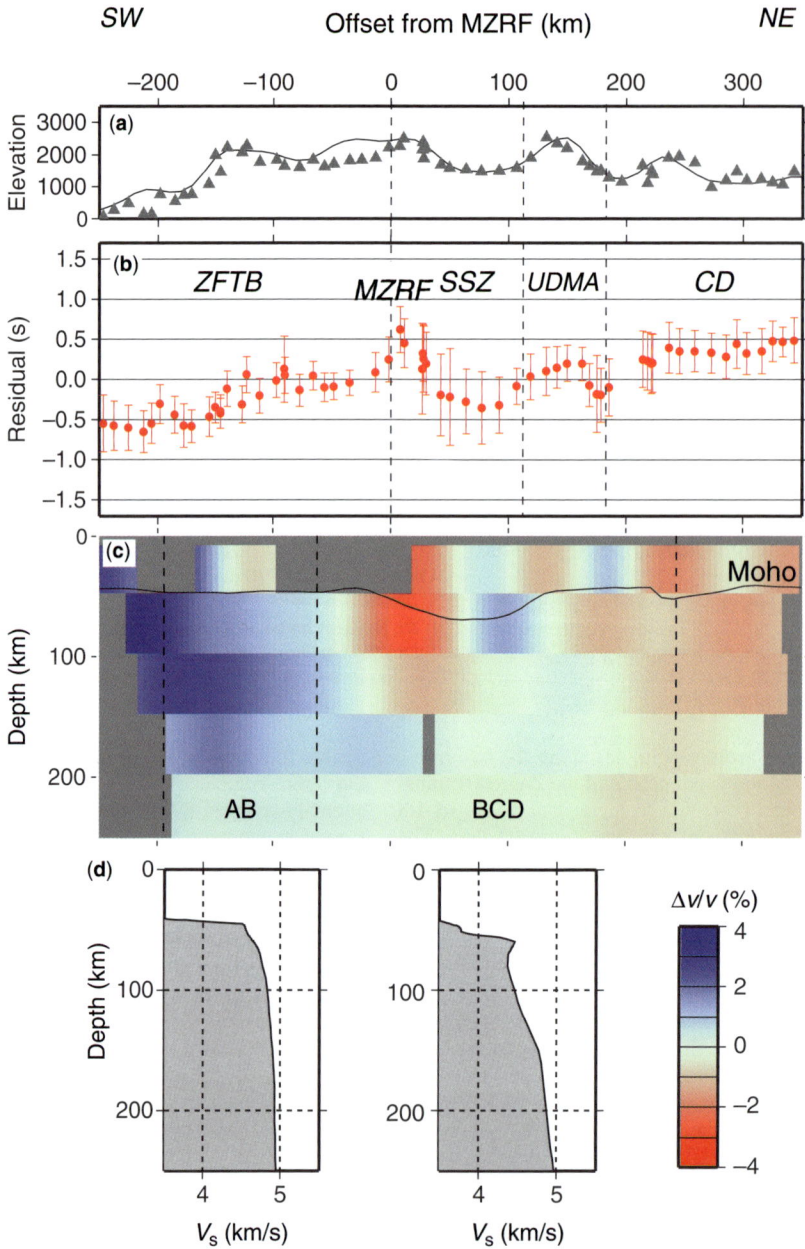

Fig. 5. P-wave tomography and local S-wave velocity models for the Central Zagros profile (Zagros01). (**a**) Topography profile and locations of seismic stations. The three vertical dashed lines show the boundaries of the main morphotectonic units. (**b**) P-traveltime relative residuals plotted for each station at the abscissa of the projection of the station onto the average profile. (**c**) Depth cross-section along the average profile in the 3D model of P-wave velocity perturbations resulting from the inversion of residuals. The continuous black line is the Moho depth profile picked from the receiver function data. The vertical dashed lines bound the areas covered by surface-wave dispersion measurements in sub-arrays AB (ZFTB) and BCD (MZRF–SSZ–UDMA region). (**d**) Mantle S-wave velocity–depth models resulting from the inversion of surface-wave dispersion curves measured in sub-arrays AB (left panel) and BCD (right panel).

Lateral variations of seismic-wave velocity in the lithosphere

Whereas receiver functions image sharp velocity discontinuities such as the Moho, lateral velocity variations are estimated from traveltime studies. Figure 5b (redrawn from Kaviani *et al.* 2007) shows the teleseismic P-wave traveltime relative residuals measured at stations of Zagros01. Residuals are plotted at the abscissa of the projection of the relevant station onto the average profile. Residuals were measured from traveltime picks of P waves on records of 111 earthquakes at teleseismic distance (25–95°). The residual curve displays both long-wavelength variations, with earlier arrivals (negative residuals) at stations in the SW than in the NE (positive residuals), and short-wavelength variations, with later arrivals in the MZRF region and earlier arrivals in the SSZ. The residuals were back-projected by the ACH inversion method to a 3D model of P-wave velocity perturbation, which explains 85% of the initial data variance. Figure 5c shows a depth cross-section in the 3D model along the average profile. Relative velocity variations are displayed only where the resolution is fair (diagonal term of the resolution matrix larger than 0.6 for the crustal layer and 0.7 for the mantle layers). The *c.* 0.9 s traveltime difference observed between the two ends of the profile projects to a strong velocity contrast in the upper mantle down to 200 km depth, with faster velocities beneath the Zagros and slower velocities beneath the UDMA and CD. Figure 5c also shows two average S-wave velocity depth models of the upper mantle estimated by Kaviani *et al.* (2007) from surface-wave dispersion measurements. They were measured in two sub-arrays of the Central Zagros transect, which cover the ZFTB (sub-array AB) and the MZRF–SSZ–UDMA areas (sub-array BCD), respectively. S-wave velocities are rather high beneath the Zagros (4.5–4.9 km s^{-1}), and they increase regularly with depth. This is not the case beneath the MZRF–SSZ–UDMA region, where a low-velocity zone (4.4 km s^{-1} minimum velocity) is detected in the shallow mantle from the Moho to 150 km depth. Therefore, we have two independent arguments to conclude on a strong velocity contrast in the lithospheric mantle beneath the Central Zagros transect. This observation agrees with the results of the regional tomography studies by Maggi & Priestley (2005) and Alinaghi *et al.* (2007), which showed that the upper mantle beneath the Iranian plateau has low P- and S-wave velocities. However, our study, conducted from a local dense station array, shows that the suture between the higher-velocity Arabian upper mantle beneath the ZFTB and the lower-velocity Iranian upper mantle is located beneath the MZRF or the SSZ regions, but

Figure 5c shows that the precise location and shape of the suture are hidden by smearing of shallow velocity anomalies.

The late arrivals observed at stations in the MZRF region project to a low-velocity anomaly concentrated at shallow depth (<100 km). As discussed by Kaviani *et al.* (2007), this low-velocity anomaly could be explained by crustal thickening. However, the high-velocity anomaly beneath the northeastern part of the SSZ in the same two upper layers is surprising, as the crust is also thickened there. To explain this paradox, Kaviani *et al.* (2007) showed that the model of crustal-scale thrust proposed by Paul *et al.* (2006) gives an acceptable fit to traveltime and surface-wave observations if it is combined with an extension of the Zagros mantle high-velocity anomaly beneath the SSZ.

Figure 6 displays the relative residual variations and P-wave tomogram along the average profile of the Zagros03 and Alborz03 arrays. Arrival times of 138 teleseismic earthquakes have been used in this inversion, which explains 83% of the initial data variance. As for Zagros01, the earliest P arrivals are observed at the southwestern end of the profile. The latest arrivals are picked at stations in the northeastern part of the SSZ and in the UDMA, and at stations on the southern flank of the Alborz. The contrast between the high-velocity upper mantle of the Arabian plate and the lower-velocity upper mantle of Central Iran explains the *c.* 1.2 s residual difference between the southwestern end of the profile and stations in the CD. Again, the location of the suture in the 50–100 km layer is blurred by a low-velocity anomaly. Unlike the Central Zagros transect, this anomaly is spatially correlated with the region of thick crust, and the thickening explains at least half of the 0.3–0.5 s difference between the residuals observed in the UDMA and in the CD. No strong crustal low-velocity anomaly is required to explain the traveltime observations in the UDMA, in contrast to the strong low- and high-velocity anomalies required in the crust beneath the MZRF and SSZ in the Zagros01 transect (Kaviani *et al.* 2007).

Discussion and conclusions

Crustal-scale thrust

The Zagros01 and Zagros03 passive seismic experiments have provided images of the lithospheric structure of the Zagros mountain belt with unprecedented resolution. Assuming reasonable crustal V_p models and a constant crustal V_p/V_s ratio of 1.8, the receiver function analysis shows that the average crustal thickness beneath the ZFTB is the same for the two transects: 42 ± 2 km. Thus, we could conclude, as Hatzfeld *et al.* (2003) did,

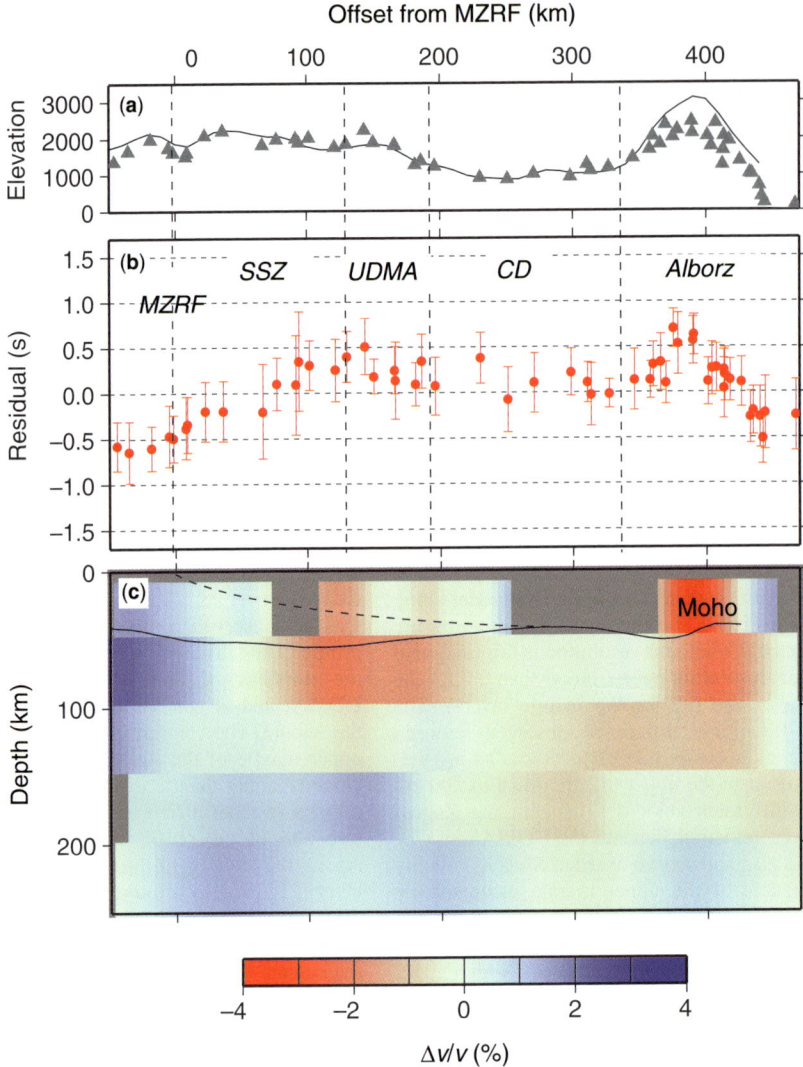

Fig. 6. (**a**) Topography profile and locations of seismic stations, (**b**) P-wave traveltime residuals and (**c**) depth profile of lateral P-wave velocity variations for the Northern Zagros profile (Zagros03).

that the crystalline crust of the ZFTB has not yet been significantly thickened by the collision, in Central and Northern Zagros. Indeed, the crystalline crust has about the same thickness (33–35 km, with 10 km of sediments) as the pre-collisional Arabian platform. However, the lack of precise information on the actual average crustal P-wave velocity and V_p/V_s ratio leads to an uncertainty as large as 5 km on the absolute Moho depth estimate, which prevents definite conclusions being drawn. Mouthereau *et al.* (2006) concluded from critical wedge modelling that basement-involved thickening and shortening of the Arabian crust is required to support the

growth of topography in the ZFTB. As the induced crustal thickening beneath the ZFTB is smaller than a few kilometres, their model is compatible with our observations.

There is no clear evidence for the existence of a major thrust fault cross-cutting the whole crust in the Zagros01 receiver function depth section. However, the station coverage in the SSZ was sparser and the seismic records of lower quality than in the Zagros03 transect. The model of crustal-scale overthrusting of the Iranian margin on the Arabian margin along the MZRF was proposed by Paul *et al.* (2006) to reconcile the Moho

depth profile with Bouguer anomaly data. However, the better-quality data of the denser Northern
Zagros experiment clearly display an intracrustal
interface with negative polarity (velocity decreasing
with increasing depth), followed closely by an interface of positive polarity. The negative-polarity
signal crosses the entire crust from the surface
exposure of the MZRF to Moho depth beneath the
southern rim of the CD. This evidence for a low-
velocity layer confirms that the MZRF is a major
crustal-scale thrust, at the suture between the
Arabian platform and the micro-blocks of Central
Iran. This interpretation also relies on geological
arguments. First, there is strong evidence that the
SSZ overlaps the highly deformed rocks of the
so-called crush zone of the MZRF region (e.g.
Stöcklin 1968; Ricou et al. 1977). Second, Agard
et al. (2005) relied on the amount of shortening,
the deformation style and the lack of high-pressure
metamorphism along the former oceanic suture in
the crush zone to suggest that the MZRF is a

major structure of crustal scale, possibly rooted to
Moho depth.

As shown by Figure 4c, crustal thickening
involves the SSZ in the Central Zagros, and the
SSZ, the UDMA and part of the CD in the Northern
Zagros. The RF migrated section of Zagros03 shows
that the MZRF interpreted as a crustal-scale thrust
roots at Moho depth (43 km) 270 km to the NE of
its surface exposure, with an average dip of 9°
(Fig. 3). If the geometry proposed for the Central
Zagros by Paul et al. (2006) is correct, the MZRF
roots at Moho depth only 170 km NE of its
surface trace, with a stronger average dip of 14°.

On the basis of tectonosedimentary evidence,
basin analysis, crustal wedge modelling and seismotectonic observations, Mouthereau et al. (2007) proposed a new crustal-scale balanced cross-section of
the Central Zagros, in which a large part of the shortening is accommodated by underplating of Arabian
crustal duplexes beneath the MZRF (Fig. 7b). Their
model implies that crustal thickening beneath the

(a) End of oceanic subduction, transition to continental collision

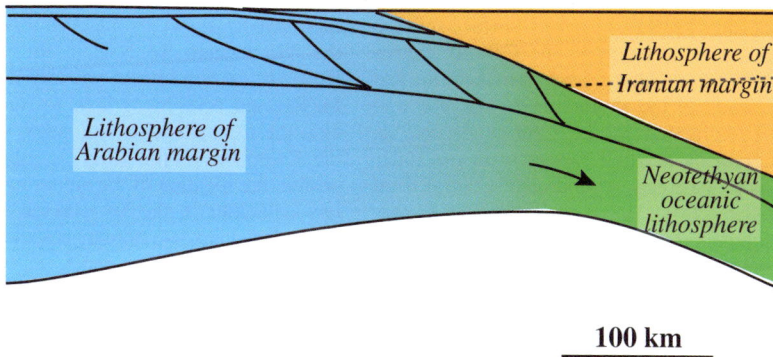

(b) Present-day structure

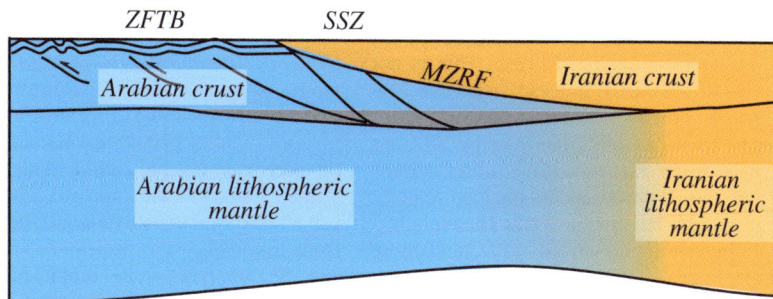

Fig. 7. Schematic model proposed for the evolution of the lithospheric structure of the Zagros from the onset of continental collision (**a**) to present time (**b**). The present-day structural model is based on the Moho depth profile and the trace of the MZRF picked from the receiver function migrated depth section of the Northern Zagros transect (Fig. 3). The surface of the crustal root shaded grey is used to estimate shortening. The mantle structure is interpreted from the P-wave tomography of Figure 6c. No vertical exaggeration.

SSZ is related to the inversion of the Arabian margin after the onset of plate coupling along the former subduction zone (Fig. 7a). This hypothesis qualitatively agrees with our crustal depth profiles, as we show that crustal thickening is located under the trace of the MRZF cross-cutting the whole crust. For a more quantitative test, we computed the surfaces of the crustal roots of the Central and Northern Zagros (depth ≥ 42 km; grey-filled area in Fig. 7b) from the two Moho depth profiles of Figures 2 and 3, and we converted these surfaces to crustal shortening estimates assuming in-plane deformation. We find a finite shortening of 59 km (23%) for the Central Zagros, and 49 km (13%) for the Northern Zagros. These values agree with the model proposed by Mouthereau *et al.* (2007), in which 75% of the total shortening (i.e. 49–58 km) is taken up by underplating of the Arabian basement beneath the SSZ.

As explained in the introduction, we installed two seismic profiles because we expected that the difference in the active deformation process between the Northern and Central Zagros would have an expression, or an origin, in the lithospheric structure. We do find that the crustal root has different widths and amplitudes, in relation to the lateral change of the MZRF dip. The surface of the crustal root is 20% larger in the Central Zagros than in the Northern Zagros, leading to the observed 10% difference in the shortening ratios. We find a stronger shortening in the Central Zagros, as expected from its location farther from the Euler rotation pole between Arabia and Central Iran (Vernant *et al.* 2004). However, because error bars on our Moho depth estimates are large (± 5 km) as a result of poor constraints on the V_p and V_p/V_s crustal models, this positive correlation cannot be taken as granted. Teleseismic traveltime tomography also documents along-range variations of crustal P-wave velocity anomalies. A significant part of the short-wavelength lateral variations of relative residuals measured along the Zagros03 transect can be explained by thickness variations of a crust with limited lateral velocity changes. This is not the case in the Zagros01 transect, where a low average crustal velocity beneath the MZRF and a high average crustal velocity beneath the SSZ are required to explain the traveltime observations. Kaviani *et al.* (2007) proposed that the low-velocity anomaly of the MZRF region could be explained by crustal thickening combined with thickening of the sedimentary wedge in front of the thrust, and the high velocity of the SSZ by superimposition of the two lower crusts by the crustal-scale thrust, combined with upper mantle high velocities. The smaller lateral velocity changes in the crust of the Northern Zagros suggest that crustal structure changes more smoothly, which is in agreement

with the greater width of the region affected by crustal-scale thrusting in relation to the smaller dip of the MZRF.

Lateral velocity contrast in the upper mantle

The traveltime study, combined for Zagros01 with surface-wave dispersion analysis, confirms that seismic velocities in the shallow mantle (50–200 km) are faster in the margin of the Arabian platform than in the Iranian micro-blocks. This contrast has already been documented at the regional scale of the Turkish–Iranian plateau from less well-resolved tomographies, and for both P- and S-wave velocities (Maggi & Priestley 2005; Alinaghi *et al.* 2007). Our study provides more precise indications on the location of the suture between the two upper mantle lids, and on the velocity contrast. Considering that the laterally heterogeneous upper mantle layer is 150 km thick, as documented by the Rayleigh-wave dispersion inversion (Fig. 5d), a lateral V_p contrast of 5–7% is required to explain the 0.9–1.2 s difference observed in the traveltime curves of the two transects. Kaviani *et al.* (2007) used independent arguments from P-wave tomography and surface-wave analysis to propose that the high-velocity upper mantle of Arabia extends toward the NE beneath the SSZ in the Central Zagros. However, strong crustal velocity anomalies in the MZRF and SSZ areas leak to mantle depth and hide the exact location and shape of the suture in the Zagros01 profile (Fig. 5c). In the Zagros03 profile, the location of the suture between high and low velocities in the 50–100 km layer is blurred by the low-velocity anomaly associated with crustal thickening. At greater depth, Figure 6c shows a smooth transition located beneath the CD at abscissa *c.* 250 km; that is, close to the abscissa where the crustal-scale thrust imaged by receiver functions reaches the Moho. Thus, our data show that the fast-velocity lithospheric mantle of Arabia extends to *c.* 140 km NE of the surface exposure of the MZRF in the Central Zagros and *c.* 250 km in the Northern Zagros, in agreement with the proposed model of crustal-scale underthrusting of the crust of Arabia by the crust of Central Iran.

Kaviani *et al.* (2007) discussed the inferences on the geodynamics that can be made from the tomographic images of the Zagros01 transect. The shear-wave absolute velocity–depth models estimated from Rayleigh-wave dispersion analysis (Fig. 5d) can be interpreted in terms of compositional and thermal constraints, which is not the case for the relative P-wave velocity tomograms. The high S-wave velocities measured in the upper mantle of the Zagros are comparable with those of a shield. The difference (0.5 km s^{-1}) between the two V_s models measured in the Zagros and in the

MZRF–SSZ–UDMA region (Fig. 5d) in the shallow mantle is probably due to a compositional change (hydrated minerals, e.g.) related to higher temperatures beneath the MZRF–SSZ–UDMA. However, Kaviani *et al.* (2007) believed that the V_s contrast is too weak to support the hypothesis of lithospheric mantle delamination proposed by Maggi & Priestley (2005).

Kaviani *et al.* (2007) explained that the V_p tomogram of Figure 5c gives no definite answer on the question of oceanic slab break-off. The image of the Neotethyan lithosphere still attached to the Arabian margin would be impossible to separate from the effect of crustal and upper mantle heterogeneity in the suture region. The V_p tomogram of Zagros03 (Fig. 6c) is less blurred by crustal structures in the upper mantle NE of the suture. In the upper two layers (0–100 km), there is no high-velocity anomaly like the one observed beneath the SSZ in the Zagros01 profile. Also, the upper mantle high-velocity anomaly that would be associated with the oceanic lithosphere attached to the Arabian margin does not appear at greater depths in Figure 6c. That is why we believe that the slab could be detached from the Zagros margin. Kaviani *et al.* (2007) gave additional elements favouring this hypothesis, including the lack of Ps conversion in the receiver function sections under the Moho beneath the UDMA and CD. Hafkenscheid *et al.* (2006) also concluded that slab detachment had occurred beneath the Zagros suture zone from a comparison of tectonic reconstructions and the mantle tomography of Bijwaard *et al.* (1998). The recent discovery in the central part of the UDMA of adakitic magmas representing potential markers of the melting at depth of oceanic crust suggests slab break-off or slab tear propagation from Eastern Anatolia to Central Iran in the last *c.* 5–10 Ma (Omrani *et al.* 2008).

Our model of the structure of the Zagros lithosphere at the onset of continental collision and at present is shown schematically in Figure 7. The crust of the Iranian micro-continent overthrusts the crust and mantle of the Arabian margin until the two plates are strongly coupled. The continuing convergence induces crustal thickening beneath the SSZ by underplating of Arabian crustal duplexes and folding and thrusting in the ZFTB. At mantle depth, the Neotethyan slab detaches from the Arabian margin; a new Moho is created (by thermal relaxation and eclogitization) beneath the MZRF crustal-scale thrust; the mantle of the Iranian lithosphere is heated by upwelling induced by slab break-off; its shallow part (50–150 km) retains higher temperature and lower seismic velocities than the neighbouring fast-velocity mantle of the Arabian platform.

The Zagros01, Zagros03 and Alborz03 experiments were funded by IIEES (International Institute of Earthquake Engineering and Seismology, Iran), and INSU–CNRS (France). The instruments belong to the French national pool of mobile seismic instruments Sismob with a complement from the British national pool Seis-UK for Alborz03 in collaboration with K. Priestley (University of Cambridge). We thank the President of IIEES, M. Ashtiany, and the Director of the Seismology Department, M. Mokhtari, for their constant support. A. K. and M. T. received fellowships from the French Embassy in Iran for their PhD studies in France. Many people participated in the field experiments. They are warmly acknowledged.

References

AGARD, P., OMRANI, J., JOLIVET, L. & MOUTHEREAU, F. 2005. Convergence history across Zagros (Iran): constraints from collisional and earlier deformation. *International Journal of Earth Sciences*, doi: 10.1007/s00531-005-0481-4.

AKI, K., CHRISTOFFERSSON, A. & HUSEBYE, E. S. 1977. Determination of 3-dimensional seismic structure of the lithosphere. *Journal of Geophysical Research*, **82**, 277–296.

ALINAGHI, A., KOULAKOV, I. & THYBO, H. 2007. Seismic tomographic imaging of P- and S-waves velocity perturbations in the upper mantle beneath Iran. *Geophysical Journal International*, **169**, 1089–1102, doi: 10.1111/j.1365-246X.2007.03317.x.

AL-LAZKI, A. I., SANDVOL, E., SEBER, D., BARAZANGI, M., TÜRKELLI, N. & MOHAMAD, R. 2004. Pn tomographic imaging of mantle lid velocity and anisotropy at the junction of the Arabian, Eurasian and African plates. *Geophysical Journal International*, **158**, 1024–1040.

ALLEN, M., JACKSON, J. & WALKER, R. 2004. Late Cenozoic reorganization of the Arabia–Eurasia collision and the comparison of short-term and long-term deformation rates. *Tectonics*, **23**, TC2008, doi: 10.1029/2003TC001530.

AUTHEMAYOU, C., CHARDON, D., BELLIER, O., MALEKZADEH, Z., SHABANIAN, E. & ABBASSI, M. R. 2006. Late Cenozoic partitioning of oblique plate convergence in the Zagros fold-and-thrust belt (Iran). *Tectonics*, **25**, TC3002, doi: 10.1029/2005TC001860.

BERBERIAN, M. & KING, G. C. P. 1981. Towards a paleogeography and tectonic evolution of Iran. *Canadian Journal of Earth Sciences*, **18**, 210–265.

BIJWAARD, H., SPAKMAN, W. & ENGDAHL, E. R. 1998. Closing the gap between regional and global travel time tomography. *Journal of Geophysical Research*, **103**, 30055–30078.

BIRD, P. 1978. Finite element modeling of lithosphere deformation: the Zagros collision orogeny. *Tectonophysics*, **50**, 307–336.

BIRD, P., TOKSÖZ, M. N. & SLEEP, N. H. 1975. Thermal and mechanical models of continent–continent convergence zones. *Journal of Geophysical Research*, **32**, 4405–4416.

BLANC, E. J. P., ALLEN, M. B., INGER, S. & HASSANI, H. 2003. Structural styles in the Zagros simple folded zone, Iran. *Journal of Structural Geology*, **160**, 401–412.

DEHGHANI, G. A. & MAKRIS, J. 1984. The gravity field and crustal structure of Iran. *Neues Jahrbuch für Geologie und Palaeontologie, Abhandlungen*, **168**, 215–229.

EVANS, J. R. & ACHAUER, U. 1993. Teleseismic velocity tomography using the ACH method: theory and application to continental-scale studies. *In*: IYER, H. M. & IRAHARA, K. (eds) *Seismic Tomography, Theory and Practice*. Chapman & Hall, London, 319–360.

GIESE, P., MAKRIS, J., AKASHE, B., RÖWER, P., LETZ, H. & MOSTAANPOUR, M. 1984. The crustal structure in Southern Iran derived from seismic explosion data. *Neues Jahrbuch für Geologie und Palaeontologie, Abhandlungen*, **168**, 230–243.

HAFKENSCHEID, E., WORTEL, M. J. R. & SPAKMAN, W. 2006. Subduction history of the Tethyan region derived from seismic tomography and tectonic reconstructions. *Journal of Geophysical Research*, **111**, B08401, doi: 10.1029/2005JB003791.

HATZFELD, D., TATAR, M., PRIESTLEY, K. & GHAFORY-ASHTYANY, M. 2003. Seismological constraints on the crustal structure beneath the Zagros mountain belt (Iran). *Geophysical Journal International*, **155**, 403–410.

HEARN, T. M. & NI, J. F. 1994. Pn velocities beneath continental collision zones: the Turkish–Iranian Plateau. *Geophysical Journal International*, **117**, 273–283.

JUDENHERC, S. 2000. *Etude et caractérisation des structures her.cyniennes à partir de données sismologiques: le cas du Massif Armoricain*. PhD thesis, Université Louis Pasteur Strasbourg I.

KAVIANI, A., PAUL, A., BOUROVA, E., HATZFELD, D., PEDERSEN, H. & MOKHTARI, M. 2007. A strong seismic velocity contrast in the shallow mantle across the Zagros collision zone (Iran). *Geophysical Journal International*, **171**, 399–410, doi: 10.1111/j.1365-246X.2007.03535.x.

LANGSTON, C. A. 1979. Structure under Mount Rainier, Washington, inferred from teleseismic body waves. *Journal of Geophysical Research*, **84**, 4749–4762.

LIGORRIA, J. P. & AMMON, C. J. 1999. Iterative deconvolution and receiver-function estimation. *Bulletin of the Seismological Society of America*, **89**, 1395–1400.

MAGGI, A. & PRIESTLEY, K. 2005. Surface waveform tomography of the Turkish–Iranian plateau. *Geophysical Journal International*, **160**, 1068–1080.

MCQUARRIE, N. 2004. Crustal-scale geometry of the Zagros fold–thrust belt, Iran. *Journal of Structural Geology*, **26**, 519–535.

MOLINARO, M., LETURMY, P., GUEZOU, J. C. & FRIZON DE LAMOTTE, D. 2005a. The structure and kinematics of the south-eastern fold–thrust belt, Iran: from thin-skinned to thick-skinned tectonics. *Tectonics*, **24**, TC3007, doi: 10.1029/2004TC001633.

MOLINARO, M., ZEYEN, H. & LAURENCIN, X. 2005b. Lithospheric structure beneath the southeastern Zagros Mountains, Iran: recent slab break-off? *Terra Nova*, **17**, 1–6.

MOUTHEREAU, F., LACOMBE, O. & MEYER, B. 2006. The Zagros Folded Belt (Fars, Iran): constraints from topography and critical wedge modelling. *Geophysical Journal International*, **165**, 336–356.

MOUTHEREAU, F., TENSI, J., BELLAHSEN, N., LACOMBE, O., DE BOISGROLLIER, T. & KARGAR, S. 2007. Tertiary sequence of deformation in a thin-skinned/thick-skinned collision belt: The Zagros Folded Belt (Fars, Iran). *Tectonics*, **26**, TC5006, doi: 10.1029/2007TC002098.

OMRANI, J., AGARD, P., WHITECHURCH, H., BENOIT, M., PROUTEAU, G. & JOLIVET, L. 2008. Arc-magmatism and subduction history beneath Zagros: new report of adakites and geodynamic consequences. *Lithos*, **106**, 380–398.

PAUL, A., KAVIANI, A., HATZFELD, D., VERGNE, J. & MOKHTARI, M. 2006. Seismological evidence for crustal-scale thrusting in the Zagros mountain belt (Iran). *Geophysical Journal International*, **166**, 227–237, doi: 10.1111/j.1365-246X.2006.02920.x.

PAUL, A., HATZFELD, D., KAVIANI, A. & TATAR, M. 2008. Lithospheric structure of the Zagros and Alborz mountain belts (Iran) from seismic imaging (abstract). *American Geophysical Union Fall Meeting, San Francisco*.

RICOU, L. E., BRAUD, J. & BRUNN, J. H. 1977. *Le Zagros*. Mémoires Hors Série, Société Géologique de France, **8**, 33–52.

ŞENGÖR, A. M. C., ÖZEREN, S., GENÇ, T. & ZOR, E. 2003. East Anatolian high plateau as a mantle-supported, north–south shortened domal structure. *Geophysical Research Letters*, **30**, 8045, doi: 10.129/2003GL017858.

SHERKATI, S. & LETOUZEY, J. 2004. Variation of structural style and basin evolution in the central Zagros (Izeh zone and Dezful Embayment), Iran. *Marine and Petroleum Geology*, **21**, 535–554.

SNYDER, D. B. & BARAZANGI, M. 1986. Deep crustal structure and flexure of the Arabian plate beneath the Zagros collisional mountain belt as inferred from gravity observations. *Tectonics*, **5**, 361–373.

STÖCKLIN, J. 1968. Structural history and tectonics of Iran: a review. *AAPG Bulletin*, **52**, 1229–1258.

TALEBIAN, M. & JACKSON, J. 2004. A reappraisal of earthquake focal mechanisms and active shortening in the Zagros mountains of Iran. *Geophysical Journal International*, **156**, 506–526.

TCHALENKO, J. S. & BRAUD, J. 1974. Seismicity and structure of the Zagros (Iran): the Main Recent Fault between 33 and 35°N. *Philosophical Transactions of the Royal Society of London*, **277**, 1–25.

VERNANT, P., NILFOROUSHAN, F. *ET AL.* 2004. Present-day crustal deformation and plate kinematics in the Middle East constrained by GPS measurements in Iran and northern Oman. *Geophysical Journal International*, **157**, 381–398.

VINNIK, L. P. 1977. Detection of waves converted from P to SV in the mantle. *Physics of the Earth and Planetary Interiors*, **15**, 39–45.

WALPERSDORF, A., HATZFELD, D. *ET AL.* 2006. Difference in GPS deformation pattern of North and Central Zagros (Iran). *Geophysical Journal International*, **167**, 1077–1088.

ZHU, L. P. & KANAMORI, H. 2000. Moho depth variation in southern California from teleseismic receiver functions. *Journal of Geophysical Research*, **105**, 2969–2980.

The kinematics of the Zagros Mountains (Iran)

D. HATZFELD[1]*, C. AUTHEMAYOU[2-4], P. VAN DER BEEK[5], O. BELLIER[2],
J. LAVÉ[6], B. OVEISI[5,7], M. TATAR[1,8], F. TAVAKOLI[1,9], A. WALPERSDORF[1] &
F. YAMINI-FARD[1,8]

[1]*Laboratoire de Géophysique Interne et Tectonophysique, CNRS, Université J. Fourier,
Maison des Géosciences, BP 53, 38041 Grenoble cedex 9, France*

[2]*Cerege-UMR CNRS 6635-Aix–Marseille Université, BP80, Europôle, Méditerranéen
de l'Arbois, 13545 Aix-en-Provence cedex 4, France*

[3]*Université Européenne de Bretagne, Brest, France*

[4]*Université de Brest, CNRS, IUEM, Domaines océaniques—UMR 6538, Place Copernic,
F-29280, Plouzané, France*

[5]*Laboratoire de Géodynamique des Chaînes Alpines, CNRS, Université J. Fourier, Maison des
Géosciences, BP 53, 38041 Grenoble cedex 9, France*

[6]*Centre de Recherches Pétrographiques et Géochimiques, 15 rue Notre Dame des Pauvres,
54501 Vandoeuvre lès Nancy, France*

[7]*Geological Survey of Iran, PO Box 13185-1494, Tehran, Iran*

[8]*International Institute of Earthquake Engineering and Seismology, PO Box 19395/3913,
Tehran, Iran*

[9]*National Cartographic Center, PO Box 13185/1684, Tehran, Iran*

Corresponding author (e-mail: denis.hatzfeld@ujf-grenoble.fr)

Abstract: We present a synthesis of recently conducted tectonic, global positioning system (GPS),
geomorphological and seismic studies to describe the kinematics of the Zagros mountain belt, with
a special focus on the transverse right-lateral strike-slip Kazerun Fault System (KFS). Both the
seismicity and present-day deformation (as observed from tectonics, geomorphology and GPS)
appear to concentrate near the 1000 m elevation contour, suggesting that basement and shallow
deformation are related. This observation supports a thick-skinned model of southwestward propagation of deformation, starting from the Main Zagros Reverse Fault. The KFS distributes
right-lateral strike-slip motion of the Main Recent Fault onto several segments located in an en
echelon system to the east. We observe a marked difference in the kinematics of the Zagros
across the Kazerun Fault System. To the NW, in the North Zagros, present-day deformation is
partitioned between localized strike-slip motion on the Main Recent Fault and shortening
located on the deformation front. To the SE, in the Central Zagros, strike-slip motion is distributed
on several branches of the KFS. The decoupling of the Hormuz Salt layer, restricted to the east
of the KFS and favouring the spreading of the sedimentary cover, cannot be the only cause of
this distributed mechanism because seismicity (and therefore basement deformation) is associated
with all active strike-slip faults, including those to the east of the Kazerun Fault System.

Mountain building is the surface expression of
crustal thickening caused by plate convergence.
Mountains are located on continental lithosphere,
which, because of its mechanical properties, generally accommodates plate convergence in a more
distributed and diffuse way than oceanic lithosphere. Because thickening stores gravitational
potential energy, it reaches a limit imposed by the

mechanical strength of the crust and lithosphere,
after which further storage of gravitational energy
is possible only by increasing the lateral size of
the mountain belt rather than its height (e.g.
Molnar & Lyon-Caen 1988). Therefore, mountain
building is a dynamic process, which, to be quanti-
fied, requires a detailed description of both the
surface kinematics and its relation with crustal

From: LETURMY, P. & ROBIN, C. (eds) *Tectonic and Stratigraphic Evolution of Zagros and Makran during the
Mesozoic–Cenozoic*. Geological Society, London, Special Publications, **330**, 19–42.
DOI: 10.1144/SP330.3   0305-8719/10/$15.00 © The Geological Society of London 2010.

deformation. In this paper, we show that shallow deformation, as evidenced by global positioning system (GPS) measurements and geomorphology, correlates well, both spatially and temporally, with basement deformation as evidenced by seismicity and topography, suggesting that they image the same mountain-building process.

The Zagros fold-and-thrust belt is located within Iran at the edge of the Arabian plate (Fig. 1). It is *c*. 1200 km long and trends NW–SE between eastern Turkey, where it connects to the Anatolian mountain belt, and the Strait of Hormuz, where it connects to the Makran subduction zone. Its width varies from *c*. 200 km in the west to *c*. 350 km in the east. The Zagros mountain belt results from convergence between Arabia and Eurasia, which has been continuous since Late Cretaceous times, with a late episode of accentuated shortening during the Pliocene–Quaternary. The Zagros is classically described in terms of longitudinal units separated by lateral discontinuities (Fig. 1). The High Zagros comprises highly deformed metamorphic rocks of Mesozoic age; it is bounded to the NE by the Main Zagros Thrust (MZT), which is the boundary with Central Iran, and to the SW by the High Zagros Fault (HZF). This is the highest part of the Zagros, with maximum elevations reaching more than 4500 m. The High Zagros overthrusts to the south the Zagros Fold Belt, which comprises a 10 km thick Palaeozoic–Cenozoic sequence of sediments. The Zagros Fold Belt is characterized by large anticlines several tens of kilometres long. Longitudinally, the Zagros is divided into two geological domains, the North Zagros (and the Dezful embayment) to the west and the Central Zagros (or Fars) to the east, separated by the north–south-trending strike-slip Kazerun Fault System that cross-cuts the entire belt. Significant differences in mechanical stratigraphy exist between the North and the Central Zagros; the sedimentary cover of the latter has been deposited on top of the infra-Cambrian Hormuz Salt layer, whereas this layer is absent in the North Zagros.

The amount of shortening between Arabia and Iran since Jurassic times, resulting from subduction of the Neotethys, is about 2000 km (McQuarrie *et al.* 2003). Ocean closure and cessation of subduction probably occurred during the Oligocene (Agard *et al.* 2005). This event is recorded by a slight decrease in the convergence velocity from 30 to 20 mm a^{-1} (McQuarrie *et al.* 2003). The total amount of shortening since the onset of continental collision is debated, depending on which marker is used to measure it. Estimates have been based on reconstructions of Late Cretaceous (Haynes & McQuillan 1974; Stöcklin 1974) to late Miocene (Stoneley 1981) strata. Shortening is accommodated differently in the North and Central Zagros because of the differing boundary conditions and pre-existing tectonics. In the North Zagros, the Main Recent Fault accommodates the lateral component of oblique convergence and may transfer some of the motion to the North Anatolian system, whereas deformation partitioning does not appear to exist in the Central Zagros.

Basement deformation

Morphotectonics and balanced cross-sections

Because the basement is decoupled from the shallow sediments by several ductile layers (e.g. the infra-Cambrian Hormuz and Miocene Gahsaran interfaces), surface deformation may not be representative of the total crustal deformation. Furthermore, deformation mechanisms may differ between the basement and the sedimentary cover because of their different mechanical properties. This view is partially supported by the fact that less than 10% of the total deformation of the Zagros (as measured at the surface) is released by seismic deformation (supposed to be related to the crustal deformation) whereas most of the deformation is seismic in other areas of Iran (Jackson & McKenzie 1988; Masson *et al.* 2005). There is no direct access to basement deformation in the Zagros because there are no basement outcrops at the surface, seismic reflection profiles do not clearly image the basement and earthquake ruptures on the reverse faults generally do not reach the surface.

An approach that implies a model assumption is to indirectly infer basement deformation from surface observations. Berberian (1995) mapped first-order changes in the stratigraphy and identified five morphotectonic units with different characteristics of folding, uplift, erosion and sedimentation. He suggested that these morphotectonic units are separated by major reverse faults affecting the basement and striking parallel to the main structures (Fig. 1). These faults are partially associated with seismicity, consistent mostly with reverse mechanisms, but the accuracy of earthquake locations (*c*. 20 km, Engdahl *et al.* 1998) does not permit mapping of active faults in detail. Moreover, some large earthquakes are not related to any of the inferred faults.

Another approach to indirectly infer crustal deformation is to compute the amount of shortening from balanced cross-sections (Blanc *et al.* 2003; McQuarrie 2004; Molinaro *et al.* 2004; Sherkati & Letouzey 2004). In this method, the different layers that constitute the sedimentary cover are supposed to only fold or fault, without internal deformation. However, the location at depth of the decoupling layers, the amount of decoupling

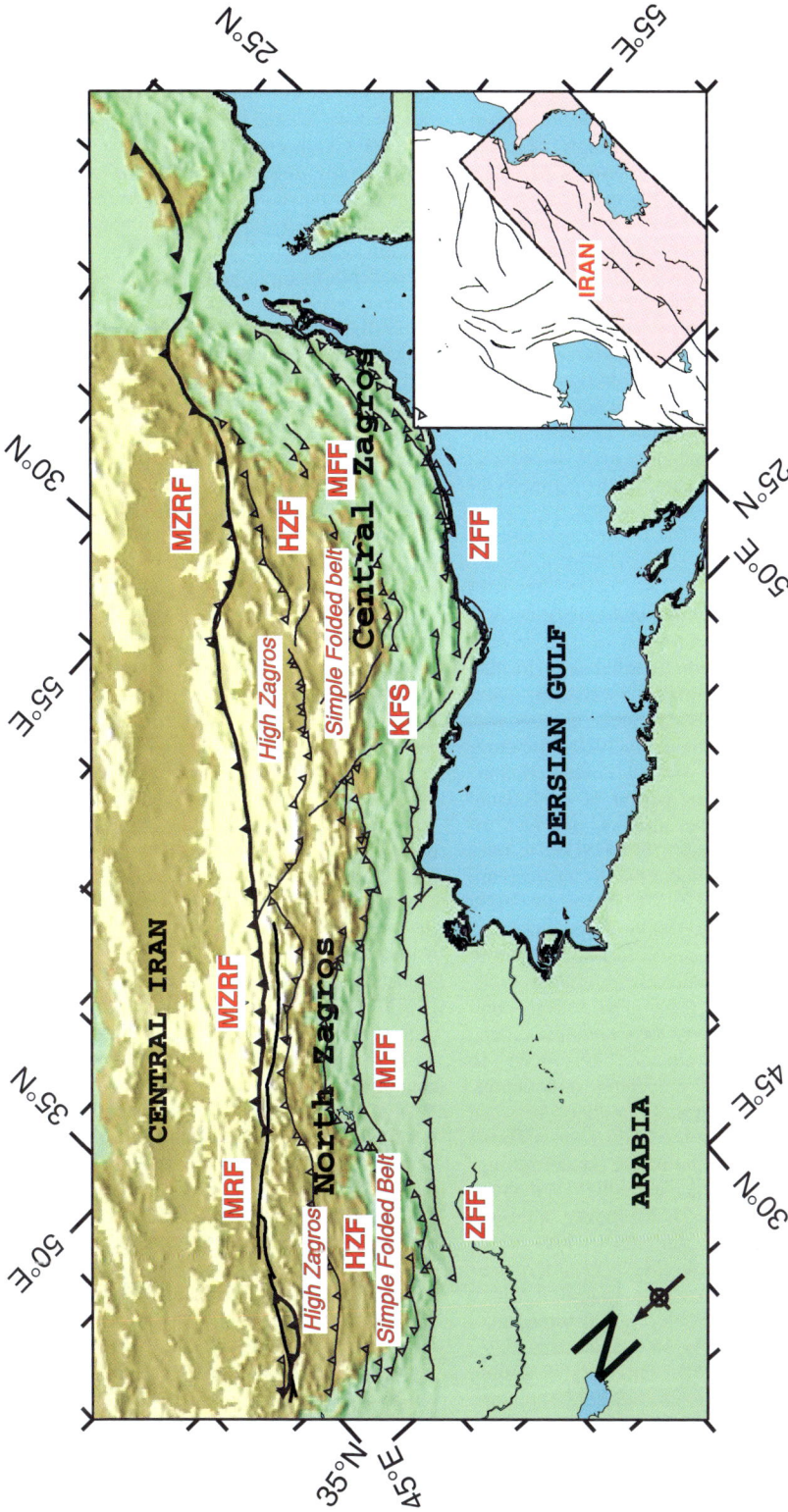

Fig. 1. Location map showing the main geographical and tectonic features of the Zagros (Iran) modified after Berberian (1995), Talebian & Jackson (2004) and Authemayou *et al.* (2006). For the faults, we use the terminology of Berberian (1995). MZRF, Main Zagros Reverse Fault; MRF, Main Recent Fault; HZF, High Zagros Fault; MFF, Main Frontal Fault; ZFF, Zagros Frontal Fault; KFS, Kazerun Fault System, which separates the North Zagros from the Central Zagros. The colours represent topography, with changes at 1000, 2000 and 3000 m levels.

related to these layers, and the relationship between folding and faulting are all complex, and solutions are generally non-unique. Usually, basement faults are assumed where unfolding creates a space problem in the core of folds. The link between surface and basement deformation is strongly debated. Some researches do not require faults in the basement (McQuarrie 2004), whereas others have proposed that deformation started in a thin-skinned mode and continued as thick-skinned deformation (Blanc *et al.* 2003; Molinaro *et al.* 2004; Sherkati *et al.* 2005). Some workers have suggested that faulting post-dates folding (Blanc *et al.* 2003; Molinaro *et al.* 2005), whereas others have proposed that basement faulting predated folding (Mouthereau *et al.* 2006). It is therefore problematic to infer basement faulting, and moreover to estimate the amount of shortening, from balanced cross-sections alone, without complete control of the geometry of the different interfaces.

Seismicity

The other way to access basement deformation is to study seismicity (Fig. 2). Two sets of data provide complementary information: earthquakes located teleseismically and earthquakes located by local networks. Teleseismically located earthquakes have been recorded since the early 1960s; the duration of the available time window is thus comparable with the usual return period of continental earthquakes. However, because of the lack of regional stations, catalogues (ISC, USGS) of teleseismically located earthquakes in Zagros are subject to large mislocations (Ambraseys 1978; Berberian 1979; Jackson 1980*b*; Engdahl *et al.* 1998, 2006). Errors in epicentre location are up to *c.* 20 km and depths are generally unreliable.

Jackson & McKenzie (1984), Ni & Barazangi (1986) and Engdahl *et al.* (2006), amongst others, filtered catalogues or relocated seismicity to improve the accuracy of epicentres and depths. The Zagros seismicity is totally confined between the Persian Gulf coast and the Main Zagros Thrust (MZT), which both limit the active (or deforming) area and exclude seismic accommodation of shortening by the MZT (Fig. 2). Moreover, although seismicity is spread over the entire width of the Zagros, the larger magnitude ($M_b > 5$) earthquakes appear to concentrate in the Zagros Fold Belt, which is an area of low ($z < 1500$–2000 m) topography (Jackson & McKenzie 1984; Ni & Barazangi 1986; Talebian & Jackson 2004). This larger seismic energy release at low elevations has been explained by differential stress owing to the gradient in topography (Jackson & McKenzie 1984; Talebian & Jackson 2004). Epicentres are not obviously correlated with geological structures or surface tectonics

(Fig. 2). Moreover, no instrumental earthquake has a magnitude M_w greater than 6.7 and, as a consequence, no co-seismic ruptures have been observed, except for one earthquake in 1990 ($M_w \approx 6.4$) located at the eastern termination of the HZF (Walker *et al.* 2005).

The only reliable depths for teleseismically located earthquakes are those computed by body-wave modelling with uncertainties in depth of ± 4 km (Talebian & Jackson 2004). In the Zagros these depth of large earthquakes is 5–19 km with a mean *c.* 11 km, suggesting that earthquakes occur in the basement below the sedimentary cover.

Most focal mechanisms computed from first-motion polarities (McKenzie 1978; Jackson & McKenzie 1984) or by body-wave modelling (Talebian & Jackson 2004) are reverse faulting with NW–SE strikes, parallel to the folding (Fig. 3). Some of these mechanisms are associated with the major faults proposed by Berberian (1995) but others are not. Most of the mechanisms are high-angle reverse faulting probably occurring in the basement at depths between *c.* 5 and 15 km; they are thus unrelated to a low-angle detachment at the base of the sedimentary layer (Fig. 3). Jackson (1980*a*) proposed that they reactivate normal faults inherited from a stretching episode affecting the Arabian platform during opening of the Tethys Ocean in the Early Mesozoic.

Strike-slip mechanisms are related to two fault systems: the north–south-trending Kazerun Fault System (KFS; comprising the Kazerun, Kareh-Bas, Sabz-Pushan and Sarvestan faults), which crosses the Zagros between 51.5°E and 54.0°E, and the Main Recent Fault (MRF), which runs parallel to the MZT and connects at its SE termination to the Kazerun Fault System. The MRF helps to accommodate the oblique shortening experienced by the North Zagros by partitioning the slip motion into pure reverse faulting and strike-slip faulting.

Early studies based on unfiltered earthquake catalogues (Nowroozi 1971; Haynes & McQuillan 1974; Bird *et al.* 1975; Snyder & Barazangi 1986) postulated that some intermediate seismicity could be related to continental subduction located NE of the MZT. However, no reliably located earthquake is located NE of the MZT (Engdahl *et al.* 1998) and no earthquakes have been located at a depth greater than 20 km in this area (Jackson & Fitch 1981; Jackson & McKenzie 1984; Maggi *et al.* 2000; Engdahl *et al.* 2006), implying that continental subduction is either aseismic or active.

Microearthquake studies complement the teleseismic information because they locate epicentres with an accuracy of a few kilometres; an order of magnitude better than teleseismic locations. On the other hand, they span a relatively short time window, which may not record the tectonic

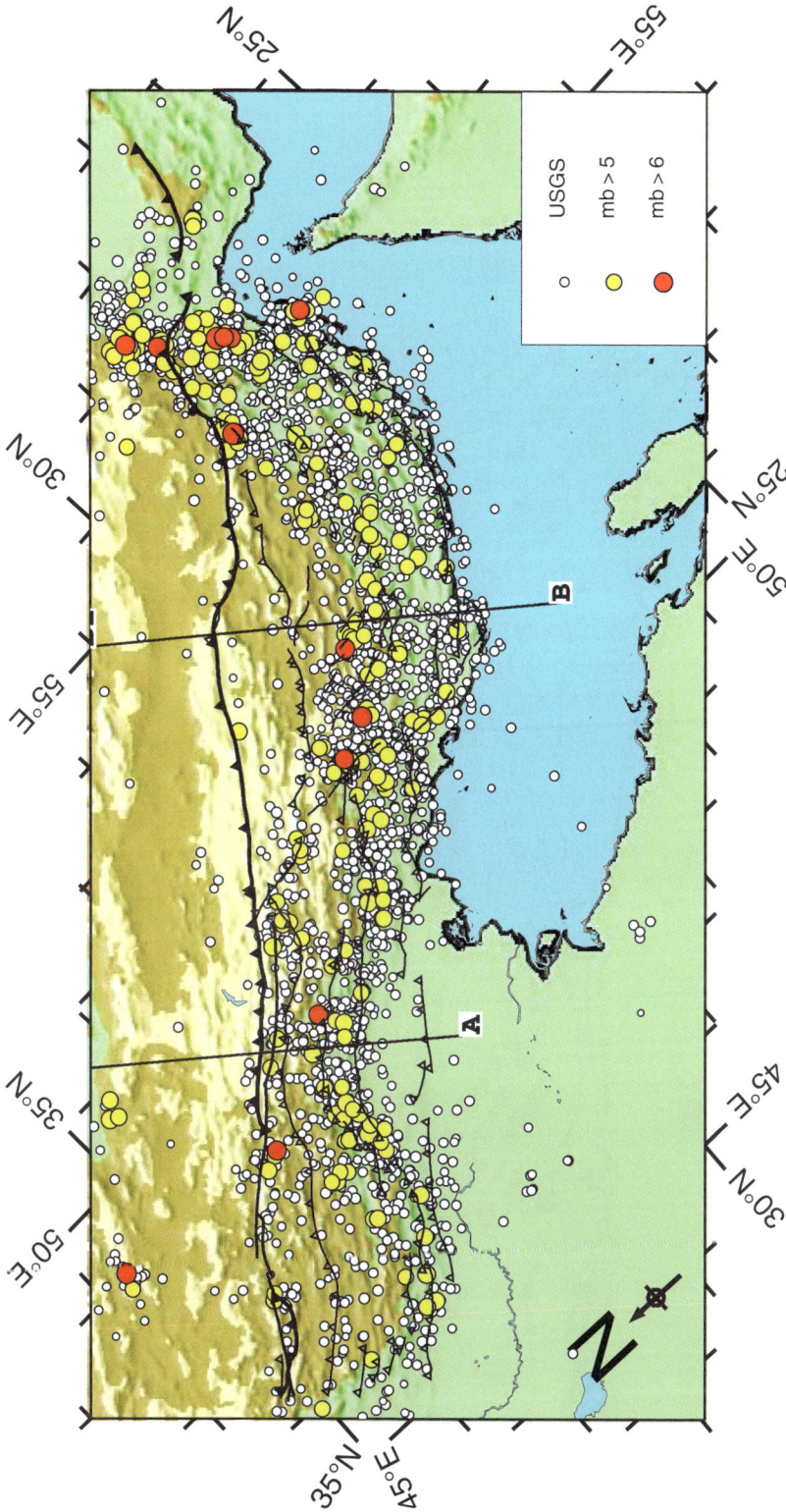

Fig. 2. Seismicity map of the Zagros based on the US Geological Survey catalogue, confirming the observation of Talebian & Jackson (2004) that seismicity (and especially large magnitude earthquakes) is restricted to the SW of the Zagros topography. Cross-sections for lines A and B are shown in Figure 11.

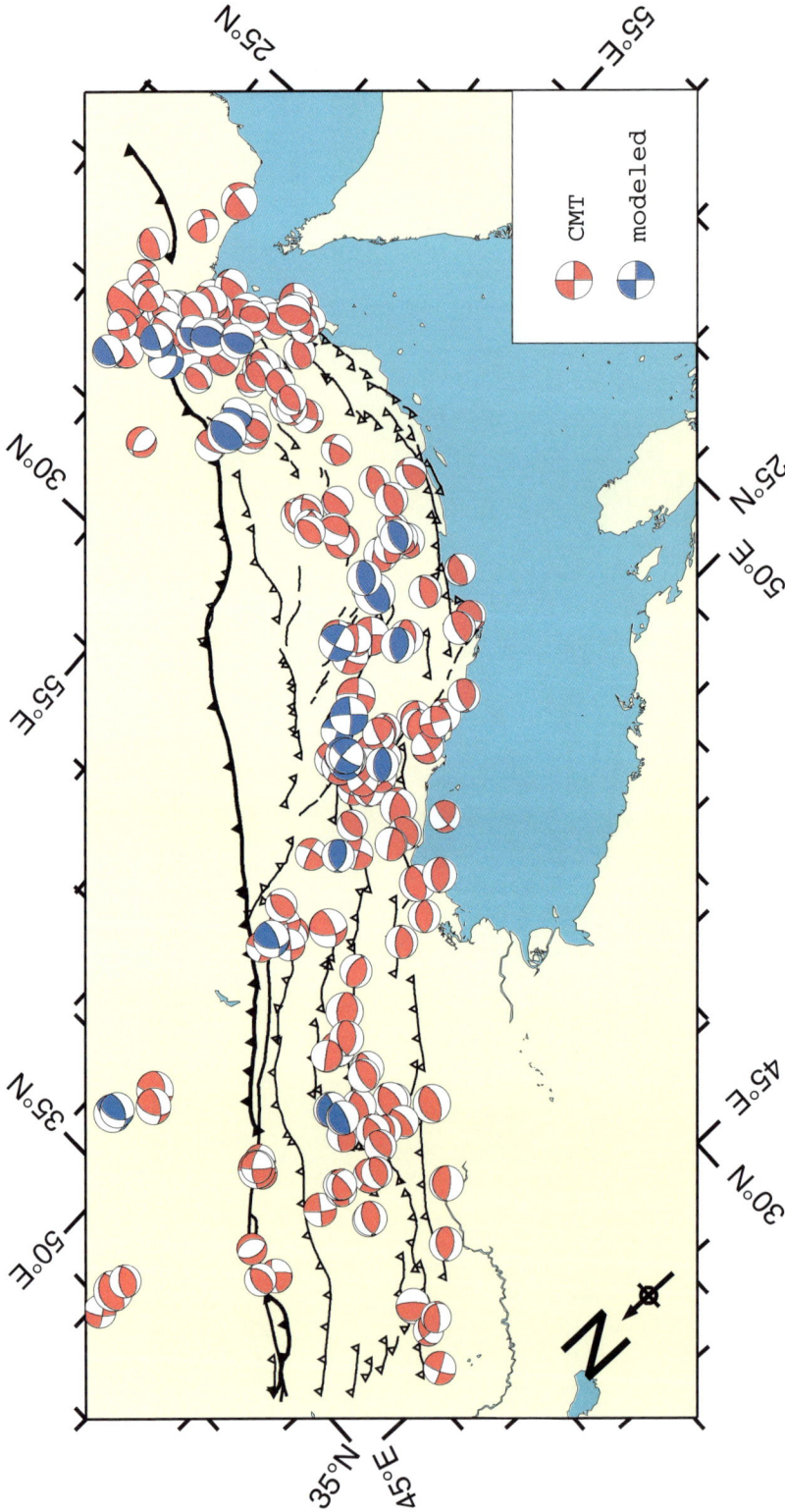

Fig. 3. Fault-plane solutions in the Zagros. Blue focal spheres are body-wave solutions modelled by Talebian & Jackson (2004) and red focal spheres are CMT solutions (http://www.seismology.harvard.edu/CMTsearch.hml). As pointed out by Talebian & Jackson (2004), most of the Zagros experiences reverse faulting, except near the MZRF and the KFS.

processes in a representative manner. Several temporary networks have been installed in the Zagros, at Qir (Savage *et al.* 1977; Tatar *et al.* 2003), Kermansha (Niazi *et al.* 1978), Bandar-Abbas (Niazi 1980; Yamini-Fard *et al.* 2007) and near the Kazerun Fault System (Yamini-Fard *et al.* 2006). Whereas earlier studies are of limited use because the small number of stations does not allow sufficient accuracy in earthquake location, more recent studies have helped to determine some aspects of the crustal structure by inverting travel-time delays of local earthquakes recorded at stations

located directly above the seismicity. Tatar *et al.* (2003) confirmed that seismicity in the Central Zagros is confined between *c.* 10 and *c.* 15 km depth, beneath the sedimentary cover and in the upper part of the basement (Fig. 4). As for the teleseismic events, no microearthquake is located north of the MZT and no earthquake is deeper than 20 km. The seismicity is not confined to the main faults, as observed at the surface, but is spread over a wider area. More interestingly, the microseismicity defines elongated NW–SE-trending lineaments parallel to the fold axes but with a different spacing,

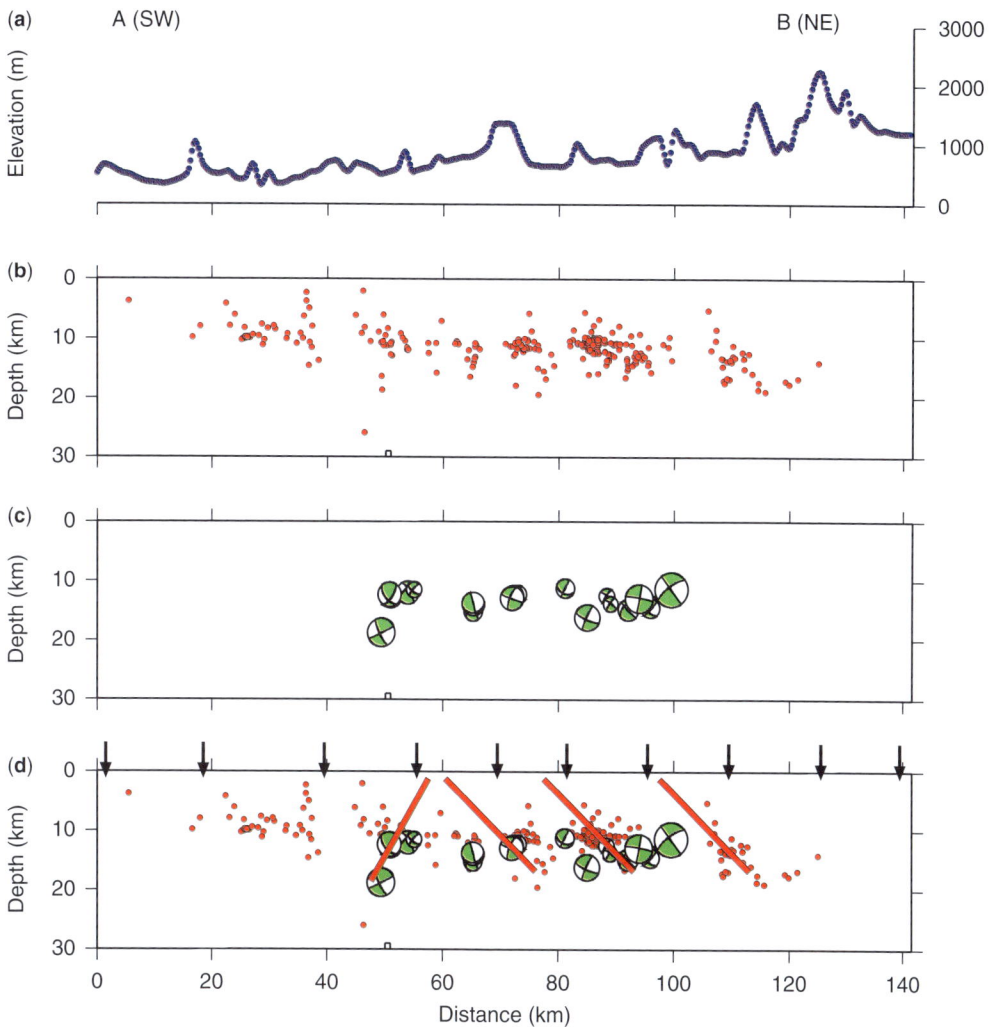

Fig. 4. SW–NE cross-section across the Central Zagros (after Tatar *et al.* 2003). (**a**) Topography. (**b**) Well-located (better than 2 km) microseismicity recorded during a 7 week period. Microseismicity is restricted to the upper basement beneath the sedimentary layer and dips slightly NE. (**c**) Fault-plane solutions (in cross-section), showing mostly reverse mechanisms. (**d**) Our interpretation of clustering possibly associated with active faults (red lines). Black arrows at the surface represent fold axes, the spacing of which is unrelated to any clustering in seismicity.

suggesting that folds and faults are not directly related. The seismicity clusters appear to dip NE (Fig. 4), supporting the model of normal-fault reactivation (Jackson 1980*a*). Focal mechanisms are consistent with NW–SE-striking reverse faults connected by NNW–SSE right-lateral strike-slip faults. The main direction of the P-axes fits well the direction of GPS shortening, suggesting that microearthquakes are the response of the crust to north–south shortening.

Two other surveys, at the intersection between the Kazerun Fault and the MRF in Borujen (Yamini-Fard *et al.* 2006) and at the transition between the Zagros collision zone and the Makran subduction zone near Bandar-Abbas (Yamini-Fard *et al.* 2007), show an interesting result. Reverse-slip focal mechanisms are confined to depths greater than 12 km along NE-dipping décollements striking perpendicular to the motion, whereas dextral strike-slip focal mechanisms are recorded at shallower depths under the trace of the MRF. This difference in mechanism with depth suggests that the upper brittle crust deforms mostly by slip (either strike-slip or reverse, depending on the orientation) on weak pre-existing faults, but that the lower crust is more pervasively weakened and accommodates the shortening by reverse faulting perpendicular to regional motion.

Surface deformation

GPS deformation

GPS measurements provide instantaneous velocities between benchmarks. Depending on the surveying procedure and on the duration of the measurements for each survey, the accuracy of the position can reach *c*. 2 mm. If the time span between two measurements is several years, and moreover if three or more measurements are available allowing some redundancy, we estimate the velocity uncertainties to be less than 2 mm a^{-1}.

Several campaigns have been conducted in the Zagros. One was part of a regional-scale survey conducted throughout Iran, with a spacing between stations larger than *c*. 150 km (Nilforoushan *et al.* 2003; Vernant *et al.* 2004; Masson *et al.* 2007), which does not provide sufficient resolution to study the deformation in great detail. However, a dozen benchmarks from this network record 6–7.5 mm a^{-1} of NNE–SSW shortening for the Zagros, which corresponds to *c*. 30% of the total convergence between Arabia and Eurasia at this longitude. The transition between the Makran subduction and the Zagros collision is clearly evidenced by the contrast in the velocities relative to Central Iran across the area.

Hessami *et al.* (2006) installed a network of 35 benchmarks covering the entire Zagros. These stations were measured during three campaigns over 3 years in 1998, 1999 and 2001. Each station was measured several times and sessions lasted 8 h. The observations of 4–6 IGS stations were included for reference. Hessami *et al.* claimed their accuracy to be 3 mm a^{-1}. The main results are that west of the Kazerun Fault shortening is accommodated by the Mountain Front Fault, whereas east of it, it seems to be accommodated 100 km north of the Mountain Front Fault and by the Main Zagros reverse Fault.

Since 1997, we installed several regional GPS networks in the Zagros (Fig. 5). These networks covered the Central Zagros (15 benchmarks), the Kazerun Fault System (11 benchmarks) and the Northern Zagros (18 benchmarks), and were measured simultaneously with several stations of the Iran Global network as well as with Iranian permanent stations. Each site was continuously observed for at least 48 h per campaign. All networks were measured a minimum of three times over a time period lasting usually 2–5 years. The data have been analysed with the GAMIT/GLOBK 10.1 software (King & Bock 2002). As many as 32 IGS stations (depending on the survey) have been included to establish the terrestrial reference frame. Final IGS orbits and corresponding Earth orientation parameters have been used. In the combination of daily solutions with the Kalman filter GLOBK, the continuous time series of daily SOPAC global solution files (IGS3 network) has been included, covering all of the measurement epoch presented here. Mean repeatability is estimated to be less than 2 mm, which yields a precision better than 2 mm a^{-1}. Details of processing procedures can be found in previous papers (Tatar *et al.* 2002; Walpersdorf *et al.* 2006; Tavakoli *et al.* 2008).

The main results (Fig. 5) show some differences from those of Hessami *et al.* (2006). As observed by those workers, the shortening component increases from NW to SE, consistent with a Arabia–Central Iran pole of rotation located at 29.8°N, 35.1°E, inferred by Vernant *et al.* (2004). However, the deformation on each side of the Kazerun Fault System is different from that proposed by Hessami *et al.* (2006). West of the Kazerun Fault System, most of the deformation is located north of the MFF, far from the Zagros Frontal Fault (ZFF). It is clearly partitioned between 4–6 mm a^{-1} of dextral strike-slip motion concentrated in the north, with probably 2–4 mm a^{-1} on the MRF alone, and 3–6 mm a^{-1} of shortening probably on the MFF. East of the Kazerun Fault, the deformation is pure shortening of 8 mm a^{-1} located along the Persian Gulf shore and associated with the

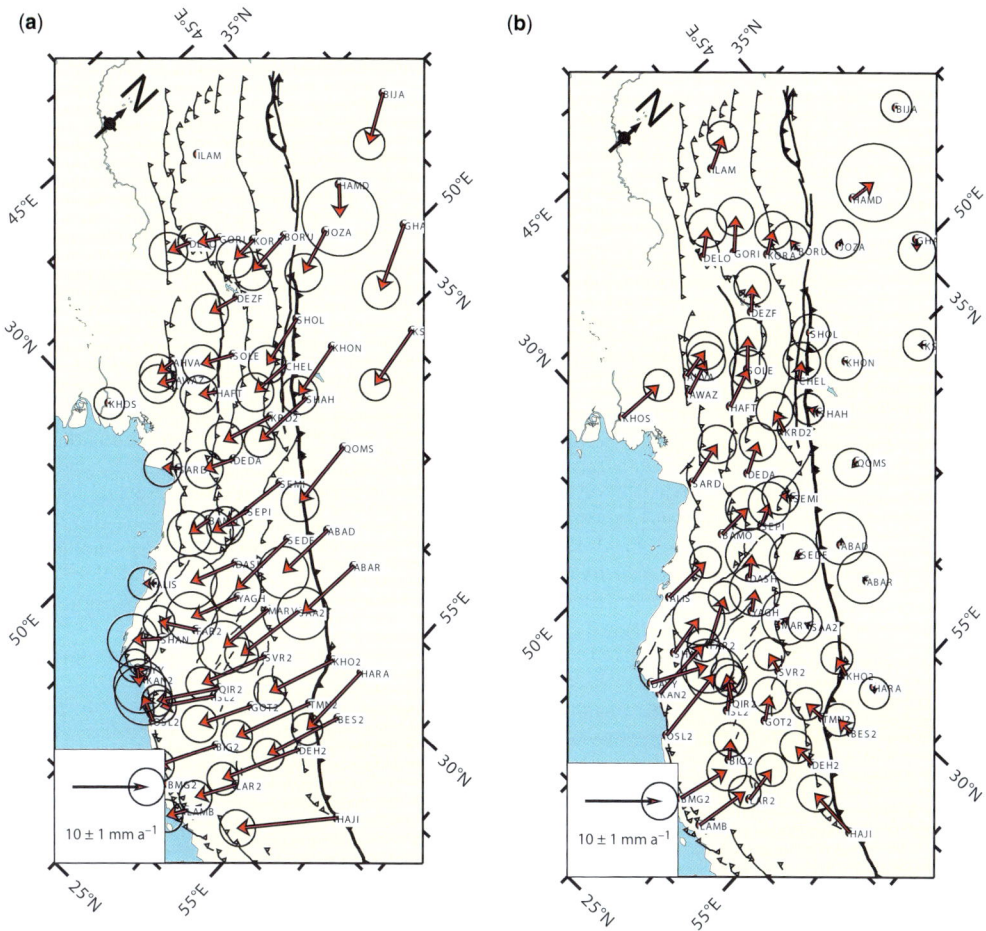

Fig. 5. GPS-detected motion of the Zagros (Tatar *et al.* 2002; Walpersdorf *et al.* 2006; Tavakoli *et al.* 2008) with 95% confidence ellipses. (**a**) Motion relative to Arabia; (**b**) motion relative to Central Iran. Deformation appears localized near the MFF. We do not observe a fan-shaped pattern in the Central Zagros, as expected from spreading of the motion as a result of the Hormuz salt layer.

ZFF. In contrast to Hessami *et al.* (2006), we do not observe significant along-strike extension (i.e. larger than 2 mm a^{-1}) between the two extremities of the Zagros. The KFS strike-slip system induces some extension oblique to the faults, but we do not observe significant along-strike extension of the Zagros associated with perpendicular shortening or thickening of the belt. This view is also evidenced by the strain rate between the benchmarks.

We computed the strain rate and rigid rotation in all triangles defined by three adjacent benchmarks, and report here the amount and direction of shortening, as well as the rotation experienced by each triangle assumed to be a rigid block (Fig. 6). GPS measurements show that most of the shortening is neither uniformly located across the belt nor located on one of the major basement faults (i.e.

MFF, ZFF) proposed by Berberian (1995). In contrast, shortening appears to be associated again with the topography and more specifically between the 1000 m elevation contour and sea level (Fig. 6a). The correlation between the gradient in topography, basement seismicity (Talebian & Jackson 2004) and instantaneous shortening rate supports the hypothesis that basement and surface deformation are related and that both propagate southwestward. Therefore, a total decoupling by the Hormuz Salt of the shallow sediments from the basement is not needed.

Finally, we observe a consistent pattern of clockwise rotation throughout the Zagros (Fig. 6b). As expected, the largest rotations are associated with the largest strain rates and follow the 1000 m elevation contour. This general rotation is probably

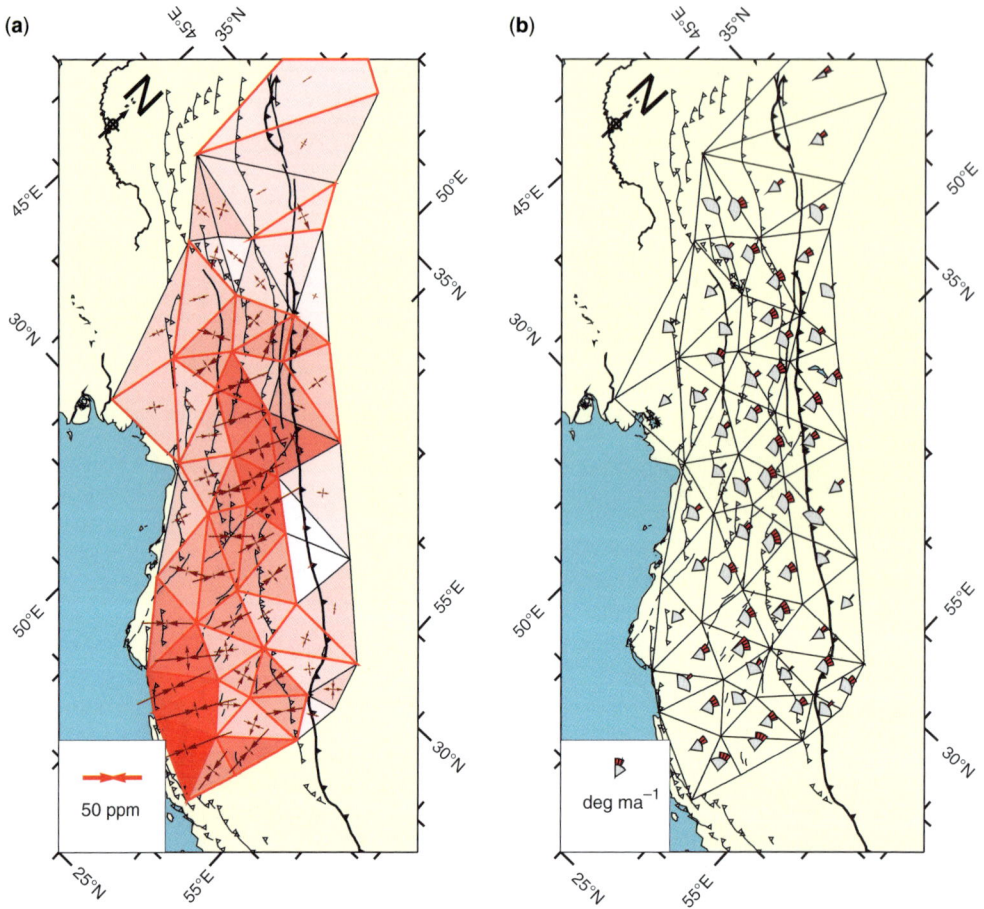

Fig. 6. (**a**) Strain rate deduced from GPS observations. Triangles are coloured as a function of the intensity of the deformation. The arrows indicate the principal strain rates. The triangles with significant deformation (exceeding the uncertainties) are surrounded by a bold line. The direction of shortening consistently trends NNE–SSW with a slight north–south rotation near the Kazerun Fault System. East of the KFS, the deformation is localized at the MFF near the Persian Gulf. West of the KFS, the deformation is localized further north, also at the MFF. In both cases it can be associated with the 1000 m topography elevation. (**b**) Rotations of triangles defined by three benchmarks. Although uncertainties are large, we observe a consistent clockwise rotation. Only two triangles located at the easternmost location show significant anticlockwise rotation. Triangles with rotations larger than $1°$ Ma^{-1} are associated with large strain and located along the MFF as is the strain.

induced by the general right-lateral transcurrent motion between Central Iran and Arabia. We do not observe larger rotation associated with the strike-slip Kazerun Fault System, nor any anti-clockwise rotation as proposed by Talebian & Jackson (2004).

Tectonics

The Zagros deformation is characterized by constant-wavelength folding, thrusting and strike-slip faulting. Models suggest that detachment folding is the main folding style (Mouthereau *et al.*

2006; Sherkati *et al.* 2005). Fold geometries vary significantly with the presence of intermediate décollements (Sherkati *et al.* 2006). Some thrusts branched on décollement levels are formed by pro-gressive fault propagation within the core of the folds. Other thrusts, associated with topographic steps, appear to be linked to basement faults. These reverse faults are generally blind. The difference in elevation of some stratigraphic marker horizons on both sides of the thrusts indi-cates 5–6 km finite vertical offset on both the MFF and the HZF (Berberian 1995; Sherkati & Letouzey 2004). The southwestward migration of sedimentary depocentres from Late Cretaceous time to Miocene collision, as well as the existence of several stages of folding, suggests that the shortening rates have varied through time (Sherkati & Letouzey 2004; Mouthereau *et al.* 2006).

In contrast to the blind reverse faults, the active traces of strike-slip faults are observable. Finite displacements on strike-slip faults are constrained by piercing points, major river offsets and fold offsets. Talebian & Jackson (2002) suggested 50 km of strike-slip offset on the MRF, which, assuming an onset 3–5 Ma ago (by analogy with the North Anatolian Fault), would require a slip rate of $10-17 \text{ mm a}^{-1}$; much larger than the GPS velocity estimate. Lateral offsets of geomorphological markers and *in situ* cosmogenic dating yield an estimated slip rate of $4.9-7.6 \text{ mm a}^{-1}$ on the MRF (Authemayou *et al.* 2009). The other strike-slip fault is the Kazerun Fault System, which we will discuss separately.

Geomorphological record of deformation

Numerous geomorphological markers such as fluvial and marine terraces occur throughout the Central Zagros and can be used to constrain fold kinematics at time scales of 10^4-10^5 years, intermediate between the instantaneous deformation recorded by GPS and seismic studies and the long-term deformation inferred from section balancing. Such markers record incremental deformation and may therefore aid in discriminating between fold models. If they can be dated sufficiently precisely they also constrain deformation rates, which can be transformed into shortening rates using an appropriate fold model.

Oveisi *et al.* (2007, 2009) studied surface deformation as recorded by marine terraces along the coastal Mand anticline, located south of the Borazjan Fault, as well as by fluvial terraces along the Dalaki and Mand rivers, which cross the northwestern Fars east of the Kazerun Fault System. Their results indicate that shortening on Late Pleistocene time scales is concentrated in the frontal part of the belt, consistent with the GPS results discussed above (Fig. 7). Three or four frontal structures appear to absorb practically all of the shortening across the Central Zagros on intermediate time scales. Immediately east of the Kazerun Fault System, the coastal Mand anticline accommodates $3-4 \text{ mm a}^{-1}$ shortening in a NE–SW direction. The Gisakan fold, located at the intersection of the Borazjan Fault and the MFF, also accommodates $2-4 \text{ mm a}^{-1}$ of shortening in the same direction. These two structures together thus account for at least 70% and possibly all of the shortening between the stable Arabian and Iranian platforms. Further to the SE, the situation is slightly more complex, with thin-skinned deformation concentrated on the Halikan fold located inboard of the MFF and only *c.* 10% ($\leq 1 \text{ mm a}^{-1}$) of the shortening taken up on the most frontal structures, such as the coastal Madar anticline.

For the active coastal anticlines, structural data as well as seismic sections preclude significant basement involvement. Instead, these anticlines evolve as open detachment or fault-propagation folds above basal (Hormuz Salt) or intermediate (Gachsaran evaporites) décollement levels. Crustal-scale shortening is fed into these structures either from the MFF or from the most internal parts of the Zagros. Active folds associated with the MFF, in contrast, do suggest basement involvement and occasional fault rupture extending to the surface, as observed at the Gisakan fold. Inboard of the MFF, minor ($<1 \text{ mm a}^{-1}$ along small-scale structures east of the Kazerun Fault) to significant (up to 5 mm a^{-1} for the Halikan anticline) amounts of shortening are absorbed by thin-skinned structures, whereas the surface expressions of major basement faults (e.g. the Surmeh Fault) provide no geomorphological evidence for recent activity.

The total amount of shortening on 10^4-10^5 years time scales, as recorded by geomorphological markers of deformation, is consistent, within error, with the GPS-derived present-day deformation rates of $8-10 \text{ mm a}^{-1}$ across the Zagros. The geomorphological data also show that deformation has been concentrated in the outboard regions of the belt, associated with the MFF and other frontal structures, during Late Quaternary times, and that both thick- and thin-skinned structures are active simultaneously.

The Kazerun Fault System

The Kazerun Fault System (KFS) separates the North Zagros from the Central Zagros (Fig. 1). It comprises several roughly north–south-trending right-lateral strike-slip faults. The Kazerun Fault itself is composed of three north–south-trending segments (Fig. 8): the Dena, Kazerun and Borazjan segments, which all terminate to the south with a north-dipping reverse fault (Authemayou *et al.* 2005, 2006). The Kazerun Fault is associated with exhumation of Hormuz Salt (Talbot & Alavi 1996) and modifies the trend of folds adjacent to it. The KFS, as well as the other north–south-trending faults, is probably inherited from a Cambrian tectonic event that affected the Arabian platform because it controls the distribution of Hormuz Salt, which is present to the east of the fault system but not to the west (Talbot & Alavi 1996; Sepehr & Cosgrove 2005). It was reactivated as early as in the Middle Cretaceous (Koop & Stoneley

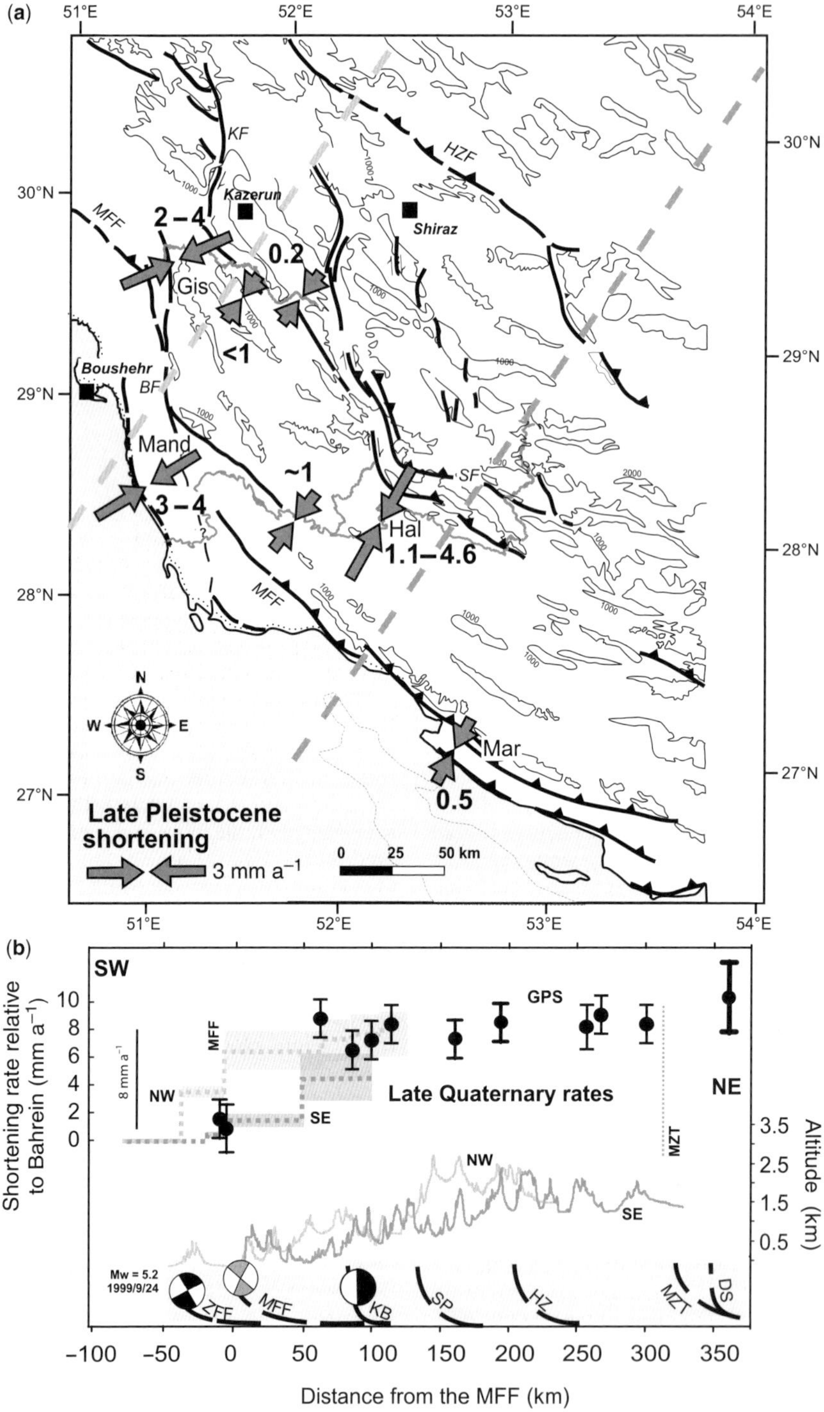

Fig. 7. Summary of the geomorphological observations of Oveisi *et al.* (2007, 2009) (**a**) Map of the Central Zagros showing the inferred shortening rates across various structures (Gis, Gisakan fold; Hal, Halikhan fold; Mand, Mand fold; Mar, Madar fold) as deduced from Late Pleistocene terrace uplift rates (wide shaded arrows, annotated with inferred rate in mm a^{-1}). This pattern should be compared with the pattern of present-day strain rates in Figure 6.

1982). The total offset along the Kazerun Fault is a matter of debate, varying from 5 km (Pattinson & Takin 1971) or 8.2 km (Authemayou et al. 2006) to 140 km (Berberian 1995), depending on the markers used to quantify strike-slip motion. This large difference in displacement results in inferred slip rates of $1-15$ mm a^{-1}. Careful mapping of the active faults and of the lateral offsets along the various segments of the fault (Fig. 9) together with precise dating of fans yields a slip rate of $c.$ $3.1-4.7$ mm a^{-1} on the Dena Fault and $1.5-3.2$ mm a^{-1} on the Kazerun Fault (Authemayou et al. 2009). The southernmost segment, the Borazjan Fault, seems to have a dominant dip-slip motion (e.g. Oveisi et al. 2009). East of the Kazerun Fault, the Kareh-Bas Fault is very active and accommodates $c.$ 5.5 mm a^{-1} of right-lateral strike slip; the Sabz-Pushan Fault in contrast looks inactive, and the Sarvestan Fault accommodates only little motion.

The onset of strike-slip motion on the Main Recent Fault is probably of Late Miocene age and therefore synchronous with the increase in shortening rate within the Zagros and the general tectonic readjustment observed throughout Iran (Allen et al. 2004). The onset of motion on both the Dena and Kazerun segments is more recent, probably $c.$ 3 Ma, and it is much younger ($c.$ $0.8-2.8$ Ma) for the Kareh-Bas Fault (Authemayou 2006; Authemayou et al. 2009).

GPS measurements of 11 benchmarks across the Kazerun Fault System (Fig. 10) allow us to infer slip rates on the various faults with uncertainties of $c.$ 2 mm a^{-1} (Tavakoli et al. 2008). The Dena and Kazerun faults accommodate $c.$ 3.5 mm a^{-1} of right-lateral strike-slip motion. The Borazjan Fault is almost inactive, but the Kareh-bas Fault also accommodates $c.$ 3.5 mm a^{-1} of right-lateral strike-slip motion. A cumulative motion of $c.$ 1.5 mm a^{-1} (within the uncertainties) affects the High Zagros Fault and the Sabz-Pushan Fault. It seems, therefore, that the motion distributes from the Main Recent Fault to the Dena and Kazerun faults, jumps to the Kareh–Bas Fault and distributes slightly on the High Zagros and Sabz–Pushan faults.

The Kazerun Fault System is seismically active (Baker et al. 1993; Berberian 1995; Talebian & Jackson 2004). Clearly, most of the seismicity and especially the largest magnitude earthquakes are located on the central segment of the Kazerun Fault (Fig. 8). The three largest ($M_s > 6$) instrumental earthquakes were located on the Kazerun segment and the Kareh-Bas and Sabz-Pushan faults. Very little activity is observed on both the Dena and Borazjan faults, and no activity is associated with either the High Zagros Fault or the Sarvestan Fault. The depth of the reliably located earthquakes associated with the KFS is 9 ± 4 km, which probably associates them with the basement. Most mechanisms are strike-slip on the Kazerun, Kareh-Bas and Sabz-Pushan faults. Reverse mechanisms are associated with the Mountain Front Fault, on both sides of the Kazerun Fault System. A few reverse mechanisms are also associated with the Borazjan segment, which suggests that it is not an active strike-slip fault but more probably a transpressive lateral ramp (e.g. Oveisi et al. 2009).

Discussion

The separation of the Zagros mountain belt into three longitudinal structural domains (sedimentary, ophiolitic and metamorphic; Ricou et al. 1977) is valid only as a first-order approximation. In a second approximation the Zagros can be divided into two main units along strike, the North Zagros and the Central Zagros (the Fars), separated by the Kazerun Fault System (Berberian 1995; Talebian & Jackson 2004). These two domains show differences in width, in the activity of bounding faults, and in the direction of folding. To further investigate the present-day kinematics of the Zagros, we need to know the relative roles of the basement (and ultimately of the lithosphere) and the surface cover. The present-day kinematics is certainly influenced by both the structure and the tectonic evolution of the fold belt, and therefore should be studied in this perspective. We thus concentrate in this discussion on the comparison of shallow and crustal deformation patterns, both spatially and in time.

Surface deformation

The coupling between surface and basement varies across the Kazerun Fault System. This variation in coupling may induce variations in the response of the surface layer to the deformation. To estimate the shortening of the North Zagros, we use the balanced cross-sections of Blanc et al. (2003) and McQuarrie (2004), because those of Sherkati &

Fig. 7. (*Continued*) BF, Borazjan Fault; HZF, High Zagros Fault; KF, Kazerun Fault; MFF, Main Frontal Fault; SF, Surmeh Fault. Light and dark grey dashed lines indicate locations of transects shown in (**b**). (b) Synthetic profiles of convergence rates (relative to stable Arabia) across the Central Zagros according to GPS and geomorphological data, compared with topographic profiles along a northwestern (light shading) and southeastern (dark shading) transect. Modified from Oveisi et al. (2009).

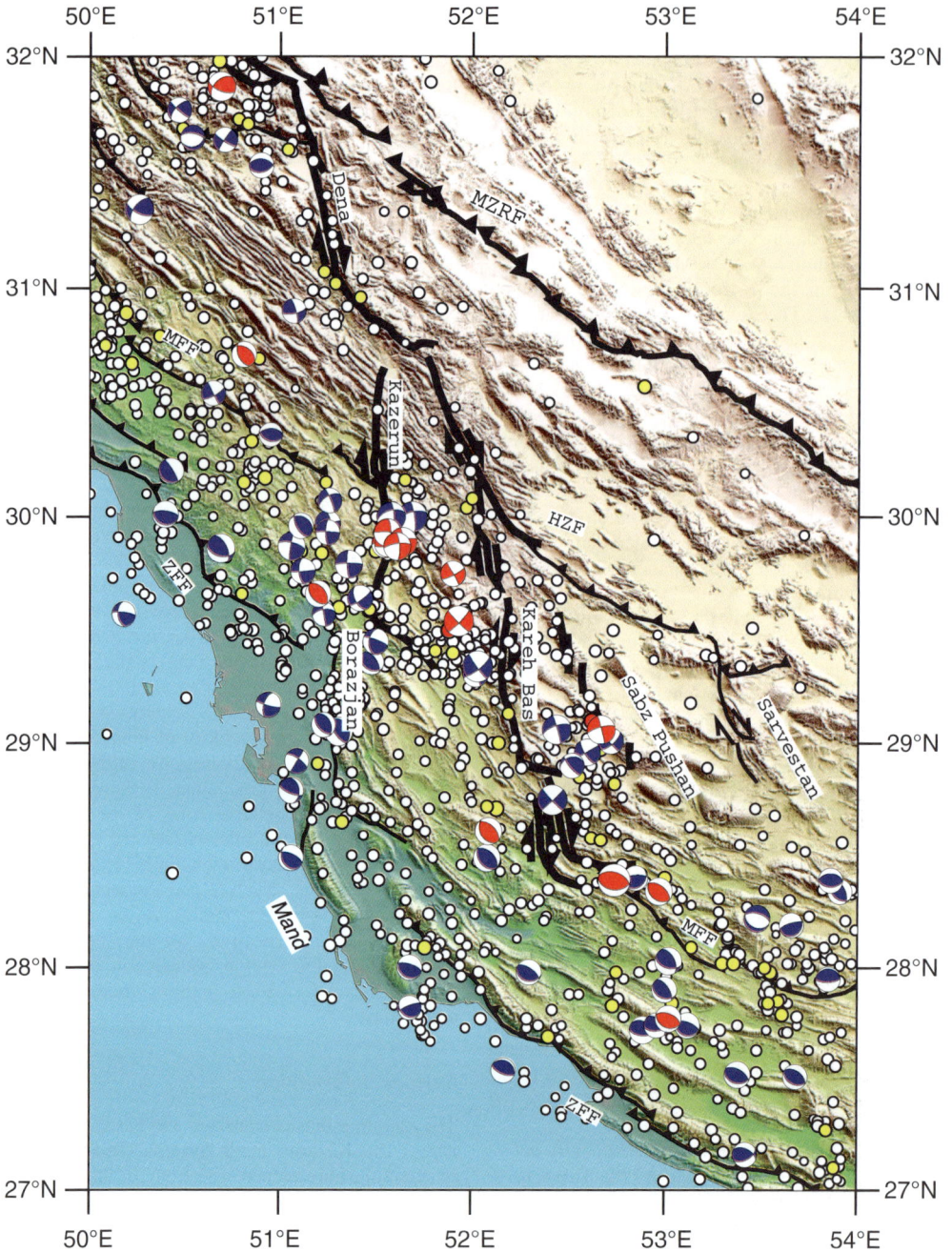

Fig. 8. Detailed seismotectonic map of the Kazerun Fault System. Bold lines indicate the active faults (Authemayou *et al.* 2006) with significant present-day motion (Tavakoli *et al.* 2007). Symbols for seismicity and focal mechanisms are as in Figures 2 and 3. The MZRF appears to be totally inactive. Most seismicity is restricted to the SW of the MFF. Seismicity is associated with the Dena, Kazerun, Kareh-Bas and Sabz-Pushan strike-slip faults.

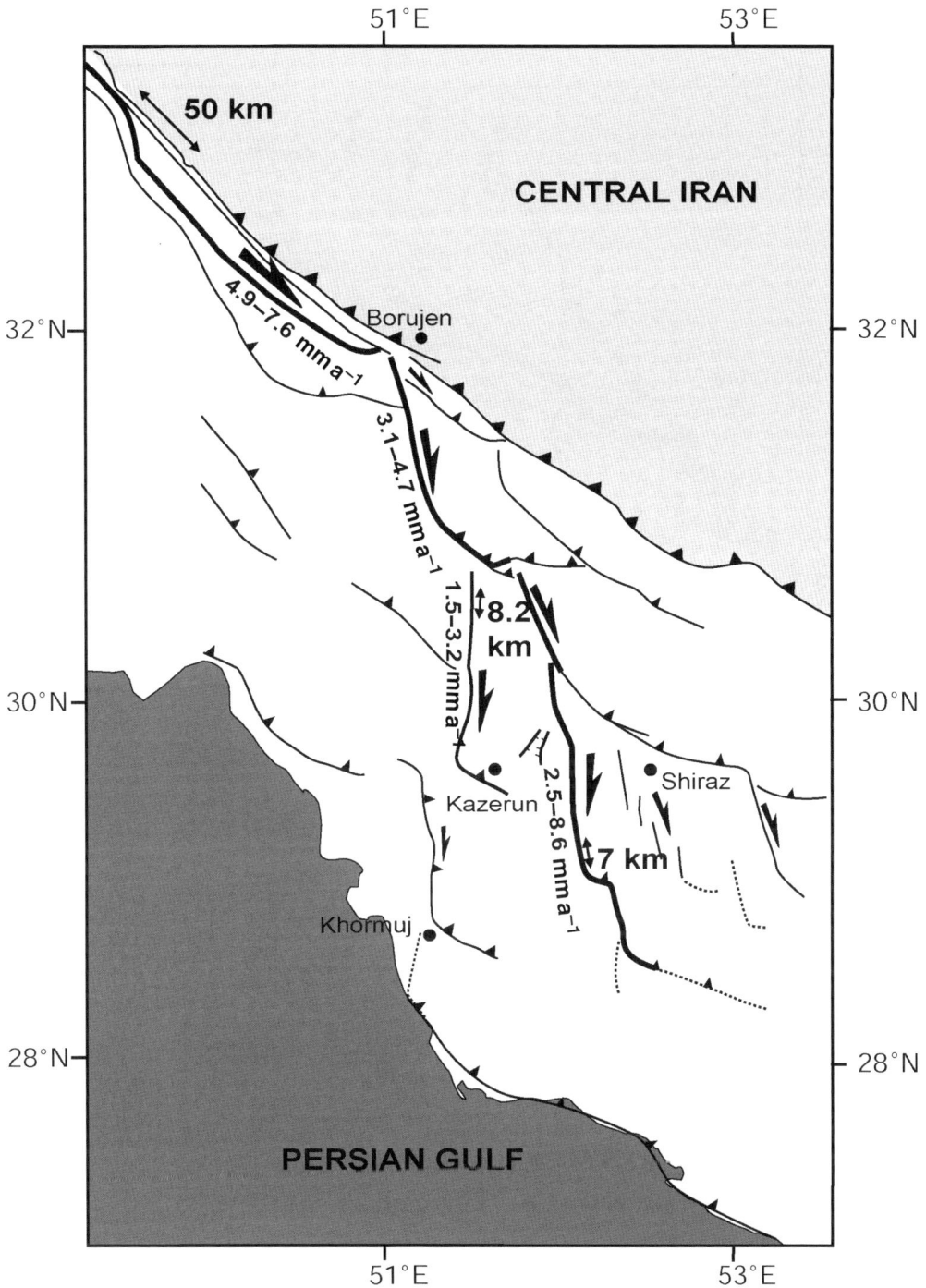

Fig. 9. Quaternary slip rate and finite horizontal displacement, showing the motion distribution from the Main Recent Fault to the Kazerun Fault System (after Authemayou *et al.* 2006).

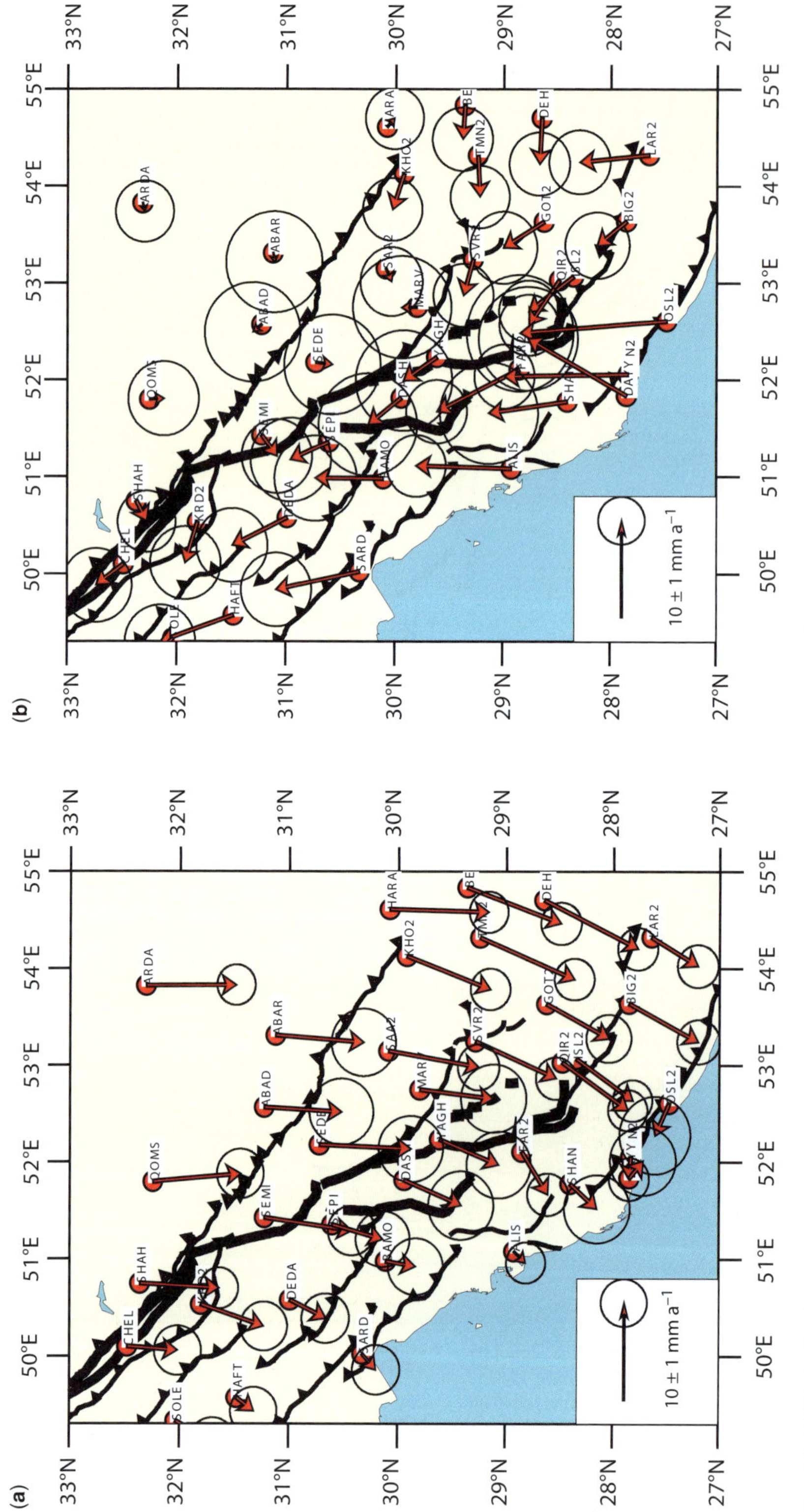

Fig. 10. GPS velocity for benchmarks located near the Kazerun Fault System (after Tavakoli *et al.* 2007). (**a**) Motion relative to Arabia; (**b**) motion relative to Central Iran.

Letouzey (2004) cross the Kazerun Fault and may not be representative of the shortening of the whole Zagros. For the Fars region, we use the cross-section of McQuarrie (2004), which is the only section that really crosses Fars, the section of Molinaro *et al.* (2004) being located at the Zagros–Makran transition. Paradoxically, the total amount of shortening is larger in the North Zagros than in Fars, both for the whole Zagros (from 57 to 85 km) and for the Zagros Fold Belt (from 35 to 50 km), even though the Fars is located further from the long-term Arabia–Central Iran pole of rotation located at 29.8°N, 35.1°E. This variation in finite shortening could be explained by an under-estimate of the displacement along the suture zone in the Central Zagros by McQuarrie (2004), or by an earlier onset of deformation in the North Zagros compared with the Central Zagros as a result of the progressive southeastward closure of the Neotethys associated with the anti-clockwise rotation of the Arabian plate.

The GPS measurements also show a difference in present-day deformation across the Kazerun Fault System (Walpersdorf *et al.* 2006). In contrast to the total shortening, the present-day shortening rates increase slightly from the North Zagros (4–6 mm a^{-1}) to the Fars (8 mm a^{-1}), consistent with the increasing distance to the pole of rotation. The strike-slip component is mostly localized on the Main Recent Fault in the North Zagros but seems to be smaller and distributed in Fars. Both in the North Zagros and in the Fars, shortening seems to be concentrated between the 1000 m elevation topography and sea level.

Geomorphological observations suggest that the folds located at the shore of the Persian Gulf are the most active structures of the Zagros. This is consistent with the GPS measurements showing that most of the present-day shortening in Fars is also accommodated at the shore. This present-day activity located at the edge of the Zagros fold belt, along the Persian Gulf shore, is consistent with the southwestward propagation of the front of the Simply Folded Belt from the Eocene (and therefore earlier than the onset of collision) to the present time (Shearman 1977; Hessami *et al.* 2001).

Basement deformation

The debate concerning thick-skinned and thin-skinned models for Zagros fold belt deformation may never find a satisfactory answer because of the lack of seismic profiles reaching the basement. The only reliably (on the base of balanced cross-sections) inferred basement reverse faults are the HZF and the MFF (Blanc *et al.* 2003; Sherkati & Letouzey 2004; Bosold *et al.* 2005) because they clearly offset the sedimentary sequence and are

controlled by seismic reflection profiles. The Zagros Frontal Fault itself generally does not propagate to the surface through the sedimentary cover, although a few surface breaks have been described (Bachmanov *et al.* 2004; Oveisi *et al.* 2009).

The seismicity associated with shortening and reverse mechanisms is mostly located in the Zagros Fold Belt (Fig. 11). Therefore neither the MZT nor the HZF is active or both are lubricated and slip aseismically. This seems true both for the North Zagros, where the only large earthquakes located north of the HZF belong to the strike-slip MRF, and for the Fars, where the seismic inactivity of these two faults is consistent with the absence of surface motion from GPS measurements across them. More precisely, the seismicity associated with reverse mechanisms is restricted to topography less than 1000 m, as pointed out by Talebian & Jackson (2004). This could be due to the gradient in topography (Talebian & Jackson 2004) but we suspect it is related to the propagation of the deformation front to the SW, as evidenced from structural studies (Sherkati & Letouzey 2004), geomorphology and GPS. The two could be linked, however, if we consider a critical-wedge model for the evolution of the Zagros Fold Belt (e.g. Mouthereau *et al.* 2006). This propagation of deformation, and therefore of the construction of topography, explains why seismicity is bounded by the Persian Gulf shore (Fig. 12), even though this shoreline has no tectonic significance and the water depth in the Persian Gulf is less than 70 m.

The relation between seismicity and surface faults differs between the North Zagros and the Fars arc (Fig. 11). In the North Zagros, seismicity is restricted to a narrow band limited by the 1000 m elevation contour, which is also the trace of the MFF. Because the topography is relatively steep, the relation between the 1000 m contour and the MFF is clear. The seismicity does not fit totally with the distribution of GPS shortening, which also affects the low topography north of the Persian Gulf. However, because GPS deformation there is controlled only by the station KHOS (Fig. 5a) and no folding or topography generation is observed in the lowland, this frontal shortening remains to be confirmed.

In Fars, seismicity is spread throughout the area between 1000 m elevation and the shore (which might be related to the MFF and the ZFF, respectively); the zone of seismicity is wider than in the North Zagros but does not encompass the entire width of the fold belt. The gradient in topography is also smoother in Fars than in the North Zagros. GPS shortening is restricted to the shore and unrelated to the high elevation.

Thus, both the seismicity and the gradient in topography (which record basement deformation)

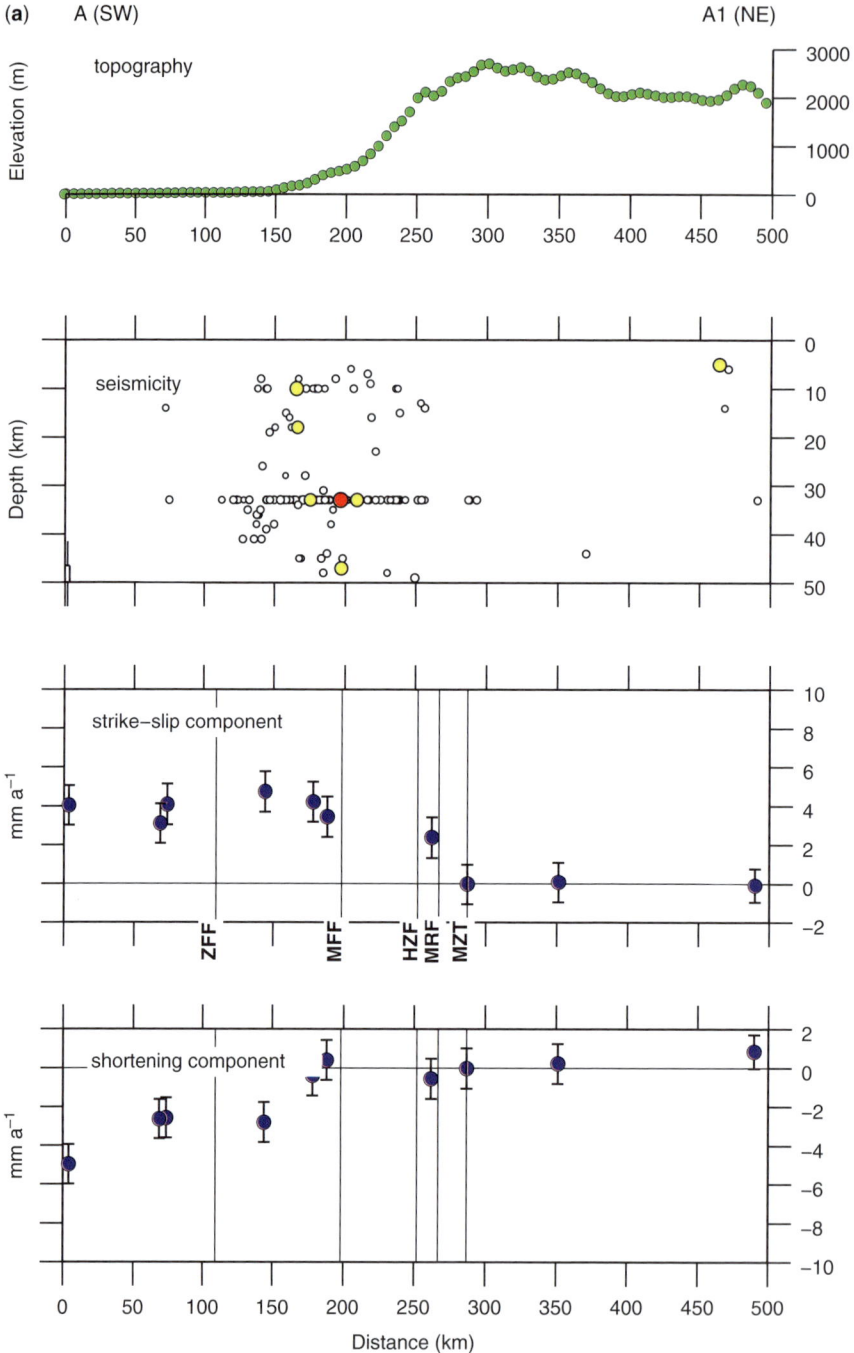

Fig. 11. (**a, b**) Cross-sections through the North and Central Zagros (see location in Fig. 2) displaying for each the topography, seismicity, present-day GPS-detected motion parallel to the mountain belt, and present-day shortening perpendicular to the mountain belt. Symbols for seismicity are as in Figure 2. The present-day motion is from GPS-determined velocities relative to Central Iran. We plot the location of the main faults (Berberian 1995). There is a strong correlation between the gradient in topography, the seismicity (relative to the basement deformation) and the shallow deformation. In the North Zagros, the strike-slip motion is concentrated near the MRF, whereas it is more distributed in the Central Zagros.

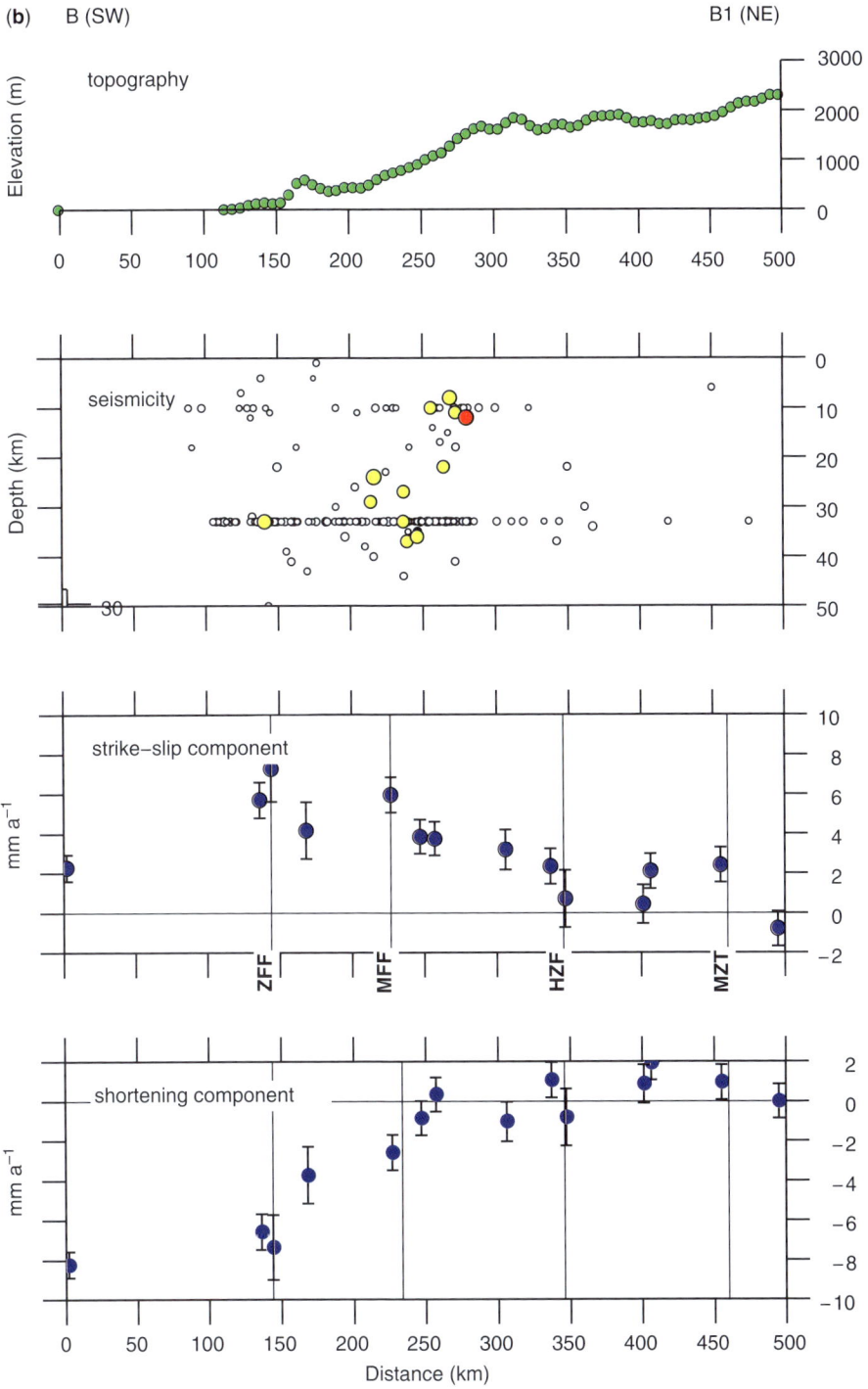

(b) B (SW)   B1 (NE)

topography

seismicity

strike−slip component

ZFF   MFF   HZF   MZT

shortening component

Distance (km)

Fig. 11.

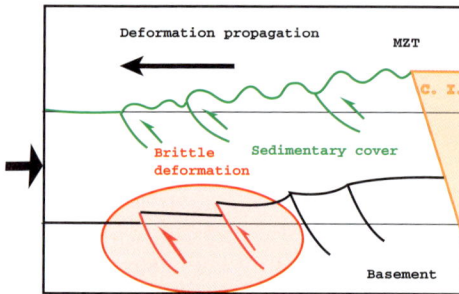

Fig. 12. Sketch summarizing our results and interpretation. C.I., Central Iran; MZT, Main Zagros Thrust. Both the shallow deformation of the sedimentary cover and the brittle deformation of the basement are associated with the gradient in topography, suggesting that they are related. Faulting in the basement is unrelated to faulting and folding in the sedimentary cover. Because we know the shallow deformation propagated southwestward with time, we suspect the basement deformation to do the same.

are correlated with the pattern of cumulative (on a million years scale) deformation. On the other hand, GPS shortening and geomorphology (which record shallow deformation) are concentrated at the front of the deformation.

Less than 10% of the total deformation is released by earthquakes. However, there is a remarkable good fit in the directions of the tensor of deformation computed from both the GPS measurements and the seismological catalogues (Masson *et al.* 2005). This deficit could mean that some faults slip aseismically. An alternative and complementary explanation is that seismicity is restricted between 10 and 15 km depth because of the thick sedimentary cover, which limits the thickness of the brittle part of the crust to 5 km only (rather than 15–18 km as usual). The stress accumulated from boundary conditions is released by seismic energy for the brittle part but also by ductile deformation both for the sedimentary cover (by folding) and by lower crustal flow. If the brittle part of the crust is 30% of the usual thickness, we expect only 30% of seismic energy release.

Significance of the Kazerun Fault System

The tectonics of the Kazerun Fault System is more complex than it looks first. The KFS is generally interpreted as an inherited fracture of an old tectonic event affecting the Arabian platform. Such inherited fractures are observed in several places in both the Zagros and the Arabian platform across the Persian Gulf, whereas we observe motion and seismicity only on part of the fractures located within the Zagros and only around the Kazerun

zone. This focusing of seismicity could be due either to a non-homogeneous state of stress within the Zagros or to the Zagros part of the Arabian platform being more brittle (it is thinner) than the remaining part.

These inherited fractures were activated during Permian and Mesozoic sedimentation, resulting in a change of the mechanical behaviour of the lithostratigraphic horizons. During collision, because the Kazerun Fault System marks the boundary of the Hormuz Salt layer in the Central Zagros, the fault plays the role of a lateral ramp for the Fars arc. A lateral ramp generally implies transpressional motion as observed along the Borazjan segment, which can be interpreted as the active part of the Kazerun Fault lateral ramp. The southward propagation of this segment can be detected by a structural study of the Mand anticline. The bending of this large coastal anticline suggests the presence of a hidden segment of the Kazerun Fault System bounding the Mand fold to the west. As the Mand anticline is a Plio-Quaternary fold, the propagation of the Kazerun Fault lateral ramp must be very recent.

If the Kazerun Fault is a lateral ramp of the Fars arc, the fault motion must be restricted to the cover. However, the seismic activity localized along the Kazerun segment implies basement faulting because earthquakes are probably located in the basement, and thus an important role for the Kazerun Fault System in the Zagros deformation.

We observe an important contrast in the style of deformation west and east of the KFS. To the west, the belt is narrow and the deformation is partitioned between the strike-slip MRF and the shortening. To the east, the belt is wider, the deformation is more localized than in the west, and the MRF spreads into several strike-slip faults that look like a large distributed en echelon system (Dena, Kazerun–HZF, HZF–Kareh-Bas–Sabz-Pushan). In fact, the Kazerun Fault System is connected to the MRF (Authemayou *et al.* 2005). Consequently, since the Pliocene, the right-lateral strike-slip motion from the MRF has been distributed onto several north–south- to NNE–SSW-trending strike-slip faults that are part of the Kazerun Fault System. The Dena Fault connects to both the Kazerun and the High Zagros faults, the High Zagros Fault connects to both the Sabz-Pushan and the Sarvestan faults, and the Kazerun Fault connects to both the Kareh-Bas and Borazjan faults. The connection between the MRF and the KFS has been attributed to the existence of inherited fractures (which were ultimately reactivated as the KFS) disturbing and stopping eastward propagation of slip on the MRF. The presence of Hormuz Salt limited to the east of the Kazerun Fault may facilitate the diffusion of deformation above a ductile layer and thus the slip motion. However, the existence of the Hormuz Salt cannot explain on its own the distribution of motion, because some of these faults (Kareh-Bas, Sabz-Pushan) are also seismically very active. Furthermore, our GPS results do not support a 'spreading' pattern of deformation for the Kazerun Fault System similar to gravity spreading as claimed by Nilforoushan & Koyi (2007) on the basis of analogue experiments. They predicted a divergent motion of the GPS vectors relative to Arabia, as reported by Hessami et al. (2006), but that does not correspond to our observations. We think that the distribution of deformation from the MRF to the Kazerun Fault System affects both the shallow sediments and the basement beneath the ductile layer.

Partitioning

Partitioning is one of the mechanisms that accommodate oblique motion (e.g. Fitch 1972). Usually, strike-slip and reverse motion occur on two parallel faults that are a few tens of kilometres apart. In continental areas, it is likely that pre-existing faults localize the deformation because they are weak (e.g. Zoback et al. 1987). It has also been proposed that a ductile layer decouples the oblique motion (Richard & Cobbold 1989) and helps partitioning. However, we observe partitioning of oblique convergence between shortening perpendicular to the belt and strike-slip motion on the MRF to the west of the Kazerun Fault System only, where the coupling between sediments and basement is strongest. Therefore, a ductile layer is probably not responsible for deformation partitioning in the North Zagros. We suspect instead that the MRF introduces a weak discontinuity that localizes strike-slip motion and, as a consequence, favours partitioning.

Vernant & Chéry (2006) designed a numerical mechanical model to explain the oblique convergence in the Zagros. They suggested low partitioning along the MRF ($1-2$ mm a^{-1}) associated with transpressionnal deformation throughout the belt. In contrast to their model predictions, GPS strike-slip motion is slightly higher ($2-4$ mm a^{-1}) and geomorphological slip rate estimates on the MRF appear to match nearly completely the strike slip component of convergence between Arabia and Central Iran. Fault kinematic measurements along the HZF, south of the MRF, indicate a transpressional regime on this fault (Malekzadeh 2007). If partitioning exists, the shortening that complements the minimum Quaternary slip rate on the MRF of $4.9-7.6$ mm a^{-1} (Authemayou et al. 2009) must be accommodated somewhere else. However, the fast slip rate along the MRF probably suggests a very weak MRF with a lower friction coefficient than adopted by Vernant & Chéry (2006), or possibly strong decoupling of the surface from the basement, rendering a model without mechanical layering somewhat irrelevant.

Conclusion

Our first conclusion is that we find, on both sides of the KFS, a good correlation between present-day surface deformation, as measured by GPS and geomorphology on one hand, and seismicity (affecting only the upper basement) and topography on the other hand (Fig. 11), suggesting that both the sedimentary cover and the basement deform together (i.e. a thick-skinned system). Because we know that deformation of the sedimentary cover propagates southwestward, we suspect basement deformation, which is required to explain the average topography, to do the same (Fig. 12). In contrast to Hessami et al. (2006), we do not observe any active shortening across the southern segment of the MZT. Thus, the reason for such propagation is probably the recent locking of the continental collision, propagating the stress away from the MZT onto inherited normal faults of the Arabian platform (Jackson 1980a). Because the strike of the belt is perpendicular to the motion of Arabia relative to Central Iran, no partitioning is required in the Central Zagros (Talebian & Jackson 2004).

The second conclusion is that the Kazerun Fault System separates the North Zagros (experiencing slip partitioning), from the Central Zagros (experiencing distributed deformation), as proposed previously. There is a good agreement between present-day deformation observed by GPS and tectonic observations, suggesting that this deformation has been stable for some time. The Kazerun Fault System distributes the strike-slip motion from the MRF onto different branches in an en echelon arrangement, from the Dena segment to the Sabz-Pushan and High Zagros faults. The presence of the decoupling Hormuz Salt layer cannot be the only reason for such distribution, because seismicity is associated with the active faults, indicating that the basement deforms in the same way. Consequently, the Kazerun Fault System affects both the sedimentary cover and the basement, playing the role of a lateral ramp of the deformation front for its southern Borazjan segment and of a 'horse-tail' termination of the MRF for its northern and central segments.

This work is part of a Franco-Iranian collaborative programme between French and Iranian scientific institutions conducted between 1997 and 2007. It has been funded by CNRS–INSU, the French Embassy in Tehran, the International Institute of Earthquake Engineering and Seismology of Iran, the Geological Survey of Iran and the National Cartographic Center of Iran. We warmly thank all the people who enthusiastically participated in the fieldwork. We are considerably indebted to all drivers of Iranian institutions who spent countless time to help in all aspects of the fieldwork. We thank M. Goraishi, M. Ghafory-Ashtiany, M. Madad and P. Vidal for encouragement and support. We benefited from numerous scientific discussions with several colleagues and especially P. Agard, J. Jackson, L. Jolivet, J. Letouzey, L.-E. Ricou and M. Talebian. The paper greatly benefited from careful reviews by R. Bendick and J. Jackson.

References

AGARD, P., OMRANI, J., JOLIVET, L. & MOUTHEREAU, F. 2005. Convergence history across Zagros (Iran): constraints from collisional and earlier deformation. *International Journal of Earth Sciences*, doi: 10.1007/s00531-005-0481-4.

ALLEN, M., JACKSON, J. & WALKER, R. 2004. Late Cenozoic re-organisation of the Arabia–Eurasia collision and the comparison of short-term and long-term deformation rates. *Tectonics*, **23**, doi: 10.1029/2003TC001530.

AMBRASEYS, N. N. 1978. The relocation of epicenters in Iran. *Geophysical Journal of the Royal Astronomical Society*, **53**, 117–121.

AUTHEMAYOU, C. 2006. *Partitionnement de la convergence oblique en zone de collision: exemple de la chaîne du Zagros (Iran)*. PhD thesis, Université Paul Cézanne, Aix–Marseille.

AUTHEMAYOU, C., BELLIER, O., CHARDON, D., MALEKZADEH, Z. & ABBASSI, M. 2005. Active partitioning between strike-slip and thrust faulting in the Zagros fold-and-thrust belt (Southern Iran). *Comptes Rendus Géosciences*, **337**, 539–545.

AUTHEMAYOU, C., CHARDON, D., BELLIER, O., MALEKZADEH, Z., SHABANIAN, E. & ABBASSI, M. R. 2006. Late Cenozoic partitioning of oblique plate convergence in the Zagros fold-and-thrust belt (Iran). *Tectonics*, **25**, article number TC3002.

AUTHEMAYOU, C., BELLIER, O., CHARDON, D., BENEDETTI, L., MALEKZADE, Z., CLAUDE, C., ANGELETTI, B., SHABANIAN, E. & ABBASSI, M. R. 2009. Quaternary slip-rates of the Kazerun and the Main Recent Faults: active strike-slip partitioning in the Zagros fold-and-thrust belt. *Geophysical Journal International*, **178**, 524–540.

BACHMANOV, D. M., TRIFONOV, V. G. *ET AL.* 2004. Active faults in the Zagros and central Iran. *Tectonophysics*, **380**, 221–241.

BAKER, C., JACKSON, J. & PRIESTLEY, K. 1993. Earthquakes on the Kazerun Line in the Zagros Mountains of Iran: strike-slip faulting within a fold-and-thrust belt. *Geophysical Journal International*, **115**, 41–61.

BERBERIAN, M. 1979. Evaluation of the instrumental and relocated epicenters of Iranian earthquakes. *Geophysical Journal of the Royal Astronomical Society*, **58**, 625–630.

BERBERIAN, M. 1995. Master blind thrust faults hidden under the Zagros folds: active basement tectonics and surface morphotectonics. *Tectonophysics*, **241**, 193–224.

BIRD, P., TOKSOZ, M. & SLEEP, N. 1975. Thermal and mechanical models of continent-continent convergence zones. *Journal of Geophysical Research*, **80**, 4405–4416.

BLANC, E. J.-P., ALLEN, M. B., INGER, S. & HASSANI, H. 2003. Structural styles in the Zagros Simple Folded Zone, Iran. *Journal of the Geological Society, London*, **160**, 401–412.

BOSOLD, A., SCHWARZHANS, W., JULAPOUR, A., ASHRAZADEH, A. R. & EHSANI, S. M. 2005. The structural geology of the High Central Zagros revisited (Iran). *Petroleum Geosciences*, **11**, 225–238.

ENGDAHL, E. R., VAN DER HILST, R. D. & BULAND, R. P. 1998. Global teleseismic earthquake relocation with improved travel times and procedures for depth determination. *Bulletin of the Seismological Society of America*, **88**, 722–743.

ENGDAHL, E. R., JACKSON, J. A., MYERS, S. C., BERGMAN, E. A. & PRIESTLEY, K. 2006. Relocation and assessment of seismicity in the Iran region. *Geophysical Journal International*, **167**, 761–778.

FITCH, T. J. 1972. Plate convergence, transcurrent faults, and internal deformation adjacent to Southeast Asia and the Western Pacific. *Journal of Geophysical Research*, **77**, 4432–4460.

HAYNES, S. J. & MCQUILLAN, H. 1974. Evolution of the Zagros suture zone, southern Iran. *Geological Society of America Bulletin*, **85**, 739–744.

HESSAMI, K., KOYI, H., TALBOT, C. J., TABASI, H. & SHABANIAN, E. 2001. Progressive unconformities within and evolving foreland fold–thrust belt, Zagros Mountains. *Journal of the Geological Society, London*, **158**, 969–981.

HESSAMI, K., NILFOROUSHAN, F. & TALBOT, C. 2006. Active deformation within the Zagros Mountains deduced from GPS measurements. *Journal of the Geological Society, London*, **163**, 143–148.

JACKSON, J. A. 1980a. Reactivation of basement faults and crustal shortening in orogenic belts. *Nature*, **283**, 343–346.

JACKSON, J. 1980b. Errors in focal depth determination and depth of seismicity in Iran and Turkey. *Geophysical Journal of the Royal Astronomical Society*, **61**, 285–301.

JACKSON, J. & FITCH, T. 1981. Basement faulting and the focal depths of the larger earthquakes in the Zagros mountains (Iran). *Geophysical Journal of the Royal Astronomical Society*, **64**, 561–586.

JACKSON, J. A. & MCKENZIE, D. 1984. Active tectonics of the Alpine–Himalayan Belt between western Turkey and Pakistan. *Geophysical Journal of the Royal Astronomical Society*, **77**, 185–264.

JACKSON, J. & MCKENZIE, D. 1988. The relationship between plate motions and seismic moment tensors and the rates of active deformation in the Mediterranean and Middle East. *Geophysical Journal of the Royal Astronomical Society*, **83**, 45–73.

KING, R. W. & BOCK, Y. 2002. *Documentation for the GAMIT analysis software, release 10.1*. Massachusetts Institute of Technology, Cambridge, MA.

KOOP, W. J. & STONELEY, R. 1982. Subsidence history of the Middle East Zagros Basin, Permian to Recent. *Philosophical Transactions of the Royal Society of London*, **305**, 149–168.

MAGGI, A., JACKSON, J. A., PRIESTLEY, K. & BAKER, C. 2000. A re-assessement of focal depth distributions in Southern Iran, the Tien Shan and Northern India; Do earthquakes really occur in the continental mantle? *Geophysical Journal International*, **143**, 629–661.

MALEKZADEH, Z. 2007. *The accommodation of the deformation from Main Recent Fault to Kazerun*. PhD thesis, Institute of Earthquake Engineering and Seismology, Tehran.

MASSON, F., CHÉRY, J., HATZFELD, D., MARTINOD, J., VERNANT, P., TAVAKOLI, F. & GHAFORY-ASTHIANI, M. 2005. Seismic versus aseismic deformation in Iran inferred from earthquakes and geodetic data. *Geophysical Journal International*, **160**, 217–226.

MASSON, F., ANVARI, M. ET AL. 2007. Large-scale velocity field and strain tensor in Iran inferred form GPS measurements: new insight for the present-day deformation pattern within NE Iran. *Geophysical Journal International*, **170**, 436–440.

MCKENZIE, D. P. 1978. Active tectonics of the Alpine–Himalayan belt: the Aegean Sea and surrounding regions. *Geophysical Journal of the Royal Astronomical Society*, **55**, 217–254.

MCQUARRIE, N. 2004. Crustal scale geometry of the Zagros fold–thrust belt, Iran. *Journal of Structural Geology*, **26**, 519–535.

MCQUARRIE, N., STOCK, J. M., VERDEL, C. & WERNICKE, B. P. 2003. Cenozoic evolution of Neotethys and implications for the causes of plate motion. *Geophysical Research Letters*, **30**, doi: 10.1029/2003GL017992.

MOLINARO, M., GUEZOU, J. C., LETURMY, P., ESHRAGHI, S. A. & FRIZON DE LAMOTTE, D. 2004. The origin of changes in structural style across the Bandar Abbas syntaxis, SE Zagros (Iran). *Marine and Petroleum Geology*, **21**, 735–752.

MOLINARO, M., LETURMY, P., GUEZOU, J.-C. & FRIZON DE LAMOTTE, D. 2005. The structure and kinematics of the south-eastern Zagros fold–thrust belt, Iran: from thin-skinned to thick-skinned tectonics. *Tectonics*, **24**, NIL42–NIL60.

MOLNAR, P. & LYON-CAEN, H. 1988. Some simple physical aspects of the support, structure, and evolution of mountain belts. *In*: CLARK, S. P., BURCHFIEL, B. C. & SUPPE, J. (eds) *Processes in Continental Lithospheric Deformation*. Geological Society of America, Special Papers, **218**, 179–207.

MOUTHEREAU, F., LACOMBE, O. & MEYER, B. 2006. The Zagros folded belt (Fars, Iran): constraints from topography and critical wedge modeling. *Geophysical Journal International*, **165**, 336–356.

NI, J. & BARAZANGI, M. 1986. Seismotectonics of the Zagros Continental Collision Zone and a Comparison with the Himalayas. *Journal of Geophysical Research*, **91**, 8205–8218.

NIAZI, M. 1980. Microearthquakes and crustal structure off the Makran coast of Iran. *Geophysical Research Letters*, **7**, 297–300.

NIAZI, M., ASUDEH, G., BALLARD, G., JACKSON, J. A., KING, G. & MCKENZIE, D. P. 1978. The depth of seismicity in the Kermansha region of the Zagros mountains (Iran). *Earth and Planetary Science Letters*, **40**, 270–274.

NILFOROUSHAN, F. & KOYI, H. A. 2007. Displacement fields and finite strains in a sandbox model simulating a fold–thrust belt. *Geophysical Journal International*, **169**, 1341–1355.

NILFOROUSHAN, F., VERNANT, P. ET AL. 2003. GPS network monitors the Arabia–Eurasia collision deformation in Iran. *Journal of Geodesy*, **77**, 411–422.

NOWROOZI, A. 1971. Seismotectonics of the Persian Plateau Eastern Turket, Caucasus, and Hindu-Kush region. *Bulletin of the Seismological Society of America*, **61**, 317–341.

OVEISI, B., LAVÉ, J. & VAN DE BEEK, P. 2007. Rates and processes of active folding evidenced by Pleistocene terraces at the central Zagros front (Iran). *In*: LACOMBE, O., LAVÉ, J., ROURE, F. & VERGÈS, J. (eds) *Thrust Belt and Foreland Basin*. Frontiers in Earth Sciences. Springer-Verlag, New York, 265–285.

OVEISI, B., LAVÉ, J., VAN DER BEEK, P., CARCAILLET, J., BENEDETTI, L., BRAUCHER, R. & AUBOURG, C. 2009. Thick- and thin-skinned deformation rates in the Zagros Simple Folded Zone (Iran) indicated by displacement of geomorphic surfaces. *Geophysical Journal International*, **176**, 627–654.

PATTINSON, R. & TAKIN, M. 1971. Geological significance of the Dezful embayment boundaries, National Iran Oil Co, Report 1166 (unpublished).

RICHARD, P. & COBBOLD, P. 1989. Structures en fleur positives et décrochements crustaux: modélisation analogique et interprétation mécanique. *Comptes Rendus de l'Académie des Sciences*, **308**, 553–560.

RICOU, L. E., BRAUD, J. & BRUNN, J. H. 1977. Le Zagros. Mémoires Hors Série, Société Géologique de France, **8**, 33–52.

SAVAGE, W. U., ALT, J. N. & MOHAJER-ASHJAL, A. 1977. Microearthquake investigations of the 1972 Qir, Iran, earthquake zone and adjacent areas. *Geological Society of America*, Abstract 9496.

SEPEHR, M. & COSGROVE, J. W. 2005. Role of the Zazerun Fault Zone in the formation and deformation Lof the Zagros Fold-Thrust Belt, Iran. *Tectonics*, **24**, 1–15.

SHEARMAN, D. J. 1977. The geological evolution of Southern Iran. *Geographical Journal*, **142**, 393–410.

SHERKATI, S. & LETOUZEY, J. 2004. Variation of structural and basin evolution in the central Zagros (Izeh zone and Dezful Embayment), Iran. *Marine and Petroleum Geology*, **21**, 535–554.

SHERKATI, S., MOLINARO, M., FRIZON DE LAMOTTE, D. & LETOUZEY, J. 2005. Detachment folding in the Central and Eastern Zagros fold-belt (Iran): salt mobility, multiple detachments and late basement control. *Journal of Structural Geology*, **27**, 1680–1696.

SNYDER, D. B. & BARAZANGI, M. 1986. Deep crustal structure and flexure of the Arabian plate beneath the Zagros collisional mountain belt as inferred from gravity observations. *Tectonics*, **5**, 361–373.

STÖCKLIN, J. 1974. Possible ancient continental margin in Iran. *In*: BURKE, C. & DRAKE, C. (eds) *Geology of Continental Margins*. Springer-Verlag, New York, 873–877.

STONELEY, R. 1981. The geology of the Kuh-e Dalneshin area of southern Iran, and its bearing on the evolution of southern Tethys. *Journal of the Geological Society, London*, **138**, 509–526.

TALBOT, C. J. & ALAVI, M. 1996. The past of a future syntaxis across the Zagros. *In*: ALSOP, G. I., BLUNDELL, D. J. & DAVISON, I. (eds) *Salt Tectonics*. Geological Society, London, Special Publications, **100**, 89–110.

TALEBIAN, M. & JACKSON, J. 2002. Offset on the Main Recent Fault of NW Iran and implications for the late Cenozoic tectonics of the Arabia–Eurasia collision zone. *Geophysical Journal International*, **150**, 422–439.

TALEBIAN, M. & JACKSON, J. 2004. A reappraisal of earthquake focal mechanisms and active shortening in the Zagros mountains of Iran. *Geophysical Journal International*, **156**, 506–526.

TATAR, M., HATZFELD, D., MARTINOD, J., WALPERSDORF, A., GHAFORI-ASHTIANY, M. & CHÉRY, J. 2002. The present-day deformation of the central Zagros (Iran) from GPS measurements. *Geophysical Research Letters*, **29**, 1927–1930.

TATAR, M., HATZFELD, D. & GHAFORY-ASHTIANY, M. 2003. Tectonics of the Central Zagros (Iran) deduced from microearthquake seismicity. *Geophysical Journal International*, **156**, 255–266.

TAVAKOLI, F., WALPERSDORF, A. *ET AL*. 2008. Distribution of the right-lateral strike-slip motion from the Main Recent Fault to the Kazerun Fault System (Zagros, Iran): evidence from present-day GPS velocities. *Earth and Planetary Science Letters*, doi: 10.1016/j.espl.2008.08.030.

VERNANT, P. & CHERY, J. 2006. Mechanical modelling of oblique convergence in the Zagros, Iran. *Geophysical Journal International*, **165**, 991–1002.

VERNANT, P., NILFOROUSHAN, F. *ET AL*. 2004. Contemporary crustal deformation and plate kinematics in the Middle East Constrained by GPS measurements in Iran and Northern Oman. *Geophysical Journal International*, **157**, 381–398.

WALKER, R., ANDALIBI, M., GHEITANCHI, M., JACKSON, J., KAREGAR, S. & PRIESTLEY, K. 2005. Seismological and field observations from the 1990 November 6 Furg (Hormozgan) earthquake: a rare case of surface rupture in the Zagros mountains of Iran. *Geophysical Journal International*, **163**, 567–579.

WALPERSDORF, A., HATZFELD, D. *ET AL*. 2006. Difference in the GPS deformation pattern of North and Central Zagros (Iran). *Geophysical Journal International*, **167**, 1077–1088.

YAMINI-FARD, F., HATZFELD, D., TATAR, M. & MOKHTARI, M. 2006. Microseismicity at the intersection between the Kazerun fault and the Main Recent Fault (Zagros-Iran). *Geophysical Journal International*, **166**, 186–196.

YAMINI-FARD, F., HATZFELD, D., FARAHBOD, A., PAUL, A. & MOKHTARI, M. 2007. The diffuse transition between the Zagros continental collision and the Makran oceanic subduction (Iran): microearthquake seismicity and crustal structure. *Geophysical Journal International*, **170**, 182–194.

ZOBACK, M. D., ZOBACK, M. L. *ET AL*. 1987. New evidence on the state of stress of the San Andreas Fault System. *Science*, **238**, 1105–1111.

The transition between Makran subduction and the Zagros collision: recent advances in its structure and active deformation

V. REGARD[1]*, D. HATZFELD[2], M. MOLINARO[3,4], C. AUBOURG[3], R. BAYER[5],
O. BELLIER[6], F. YAMINI-FARD[7], M. PEYRET[5] & M. ABBASSI[7]

[1]LMTG, Université de Toulouse–CNRS–IRD–OMP, 14 av. E. Belin, 31400 Toulouse, France

[2]LGIT, Maison des Géosciences, BP 53, 38400 Grenoble Cedex 9, France

[3]Laboratoire de Tectonique, CNRS, Université de Cergy-Pontoise, Cergy, France

[4]Present address: Shell International Exploration and Production B.V., Rijswijk, Netherlands

[5]Géosciences Montpellier, UMR 5243–CC 60, Université Montpellier 2, Place E. Bataillon,
34095 Montpellier cedex 5, France

[6]CEREGE, Aix–Marseille Université–CNRS & IRD, Europôle Méditerranéen Arbois,
F-13545 Aix-en-Provence, France

[7]International Institute of Earthquake Engineering and Seismology, Tehran, Iran

*Corresponding author (e-mail: regard@lmtg.obs-mip.fr)

Abstract: SE Iran is the site of a rare case of young transition between subduction and collision. We have synthesized recent results in geodesy, tectonics, seismology and magnetism to help understand the structure and kinematics of the Zagros–Makran transition. Surface observations (tectonics, magnetism and geodesy) indicate a transpressive discontinuity consisting of several faults striking obliquely to the convergent plate motion, whereas deeper observations (seismology) support a smooth transition across the fault system. No lithospheric transform fault has been created, although the transition already behaves like a major boundary in terms of tectonic style, seismic structure, lithology and magnetism. The Zendan–Minab–Palami fault system consists of several faults that accommodate a transpressive tectonic regime. It is the surface expression of a southward propagation of the north–south-trending right-lateral strike-slip fault system of Jiroft–Sabzevaran. Within each system the numerous faults coalesce into a single, lithospheric, wrench fault.

The spatial transition between subduction and continental collision is by itself unstable and often a transform fault will develop to accommodate the differences in tectonic setting, as the Chaman Fault in central Asia does (Lawrence et al. 1992). Interestingly, the Hormuz Strait area in Iran (26.5°N, 56.5°E, Fig. 1) displays such a setting, but in a juvenile stage. At this point Arabia converges northward with Eurasia at a velocity of 23–25 mm a^{-1} according to global positioning system (GPS) measurements (Bayer et al. 2003, 2006; McClusky et al. 2003; Vernant et al. 2004; Masson et al. 2007). The Arabian plate is oceanic to the east in the Oman Gulf whereas it is continental to the west in the Arabian platform (Fig. 1). As expected, the Arabian and Eurasian continental plates collide to the west, forming the NW–SE-striking Zagros fold-and-thrust belt (ZFTB), which is a continental accretionary prism within the Arabian plate and accommodates about 10 mm a^{-1}

of NNE–SSW-trending shortening (Alavi 1994; Talebian & Jackson 2002; Tatar et al. 2002; Blanc et al. 2003). To the east, Arabia subducts under Iran, resulting in an extensive accretionary prism, of which the east–west-striking Makran belt is the emerged portion (Byrne et al. 1992; McCall 1997; Kopp et al. 2000). Between these regions, the structures in the Hormuz area define a curved structure connecting the Main Zagros Thrust (MZT) suture to the Makran frontal thrust (Fig. 1); this curved structure probably represents the newly formed transform fault. These structures connecting the Zagros and Makran mountain belts trend north–south to NNW–SSE, and are therefore highly oblique to the north–south convergence velocity, with an expected transpressive character.

This area has been studied for some time by geologists and geophysicists. The latter have described a sharp boundary between the eastern and western domains called the 'Oman Line'. Despite

From: LETURMY, P. & ROBIN, C. (eds) Tectonic and Stratigraphic Evolution of Zagros and Makran during the Mesozoic–Cenozoic. Geological Society, London, Special Publications, **330**, 43–64.
DOI: 10.1144/SP330.4   0305-8719/10/$15.00 © The Geological Society of London 2010.

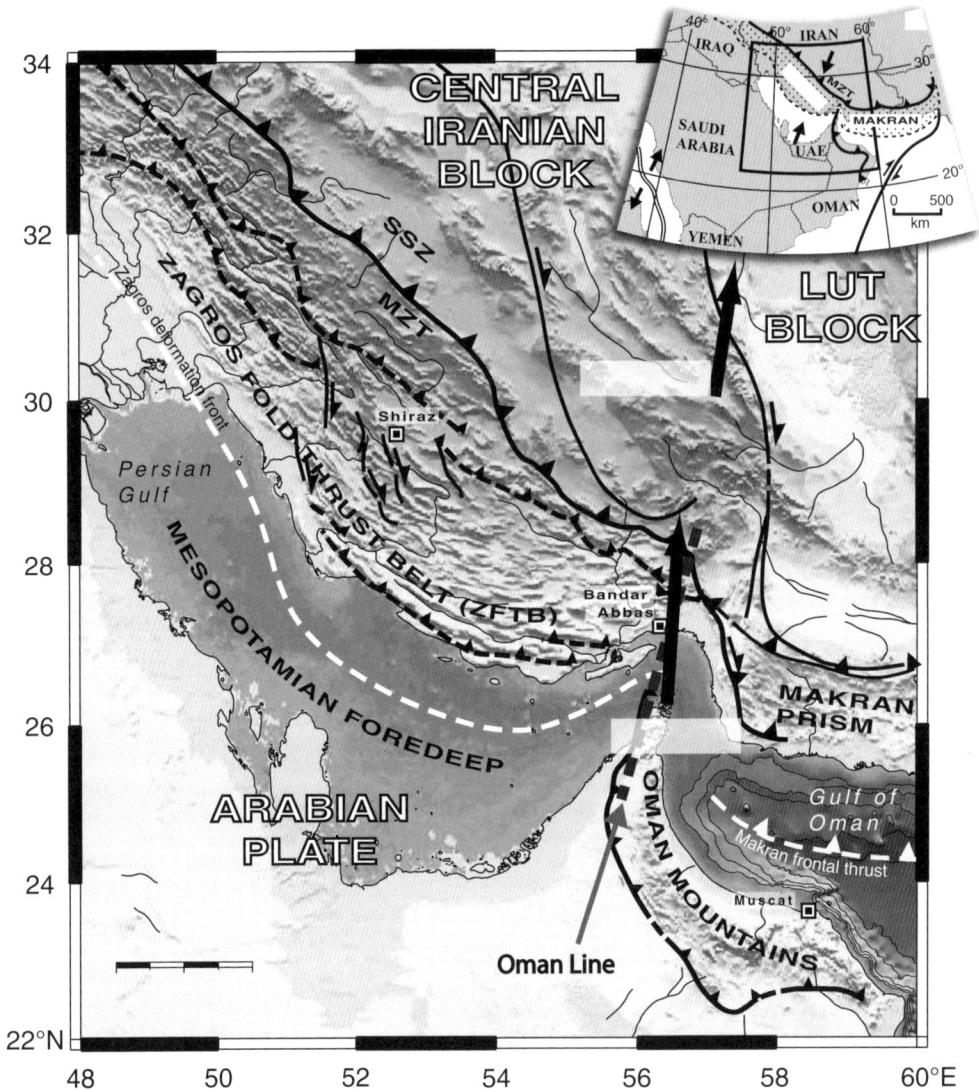

Fig. 1. Map of South Iran–North Arabia (modified after Molinaro *et al.* 2005*a*). The arrows represent the velocity relative to Eurasia (Vernant *et al.* 2004). It should be noted that the convergence accommodated through the Zagros Fold-and-Thrust Belt (ZFTB) is only *c.* 9 mm a^{-1}, and there is no value currently available for the part of the Arabia–Eurasia convergence accommodated through the Makran between Arabia and Lut [except using the Vernant *et al.* (2004) results with some assumptions]. MZT, Main Zagros Thrust; SSZ, Sanandaj–Sirjan Zone.

the associated low seismicity its sharpness was interpreted as evidence of a transform fault (Kadinsky-Cade & Barazangi 1982).

To provide a better understanding of this newly evolving transform fault system, the area has been recently studied using various geological and geophysical techniques, in the framework of an Iranian–French collaboration, including palaeomagnetism (Aubourg *et al.* 2004, 2008; Smith *et al.*

2005), Tertiary and active tectonics (Molinaro *et al.* 2004, 2005*a*; Regard *et al.* 2004, 2005*a*), seismology (Yamini-Fard 2003; Yamini-Fard & Hatzfeld 2008), and geodesy (Vernant *et al.* 2004; Bayer *et al.* 2006; Masson *et al.* 2007). The purpose of this paper is to provide a synthesis of these acquired data and to discuss this subduction–collision transition. Thus, after introducing the general setting of the area, the present-day and recent deformation, and the upper crustal structure from balanced cross-sections, we discuss the deep crustal structure revealed by seismological studies and finally present the actual deformation rates from GPS and geodesy.

Geological setting

Zagros

The Zagros mountain belt is a NW–SE-trending fold-and-thrust belt, consisting of a 6–15 km thick sedimentary pile that overlies Precambrian metamorphic basement (McCall et al. 1985; McCall 1997). The sedimentary cover can be divided into three successive sequences. At its base, it is composed of thick late Precambrian evaporitic deposits (the so-called 'Hormuz Salt'), which constitute the main regional décollement for most of the larger folds within the Zagros fold-and-thrust belt (ZFTB). This layer is the origin of numerous salt diapirs that have pierced the overlying sedimentary cover and risen to the surface. A c. 4000 m thick Cambrian to Eocene sequence forms the so-called Competent Group. Apart from the initial Cambrian–Carboniferous clastic formations, the majority of this group until the Upper Cretaceous units consists of massive platform carbonate rocks (James & Wynd 1965; Faure-Muret & Choubert 1971; Szabo & Kheradpir 1978; Sharland et al. 2001). The remainder of the stratigraphic sequence is represented by the Miocene to Recent clastic sediments of the Incompetent Group. These molasse-type sediments, derived from the uplift and erosion of the Zagros Mountains, show a typical coarsening-up evolution from marine-to-continental clastic deposits to coarse proximal conglomerates at the top (James & Wynd 1965; Edgell 1996; Hessami et al. 2001).

Makran

The Makran accretionary wedge stretches from Iran to central Pakistan and extends off the south coast of this region (Schluter et al. 2002). It has been formed by the subduction of the oceanic portion of the Arabian plate beneath Eurasia and is built up by sediments scraped off the Arabian plate since the early Tertiary (Berberian & King 1981; Harms et al. 1984; Kopp et al. 2000). Subduction was probably initiated during Palaeocene time (Platt et al. 1988) and accretion started during Eocene time (Byrne et al. 1992). The modern Makran accretionary prism has developed since the Late Miocene (Platt et al. 1985, 1988), and is still propagating seaward at a rate of c. 10 mm a^{-1} (White 1982). Two features make this accretionary wedge unusual: (1) the sediment thickness on top of the

oceanic crust is extremely high (at least 6 km); (2) the dip angle of subduction is extremely low (c. 5°, Jacob & Quittmeyer 1979; Byrne et al. 1992; Carbon 1996).

Zagros–Makran transition

The Zagros and Makran domains are both bounded to the north by a continuous ophiolitic belt along the Main Zagros reverse fault and its eastern continuation along the Makran Thrust (Fig. 1) (McCall 1997). South of this suture, the Zagros and Makran regions behave differently, highlighting how the subduction and collision settings differ. Whereas the convergence velocity accommodated by the Zagros collision increases progressively from NW to SE, the transition from the collision to the Makran subduction is marked by a jump from 9 ± 2 to c. 19 ± 2 mm a^{-1} (Vernant et al. 2004; Masson et al. 2007). The Zagros–Makran transition is thus expected to have a wrench motion of at least some 10 mm a^{-1}. During the late Cenozoic, Arabia and Eurasia continuously converged (Fig. 2; McQuarrie et al. 2003). Palinspastic reconstructions show how this continuous convergence was accommodated in the Zagros and Makran during the last 30 Ma (Fig. 2), suggesting that the Zagros–Makran transition zone must have been formed during the last c. 15 Ma.

The transition takes place at the front of the mountainous Musandam peninsula (Hormuz Strait, Fig. 1). Formed during the Late Cretaceous, this mountain range not only magnifies the differences between the continent and the ocean, but also acts as a heterogeneity within the colliding Arabia plate. It can be traced up to the Hormuz Strait, in a seismic profile running from Qeshm to Minab (Ross et al. 1986). Such a set-up suggests that the Oman peninsula may have interfered with the Zagros collision and be partly responsible for the curved shape of the Zagros–Makran transition (Ricou et al. 1977; Kadinsky-Cade & Barazangi 1982; Aubourg et al. 2008). Several other mechanisms have been proposed to explain this curved shape of the Fars Arc: (1) rotation of basement faults (Hessami et al. 2001); (2) a change in the style of accommodation of shortening from head-on shortening in the SE to oblique shortening to the NW of the ZFTB, observed by Talebian & Jackson (2004); (3) a Jura style of deformation in which the arcuate shape is controlled by the progressive lateral pinch-out of the basal Hormuz evaporites upon which the Zagros folds detach (Molinaro et al. 2005a).

The transition between the Zagros and Makran is also often described in the literature as the 'Oman Line' trending N20°E and running northward from the Musandam peninsula (Kadinsky-Cade &

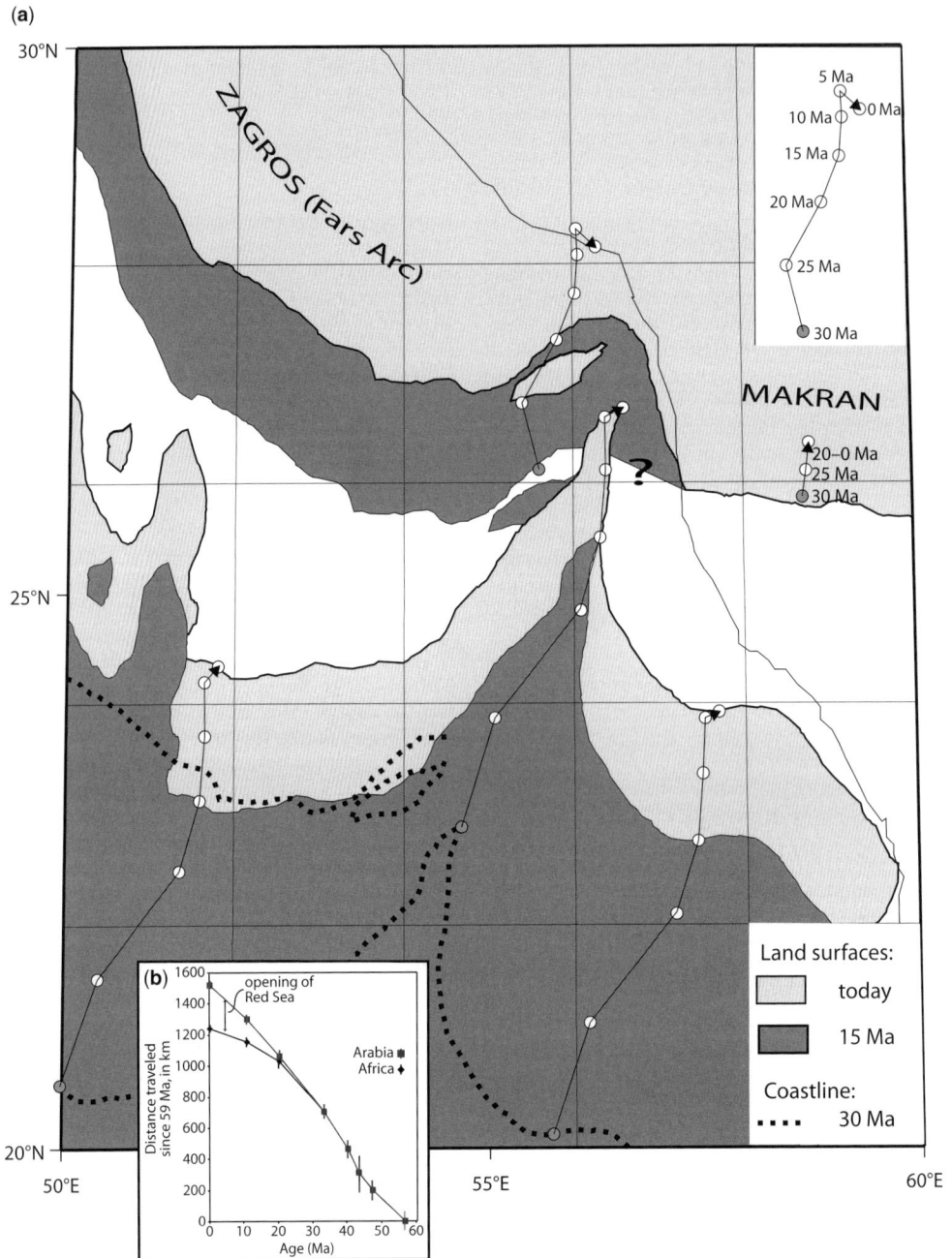

Fig. 2. (**a**) Palinspastic reconstructions around the study area for the last 30 Ma, with plates moving relative to Eurasia. The segmented lines with circles represent the displacement trajectories (see inset for stages). The reconstruction is according to the ODSN Plate Tectonic Reconstruction Service (http://www.odsn.de/odsn/services/paleomap/paleomap.html). Reconstructions are based on block rotations calculated from magnetic anomalies (Hay *et al.* 1999; Soeding 1999). The ancient Makran shoreline cannot be estimated as it is formed by an accretionary prism under construction; the motion-segmented arrow is estimated by the Lut block reconstructed position. (**b**) Graph showing the position of a reference point (38°N, 48°E) representing Arabia with respect to Eurasia v. time, both including (grey boxes) and excluding (black diamonds) rotation describing the opening of the Red Sea, modified after McQuarrie *et al.* (2003).

Barazangi 1982). It separates the continent to the east from the Oman Gulf oceanic crust to the west; it represents an inactive transform zone inherited from the Neotethys ocean opening (White & Ross 1979). This transition marks the boundary between a region of high seismicity located in the NW (the Zagros domain) and a region of low seismicity to the east, and is highlighted by the Musandam peninsula trend. Formerly the Oman Line was suggested to represent the transition between the Zagros and Makran (Kadinsky-Cade & Barazangi 1982), but it is now thought that this role is played by the Minab–Zendan Fault system, whose southern prolongation has been seen on seismic profiles from the Oman Gulf (Ravaut et al. 1998).

The multidisciplinary study presented here aims at better determining the structure and kinematics

of this key area. In particular, the Zendan–Minab Fault system (Fig. 3) is studied to help us understand why, as it seems to play a major role in the accommodation of the deformation, the associated seismicity is low. We also discuss the deformation related to this zone and see if it could be the result of the indentation by the Musandam peninsula.

Results

Deformation pattern

Tectonics. Using satellite images, and structural and geomorphological field observations, Regard et al. (2004) illustrated the study area's present-day deformation pattern accommodated by faulting. The study area shows a distributed deformation

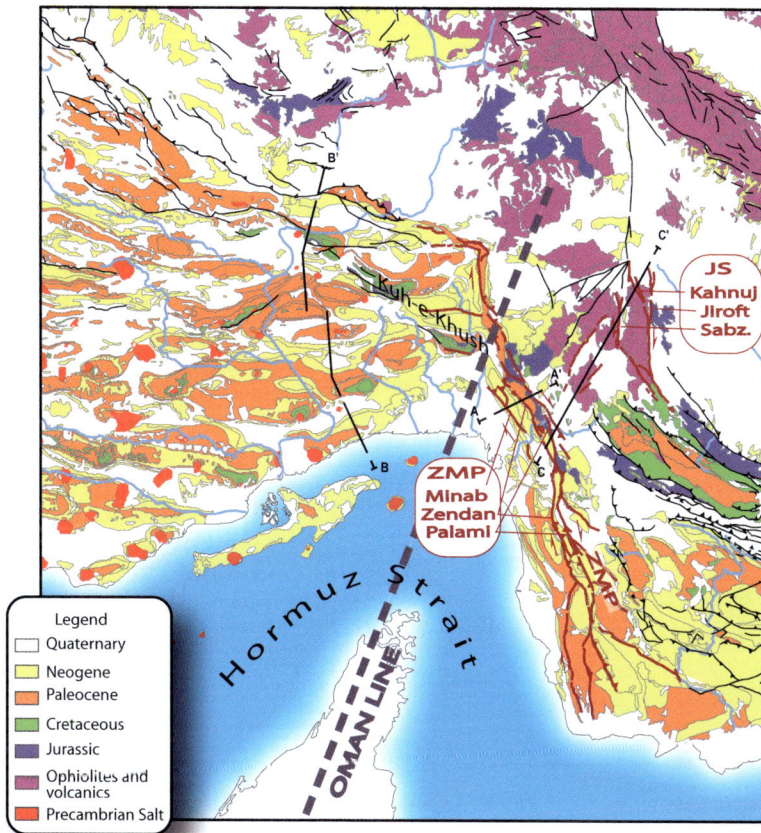

Fig. 3. Geological map of the study area, modified from the 1:1 000 000 geological map of Iran. Faults in black are those from the original drawing; faults in red are the active faults mapped by Regard et al. (2004), which are distributed in two fault systems: (1) the Jiroft, Sabzevaran (abbreviated as Sabz.) and Kahnuj faults in the Jiroft–Sabzevaran fault system (JS); (2) the Minab, Zendan and Palami faults in the Zendan–Minab–Palami fault system (ZMP). A–A' and B–B' are the cross-sections (Fig. 5) from Molinaro et al. (2004). C–C' is the local tomography cross-section (Fig. 8), from Yamini-Fard et al. (2007).

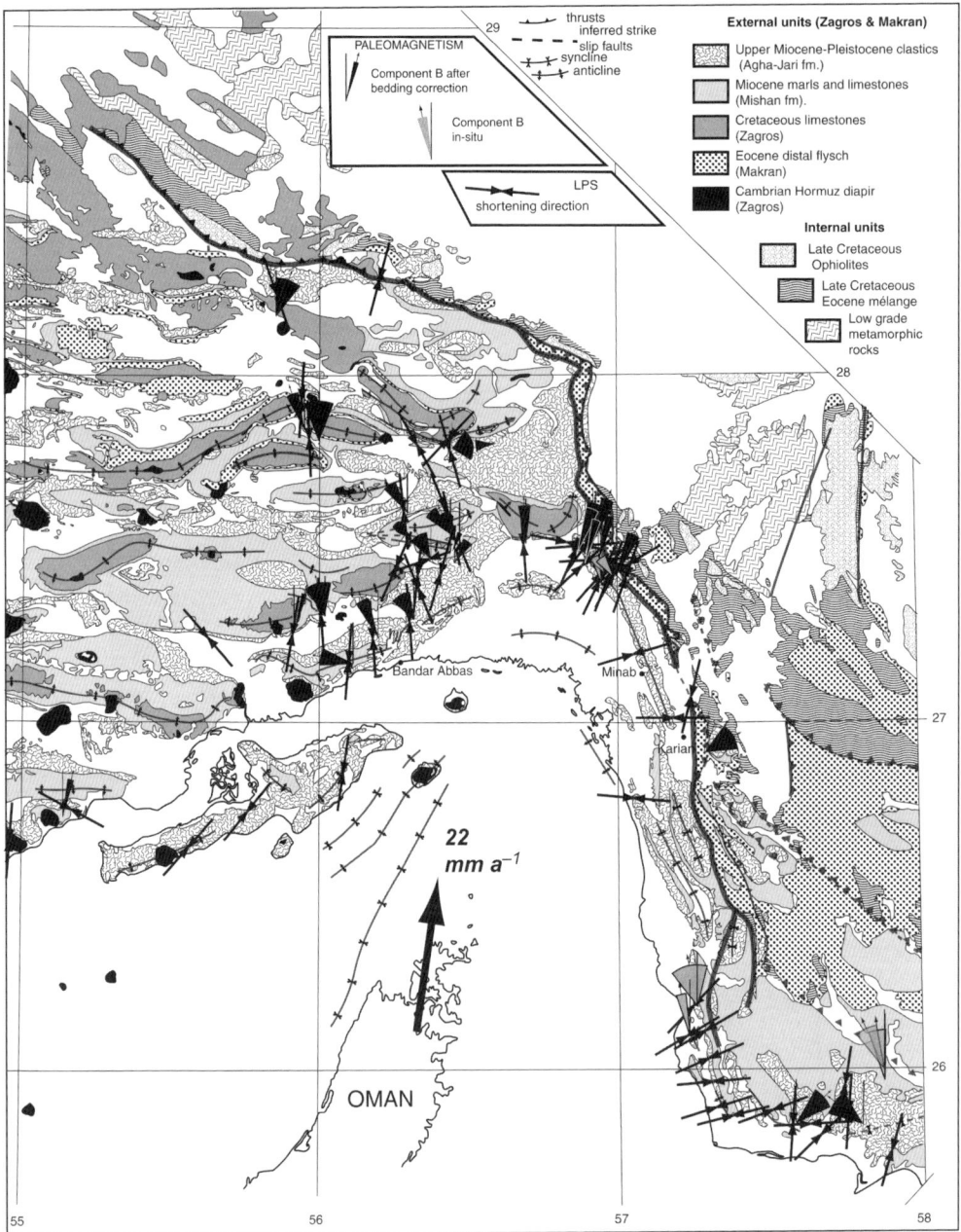

Fig. 4. Summary of magnetic studies on the Fars Arc and western Makran (Bakhtari *et al.* 1998; Aubourg *et al.* 2004; Smith *et al.* 2005).

pattern covering a wide domain (Fig. 3). Six north–south- to NW–SE-trending major faults were identified, each displaying clear evidence of late Quaternary reverse right-lateral slip. They constitute two fault systems. The first one encompasses, from west to east, the Minab, Zendan and Palami faults (the ZMP fault system; Fig. 3). The Zendan Fault corresponds to the lithological boundary between the Zagros and Makran. The second one, the Jiroft–Sabzevaran (JS) fault system, comprises the Jiroft, Kahnuj and Sabzevaran north–south-trending faults (Fig. 3). The ZMP fault system transfers the Zagros (continental prism) deformation to the Makran accretionary prism, whereas the JS fault system transfers some motion northward to the Alborz–Kopet Dagh convergence zone in northern Iran (Figs 1 & 3). Tectonic study and fault slip vector analyses indicate that two distinct tectonic regimes have occurred successively since the Miocene within a consistent regional NE–SW-trending compression: (1) a late Miocene to Pliocene tectonic regime characterized by partitioning between reverse faulting and en echelon folding; (2) a NE–SW-trending σ_1 axis transpressional regime homogeneously affecting the region since the late Pliocene (Regard et al. 2004). The change is contemporaneous with a major regional tectonic reorganization (Allen et al. 2004). This study provides evidence of active deformation that is not localized, but is distributed across a wide zone. It accommodates the convergence and transfers it from collision to subduction by transpressional tectonics without any partitioning process in the present-day period.

Magnetism. Several studies using magnetic fabric data were conducted in the Fars Arc and in the eastern Makran (Bakhtari et al. 1998; Aubourg & Robion 2002; Aubourg et al. 2004; Smith et al. 2005). In the Agha-Jari formation (Upper Miocene–Pleistocene silicoclastic rocks), the magnetic foliation is parallel to the bedding, whereas the magnetic lineation strikes perpendicular to the horizontal shortening or layer-parallel shortening (LPS). Several studies from other thrust belts (e.g. in the Pyrenees, Averbuch et al. 1992; Pares et al. 1999) suggest that the magnetic lineation records LPS that occurs before any detectable folding.

Near our study area, Aubourg et al. (2004) compared the magnetic fabric data with P-axis earthquake focal mechanisms and found a good agreement west of the Zagros–Makran transition, contrasting with significant differences in the east. Data collected during the last 10 years are summarized in Figure 4. To the west, in the Fars Arc, the general trend is a shortening direction (LPS) close to the GPS convergence direction (north–south), with slight anticlockwise rotations. Just west of and within the Minab–Zendan fault system, the shortening direction is roughly normal to the structures whereas clockwise rotations are recorded. To the east (within the Makran), data are sparse and do not show a clear signal, despite much LPS directed parallel to the north–south convergence direction.

Smith et al. (2005) and Aubourg et al. (2008) measured palaeomagnetic data in the Agha-Jari formation (Mio-Pliocene) to document vertical-axis block rotations. The pre-tilting palaeomagnetic component B documents the rotation that occurred between the age of Agha-Jari Fm. and recent time

(Fig. 4). At first glance, it is apparent that clockwise and counterclockwise rotations of small magnitude (typically less than 20°) occurred respectively in the eastern and western Zagros–Makran transition zone. If the block rotation is removed, the LPS directions are in good agreement with the overall convergence direction. This suggests that the present shape of the Zagros–Makran transition zone, oblique to the convergence direction, has been acquired only recently (since Mio-Pliocene time).

Together, the palaeomagnetic and magnetic fabric data (LPS) document the roughly north–south convergence, except near the Zagros–Makran transition zone, where the structures have experienced recent clockwise rotations that may continue at present (Fig. 4). These rotations agree with a wrench zone that is not localized on a single structure but extends over an area that is some tens of kilometres wide. The Fars Arc exhibits slight anticlockwise rotations, suggesting it has experienced some indentation since Middle Miocene time (Fig. 4).

Balanced cross-sections

Molinaro et al. constructed two cross-sections on each side of the Zagros–Makran transition (Molinaro et al. 2004, 2005a). The contribution of that work is not only to provide quantitative shortening values but also to reveal how the deformation is accommodated. The section on the eastern side of the transition indicates some 6 km of shortening in front of the Zendan Fault, affecting only the Tertiary cover, over a décollement c. 6 km deep (Fig. 5; section AA′ in Fig. 3). The shortening is measured perpendicular to the Zendan Fault (i.e. along a N150°E trend). In the SE Zagros, the Bandar Abbas–Hadjiabad section (BB′ in Fig. 3) displays two different north–south shortening values of 10 and 45 km, respectively, for the basement and the cover (Fig. 5). In particular, with this section Molinaro et al. documented two main steps in the evolution of the Zagros fold–thrust belt. In the first (Mio-Pliocene) stage the deformation was thin-skinned in style, with a décollement at c. 8–9 km depth. In the second (Pliocene to Recent) shortening stage, the basement must be involved through major thrust faults, inferred from focal mechanisms and observation of steps in the general topography and structural elevation of the Zagros mountains. These faults ramp up through the cover and cut the former folds obliquely; the best example is the Kuh-e-Khush (Fig. 2). This obliquity could be due to rotation of folds, reactivation of old basement structures or stress rotation. Geographically, to the west, in the Zagros, deformation involves the entire crust, with folding in the cover controlled by the basal Hormuz Salt layer and

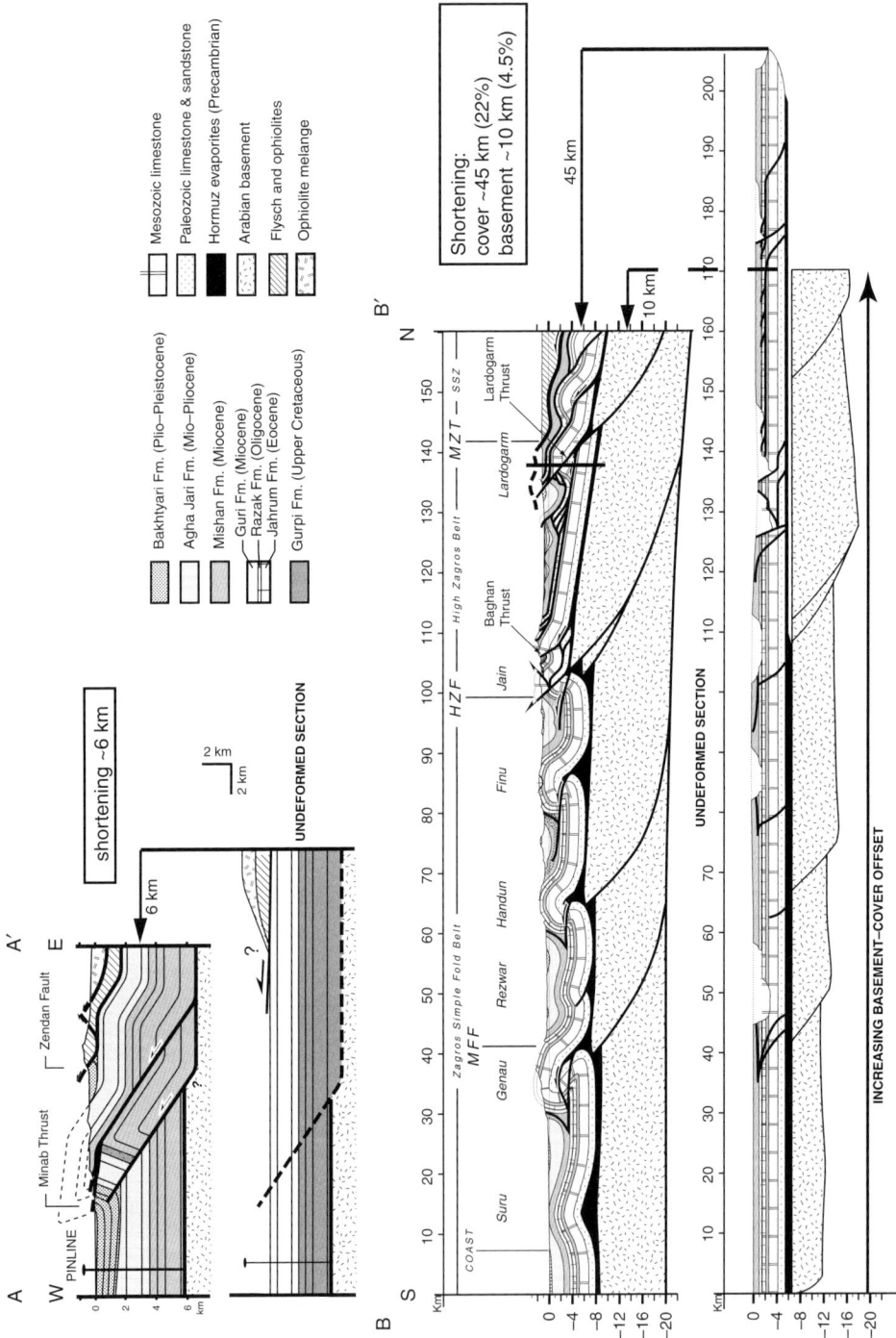

Fig. 5. Balanced cross-sections, from Molinaro *et al.* (2004, 2005*a*). (See Fig. 3 for profile location.) A–A′, cross-section through the Minab–Zendan fault system; B–B′, cross-section from Bandar Abbas to Hajiabad, southeastern Zagros. MFF, Main Front Fault; HZF, High Zagros Fault; MZT, Main Zagros Thrust; SSZ, Sanandaj–Sirjan Zone.

major thrusting in the basement. To the east, on the other hand, folding is controlled by a décollement located at 6 km depth and the basement does not appear to be involved in the deformation.

Seismotectonics and deep structure

Seismotectonics. Yamini-Fard *et al.* (2007) installed a dense seismological mobile network of 24 stations

for 8 weeks in 1999, in the area of Bandar Abbas, to locate precisely the microseismicity of the area and calculate focal mechanisms (Fig. 6). In addition, they carried out a tomographic traverse, with 25 stations from Hadjiabad to Minab and the Makran ranges (Fig. 7).

The microearthquake distribution around the transition between the Zagros continental collision and the Makran subduction is restricted to the west

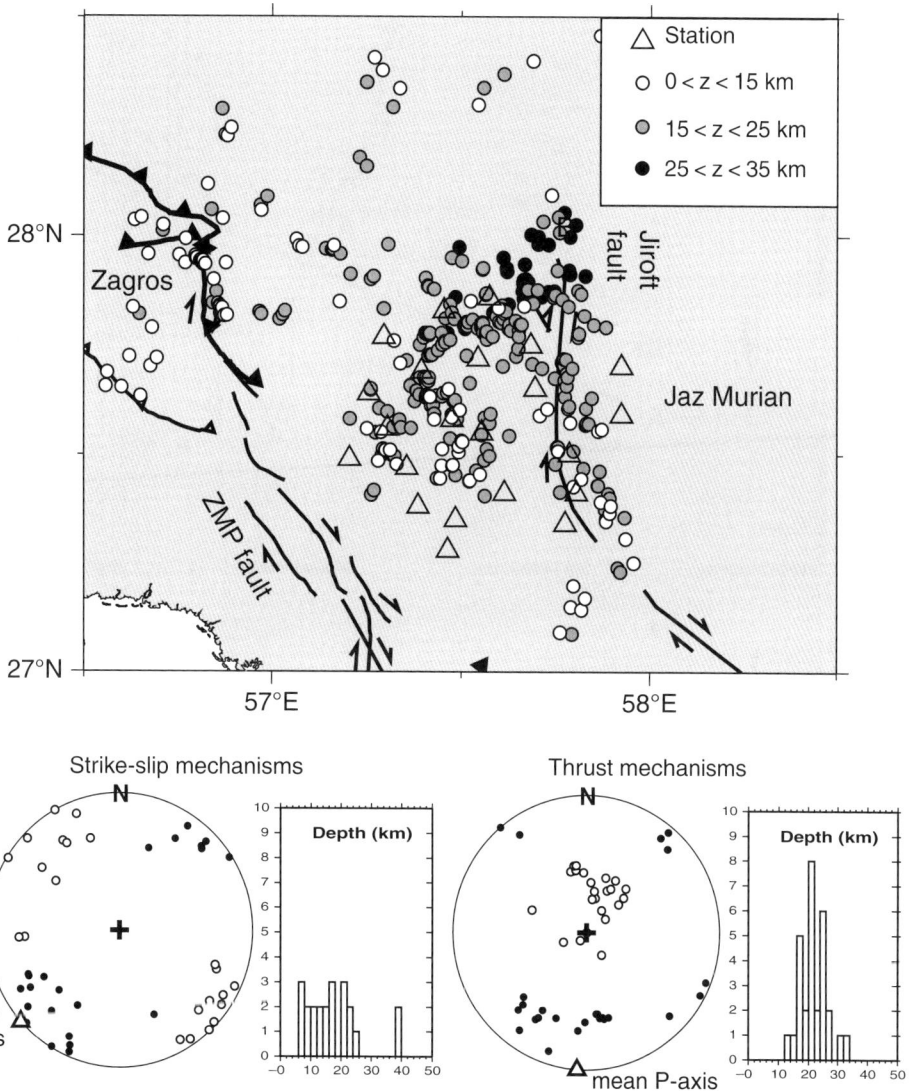

Fig. 6. Map of the microseismicity recorded for the period 17 November 1999 to 6 January 2000 by Yamini-Fard (2003) and Yamini-Fard *et al.* (2007). They computed 59 focal mechanisms that can be divided into two groups in terms of their mechanism: strike-slip (horizontal T-axis) or thrusting (vertical T-axis). These lead, respectively, to a *c.* N45°E-trending and roughly north-trending mean P-axis. Interestingly, thrust mechanisms occur at depths (15–35 km) greater than strike-slip mechanisms (5–25 km); filled and open circles are P- and T-axes, respectively (redrawn after Yamini-Fard *et al.* 2007).

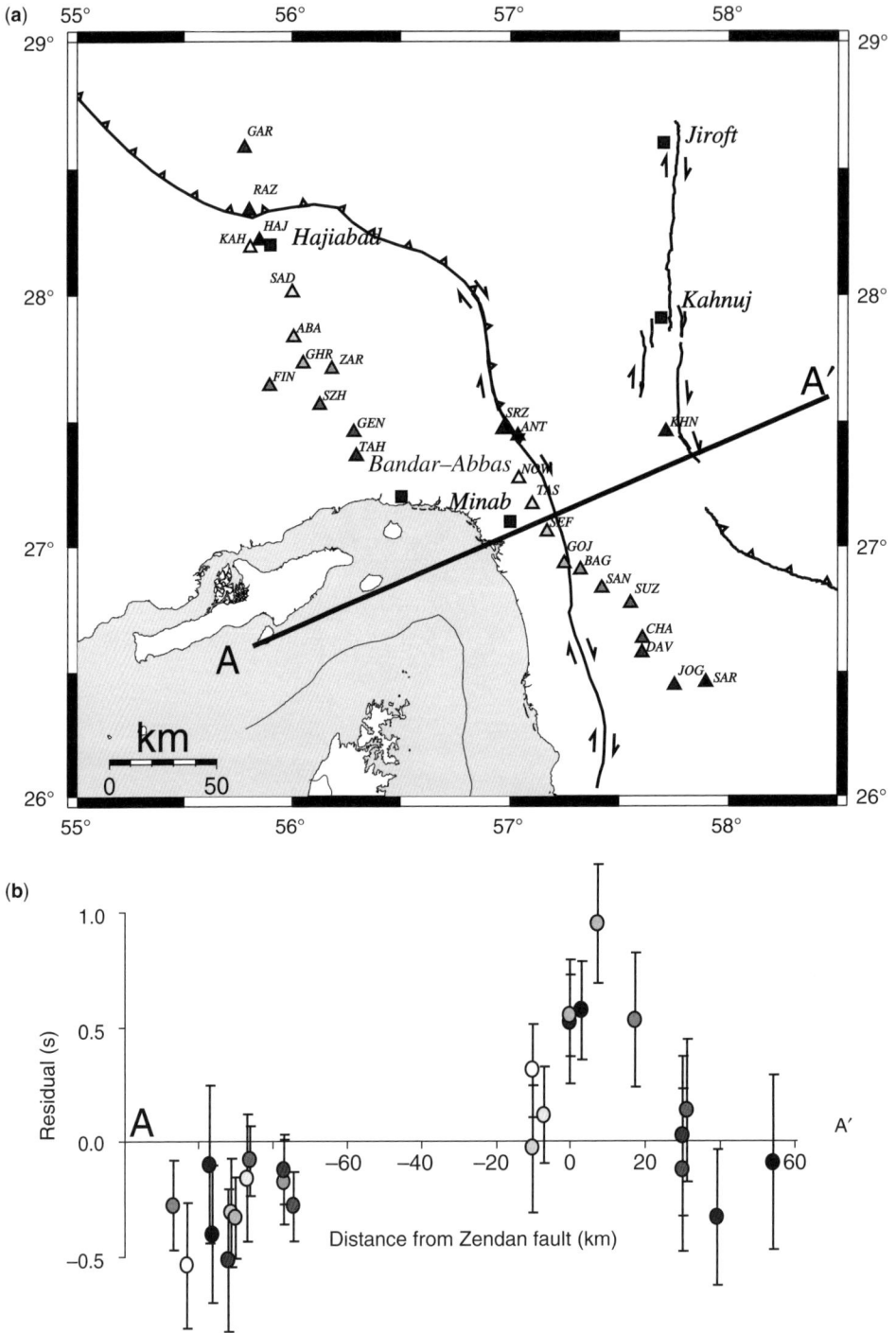

Fig. 7. Teleseismic travel-times residuals at stations along the Minab profile. (**a**) Location of seismological stations. (**b**) Residual values, given by reference to the average residual (in s) v. eastward distance to the Zendan Fault. Redrawn from Yamini-Fard (2003) and Yamini-Fard & Hatzfeld (2008). Corresponding symbol shades are used in the map and the residual graph.

of the Jaz Murian depression and the Jiroft Fault (Fig. 6). No earthquakes seem to be related to the Zendan–Minab–Palami fault system. Most of the shallow seismicity is related either to the Zagros mountain belt, located to the west, or to the Sabzevaran–Jiroft fault system, located to the north. The depth distribution of the microearthquakes increases northeastward to an unusual value for the Zagros of 40 km. Two dominant types of focal mechanisms are observed in this region: low-angle thrust mostly restricted to the lower crust (at depths between 15 and 30 km) and strike-slip at shallow depth (10–20 km, Fig. 6). Both are consistent with north–south to NE–SW shortening.

Crustal seismic structure. Teleseismic P-wave travel times, from *c*. 50 earthquakes, were calculated along a profile that crosses the Minab–Zendan fault system at three locations. The orientation of the profile was governed by accessibility and safety of the road. We observe a large delay in the travel time residual every time the profile crosses the ZMP fault system (Fig. 7). The inversion of the travel time residuals performed by Yamini-Fard (2003) highlights perturbations relative to a homogeneously layered velocity structure. In the first

layer (associated with the crust), two low-velocity anomalies are related to the ZMP fault zone. The deeper layers related to the upper mantle show a slow velocity in the west relative to a fast velocity in the east of the ZMP fault system. If we project the residual on a profile perpendicular to the ZMP fault system as a function of the distance to the ZMP fault system (Fig. 7), we see a clear offset of up to 1 s at the location of the ZMP fault system. For stations located on the Makran block, we observe a decrease with the distance to the ZMP fault system, which is also related to a southeastward orientation of the profile. Because the crustal structure is 3D, and the slab is dipping northward beneath the Makran, it is likely that the offset decreases southward. The strong offset exactly related to the ZMP fault system, however, indicates a strong velocity contrast between the Zagros and Makran.

Yamini-Fard (2003) also computed teleseismic receiver functions. The P–S converted phases usually image the Moho discontinuity well. Their interpretation is not straightforward: the receiver function data are complex and cannot be easily used to draw a Moho profile. This complexity probably comes from the departure of the flat-layered

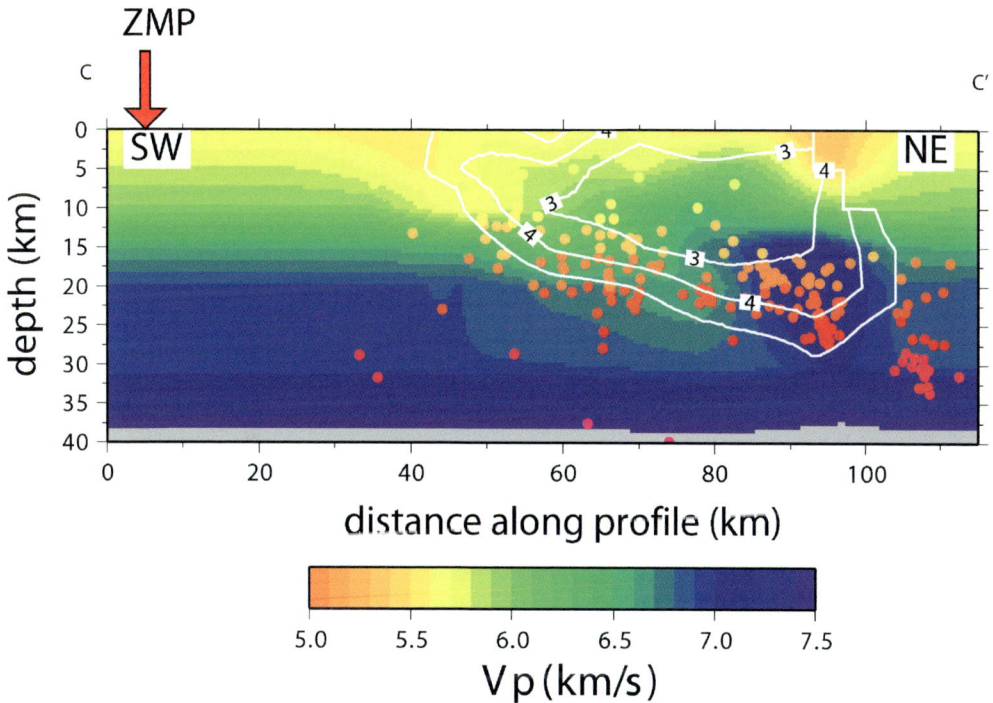

Fig. 8. Cross-section of the 3D velocity structure trending SW–NE computed from local travel-time residuals (see Fig. 3 for location.) Results are reliable for a spread function <5 (white contour). The hypocentres are reported. There is a clear indication of a northward dipping anomaly related to the seismicity (Yamini-Fard *et al.* 2007).

structure that is used as an initial model and generates complex wave propagations in the crust.

Finally, Yamini-Fard *et al.* (2007) computed a local 3D crustal velocity structure and relocated simultaneously the local earthquakes inverting the travel times of local seismicity (Yamini-Fard *et al.* 2007). The resulting velocity structure suggests a high-velocity body dipping northeastward (Fig. 8). An important result is that no seismicity appears to be associated with the Zendan–Minab–Palami fault system, suggesting that the transition between the Zagros collision and the Makran subduction is not associated with a sharp transform fault. Instead, it is associated with a progressive transition located in the lower crust. The shallow right-lateral strike-slip faulting is the response of the upper crust to the shortening. This 'partitioning' in depth is probably related to the difference in the strength of the upper and lower crusts.

Modern kinematics

GPS. Vernant *et al.* (2004; updated by Masson *et al.* 2007) used a network of 27 GPS sites to establish the current large-scale deformation rates within Iran. The network was measured three times, during September 1999, October 2001 and October 2005. This work provided a useful overview of Iranian geodynamics. The researchers then divided Iran into various rigid blocks. They concluded that, to the west, the Zagros is undergoing 9 ± 2 mm a^{-1} of shortening, in a direction close to north–south. To the east, the Makran subduction accommodates 19 ± 2 mm a^{-1} of N20°E-trending convergence. The Makran–Zagros transition zone should thus accommodate some 11 ± 2 mm a^{-1} of right-lateral movement.

Bayer *et al.* (2006) used a denser network focused on the Zagros–Makran transition zone. It consisted of 15 stations, with an average separation of *c.* 60 km, which were measured in 2000 and 2002. The GPS-derived velocity field can be expressed in various useful frames: fixed Eurasia, fixed Central Iran or fixed Arabia (Fig. 9). When expressed in a Central Iran fixed frame, the velocity directions are more or less parallel to the convergence velocity. There is no clear divergence from the Musandam peninsula (Oman), contrary to what would be expected if Musandam was acting as an indentor, as proposed by Kadinsky-Cade & Barazangi (1982; Fig. 9).

Assuming a rigid block model, Bayer *et al.* (2006) computed the motion accommodated by the Minab–Zendan–Palami fault system. They found 15 mm a^{-1} and 6 mm a^{-1} for the motions parallel or perpendicular to the direction of the fault (i.e. N160°E; Fig. 9). In addition, they estimated the strike-slip motion of the Jiroft–Sabzevaran

fault system (north–south-trending) to be 3.1 ± 2.5 mm a^{-1}. The Minab–Zendan–Palami fault system motion is estimated in its southern part, where the entire Zagros–Makran motion must be accommodated (Regard *et al.* 2004), whereas in its northern part, Zagros–Makran transition zone deformation is distributed over the two fault systems. The 15 mm a^{-1} MZP fault system displacement rate must therefore encompass the 3.1 ± 2.5 mm a^{-1} of the Jiroft–Sabzevarn fault system calculated by Regard *et al.* (2005*a*; see below).

Geodetic data show that almost all the convergence is accommodated in the east by the Makran subduction zone, whereas only half of it is accommodated in the west by the Zagros. The Zagros–Makran transition zone clearly accommodates *c.* 15 mm a^{-1} of differential motion; the transpressive character of the transition zone is because of the fault obliquity relative to the overall plate convergence direction. Geodesy does not provide evidence of a rigid indentation of the Musandan peninsula into Iran, which would be represented by a velocity field pattern divergent from the Musandam peninsula.

Tectonics and geomorphology. As described above, Regard *et al.* (2004) provided evidence of two fault systems accommodating the relative velocities in the northern part of the study area (whereas the results of Bayer *et al.* (2006) concern the southern part of the system). Tectonic and geomorphological analyses combined with cosmogenic nuclide-dating (^{10}Be) have revealed a total right-lateral slip rate of 4.7 ± 2.0 mm a^{-1} to 6.3 ± 2.3 mm a^{-1} for the ZMP fault system, depending on the ages of offsets, and 5.7 ± 1.7 mm a^{-1} for the JS fault system (table 3 of Regard *et al.* 2005*a*; see also Regard *et al.* 2006). Regard *et al.* evaluated the total motion accommodated across the area to be 11.3 ± 3.9 mm a^{-1} or 13.1 ± 4.3 mm a^{-1} in a direction N10 $\pm$ 20°E.

The total shortening at the west front of the ZMP fault system is estimated to be *c.* 6 km (Fig. 4) since the Mio-Pliocene (*c.* 5 Ma, Molinaro *et al.* 2005*a*), and implies an average shortening rate of the order of 1 mm a^{-1}.

Discussion

The work presented here is a compilation of data collected by various means. It gives a unique insight into the current dynamics of a subduction–collision transition zone. In particular, important questions arise, such as: What is the lithospheric structure? Where is the main structure? Does the Musandam peninsula act as an indentor? To give a coherent view of this transition we first describe what we do know about its structure, then try to

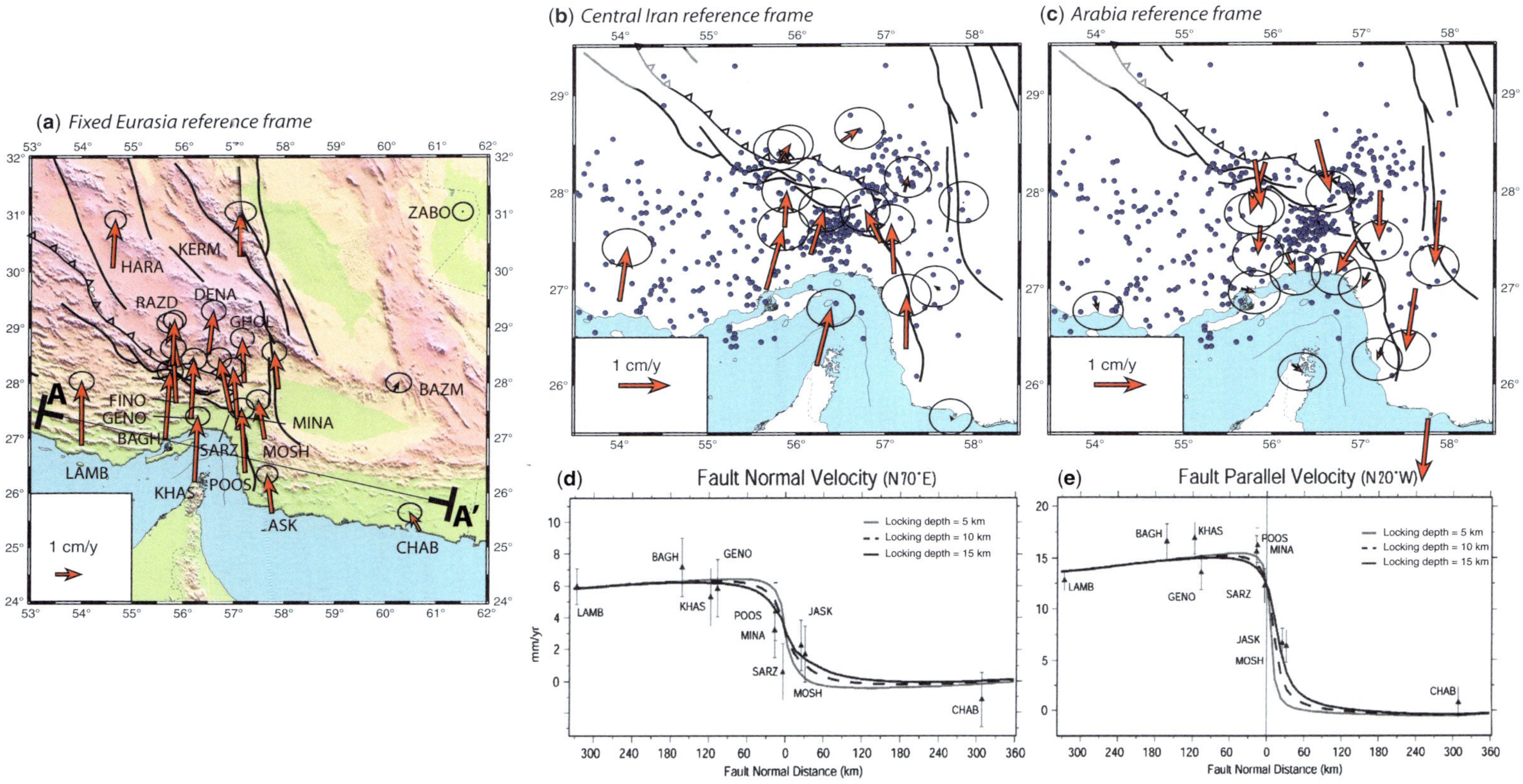

Fig. 9. GPS data for the study area (Bayer *et al.* 2006). (**a**) GPS velocities and their 95% confidence ellipses in a fixed Eurasia reference frame. (**b**, **c**) Detailed view of the study area in, respectively, a fixed Central Iran reference frame and a fixed Arabia reference frame; dots represent instrumental seismicity. (**d**, **e**) GPS velocity profiles, normal and perpendicular, respectively, to the Zendan–Minab fault trend [A–A′ in (a)]. Lines refer to best-fit deformation with a block model (see Bayer *et al.* 2006, for details).

assess the question of its modern kinematics and the timing of its set-up before inferring its future through comparison with laboratory experiments and other subduction–collision transitions.

Crustal and lithospheric structure

Before Arabia and Iran began to collide, probably in the late Oligocene, the area was occupied by a continuous subduction zone. The subduction of continental Arabia to the west led to a collision, whereas to the east the oceanic part of the Arabian plate is still subducting at present. Under the Zagros collision zone the plate probably broke and the oceanic part may have sunk deep into the mantle (Molinaro *et al.* 2005*b*). Further to the east, in the study area, it is difficult to know if the nearby Zagros underwent such a slab break-off or if the subducted slab is still attached. In particular, the transitional area is wide, which suggests a gentle transition at depth, compatible with a continuous deep slab (Regard *et al.* 2005*a, b*). Closer to the surface, seismological data display a clear view of a NE-dipping surface (Fig. 8). This plane dips *c.* 15° and is likely to originate at the surface near the ZMP fault system (Fig. 8; Yamini-Fard *et al.* 2007). This plane is associated with significant microseismicity with NE–SW-trending P-axes, showing thrusting at depth (Fig. 6; Yamini-Fard *et al.* 2007). It would thus correspond to an active crustal-scale thrust, separating a Zagros-related part in its footwall from Makran formations in its hanging wall.

The system tectonics appears more complicated in the uppermost part. It is dominated by folds and faults organized in two fault systems: (1) the ZMP fault system to the SW; (2) the JS fault system to the NE. (1) The NNW–SSE-trending ZMP fault system is associated with en echelon folding and constitutes the lithological boundary between the Zagros and Makran (Regard *et al.* 2004). It presents high-velocity anomalies (1 s residuals, Fig. 7) (Yamini-Fard 2003). It is made up of numerous and highly segmented faults. The system's western faults dip ENE whereas the eastern ones dip WSW, giving a flower-structure-like superficial organization (Regard *et al.* 2004). This system could act as a developing crustal-scale strike-slip fault, with infilling by dense material, but these observations do not agree with the balanced cross-section, which implies an 8 km deep décollement surface that should extend a couple of kilometres eastward from the ZMP fault system. This apparent discrepancy will be resolved in the discussion below. (2) The north–south-trending Jiroft–Sabzevaran fault system does not appear to be marked by any seismicity alignment although the local seismicity level is high (Yamini-Fard 2003). The faults are strike-slip with a small component of vertical motion. The fault system is partly linked to the south with the tectonic Makran northern boundary, south of the Jaz Murian depression, and partly to the ZMP fault system (Regard *et al.* 2004). To the north the system seems to continue northward to the Nayband and Gowk faults, which mark the boundary between Central Iran and the Lut Block (Walker & Jackson 2002).

Surface structural setting and modern kinematics (map view)

The current deformation at the surface and close to the fault systems is shown by seismology and active tectonics to be caused by a NE–SW-trending main compressional direction (Regard *et al.* 2004; Yamini-Fard *et al.* 2007). Active tectonics also indicates that convergence is accommodated nearly equally by the two fault systems, the deformation being distributed within each fault system instead of being localized (Regard *et al.* 2005*a*).

On a wider scale, the palaeomagnetic data show that rotations occurred both clockwise to the east and anticlockwise to the west of the transition zone. This could indicate an indentor role of the Musandam peninsula (Aubourg *et al.* 2004, 2008). On the other hand, the GPS velocity field does not show any divergence away from the 'indentor', and this would suggest that there is no rigid indentation in the area (Figs 9 & 10). Tectonic observations also do not favour the Musandam peninsula indentor hypothesis, as they indicate that the motion between Arabia and Makran is accommodated only by one transcurrent fault system (the ZMP fault system), on which all the relative motion is accommodated, as indicated by the similarity in GPS and tectonic velocities.

Indeed, the global motion accommodated by the transition zone fault systems is evaluated by GPS to be 15 mm a^{-1}, close to the evaluation from tectonics of 11–13 mm a^{-1}. The agreement between GPS and tectonics is not so good on the part accommodated by the Jiroft–Sabzevaran fault system, the transcurrent motion of which is evaluated to be 3.1 ± 2.5 mm a^{-1} or 5.7 ± 1.7 mm a^{-1}, respectively (Fig. 10). The discrepancy comes from the loss of the elastic component away from the network, the easternmost station of which is close to the Jiroft Fault, as highlighted by recent results (Peyret *et al.* 2009).

Two-stage post-Miocene evolution

Some of the studies presented here highlight a two-stage scenario for the recent (since Miocene times) evolution of the area. First, magnetic studies

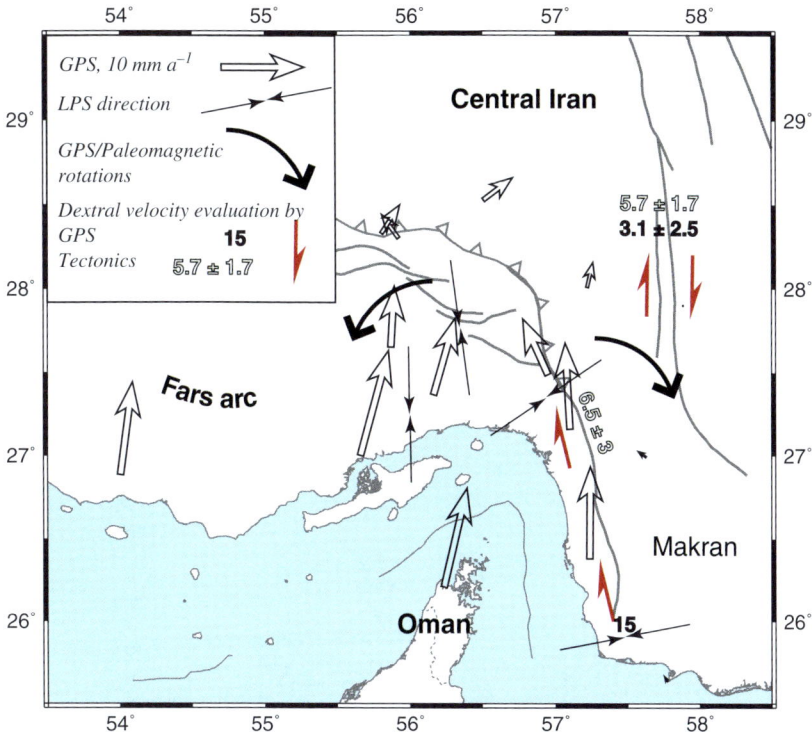

Fig. 10. Summary of observations at the surface. GPS results are indicated in a fixed Central Iran reference frame.

indicate that the transition shape, oblique to the convergence, has been acquired recently, since the Mio-Pliocene. Before the transition structural set-up, the deformation recorded by magnetism was roughly coherent with the overall plate convergence. Superimposed on this, some moderate rotations, clockwise to the east and anticlockwise to the west, may account for a slight indentation by the Musandam peninsula. Second, a major change in tectonics was also recorded. As emphasized above, the balanced cross-section of the Minab fold proposed by Molinaro et al. (2004) appears not to be compatible with modern kinematics and structure. The Minab fold may have been formed during a former deformation stage. In addition, palaeostress determinations by Regard et al. (2004) show a stress-orientation change in the Pliocene. Those workers suggested that it corresponds to a change from a partitioned convergence accommodation through folds and reverse faults (ZMP fault system) to non-partitioned convergence accommodation through the ZMP transpressional faulting. The best way to combine these results (tectonics, structure and kinematics) is to assume that the flat thrust shown by balanced cross-sections has been cut through by a more vertical fault, possibly evolving to a flower

structure near the surface. This fault is likely to carry exotic slices such as the Palami range lying between the Palami and Zendan faults.

Comparisons with other settings, both in the laboratory and in nature

Cotton & Koyi (2000) proposed a sandbox experiment that interestingly mimics the study area although it was not intended to reproduce it (Fig. 11). The experiment comprises a sandbox undergoing north–south compression; the sand is lying over a frictional substrate in the western part whereas it is underlain by a ductile level in its eastern part (Fig. 11). Although the experiment was designed for basin tectonics, we could interpret it in terms of lithospheric structure. The frictional v. ductile substrate should be compared with the collision v. subduction convergence settings. In this experiment, the eastern range (over ductile substrate) extends further to the south than the western range. Between the two ranges a transfer zone is produced, which is formed by two north–south-trending strike-slip faults: (1) a southwestern one connecting the deformation zone of the western

(a)

(b)

(c)

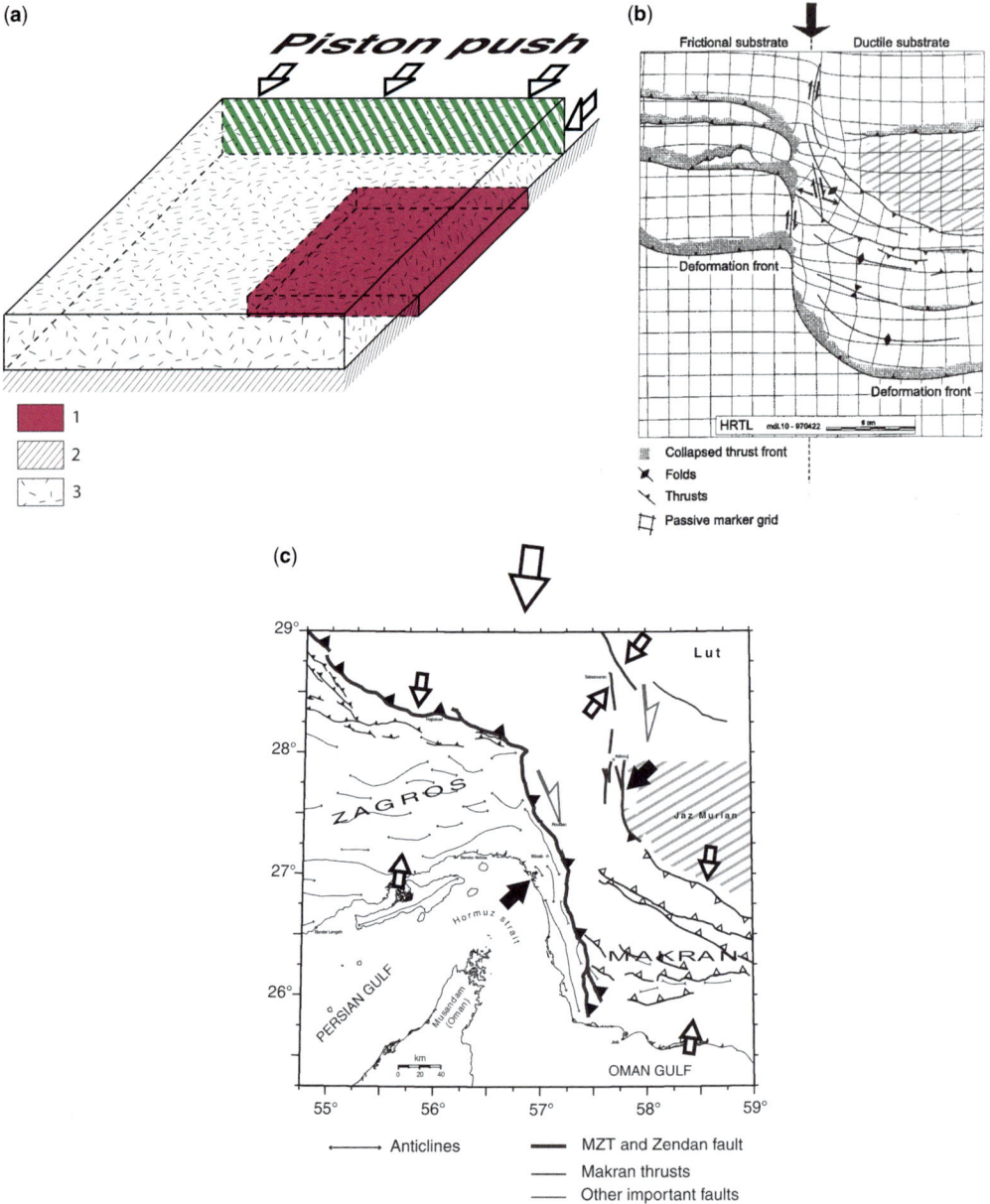

Fig. 11. Sandbox experiment by Cotton & Koyi (2000). (**a**) The setup consists of a box filled with sand (3) over a rigid basement (2), which is partly replaced by a low-viscosity layer (1). The sand layers experience convergence owing to piston push. (**b**) The experimental result shows that the structures propagate further over the viscous basement than over the rigid basement; a transition zone is created. (**c**) Structural scheme of our study area, with strain arrows, which resembles the experimental result (b).

range (over a frictional substrate) to the frontal thrust of the eastern one; (2) a northern one that connects the northern part of the experiment with the backstop of the eastern range (Fig. 11). Comparison of this scheme with the structural setting of the Zagros–Makran transition zone highlights many similarities; in particular, the faults (1) and (2) probably represent the ZMP and JS fault systems, respectively (Fig. 11). However, the experiment cannot explain the curvature of the Fars Arc (southeastern

Zagros); in the experiment the range over the frictional substrate has a linear trend (Fig. 11). A possible explanation for this is that in the Fars Arc the curvature is due to the underthrusting of the Musandam peninsula.

Another transition between subduction and collision can be found at the other end of the Makran. Some of us have already initiated discussion on this subject (Regard *et al.* 2005a). There, the transfer zone is formed by three single faults, the Chaman, Ghazaband and Ornach-Nal faults (Fig. 12). The system connects the Makran accretionary prism to the Pamirs (Panjshir Fault) (Lawrence *et al.* 1992). The overall system is thought to accommodate between 25 and 35 mm a^{-1} according to geological evidence (Beun *et al.* 1979), or 40 mm a^{-1} according to the NUVEL-1 model (DeMets *et al.* 1990). The most important part of the deformation is transmitted to the inner Makran and there is no evidence of active deformation transmission to the frontal, offshore, Makran thrust. The transition zone is more mature than the Zagros–Makran transition zone. Indeed, it is known to be much older (20–25 Ma instead of 5 Ma or less) and its length (more than 1000 km) implies a considerable slab stretching such that the slab must no longer be continuous between the Makran and Pamirs. Consequently, this system could be viewed as a possible future for our study area. In particular, it is noteworthy that the accommodation by two or three disconnected systems characterizes the transition zone as well as in the Cotton & Koyi (2000) experiment and at the Zagros–Makran transition (Figs 11 & 12). Some partitioning also occurs; for example, between the Chaman strike-slip fault and the Sulaiman fold belt (Lawrence *et al.* 1992; Davis & Lillie 1994). In turn, the maturity is expressed by the localization of the deforming structures, which are long, low-segmented faults in contrast to the highly segmented fault systems in the Zagros–Makran area (Regard *et al.* 2005a).

Synthesis

The evolution of the Zagros–Makran transition zone has probably been influenced by an inherited structural setting. For a better understanding of the way it has recently evolved, a simplified scenario is proposed in Figure 13. The way in which this area is deforming is comparable with both the Makran–Pamir transition and the laboratory experiment. Two characteristic fault systems are found to the SW (Zendan–Minab–Palami fault system) and NE (Jiroft–Sabzevaran fault system). To the east the Makran deformation shows the same pattern as in the experiment and at the eastern Makran boundary, but to the west, the Fars Arc shape disagrees with the experiment. We hypothesize that this is

due to the complexity introduced by the Musandam peninsula, which is topographically much higher than the surrounding parts of Arabia.

The first fault system (the ZMP fault system) currently trends N160°E, oblique to the convergence. It is made up of three highly segmented faults that are the possible expression at the surface of a flower structure that changes to an oblique thrust at its northern boundary. Seismology indicates that it is correlated with a strong discontinuity at depth and that near its northern termination its deep structure is a NE-dipping plane. In this scheme some 6 km structure-normal shortening occurs that could be the result of a former strain distribution or of a slight deformation partitioning, with a frontal fault accommodating up to 5 mm a^{-1} of shortening, as indicated by the recent results of Peyret *et al.* (2009). The 6 km deep décollement proposed by Molinaro *et al.* (2004) may therefore now branch at depth to the flower structure. We propose that its central fault (Zendan) represents the main boundary between the Zagros and Makran.

The second fault system (the JS fault system) trends north–south. It probably connects to the north to a well-known strike-slip system bounding the Lut block to the west. To the south its deformation probably transfers partly to the northern Makran tectonic boundary and partly to the ZMP fault system, as suggested by the laboratory experiment of Cotton & Koyi (2000).

Of the *c.* 19 mm a^{-1} differential convergence rate, some 3–6 mm a^{-1} are accommodated by the JS fault system and *c.* 6–7 mm a^{-1} in the ZMP fault system at the Minab latitude, increasing to *c.* 15 mm a^{-1} to the south; this increase corresponds to the progressive deformation transmission from the JS fault system to the ZMP fault system. It should be noted that if one part of the JS-accommodated deformation is transmitted to the Makran northernmost thrust, the 15 mm a^{-1} strike-slip deformation observed to the south of the ZMP fault system is not fully explained by addition of the ZMP and JS motions (maximum 13 mm a^{-1}). This discrepancy is not yet resolved.

The data presented here provide evidence of post-Miocene system evolution (Fig. 13). The overall oblique structure is from a recent (less than 5 Ma) setting as indicated by magnetism and tectonics. The set-up time is contemporaneous with a major change in the Middle East tectonics and with Zagros topography-building initiation (Allen *et al.* 2004). During this set-up, the Zagros and Makran, which were originally continuous, differentiated. Interestingly, this differentiation occurred at the time when deformation in the Zagros passed from thin-skinned to thick-skinned (Molinaro *et al.* 2004); this is very different from the Makran,

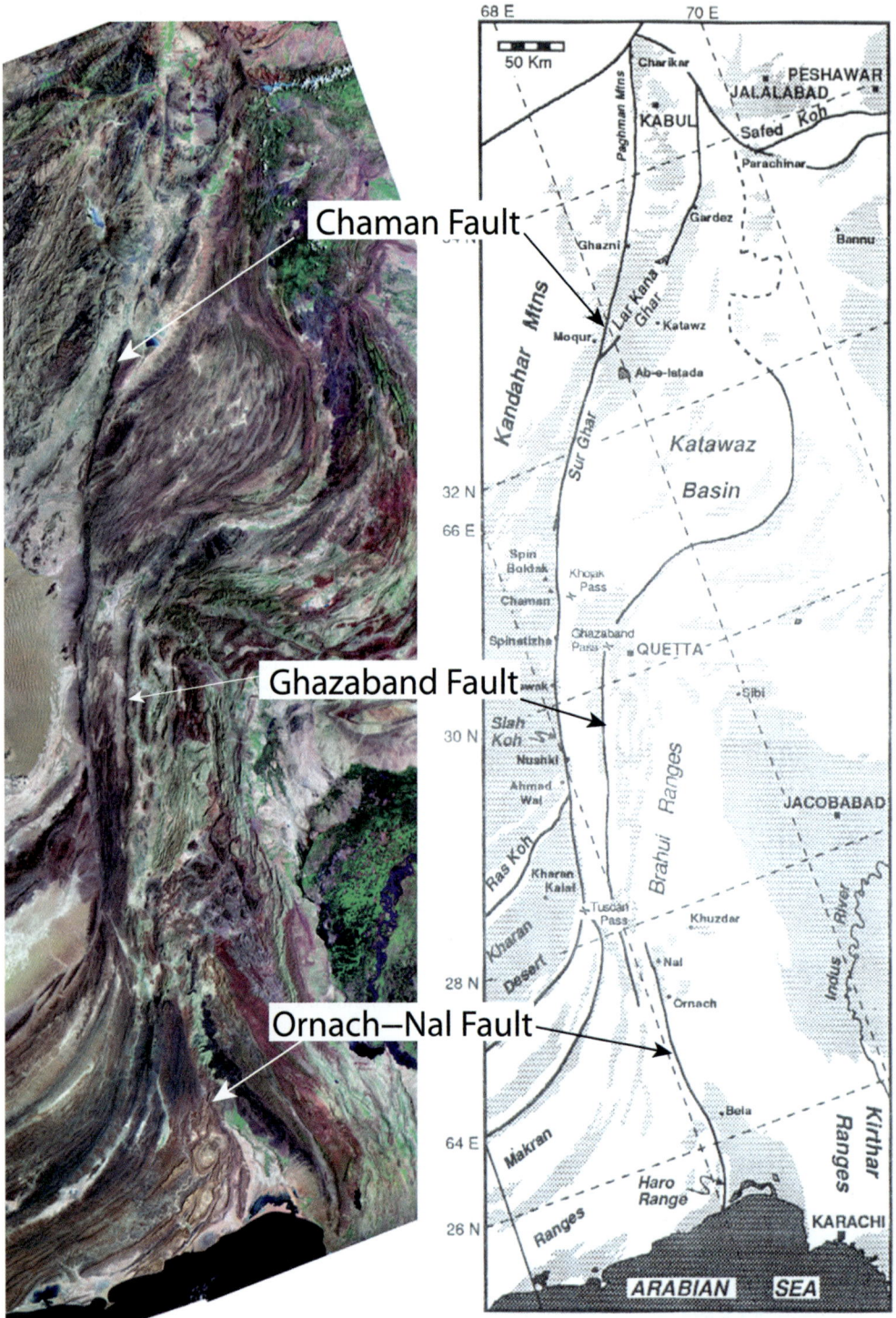

Fig. 12. Landsat image of the Chaman Fault (bands 7, 4 and 2) and its tectonic interpretation, after Lawrence *et al.* (1992). It should be notes that the transfer zone is formed by three continuous but disconnected faults: the Chaman, Ghazaband and Ornach–Nal faults.

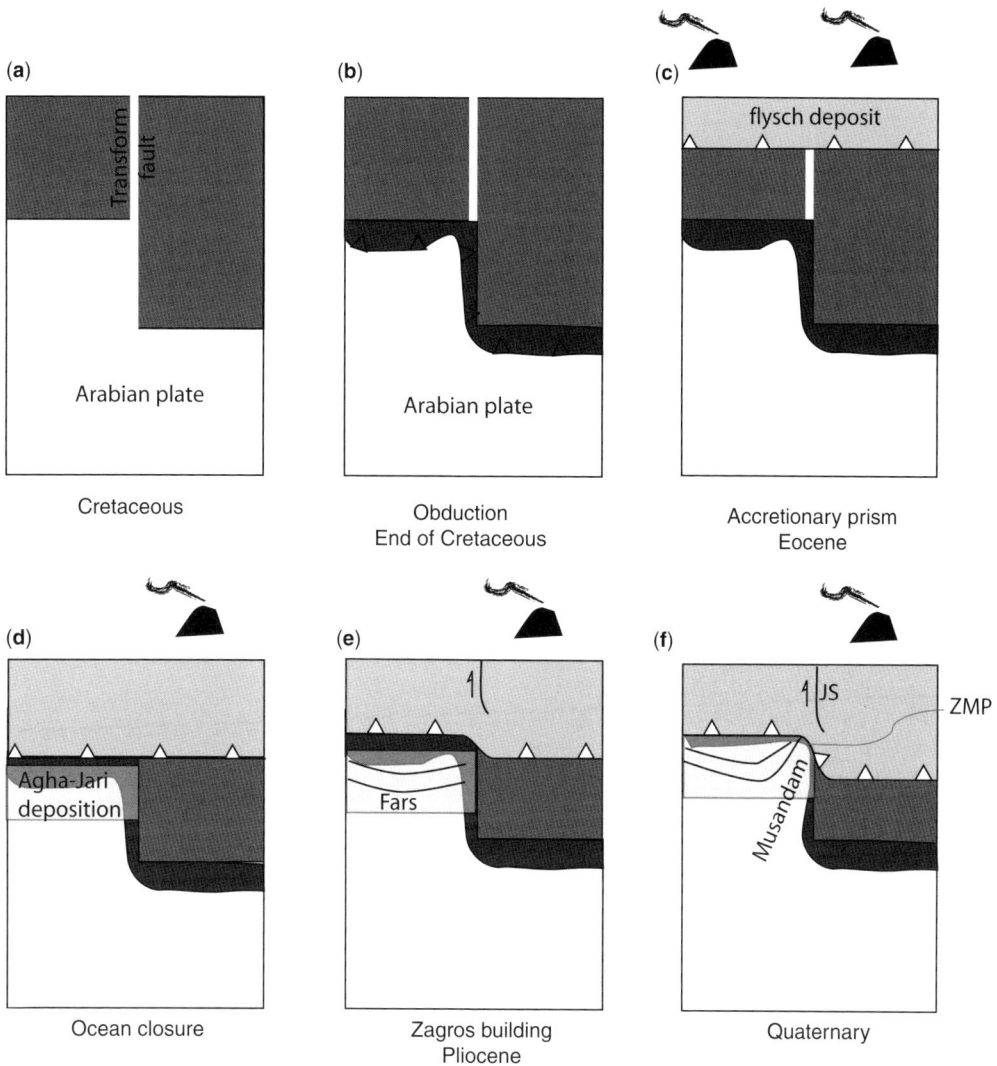

Fig. 13. Sketch of the study area history. (**a**) Initial setting: a former passive margin south of the Tethys, cut by a transform fault. (**b**) Obduction occurred near the end of the Cretaceous. (**c**) The Tethys northern margin had been subducting for some time when the accretionary prism began to be built (Eocene). (**d**) After ocean closure and emergence, the Agha-Jari formation was deposited. (**e**) Zagros mountain building during the Pliocene; onset of Zagros–Makran syntaxis. (**f**) Present setting: JS and ZMP are, respectively, the Jiroft–Sabzevaran and Zendan–Minab–Palami fault systems. The Musandam topographic high causes the Fars Arc curvature.

where only the sediment cover is scraped off. It is tempting to relate the onset of the Zagros–Makran transition zone to the initiation of thin-skinned tectonic shortening in the Zagros, which from then on was very different from the Makran tectonic style, where only the sediments overlying the oceanic crust are affected by deformation. The faults and folds forming the system are currently in a young stage and they sometimes undergo some change, as indicated by stress tensor orientation changes. At present the system is divided into two fault systems, which will endure, whereas their internal organization will simplify to localize in a single and continuous structure, although some partitioning is likely to continue. Obviously, such a transition is a lithospheric-scale deformation zone.

Conclusion

This review of studies on the Zagros–Makran transition zone clarifies many points, giving a coherent overview of its structural setting and behaviour.

(1) The location of this transition is dictated by the past. Indeed, it corresponds to a coastline offset by a transform fault at the time of ocean opening (Fig. 13a).

(2) There is no evidence that currently the two deformation zones, the Zagros and Makran, are connected by a transform fault. Numerous faults accommodate the deformation, organized in two fault systems. Some of the faults could be of crustal extent.

(3) The current stress state is transpressional with a NE–SE-trending σ_1; the associated strain is strike-slip with some transpressional component.

(4) An important change in tectonic style occurred at some time in the Mio-Pliocene, contemporaneous with other changes widely recorded in the Middle East, and in the Zagros in particular. This may indicate the initiation of the zone as a transform zone between the Zagros and Makran, which from then on evolved differently (Fig. 13b–d). This initiation may be closely related to the change from thin-skinned to thick-skinned tectonics at the same time in the Zagros, whereas in Makran tectonics remained unchanged.

(5) At depth, a northeastward dipping plane is linked to the Zagros; this is probably an underthrust slice. This plane seems to connect to the Zendan Fault at the surface.

(6) An important clue is whether Musandam (Oman) acts as an indentor. Our conclusion is that there is no first-order indentation (Fig. 13e). However, it could be a second-order indentation explaining the modern Fars Arc curvature (Fig. 13f).

(7) Laboratory experiments and a real-world analogue help us to imagine the future of such a transform zone: the fault zones are likely to simplify in locating the deformation in a single low-segmented fault. There is no clear evidence at present, however, that the fault zones will coalesce into a single one.

The tectonic study is indebted to the ISIS programme for SPOT satellite image acquisition (©CNES 2004 to 2007, distribution SPOT images S. A.). We wish to thank R. Walker and an anonymous reviewer for their constructive reviews. This work benefited from a 10 year collaboration between Iranian and French scientists, who all participated in data collection and discussions.

References

ALAVI, M. 1994. Tectonics of the Zagros orogenic belt of Iran: new data and interpretations. *Tectonophysics*, **229**, 211–238.

ALLEN, M., JACKSON, J. & WALKER, R. 2004. Late Cenozoic reorganization of the Arabia–Eurasia collision and the comparison of short-term and long-term deformation rates. *Tectonics*, **23**, TC2008, doi: 10.1029/2003TC001530.

AUBOURG, C. & ROBION, P. 2002. Composite ferromagnetic fabrics (magnetite, greigite) measured by AMS and partial AARM in weakly strained sandstones from western Makran, Iran. *Geophysical Journal International*, **151**, 729–737.

AUBOURG, C., SMITH, B. *ET AL.* 2004. Post-Miocene shortening pictured by magnetic fabric across the Zagros–Makran syntaxis. *In*: SUSSMAN, A. J. & WEIL, A. B. (eds) *Orogenic Curvature: Integrating Paleomagnetic and Structural Analyses*. Geological Society of America, Special Papers, **383**, 17–40.

AUBOURG, C., SMITH, B., BAKHTARI, H. R., GUYA, N. & ESHRAGHI, A. 2008. Tertiary block rotations in the Fars Arc (Zagros, Iran). *Geophysical Journal International*, **173**, 659–673.

AVERBUCH, O., FRIZON DE LAMOTTE, D. & KISSEL, C. 1992. Magnetic fabric as a structural indicator of the deformation path within a fold–thrust structure: a test case from the corbières (NE Pyrenees, France). *Journal of Structural Geology*, **14**, 461–474.

BAKHTARI, H. R., DE LAMOTTE, D. F., AUBOURG, C. & HASSANZADEH, J. 1998. Magnetic fabrics of Tertiary sandstones from the Arc of Fars (Eastern Zagros, Iran). *Tectonophysics*, **284**, 299–316.

BAYER, R., SHABANIAN, E. *ET AL.* 2003. Active deformation in the Zagros–Makran transition zone inferred from GPS measurements in the interval 2000–2002 (abstracts EGS 2003). *Geophysical Research Abstracts*, **5**, 05891.

BAYER, R., CHERY, J. *ET AL.* 2006. Active deformation in Zagros–Makran transition zone inferred from GPS measurements. *Geophysical Journal International*, **165**, 373–381.

BERBERIAN, M. & KING, G. C. P. 1981. Towards a paleogeography and tectonic evolution of Iran. *Canadian Journal of Earth Sciences*, **18**, 210–265.

BEUN, N., BORDER, P. & CARBONNEL, J. 1979. Premières données quantitative relatives au coulissage du décrochement de Chaman (Afghanistan du sud-est). *Comptes Rendus de l'Académie des Sciences*, **288**, 931–934.

BLANC, E. J.-P., ALLEN, M. B., INGER, S. & HASSANI, H. 2003. Structural styles in the Zagros Simple Folded Zone, Iran. *Journal of the Geological Society, London*, **160**, 401–412.

BYRNE, D. E., SYKES, L. R. & DAVIS, D. M. 1992. Great thrust earthquakes and aseismic slip along the plate boundary of the Makran subduction zone. *Journal of Geophysical Research*, **97**, 449–478.

CARBON, D. 1996. *Tectonique post-obduction des montagnes d'Oman dans le cadre de la convergence Arabie–Iran*. PhD thesis, Université Montpellier II.

COTTON, J. T. & KOYI, H. A. 2000. Modeling of thrust fronts above ductile and frictional detachments: Application to structures in the Salt Range and Potwar Plateau, Pakistan. *Geological Society of America Bulletin*, **112**, 351–363.

DAVIS, D. M. & LILLIE, R. J. 1994. Changing mechanical response during continental collision: active examples from the foreland thrust belts of Pakistan. *Journal of Structural Geology*, **16**, 21–34.

DEMETS, C., GORDON, R. G., ARGUS, D. F. & STEIN, S. 1990. Current plate motions. *Geophysical Journal International*, **101**, 425–478.

EDGELL, H. S. 1996. Salt tectonism in the Persian Gulf Basin. *In*: ALSOP, J. L., BLUNDELL, D. J. & DAVISON, I. (eds) *Salt Tectonics*. Geological Society, London, Special Publications, **100**, 129–151.

FAURE-MURET, A. & CHOUBERT, G. 1971. Aperçu de l'évolution structurale de l'Iran. *In*: CHOUBERT, G. (ed.) *Tectonique de l'Afrique*. Sciences de la Terre, **6**, 141–151.

HARMS, J. C., CAPPEL, H. N. & FRANCIS, D. C. 1984. The Makran coast of Pakistan: its stratigraphy and hydrocarbon potential. *In*: HAQ, B. U. & MILLIMAN, J. D. (eds) *Marine Geology and Oceanography of Arabian Sea and Coastal Pakistan*. Van Nostrand Reinhold, New York, 3–26.

HAY, W. W., DECONTO, R. *ET AL.* 1999. Alternative global Cretaceous paleogeography. *In*: BARRERA, E. & JOHNSON, C. (eds) *The Evolution of Cretaceous Ocean/Climate Systems*. Geological Society of America, Special Papers, **332**, 1–47.

HESSAMI, K., KOYI, H. A. & TALBOT, C. J. 2001. The significance of strike-slip faulting in the basement of the Zagros fold and thrust belt. *Journal of Petroleum Geology*, **24**, 5–28.

JACOB, K. H. & QUITTMEYER, R. L. 1979. The Makran region of Pakistan and Iran: Trench–arc system with active plate subduction. *In*: FARAH, A. & DE JONG, K. A. (eds) *Geodynamics of Pakistan*. Geological Survey of Pakistan, Quetta, 305–317.

JAMES, G. A. & WYND, J. G. 1965. Stratigraphic nomenclature of Iranian oil consortium agreement area. *AAPG Bulletin*, **49**, 2162–2245.

KADINSKY-CADE, K. & BARAZANGI, M. 1982. Seismotectonics of Southern Iran: the Oman Line. *Tectonics*, **1**, 389–412.

KOPP, C., FRUEHN, J., FLUEH, E. R., REICHERT, C., KUKOWSKI, N., BIALAS, J. & KLAESCHEN, D. 2000. Structure of the Makran subduction zone from wide angle and reflection seismic data. *Tectonophysics*, **329**, 171–191.

LAWRENCE, R. D., KHAN, S. H. & NAKATA, T. 1992. Chaman Fault, Pakistan–Afghanistan. *Annales Tectonicae Special Issue*, **VI** (Supplement), 196–223.

MASSON, F., ANVARI, M. *ET AL.* 2007. Large-scale velocity field and strain tensor in Iran inferred from GPS measurements: new insight for the present-day deformation pattern within NE Iran. *Geophysical Journal International*, **170**, 436–440.

MCCALL, G. J. H. 1997. The geotectonic history of the Makran and adjacent areas of southern Iran. *Journal of Asian Science*, **15**, 517–531.

MCCALL, G. J. H., MORGAN, K. H. *ET AL.* 1985. Minab quadrangle map 1:250 000 and explanatory text. Geological Survey of Iran, Tehran.

MCCLUSKY, S. M., REILLINGER, R., MAHMOUD, S., BEN SARI, D. & TEALEB, A. 2003. GPS constraints on Africa (Nubia) and Arabia plate motions. *Geophysical Journal International*, **155**, 126–138.

MCQUARRIE, N., STOCK, J. M., VERDEL, C. & WERNICKE, B. P. 2003. Cenozoic evolution of

Neotethys and implications for the causes of plate motions. *Geophysical Research Letters*, **30**, 2036, doi: 10.1029/2003GL017992.

MOLINARO, M., GUEZOU, J. C., LETURMY, P., ESHRAGHI, S. A. & FRIZON DE LAMOTTE, D. 2004. The origin of changes in structural style across the Bandar Abbas syntaxis, SE Zagros (Iran). *Marine and Petroleum Geology*, **21**, 735–752.

MOLINARO, M., LETURMY, P., GUEZOU, J. C., FRIZON DE LAMOTTE, D. & ESHRAGHI, S. A. 2005*a*. The structure and kinematics of the southeastern Zagros fold–thrust belt, Iran: from thin-skinned to thick-skinned tectonics. *Tectonics*, **24**, TC3007.

MOLINARO, M., ZEYEN, H. & LAURENCIN, X. 2005*b*. Lithospheric structure beneath the south-eastern Zagros Mountains, Iran: recent slab break-off? *Terra Nova*, **17**, 1–6.

PARES, J. M., VAN DER PLUIJM, B. A. & DINARES-TURELL, J. 1999. Evolution of magnetic fabrics during incipient deformation of mudrocks (Pyrenees, northern Spain). *Tectonophysics*, **307**, 1–14.

PEYRET, M., BAYER, R. *ET AL.* 2009. Present-day strain distribution across the Minab-Zendan-Palami fault system from dense GPS transects. *Geophysics Journal International*, **179**, 751–762, doi: 10.111/j.1365-246X.2009.04321x.

PLATT, J. P., LEGGETT, J. K., YOUNG, J., RAZA, H. & ALAM, S. 1985. Large-scale underplating in the Makran accretionary prism, southwest Pakistan. *Geology*, **13**, 507–511.

PLATT, J. P., LEGGETT, J. K. & ALAM, S. 1988. Slip vectors and fault mechanics in the Makran accretionary wedge, southwest Pakistan. *Journal of Geophysical Research*, **93**, 7955–7973.

RAVAUT, P., CARBON, D., RITZ, J. F., BAYER, R. & PHILIP, H. 1998. The Sohar Basin, Western Gulf of Oman: description and mechanisms of formation from seismic and gravity data. *Marine and Petroleum Geology*, **15**, 359–377.

REGARD, V., BELLIER, O. *ET AL.* 2004. Accommodation of Arabia–Eurasia convergence in the Zagros–Makran transfer zone, SE Iran: a transition between collision and subduction through a young deforming system. *Tectonics*, **23**, TC4007.

REGARD, V., BELLIER, O. *ET AL.* 2005*a*. Cumulative right-lateral fault slip rate across the Zagros–Makran transfer zone: role of the Minab–Zendan fault system in accommodating Arabia–Eurasia convergence in southeast Iran. *Geophysical Journal International*, **162**, 177–203.

REGARD, V., FACCENNA, C., MARTINOD, J. & BELLIER, O. 2005*b*. Slab pull and indentation tectonics: insights from 3D laboratory experiments. *Physics of the Earth and Planetary Interiors*, **149**, 99–113.

REGARD, V., BELLIER, O. *ET AL.* 2006. [10]Be dating of alluvial deposits from Southeastern Iran (the Hormoz Strait area). *Palaeogeography, Palaeoclimatology, Palaeoecology*, **242**, 36–53.

RICOU, L.-E., BRAUD, J. J. & BRUNN, J. H. 1977. Le Zagros. *In*: *Livre à la memoire de Albert F. de Lapparent (1905–1975) consacre aux Recherches géologiques dans les chaînes alpines de l'Asie du Sud-Ouest*. Mémoires hors série Société Géologique de France, **8**, 33–52.

Ross, D. A., Uchupi, E. & White, R. S. 1986. The geology of the Persian Gulf–Gulf of Oman region: a synthesis. *Reviews of Geophysics*, **24**, 537–556.

Schluter, H. U., Prexl, A., Gaedicke, C., Roeser, H., Reichert, C., Meyer, H. & von Daniels, C. 2002. The Makran accretionary wedge: sediment thicknesses and ages and the origin of mud volcanoes. *Marine Geology*, **185**, 219–232.

Sharland, P. R., Archer, R. *et al.* 2001. *Arabian Plate Sequence Stratigraphy*. GeoArabia, Bahrain.

Smith, B., Aubourg, C., Guezou, J. C., Nazari, H., Molinaro, M., Braud, X. & Guya, N. 2005. Kinematics of a sigmoidal fold and vertical axis rotation in the east of the Zagros–Makran syntaxis (southern Iran): Paleomagnetic, magnetic fabric and microtectonic approaches. *Tectonophysics*, **411**, 89–109.

Soeding, E. 1999. ODSN Plate Tectonic Reconstruction Service. World Wide Web Address: http://www.odsn.de/odsn/services/paleomap/paleomap.html.

Szabo, F. & Kheradpir, A. 1978. Permian and Triassic stratigraphy, Zagros Basin, southwest Iran. *Journal of Petroleum Geology*, **1**, 57–82.

Talebian, M. & Jackson, J. 2002. Offset on the Main Recent Fault of NW Iran and implications for the late Cenozoic tectonics of the Arabia–Eurasia collision zone. *Geophysical Journal International*, **150**, 422–439.

Talebian, M. & Jackson, J. 2004. A reappraisal of earthquake focal mechanisms and active shortening in the Zagros mountains of Iran. *Geophysical Journal International*, **156**, 506–526.

Tatar, M., Hatzfeld, D., Martinod, J., Walpersdorf, A., Ghafori-Ashtiany, M. & Chery, J. 2002. The present-day deformation of the central Zagros from GPS measurements. *Geophysical Research Letters*, **29**, doi: 10.1029/2002GL015427.

Vernant, P., Nilforoushan, F. *et al.* 2004. Contemporary crustal deformation and plate kinematics in Middle East constrained by GPS Measurements in Iran and Northern Oman. *Geophysical Journal International*, **157**, 381–398.

Walker, R. & Jackson, J. 2002. Offset and evolution of the Gowk fault, SE Iran; a major intra-continental strike-slip system. *Journal of Structural Geology*, **24**, 1677–1698.

White, R. S. 1982. Deformation of the Makran accretionary sediment prism in the Gulf of Oman (north-west Indian Ocean). *In*: Leggett, J. K. (ed.) *Trench and Fore-Arc Geology: Sedimentation and Tectonics on Modern and Ancient Active Plate Margins*. Geological Society, London, Special Publications, **10**, 357–372.

White, R. S. & Ross, D. A. 1979. Tectonics of the Western Gulf of Oman. *Journal of Geophysical Research*, **84**, 3479–3489.

Yamini-Fard, F. 2003. *Sismotectonique et structure lithosphérique de deux zones de transition dans le Zagros (Iran): la zone de Minab et la zone de Qatar–Kazerun*. PhD thesis, Université J. Fourier, Grenoble I.

Yamini-Fard, F., Hatzfeld, D., Farahbod, A. M., Paul, A. & Mokhtari, M. 2007. The diffuse transition between the Zagros continental collision and the Makran oceanic subduction (Iran): microearthquake seismicity and crustal structure. *Geophysical Journal International*, **170**, 182–194.

Yamini-Fard, F. & Hatzfeld, D. 2008. Seismic structure beneath Zagros–Makran transition zone (Iran) from teleseismic study: seismological evidence for underthrusting and buckling of the Arabian plate beneath central Iran. *Journal of Seismology and Earthquake Engineering*, **10**, 11–24.

Mesozoic extensional brittle tectonics of the Arabian passive margin, inverted in the Zagros collision (Iran, interior Fars)

PAYMAN NAVABPOUR[1]*, JACQUES ANGELIER[2]† & ERIC BARRIER[3]

[1]*Universität Jena, Institut für Geowissenschaften, Burgweg 11, D-07749 Jena, Germany*

[2]*Géoscience Azur, La Darse, BP 48, 06235 Villefranche-Sur-Mer Cedex, France*

[3]*Université Pierre et Marie Curie, Case 129, 75252 Paris Cedex 05, France*

*Corresponding author (e-mail: payman.navabpour@gmail.com)

†Deceased

Abstract: The present Zagros mountain belt of SW Iran is known to be the former NE Arabian passive continental margin of the southern Neo-Tethyan basin, which originated by Permian–Triassic rifting, and has a late Cenozoic collisional imbricate structure. We carried out brittle tectonic analyses of syndepositional normal fault slip data in the High Zagros Belt of the Fars Province to reconstruct the extensional deformation of the passive margin during the Mesozoic era in terms of stress tensor inversion. This reconstruction revealed two main directions of extension, developing from a north–south margin-oblique trend to a NE–SW margin-perpendicular one. Considering the basement structures and the existence of the basal Infracambrian salt detachment, we infer that a transtensional extension could have initiated two major periods of crustal stretching: a Permian–Triassic thick-skinned phase with the basement faults developing in an oblique rifting, and a Mesozoic thin-skinned phase with the sedimentary cover being affected by successive extensional structures and block tilting. This extensional tectonic history probably continued during the early Tertiary period, prior to the continental collision. Fault slip geometries and structural patterns of both the Mesozoic extension and the late Cenozoic compression indicate inversion of the inherited structures in the Zagros collision during the subsequent thin- and thick-skinned stages of crustal shortening.

The Zagros fold-and-thrust belt of Iran is a result of the Alpine orogenic events (Ricou *et al.* 1977) in the Alpine–Himalayan mountain range. It extends in a NW–SE direction from eastern Turkey to the Strait of Hormuz in southern Iran (Fig. 1). The High Zagros Belt (HZB, also known as the Imbricate Zone) marks the northeastern part of the former Arabian passive margin, which is separated from the Iranian plate along the Main Zagros Thrust (MZT) and the Main Recent Fault (MRF; Braud 1971; Tchalenko & Braud 1974; Berberian 1995). The two major parallel domains of the Sanandaj–Sirjan Metamorphic Zone (SSMZ) and the Urumieh–Dokhtar Magmatic Arc (UDMA), to the NE of the MZT, are presumed to be the result of NE-dipping subduction of the Neo-Tethyan oceanic crust beneath the Iranian continental active margin (Berberian & King 1981). The subduction process started in the late Jurassic period (Stampfli & Borel 2002) and terminated with the continental collision between the Arabian and Iranian plates in Oligocene–Miocene times (Sherkati & Letouzey 2004; Agard *et al.* 2005).

SW of the HZB, the belt is characterized by the major morphotectonic units of the Zagros Simply Fold Belt (ZSFB), Dezful Embayment and the Persian Gulf–Mesopotamian Lowland. The High Zagros Fault (HZF) marks the ZSFB–HZB boundary, with a concentration of seismic activity mainly to the SW (Talebian & Jackson 2004). From NW to SE, the ZSFB itself is divided into major sub-areas of the Lurestan Area, Izeh Zone and the Fars Arc, by the transverse strike-slip faults of the Balarud, Izeh and Kazerun Fault Zones, respectively. The Mountain Front Fault (MFF) bounds the ZSFB to the SW and separates it from the Dezful Embayment. The Zagros Front separates the Persian Gulf–Mesopotamian basin from the entire mountain belt. The presence of all these major belt-parallel and belt-oblique fault zones (Fig. 1), that are known to be pre-existing deep-seated crustal structures, facilitated different morphotectonic developments of the belt segments and highlighted significant differences in earlier palaeostratigraphic history and structural style (Berberian 1995; Sepehr & Cosgrove 2004).

Although the tectonostratigraphic evolution and deformation styles of the Zagros mountain belt have been studied in detail from stratigraphical, structural, seismic and geophysical points of view,

From: LETURMY, P. & ROBIN, C. (eds) *Tectonic and Stratigraphic Evolution of Zagros and Makran during the Mesozoic–Cenozoic*. Geological Society, London, Special Publications, **330**, 65–96.
DOI: 10.1144/SP330.5   0305-8719/10/$15.00 © The Geological Society of London 2010.

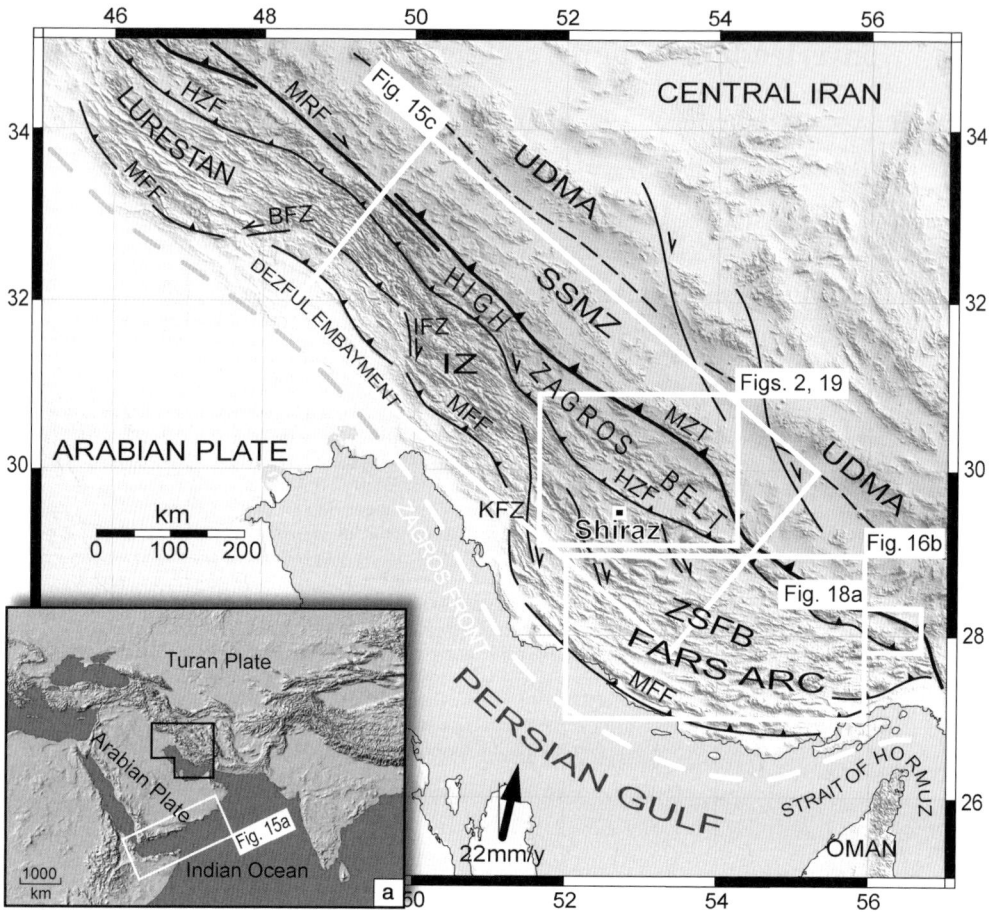

Fig. 1. Index map of the study area within the Zagros fold-and-thrust belt (from Navabpour *et al.* 2007*a*), showing the structural domains in SW Iran. Topography is from GTOPO30. UDMA, Urumieh–Dokhtar Magmatic Arc; SSMZ, Sanandaj–Sirjan Metamorphic Zone; MZT, Main Zagros Thrust; MRF, Main Recent Fault; HZF, High Zagros Fault; ZSFB, Zagros Simply Folded Belt; MFF, Mountain Front Fault; IZ, Izeh Zone; BFZ, Balarud Fault Zone; IFZ, Izeh Fault Zone; KFZ, Kazerun Fault Zone. Black arrow indicates GPS convergence vector from Vernant *et al.* (2004). Inset (**a**) is regional location map.

almost no attempt has been made to analyse the evolution of the HZB during the tectonic history of the NE Arabian Passive Margin (NEAPM) in terms of brittle tectonics. In this paper, we aim at deciphering the Mesozoic extensional brittle tectonics of the HZB of interior Fars (Fig. 1). Our study is primarily based on systematic stress inversion of brittle tectonic data with particular attention to the Mesozoic syndepositional faulting that reflects deep-seated extensional structures during the history of the NEAPM. However, the extensional history of the HZB is not limited to the Mesozoic era. We also observed Cenozoic normal faults associated with tension cracks that mainly developed as a function of the Cenozoic

compressional deformation in this region (see Navabpour *et al.* 2007*a*).

Geological setting

The HZB is the most uplifted and eroded part of the Zagros mountain belt. Folds and thrusts trending NW–SE are the most common tectonic features of the HZB in the study area (Fig. 2). These features change into a highly imbricate structure to the NE (Molinaro *et al.* 2004, 2005*a*). Most of the affected rock units are shelf and open marine limestones and marls of the Mesozoic Arabian passive margin. Towards the SW, in the ZSFB, large folds have

Fig. 2. Geological map of the study area in the interior Fars (from Navabpour *et al.* 2007*a*), showing the position of the sites where brittle structures were collected and measured. The map is mainly redrawn and simplified based on the geological map of NIOC (1977) with some changes based on NIOC (1979) and GSI (1985, 1990). The map is based on the major difference in lithology from the Palaeozoic and metamorphic rocks of the SSMZ to the Mesozoic rocks of the HZB and the Tertiary rocks of the ZSFB. AA′, location of cross-section in the map. Inset (**a**) shows the strike rose distribution of the folded strata ($n = 136$) with an average trend of N123°. Stratigraphic details are given in Figure 3. Location is shown in Figure 1; abbreviations are as in Figure 1.

Age		SW	**Fars Stratigraphic Units**	NE	Tectonic history
Ma	Geological				

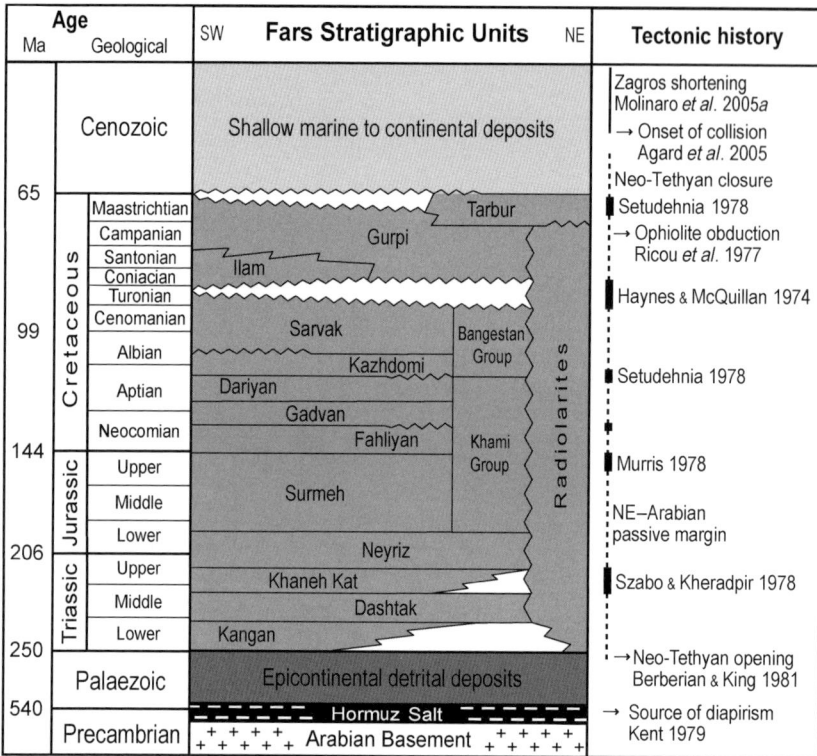

Fig. 3. Stratigraphic chart of the Fars region modified and simplified from Beydoun *et al.* (1992) and Motiei (1993). Absolute ages are from Palmer & Geissman (1999). Summary of tectonic history is added for a more complete account. Bold black lines show the various reported unconformities during the time of existence of the NEAPM (dashed line).

affected the shallow water carbonates and the clastic deposits of the Tertiary period, which belong to the platform domain of the Arabian margin. These two major domains are separated by the HZF (Figs 1 & 2). In the study area, the metamorphic rocks of the SSMZ thrust over the HZB along the MZT, which is expected to be the surface trace of the Arabia–Eurasia suture zone (Berberian & King 1981; Paul *et al.* 2003, 2006; Agard *et al.* 2005, 2006). In addition, a narrow belt of ophiolite and radiolarite nappe is situated between the MZT and the HZB to the east of the area (Fig. 2). This thrust sheet of ophiolite and radiolarite is considered as a remnant of the Neo-Tethyan oceanic crust (Ricou 1968).

Stratigraphy

The entire Zagros belt is formed by more than 10 km thickness of Palaeozoic–Cenozoic sedimentary layers, detached from metamorphic continental basement above the thick Infracambrian Hormuz Salt Formation (Falcon 1974) (Figs 2 & 3). Geological evidence suggests that Iran was a part of the Afro-Arabian continental platform, at least from the

late Precambrian to late Palaeozoic times, with epicontinental detrital deposits (Stocklin 1968; Nabavi 1976). During the Permian period, the Arabian foreland (i.e. the HZB) gradually subsided and the sea invaded most of the area. This event is marked by a significant change in sedimentation over the Arabian foreland, from dominantly Palaeozoic clastic sediments to the carbonate series of Permian, Mesozoic and Tertiary times (Powers 1968; see also Fig. 3). There is evidence of volcanic activity associated with the Permian–Triassic rifting in the HZB, where a few amygdaloidal basaltic flows of Permian age crop out (Thiele *et al.* 1968; Berberian 1977). In the NE of the HZB, the late Triassic black marls of the Arabian foreland are covered by a thick sequence of Jurassic–Cretaceous deep-sea red radiolarian chert and siliceous limestone (Ricou 1974, 1976) that belongs to the Neo-Tethyan oceanic environment (Berberian & King 1981; Ricou 1994).

To the SW, the Arabian foreland (i.e. the Zagros basin) steadily subsided along inherited faults during the Mesozoic era. The marine carbonate sedimentary regime persisted during the early Triassic period, with deposition of the Kangan Formation (Szabo & Kheradpir 1978; see also Fig. 3).

Regressive conditions then occurred in the middle Triassic period, resulting in deposition of the eva- porites of the Dashtak Formation (Berberian & King 1981). In the HZB, an unconformity is present between the middle Triassic and Jurassic beds. Above the unconformity, the Liassic terrige- nous clastic deposits and the transitional terrigenous to open marine sediments of the Neyriz Formation accumulated (Szabo 1977; Szabo & Kheradpir 1978). Overlying the Neyriz Formation, the marine limestones and marls of the Khami Group (Middle Jurassic–Aptian; Fig. 3) reveal a steadily subsiding basin and continuous sedimentation in the interior Fars, during middle Jurassic–early Cre- taceous times (James & Wynd 1965; Setudehnia 1978).

In the interior Fars, the siltstones and iron- stained glauconitic sandstones found in the upper parts of the Fahliyan and Dariyan Formations indi- cate periods of regression, emergence and erosion in the Neocomian and Late Aptian, respectively (James & Wynd 1965; Setudehnia 1978). The sedi- mentation continued with the shallow marine shales and carbonates of the Bangestan Group (Albian– Turonian; Fig. 3). The conglomerates, breccia, fer- ruginous materials and a weathered zone on top of the Sarvak Formation indicate tectonic activity in most of the interior Fars in the Late Turonian (Berberian & King 1981). East of Arsanjan, a tran- sition occurs between the iron-stained Turonian carbonate platform and deep-sea radiolarian sedi- ments towards the NE (Fig. 2). A submarine channel filled with turbiditic carbonate breccia cut through the radiolarite unit (Haynes & McQuillan 1974). The late Cretaceous sedimentation in most parts of the Zagros basin usually began with neritic carbonates of the Ilam Formation (Santonian–Early Campanian). This carbonate sedimentation was followed by deeper water conditions, with marls and shales of the Gurpi Formation (Campanian– Maastrichtian), which covered almost the entire Zagros basin. In the late Maastrichtian, a general regression created a major Cretaceous–Cenozoic unconformity throughout the Zagros (James & Wynd 1965; Setudehnia 1978).

The Cenozoic era was characterized by a regional change in sedimentary facies, from an open marine to a continental environment. During this era, the Palaeocene–Eocene neritic shales and marls of the Pabdeh Formation changed to the Eocene– Oligocene shallow water limestones of the Jahrom and Asmari Formations and were covered by the Miocene evaporites and red beds of the Gachsaran– Agha Jari Formations and the Pliocene–Pleistocene molasse-type Bakhtyari conglomerates. All these facies changes involved angular unconformities that are thought to be the result of folding and erosion that originated from the Tertiary Neo- Tethyan closure process and collision between the Arabian and Iranian continental plates (Berberian & King 1981; Hessami et al. 2001).

Structures and tectonic setting

Most of the linear facies boundaries trending NW– SE and the related different subsidence rates during the Mesozoic era were the result of normal faulting parallel to the NEAPM (Berberian & King 1981; Koop & Stoneley 1982). This is supported by the idea that the present-day seismicity of the Zagros mountain belt is a result of inverse reactivation on pre-existing NE-dipping normal faults during the Cenozoic continental collision between the Arabian and Iranian plates and the following plate conver- gence (Jackson 1980; Jackson & McKenzie 1984). It has also been shown that the conglomerates on top of the Sarvak Formation in SE Zagros resulted from normal faulting within the sedimentary basin (Stoneley 1990).

According to Chauvet et al. (2004), syndeposi- tional normal faults are observed in the Permian car- bonate platform of Oman, south of the Hormuz Strait. Conjugate normal faults with metre-scale offsets and block tilting associated with extensional structures in that region indicate a north–south direction of extension. Those researchers have shown that the tilted structures, which were sealed by an unconformity associated with dolomitization, reveal a middle Permian extensional phase in the Arabian margin during the first stage of the Neo- Tethyan rifting. The analysis of seismic profiles revealed that the growth of the Permian–Triassic strata was related to the hanging wall of a buried normal fault in the Dezful Embayment (Sepehr & Cosgrove 2004), evidencing a Permian–Triassic extension in the Zagros. This evidence suggested that at that time the ZSFB was the domain of a fault- controlled horst-and-graben structure with a greater subsidence rate as compared with the HZB (Sepehr & Cosgrove 2004).

The relatively stable oceanic environment of the NEAPM was affected by a southwestward obduction of the Neo-Tethyan ophiolite–radiolarite nappe (the Semail Ocean of Stampfli 2000) in the late Cretaceous period (Béchennec et al. 1990; Breton et al. 2004). In the Neyriz area, the shallow water reef limestones of the Upper Campanian– Maastrichtian Tarbur Formation covered the ophio- litic thrust stack unconformably (Ricou 1974). The obduction of the ophiolites was thus thought to have occurred during the Late Santonian–Early Campanian (Berberian & King 1981), as the deposi- tion of the Tarbur limestones postdates the emplace- ment of the ophiolite–radiolarite nappe. However, Stoneley (1990) believed that the emplacement of the ophiolites in the HZB was due to a change in the position of oceanic spreading axis to the south, uplifting of the Arabian shelf margin and a subsequent southwestward gravitational sliding of the oceanic deposits over the shelf. He thus inferred that the emplacement of the ophiolites in the HZB differs from the tectonic compressive obduction that occurred in Oman.

The present-day morphotectonic features of the Zagros mountain belt (Berberian 1995) are a consequence of the late Cenozoic Neo-Tethyan closure and continental collision under a north–south plate convergence (McQuarrie *et al.* 2003; Agard *et al.* 2005). The angular unconformities and facies changes within the Cenozoic sedimentary sequence show that the folding process, which was initiated in the HZB in late Oligocene–early Miocene times (Sherkati *et al.* 2005), was a result of separate tectonic episodes during which the deformation front propagated from NE to SW (Hessami *et al.* 2001). In NW Zagros, the north–south plate convergence is partitioned into a belt-parallel right-lateral slip along the MRF and a belt-perpendicular shortening across the ZSFB (Talebian & Jackson 2004; Authemayou *et al.* 2006). In SE Zagros, structural studies have indicated that the belt has undergone two distinct thin- and thick-skinned north–south crustal shortening episodes since the Miocene epoch (Molinaro *et al.* 2005a; Sherkati *et al.* 2005; Mouthereau *et al.* 2007). This idea is consistent with the recent seismicity of the basement reverse faults under an overall north–south compression (Talebian & Jackson 2004). Detailed brittle tectonic reconstructions have shown that the late Cenozoic deformation involve an anticlockwise reorientation of the compressive stress trend, from a pre-folding NE–SW direction to a post-folding north–south one, associated with a more distributed strike-slip reactivation of the earlier reverse faults across the HZB of interior Fars (Navabpour *et al.* 2007a).

Brittle tectonic analysis and main extensional events

Not only do the tectonic events affecting the HZB differ in nature (compression, extension, wrench faulting), but they also show contrasting stress fields in the interior Fars. We used systematic brittle tectonic analyses, especially in terms of stress tensor inversion, to decipher the various extensional tectonic episodes that have affected the NEAPM during the Mesozoic era. The common steps in these analyses involve data collection in the field, data separation, age recognition, computation of the stress fields and finally characterization and classification of the various events. For most of the brittle structures that we studied, we collected fault slip data (oblique-slip, normal, strike-slip and reverse). These data include the orientation of fault surfaces, slickenside lineation and sense of motion indicators

where available. Tension cracks were also abundant, but there are few data on pressure-solution seams. The bedding attitudes of the sedimentary units were recorded at all sites to reconstruct the initial pre-folding attitudes of strata and related brittle structures where appropriate.

Palaeostress reconstructions

Our palaeostress determinations are based on systematic use of the direct inversion method (Angelier 1984, 1990, 2002). Where data distribution did not allow a reliable stress tensor determination, we used the right dihedra method (Angelier & Mechler 1977). A widespread difficulty in palaeostress determinations was related to the heterogeneity of the datasets in the study area, which is often a result of two or more successive tectonic events involving polyphase brittle deformation. In such cases, computing an average stress regime was meaningless, and data separation into more mechanically homogeneous subsets was compulsory. These subsets fit the different stress states and may reveal distinct tectonic events. Various criteria are used in this relation, including crosscutting relationships between different brittle structures and successive striae observed on fault surfaces that indicate fault reactivation under a new stress field. In addition, the sedimentary layers have been subjected to faulting during both the basin sedimentation of the passive margin and the folding process of the collision. It was thus indispensable to pay close attention to geometrical and relative relationships between folds, brittle structures and syndepositional evidence. Particular attention was paid to the stratigraphic ages of affected rocks and the unconformities revealed by geological mapping (NIOC 1977, 1979; GSI 1985, 1990, 1996a, b, 1999, 2000a–d, 2001a, b). Such information led to more accurate dating of syndepositional tectonic events. The succession of compressional and extensional stress fields during the late Cenozoic collisional history of the HZB has already been discussed elsewhere (Authemayou *et al.* 2006; Navabpour *et al.* 2006, 2007a, b). Below, we present only the relevant brittle tectonic data used to reconstruct the various extensional events within the sedimentary basin of this area.

In the reconstruction of palaeostress directions, consideration of possible block rotation is crucial. In the absence of palaeomagnetic data, horizontal block rotations around vertical axes could not be reconstructed. Geologically, there is no evidence of block rotation related to the mountain-building process, because the trends of the late Cenozoic folds remained consistent throughout the HZB in the study area, as suggested by the rose diagram of the folded strata (Fig. 2a). Both the MZT and the HZF are thought to have acted as reverse faults with no strike-slip movement during the late Cenozoic shortening, after the onset of collision (Molinaro *et al.* 2005*a*) under a general NE–SW direction of compression (Navabpour *et al.* 2007*a*). The earthquake focal mechanisms along the HZF invariably show reverse faulting (see Talebian & Jackson 2004) on pre-existing high-angle faults (Yamini-Fard *et al.* 2006) and thus do not confirm a major active strike-slip movement that could have induced large block rotations across the HZB of interior Fars even under the present-day north–south oblique convergence. Moreover, the consistency of the stress directions obtained from each brittle tectonic event (Navabpour *et al.* 2007*a*) suggests that no significant block rotation has occurred since the Miocene compression, at least in the study region, where sites are densely distributed. However, further to the south, a possible horizontal rotation of the sedimentary cover is evidenced by magnetic fabric studies of the late Tertiary sandstones in the ZSFB of the Fars Arc (Bakhtari *et al.* 1998; Aubourg *et al.* 2004). We thus infer that no major horizontal rotation around vertical axis has occurred in the study area, except the well-documented rotation of Arabia as a whole, which is revealed by plate kinematic reconstructions (e.g. Dercourt *et al.* 2000; McQuarrie *et al.* 2003). On the other hand, all the orientations referred to hereafter are corrected by back-tilting the strata to their initial horizontal attitudes to reconstruct the Mesozoic extensional events that gave way to deep-seated vertical block rotation around horizontal axes and related syndepositional normal faulting.

A summary of palaeostress reconstructions is provided for a more complete account (Table 1), in which data for each stress regime are presented based on their geographical positions and the stratigraphic age of the affected rocks. The geological situation of each site is indicated in the geological map (Fig. 2). As mentioned above, the relative relationships of different normal faults and syndepositional evidence are crucial. For this reason, these data are presented separately in the Table 1. The bedding attitude of each site is given to allow reconstruction of the initial horizontal attitudes of the related brittle structure (BS) with respect to local tilted strata. Calculated stress fields are defined by the direction and plunge of the principal stress axes (σ_1, σ_2 and σ_3), method of analysis (M), number of data (N), ratio of the stress magnitude differences (Φ) and the reduced misfit angle (α), which is acceptable for $\alpha < 25°$. In addition, a quality estimator (Q) is attached to each determination, because local results of the stress calculations cannot be regarded as equal in value. This quality estimator provides a multivariable evaluation between different stress solutions that strictly depends on quantitative (such

as the number of acceptable data and the average misfit level) and semi-quantitative and qualitative aspects (such as the confidence level and accuracy of fault slip measurements). The reader is referred to Angelier (1984, 1990, 2002) for further details on the inversion parameters.

Identification of the Mesozoic extensions

To reconstruct the Mesozoic extensional tectonic history of the NEAPM we used syndepositional evidence within each stratigraphic sequence during which the normal faults have been developed. Two typical main features of such evidence are the sedimentary wedges on the hanging wall of normal faults (Nottvedt *et al.* 1995) and the tilted blocks bounded by normal faults subjected to local erosion and sealed by younger sedimentary layers (Leeder & Gawthorpe 1987). In the first case, during the development of listric normal faults within a sedimentary basin, the hanging-wall block tilts towards the footwall of the fault so that an asymmetric trough is generated. Continuing deposition then fills this asymmetric trough, creating a syntectonic sedimentary wedge. We used the presence of such sedimentary wedges to detect syndepositional normal faults (e.g. site 43; Fig. 4a). In the second case, an extensional episode creates a bookshelf structure with multiple tilted blocks and normal faults that is subsequently covered and sealed by younger sedimentary layers. The time span between the older and younger sediments constrains the age of the extensional event (e.g. site 45; Fig. 4b).

A difficulty in determining the age of extension arises in the absence of syndepositional evidence. In such a case, a systematic back-tilting can be applied for tilted strata (e.g. site 15; Fig. 5a), revealing the pre-tilt attitude of corresponding stress axes. Ambiguities may, however, occur. For instance, it may be difficult to distinguish between pre-tilt normal faults (indicating an early extension) and post-tilt reverse faults (related to a late compression), where the layers have been tilted by folding to a subvertical attitude. A set of conjugate-like normal faults that developed in the sedimentary basin can be observed as conjugate reverse faults at the present-day attitude in the nearly vertical fold flanks. However, a comparison between sites with a variety of stratal dips across the fold structure usually solves the ambiguity (Fig. 5b). Back-tilting the strata will in this case yield a consistent pre-fold attitude for the relevant extension (Fig. 5c). Despite these possibilities, some uncertainties remain where strata are horizontal (unfolded). In such cases, it may be almost impossible to determine the age of the extension by direct observation (except for the maximum age indicated by dating of the affected sedimentary rock). Because of these limitations,

Table 1. *Summary of geological position of brittle structures and reconstructed extensional stress regimes*

Stress regime	Location	Coordinates Long. (E)	Coordinates Lat. (N)	Stratigraphic age	Bedding Strike/dip	Site no.	BS	M	N	σ_1 Dir.	σ_1 Plunge	σ_2 Dir.	σ_2 Plunge	σ_3 Dir.	σ_3 Plunge	Φ	α	Q
Mesozoic syndepositional extensional regimes (Fig. 7)	NE Marvdasht	52.89	29.99	Alb.–Cenom.	014/13W	01	N*	DI	19	282	60	125	28	030	10	0.49	15	A
	NW Arsanjan	53.19	30.10	Aptian	167/10W	17	N*	DI	8	026	65	120	02	211	25	0.45	20	B
	E Arsanjan	53.45	29.92	Alb.–Cenom.	137/05E	19	N	DI	8	010	66	122	09	216	22	0.34	19	B
	E Kazerun	51.90	29.65	Camp.–Maas	149/12W	24	N*	DI	5	264	85	031	03	122	04	0.47	9	A
	NW Arsanjan	53.15	30.09	Aptian	124/33S	35	N*	RD	8	284	76	101	15	191	01	0.48		A
	N Marvdasht	52.91	30.11	Aptian	100/24N	36	N*	DI	10	343	79	098	05	189	01	0.46	11	A
	NW Arsanjan	53.20	30.10	Aptian	138/31W	41	N*	RD	26	331	85	120	04	210	03	0.61		A
	NE Ahmadabad	52.79	30.46	M.–L. Jurassic	032/35E	43	N*	DI	10	151	63	033	13	297	23	0.44	12	A
	NE Ahmadabad	52.77	30.44	M.–L. Jurassic	168/17E	44	N*	DI	10	144	78	272	07	003	09	0.50	9	A
	S Ahmadabad	52.67	30.22	Cenomanian	087/07S	45	N	RD	20	187	72	293	05	024	17	0.48		A
	E Arsanjan	53.45	29.92	Cenomanian	Horizontal	52	N	RD	5	139	86	312	04	042	01	0.49		B
	NW Marvdasht	52.59	30.10	Cenomanian	136/16E	55	N*	DI	10	037	66	269	15	174	18	0.50	5	A
	S Arsanjan	53.26	29.67	Cenomanian	113/34S	59	N*	DI	13	227	73	099	10	006	13	0.47	15	A
	S Eghlid	52.59	30.75	E. Jurassic	092/26N	63	N*	DI	8	359	77	105	04	195	13	0.47	17	B
	S Neyriz	54.19	29.18	Jurassic	096/43S	76	N*	RD	6	295	59	101	31	194	06	0.61		B
	S Neyriz	54.19	29.18	Jurassic	096/43S	76	N*	RD	2	194	00	284	64	104	27	0.63		C
	NE Sarvestan	53.59	29.45	M.–L. Triassic	178/33W	77	N*	DI	4	330	74	101	11	193	12	0.47	12	A
	S Neyriz	54.11	29.18	Jurassic	104/85N	79	N*	DI	8	001	77	266	01	176	13	0.50	9	A
	SW Tashk	53.44	29.66	E. Cretaceous	127/17N	87	N*	DI	25	055	77	271	10	180	07	0.51	11	A
Back-tilted pre-Cenozoic folding extensional regimes (Fig. 8)	N Sepidan	51.97	30.30	Cretaceous	100/26N	04	N*	DI	5	344	66	094	09	187	23	0.48	22	B
	W DashteArjan	51.93	29.63	Camp.–Maas.	028/03E	13	N	DI	13	335	81	121	08	211	05	0.44	13	A
	N Kazerun	51.66	30.02	Camp.–Maas.	142/63E	14	N*	DI	40	227	70	135	01	045	20	0.51	10	A
	SW Ahmadabad	52.58	30.39	Cenomanian	117/20N	18	N*	DI	8	065	67	272	20	179	09	0.52	13	A
	NE Marvdasht	52.93	30.11	Cenomanian	102/73N	40	N*	RD	5	246	55	116	24	014	24	0.53	25	B
	N Ghaderabad	53.30	30.33	Aptian	123/20N	49	N*	DI	15	229	63	105	16	008	21	0.37	19	B
	NW Arsanjan	53.20	30.10	Aptian	150/20W	54	N*	RD	11	330	74	123	15	215	07	0.73		B
	SW Arsanjan	53.16	29.85	M.–L. Jurassic	123/12N	58	N*	DI	13	234	79	121	04	030	10	0.44	22	B
	NE Marvdasht	52.94	29.95	Albian	135/66N	61	N*	RD	4	204	66	311	07	044	23	0.51		B
	SW Eghlid	52.53	30.81	E. Jurassic	141/37E	64	N*	DI	9	090	67	291	22	198	08	0.52	20	B
	W Neyriz	53.89	29.25	Cenomanian	125/43N	81	N*	DI	9	097	84	300	06	210	02	0.52	11	A

BS, brittle structures: N, normal faults (where * indicates back-tilted strata). M, method of stress determination (RD, right dihedra; DI, direct inversion); N, number of fault slip data; Φ, ratio of stress magnitude differences [$\Phi = (\sigma_2 - \sigma_3)/(\sigma_1 - \sigma_3)$]; α, average angle between observed slip and computed shear in degrees (acceptable for $\alpha < 25°$); Q, quality of stress calculation (A, high; B, moderate; C, low).
Bedding attitudes and stress axes are given after magnetic north correction. Site location is shown in Figure 2.

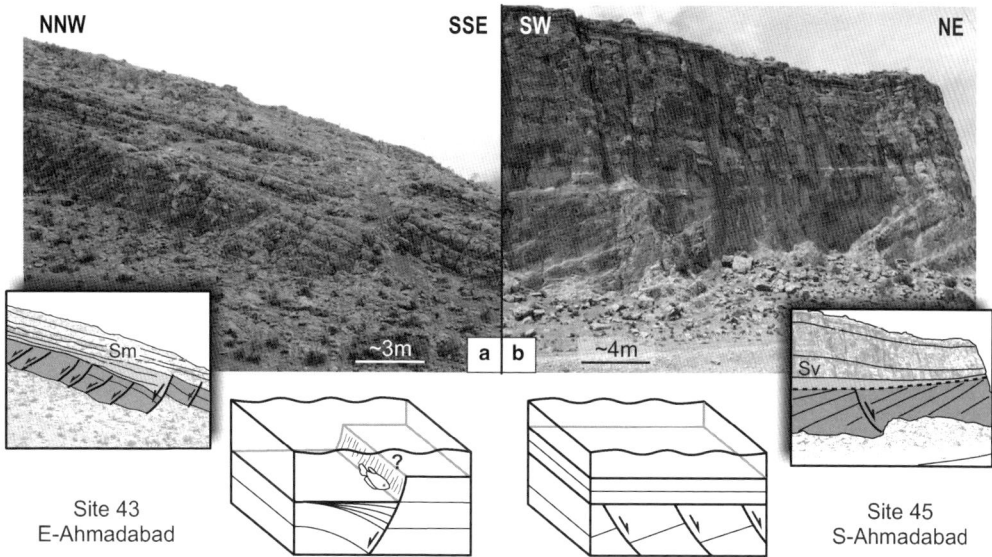

Fig. 4. Syndepositional extensional structures buried within the Mesozoic stratigraphic sequence of the HZB.
(a) Sedimentary wedge in the hanging wall of a palaeo-submarine normal fault observed at site 43, revealing active extension during deposition of the Surmeh Formation (Sm) in the middle–late Jurassic period. (b) Tilted blocks along a normal fault within the Sarvak Formation (Sv) sealed by younger sedimentary layers of the same formation at site 45, indicating an intraformational angular unconformity (dashed line) during a middle Cretaceous phase of extension. In the insets, faulted key-beds are shown in dark grey whereas sedimentary wedges and unaffected beds are shown in light grey. Small 3D schemes of deformation are added for more detail. It should be noted that in each photograph the entire outcrop is formed of the same rock unit.

we paid special attention to syndepositional deformations that could be observed in the field.

For a complete account, we have provided some of the field examples of the Mesozoic syndepositional extensional structures, together with the stress tensors analysed after back-tilting the corresponding fault slip data (Fig. 6). For instance, at site 01, the upper layers of the Sarvak Formation show different thicknesses involved with normal faulting (Fig. 6a). This situation indicates that faulting was active during the Cenomanian. The stress tensor inversion of the normal faults at this site reveals a NE–SW direction for the extensional σ_3 stress axis. At sites 17, 19 and 59, the faulted key-beds of the Sarvak and Dariyan Formations are sealed with unaffected upper layers of the same formations (Fig. 6b–d), indicating extensional tectonic activity during the deposition of these formations during the Cenomanian and Aptian, respectively. The stress tensor inversion of the corresponding normal faults reveals two NE–SW and north–south directions of extension. At sites 36 and 55, a sedimentary wedge is observed above the major normal faults within the Sarvak Formation (Fig. 6e, f), characterizing a north–south direction of extension. At site 77, the Jurassic limestones of the Surmeh Formation seal tilted blocks of the

Upper Triassic Khaneh Kat Formation (Fig. 6g), indicating a north–south direction of extension. A syndepositional north–south direction of extension is also evidenced by normal faults within the Jurassic radiolarite layers of the Neo-Tethyan basin at site 79. The sedimentary layers at this site are folded and tilted to a subvertical present-day attitude (Fig. 6h), presenting conjugate reverse faults. However, after back-tilting the strata to a pre-fold horizontal situation, the unaffected uppermost layers agree with a syndepositional normal faulting.

Numerous outcrops throughout the study area revealed syndepositional Mesozoic normal faulting. Azimuthal distribution of the calculated stress axes for the corresponding fault slip data reveals two major north–south and NE–SW trends for the extensional σ_3 stress axes after back-tilting correction (Fig. 7). The average trends of extension are N006° and N034°, respectively. These directions were reconstructed based on consideration of 85% of the Mesozoic syndepositional structures that could be identified as normal faults striking east–west and NW–SE, respectively. We also measured consistent brittle structures at some sites where no evidence of syntectonic sedimentation could be identified. Azimuthal distribution of calculated stress axes for the corresponding back-tilted fault

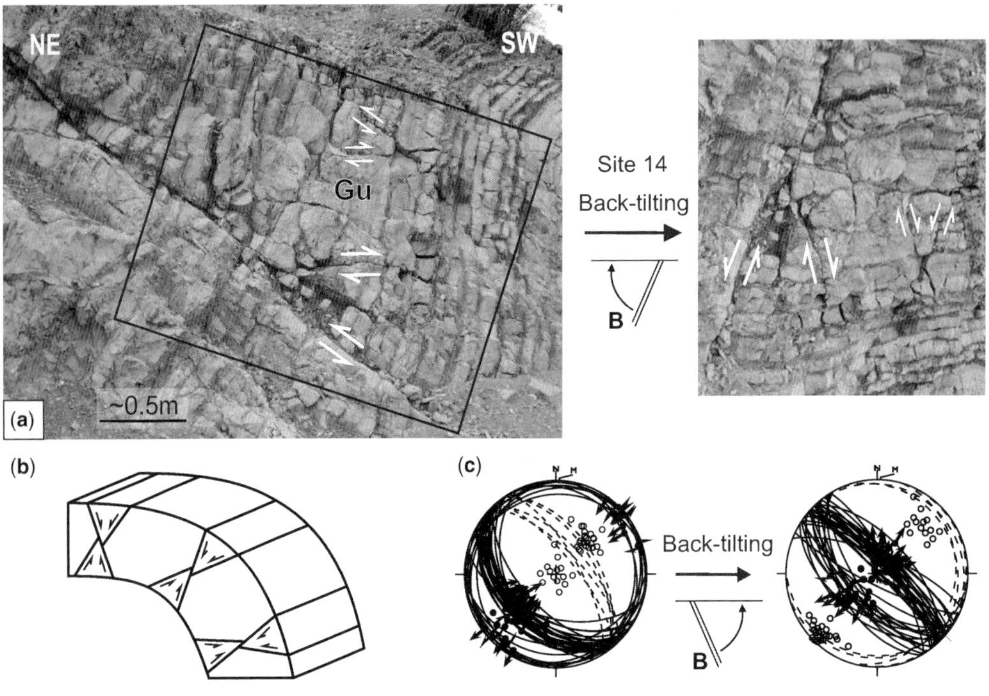

Fig. 5. Pre-folding conjugate normal faults within the upper Cretaceous sedimentary layers, indicating an extension that occurred between the late Cretaceous sedimentation and the middle Cenozoic folding. (**a**) Conjugate reverse faults in the present-day folded strata of the Gurpi Formation (Gu) at site 14 (left), interpreted as pre-folding conjugate normal faults by back-tilting the strata to an initial horizontal attitude (right). (**b**) Schematic illustration of pre-folding conjugate normal faults in different positions within folded strata, characterizing a uniform pre-fold extension. (**c**) Back-tilting presentation for the fault slip data of site 14. In the stereoplots, faults are shown by great circles, including the striae as indicated by filled circles and slip arrows. Dashed great circles are bedding planes. Isolated filled and open circles are poles to bedding and faults, respectively. Stereoplots are Schmidt (equal area) projections of the lower hemisphere, including the position of magnetic north (marked by small M).

slip data reveals the same north–south and NE–SW directions for the extensional σ_3 stress axes, with similar average trends of N007° and N036°, respectively, in a pre-fold attitude (Fig. 8). Such a similarity in the results led us to infer that even in the absence of syntectonic sedimentation these pre-folding normal faults can also be considered as Mesozoic brittle structures.

The regional relationship between these two directions of extension and the general trend of the fold-and-thrust belt indicates that the north–south extension is oblique to the present NW–SE structural grain of the belt, whereas the NE–SW extension is perpendicular to it (Figs 7 & 8). A comparison between the two extensional stress trends shows that the north–south direction prevails among the syndepositional extensions (Fig. 7a). In contrast, the NE–SW direction dominates among the pre-folding extensions with no syntectonic sedimentation (Fig. 8a). This could be because in many

cases the syndepositional normal faults probably reveal an older extension (see Table 1). The brittle structures related to these two major directions of extension are distributed throughout the entire study area in the HZB of interior Fars and each direction cannot be referred to a specific area. Examining the age of the affected rocks reveals that both populations of the normal faults developed partly contemporaneously during the Mesozoic era and do not belong to a distinct time span.

A minority of the Mesozoic syndepositional brittle structures (15%) comprise normal faults trending NNE–SSW, indicating a NW–SE extensional direction (sites 24, 43 and 76 in Fig. 7). Such an association between synchronous major and minor orthogonal normal faults can be interpreted as indicating the existence of cross-faults that separate different lateral blocks in the hanging wall of the major normal faults (Laubach & Marshak 1987). The coeval activity of the two orthogonal fault sets can also be interpreted as indicating the existence of a regional biaxial extension, creating chocolate-tablet structures. Such a mechanism has already been identified by field evidence of normal faulting (Angelier & Bergerat 1983) and numerical modelling (Hu & Angelier 2004). However, such evidence of syndepositional NW–SE Mesozoic extension is very sparse in the study area relative to the other major north–south and NE–SW extensions and cannot be interpreted as indicating a distinct tectonic event.

Geometry of the Mesozoic structures

In the study area, the two calculated Mesozoic directions of extension are obtained from two sets of normal faults, displaying consistent average strikes of c. N098° and c. 135°, resulting in an acute angle of c. 37° between these two trends (Fig. 9a). This combination of differently trending normal faults is well demonstrated by major structures in the geological map of Arsanjan, to the NE of Shiraz (Fig. 9b; GSI 2001a). These major normal structures are situated in the transition area between the Turonian iron-stained carbonate platform of the NEAPM and the deep-sea radiolarian sediments of the Neo-Tethyan oceanic basin (see Haynes & McQuillan 1974). Some of the normal faults of these two subsets have preserved their initial normal situations whereas others have been reactivated as reverse (Fig. 9c) and/or strike-slip faults (Fig. 9d) (see also Navabpour et al. 2006, 2007a). More precise regional fault slip observations were made to determine the possible relationship between these structures. It was thus revealed that successive striae are well preserved on some of the NW–SE faults, indicating two separate dip-slip and oblique-slip normal vectors (Fig. 10a, b). Further data analysis led us to separate and distinguish three kinds of fault slip movements (Fig. 10c): (1) dip-slip normal faults trending east–west; (2) dip-slip normal faults trending NW–SE; (3) oblique-slip sinistral normal faults trending NW–SE.

In the first case, the normal faults with east–west strikes are observed as conjugate brittle structures that dip north and south. Dip-slip fault striations with rake angles (R) greater than 70° on these fault planes indicate a north–south extensional stress regime, with horizontal σ_3 and σ_2 stress axes and vertical σ_1 axis (Fig. 10c, top). We could not find other successive normal striations on these fault surfaces. In the second case, dip-slip fault striations with $R > 70°$ also exist on conjugate NW–SE normal faults, indicating a NE–SW extensional stress regime with the same attitudes for the stress axes but different trends for σ_3 and σ_2 axes; most of the faults (c. 70%) dip to the NE (Fig. 10c, middle). The third case resembles the second one,

but with oblique striae ($R < 70°$) on the same faults, indicating successive slips. The fault slip data thus show an oblique-slip sinistral normal faulting, indicating a transtensional stress regime with an approximately north–south direction of extension; most of the faults (c. 75%) dip to the NE, as before (Fig. 10c, bottom; note the inclined attitudes of σ_1 and σ_2 stress axes in the transtensional stress regime). Although the three fault slip subsets are separated based on stress tensor analysis of successive fault striations, the age relationship between the different normal faults remains unclear. In comparisons between sites, shifts in the chronology commonly occur. At some sites, the dip-slip movement occurred before the oblique-slip one (Fig. 10a), but at some other sites the relative chronology is opposite (Fig. 10b).

A further and major problem in our study resulted from the existence of local extensions induced by folding during the late Cenozoic era. For this reason, our conclusions regarding the Mesozoic extension were drawn from only the Mesozoic outcrops where evidence of syntectonic sedimentation could be found. We could not find systematic sedimentary evidence associated with normal faulting revealing a consistent regional syndepositional extension during the Cenozoic basin evolution of the study area. However, some local syndepositional normal faulting affected the Eocene Pabdeh marls in the southern rim of the HZB of interior Fars (North Sepidan, Fig. 11), indicating a possible extensional activity of the HZF during the early Tertiary period. Unfortunately, the related brittle structures were too few to allow an acceptable stress tensor analysis and to draw a firm conclusion. Our field observation suggests a NNW–SSE strike for this faulting, indicating an ENE–WSW direction of extension. Further studies are needed to evaluate the importance and significance of such an early Tertiary extension.

Age of the extensional events

Almost all the evidence suggests that the first stage of the Neo-Tethyan opening started by a rifting process between the Iranian and Arabian continental plates during Permian–Triassic times. This extensional event is documented by some Permian amygdaloidal basaltic flows in the HZB (Thiele et al. 1968; Berberian 1977). The seismic profile analyses across the Dezful Embayment have revealed that the Permian–Triassic sedimentation was related to the hanging wall of a normal fault (Fig. 12a; Sepehr & Cosgrove 2004), characterizing a synrift structure. Syndepositional conjugate normal faults are also observed in the Permian carbonate platform of Oman, south of the Hormuz Strait, where the tilted

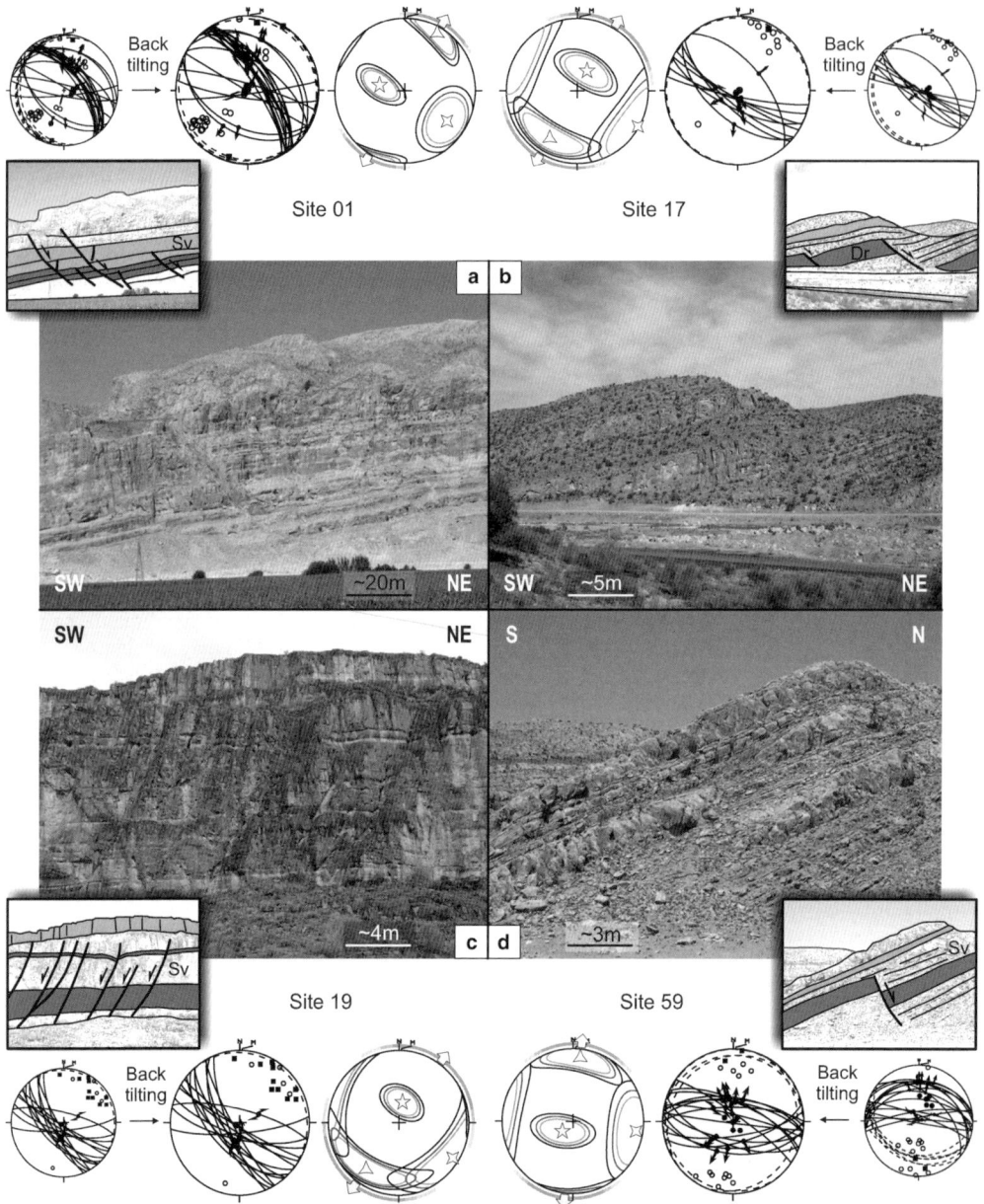

Fig. 6. Field evidence of syndepositional faulting within the Mesozoic sedimentary sequence, based on the situations observed at sites 01, 17, 19, 36, 55, 59, 77 and 79, indicating normal fault blocks associated with thickness change of coeval sediments (**a**) that are covered by non-faulted strata (**b, c, d, g** and **h**) and syntectonic sedimentary wedges (**e** and **f**). The related fault slip patterns and stress tensor determinations are included, indicating two major north–south and NE–SW direction of extension. Dr and Kn, Dariyan and Khaneh Kat Formations; Rd, radiolarite unit (note the unaffected layers on the right side of the photograph at site 79). In the stereoplots, small filled squares are poles to the tension cracks; stars with 5, 4 and 3 points mark the principal stress axes of σ_1, σ_2 and σ_3 respectively, with their 60, 75 and 90% confidence ellipses. Large open arrows show directions of extension with azimuthal confidence in grey curves. Other symbols and descriptions as in Figures 4 and 5.

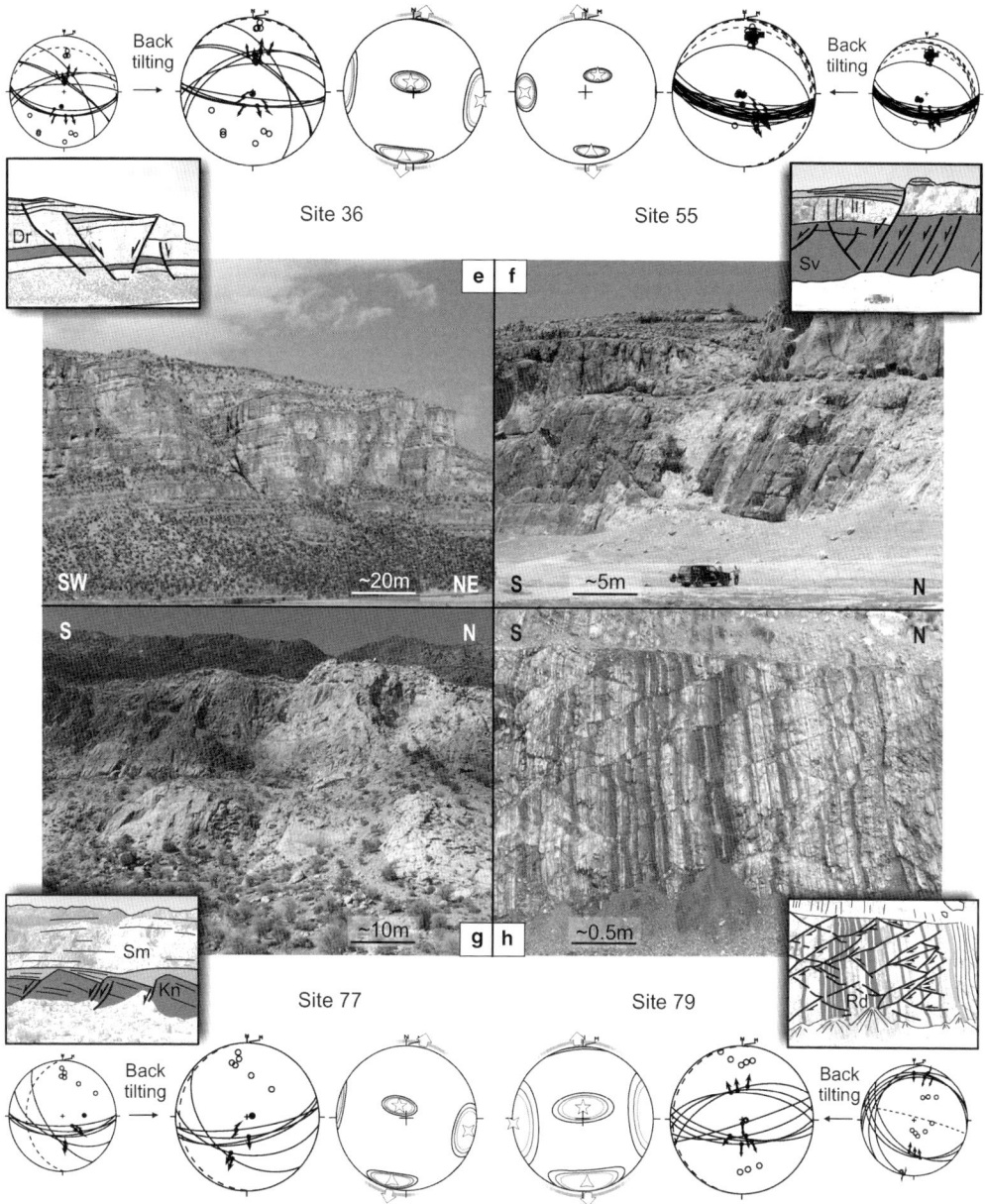

Fig. 6. (*Continued*)

blocks are sealed by an unconformity associated with dolomitization (Chauvet *et al.* 2004). These structures indicate a north–south direction of extension in the middle Permian period. We cannot discuss the Permian and Triassic extensions in the HZB, because the outcrops were scattered and not good enough for data collection. However, a late Triassic extension could be identified in the

Khaneh Kat Formation, east of Shiraz (site 77, Fig. 6).

Leeder & Gawthrope (1987) have discussed the tectonostratigraphic evolution of a half-graben structure in coastal environments. They have shown how the uplifted footwall and downthrown hanging wall of the tilted blocks can be related to local erosion and facies changes in shallow-water

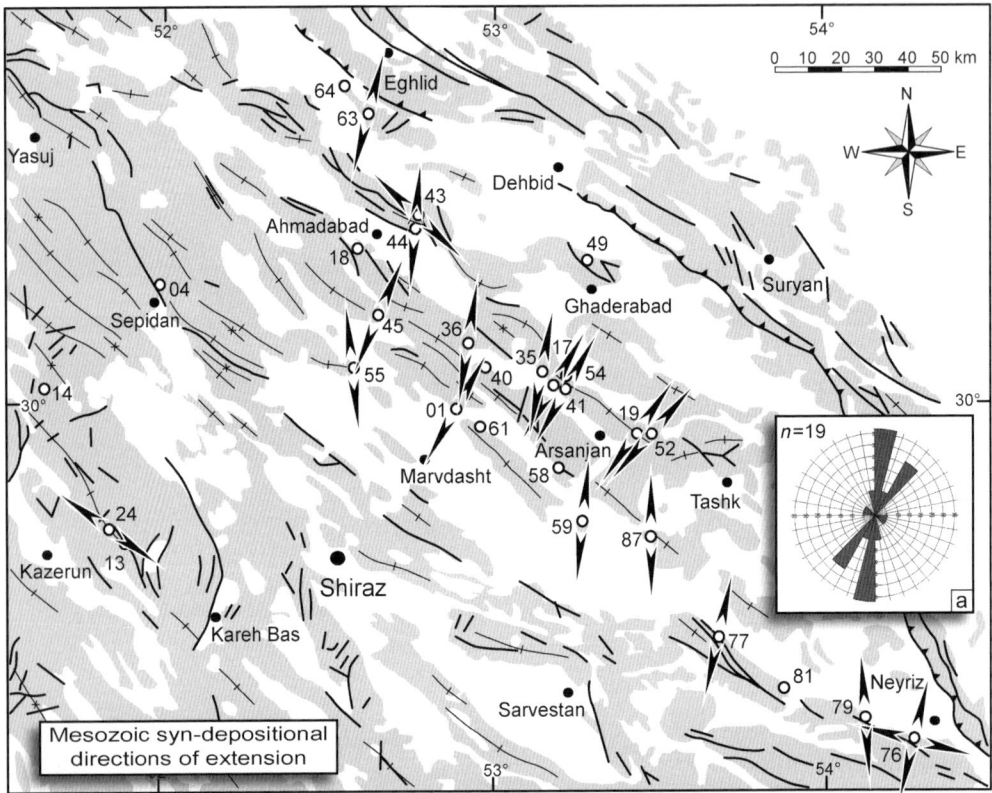

Fig. 7. Distribution map of the reconstructed σ_3 stress axes (pairs of divergent black arrows), for extensional stress regimes obtained from the Mesozoic syndepositional normal faults throughout the study area. Inset (**a**) shows a rose diagram of σ_3 stress axes ($n = 19$), indicating two distinct north–south and NE–SW directions of extension with average trends of N006° and N034°. Geological details are shown in Figures 2 and 3, and listed in Table 1.

conditions, respectively. Such a condition has already been reported from the top of the Sarvak Formation in SE Zagros at Kuh-e-Faraghun, where wedges of conglomerates were derived from adjacent normal fault scarps and were sealed by the overlying Palaeocene–Eocene sediments (Fig. 12b; Stoneley 1990). In our study area, we observed intraformational angular unconformities within the Sarvak Formation along approximately the middle axis of the HZB of interior Fars, associated with block tilting, syndepositional normal faults and brecciated facies in the upper parts of the Sarvak Formation (Figs 4b and 12c–f). This evidence can be correlated especially along the major normal fault blocks of the Turonian–Early Coniacian in the Arsanjan area, reconstructed by geological observations of Stoneley (1981).

Most of the syndepositional normal faulting and tilted blocks that we could determine occurred in the Middle–Upper Triassic (Khaneh Kat Fm), Middle–Upper Jurassic (Surmeh Fm), Aptian

(Dariyan Fm), Cenomanian–Turonian (Sarvak Fm) and Campanian–Maastrichtian (Gurpi Fm) sedimentary rock units (see Table 1). As mentioned above, there are some reported Mesozoic angular unconformities from sedimentary sequences of Late Triassic (Szabo 1977; Szabo & Kheradpir 1978), Late Jurassic (Murris 1978), Aptian–Albian (James & Wynd 1965; Setudehnia 1978), Late Turonian (Haynes & McQuillan 1974) and Late Maastrichtian (James & Wynd 1965; Setudehnia 1978) age (see Fig. 3). Most of these unconformities have been reported from the HZB and were associated with intraformational conglomerate, breccia, glauconitic sandstone and iron-stained dolomitic shelf carbonate.

A good correlation thus is obvious between the angular unconformities and syndepositional extensional structures within the Mesozoic stratigraphic sequence of the HZB, when we examine their ages in a correlation chart (Fig. 13). This correlation becomes more acceptable when is compared with

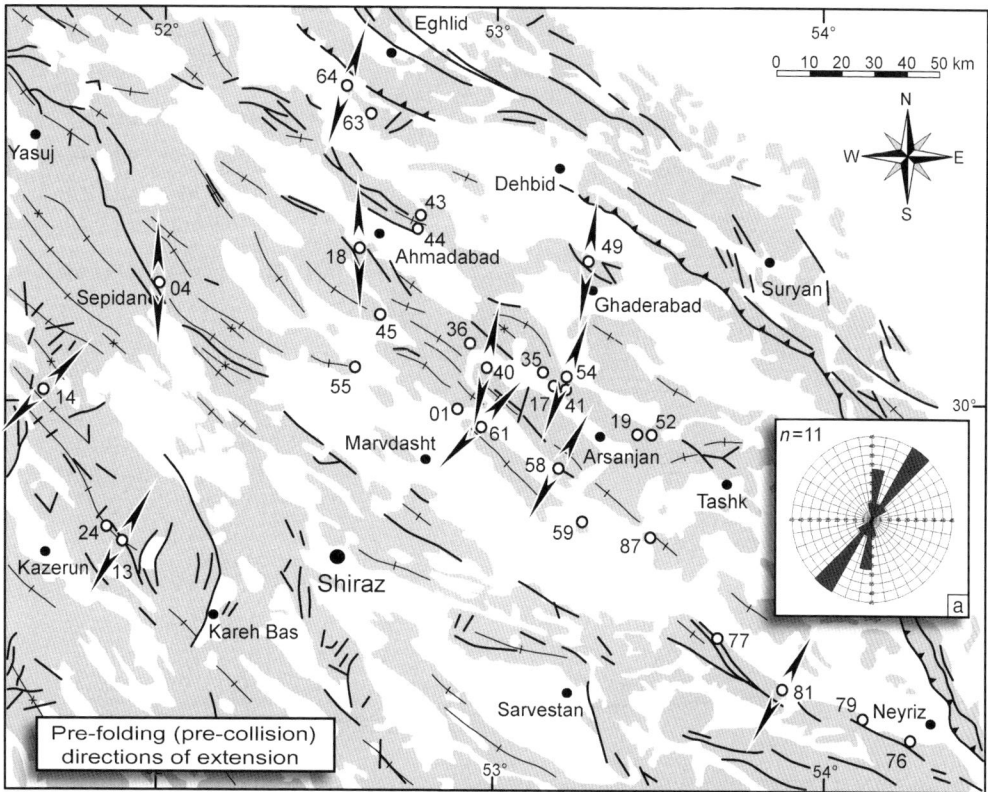

Fig. 8. Distribution map of the reconstructed σ_3 stress axes for extensional stress regimes obtained from back-tilted Mesozoic strata with no evidence of syntectonic sedimentation, indicating extensions that occurred before the late Cenozoic folding. Inset (**a**) shows a rose diagram of σ_3 stress axes ($n = 11$), indicating two distinct north–south and NE–SW directions of extension with average trends of N007° and N036°, similar to that of the syndepositional structures (see Fig. 7). Arrows as in Figure 7. Geological detail are shown in Figures 2 and 3, and listed in Table 1.

the subsidence rates of the ZSFB of Fars Arc, based on formation isopach maps. According to Bordenave & Hegre (2005), an average background apparent subsidence rate of c. 10 m Ma^{-1} within the ZSFB of Fars Arc was increased to about 30, 25, 60 and 20 m Ma^{-1} during the deposition of the Khaneh Kat, Dariyan, Sarvak and Gurpi–Pabdeh Formations, respectively (Fig. 13). However, the successive angular unconformities of the HZF were not reported from the ZSFB. This situation indicates that the inner parts of the NEAPM were more stable to the SW of the HZF than to the NE. We thus infer that most of the reported unconformities within the Mesozoic sedimentary sequence of the HZB could have been related to a contribution of both the block tilting and the basin deposition or erosion processes in the northeastern rim of the NEAPM, suggesting successive episodes of extension that have been recorded as brittle structures during the continuous subsidence. This aspect of

unconformities related to block tilting was also illustrated in the North Sea basin (Kyrkjebo *et al.* 2004) and in the Neo-Tethyan passive margin of both the North Anatolian Palaeorift (Kocyigit & Altiner 2002) and Oman (Chauvet *et al.* 2004).

According to our data, there is little evidence of a syndepositional normal faulting within the Eocene marls of the Pabdeh Formation (see Figs 11 & 13). It has been reported that outcrops of the Eocene rocks have been subjected to subaerial weathering (James & Wynd 1965; Motiei 1993), because of uplifting and folding along the northeastern rim of the ZSFB during the late Eocene–early Oligocene epochs (Hessami *et al.* 2001). New field observations confirmed that the onset of collision occurred in the late Oligocene–early Miocene epochs (Agard *et al.* 2005, 2006) and that the folding started in the early Miocene epoch (Sherkati *et al.* 2005). A dramatic increase in subsidence rate, to more than 140 m Ma^{-1}, is evidenced in the

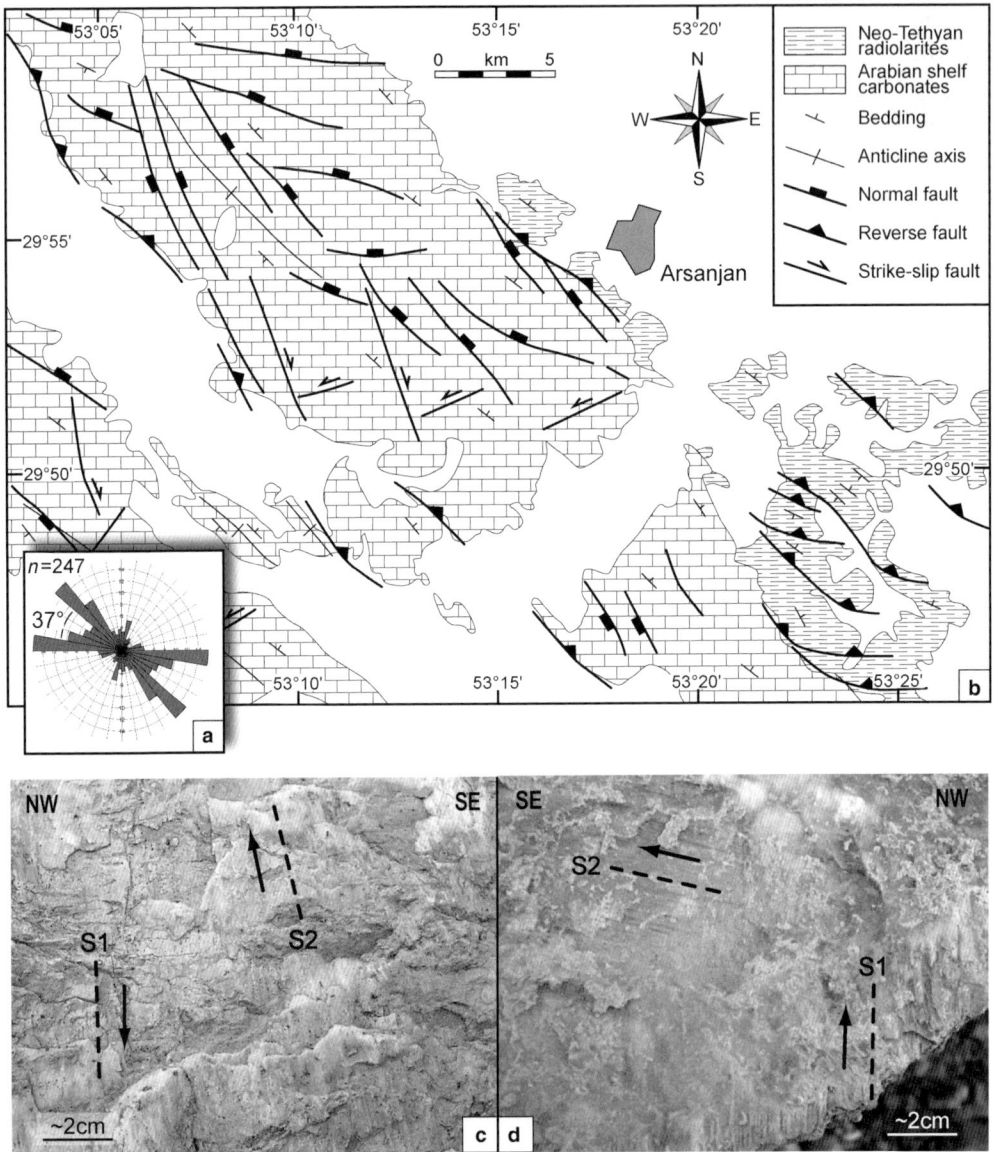

Fig. 9. Regional attitudes of the Mesozoic normal faults. (**a**) Rose distribution of the strikes of the Mesozoic normal faults throughout the study area (*n* = 247), revealing two main *c.* N098° and *c.* N135° trends at an acute angle of *c.* 37°. (**b**) Structural map of the Arsanjan area (redrawn from GSI 2001*a*), indicating two east–west and NW–SE strikes of the Mesozoic major normal fault sets that dip mainly to the north and NE throughout a gentle anticline. Location is shown as in Figure 2. (**c**) Reverse reactivation on a pre-existing Mesozoic normal fault within the Gurpi Formation. S1 indicates a normal fault slip vector that is characterized by old striae grooved on the fault surface; S2 indicates a reverse fault slip vector that is characterized by fresh calcite steps. (**d**) Successive reverse and strike-slip reactivation on a high-angle fault within the Sarvak Formation, S1 then S2 respectively, indicating a change in stress state during the late Cenozoic collisional process (see Navabpour *et al.* 2006, 2007*a*).

Oligocene epoch during the deposition of the Asmari Formation in the ZSFB of the Fars Arc (Fig. 13; Bordenave & Hegre 2005). Considering these ages, it may also be inferred that the subaerial weathering of the Eocene rocks could have been a result of extensional tectonics and block tilting, as

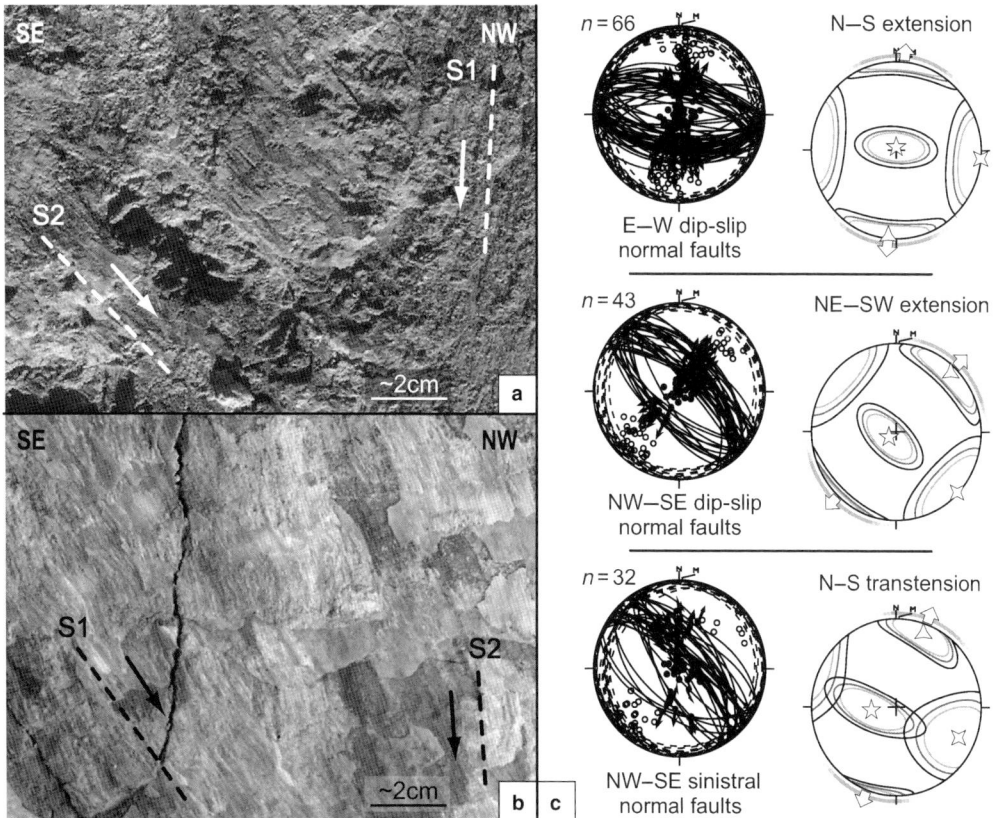

Fig. 10. Major normal fault slip geometries within the Mesozoic units of the HZB. (**a, b**) Successive striae on a fault plane, S1 then S2, indicate two different relative fault slip chronologies. This situation implies that no systematic succession of movements can be characterized between dip-slip and oblique-slip sinistral normal fault slips. (**c**) Three normal fault geometries separated based on different strikes, rakes of striae and stress tensor determination at different sites throughout the study area: east–west normal faults indicating a north–south extension (top), NW–SE normal faults indicating a NE–SW extension (middle), and NW–SE sinistral normal faults indicating a transtensional north–south extension (bottom). Number of fault slip data, *n*, shown at top left corner of each diagram. Symbols of stereoplots as in Figures 5 and 6.

is described above for the Mesozoic units. Such an interpretation would indicate that the Mesozoic syndepositional history of extension and block tilting of the NEAPM probably continued during the early Cenozoic era, prior to the continental collision. Further studies would be essential to indicate if such view is correct. After the onset of collision, the angular unconformities of the middle–late Cenozoic stratigraphic sequence characterize the different folding episodes of the Zagros basin (Sherkati *et al.* 2005).

Discussion

In contrast to brittle tectonic analysis in the tabular areas of plate interiors, our study in the HZB required careful consideration of relationships between the brittle structures and folds. This interaction between brittle structures and folds results in structural complexity. In particular, the number of stress regimes reconstructed in the present-day structure is larger than the number of tectonic events that occurred in the Mesozoic history of the HZB, because of the reactivation of inherited faults and the presence of brittle structures related to the late Cenozoic compressional stress fields. However, numerous observations of syndepositional normal faulting related to the Mesozoic history of the study area facilitated deciphering of the tectonic relative chronology. Similar aspects of syndepositional normal faulting have already been highlighted for the HZB of Kermanshah (Angelier *et al.* 2006), the HZB of

Fig. 11. Syndepositional evidence of normal faulting in the Eocene Pabdeh marls (Pd) in the southern rim of the HZB (N Sepidan; see Fig. 2 for location map), indicating a possible extensional activity of the HZF during the early Tertiary period, prior to the continental collision.

interior Fars (Navabpour *et al.* 2006) and in the northern North Sea along the Norwegian continental shelf (Nottvedt *et al.* 1995).

The brittle tectonic analysis and palaeostress reconstructions presented in this study bring new approaches to the study of the tectonic evolution of the former NEAPM in the Zagros collisional belt. However, questions remain regarding some natural limitations of the methods. For instance, small outcrops of the Palaeozoic and Triassic sedimentary rocks do not provide sufficient information about the Permian–Triassic rifting and oceanic opening process that are recorded within the NEAPM, as was reported from the Permian carbonate platform of Oman by Chauvet *et al.* (2004). The quality of fault slip striae preserved on the Mesozoic faults is not often good enough to allow identification of clear relative chronologies between the various Mesozoic tectonic events. In addition, Cenozoic outcrops are few in the HZB of interior Fars, which makes identification of possible syndepositional tectonic activities in the early Cenozoic sedimentary basin difficult.

In this section, we intend to discuss the Mesozoic extensional stress regimes and to reconstruct a possible structural model for the Mesozoic brittle tectonic evolution of the HZB, considering regional tectonic aspects. To this end, we considered the stratigraphic ages of syndepositional normal faulting (see Figs 6 & 7) and the maximum stratigraphic ages that can be assumed for the reconstructed prefolding extensional events (see Figs 5 & 8; see also Table 1 for a complete account). It was thus possible to plot the age of the events as a function of the

extensional directions revealed by the inversion of fault slip data (Fig. 14). The resulting distribution clearly shows the two main north–south and NE–SW directions of extension (Fig. 14a), as already illustrated by the rose diagrams of reconstructed σ_3 stress axes based on the Mesozoic and prefolding normal fault slip data (see Figs 7a & 8a). It also shows that these two directions of extension correspond to roughly different time intervals, *c.* 240–90 Ma for the north–south extensional trend and 180–60 Ma for the NE–SW extensional trend (Fig. 14b). It is interesting to note that both the directions of extension coexisted during the period of 180–90 Ma (marked as the light grey area in Fig. 14b).

This approach indicates a possible gradual regional change in direction of extension from north–south to NE–SW. In other words, it seems that the direction of extension has changed slightly from a margin-oblique trend to a margin-perpendicular one, during the Mesozoic history of the NEAPM. First, margin-oblique extension occurred during the Triassic–early Jurassic period. The margin-perpendicular extension then prevailed over the area during the late Jurassic–Cretaceous period, whereas the margin-oblique one continued until the middle Cretaceous period. The existence of such a continuous extension and a possible gradual change in stress direction during the Mesozoic era, as well as the similarities between the trends of the corresponding normal faults and the late Cenozoic reverse structures, encouraged us to search for possible explanations.

Existence of an oblique crustal stretching?

The geometry of fault systems of a rift zone reflects whether it evolved in response to orthogonal or oblique divergence of its flanking stable blocks (Tron & Brun 1991). A classical example of oblique crustal stretching is the western branch of the East African rift system (Rosendahl *et al.* 1992). It should also be kept in mind that during the evolution of a rift the stress regime governing its development may significantly change. The structural pattern of the northern margin of the Aden Gulf presents two differently oriented normal fault systems that were generated on the continental margin under an oblique extensional stress regime (Huchon & Khanbari 2003). The strikes of the normal faults are parallel to the trend of both the oceanic spreading axis and the continental margin, indicating an acute angle of about 40° between the fault trends (Fig. 15a). Creating a mirror view of the Aden Gulf structures (Fig. 15b) and comparing it with the Mesozoic extensional structures of the HZB (Fig. 15c), the question of the history of the NEAPM arises: has it undergone two differently

Fig. 12. Aspects of syndepositional extensional structures within the Zagros basin. (**a**) Permian–Triassic strata (in grey) that are involved with the hanging wall of a buried normal fault inverted during the late Cenozoic folding, based on seismic profile analysis across one of anticlines of the Dezful Embayment (redrawn from Sepehr & Cosgrove 2004; see Fig. 1 for location map). (**b**) Intraformational conglomerate and breccia (in grey) related to normal faulting within the sedimentary basin of the Sarvak (Sv) and Gurpi (Gu) Formations in SE Zagros at Kuh-e-Faraghun during the Turonian–Coniacian (redrawn from Stoneley 1990; see Fig. 18a for location map). (**c, d**) Intraformational angular unconformities (dashed lines) associated with normal faulting within the Sarvak Formation. (**e**) Extensional block tilting associated with syndepositional sedimentary wedge (in grey) within the Sarvak Formation. (**f**) Brecciated facies of the Sarvak Formation mainly observed at the upper parts above the intraformational unconformities. (See Fig. 2 for localities.)

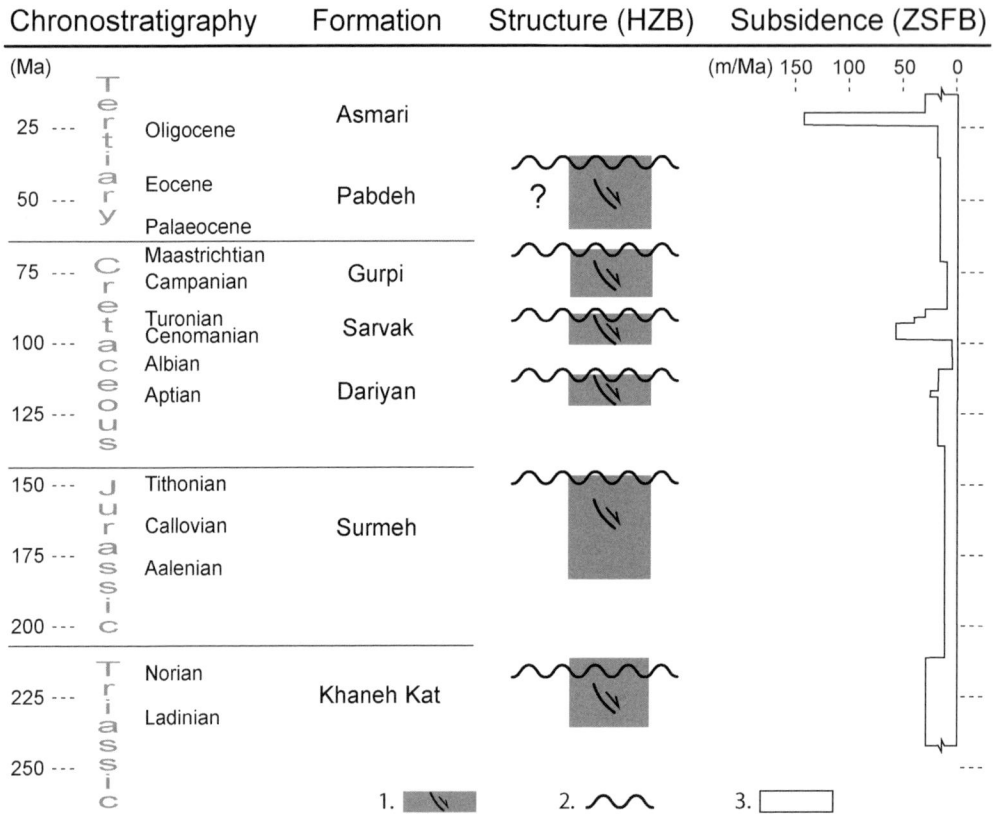

| Chronostratigraphy | Formation | Structure (HZB) | Subsidence (ZSFB) |

Fig. 13. Correlation chart for syndepositional normal faulting and unconformities of the HZF of interior Fars and apparent subsidence rates of the ZSFB of the Fars Arc for the Mesozoic–early Cenozoic time interval. 1, Rock unit associated with syndepositional normal faulting; 2, angular unconformity (references in the text and Fig. 3); 3, apparent subsidence rate (i.e. the ratio between thickness of the sedimentary rock unit and the duration of its deposition) from Bordenave & Hegre (2005).

oriented far-field crustal extensions, or have the structures resulted from different local phenomena under a single far-field extension? If we assume that two differently oriented far-field extensional tectonic events have affected the region, each related structure should be referable to a specific time span, and a clear chronology should emerge between these extensional tectonic regimes. A general change in the direction of extension certainly occurred during the Mesozoic era, but as shown by our data this change was gradual and both the different normal fault movements and the related north–south and NE–SW extensional directions occurred contemporaneously during middle Jurassic–middle Cretaceous times (Fig. 14b).

On the other hand, the Mesozoic syndepositional extensional brittle structures of the HZB of interior Fars developed within the sedimentary cover of the NEAPM after the Permian–Triassic rifting. Some additional information on the basement structures

is necessary to simulate a possible structural evolution of the Permian–Triassic rifting. In most cases, the direction of a single segment of an oceanic ridge does not change significantly with respect to the corresponding continental margin segment during sea-floor spreading, because of the rigidity of the oceanic crust. A study of a frozen mantle flow structure in the Neyriz ophiolitic complex revealed the existence of a palaeo-oceanic spreading centre with a reconstructed axis trending N105°, compatible with the geometry and orientation of harzburgite foliations and lineations and sheeted dykes (Fig. 15d; Nadimi 2002). Although such a structure could have been involved with differential changes in direction during the late Cretaceous ophiolite obduction, it strikes almost parallel to the strike of syndepositional normal faults within both the Neyriz radiolarites and the Mesozoic sedimentary sequence of the NEAPM that revealed the north–south margin-oblique extension throughout the study area (see

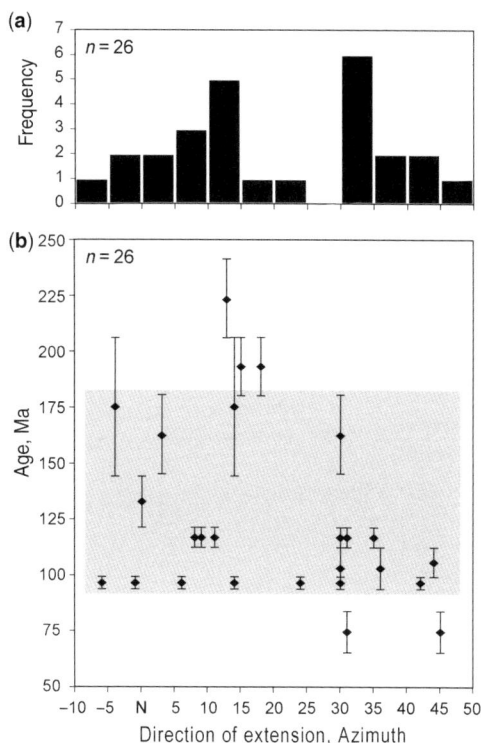

Fig. 14. Relationships between age, trend and frequency of the Mesozoic extensions. (**a**) Frequency–direction relationship for the extensional stress trends ($n = 26$), representing the main north–south and NE–SW directions (see Fig. 7). (**b**) Age–direction relationship for the extensional events, suggesting a general change in the direction from roughly north–south to NE–SW during the Mesozoic era. Vertical bars show the age range of the corresponding lithostratigraphic unit in which normal faulting was observed. The time overlap of the two directions for the period of 180–90 Ma ago is marked by the light grey area.

Figs 6, 7 & 9a). This situation indicates that no major differential change in structural orientation occurred during the late Cretaceous ophiolite obduction, suggesting a possible obliquity between the Neo-Tethyan palaeo-oceanic ridge and the NEAPM.

The existence of such a possible obliquity led us to infer that the coeval bidirectional Mesozoic extension could have originated by gravitational reactivation of similar structures within both the Palaeozoic sedimentary sequence and the continental basement similar to that of the Aden Gulf, characterizing a Permian–Triassic oblique crustal stretching. By adopting the results of experimental oblique extensional modelling of Tron & Brun (1991) for the present-day situation of the observed extensional structures in the HZB of interior Fars

(Fig. 15c), we assume an oblique rifting under a far-field north–south extensional stress trend at a stretching angle of 60° (Fig. 16a). In such a transtensional condition, two sets of normal faults could have been formed within the stretched continental crust parallel to both the divergent flanks and the oblique spreading axis, trending NW–SE and east–west respectively within the present-day geographical reference frame, with an acute angle of 30° between the fault trends. Whereas Mesozoic normal faults are well documented within the sedimentary cover of the HZB of interior Fars (Navabpour et al. 2006; this study), there is unfortunately no seismic evidence to reveal the existence of such structures within the basement of the study area. However, further to the south, considerable seismic activity throughout the ZSFB of the Fars Arc has revealed the existence of active reverse faults (Berberian 1995; Gillard & Wyss 1995) within the basement (Maggi et al. 2000; Talebian & Jackson 2004; Engdahl et al. 2006) (Fig. 16b; Table 2), based on consideration of earthquake focal mechanism solutions.

Adopting a half-graben structure for the NEAPM, we considered the NE-dipping nodal planes of the earthquake focal mechanisms as fault planes of the inherited rift structures. A statistical rose analysis of these fault planes (Fig. 16c) reveals two aspects that deserve attention. First, the strikes of the reverse faults indicate two major c. N085° and c. N120° trends for the basement inherited brittle structures, with an acute angle of c. 35° between these faults similar to that is expected from the oblique stretching model. Second, most of these faults show an average dip angle of c. 60°, characterizing high-angle reverse faults (Berberian 1995), which is consistent with the usual continental normal faulting geometry. Therefore, we prefer to interpret the structural pattern of the present-day basement reverse faults together with the obliquity of the Neyriz palaeo-oceanic spreading centre as sufficient evidence supporting the Permian–Triassic oblique stretching model for the Neo-Tethyan opening. The consistency between the strikes of both the basement reverse faults of the ZSFB of the Fars Arc and the Mesozoic normal faults within the sedimentary cover of the HZB of interior Fars also provides indirect confirmation that no major horizontal block rotation has occurred between these domains during the evolution of the NEAPM and the subsequent Zagros collision.

Possible scenario for the Mesozoic extensions

A possible scenario has already been proposed for the oblique opening of the Aden Gulf, by Huchon & Khanbari (2003). Those workers have shown that the NE–SW extensional stress related to the

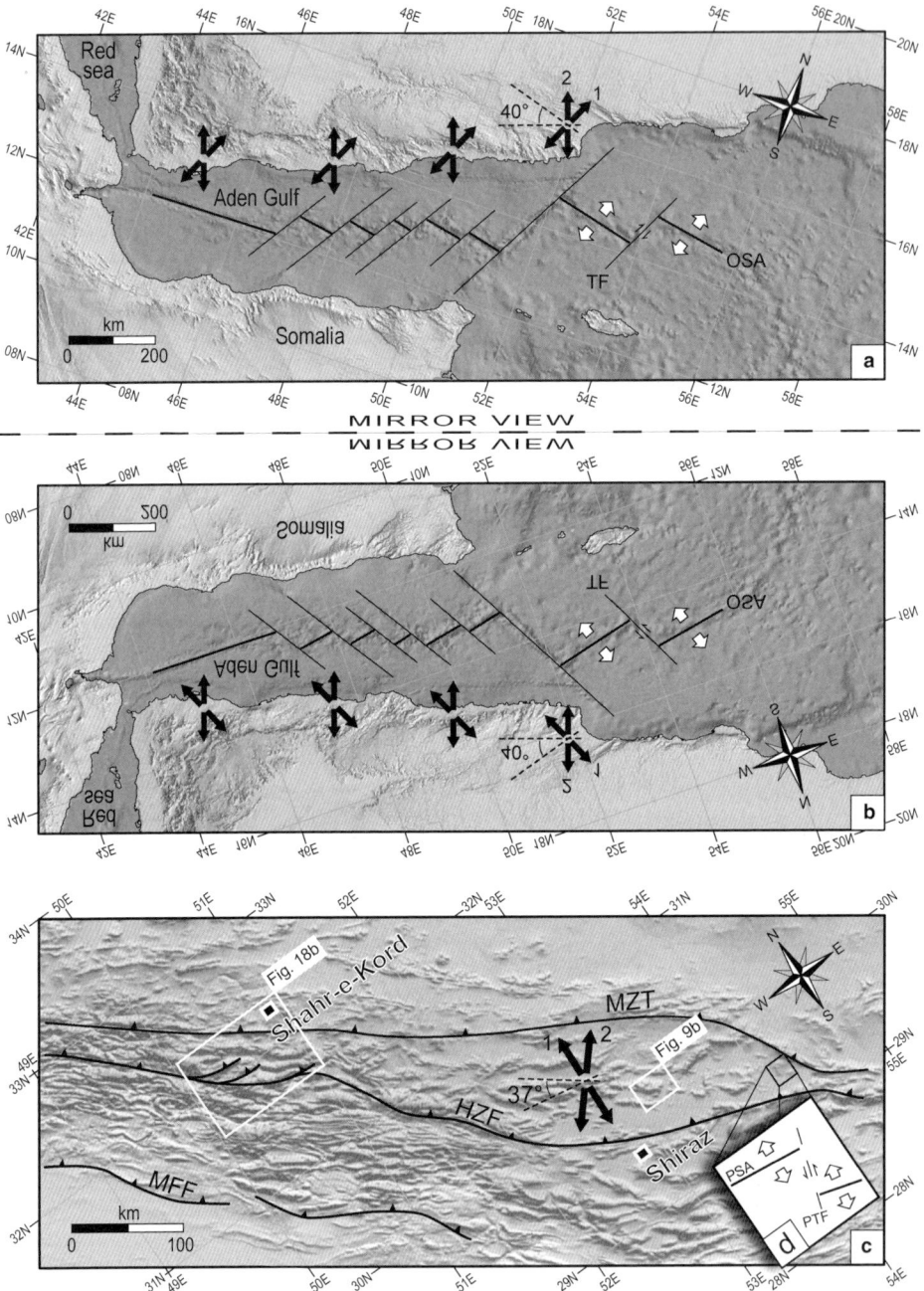

Fig. 15. Recent and ancient oblique extensional tectonics of the Arabian continental margins. (**a**) Recent active oblique opening of the Aden Gulf (redrawn from Huchon & Khanbari 2003). (**b**) Mirror view of (**a**) to be compared with (**c**). (**c**) Regional view of the reconstructed Mesozoic extensional trends of the former NEAPM in the HZB (see also Figs 7–9). (**d**) Inferred palaeo-oceanic spreading axes of the Neo-Tethyan basin within the ophiolites of the Neyriz area (redrawn from Nadimi 2002). OSA, oceanic spreading axis; TF, transform fault; PSA, palaeo-spreading axis; PTF, palaeo-transform fault. White and black pairs of arrows show far-field and near-field stress components, respectively. Numbers refer to relative chronology. Dashed lines show trends of normal faults. The acute angle between the two normal fault trends indicates a possible obliquity between the PSA and the former NEAPM. Location is shown in Figures 1 and 2.

active spreading axes of the Indian Ocean is oblique to the diverging east–west trend of the southern Arabian continental passive margins. This extensional trend is preserved on the continental margins as normal faults trending NW–SE and is replaced by a younger trend of an almost north–south extension perpendicular to the trend of the continental margins (see Fig. 15a). The north–south direction of extension was thus interpreted as the result of an eastward propagation of the oceanic spreading axis into the continental crust, following a mechanism similar to that of a tension crack but at lithospheric scale, which characterizes a rift-to-drift transition. Such a scenario may be sufficient to describe a similar process for evolution of the brittle structures within the Palaeozoic–Triassic sedimentary sequence during the Permian–Triassic Neo-Tethyan oblique opening between the Arabian and Iranian continental plates, as proposed earlier. However, there is still debate on the interpretation of the different syndepositional normal fault slip geometries of the HZB of interior Fars, especially the oblique-slip sinistral normal fault movements, during the Jurassic–Cretaceous history of the NEAPM long after the creation of the Neo-Tethyan oceanic environment. The structural pattern of the major Mesozoic normal faults could have been produced by gravitational reactivation of similar inherited structures within either the basement and/or the underlying Palaeozoic sedimentary sequence over the basal Infracambrian Hormuz Salt detachment (see Fig. 3).

It is known that at shallow levels the structural style of a rift is strongly influenced by the lithological composition of the down-faulted pre-rift and synrift sediments. Evaporitic rocks can give way to the development of multilevel extensional detachment faulting (Jarrige et al. 1990; Jarrige 1992) and can act as sole-out levels for secondary fault systems developing in response to gravitational instability (Nottvedt et al. 1995). Recent analogue modelling studies of Soto et al. (2007) indicated how different basin-bounding fault geometries and thickness of a viscous layer within the brittle pre-rift sequence influence the deformation and sedimentary basin patterns related to the half-graben extension. For a long time, it has been known that the Intracambrian Hormuz Salt Formation has behaved as a detachment layer controlling different structural styles of the folded sedimentary cover with respect to the faulted basement (Colman-Sadd 1978; Talbot 1979; Koyi et al. 2000; Sherkati et al. 2005; Sepehr et al. 2006). Such a thick detachment layer could have played an important role in the extensional Mesozoic history of the NEAPM as well. The relatively high subsidence rate (i.e. c. 60 m Ma^{-1}) in the Fars Arc during the Cenomanian, which is marked by the brecciated facies and

intraformational angular unconformities of the Sarvak Formation in the HZB (Stoneley 1981, 1990; this study), is attributed to readjustment of faulted blocks and gravitational movements on the basal salt detachment (Bordenave & Hegre 2005). This is consistent with the idea that the first gentle movements within the sedimentary cover, predating the development of the major late Cenozoic Zagros structures, probably triggered the salt plugs to well up (Player 1969; Kent 1970, 1979; Koyi 1988; Talbot & Alavi 1996). Regional extension is believed to be an important initiator of salt diapirs during thin- and thick-skinned stretching (Vendeville & Jackson 1992). It seems likely that the salt plugs pierced the thinned Palaeozoic sedimentary sequence along the HZF during the Permian–Triassic extension (Sepehr & Cosgrove 2004).

We infer that during the Permian–Triassic oblique continental stretching, the initial east–west normal faults could have been produced within both the basement and the Palaeozoic sedimentary cover under a north–south extensional stress field in the future HZB close to the rift axis, characterizing a thick-skinned extension (Fig. 17a). These east–west normal faults could be associated with the NW–SE normal faults further to the south in the future ZSFB on the NEAPM, as evidenced by the present-day seismicity of the basement structures. Creation of the oceanic environment to the NE of the Arabian margin together with the role of the Hormuz Salt detachment then provided conditions for the sedimentary cover to undergo continuous extensional block tilting on the inherited east–west margin-oblique normal faults, with some localized oblique-slip and margin-parallel faulting, during Triassic–Jurassic times (Fig. 17b). The extensional tectonic regime then gradually changed and the NE-dipping NW–SE margin-parallel normal faults prevailed over the future HZB, deeply affecting the entire sedimentary cover to the basal detachment, characterizing a dominant thin-skinned extension. Such continuous extensional tectonics over a salt detachment layer has already been discussed for the structural evolution of the West African passive margin after the South Atlantic rifting (Marton et al. 2000). Meanwhile, the reactivation of the preliminary inherited east–west normal faults within the Palaeozoic rocks (possibly as a result of the Mesozoic sedimentary load) could have induced local north–south extensional stress trends beneath the Mesozoic sedimentary sequence. These localized north–south extensions could then result in initiation of the syndepositional earth–west normal faults and some sinistral reactivation on the NW–SE normal faults within the overlying sedimentary layers during the Cretaceous history of the NEAPM (Fig. 17c). Interestingly, the higher frequency of

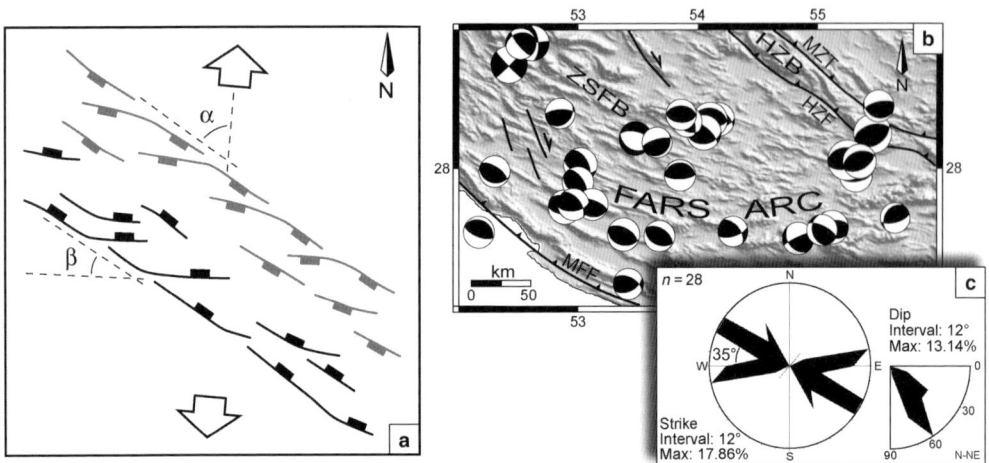

Fig. 16. Structural model of an oblique extension and basement structures of the Zagros. (**a**) Schematic illustration of a classical transtensional extension with a stretching angle of $\alpha = 60°$ (redrawn from Tron & Brun 1991). Adopting this model for the present-day geographical reference of the Zagros suggests an oblique half-graben structure composed of north- to NE-dipping normal faults (highlighted as black faults in the figure) trending east–west and NW–SE with an acute angle of $\beta = 30°$ between the fault trends. (**b**) Basement structures of the ZSFB of the Fars Arc revealed by earthquake focal mechanisms, most of which indicate reverse faulting (based on CMT database for 1976–2007 time interval; see Table 2), envisioned as inversion on the earlier normal faults (e.g. Jackson 1980). Considering the half-graben structure of the former NEAPM, most of the north- to NE-dipping nodal planes indicate high-angle (*c.* 60°) reverse faults (**c**) trending *c.* N085° and *c.* N120° with an acute angle of *c.* 35° between the fault trends. Location is shown in Figure 1.

the NE-dipping normal faults with respect to the SW-dipping ones (see Fig. 10c) characterizes the half-graben structure of the sedimentary cover of the NEAPM during the Mesozoic era, which is also evidenced by a micro-earthquake survey of the present-day active reverse faults within the basement of the ZSFB of the Fars Arc (see Tatar *et al.* 2004). It should be noted that the tectonic evolution may significantly differ in the NW Zagros, where the thickness and plasticity of the equivalent Infracambrian facies were probably insufficient to behave as a widespread tectonically active detachment layer (e.g. Kent 1970, 1979; Talbot & Alavi 1996).

Inherited extension and collisional inverted structures

For a long time, it has been envisioned that the present-day seismic activities of the Zagros reverse faults is due to inverse reactivation on pre-existing NE-dipping normal faults that were produced during the earlier Permian–Triassic rifting process (Jackson 1980; Jackson & McKenzie 1984). An integrated lithospheric modelling revealed a stretched continental crust with a thickness of *c.* 42 km beneath the Arabian Platform decreasing to *c.* 37 km beneath the Persian Gulf, as evidence for the Permian–Triassic

rifting (Molinaro *et al.* 2005*b*), based on a combined interpretation of gravity, geoid, topography and isotherm curves. The recent studies on the structural styles and balancing cross-sections suggested that the belt has undergone two distinct episodes of crustal shortening across SE Zagros since the late Cenozoic collision (Molinaro *et al.* 2005*a*; Sherkati *et al.* 2005). First, during a Miocene–Pliocene thin-skinned phase, the main folded structures developed within the sedimentary cover. Second, during a Pliocene–Recent thick-skinned phase, the basement faults were reactivated as thrusts and created out-of-sequence structures, some of which cut obliquely through the earlier folded structures of the sedimentary cover. It is shown that the oblique out-of-sequence structures are characterized by major east–west reverse faults, which branch from the HZF in the HZB of SE Zagros in the Hadjiabad region (Fig. 18a; Molinaro *et al.* 2005*a*). This structural pattern can also be observed in the HZB of the Shahr-e-Kord region, to the NW of our study area (Fig. 18b). Ricou (1974) has already considered the east–west reverse faults of the HZB as reverse splays of the HZF, indicating transpressional deformation under the overall recent north–south compression associated with the right-lateral slip of the MRF (Authemayou *et al.* 2006). However, as mentioned elsewhere, the HZF is associated with a

Table 2. *Earthquake source parameters of the ZSFB of the Fars Arc*

Date	Time	Long. (E)	Lat. (N)	D	M_w	NP1 Strike	NP1 Dip	NP1 Rake	NP2 Strike	NP2 Dip	NP2 Rake	F	P Dir.	P Plunge	N Dir.	N Plunge	T Dir.	T Plunge
1977.10.19	06:35	55.12	27.57	15	5.5	117	41	120	259	56	66	R	006	08	273	19	117	69
1983.02.18	07:40	53.85	27.95	15	5.2	272	20	94	88	70	89	R	179	25	088	01	356	65
1984.12.22	16:05	53.68	27.51	49	5.1	115	41	84	303	49	95	R	029	04	120	04	255	85
1985.02.02	20:52	53.48	28.22	22	5.5	114	32	−81	284	58	−95	N	179	76	287	05	018	13
1985.08.07	15:43	52.89	27.72	15	5.6	303	39	116	91	55	70	R	195	08	103	16	311	72
1986.05.02	03:18	53.02	28.03	15	5.2	107	47	57	331	52	121	R	040	02	131	24	304	66
1986.05.03	10:37	53.00	27.90	15	5.2	111	33	60	325	62	108	R	042	15	136	16	270	68
1987.05.12	07:15	55.32	27.95	15	5.5	278	34	104	80	57	80	R	177	11	086	08	321	76
1990.11.06	18:45	55.25	28.06	15	6.6	274	37	107	73	55	77	R	172	09	080	10	302	76
1991.05.22	16:29	55.43	27.04	15	5.4	98	47	66	311	48	114	R	025	00	115	18	293	72
1992.05.19	12:24	55.35	28.05	15	5.6	254	40	99	63	51	83	R	158	05	067	06	291	82
1993.03.29	15:20	52.30	27.98	40	5.2	104	28	72	305	64	99	R	028	18	120	08	234	70
1993.07.09	10:29	55.51	28.45	23	5.2	110	26	120	257	68	76	R	357	22	262	13	144	65
1994.03.01	03:49	52.42	28.75	17	6.1	136	85	−176	46	86	−5	S	001	06	190	84	091	01
1994.03.30	19:55	52.60	28.96	33	5.4	148	71	177	239	87	19	S	012	11	248	71	105	15
1995.01.24	04:14	55.65	27.64	15	5.0	217	31	56	75	64	109	R	151	17	247	17	018	65
1996.05.24	06:36	53.12	27.74	15	5.2	107	22	88	289	68	91	R	019	23	109	01	201	67
1997.05.05	15:11	53.42	27.16	15	5.1	296	52	128	64	52	52	R	180	00	090	29	270	61
1997.10.03	11:28	54.84	27.49	15	5.3	142	33	164	246	81	58	R	001	29	251	32	124	44
1998.11.13	13:01	53.38	27.52	33	5.4	103	35	78	298	56	98	R	022	11	113	07	235	77
1999.04.30	04:20	52.96	27.74	45	5.2	321	53	134	82	55	47	R	201	01	110	34	293	56
2000.03.01	20:06	52.85	28.40	15	5.0	49	26	55	267	69	106	R	345	23	081	14	201	63
2001.04.13	01:04	55.04	27.55	26	5.1	166	34	55	295	67	64	R	044	18	306	24	168	60
2003.07.10	17:06	54.10	28.35	15	5.8	277	33	93	93	57	88	R	185	12	094	02	355	78
2003.07.10	17:40	54.05	28.26	15	5.7	83	34	49	310	65	114	R	022	17	119	21	257	62
2003.10.24	05:58	53.91	28.34	33	5.0	128	39	70	333	54	105	R	052	08	144	12	290	75
2003.11.28	23:19	53.66	28.19	33	5.0	43	19	60	255	74	100	R	337	28	072	09	179	60
2003.12.15	22:57	53.86	28.39	15	5.1	272	43	90	92	47	90	R	182	02	092	00	001	88
2004.01.14	16:58	52.17	27.54	12	5.2	297	25	95	112	65	88	R	203	20	113	02	017	70
2005.08.09	05:09	52.52	28.90	16	5.1	257	30	31	139	75	117	R	209	25	312	26	081	53
2006.09.10	08:57	54.29	27.54	23	5.0	321	52	147	73	65	43	R	194	08	098	41	293	48

D, centroid depth (km). Nodal planes are shown by their strike and dip with the rake of slip vector. F, fault type (R, reverse; S, strike-slip; N, normal) P, N and T are pressure, neutral and tension axes, which are indicated by their direction and plunge. Source: Harvard CMT solution (http://www.seismology.harvard.edu). Site location is shown in Figure 16b.

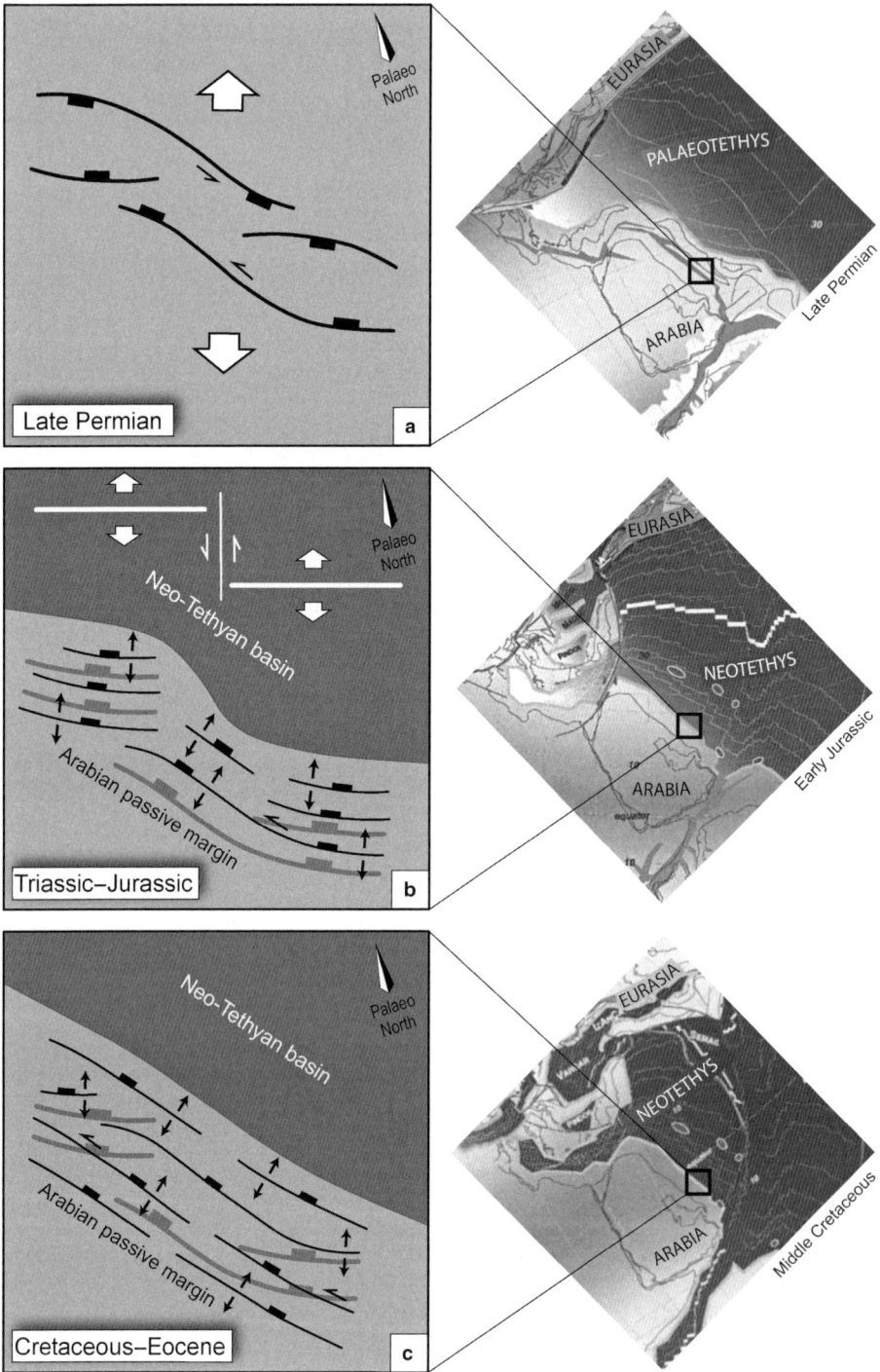

Fig. 17. Schematic illustration of a possible scenario for the Mesozoic extensional brittle tectonic evolution of the NEAPM. (**a**) Inferred Permian–Triassic oblique stretching that initiated major east–west normal faults within both the continental basement and the Palaeozoic sedimentary sequence close to the rift axis. (**b**) Continuation of margin-oblique faulting and extensional block tilting on the inherited normal faults, with some localized margin-parallel and

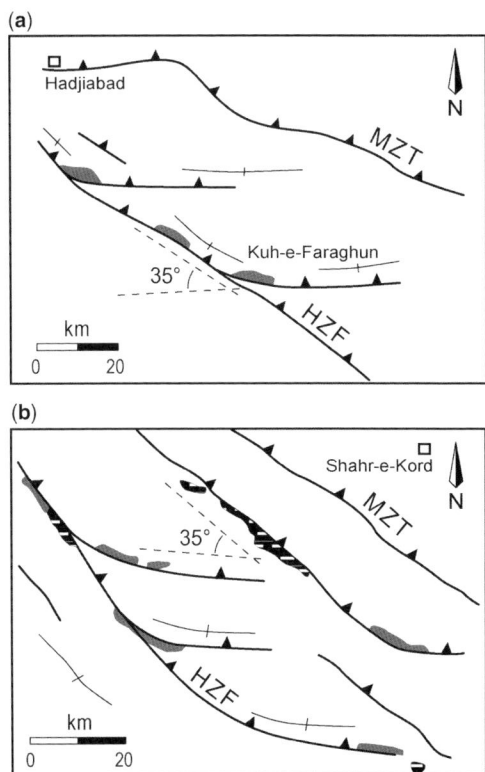

Fig. 18. Major reverse fault pattern of the HZB.
(a) Out-of-sequence structures of the Hadjiabad area (after Molinaro *et al.* 2005*a*), indicating east–west reverse faults branching from the HZF at an acute angle of *c.* 35°, associated with the Permian–Triassic outcrops. Location is shown in Figure 1. (b) Structural map of the Shahr-e-Kord area (after NIOC 1975), indicating similar reverse fault pattern, associated with the Hormuz Salt plugs and Palaeozoic outcrops. Location is shown in Figure 15c. Description of outcrops as in Figure 2. Abbreviations as in Figure 1.

high degree of uplifting to the NE, separating the Mesozoic outcrops of the HZB from the Cenozoic ones of the ZSFB (see Fig. 2). The presence of outcrops of Palaeozoic sedimentary rocks, together with the existence of numerous Hormuz Salt plugs along both the HZF (e.g. Sepehr & Cosgrove 2004) and the corresponding east-west reverse faults throughout the HZB (Fig. 18a, b) imply involvement of the deep-seated basement structures in such a deformation.

We acknowledge that the east–west reverse faults of the HZB can be interpreted as reverse splays of the HZF as a result of a right-lateral transpressional deformation especially in the Shahr-e-Kord region. However, such a high right-lateral movement similar to the MRF was not observed in the field along the HZF, and the focal mechanisms determined for the earthquakes along the HZF invariably show basement reverse faulting (see Talebian & Jackson 2004) on pre-existing high-angle faults (Yamini-Fard *et al.* 2006). The angle between the trends of these two reverse fault sets is about 35°, similar to that already observed between both the Mesozoic normal faults of the HZB and the basement reverse faults of the ZSFB of Fars Arc (see Figs 9a, 15c, 16c & 18a, b). We could identify the inverse reactivation on some of the Mesozoic normal faults, as is characterized by the existence of both normal and reverse fault slip striations on the same fault planes at different sites (see Fig. 9c, d). Consequently, we prefer to interpret these reverse structures as the result of an inversion on the pre-existing deep-seated normal faults (i.e. the old half-graben structure of the NEAPM), as was suggested by Sepehr & Cosgrove (2004) based on interpretation of a seismic profile across one of the anticlines of the Dezful Embayment (see Fig. 12a). Numerical modelling of crustal stretching has already suggested that inversion structures depend strongly on the inherited extensional geometry during the subsequent contraction (Buiter *et al.* 2002). This evidence led us to infer that the present-day structural styles of the HZB, which developed during the late Cenozoic collision and the subsequent crustal shortening, are compatible with, and partly controlled by, the inherited Mesozoic structures. In other words, the pre-existing Mesozoic structures could have played an essential basic role in the late Cenozoic structural evolution, under differently oriented compressional stress trends (see Navabpour *et al.* 2006, 2007*a*).

Therefore, the Miocene–Pliocene thin-skinned shortening could have occurred by roll-back on the pre-existing NW–SE thin-skinned extensional brittle structures (i.e. the normal faults of the sedimentary cover; Fig. 19a) in the HZB, under a general NE–SW compressional stress trend during the first stages of continental collision. During this period, the MZT was the location of a southwestward active thrusting of the SSMZ over the HZB (Braud 1971; Agard *et al.* 2005). The inherited

Fig. 17. (*Continued*) oblique-slip faulting during the Triassic–Jurassic period. (**c**) Gradual change in the extensional tectonic regime, prevailing NW–SE margin-parallel faults with reactivation of the inherited margin-oblique faults inducing some oblique-slip normal movements during the Cretaceous–early Tertiary period. Light and dark grey backgrounds indicate continental and oceanic crust, respectively. Inherited faults and oceanic structures are marked as grey and white lines, respectively. Location maps from Stampfli & Borel (2002). Palaeo-north directions from Gaetani *et al.* (2003) after Dercourt *et al.* (2000). Arrows as in Figure 15.

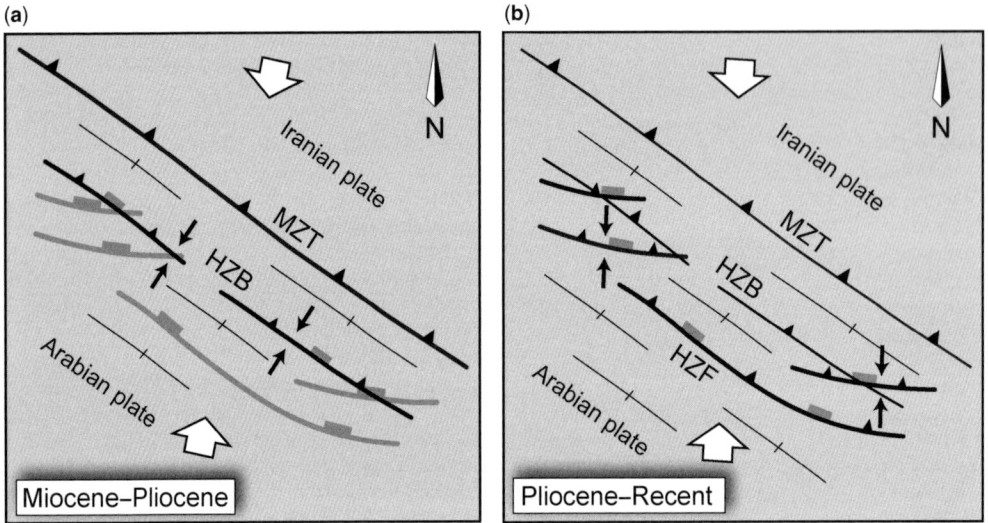

Fig. 19. Inversion of the inherited extensional structures during the late Cenozoic shortening of the Zagros in the HZB of interior Fars. (**a**) Thin-skinned inversion on the margin-parallel normal faults of the sedimentary cover. (**b**) Thick-skinned inversion on the margin-oblique normal faults of the basement. In each stage, bold black lines mark active reverse faults. Pairs of black arrows indicate direction of compression (from Navabpour *et al.* 2007*a*). Pairs of white arrows indicate direction of plate convergence (from McQuarrie *et al.* 2003). Other descriptions as in Figure 17. Location is shown in Figure 1; abbreviations as in Figure 1.

normal faults within the Mesozoic units could have facilitated deep ramps to develop and transfer the reverse displacement, detached from the basement by the basal Hormuz Salt horizon, to an upper flat detachment situated within the Upper Cretaceous Gurpi marls, producing the late Tertiary imbricate structure of the HZB (Molinaro *et al.* 2005*a*). The Pliocene–Recent thick-skinned shortening could have taken place by roll-back on the inherited east–west thick-skinned extensional brittle structures (i.e. the normal faults of the basement; Fig. 19b), under the recent north–south compressional stress trend, creating the out-of-sequence high-angle reverse faults that cut obliquely through the earlier structures of the overlying sedimentary cover of the HZB. During this period, the activity of the MZT had ceased, as shown by the Upper Pliocene Bakhtyari deposits sealing the corresponding structures (Gidon *et al.* 1974*a*, *b*), and the HZF is known to have a significant present-day seismic reverse activity in the HZB (Yamini-Fard 2003; Yamini-Fard *et al.* 2006). Further south, in the ZSFB, the basement active reverse faulting is characterized by major topographic steps at the surface of the sedimentary cover (Mouthereau *et al.* 2007) and highlighted by the numerous present-day earthquakes, with focal mechanisms indicating an average N009° trend for the recent compressive stress axes across the Fars Arc (Zamani 2009).

Conclusion

According to our analysis of brittle structures and based on the field evidence of syndepositional normal faulting in the HZB of interior Fars, we can reconstruct an extensional tectonic model for the Mesozoic history of the NEAPM. The reconstruction of tectonic palaeostress regimes and the corresponding structural patterns revealed the existence of two major margin-oblique and margin-parallel Mesozoic normal fault systems within the sedimentary cover. Further evidence of the existence of a possible Neo-Tethyan oblique palaeo-oceanic spreading axis together with the structures of the basement suggest that the Mesozoic structural pattern of the NEAPM could have been originated by gravitational reactivation on similar structures within the basement, which were inherited from an oblique oceanic opening. In this model, it is assumed that an oblique stretching process initiated thick-skinned extensional brittle structures as deep-seated normal faults within both the continental basement and the Palaeozoic sedimentary cover during the Permian–Triassic period. The deformation then continued during the Mesozoic era, commonly inducing extensional block tilting along the major inherited normal faults above the basal Infracambrian salt detachment, characterizing a thin-skinned extension.

Despite the major differences in region, age and orientation, a striking similarity exists between the proposed extensional tectonic evolution during the Neo-Tethyan opening in the NEAPM and the present-day extension in the Aden Gulf of the south Arabian continental passive margin. In both cases, the continental break-up was oblique in type, involving both the normal and transform fault segments of the stretching basement. The initial direction of extension was thus close to that of the plate divergence, oblique to the forthcoming continental margins, initiating deep-seated margin-oblique normal faults within the continental basement. During the next steps of oceanic opening, the direction of extension became nearly perpendicular to the continental margin, as local extensional tectonic processes prevailed by margin-parallel normal faults within the sedimentary cover. We infer that such resemblances in structural evolution reflect a similar transition from margin-oblique to margin-perpendicular extension as an oblique oceanic opening continues.

Although more detailed brittle tectonic analysis is needed at the scale of the whole Zagros belt, our reconstruction also highlights the role of the inherited extensional brittle structures of both basement and the sedimentary cover in the subsequent structural evolution characterized by the late Cenozoic collision and crustal shortening. Our fault slip data analysis strongly suggests that the inherited normal faults within the sedimentary sequence of the Arabian margin were reactivated as reverse faults during the folding process of the Miocene–Pliocene thin-skinned shortening. In the next stage of late Cenozoic compression, the out-of-sequence reverse structures related to the Pliocene–Recent thick-skinned shortening involved inversion of the inherited basement normal faults. The reactivated brittle structures were a function of both the inherited normal fault patterns and the late Cenozoic compressive stress orientations. Such a tectonic evolution, however, should not be extrapolated to other regions of the Zagros, where the basal salt detachment is absent or is of lesser importance.

We thank the Geological Survey of Iran (GSI) for providing vehicles and financial support during the extensive fieldwork in Iran. Trips to Iran were supported by the Middle East Basin Evolution (MEBE) project. Valuable comments by A. Zanchi and O. Bellier resulted in significant improvement of the manuscript, and M. Saadat proofread the manuscript; their help is gratefully acknowledged. The senior author thanks the Cultural Service of the French Embassy in Tehran and the French Ministry of Foreign Affairs for thesis grant through the CNOUS–CROUS, enabling him to study in the University of Nice–Sophia Antipolis.

References

AGARD, P., OMRANI, J., JOLIVET, L. & MOUTHEREAU, F. 2005. Convergence history across Zagros (Iran): constraints from collisional and earlier deformation. *International Journal of Earth Sciences*, **94**, 401–419, doi: 10.1007/s00531-005-0481-4.

AGARD, P., MONIÉ, P. *ET AL*. 2006. Transient, synobduction exhumation of Zagros blueschists inferred from *P–T*, deformation, time, and kinematic constraints: Implications for Neotethyan wedge dynamics. *Journal of Geophysical Research*, **111**, B11401, doi: 10.1029/2005JB004103.

ANGELIER, J. 1984. Tectonic analyses of fault slip data sets. *Journal of Geophysical Research*, **89**, 5835–5848.

ANGELIER, J. 1990. Inversion of field data in fault tectonics to obtain the regional stress III. A new rapid direct inversion method by analytical means. *Geophysical Journal International*, **103**, 363–376.

ANGELIER, J. 2002. Inversion of earthquake focal mechanisms to obtain the seismotectonic stress IV. A new method free of choice among nodal planes. *Geophysical Journal International*, **150**, 588–609.

ANGELIER, J. & BERGERAT, F. 1983. Systèmes de contrainte et extension intra-continentale. Colloque 'Rifts et Fossés Anciens', Marseille. *Bulletins du Centre de Recherches Exploration–Production ELF-Aquitaine*, **7**, 137–147.

ANGELIER, J. & MECHLER, P. 1977. Sur une méthode graphique de recherche des contraintes principales également utilisable en tectonique et en séismologie: la méthode des dièdres droits. *Bulletin de la Société Géologique de France*, **7**, 1309–1318.

ANGELIER, J., BARRIER, E. & NAVABPOUR, P. 2006. Brittle tectonic analyses near Kermanshah and Shiraz–Neyriz, SW-Zagros, Iran: mesozoic extension across the Arabian margin. *The Middle East Basin Evolution Conference Abstracts, December 2006, University of Milan, Milan*, 7–8.

AUBOURG, C., SMITH, B. *ET AL*. 2004. Post-Miocene shortening pictured by magnetic fabric across the Zagros–Makran syntaxis. *In: Orogenic Curvatures*. Geological Society of America, Special Papers, **383**, 17–40.

AUTHEMAYOU, C., CHARDON, D., BELLIER, O., MALEK-ZADEH, Z., SHABANIAN, E. & ABBASSI, M. R. 2006. Late Cenozoic partitioning of oblique plate convergence in the Zagros fold-and-thrust belt (Iran). *Tectonics*, **25**, TC3002, doi: 10.1029/2005TC001860.

BAKHTARI, H., FRIZON DE LAMOTTE, D., AUBOURG, C. & HASSANZADEH, J. 1998. Magnetic fabric of Tertiary sandstones from the Arc of Fars (Eastern Zagros, Iran). *Tectonophysics*, **284**, 299–316.

BÉCHENNEC, F., LE MÉTOUR, J., RABU, D., BOURDILLON-JEUDY-DE-GRISSAC, C., DE WEVER, P., BEURRIER, M. & VILLEY, M. 1990. The Hawasina Nappes: stratigraphy, palaeogeography and structural evolution of a fragment of the south-Tethyan passive continental margin. *In:* ROBERTSON, A. H. F., SEARLE, M. P. & RIES, A. C. (eds) *The Geology and Tectonics of the Oman Region*. Geological Society, London, Special Publications, **49**, 213–223.

BERBERIAN, M. 1977. Three phases of metamorphism in Hajiabad quadrangle map; a paleotectonic discussion. *In*: BERBERIAN, M. (ed.) *Contribution to the seismotectonics of Iran (Part III)*. Geological Survey of Iran, **40**, 239–260.

BERBERIAN, M. 1995. Master 'blind' thrust faults hidden under the Zagros folds: active basement tectonics and surface morphotectonics. *Tectonophysics*, **241**, 193–224.

BERBERIAN, M. & KING, G. C. P. 1981. Towards a paleogeography and tectonic evolution of Iran. *Canadian Journal of Earth Sciences*, **18**, 210–265.

BEYDOUN, Z. R., HUGHES CLARKE, M. W. & STONELEY, R. 1992. Petroleum in the Zagros basin: A Late Tertiary foreland basin overprinted onto the outer edge of a vast hydrocarbon-rich Palaeozoic–Mesozoic passive margin shelf. *AAPG Bulletin*, **55**, 309–339.

BORDENAVE, M. L. & HEGRE, J. A. 2005. The influence of tectonics on the entrapment of oil in the Dezful Embayment, Zagros Fold Belt, Iran. *Journal of Petroleum Geology*, **28**, 339–368.

BRAUD, J. 1971. La nappe du Kuh-e-Garun (région de Kermanshah, Iran), chevauchement de l'Iran Central sur le Zagros. *Bulletin de la Société Géologique de France*, **13**, 416–419.

BRETON, J. P., BÉCHENNEC, F., LE MÉTOUR, J., MOEN-MAUREL, L. & RAZIN, P. 2004. Eoalpine (Cretaceous) evolution of the Oman Tethyan continental margin: insights from a structural field study in Jabal Akhdar (Oman Mountains). *GeoArabia*, **9**, 41–58.

BUITER, S. J. H., HUISMANS, R. S., BEAUMONT, C. & PFIFFNER, O. A. 2002. Self-consistent numerical models of extensional basin formation and subsequent contractional inversion. http://www.geodynamics.no/BUITER/buiterrm02A.pdf

CHAUVET, F., BASILE, C. & DUMONT, T. 2004. Tectonic remnants of a Permian rifting in Oman autochthon (Saih Hatat and Jabal Akhdar windows). http://www.cosis.net/abstracts/RSTGV/00361/RSTGV-A-00361.pdf

COLMAN-SADD, S. P. 1978. Fold development in Zagros simply folded belt, southwest Iran. *AAPG Bulletin*, **62**, 984–1003.

DERCOURT, J., GAETANI, M. *ET AL.* 2000. *Atlas Peri-Tethys, Palaeogeographical Maps*. CCGM/CGMW, Paris.

ENGDAHL, E. R., JACKSON, J. A., MYERS, S. C., BERGMAN, E. A. & PRIESTLEY, K. 2006. Relocation and assessment of seismicity in the Iran region. *Geophysical Journal International*, **167**, 761–778, doi: 10.1111/j.1365-246X.2006.03127.x.

FALCON, N. L. 1974. Southern Iran: Zagros Mountains. *In*: SPENCER, A. (ed.) *Mesozoic–Cenozoic Orogenic Belts*. Geological Society, London, Special Publications, **4**, 199–211.

GAETANI, M., DERCOURT, J. & VRIELYNCK, B. 2003. The Peri-Tethys Programme: achievements and results. *Episodes*, **26**, 79–93.

GIDON, M., BERTHIER, F., BILLIAULT, J.-P., HALBRONN, B. & MAURIZOT, P. 1974*a*. Charriage et mouvements synsédimentaires tertiaires dans la région de Borudjerd (Zagros, Iran). *Comptes Rendus de l'Académie des Sciences, Série D*, **278**, 421–424.

GIDON, M., BERTHIER, F., BILLIAULT, J.-P., HALBRONN, B. & MAURIZOT, P. 1974*b*. Sur les caractères

et l'ampleur du coulissement de la 'Main Fault' dans la région de Borudjerd–Dorud (Zagros oriental, Iran). *Comptes Rendus de l'Académie des Sciences, Série D*, **278**, 701–704.

GILLARD, D. & WYSS, M. 1995. Comparison of strain and stress tensor orientation: Application to Iran and southern California. *Journal of Geophysical Research*, **100**, 22197–22213.

GSI 1985. *Geological Map of Neyriz*. Quadrangle **H11**, 1:250 000. Geological Survey of Iran, Tehran.

GSI 1990. *Geological Map of Eqlid*. Quadrangle **G10**, 1:250 000. Geological Survey of Iran, Tehran.

GSI 1996*a*. *Geological Map of Saadatshahr*. Quadrangle **6650**, 1:100 000. Geological Survey of Iran, Tehran.

GSI 1996*b*. *Geological Map of Neyriz*. Quadrangle **6848**, 1:100000. Geological Survey of Iran, Tehran.

GSI 1999. *Geological Map of Shiraz*. Quadrangle **6549**, 1:100 000. Geological Survey of Iran, Tehran.

GSI 2000*a*. *Geological Map of Sivand*. Quadrangle **6550**, 1:100 000. Geological Survey of Iran, Tehran.

GSI 2000*b*. *Geological Map of Eghlid*. Quadrangle **6551**, 1:100 000. Geological Survey of Iran, Tehran.

GSI 2000*c*. *Geological Map of Abadeh-e-Tashk*. Quadrangle **6649**, 1:100 000. Geological Survey of Iran, Tehran.

GSI 2000*d*. *Geological Map of Suryan*. Quadrangle **6750**, 1:100 000. Geological Survey of Iran, Tehran.

GSI 2001*a*. *Geological Map of Arsanjan*. Quadrangle **6649**, 1:100 000. Geological Survey of Iran, Tehran.

GSI 2001*b*. *Geological Map of Dehbid*. Quadrangle **6651**, 1:100 000. Geological Survey of Iran, Tehran.

HAYNES, S. J. & MCQUILLAN, H. 1974. Evolution of the Zagros suture zone, southern Iran. *Geological Society of America Bulletin*, **85**, 739–744.

HESSAMI, K., KOYI, H. A., TALBOT, C. J., TABASI, H. & SHABANIAN, E. 2001. Progressive unconformities within an evolving foreland fold–thrust belt, Zagros Mountain. *Journal of the Geological Society, London*, **158**, 969–981.

HU, J.-C. & ANGELIER, J. 2004. Stress permutations: Three-dimensional distinct element analysis accounts for a common phenomenon in brittle tectonics. *Journal of Geophysical Research*, **109**, B09403, doi: 10.1029/2003JB002616.

HUCHON, PH. & KHANBARI, KH. 2003. Rotation of the syn-rift stress field of the northern Gulf of Aden margin, Yemen. *Tectonophysics*, **364**, 147–166.

JACKSON, J. 1980. Reactivation of basement faults and crustal shortening in orogenic belts. *Nature*, **283**, 343–346.

JACKSON, J. & MCKENZIE, D. P. 1984. Active tectonics of the Alpine–Himalayan belt between western Turkey and Pakistan. *Geophysical Journal of the Royal Astronomical Society*, **77**, 185–264.

JAMES, G. A. & WYND, J. G. 1965. Stratigraphic nomenclature of Iranian Oil Consortium Agreement area. *AAPG Bulletin*, **49**, 2182–2245.

JARRIGE, J.-J. 1992. Variation of extensional fault geometry related to detachment surfaces within sedimentary sequences and basement. *Tectonophysics*, **215**, 161–166.

JARRIGE, J.-J., OTT D'ESTEVOU, P., BUROLLET, P. F., MONTENAT, C., PRAT, P., RICHERT, J.-P. & THIRIET, J.-P. 1990. The multistage tectonic evolution of the Gulf of Suez and Northern Red Sea continental rift from field observations. *Tectonics*, **9**, 441–465.

KENT, P. E. 1970. The salt plugs of the Persian Gulf region. *Leicester Literary and Philosophical Society Transactions*, **64**, 56–88.

KENT, P. E. 1979. The emergent Hormoz salt plugs of southern Iran. *Journal of Petroleum Geology*, **2**, 117–144.

KOCYIGIT, A. & ALTINER, D. 2002. Tectonostratigraphic evolution of the North Anatolian Palaeorift (NAPR): Hettangian–Aptian passive continental margin of the northern Neo-Tethys, Turkey. *Turkish Journal of Earth Sciences*, **11**, 169–191.

KOOP, W. J. & STONELEY, R. 1982. Subsidence history of the Middle East Zagros Basin, Permian to recent. *Philosophical Transactions of the Royal Society of London, Series A*, **305**, 149–168.

KOYI, H. A. 1988. Experimental modelling of role of gravity and lateral shortening in Zagros mountain belt. *AAPG Bulletin*, **72**, 1381–1394.

KOYI, H., HESSAMI, KH. & TEIXELL, A. 2000. Epicenter distribution and magnitude of earthquakes in fold–thrust belts: insights from sandbox models. *Geophysical Research Letters*, **27**, 273–276.

KYRKJEBO, R., GABRIELSEN, R. H. & FALEIDE, J. I. 2004. Unconformities related to the Jurassic–Cretaceous synrift–postrift transition of the northern North Sea. *Journal of the Geological Society, London*, **161**, 1–18.

LAUBACH, S. E. & MARSHAK, S. 1987. Fault patterns generated during extensional deformation of crystalline basement, NW Scotland. *In*: COWARD, M. P., DEWEY, J. F. & HANCOCK, P. L. (eds) *Continental Extensional Tectonics*. Geological Society, London, Special Publications, **28**, 495–499.

LEEDER, M. R. & GAWTORPE, R. L. 1987. Brittle modes of foreland extension. *In*: COWARD, M. P., DEWEY, J. F. & HANCOCK, P. L. (eds) *Continental Extensional Tectonics*. Geological Society, London, Special Publications, **28**, 139–152.

MAGGI, A., JACKSON, J. A., PRIESTLEY, K. & BAKER, C. 2000. A re-assessment of focal depth distributions in southern Iran, Tien Shan and northern India: do earthquakes really occur in the continental mantle? *Geophysical Journal International*, **143**, 629–661.

MARTON, L. G., TARI, G. C. & LEHMANN, C. T. 2000. Evolution of the Angolan Passive Margin, West Africa, with emphasis on post-salt structural styles. *In*: MOHRIAC, W. & TALWANI, M. (eds) *Atlantic Rifts and Continental Margins*. American Geophysical Union, Monograph series, **115**, 129–149.

MCQUARRIE, N., STOCK, J. M., VERDEL, C. & WERNICKE, B. P. 2003. Cenozoic evolution of Neotethys and implications for the causes of plate motions. *Geophysical Research Letters*, **30**, 2036, doi: 10.1029/2003GL017992.

MOLINARO, M., GUEZOU, J. C., LETURMY, P., ESHRAGHI, S. A. & FRIZON DE LAMOTTE, D. 2004. The origin of changes in structural style across the Bandar Abbas syntaxis, SE Zagros (Iran). *Marine and Petroleum Geology*, **21**, 735–752.

MOLINARO, M., LETURMY, P., GUEZOU, J.-C., FRIZON DE LAMOTTE, D. & ESHRAGHI, S. A. 2005a. The structure and kinematics of the southeastern Zagros fold–thrust belt, Iran: from thin-skinned to thick-skinned tectonics. *Tectonics*, **24**, TC3007, doi: 10.1029/2004TC001633.

MOLINARO, M., ZEYEN, H. & LAURENCIN, X. 2005b. Lithospheric structure underneath the south-eastern Zagros Mountains, Iran: recent slab break-off? *Terra Nova*, **17**, 1–6.

MOTIEI, H. 1993. Stratigraphy of Zagros. *In*: HUSHMANDZADEH, A. (ed.) *Treatise on the Geology of Iran*. Geological Survey of Iran, Tehran [in Farsi].

MOUTHEREAU, F., TENSI, J., BELLAHSEN, N., LACOMBE, O., DE BOISGROLLIER, T. & KARGAR, S. 2007. Tertiary sequence of deformation in a thin-skinned/thick-skinned collision belt: The Zagros Folded Belt (Fars, Iran). *Tectonics*, **26**, TC5006, doi: 10.1029/2007TC002098.

MURRIS, R. J. 1978. *Hydrocarbon habitat of the Middle East*. Shell International Petroleum Maalschappij B.V., Exploration and Production, **179**.

NABAVI, M. 1976. *An Introduction to the Iranian Geology*. Geological Survey of Iran, Tehran [in Farsi].

NADIMI, A. 2002. Mantle flow patterns at the Neyriz paleo-spreading center, Iran. *Earth and Planetary Science Letters* **203**, 93–104.

NAVABPOUR, P., ANGELIER, J. & BARRIER, E. 2006. Paleostress and brittle tectonic evolution from continental passive margin to continental collision: a case study of the High Zagros Belt of interior Fars, Iran. *The Middle East Basin Evolution Conference Abstracts, December 2006*, University of Milan, Milan, 34–37.

NAVABPOUR, P., ANGELIER, J. & BARRIER, E. 2007a. Cenozoic post-collisional brittle tectonic history and stress reorientation in the High Zagros Belt (Iran, Fars Province). *Tectonophysics*, **432**, 101–131, doi: 10.1016/j.tecto.2006.12.007.

NAVABPOUR, P., ANGELIER, J. & BARRIER, E. 2007b. Shift of deformation partitioning from active to passive margin in continental collision: a case study of W Zagros, Iran. *The Middle East Basin Evolution Conference Abstracts, December 2007*, University of Pierre and Marie Curie, Paris, C-32.

NIOC 1975. *Geological Map of Iran*. Sheet **4** South–West Iran, 1:1 000 000. National Iranian Oil Company, Tehran.

NIOC 1977. *Geological Map of Iran*. Sheet **5** South–Central Iran, 1:1 000 000. National Iranian Oil Company, Tehran.

NIOC 1979. *Geological Map of Shiraz*. Quadrangle **G11**, 1:250 000. National Iranian Oil Company, Tehran.

NOTTVEDT, A., GABRIELSEN, R. H. & STEEL, R. J. 1995. Tectonostratigraphy and sedimentary architecture of rifted basins, with reference to the northern North Sea. *Marine and Petroleum Geology*, **12**, 881–901.

PALMER, A. R. & GEISSMAN, J. 1999. *Geological Time Scale*. Geological Society of America, **CTS004**, http://www.geosociety.org/science/timescale/timescl-1999.pdf.

PAUL, A., KAVIANI, A., HATZFELD, D. & MOKHTARI, M. 2003. Lithospheric structure of the Central Zagros from seismological tomography. *4th International Conference of Earthquake Engineering and Seismology, May 2003*, University of Tehran, Tehran. p. 8.

PAUL, A., KAVIANI, A., HATZFELD, D., VERGNE, J. & MOKHTARI, M. 2006. Seismological evidence for crustal-scale thrusting in the Zagros mountain belt (Iran). *Geophysical Journal International*, **166**, 227–300.

PLAYER, R. A. 1969. *The Hormoz salt plugs of southern Iran.* PhD thesis, University of Reading.

POWERS, R. W. 1968. *Saudi Arabia (excluding Arabian Shield).* Centre National de la Recherche Scientifique, Lexique Stratigraphique International, III Asie, Fascicule 10b1.

RICOU, L. E. 1968. Sur la mise en place au Crétacé supérieur d'importantes nappes a radiolarites et ophiolites dans les monts Zagros (Iran). *Comptes Rendus de l'Académie des Sciences, Série D,* **267**, 2272–2275.

RICOU, L. E. 1974. *L'évolution géologique de la région de Neyriz (Zagros iranien) et l'évolution structurale des Zagrides.* Thèse d'état, Université d'Orsay.

RICOU, L. E. 1976. *Evolution Structurale des Zagrides. La Région Clef de Neyriz (Zagros Iranien).* Mémoire de la Société Géologique de France, **125**.

RICOU, L. E. 1994. Tethys reconstructed: plates, continental fragments and their boundaries since 260 Ma from Central America to Southeastern Asia. *Geodinamica Acta,* **7**, 169–218.

RICOU, L. E., BRAUD, J. & BRUNN, J. H. 1977. *Le Zagros.* Mémoire Hors-série de la Société Géologique de France, **8**, 33–52.

ROSENDAHL, B. R., KILEMBE, E. & KACZMARICK, K. 1992. Comparison of the Tanganyika, Malawi, Rukwa and Turkana rift basins from analyses of seismic reflection data. *Tectonophysics,* **213**, 235–256.

SEPEHR, M. & COSGROVE, J. W. 2004. Structural framework of the Zagros Fold–Thrust Belt, Iran. *Marine and Petroleum Geology,* **21**, 829–843.

SEPEHR, M., COSGROVE, J. & MOIENI, M. 2006. The impact of cover rock rheology on the style of folding in the Zagros fold–thrust belt. *Tectonophysics,* **427**, 265–281.

SETUDEHNIA, A. 1978. The Mesozoic sequence in southwest Iran and adjacent areas. *Journal of Petroleum Geology,* **1**, 3–42.

SHERKATI, S. & LETOUZEY, J. 2004. Variation of structural style and basin evolution in the central Zagros (Izeh zone and Dezful Embayment), Iran. *Marine and Petroleum Geology,* **21**, 535–554.

SHERKATI, S., MOLINARO, M., FRIZON DE LAMOTTE, D. & LETOUZEY, J. 2005. Detachment folding in the central and eastern Zagros fold-belt (Iran): salt mobility, multiple detachment and final basement control. *Journal of Structural Geology,* **27**, 1680–1696.

SOTO, R., CASAS-SAINZ, A. M. & DEL RIO, P. 2007. Geometry of half-grabens containing a mid-level viscous décollement. *Basin Research,* **19**, 437–450, doi: 10.1111/j.1365-2117.2007.00328.x.

STAMPFLI, G. 2000. Tethyan oceans. *In*: BOZKURT, E., WINCHESTER, J. A. & PIPER, J. D. A. (eds) *Tectonics and Magmatism in Turkey and the Surrounding Area.* Geological Society, London, Special Publications, **173**, 1–23.

STAMPFLI, G. & BOREL, G. D. 2002. A plate tectonic model for the Palaeozoic and Mesozoic constrained by dynamic plate boundaries and restored synthetic oceanic isochrones. *Earth and Planetary Science Letters,* **196**, 17–33.

STOCKLIN, J. 1968. Structural history and tectonics of Iran; a review. *AAPG Bulletin,* **52**, 1229–1258.

STONELEY, R. 1981. The geology of the Kuh-e Dalneshin area of Southern Iran and its bearing on the evolution of Southern Tethys. *Journal of the Geological Society, London,* **138**, 509–526.

STONELEY, R. 1990. The Arabian continental margin in Iran during the Late Cretaceous. *In*: ROBERTSON, A., SEARLE, M. & RIES, A. (eds) *The Geology and Tectonics of the Oman Region.* Geological Society, London, Special Publications, **49**, 787–795.

SZABO, F. 1977. Permian and Triassic stratigraphy of Fars area, southwest Iran. *2nd Iranian Geological Symposium, Iranian Petroleum Institute, Tehran,* 308–341.

SZABO, F. & KHERADPIR, A. 1978. Permian and Triassic stratigraphy, Zagros basin, southwest Iran. *Journal of Petroleum Geology,* **1**, 57–82.

TALBOT, C. J. 1979. Fold trains in a glacier of salt in southern Iran. *Journal of Structural Geology,* **1**, 5–18.

TALBOT, C. J. & ALAVI, M. 1996. The past of a future syntaxis across the Zagros. *In*: ALSOP, G. I., BLUNDELL, D. J. & DAVISON, I. (eds) *Salt Tectonics.* Geological Society, London, Special Publications, **100**, 89–109.

TALEBIAN, M. & JACKSON, J. 2004. A reappraisal of earthquake focal mechanisms and active shortening in the Zagros mountains of Iran. *Geophysical Journal International,* **156**, 506–526.

TATAR, M., HATZFELD, D. & GHAFORY-ASHTIANY, M. 2004. Tectonics of the Central Zagros (Iran) deduced from microearthquake seismicity. *Geophysical Journal International,* **156**, 255–266.

TCHALENKO, J. S. & BRAUD, J. 1974. Seismicity and structure of the Zagros: the Main Recent Fault between 33° and 35°N. *Philosophical Transactions of the Royal Society of London, Series A,* **277**, 1–25.

THIELE, O., ALAVI, M., ASSEFI, R., HUSHMANDZADEH, A., SEYED-EMAMI, K. & ZAHEDI, M. 1968. *Explanatory text of the Golpaygan quadrangle map.* Quadrangle **E7**. Geological Survey of Iran, Tehran.

TRON, V. & BRUN, J.-P. 1991. Experiments on oblique rifting in brittle–ductile systems. *Tectonophysics,* **188**, 71–84.

VENDEVILLE, B. C. & JACKSON, M. P. 1992. The rise of diapirs during thin-skinned extension. *Marine and Petroleum Geology,* **9**, 331–353.

VERNANT, P., NILFOROUSHAN, F. *ET AL.* 2004. Present-day crustal deformation and plate kinematics in the Middle East constrained by GPS measurements in Iran and northern Oman. *Geophysical Journal International,* **157**, 381–398.

YAMINI-FARD, F. 2003. *Sismotectonique et structure lithosphérique de 2 zones de transition dans le Zagros (Iran): La zone de Minab et la zone de Qatar–Kazerun.* PhD thesis, Université Joseph Fourier, Grenoble.

YAMINI-FARD, F., HATZFELD, D., TATAR, M. & MOKHTARI, M. 2006. Microearthquake seismicity at the intersection between the Kazerun fault and the Main Recent Fault (Zagros, Iran). *Geophysical Journal International,* **166**, 186–196, doi: 10.1111/j.1365-246X.2006.02891.x.

ZAMANI, B. 2009. *State of tectonic stress in the Iranian crust, indicated by focal mechanisms of earthquackes.* PhD thesis, University of Shiraz [in Farsi].

New magnetic fabric data and their comparison with palaeostress markers in the Western Fars Arc (Zagros, Iran): tectonic implications

CHARLES AUBOURG[1]*, BRIGITTE SMITH[2], ALI ESHRAGHI[3], OLIVIER LACOMBE[4], CHRISTINE AUTHEMAYOU[5], KHALED AMROUCH[4], OLIVIER BELLIER[6] & FRÉDÉRIC MOUTHEREAU[4]

[1]*Géosciences & Environnement Cergy, Université Cergy Pontoise, CNRS, 5, mail Gay Lussac, Neuville-sur-Oise, 95031 Cergy, France*

[2]*Géosciences Montpellier, Université de Montpellier 2, CNRS, Place Eugène Bataillon, 34095 Montpellier, France*

[3]*Geological Survey of Iran, Tehran, Iran*

[4]*Laboratoire de Tectonique, Université P. et M. Curie-Paris 6, CNRS, Paris, France*

[5]*Laboratoire Domaines Océaniques, CNRS, Institut Universitaire Européen de la Mer, Université de Brest, Place Nicolas Copernic, 29280 Plouzane, France*

[6]*CEREGE – UMR CNRS 6635 – Aix – Marseille Université, BP 80, Europôle, Méditerranéen de l'Arbois, 13545 Aix-en-Provence, Cedex 4, France*

Corresponding author (e-mail: aubourg@u-cergy.fr)

Abstract: The Zagros Simply Folded Belt (ZSFB) is an active fold-and-thrust belt resulting from the still continuing continental collision between the Arabian plate and the Iranian plate, which probably started in the Oligocene. The present-day shortening (N25°) is well documented by focal mechanisms of earthquakes and global positioning system (GPS) surveys. We propose in this study a comparison of published palaeostress markers, including magnetic fabric, brittle deformation and calcite twinning data. In addition, we describe the magnetic fabric from Palaeocene carbonates (10 sites) and Mio-Pliocene clastic deposits (15 sites). The magnetic fabrics are intermediate, with magnetic foliation parallel to the bedding, and a magnetic lineation mostly at right angles to the shortening direction. This suggests that the magnetic fabric retains the record of an early layer-parallel shortening (LPS) that occurred prior to folding. The record of LPS allows the identification of originally oblique folds such as the Mand Fold, which have developed in front of the Kazerun Fault. The shape parameter of the magnetic fabric indicates a weak strain compatible with the development of detachment folds in the ZSFB. The palaeostress datasets, covering the Palaeocene to Pleistocene time interval, support several folding episodes accompanied by a counter-clockwise rotation of the stress field direction. The Palaeocene carbonates in the ZSFB record a N47 LPS during early to middle Miocene detachment folding in the High Zagros Belt (HZB). The Mio-Pliocene clastic deposits recorded a N38 LPS prior to and during detachment folding within the ZSFB at the end of the Miocene–Pliocene. Similarly, fault slip and calcite twin data from the ZSFB also support a counter-clockwise rotation from NE to N20 between the pre-folding stage and the late rejuvenation of folds. This counter-clockwise trend of palaeostress data agrees with fault slip data from the HZB. During the late stage of folding in the ZSFB, the Plio-Quaternary palaeostress trends are consistently parallel to the present-day shortening direction.

In the Zagros active fold-and-thrust belt, the present-day stress and strain fields are now well constrained by geodetic and seismic data (Tatar et al. 2002; Talebian & Jackson 2004; Hessami et al. 2006; Lacombe et al. 2006; Walpersdorf et al. 2006). It is, however, also important to elucidate the stress pattern in the different stages of fold-and-thrust belt formation, from layer-parallel shortening (LPS) to folding and thrusting. Palaeostress or -strain can be determined by several means, including analyses of magnetic fabric, striated microfaults and calcite twinning.

From: LETURMY, P. & ROBIN, C. (eds) *Tectonic and Stratigraphic Evolution of Zagros and Makran during the Mesozoic–Cenozoic*. Geological Society, London, Special Publications, **330**, 97–120.
DOI: 10.1144/SP330.6   0305-8719/10/$15.00 © The Geological Society of London 2010.

The objectives of this work are the following. We will first characterize the general pattern of the magnetic fabric in the western Fars Arc, based on new data combined with previously published data. Then we will compare this information with the Late Cenozoic palaeostress data deduced from analyses of both small-scale deformation recorded in the field and calcite twinning. Finally, we will integrate the Late Cenozoic stress pattern in a comprehensive scheme of the tectonic evolution of the Fars.

Palaeostress markers in fold-and-thrust belts

Magnetic fabric is analysed from the measurement of standard cores of *c.* 10 cm^3. In essence, the magnetic fabric averages the 3D preferred orientation of billions of magnetic grains with anisotropy as small as 0.1% (Hrouda 1982). In unmetamorphosed rocks from fold-and-thrust belts, the magnetic fabric generally integrates the record of burial and subsequent deformation (Graham 1966; Hrouda 1982; Borradaile 1987) (Fig. 1). Numerous studies in fold-and-thrust belts have shown that magnetic fabric, when measured with the anisotropy of low-field magnetic susceptibility (AMS), can be used successfully as a record of layer-parallel shortening (LPS), thus behaving as a good proxy for strain (Graham 1966; Kissel *et al.* 1986; Averbuch *et al.* 1992; Hirt *et al.* 1995; Aubourg *et al.* 1997; Parés *et al.* 1999). Generally, the magnetic lineation from AMS (K_1) lies at right angles to the LPS direction whereas the magnetic foliation (the plane containing AMS K_1 and K_2 axes) remains parallel to the bedding. This fabric is labelled 'intermediate fabric' according to the nomenclature proposed by

Averbuch *et al.* (1992). When rocks are more strained, the bedding-related magnetic foliation is progressively lost and a tectonic-related magnetic foliation may develop. This fabric is called 'tectonic fabric'. It should be noted that tectonic-related magnetic foliation can develop without its counterpart being visible in the field (such as a cleavage). Several pioneering studies envisaged the quantitative issue of AMS by using appropriate parameters (see the review by Borradaile 1987). However, it appears that, in addition to strain, the nature of magnetic carriers of AMS controls also the magnitude of AMS parameters (Rochette *et al.* 1992; Hrouda *et al.* 1993). Despite this complication, the shape parameter *T* (Jelinek 1981) may provide a valuable indication of the progressive loss of bedding-parallel magnetic foliation during the imprint of LPS in sedimentary rocks from fold-and-thrust belts (Parés *et al.* 1999).

The occurrence of intermediate or tectonic magnetic fabric is apparently dependent on the efficiency of the décollement level and the nature of the sedimentary rocks (Frizon de Lamotte *et al.* 2002). Several researchers observed that claystones, carbonates and clastic deposits have developed distinct magnetic fabrics in response to similar strain history (Bakhtari *et al.* 1998; Sagnotti *et al.* 1998). It is easier to develop a tectonic fabric in carbonates and clastic deposits compared with claystones, where magnetic foliation is strongly controlled by the bedding. When the décollement level is frictionless, as it may be in salt-based thrust belts, intermediate LPS fabric is dominant (Parés *et al.* 1999; Kanamatsu *et al.* 2001). However, when the décollement level has high friction, tectonic LPS fabrics are likely to develop (see discussion by Robion *et al.* 2007). During folding, we may distinguish between detachment folds and fault-related folds.

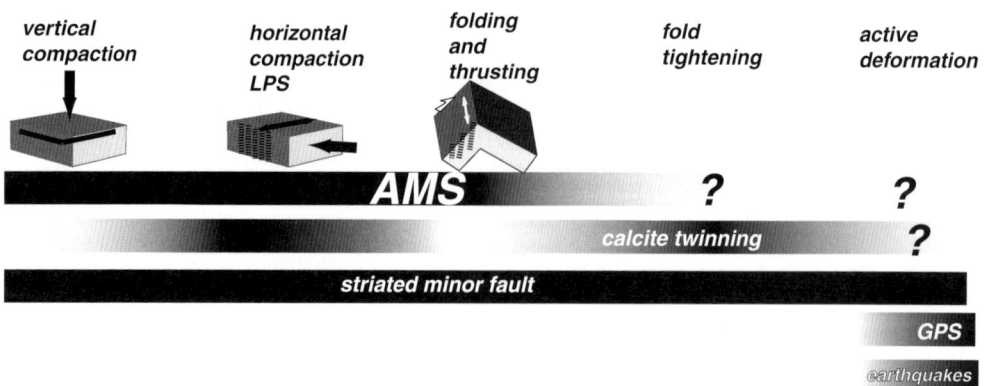

Recovering strain and stress from a fold-and-thrust belt

Fig. 1. Record of palaeostress in fold-and-thrust belt by several techniques, all used in this study. Dark shading indicates the timing of the palaeostress record.

In the case of a detachment fold, the strain is weak and the LPS-related magnetic fabric is generally preserved (Aubourg et al. 2004). In contrast, when the strain is more pronounced, as it is in a fault-propagation fold, a fold-related magnetic fabric can develop (Saint-Bezar et al. 2002) (Fig. 1). The strain imprint by AMS in the later stage of fold-and-thrust belt formation as fold tightening and active deformation occurs is not yet well documented. Hamilton et al. (2004) envisaged the record of post-folding strain by AMS in thrust belts.

The analysis of small-scale brittle deformation is performed directly in the field. This consists of inverting fault slip data into stress tensors representative of the fault population (Angelier 1990; Mercier et al. 1991). Generally, several tens of striated minor faults are analysed to compute a palaeostress tensor, including the orientations of the three principal stress axes σ_1, σ_2 and σ_3 and the stress ellipsoid shape ratio Φ defined as $\Phi = (\sigma_2 - \sigma_3)/(\sigma_1 - \sigma_3)$. This now classical approach has given rise to many theoretical developments and successful applications over the last 30 years, so there is no need to enter into much detail here [refer to Authemayou et al. (2006) and Lacombe et al. (2006) for basic assumptions, limitations and references on this technique]. To provide time constraints on palaeostress data, several criteria are used such as the age of faulted rocks, the eventual superimposition of striations along the fault plane and the orientation of palaeostress axes with respect to bedding, in addition to evidence of syntectonic sedimentation when available. In favourable situations, it is possible to recover the whole palaeostress story of the thrust belt, from burial to active deformation (Fig. 1).

Mechanical e-twinning readily occurs in calcite deformed at low temperature (Burkhard 1993). Calcite twinning requires a low critical resolved shear stress (CRSS), which depends on grain size (Rowe & Rutter 1990) and internal twinning strain, and has only a slight sensitivity to temperature, strain rate and confining pressure; thus calcite twinning fulfils most of the requirements for palaeopiezometry (Lacombe 2007). In this paper, we used Etchecopar's method of inverting calcite twin data (Etchecopar 1984; see details given by Lacombe 2001, 2007). This method applies to small twinning strain that can be approximated by coaxial conditions, so orientation of twinning strain can be correlated with palaeostress orientation (Burkhard 1993). Calcite twinning analysis is performed optically under a U-stage microscope. From mutually perpendicular thin-sections, tens of calcite grains are analysed for a sample at a given site, from host rock matrix and/or veins. The inversion process takes into account both the twinned and the untwinned planes, the latter being those of the

potential e-twin planes that never experienced a resolved shear stress of sufficient magnitude to cause twinning. The inverse problem consists of finding the stress tensor that best fits the distribution of twinned and untwinned planes. As for fault slip data, the orientations of the three principal stresses σ_1, σ_2, and σ_3 are calculated, together with the Φ ratio, but in addition the peak differential stress $(\sigma_1 - \sigma_3)$ is also computed. If more than c. 30% twinned planes in a sample are not explained by a unique stress tensor, the inversion process is repeated with the uncorrelated twinned planes and the whole set of untwinned planes. Where polyphase deformation has occurred, this process provides an efficient way of separating superimposed twinning events. The stress inversion technique is to date the only technique that allows simultaneous calculation of principal stress orientations and differential stress magnitudes from a set of twin data, and that therefore allows differential stress magnitudes to be related unambiguously to a given stress orientation and stress regime (Lacombe 2007). To date the palaeostress tensor derived from calcite twinning data we have to take into account the age of rocks, the various generations of calcite veins and the orientation of palaeostress axes with respect to bedding. It is generally assumed that calcite twinning records early LPS (Craddock & van der Pluijm 1999) but several studies have also reported the potential of calcite twinning to record late-stage fold tightening strain (Fig. 1; Harris & van der Pluijm 1998; Lacombe 2001; Lacombe et al. 2007).

Geological setting

The Zagros belt is one of the youngest continental collision belts, resulting from the convergence between the Arabian and the Iranian plates (Fig. 2a). The subduction started in the late Jurassic and the continental collision began by the Late Oligocene–Miocene. Geodetic data show that about one-third of the total c. 22 mm a^{-1} present shortening between Arabia and Eurasia is accommodated in the external part of the Zagros (Fig. 2b). The Zagros comprises two major NW–SE trending structural zones, the High Zagros Belt (HZB) and the Zagros Simply Folded Belt (ZSFB) (Fig. 2c). They are bounded by two major thrusts: the Main Zagros Thrust (MZT), which is the inactive suture between the Arabian plate and the Iranian plate (Ricou et al. 1977; Berberian 1995), and the High Zagros Thrust (HZT), which marks the NE boundary of the Arabian passive palaeomargin. We limit the presentation of these units to the central Zagros, bracketed between the longitudes 51° and 55°E, where the ZSFB corresponds to the Fars Arc (Fig. 2c). In this study we use the fault

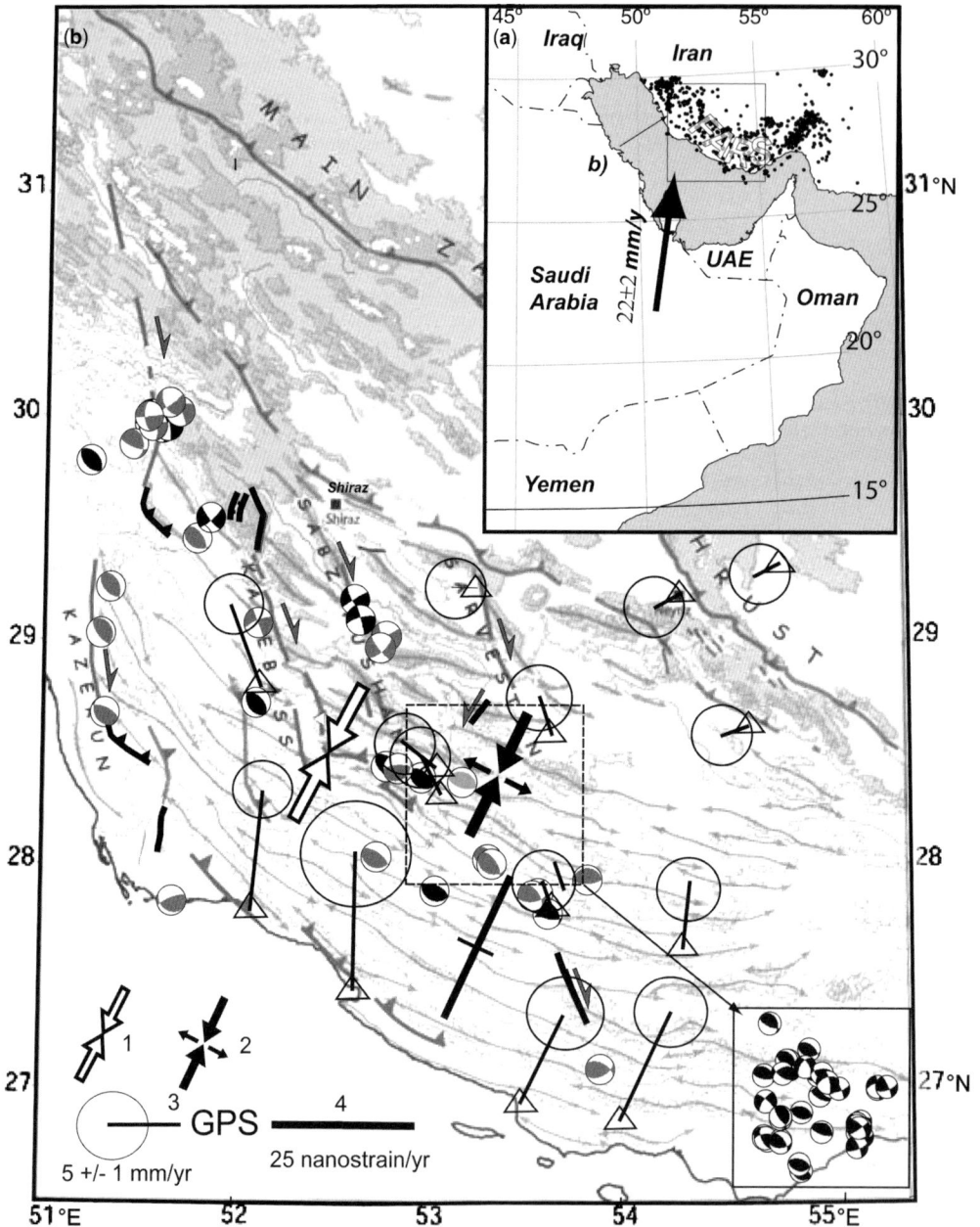

Fig. 2. Present-day deformation and structure of the Fars Arc. (**a**) View of the Arabian plate and the Zagros, marked by a pervasive seismicity. (**b**) Present-day deformation as indicated by seismicity and GPS displacement. Shortening directions inferred from the inversion of earthquake focal mechanisms. 1 and 2, current compressional trend derived from moderate earthquakes and microearthquakes, respectively (Lacombe *et al.* 2006). 3 and 4, GPS velocity field relative to central Iran and related strain rate, respectively (Walpersdorf *et al.* 2006). (**c**) Main tectonic structures of the High Zagros Belt and the Zagros Simply Folded Belt in the setting of Arabian and Iranian plate convergence. The disposition of major blind thrusts (High Zagros Fault, Mountain Front Fault, and Zagros Front Fault) along the Kazerun Fault (KZ) in the western Fars Arc and their intersection in the Eastern Fars Arc should be noted. K, Karehbass Fault; ZMS, Zagros–Makran Syntaxis. We indicate the location of the Mand and Minab anticlines at the western and eastern tips of the Fars Arc, respectively. The magnetic fabric of these anticlines is shown in Figure 6.

(c)

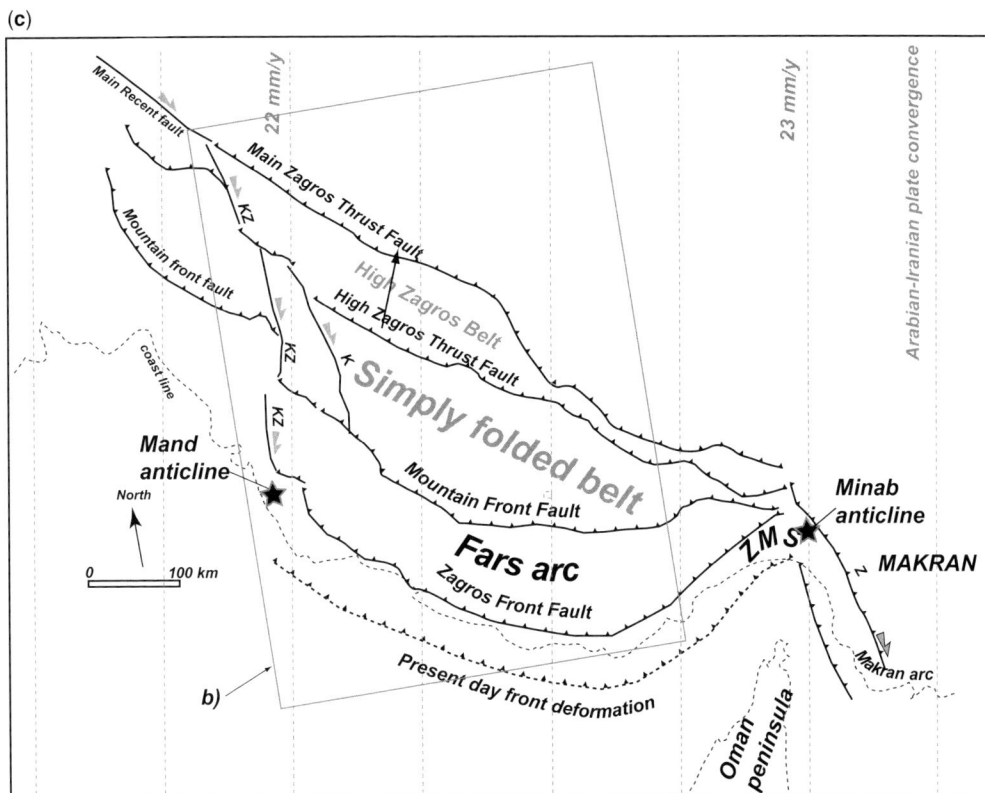

Fig. 2.

nomenclature proposed by Sepehr & Cosgrove (2004) & Sherkaty & Letouzey (2004). The ZSFB is bounded by the High Zagros Thrust to the NE and the Zagros Front Fault to the SW. It should be noted that a blind and active thrust, the Mountain Front Fault (MFF) is localized in the intermediate part of ZSFB. The MFF coincides approximately with the 1500 m topographic contour map and major zone of seismicity (Sepehr & Cosgrove 2004).

The HZB is the most uplifted (up to 4000 m) and eroded part of the Zagros mountain belt. However, its present-day seismic activity is low (Talebian & Jackson 2004). The main folding stage started during the Late Oligocene–Early Miocene and ended in the late Miocene (Sherkati et al. 2005). Navabpour et al. (2007) documented successive palaeostress fields in the central HZB (Shiraz area) using tectonic analysis of small-scale brittle deformation. They reported a c. 50° counter-clockwise rotation of the palaeostress field from the Late Oligocene–Early Miocene (c. N53°) to Quaternary (c. N2°). They proposed that this counter-clockwise rotation of the palaeostress field reflects large-scale

plate kinematic changes (McQuarrie et al. 2003). The reconstructed compressional trends are reported in Figure 2b.

In contrast to the inactive HZB, the ZSFB is seismically active (Berberian 1995) and it concentrates c. 40% of the convergence between the Arabian and Iranian plates (Tatar et al. 2002). Most of the earthquakes take place in the upper part of the basement (at 11–15 km depth) (Tatar et al. 2004). A large part (c. 95%) of the deformation is thus accommodated aseismically by creep on faults and folding in the cover (Masson et al. 2005). Using balanced cross sections, several studies bracketed the total shortening between 25 and 37 km in the Fars Arc (Blanc et al. 2003; Molinaro et al. 2003, 2004a, b; McQuarrie 2004; Sherkati & Letouzey 2004; Sherkati et al. 2006; Mouthereau et al. 2007a, b). Both basement and the c. 10 km of Palaeozoic to Cenozoic cover are involved in collisional shortening. Folding is still continuing, as Quaternary fold growth is commonly observed along the coast of the Fars Arc (Homke et al. 2004; Oveisi et al. 2007). The Kazerun Fault (Berberian 1995) laterally bounds the Fars Arc along its SW margin. The Kazerun Fault has been studied by Authemayou *et al.* (2006), and its satellite faults (Fig. 2b), the Karehbass, Sabz Pushan and Sarvestan faults, by Berberian (1995). It constitutes a system of dextral strike-slip faults along which the cumulative right lateral shear reaches 6 mm a^{-1} (Authemayou *et al.* 2006; Walpersdorf *et al.* 2006).

Based on geodetic data (Walpersdorf *et al.* 2006) and inversion of earthquake focal mechanisms (Lacombe *et al.* 2006), the present-day shortening or compression directions are well constrained in the Northern and Central ZSFB. These directions are parallel and trend *c.* N25. This direction is slightly oblique to the *c.* N10° lithospheric convergence deduced from the Global Iran geodetic data (Vernant *et al.* 2004). This obliquity reflects a partitioning of oblique convergence in the western part of Fars Arc (Talebian & Jackson 2004; Authemayou *et al.* 2006).

Small-scale brittle deformation (Authemayou *et al.* 2006; Lacombe *et al.* 2006; Navabpour *et al.* 2007), calcite twinning (Lacombe *et al.* 2007) and AMS analyses (Bakhtari *et al.* 1998; Aubourg *et al.* 2004) provide a comprehensive pattern of Cenozoic palaeostress and strain data. The main trend of shortening or compression derived from these data is indicated in Figure 2b. It can be seen that a counter-clockwise rotation of the shortening direction is recorded in the Fars Arc from the Middle Miocene to the present. The palaeostress pattern will be discussed in the light of new AMS data.

Aubourg *et al.* (2008) proposed a first pattern of block rotations using palaeomagnetic data in the western Fars Arc. Both counter-clockwise and clockwise rotations have been reported mainly in the Agha-Jari Fm. (Fig. 3a). Although the dominant sense of rotation is clockwise, this first block rotation pattern does not support or rule out the various models of block rotations proposed by several researchers in the western Fars Arc (Bakhtari *et al.* 1998; Talebian & Jackson 2004; Molinaro *et al.* 2005; Authemayou *et al.* 2006; Lacombe *et al.* 2006; Navabpour *et al.* 2007).

New AMS results

Sampling

We sampled 10 sites in the limestone and marly limestone levels of the Palaeocene–Eocene carbonates of the Pabdeh Fm., four sites in the Miocene marly limestones of the Razk Fm., and 11 sites in the Mio-Pliocene clastic deposits of the Mishan and Agha-Jari Fms., from the western part of the Fars Arc, in the ZSFB (Fig. 3a). We cored rocks using a portable drilling machine and determined the geographical orientation using both magnetic compass and sun angles. Table 1 provides information about sampling, rock formation and bedding orientation. In Figure 3b, we indicate the sites in the stratigraphic column. We sampled one site in the Gurpi Fm., nine in the Padbeh Fm., four in the Razak Fm., two in the Mishan Fm. and nine in the Agha-Jari Fm. (Fig. 3). It should be noted that the four sites from the Mand anticline (8K–11K) were sampled in the Pliocene coastal Lahbari member (upper Agha-Jari Fm.). This may represent therefore the youngest deformation recorded by AMS. For palaeomagnetic and magnetic fabric investigations, we selected the finest-grained formations and sampled the paired limbs of anticlines or synclines. The Mio-Pliocene formations are contemporaneous with the folding events, especially the Agha-Jari Fm., where intraformational growth strata are commonly observed (Berberian & King 1981; Hessami *et al.* 2001; Homke *et al.* 2004; Sherkati & Letouzey 2004; Sherkati *et al.* 2005; Lacombe *et al.* 2006). All rocks sampled are weakly strained. Apart from small-scale brittle deformation (joints and faults), we never observed penetrative strain such as cleavage.

The Palaeocene sites are mainly located in the northern part of the ZSFB (north of 35°N latitude) on both sides of the Kazerun Fault (Fig. 3a). The Mio-Pliocene sites are in the southwestern part of the Fars Arc near the Kazerun, Karebass and Sabz-Pushan faults. Sites 1K, 2K, 5K (Palaeocene) and sites 6K and 7K (Mio-Pliocene) are close to the same Kazerun Fault segment. Sites 8K to 11K are in the Mand anticline, which is developed in front of the southernmost thrust termination of the Kazerun–Borazjan Fault (Authemayou *et al.* 2006; Sherkati *et al.* 2006; Oveisi *et al.* 2007). Sites 15K, 16K and 21K are situated in front of the southern termination of the Karebass Fault, where Permo-Triassic rocks are exhumed (Talebian & Jackson 2004). Sites 14K, 17K, 18K and 22K are along the Sabz-Pushan Fault. Only sites 19K and 20K are away from identified strike-slip faults.

AMS data

General behaviour. We measured the anisotropy of low-field magnetic susceptibility (AMS) of 327 standard oriented cores (*c.* 10 cm^3) using an Agico KLY-3S system. We processed the AMS data using standard tensorial Jelinek statistics (Jelinek 1978). Mean AMS data are compiled in Table 1. We plot in equal-area stereoplots the principal axes of the anisotropy ellipsoid ($K_1 \geq K_2 \geq K_3$) for each site, and the density diagrams of the K_1 and K_3 axes for two chronological groups of samples: the Mio-Pliocene and the Palaeocene rocks. We used three systems of coordinates: (1) the geographical coordinates (GC); (2) the stratigraphic coordinates (SC);

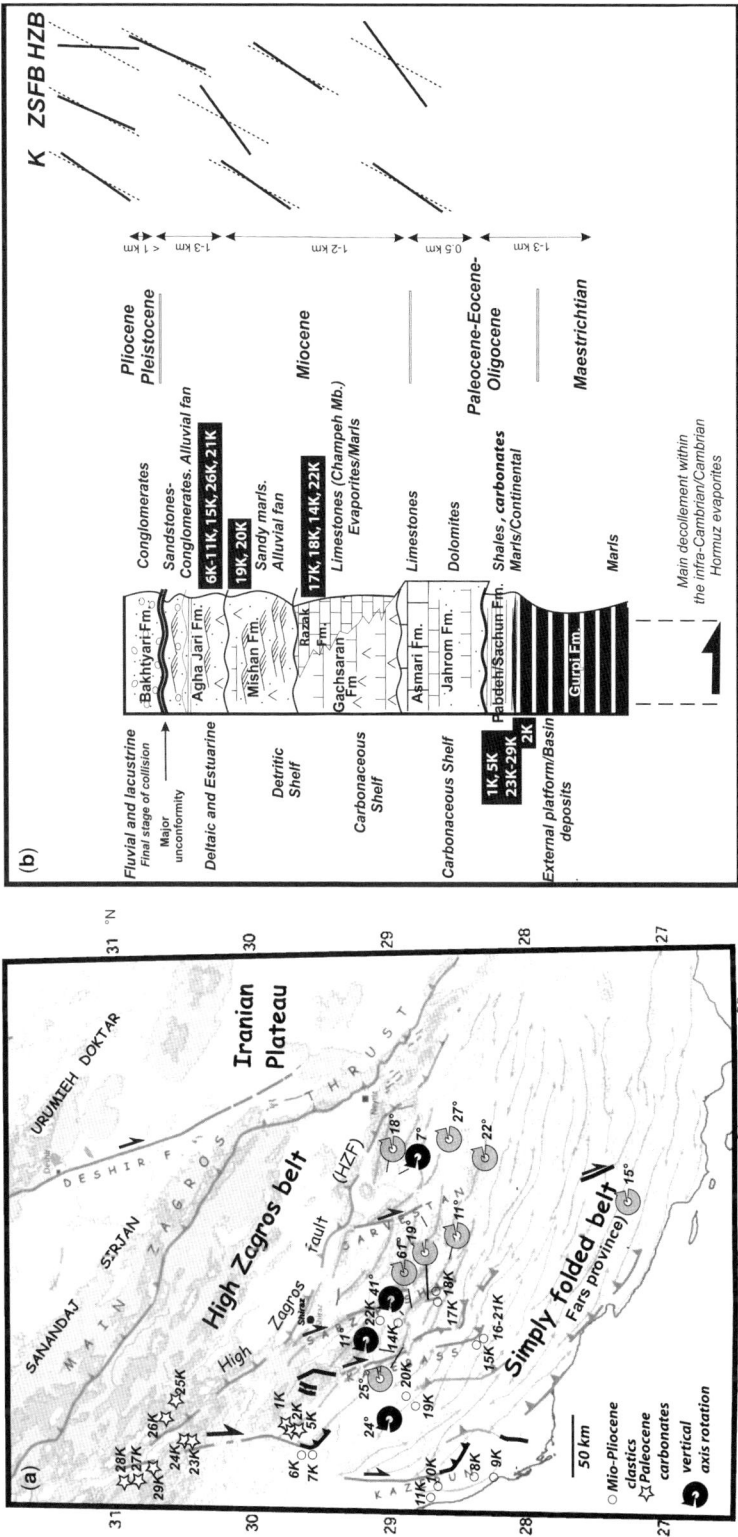

Fig. 3. (a) Location of the sites sampled. The vertical-axis rotation derived from palaeomagnetic data (Aubourg et al. 2008) is also indicated. (b) Stratigraphic column with location of sites. Also indicated is the shortening direction derived from brittle deformation analyses from the Kazerun Fault (K: Authemayou et al. 2006), Zagros Simply Fold Belt (ZSFB; Lacombe et al. 2006) and High Zagros Belt (HZB; Navabpour et al. 2007). Dashed lines represent the present-day shortening direction from GPS data.

Table 1. *Data for sites grouped by fold*

Site	Fold	Fm.	Age	Lithology	S_0	Longitude	Latitude	n	AMS Scalar data			GC				SC	
									K_m	P'	T	K_1	e_1/e_2	K_3	e_1/e_2	K_1	K_3
1 K	Chowgan	Padbeh	Paleocene	carbonates	290N25	51°37.01′	29°48.10′	16	26	1.014	0.64	338-20	27-9	222-14	9-9	161-0	253-76
2 K		Gurpi	Eocene	marls	192W12	51°36.55′	29°47.65′	12	39	1.018	0.62	311-5	11-9	78-82	10-8	131-6	323-84
5 K		Padbeh	Paleocene	carbonates	170W19	51°35.27′	29°47.06′	8	15	1.016	−0.17	226-40*	23-6	318-2*	12-7	232-24	138-9*
6 K	Rudak	Agha Jhari	Mio-Pliocene	red sandstones	256NW12	51°26.34′	29°37.68′	12	577	1.074	0.77	342-8	10-5	150-81	6-2	162-4	18-86
7 K		Agha Jhari	Mio-Pliocene	red sandstones	191W53	51°26.96′	29°37.83′	13	251	1.039	0.72	328-3	10-5	222-80	6-4	328-3	222-80
8 K	Mand	Lahbari	Mio-Pliocene	red sandstones	316E25	51°17.30′	28°25.81′	12	355	1.039	0.71	120-12	12-8	235-65	8-4	116-4	316-86
9 K		Lahbari	Mio-Pliocene	red sandstones	134W20	51°16.84′	28°15.78′	13	310	1.063	0.89	296-4	6-3	36-69	3-1	115-3	320-87
10 K		Lahbari	Mio-Pliocene	red sandstones	20E15	51°12.20′	28°41.81′	12	249	1.044	0.78	131-10	17-8	284-79	11-8	311-4	153-86
11 K		Lahbari	Mio-Pliocene	red sandstones	190W15	51°07.36′	28°41.48′	14	558	1.001	0.80	301-7	7-2	76-4	3-2	121-7	316-83
21 K	Daryau	Agha Jhari	Mio-Pliocene	red sandstones	300NE34	52°24′26.6″	28°18′46.6″	15	670	1.074	0.44	121-4	5-2	216-55	5-2	119-2	240-86
15 K		Agha Jhari	Mio-Pliocene	red sandstones	125SW40	52°22′42.7″	28°20′19.6″	15	324	1.024	−0.35	302-3	8-2	211-21	27-5	122-1	31-19
16 K		Agha Jhari	Mio-Pliocene	red sandstones	295NE25	52°24′25.4″	28°18′51.4″	16	443	1.049	0.24	121-0	7-6	212-51	11-7	302-1	210-83
17 K	Qir	Razak	Miocene	red sandstones	300NE41	52°41′03.1″	28°51′19.0″	12	93	1.030	0.71	115-11	10-3	216-46	4-2	110-2	315-88
18 K		Razak	Miocene	red sandstones	120SW46	52°42′32.6″	28°42′06.5″	11	72	1.034	0.73	136-10	15-3	36-41	5-3	319-2	71-86
19 K	Bushgan	Mishan	Miocene	marls	140SW15	51°49′26.9″	28°53′24.3″	14	586	1.087	0.59	151-7	5-2	35-76	4-2	152-0	277-89
20 K		Mishan	Miocene	marls	305NE15	51°52′14.4″	28°54′36.3″	12	539	1.121	0.80	143-3	7-2	244-76	3-2	142-7	324-84
14 K	Amirabad	Razak	Miocene	marls	290NE52	52°33′28.1‴	28°57′49.6″	13	91	1.027	0.71	288-3	6-5	195-50	6-3	290-2	47-86
22 K		Razak	Miocene	marls	127SW57	52°38′36.3‴	29°03′49.7″	14	59	1.030	0.62	286-37	4-3	39-28	3-2	266-2	72-88
23 K	Darihsk-East	Padbeh	Paleocene	carbonates	130SW35	51°32′32.7‴	30°31′40.4″	12	25	1.018	0.80	268-14	32-6	29-66	11-2	87-11	242-78
24 K		Padbeh	Paleocene	carbonates	307N26	51°34′01.0‴	30°32′11.7″	12	27	1.012	0.87	360-16	90-11	213-71	10-4	182-5	48-84
25 K	Vasag	Padbeh	Paleocene	carbonates	326NE75	51°40′18.2″	30°33′08.2″	17	14	1.015	0.73	148-7	17-5	241-25	5-5	140-4	30-79
26 K		Padbeh	Paleocene	carbonates	326NE23	51°39′09.3‴	30°34′25.0″	12	20	1.014	0.51	339-5	32-2	240-62	4-2	136-4	308-86
27 K	Darihsk-West	Padbeh	Paleocene	carbonates	152SW40	51°19′26.5‴	30°43′17.2″	14	10	1.016	0.62	167-13	90-6	62-52	24-5	353-1	239-88
28 K		Padbeh	Paleocene	carbonates	265N15	51°19′25.5‴	30°45′16.8″	13	48	1.025	0.66	301-5	7-3	192-75	3-2	121-4	271-86
29 K		Padbeh	Paleocene	carbonates	310N28	51°24′45.5′	30°40′54.4″	13	106	1.043	0.80	320-11	10-2	209-62	3-2	324-5	128-85

The formation (Fm.), stratigraphic age and lithology are indicated. Latitude and longitude locate the sampling sites. For bedding (S_0), numbers give strike (right-hand rule), dip direction, dip. n, number of samples measured. AMS scalar parameter. K_m, mean magnetic susceptibility, where $K_m = (K_1 + K_2 + K_3)/3$; 10^{-6} SI. P', corrected degree of anisotropy. $P = \exp\sqrt{2[(\eta_1 - \eta_m)^2 + (\eta_2 - \eta_m)^2 + (\eta_3 - \eta_m)^2]}$, where $\eta_i = \ln K_i$ and $\eta_m = (\eta_1 + \eta_2 + \eta_3)/3$. T, shape of AMS ellipsoid. $T = 2(\eta_1 - \eta_2)/(\eta_2 - \eta_3) - 1$. AMS direction K_1 and K_3 (declination/inclination) with their geographical coordinates (GC) and stratigraphic coordinates (SC). Confidence angles e_1/e_2 from Jelinek statistics (Jelinek 1978) are provided for geographical system of coordinates. For K_1, e_1 and e_2 refer to planes K_1–K_2 and K_1–K_3. For K_3, e_1 and e_2 refer to planes K_1–K_3 and K_2–K_3.

(3) the bedding strike coordinates (BSC). Rotating from GC to SC coordinates consists in untilting the AMS directions around the local bedding strike by an amount equal to the dip angle. In the BSC coordinates, a further vertical axis rotation is applied, so that all the local bedding strikes are rotated (clockwise or counter-clockwise) onto an arbitrary north reference direction, by the smallest angle between local strike and north (Aubourg *et al.* 2004). Quantitative information about AMS is provided by standard AMS parameters (Tarling & Hrouda 1993). K_m is the bulk magnetic susceptibility, P' is the degree of anisotropy, and T is the shape parameter. The definition and the mean values of these parameters are given in Table 1. K_m is a measure of magnetic grain concentration, including paramagnetic and ferromagnetic *sensu lato* minerals. P' is proportional to the degree of the magnetic grains' preferred orientation. It is dependent upon strain record, magnetic mineralogy and lithology (Rochette *et al.* 1992). When magnetic mineralogy and lithology are constant, P' is thus indicative of the degree of deformation. The shape factor T is also dependent upon strain, magnetic mineralogy and lithology. However, in thrust belts, the pattern of T from oblate $(+1)$ to prolate (-1) is a good indication of increasing strain record (Averbuch *et al.* 1992; Parés *et al.* 1999; Aubourg *et al.* 2004; Robion *et al.* 2007).

To obtain an overall picture of the AMS data, we first show the density diagram of the AMS K_1 and K_3 axes in the three orientation systems (Fig. 4). In geographical coordinates, the K_3 axes are spread along a direction roughly perpendicular to the main fold axis trend. In the Mio-Pliocene formation, the main trend of K_3 axes is around *c.* N35°. In the Palaeocene formation, it is *c.* N45°. The magnetic lineations are subhorizontal. They are well grouped around *c.* N120° and *c.* N310° for the Mio-Pliocene and Palaeocene formations, respectively. In stratigraphic coordinates (Fig. 4), the K_3 are centred on the vertical axis in both the Mio-Pliocene and Palaeocene rocks, indicating that magnetic foliation is mostly parallel to bedding. The maximum density of magnetic lineations is at *c.* N120° and *c.* N130° for the Mio-Pliocene and Palaeocene rocks, respectively. It should be noted, however, that magnetic lineations are scattered around the horizontal plane, and in the NW and SE quadrants. For the Mio-Pliocene rocks, when the bedding strikes are transferred onto the north reference direction (BSC, Fig. 4a), the first maximum density of magnetic lineation is parallel to this direction, but a secondary maximum lies around N140°. In the Palaeocene formations, the magnetic lineations are much more scattered. The first maximum is observed at *c.* N35° (i.e. oblique to the bedding strike) but secondary maxima can also be seen around N10° and N150°.

The overall features of the magnetic fabric (i.e. magnetic foliation parallel to bedding and magnetic lineation parallel to the fold axis) indicate that magnetic fabric is essentially intermediate. The magnitude of the anisotropy parameters, P', as a function of the shape parameter, T, is also shown in Figure 4. The highest anisotropy factors are found in the clastic rocks, particularly in the Mishan black marls, and the lowest in the Palaeocene and Early Miocene carbonates. On average, the Mio-Pliocene clastic deposits and Palaeocene carbonates have values of $P' = 1.06 \pm 0.03$ and $P' = 1.02 \pm 0.01$, respectively. Bakhtari *et al.* (1998) observed the same difference of P' values. For a large majority of sites, the shape parameter is positive, indicating an oblate shape of AMS ellipsoid. On average, the Mio-Pliocene clastic deposits and Palaeocene carbonates have values of $T = 0.57 \pm 0.32$ and $T = 0.52 \pm 0.35$, respectively. Only sites 5K, 15K and 16K show negative values of T. At site 5K, the magnetic fabric, as we will see below, is inverse, which means that there is an exchange of AMS axes and an inverse trend of T value (Rochette *et al.* 1992). At sites 15K and 16K, we will see that there is a loss of bedding-related magnetic foliation as a result of a larger strain imprint, leading to a tectonic magnetic fabric. As a whole, the regional observations of magnetic fabric indicate therefore that the magnetic fabric is intermediate, with a magnetic lineation developing more or less parallel to the strike of the bedding.

Magnetic fabric at the fold scale. We now examine the magnetic fabric fold by fold (Fig. 5). At Mio-Pliocene sites, all magnetic foliations are parallel to bedding except at site 15K. The magnetic lineation is parallel to the strike of bedding and to the fold axis at folds 15K–16K–21K and 17K–18K. However, some magnetic lineations develop also oblique to the strike of bedding at folds 6K–7K, 8K–11K and 19K–20K and at site 22K. As a reference frame, we plot the shortening direction derived from geodetic data and inversion of earthquake focal mechanisms (N25°). We compare this direction with the AMS shortening direction (ASD). ASD is the strike of the vertical plane containing K_2 and K_3 after bedding correction. After bedding correction, we note that ASD, within the 95% uncertainty of mean magnetic lineation, is parallel to the present-day shortening direction at folds 8K–11K, 15K–16K–21K and 17K–18K. In contrast, ASD is rotated clockwise at folds 19K–20K and 6K–7K, and counter-clockwise at fold 14K–22K with respect to the present-day shortening direction. At Palaeocene sites, the magnetic foliation is mainly parallel to bedding. It should be noted, however, that the fabric is inverse at site 5K, where K_2 is close to the pole of bedding.

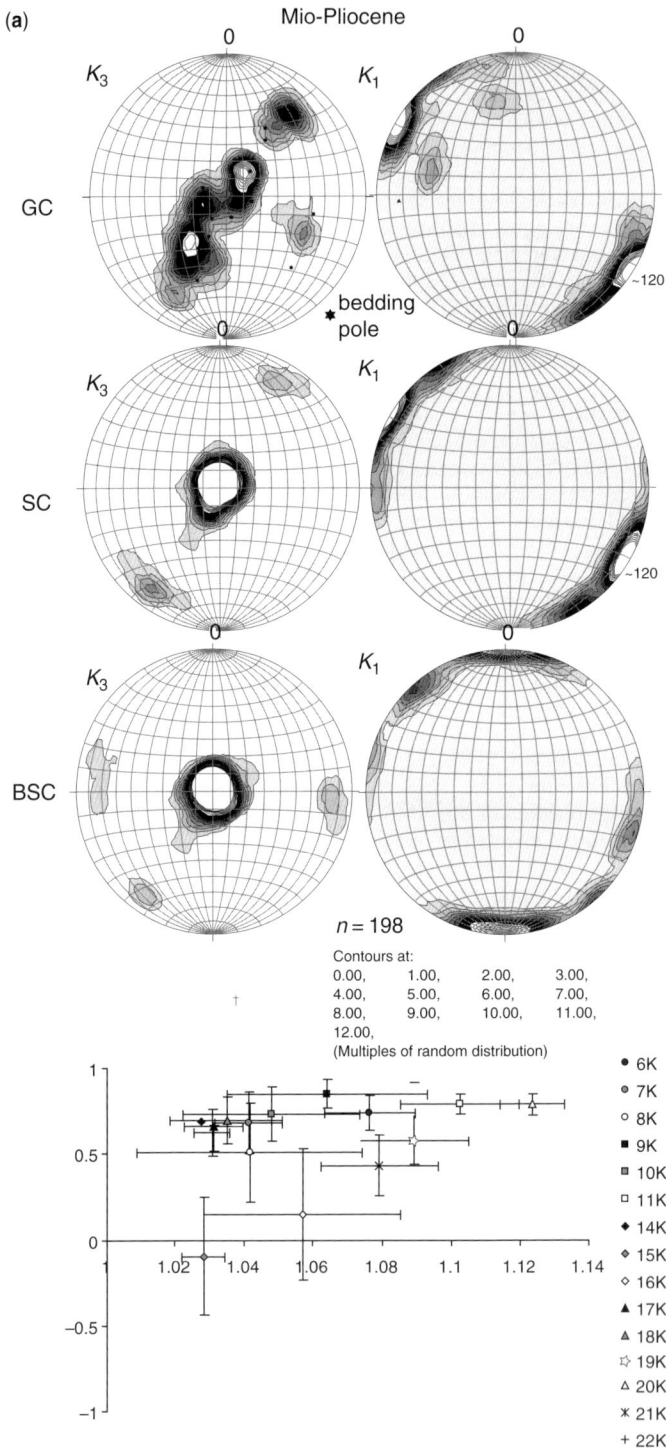

Fig. 4. AMS density diagrams produced using StereoNet. Stereographic projection in the lower hemisphere. GC, geographical coordinate; SC, stratigraphic coordinates; BSC, bedding strike coordinate. P' v. T and their standard deviation are shown (see Table 1 for definition of P' and T).

(b)

Palaeocene

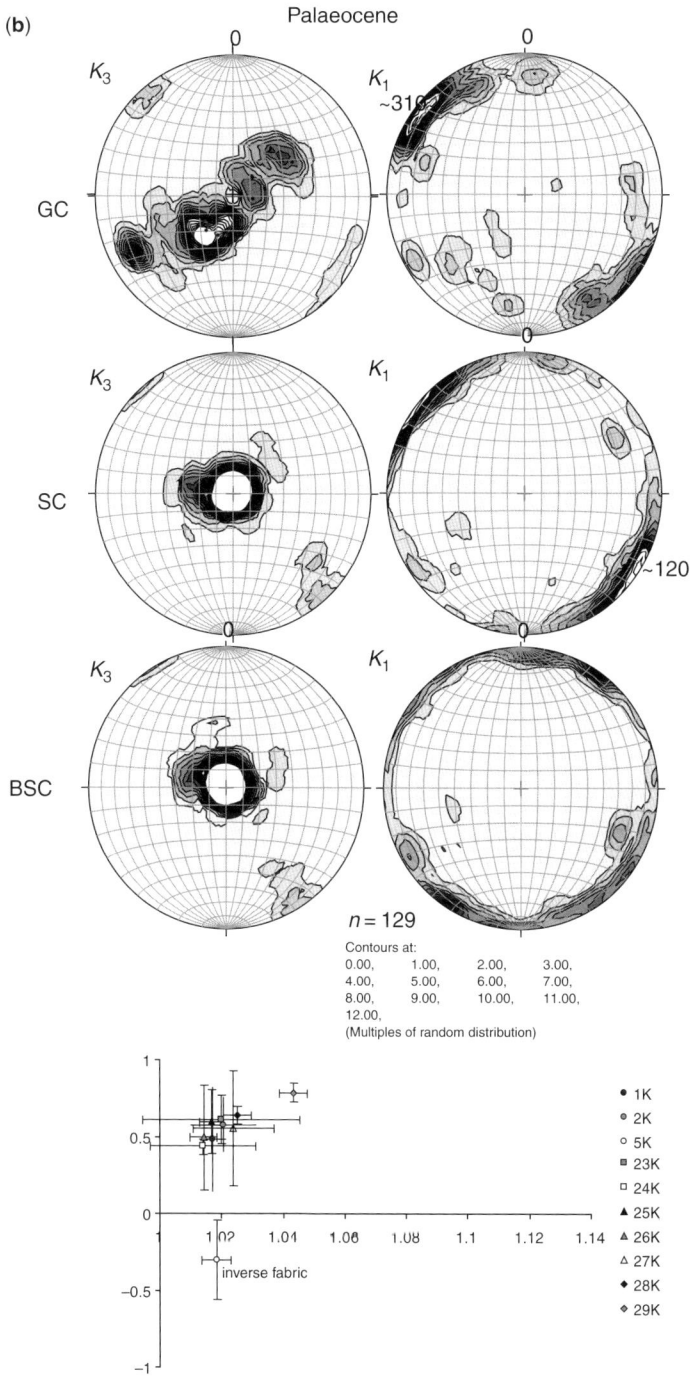

n = 129

Contours at:
0.00,	1.00,	2.00,	3.00,
4.00,	5.00,	6.00,	7.00,
8.00,	9.00,	10.00,	11.00,
12.00,			

(Multiples of random distribution)

- • 1K
- ◉ 2K
- ○ 5K
- ▪ 23K
- ▫ 24K
- ▲ 25K
- ▴ 26K
- △ 27K
- ◆ 28K
- ◇ 29K

inverse fabric

Fig. 4.

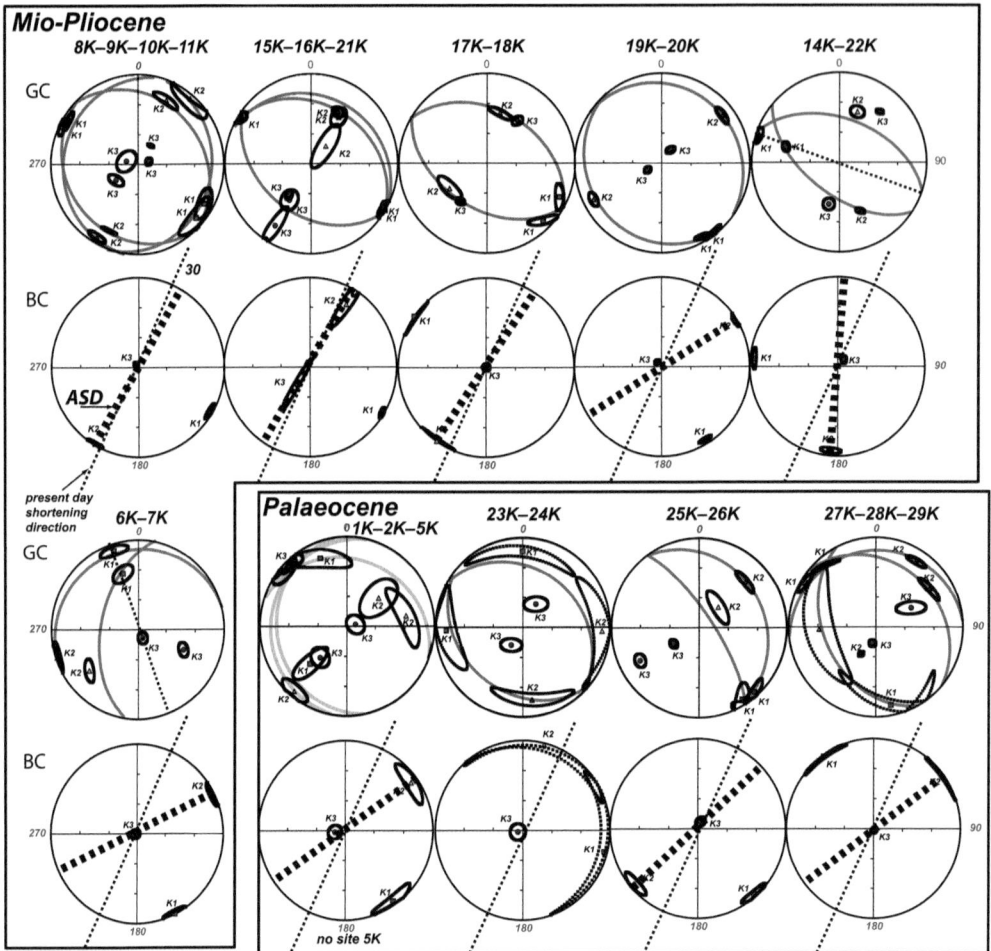

Fig. 5. AMS principal axes for the various folds, using the same conventions as in Figure 4. AMS K_1 (squares) K_2 (triangles) and K_3 (circles) are plotted with their confidence ellipse from the Jelinek's statistics (Jelinek 1978). ASD is the AMS shortening direction deduced from the K_2-K_3 vertical plane in stratigraphic coordinates. The present-day shortening direction is inferred from GPS (Walpersdorf *et al.* 2006).

Contrary to what observed in Plio-Miocene clastic deposits, the magnetic lineations are not as well defined in Palaeocene carbonates (see confidence angles in Table 1) because the magnetic fabric is dominantly of sedimentary origin. This is particularly true at fold 23K–24K, where the scatter of magnetic lineation is too large for any interpretation. However, the magnetic lineations are sufficiently accurate for interpretation in the other folds. ASDs are rotated clockwise with respect to the present-day shortening direction (Fig. 5).

We focus our attention to two specific folds: the Mand anticline, where oblique magnetic lineations are observed, and the syncline that develops in front of the Daryau anticline, where tectonic magnetic foliation is observed.

The Mand anticline provides a good example where the magnetic lineations are strongly oblique to the local strike of the bedding (Fig. 6). It is interesting to compare the Mand fold with its counterpart from the eastern Fars Arc: the Minab fold (Molinaro *et al.* 2004a; Smith *et al.* 2005). In both folds, AMS is measured from similar red sandstones. The location of these two folds is indicated in Figure 2c. The Landsat pictures of these two folds with the major faults, AMS data, and ASD are shown in Figure 6. The Mand anticline, with its symmetrical *c.* 20° limbs, is a detachment fold that develops above the Cambrian Hormuz décollement level (Sherkati *et al.* 2006; Oveisi *et al.* 2007). The Minab anticline, with steep and asymmetrical limbs, is a fault propagation fold above a shallower

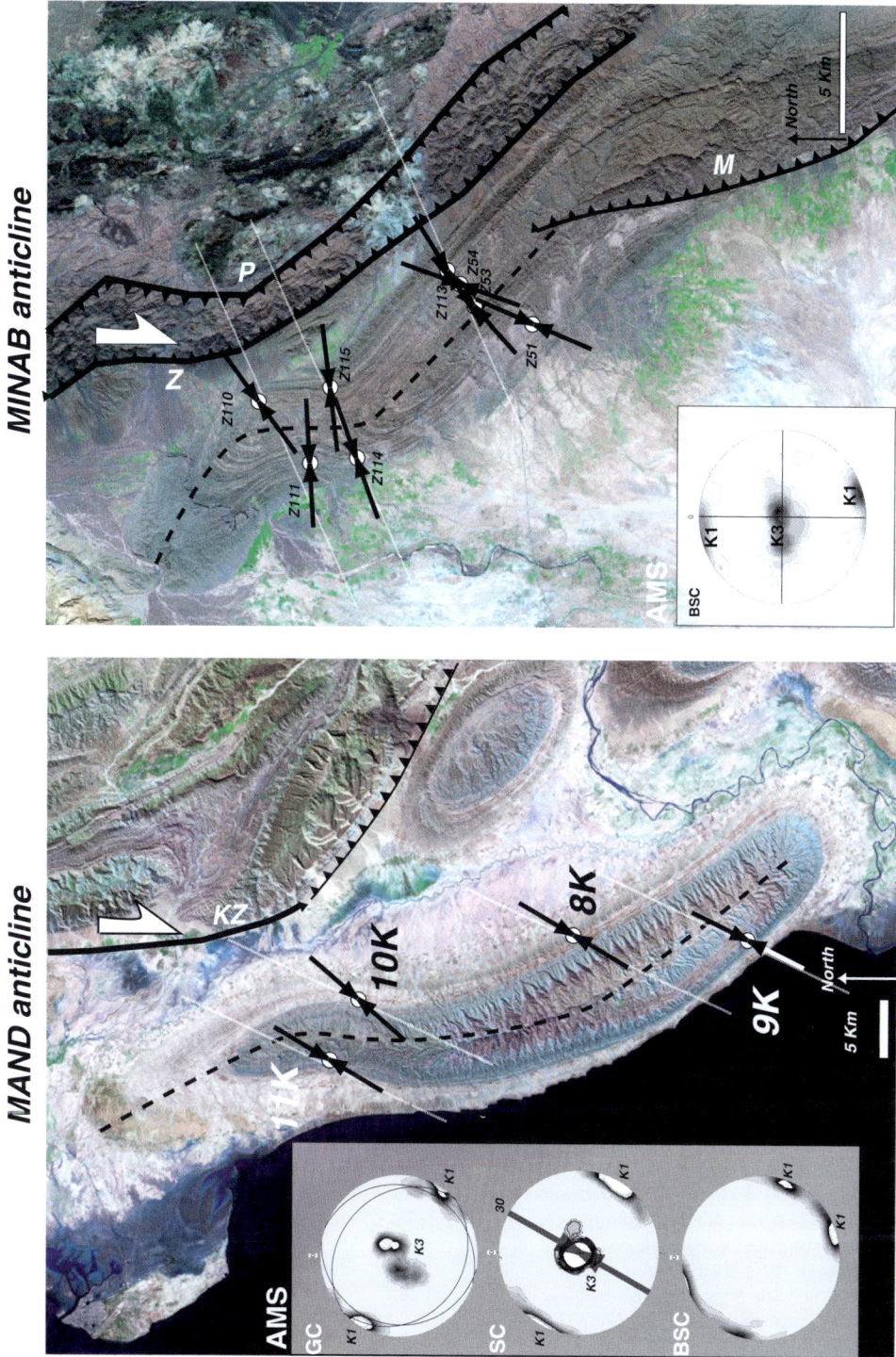

Fig. 6. Magnetic fabric in the Mand and Minab anticlines. (See Fig. 2b for location of these two folds). The Mand anticline is a detachment fold whereas the Minab anticline is a fault propagation fold. Black arrows, AMS shortening direction. It should be noted that the ASD is consistent in the Mand anticline but follows the strike of the bedding in the Minab anticline. White dashed line, present-day shortening direction inferred from GPS. KZ, Kazerun Fault; P, Palami Fault; Z, Zendan Fault; M, Minab Fault.

décollement level at 6 km depth (Molinaro *et al.* 2004*a*). In contrast to the Mand fold, where rocks are weakly strained, the deformation in the Minab anticline is much more pronounced (kink folds, jointing, spaced cleavage) (Aubourg *et al.* 2004; Molinaro *et al.* 2004*a*). The Mand and Minab folds show a distinct torsion in their northern part (Fig. 6). They are bounded to the NE by transpressive faults; the Kazerun Fault in the western Fars Arc (Authemayou *et al.* 2006), and the Zendan Fault in the eastern Zagros Makran Syntaxis (Regard *et al.* 2003). The Minab anticline is thrusted by the Zendan transpressive fault along its northeastern margin (Regard *et al.* 2003).

The AMS pattern is different in the two folds. Whereas the magnetic foliation is parallel to the bedding in the Mand fold, the K_3 axes are slightly scattered along the strain direction in Minab fold (Fig. 6). Aubourg *et al.* (2004) interpreted this pattern as the record of a more intense strain compatible with field observation. The magnetic lineations form a distinguishable pattern in the two anticlines. In the Minab fold, the magnetic lineation follows the change of bedding strike, whereas the palaeomagnetic data demonstrate that the bedding strike torsion is primary (Smith *et al.* 2005). In contrast, the magnetic lineation in the Mand fold remains remarkably constant, despite a significant change (*c.* 50°) of bedding strike azimuth (Table 1). As a result, the magnetic lineations group better in bedding strike coordinates in the Minab fold, whereas magnetic lineations split into two groups in the Mand fold, indicating a moderate (*c.* 20°) and a strong (*c.* 70°) counter-clockwise rotation of the magnetic lineations with respect to the local fold axis strike (Fig. 6). In the Mand fold, the ASD is remarkably parallel to the present-day shortening direction. In the Minab fold, there is also a rather good consistency between ASD and the present-day shortening direction derived from geodetic data (Bayer *et al.* 2006).

Smith *et al.* (2005) proposed that the Minab fold developed above an inherited north–south-trending tectonic structure. Because the palaeostress pattern (AMS and brittle deformation) follows the strike of the bedding, this implies that stresses deviated near the north–south inherited structure, probably before the fault propagation fold development. In the Mand fold, the interpretation is different because there is no deviation of the palaeostress in relation to the strike of the bedding. Our data suggest that strain does not deviate because of inherited structure, if there is any. The obliquity of magnetic lineation with respect to the bedding strike strongly supports the Mand anticline being an oblique fold.

Throughout the western Fars Arc, the intermediate magnetic fabric as developed in the Mand anticline is the rule (Bakhtari *et al.* 1998; Aubourg *et al.* 2004). Nevertheless, a tectonic fabric is observed at sites 15K and 16K (Fig. 7d). These sites are located in a tight syncline, which developed in front of the Daryau anticline where overturned dips of the Guri Fm. are mapped in the core of the anticline (Fig. 7a). This is one of the rare places in ZSFB where overturned dips are identified. The Agha-Jari rocks are, however, weakly strained and only small-scale brittle deformation is observed (Fig. 7b). To better understand the situation of sites 15K and 16K–21K, we sketch the cross-section of the Daryau anticline (Fig. 7c) together with the three axes of the AMS ellipsoids for sites 15K and 16K (Fig. 7d). Some of the magnetic foliations are oblique to bedding and the K_3 axes are spread along a direction perpendicular to the fold trend. It is thus likely that sites 15K and 16K record a tectonic imprint related to the late thrusting stage of the Daryau anticline. Consistent with this, the ASD are parallel to the present-day shortening direction (Fig. 7a).

Interpretation of AMS results

Bakhtari *et al.* (1998) reported that about 55% of magnetic fabrics are intermediate with magnetic foliation parallel to bedding in the Fars Arc. Apart from sites 5K, 15K and 16K, all the sites studied in the present study display intermediate magnetic fabric (*c.* 92% of sites). In the western Fars Arc, we propose that the acquisition of magnetic fabric is mainly coeval with LPS. There are, however, some folds in which a tectonic magnetic foliation can develop simultaneously with folding and faulting near major active faults (Aubourg *et al.* 2004). One striking result of initial AMS studies in the Fars Arc is the recognition of magnetic lineation oblique to the fold axis. Bakhtari *et al.* (1998) observed at the scale of the western and central Fars Arc a *c.* 15° counter-clockwise obliquity between the magnetic lineation and the fold axis. Aubourg *et al.* (2004) reported both clock-wise and counter-clockwise obliquity larger than 30° at *c.* 40% of the sites in the eastern Fars Arc and Zagros–Makran Syntaxis. In the Mio-Pliocene rocks of the present study, the magnetic lineation K_1 is generally parallel to the fold axis trend, except in the Mand anticline (sites 8K–11K), in fold 6K–7K and at site 22K (Fig. 5), where K_1 is rotated counter-clockwise relative to the fold axis direction. At Palaeocene sites, it is difficult to determine an obliquity fold by fold, because the dispersion of the K_1 directions between the two opposite limbs of the folds is large. This is illustrated by the density diagram, where several maxima can be seen in bedding strike coordinates. We note, however, that the first maximum density

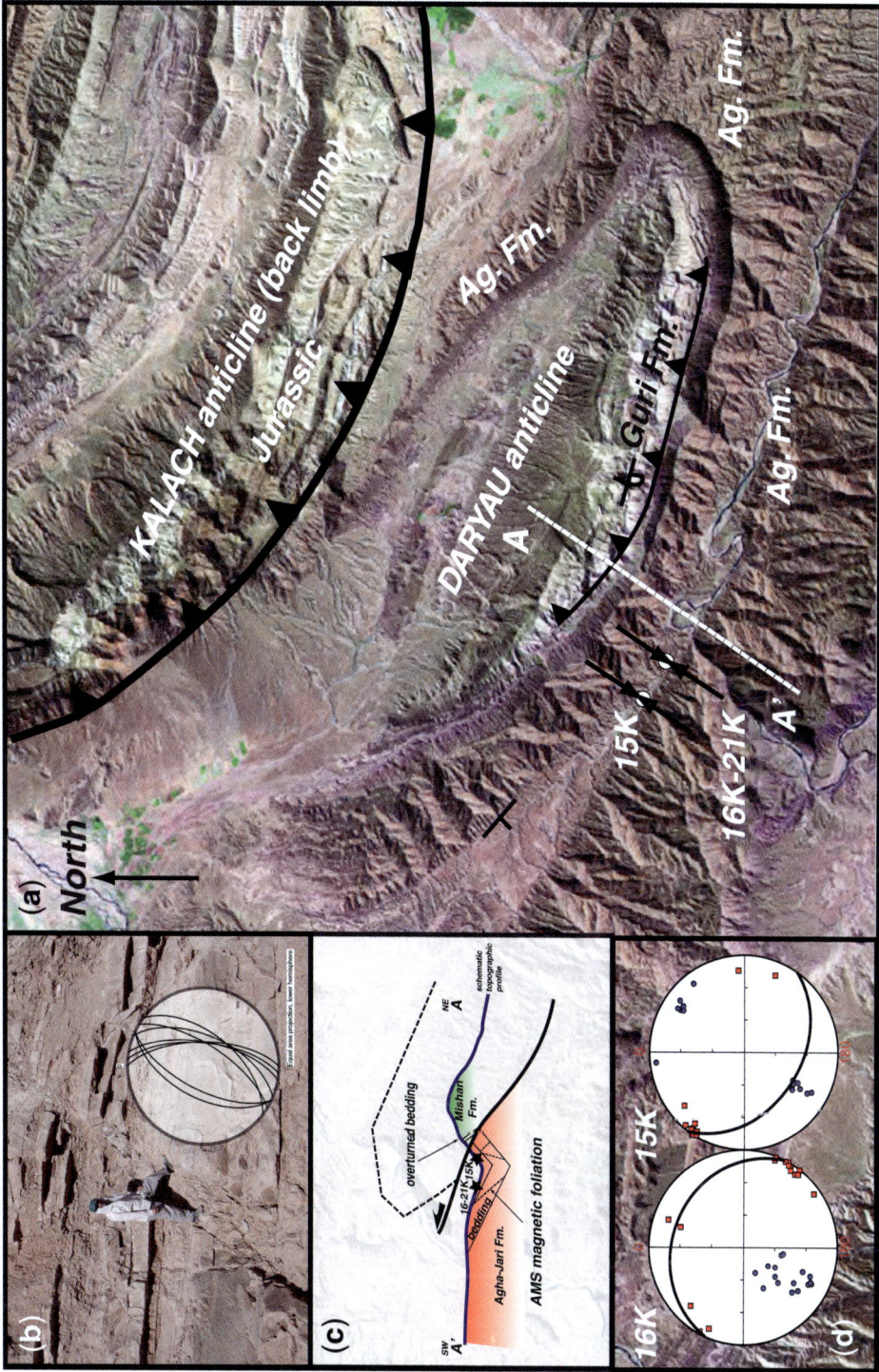

Fig. 7. Magnetic fabric in the footwall of the Daryau thrust. (**a**) Landsat picture. Sites 15K, 16K and 21K are sampled in Agha-Jari Fm (Ag. Frm). ASD is plotted (black arrows). A–A', line of the cross-section shown in (**c**). (**b**) Photograph of Agha-Jari Fm. at site 16K. We see conjugate normal faults compatible with c. N120 extension. (**c**) Schematic cross-section. The attitude of magnetic foliation is sketched with respect to the bedding. Same convention as in Figure 5. Only axes K_1 (squares) and K_3 (circles) are plotted. Magnetic fabric. (**d**) Magnetic fabric. Same convention as in Figure 5. Only axes K_1 (squares) and K_3 (circles) are plotted. Magnetic fabric is here tectonic, as shown by the incipient loss of bedding-parallel magnetic foliation at site 16K and the appearance of a tectonic magnetic foliation at site 15K.

of magnetic lineations is rotated clockwise with respect to the bedding strike.

Two mechanisms can explain the obliquity of the magnetic lineation with respect to the fold trend or bedding strike. A fold can develop obliquely above an inherited structure (Frizon de Lamotte *et al.* 1995; Smith *et al.* 2005). This is comparable with the 'forced' folds above blind thrusts as proposed in the ZSFB by several workers (Cosgrove & Ameen 2000; Sattarzadeh *et al.* 2000). Another plausible mechanism to explain the obliquity of magnetic lineation that the shortening direction changed between the onset of LPS and folding. We sketch in Figure 8 a two-step deformation phase: LPS followed by folding–faulting systems. We show in this system a blind fault, where forced folds can develop during folding. Between the two events, the shortening direction is rotated counter-clockwise. Our two-step model therefore combines the two mechanisms, oblique fold and rotation of the shortening direction. During LPS, we assume a regular imprint of a magnetic fabric as a result of stress (Fig. 8a). During folding and faulting (Fig. 8b), forced and frontal folds develop along and away from the fault, but in our hypothesis, the LPS-related magnetic fabric is preserved. In this model, we observe two kinds of oblique magnetic lineations. For the forced folds along the blind fault, we note a counter-clockwise obliquity of magnetic lineation with respect to the bedding (Fig. 8b). In contrast, for the frontal folds, we see a clockwise obliquity of magnetic lineation. According to this simple model, the *c.* 15° counter-clockwise obliquity evidenced both by Bakhtari *et al.* (1998) and the present dataset in the Agha-Jari Fm. may be explained either by forced folds above north–south-trending blind faults or by a clockwise rotation of

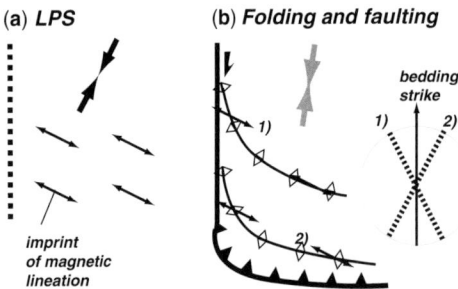

Fig. 8. Model of the development of oblique magnetic lineation. (**a**) Imprint of LPS by magnetic fabric. (**b**) Folding and faulting. Some folds develop oblique to the shortening direction (grey arrow). It should be noted that the shortening direction is rotated counter-clockwise with respect to LPS. Some magnetic lineations are oblique to the fold axis. As a result, magnetic lineations are not parallel to the bedding strike.

the shortening direction through time, or a combination of both. It should be noted that Bakhtari *et al.* (1998) advocated a block rotation mechanism to explain the obliquity of magnetic lineations. Continuing in the frame of this model, we attempt to explain the oblique magnetic lineations observed at the Mand anticline. Rotation of the shortening direction is unlikely in the Mand anticline as the Lahbari Mb. is the youngest formation of Pliocene age sampled in this study. Thus, to account for the counter-clockwise obliquity of magnetic lineation with respect to the fold axis, the Mand anticline is probably a forced fold above a blind segment of a north–south-trending fault, which may be a possible southward extension of the Kazerun Fault as was suggested by Authemayou *et al.* (2006).

Discussion

Comparison of palaeostress markers

In the western Fars Arc, we have the opportunity to compare present-day shortening direction with palaeostress or palaeostrain data, derived from magnetic fabric, small-scale brittle deformation and calcite twinning analyses. These data are reported in Figure 9.

AMS data. The AMS shortening direction (ASD) derived from this study and Bakhtari *et al.* (1998) (Fig. 9a) provides essentially a picture of LPS, prior to folding. We see that the ASD pattern is rather homogeneous, apart from local deviations caused by uncertainties of magnetic lineations (see Table 1) or tectonic complication such as block rotation (see Fig. 3a). From the western to central Fars Arc, ASD swings from a NE–SW to a north–south direction as previously observed by Bakhtari *et al.* (1998). For the new set of AMS data confined to the western Fars Arc, there is an overall good agreement between ASD and present-day shortening direction in the Mio-Pliocene formations to the south, and a systematic *c.* 10° clockwise deviation for the Palaeocene formations (Figs 5 & 9a). When combining AMS data from the present study and from Bakhtari *et al.* (1998) restricted to the western Fars Arc, we obtain an LPS direction at N47° ± 13° (eight sites) for the Palaeocene formations and N38° ± 32° (31 sites) for the Mio-Pliocene formations.

Fault slip data. We report the palaeostress σ_1 trends (Fig. 9b) resulting from the inversion of the small-scale brittle deformation from two studies (Authemayou *et al.* 2006; Lacombe *et al.* 2006). As noted in the introduction, small-scale brittle deformation can record all steps of deformation during the formation of a thrust belt (Fig. 1). The

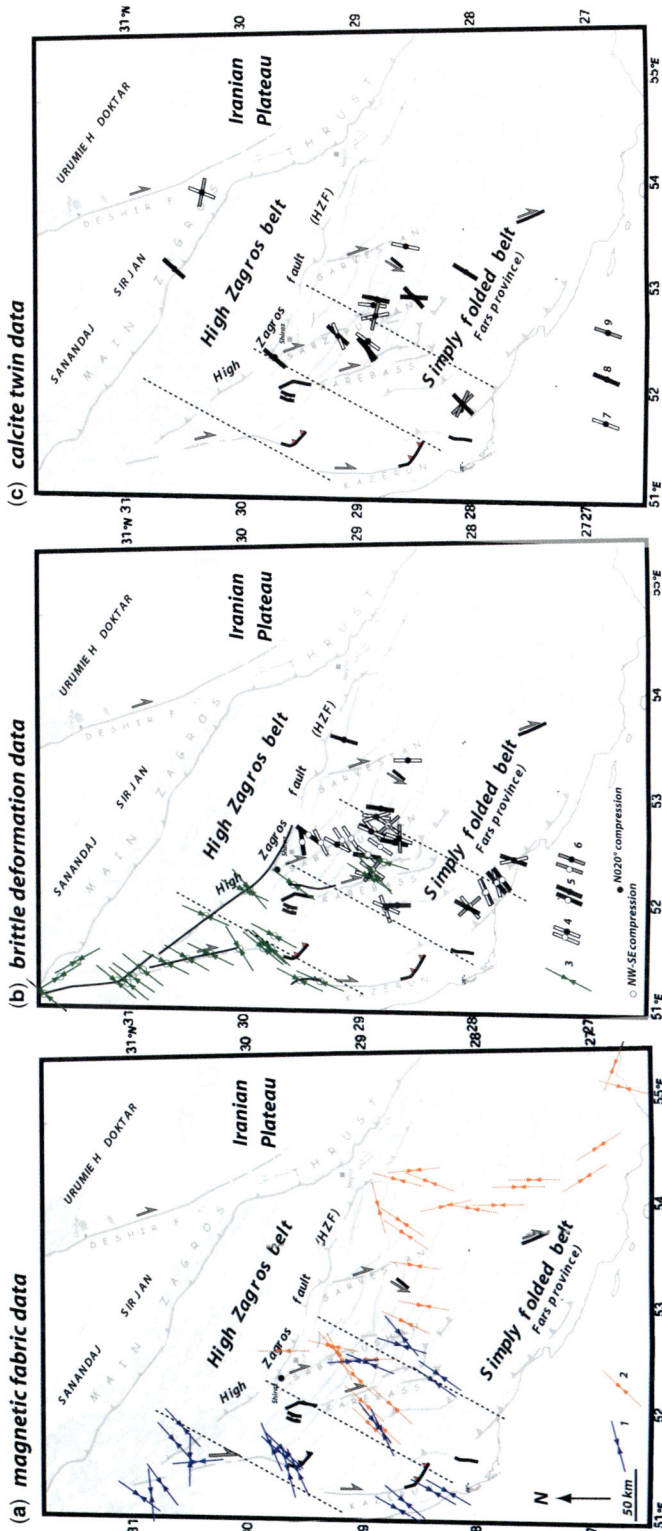

Fig. 9. Palaeostress map. (**a**) AMS shortening direction. 1, From this study; 2, from Bakhtari *et al.* (1998). (**b**) Brittle deformation data. 3, Compressional trends from Authemayou *et al.* (2006); 4–6, compressional or extensional trends from Lacombe *et al.* (2006) (4, compressional trend, strike-slip regime; 5, compressional trend, compressional regime; 6, extensional trend, extensional regime). (**c**) Calcite twin data (Lacombe *et al.* 2007). 7, Compressional trend, strike-slip regime; 8, compressional trend, compressional regime; 9, extensional trend, extensional regime. Dashed lines represent the present-day shortening direction inferred from earthquake focal mechanisms (Lacombe *et al.* 2006).

identification and separation of successive generations of faults and related stress regimes is based on both mechanical incompatibility between fault slips (single misfits of fault slips with the computed stress tensors) and relative chronology observations (e.g. superimposed striations on fault surfaces, crosscutting relationships between faults). To establish a time distribution of tectonic regimes, dating of the brittle structures also requires stratigraphic information about the age of the deformed units and/or evidence of syndepositional tectonism. Particular attention was also paid to horizontal-axis rotations of rock masses as a result of folding. During folding, several cases deserve consideration, because faults may have formed before, during or after folding. For instance, pre-folding strike-slip faults, a common feature in the SFB, can be unambiguously identified by the attitude of the striations,

which always lie within the bedding regardless of the strata attitude, and thus have to be interpreted in their back-rotated attitude. Following Anderson (1951), it is assumed that away from major fault zones one of the three principal stress axes of a tensor is generally vertical. If a fault set formed before folding and was secondarily tilted with the bedding, the tensor calculated on this set does not display a vertical axis. Instead, one of the stress axes is generally found to be perpendicular to bedding, whereas the two others lie within the bedding plane. In such a case, the fault system is interpreted after back tilting to its initial position. Within a heterogeneous fault population this geometrical reasoning allows separation of data subsets based on their age relative to fold development (Fig. 10). In the case of the very simple geometry and cylindrical character of folds in the Zagros

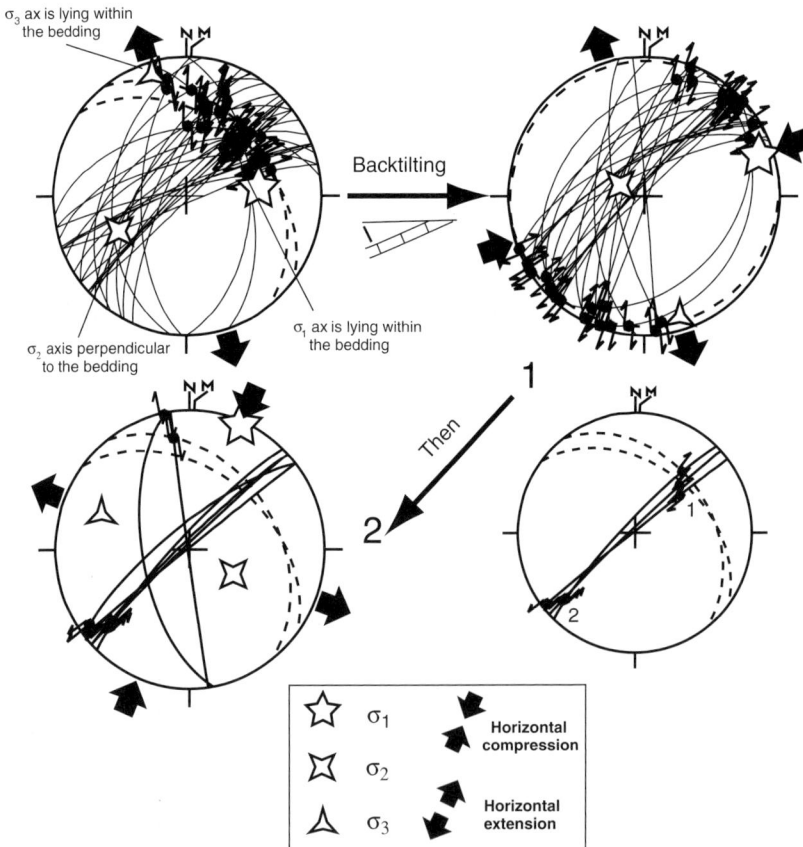

Fig. 10. Example of chronological relationships between faulting related to ENE–WSW compression and faulting related to N020° compression in limestones from the Champeh Member of the Gachsaran Fm. Stereodiagrams in the left part of the figure show striated microfaults in their current attitude (tilted strata), in contrast to the upper right diagram, in which faults related to ENE–WSW compression have been backtilted with bedding. The first strike-slip system predates folding as revealed by the attitude of principal stress axes and striations with respect to bedding; the N020° compression reactivates some faults consistent with the former stress regime and post-dates folding.

SFB, this criterion is of primary importance for establishing a relative chronology. This criterion is further combined with dating of fold development using unconformities and growth strata within synorogenic deposits. The chronology inferred in this way is usually confirmed by identification of superimposed striations on reactivated fault surfaces where observable; it therefore reliably reflects the local succession of faulting events and related stress regimes.

We discuss first the palaeostress pattern in the area of the Karebass and Sabz Pushan faults (Lacombe *et al.* 2006) and then the palaeostress pattern along the Kazerun Fault (Authemayou *et al.* 2006).

For the Karebass and Sabz Pushan faults, the first faulting event is marked by reverse and strike-slip faults and is related to a compressional trend striking NE–SW to ENE–WSW on average. When bed tilting is sufficiently steep to prevent uncertainties, it can be unambiguously identified as having occurred mainly before folding, but also sometimes during and after folding. Although these observations were not always made together at all sites, this compression is clearly associated with the main folding phase at the regional scale.

A more recent faulting event related to a N20° compression has been distinguished from the previous NE–SW compression. At sites where both compressions have been recognized, superimposed striations on fault surfaces and considerations of fault v. bedding attitudes suggest that the N20° compression-related faulting episode postdates that related to the NE–SW compression. We show an example of a superimposed record of pre-tilting and post-tilting deformation (Fig. 10). In this example, some NE–SW-trending strike-slip faults show two striations, the horizontal ones indicating left-lateral motion cutting the NE-dipping ones consistent with right-lateral motion. Additionally, the latter lie within the bedding as the σ_1 and σ_3 axes of the computed stress tensor, whereas the σ_2 axis is perpendicular to bedding. This means that faulting related to the NE–SW compression occurred first, mainly before local folding, and must be interpreted as the earliest event. It predates fault development related to the N20° compression, which clearly occurred after folding. At most sites, faulting related to the N20° compression postdates folding. However, folding may have continued and probably ended during this faulting event, at least locally.

For the Kazerun Fault, inversions of the fault slip data indicate a strike-slip regime with a N35–40°E-trending σ_1 with a mean value of N36° $\pm$ 15°. Stress tensors determined within Mesozoic rocks are consistent within 10° with the stress tensors deduced from inversion of data collected within Pliocene to Quaternary sediment

(N27°). Consequently, no significant change of σ_1 trend has been observed along the Kazerun system. This constancy could be attributed to the reorientation and partitioning mechanisms near this major structural boundary.

Calcite twin data. We plot the main compressional trends σ_1 resulting from the inversion of calcite twinning data collected in both fold-related veins and host rocks with ages ranging from Late Cretaceous to Middle Miocene (Lacombe *et al.* 2007) (Fig. 9c). The relationship between the palaeostress axes and bedding indicates that calcite twinning mainly recorded the late stage of folding (fold tightening) rather than the LPS. As is generally the case for fault slip data and for earthquake focal mechanisms (Lacombe *et al.* 2006), the stress regime is either truly compressional (vertical σ_3 axis) or strike-slip (vertical σ_2 axis), without any obvious regional variation and chronology in the results (except close to the Kazerun Fault). It should be noted that some samples also reveal a component of fold-parallel extension. On average, σ_1 trends N25 $\pm$ 15° (Lacombe *et al.* 2007, on 16 analyses). We observe, however, some departures from the N25 trend. These results are in good agreement with the post-folding palaeostress data of Lacombe *et al.* (2006). In addition, the estimated pre-folding (although few) and post-folding differential stress magnitudes obtained from the twinning analysis are low and, to a first approximation, they are constant across the Zagros Simply Folded Belt. This led Lacombe *et al.* (2007) to propose that most of the folds in the ZSFB formed under low differential stresses and resulted from buckling of the detached Zagros cover, as fault-related folding would be expected to have occurred under higher differential stresses owing to friction on the ramps. The overall constant wavelength of folds, their nearly coeval development and hence the first-order absence of clear propagation of deformation across the SFB, and their rapid growth rates also support buckling of the cover (Mouthereau *et al.* 2006). This is in line with the value of the AMS shape parameter T, which is generally larger than 0.5 (see Table), indicating for a weak input of strain and limited internal deformation. Therefore both independent calcite twinning and AMS approaches support the hypothesis that most folds in the ZSFB are mainly detachment folds (Falcon 1961; Molinaro *et al.* 2003; Sherkati *et al.* 2005; Mouthereau *et al.* 2006).

Integrating palaeostress data

Several workers agree on a two-stage model of formation of the Zagros thrust-and-fold belt since the Miocene (Molinaro *et al.* 2004, 2005; Sherkati *et al.* 2005, 2006; Mouthereau *et al.* 2007*a, b*).

The first stage is a wide detachment-folding phase (or buckling phase) in both the High Zagros Belt and the Zagros Simply Fold Belt (Mouthereau *et al.* 2007*a*), as a result of the decoupling of the sedimentary cover above the 1 km thick Eo-Cambrian Hormuz Salt Fm. The precise age of this tectonic phase is still debated. It is considered to start in the Early Miocene and finish at the end of Middle Miocene in the HZB, based on the observation of unconformities (Navabpour *et al.* 2007). In the ZSFB, the buckling phase occurred during the Late Miocene to Pliocene, based on the observation of growth strata in the Upper Agha-Jari Fm. (Homke *et al.* 2004; Sherkati *et al.* 2005; Lacombe *et al.* 2006). The second folding stage occurred mainly in the ZSFB after the deposition of the Bakhtyari Fm. It consists mostly of fold reactivation and generalized blind basement faulting. This tectonic phase is still active at present throughtout the ZSFB, although the Quaternary deformation is essentially localized in the frontal folds (Oveisi *et al.* 2007).

Although the above-mentioned studies agree on the broad lines of the folding history and with the major present-day role of the basement, it must be mentioned that there is no general agreement on the timing of these two tectonic phases, nor on the time when the basement was first involved in the Zagros deformation. In the present study, we attempt to combine all the palaeostress and palaeostrain data, including those of Navabpour *et al.* (2007) for the HZB, in a comprehensive scheme of a two-stage formation of the High Zagros Belt and the Zagros Simply Folded Belt (Fig. 11).

During the buckling stage in the HZB, our data support the idea that rocks older than the Mishan Fm. and Agha-Jari Fm. of the ZSFB experienced both LPS and faulting. This is the first step of our model in Figure 11a. We plot Early to Middle Miocene data of Navabpour *et al.* (2007), pre-folding fault slip data from Lacombe *et al.* (2006), and Palaeocene AMS data (this study) for this first buckling step. It should be noted that pre-folding fault slip data are here interpreted as older than in the previous interpretation by Lacombe *et al.* (2006). It can be seen that there is a good agreement between all the data, indicating a main NE shortening direction. When the buckling stage occurred in the ZSFB during the Late Miocene (Fig. 11b), the AMS data from the Agha-Jari Fm., calcite twin palaeostress data (2007), late Miocene to Early Pliocene palaeostress data from the HZB (Navabpour *et al.* 2007), and syn- to post-folding palaeostress data from the ZSFB (Lacombe *et al.* 2006) are all consistent with a shortening direction between

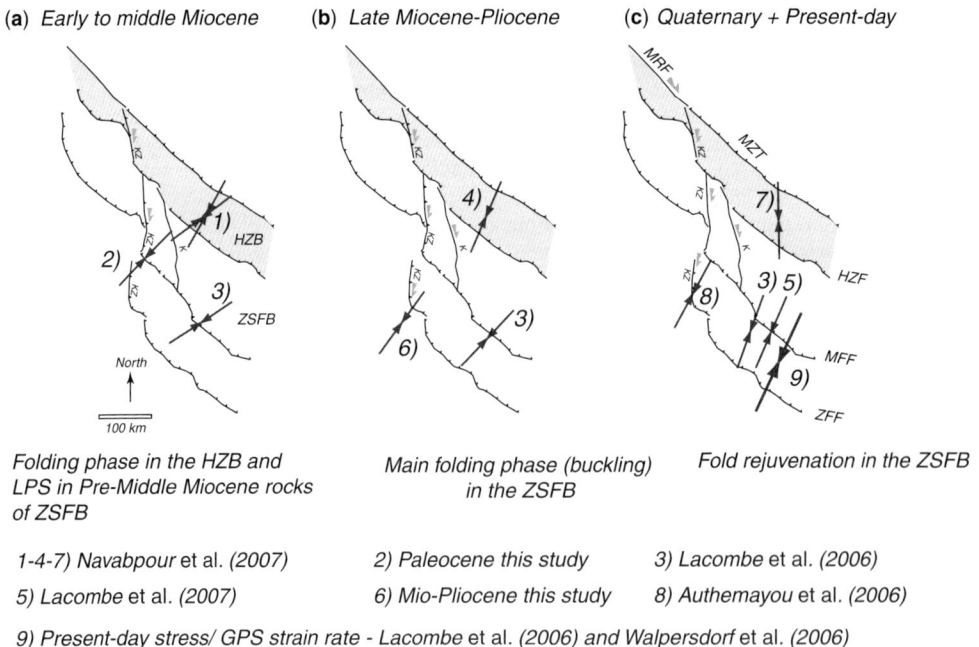

(a) *Early to middle Miocene*   **(b)** *Late Miocene-Pliocene*   **(c)** *Quaternary + Present-day*

Folding phase in the HZB and LPS in Pre-Middle Miocene rocks of ZSFB

Main folding phase (buckling) in the ZSFB

Fold rejuvenation in the ZSFB

1-4-7) Navabpour et al. (2007)   2) Paleocene this study   3) Lacombe et al. (2006)

5) Lacombe et al. (2007)   6) Mio-Pliocene this study   8) Authemayou et al. (2006)

9) Present-day stress/ GPS strain rate - Lacombe et al. (2006) and Walpersdorf et al. (2006)

Fig. 11. Tectonic scenario of folding in the Zagros belt and related palaeostress data. (**a**) Detachment folding in the HZB and coeval LPS–faulting in the ZSFB; (**b**) detachment folding in the ZSFB; (**c**) late fold rejuvenation in the ZSFB; only faulting occurred in the HZB. KZ, Kazerun Fault; K, Karebas Fault; HZF, High Zagros Fault; MFF, Mountain Front Fault; ZFF, Zagros Front Fault; MZT, Main Zagros Thrust; MRF, Main Recent Fault.

N25 and N45, although a counter-clockwise rotation of shortening is locally suspected using fault slip data (Lacombe *et al.* 2006). For the second folding phase, which consists mainly of fold rejuvenation, we plot (Fig. 11c) the recent palaeostress recorded by Quaternary sediments along the Kazerun Fault (Authemayou *et al.* 2006), Late Pliocene to recent fault slip data in the HZB (Navabpour *et al.* 2007), inversion of focal mechanisms of earthquakes from the western Fars Arc (Lacombe *et al.* 2006) and the GPS-derived shortening trend (Walpersdorf *et al.* 2006). All these data reveal that the shortening direction remains more or less at N20–N30 in the ZSFB, but instead trends north–south in the HZB.

Our model therefore shows that it is possible to fit the palaeostress data from the HZB and the ZSFB within the uncertainties of the dataset and the history of deformation. A counter-clockwise rotation of at least 10° of the shortening direction between the first stage of folding in the HZB (buckling) and the second stage of folding in the ZSFB (fold rejuvenation) is likely. This rotation is possibly due to far-field geodynamic constraints as discussed by Navabpour *et al.* (2007), or to clockwise block rotations close to the set of right-lateral strike-slip faults bounding the Fars Arc to the west as initially stated by Bahktari *et al.* (1998) and developed by Lacombe *et al.* (2006). Whatever its origin, the counter-clockwise rotation of the shortening direction has consequences for the palaeostress regime during faulting, especially along these major strike-slip faults. Although this tendency is not clear throughout the ZSFB, fault slip data collected from the HZB and along the Kazerun Fault support a temporal evolution from reverse to strike-slip regimes (Authemayou *et al.* 2006; Navabpour *et al.* 2007). This is consistent with a counter-clockwise rotation of the shortening direction, and this may explain how an initial thrust trending NW–SE during the first stage of folding in the HZB evolved into a transpressive fault during the second folding stage in the ZSFB and in the faulting stage in the HZB.

Conclusion

To recover the imprint of palaeostress and -strain during the development of the Zagros Simply Folded Belt, we have performed an integrated study of palaeostress data obtained by different techniques, including magnetic fabric, fault slip and calcite twinning data. The magnetic fabric from Palaeocene carbonates and Mio-Pliocene clastic deposits retains the record of the layer-parallel shortening (LPS) that occurred prior to folding. We propose that the Palaeocene carbonates record a N47° ± 13° LPS during early to middle Miocene detachment folding in the High Zagros Belt. Before or during later (late Miocene–Pliocene) detachment folding in the ZSFB, the Mio-Pliocene clastic deposits recorded the N38° ± 32° LPS. Fault slip and calcite twinning data recorded a two-stage folding in the ZSFB: first the detachment folds and then a reactivation of folds. All the techniques suggest a counter-clockwise rotation of the shortening direction from NE to N20° between the onset of the detachment-fold phase of the HZB and the late stage of the detachment-fold phase of the ZSFB.

This work was funded by a DYETI programme led by D. Hatzfeld. The Geological Survey of Iran, thanks to Dr M. R. Ghassemi, provided invaluable help in logistics and science. We have benefited from constructive discussion with members of the DYETI and MEBE groups. D. Frizon de Lamotte is particularly thanked for reviewing the initial manuscript.

References

ANDERSON, E. M. 1951. *The Dynamics of Faulting.* Oliver & Boyd, White Plains, NY.

ANGELIER, J. 1990. Inversion of field data in fault tectonics to obtain the regional stress—III. A new rapid direct inversion method by analytical means. *Geophysical Journal International*, **103**, 363–376.

AUBOURG, C., FRIZON DE LAMOTTE, D., POISSON, A. & MERCIER, E. 1997. Magnetic fabrics and oblique ramp-related folding. A case study from the Western Taurus (Turkey). *Journal of Structural Geology*, **19**, 1111–1120.

AUBOURG, C., SMITH, B. *ET AL.* 2004. Post-Miocene shortening pictured by magnetic fabric across the Zagros–Makran syntaxis. *In:* SUSSMAN, A. B. (ed.) *Orogenic Curvature: Integrating Palaeomagnetic and Structural Analyses.* Geological Society of America, Special Papers, **383**, 17–40.

AUBOURG, C., SMITH, B., BAKHTARI, H., GUYA, N. & ESHRAGHI, A. R. 2008. Tertiary block rotations in the Fars Arc (Zagros, Iran). *Geophysical Journal International*, **173**, 659–673.

AUTHEMAYOU, C., CHARDON, D., BELLIER, O., MALEKZADEH, Z., SHABANIAN, E. & ABBASSI, M. R. 2006. Late Cenozoic partitioning of oblique plate convergence in the Zagros fold-and-thrust belt (Iran). *Tectonics*, **25**, TC3002.

AVERBUCH, O., FRIZON DE LAMOTTE, D. & KISSEL, C. 1992. Magnetic fabric as a structural indicator of the deformation path within a fold–thrust structure: a test case from the Corbières (NE Pyrenees, France). *Journal of Structural Geology*, **14**, 461–474.

BAKHTARI, H., FRIZON DE LAMOTTE, D., AUBOURG, C. & HASSANZADEH, J. 1998. Magnetic fabric of Tertiary sandstones from the Arc of Fars (Eastern Zagros, Iran). *Tectonophysics*, **284**, 299–316.

BAYER, R., CHERY, J. *ET AL.* 2006. Active deformation in Zagros–Makran transition zone inferred from GPS measurements. *Geophysical Journal International*, **165**, 373–381.

BERBERIAN, M. 1995. Master 'blind' thrust faults hidden under the Zagros folds: active basement tectonics and surface morphotectonics. *Tectonophysics*, **241**, 193–224.

BERBERIAN, M. & KING, G. C. P. 1981. Towards a paleo-geography and tectonic evolution of Iran. *Canadian Journal of Earth Sciences*, **18**, 210–265.

BLANC, E. J.-P., ALLEN, M. B., INGER, S. & HASSANI, H. 2003. Structural styles in the Zagros Simply Folded Zone, Iran. *Journal of the Geological Society, London*, **160**, 401–412.

BORRADAILE, G. J. 1987. Anisotropy of magnetic susceptibility: rock composition versus strain. *Tectonophysics*, **138**, 327–329.

BURKHARD, M. 1993. Calcite twins, their geometry, appearance and significance as stress–strain markers and indicators of tectonic regime: a review. *Journal of Structural Geology*, **15**, 351–368.

COSGROVE, J. W. & AMEEN, M. S. 2000. A comparison of the geometry, spatial organisation and fracture patterns associated with forced folds and buckle folds. *In*: COSGROVE, J. W. & AMEEN, M. S. (eds) *Forced Folds and Fractures*. Geological Society London, Special Publications, **169**, 7–21.

CRADDOCK, J. P. & VAN DER PLUIJM, B. 1999. Sevier–Laramide deformation of the continental interior from calcite twinning analysis, west–central North America. *Tectonophysics*, **305**, 275–286.

ETCHECOPAR, A. 1984. *Etude des états de contraintes en techonique cassante et simulation des déformations plastiques*, PhD thesis, Montpellier University, France, 270.

FALCON, N. L. 1961. Major earth-flexuring in the Zagros mountains of south-west Iran. *Quarterly Journal of the Geological Society*, **117**, 367–376.

FRIZON DE LAMOTTE, D., GUEZOU, J.-C. & AVERBUCH, O. 1995. Distinguishing lateral folds in thrust-systems; examples from Corbières (SW France) and Betic Cordillieras (SE Spain). *Journal of Structural Geology*, **17**, 233–244.

FRIZON DE LAMOTTE, D., SOUQUE, C., GRELAUD, S. & ROBION, P. 2002. Early record of tectonic magnetic fabric during inversion of a sedimentary basin. Short review and examples from the Corbières transfer zone (France). *Bulletin de la Sociéte Géologique de France*, **173**, 461–469.

GRAHAM, J. W. 1966. Significance of magnetic anisotropy in Appalachian sedimentary rocks. *In*: STEINHART, J. S. & SMITH, T. J. (eds) *The Earth Beneath the Continents*. American Geophysical Union, Geophysical Monograph, **10**, 627–648.

HAMILTON, T. D., BORRADAILE, G. J. & LAGROIX, F. 2004. Sub-fabric identification by standardization of AMS: an example of inferred neotectonic structures from Cyprus. *In*: MARTIN-HERNANDEZ, F., LUNEBURG, C. M., AUBOURG, C. & JACKSON, M. (eds) *Magnetic Fabric: Methods and Applications*. Geological Society, London, Special Publications, **238**, 527–540.

HARRIS, J. H. & VAN DER PLUIJM, B. A. 1998. Relative timing of calcite twinning strain and fold–thrust belt development: Hudson Valley fold–thrust belt. New York, USA. *Journal of Structural Geology*, **20**, 21–31.

HESSAMI, K., KOYI, H. A., TALBOT, C. J., TABASI, H. & SHABANIAN, E. 2001. Progressive unconformities within an evolving foreland fold–thrust belt, Zagros Mountains. *Journal of the Geological Society, London*, **158**, 969–981.

HESSAMI, K., NILFOROUSHAN, F. & TALBOT, C. J. 2006. Active deformation within the Zagros Mountains deduced from GPS measurements. *Journal of the Geological Society, London*, **163**, 143–148.

HIRT, A. M., EVANS, K. F. & ENGALDER, T. 1995. Correlation between magnetic anisotropy and fabric for Devonian shales on the Appalachian plateau. *Tectonophysics*, **247**, 121–132.

HOMKE, S., VERGES, J., GARCES, M., EMAMI, H. & KARPUZ, R. 2004. Magnetostratigraphy of Miocene–Pliocene Zagros foreland deposits in the front of the Push-e Kush Arc (Lurestan Province, Iran). *Earth and Planetary Science Letters*, **225**, 397–410.

HROUDA, F. 1982. Magnetic anisotropy of rocks and its application in geology and geophysics. *Geophysical Surveys*, **5**, 37–82.

HROUDA, F., PROS, Z. & WOHLGEMUTH, J. 1993. Development of magnetic and elastic anisotropies in slates during progressive deformation. *Physics of the Earth and Planetary Interiors*, **77**, 251–265.

JELINEK, V. 1978. Statistical processing of anisotropy of magnetic susceptibility mesured on group of specimen. *Studia Geophysica et Geodaetica*, **22**, 50–62.

JELINEK, V. 1981. Characterization of the magnetic fabric of the rocks. *Tectonophysics*, **79**, 63–67.

KANAMATSU, T., HERRERO-BERVERA, E. & ASAHIKO, T. 2001. Magnetic fabric of soft-sediment folded strata within a neogene accretionary complex, the Miura group, Central Japan. *Earth and Planetary Science Letters*, **187**, 333–343.

KISSEL, C., BARRIER, E., LAJ, C. & LEI, T.-Q. 1986. Magnetic fabric in 'undeformed' marine clays from compressional zones. *Tectonics*, **5**, 769–781.

LACOMBE, O. 2001. Palaeostress magnitudes associated with development of mountain belts: Insights from tectonic analyses of calcite twins in the Taiwan Foothills. *Tectonics*, **20**, 834–849.

LACOMBE, O. 2007. Comparison of palaeostress magnitudes from calcite twins with contemporary stress magnitudes and frictional sliding criteria in the continental crust: Mechanical implications. *Journal of Structural Geology*, **29**, 86–99.

LACOMBE, O., MOUTHEREAU, F., KARGAR, S. & MEYER, B. 2006. Late Cenozoic and modern stress fields in the western Fars (Iran): implications for the tectonic and kinematic evolution of central Zagros. *Tectonics*, **25**, TC1003.

LACOMBE, O., AMROUCH, K., MOUTHEREAU, F. & DISSEZ, L. 2007. Calcite twinning constraints on late Neogene stress patterns and deformation mechanisms in the active Zagros collision belt. *Geology*, **35**, 263–266.

MASSON, F., CHERY, J., HATZFELD, D., MARTINOD, J., VERNANT, P., TAVAKOLI, F. & GHAFORY-ASHTIANI, M. 2005. Seismic versus aseismic deformation in Iran inferred from earthquakes and geodetic data. *Geophysical Journal International*, **160**, 217–226.

MCQUARRIE, N. 2004. Crustal scale geometry of the Zagros fold–thrust belt, Iran. *Journal of Structural Geology*, **26**, 519–535.

MCQUARRIE, N., STOCK, J. M., VERDEL, C. & WERNICKE, B. P. 2003. Cenozoic evolution of Neotethys and implications for the causes of plate motions. *Geophysical Research Letters*, **30**, 2036, doi: 10.1029/2003GL017992.

MERCIER, J.-L., CAREY-GAILHARDIS, E. & SÉBRIER, M. 1991. Palostress determinations from fault kinematics: Application to the Neotectonics of the Himalayas–Tibet and the central Andes. *Philosophical Transactions of the Royal Society of London*, **337**, 41–52.

MOLINARO, M., GUEZOU, J.-C., AUBOURG, C., LETURMY, P. & ESHRAGHI, S. A. 2003. Structural style across the Bandar Abbas syntaxis, SE Zagros: facts and factors of a subduction to collision transition. *EGS–AGU–EUG Joint Assembly, Nice.*

MOLINARO, M., GUEZOU, J. C., LETURMY, P., ESHRAGHI, S. A. & FRIZON DE LAMOTTE, D. 2004a. The origin of changes in structural style across the Bandar Abbas syntaxis, SE Zagros (Iran). *Marine and Petroleum Geology*, **21**, 735–752.

MOLINARO, M., LETURMY, P., GUEZOU, J.-C. & FRIZON DE LAMOTTE, D. 2004b. The structure and kinematic evolution of the south-eastern Zagros Mountains, Iran. *EGU 1st General Assembly, Nice.*

MOLINARO, M., LETURMY, P., GUEZOU, J.-C., FRIZON DE LAMOTTE, D. & ESHRAGHI, S. A. 2005. The structure and kinematics of the southeastern Zagros fold–thrust belt, Iran: From thin-skinned to thick-skinned tectonics. *Tectonics*, **24**, TC3007.

MOUTHEREAU, F., LACOMBE, O. & MEYER, B. 2006. The Zagros folded belt (Fars, Iran): constraints from topography and critical wedge modelling. *Geophysical Journal International*, **165**, 336–356.

MOUTHEREAU, F., TENSI, J., BELLAHSEN, N., LACOMBE, O., DE BOISGROLLIER, T. & KARGHAR, S. 2007a. Tertiary sequence of deformation in a thin-skinned/thick-skinned collision belt: the Zagros Folded Belt (Fars, Iran). *Tectonics*, **26**, TC5006.

MOUTHEREAU, F., LACOMBE, O., TENSI, J., BELLAHSEN, N., KARGAR, S. & AMROUCH, K. 2007b. Mechanical constraints on the development of the Zagros Folded Belt. *In:* LACOMBE, J. L. O., VERGÉS, J. & ROURE, F. (eds) *Thrust Belts and Foreland Basins: From Fold Kinematics to Hydrocarbon Sytems.* Frontiers in Earth Sciences. Springer, New York, 247–266.

NAVABPOUR, P., ANGELIER, J. & BARRIER, E. 2007. Cenozoic post-collisional brittle tectonic history and stress reorientation in the High Zagros Belt (Iran, Fars Province). *Tectonophysics*, **432**, 101–131.

OVEISI, B., LAVÉ, J. & VAN DER BEEK, P. A. 2007. Rates and processes of active folding evidenced by Pleistocene terraces at the central Zagros front (Iran). *In:* LACOMBE, O., LAVÉ, J., ROURE, F. & VERGÈS, J. (eds) *Thrust Belts and Foreland Basins.* Frontiers in Earth Sciences. Springer, New York, 265–285.

PARÉS, J. P., VAN DER PLUIJM, B. A. & DINARÈS-TURELL, J. 1999. Evolution of magnetic fabric during incipient deformation of mudrocks (Pyrenees, Northern Spain). *Tectonophysics*, **307**, 1–14.

REGARD, V., BELLIER, O., THOMAS, J. C., ABBASSI, M. R. & MERCIER, J. L. 2003. Tectonics of a lateral transition between subduction and collision: the Zagros–Makran transfer deformation zone (SE Iran). *In: EGS–AGU–EUG Joint Assembly, Nice.*

RICOU, L., BRAUD, J. & BRUNN, J. H. 1977. *Le Zagros.* Mémoires Hors Série de la Société Géologique de France, **8**, 33–52.

ROBION, P., GRELAUD, S. & FRIZON DE LAMOTTE, D. 2007. Pre-folding magnetic fabrics in fold-and-thrust belts: Why the apparent internal deformation of the sedimentary rocks from the Minervois basin (NE Pyrenees, France) is so high compared to the Potwar basin (SW Himalaya, Pakistan). *Sedimentary Geology*, **196**, 181–200.

ROCHETTE, P., JACKSON, J. & AUBOURG, C. 1992. Rock magnetism and the interpretation of ansisotropy of magnetic susceptibility. *Review of Geophysics*, **30**, 209–226.

ROWE, K. J. & RUTTER, E. H. 1990. Palaeostress estimation using calcite twinning: experimental calibration and application to nature. *Journal of Structural Geology*, **12**, 1–17.

SAGNOTTI, L., SPERENZA, F., WINKLER, A., MATTEI, M. & FUNICIELLO, R. 1998. Magnetic fabric of clay sediments from the external northern Apennines (Italy). *Physics of the Earth and Planetary Interiors*, **105**, 73–93.

SAINT-BEZAR, B., HEBERT, R. L., AUBOURG, C., ROBION, P., SWENNEN, R. & FRIZON DE LAMOTTE, D. 2002. Magnetic fabric and petrographic investigation of hematite-bearing sandstones within ramp-related folds: examples from the South Atlas Front (Morocco). *Journal of Structural Geology*, **24**, 1507–1520.

SATTARZADEH, Y., COSGROVE, J. W. & VITA-FINZI, C. 2000. The interplay of faulting and folding during the evolution of the Zagros deformation belt. *In:* COSGROVE, J. W. & AMEEN, M. S. (eds) *Forced Folds and Fractures.* Geological Society, London, Special Publications, **169**, 187–196.

SEPEHR, M. & COSGROVE, J. W. 2004. Structural framework of the Zagros Fold–Thrust, Iran. *Marine and Petroleum Geology*, **21**, 829–843.

SHERKATI, S. & LETOUZEY, J. 2004. Variation of structural style and basin evolution in the central Zagros (Izeh Zone and Dezful Embayment, Iran). *Marine and Petroleum Geology*, **21**, 535–554.

SHERKATI, S., MOLINARO, M., FRIZON DE LAMOTTE, D. & LETOUZEY, J. 2005. Detachment folding in the central and eastern Zagros fold-belt (Iran): salt mobility, multiple detachment and final basement control. *Journal of Structural Geology*, **27**, 1680–1696.

SHERKATI, S., LETOUZEY, J. & FRIZON DE LAMOTTE, D. 2006. Central Zagros fold–thrust belt (Iran): new insights from seismic data, field observation, and sandbox modeling. *Tectonics*, **25**, TC4007, doi: 10.1029/2004TC001766.

SMITH, B., AUBOURG, C., GUÉZOU, J. C., NAZARI, H., MOLINARO, M., BRAUD, X. & GUYA, N. 2005. Kinematics of a sigmoidal fold and vertical axis rotation in the east of the Zagros–Makran syntaxis (Southern

Iran): palaeomagnetic, magnetic fabric and microtectonic approaches. *Tectonophysics*, **411**, 89–109.

TALEBIAN, M. & JACKSON, J. 2004. A reappraisal of earthquake focal mechanisms and active shortening in the Zagros mountains of Iran. *Geophysical Journal International*, **156**, 506–526.

TARLING, D. H. & HROUDA, F. 1993. *The Magnetic Anisotropy of Rocks*. Chapman & Hall, London.

TATAR, M., HATZFELD, D., MARTINOD, J., WALPERSDORF, A., GHAFORI-ASHTIANY, M. & CHÉRY, J. 2002. The present-day deformation of the central Zagros from GPS measurements. *Geophysical Research Letters*, **29**, 19–27.

TATAR, M., HATZFELD, D. & GHAFORY-ASHTIANY, M. 2004. Tectonics of the Central Zagros (Iran) deduced from microearthquake seismicity. *Geophysical Journal International*, **156**, 255–266.

VERNANT, P., NILFOROUSHAN, F. *ET AL.* 2004. Present-day crustal deformation and plate kinematics in the Middle East constrained by GPS measurements in Iran and northern Oman. *Geophysical Journal International*, **157**, 381–398.

WALPERSDORF, A., HATZFELD, D. *ET AL.* 2006. Difference in the GPS deformation pattern of North and Central Zagros (Iran). *Geophysical Journal International*, **167**, 1077–1088.

Structure, timing and morphological signature of hidden reverse basement faults in the Fars Arc of the Zagros (Iran)

PASCALE LETURMY*, MATTEO MOLINARO[1] & DOMINIQUE FRIZON DE LAMOTTE

Université Cergy-Pontoise, Département des Sciences de la Terre et Environnement, F-95000 Cergy-Pontoise, France

[1]*Present address: Shell International Exploration and Production BV, Riswijk, Netherlands*

Corresponding author (e-mail: pascale.leturmy@u-cergy.fr)

Abstract: In the Zagros Fold–Thrust Belt (ZFTB) of Iran it is firmly established that the basement is involved in the deformation. The strongest line of evidence for this assertion comes from the relatively intense mid-crustal seismic activity. On one hand, the main basement structures such as the Main Zagros Fault (MZT) and High Zagros Fault (HZF) reach the surface and are therefore well identified. On the other hand, basement faults south of the HZF are hidden by sedimentary cover and their location is uncertain. In the Eastern Zagros, basement control on surface structures occurred only at a late stage of the tectonic evolution. In other words, the current thick-skinned style of Zagros deformation succeeded a more general thin-skinned phase of orogeny. This chronology is particularly well illustrated by spectacular interference patterns, in which early detachment folds are cut by late oblique basement faults. We present a combined morphological and structural analysis of such structures, and we explore their impact on the river network. We confirm that basement involvement occurred at a late stage of deformation and we show that thick-skinned deformation progressively propagated towards the foreland. An overview of basement steps throughout the Zagros based on published cross-sections allows us to conclude that although basement deformation is concentrated on two major faults in the Central Zagros, it is distributed on several segmented faults in the Fars Arc. This segmentation increases to the SE towards the so-called Oman line and the transition to the Makran accretionary prism.

Within foreland fold–thrust belts, shortening is often accommodated within the crystalline basement rocks underlying the sedimentary cover by major reverse faulting (Woodward 1988; Narr & Suppe 1994; Mitra & Mount 1998). Basement thrusting has been recognized in numerous fold–thrust belts around the world and is commonly associated with irregular topographic and structural uplift and alignments of seismic activity (Jackson 1980; Jackson & Fitch 1981; Molnar & Chen 1983; Talebian & Jackson 2004). However, as basement faults are hidden from view beneath the more recent sedimentary units and rarely reach the surface, it is often difficult to obtain an accurate picture of their geometry and their relationships with the structures in the overlying cover. In many cases even first-order information such as their location and trends are controversial. As a consequence, the identification and study of basement faults in foreland settings has often relied on indirect information such as seismic data when they exist, seismicity, potential field (gravity and magnetic) measurements, and anomalies in topography and drainage network patterns.

In the Zagros Fold–Thrust Belt (ZFTB) of Iran (Fig. 1) it is firmly established that the basement is involved in the deformation. Evidence for this comes essentially from the relatively intense mid-crustal seismic activity recorded in the Zagros. Focal mechanisms determined for the seismic events show that the majority of basement faults are of reverse type (Fig. 1b). This is supported by the fact that the basement faults are also associated with major vertical steps in the general topography and a vertical shift of the syncline base level in the hanging wall of basement faults. Based on these two criteria, seismicity and topography, Berberian (1995) proposed a map of the main active basement faults in the ZFTB. Other maps have been proposed based upon analysis of satellite imagery (Fürst 1990; Hessami *et al.* 2001; Authemayou *et al.* 2006). However, because the basement faults in Zagros rarely reach the surface, their overall geometry and lateral extent, and in many cases their precise location, remain a matter of debate.

Basement faults are thought to have played a role in the development of the Arabian palaeo-margin since the Late Precambrian and the deposition of

From: LETURMY, P. & ROBIN, C. (eds) *Tectonic and Stratigraphic Evolution of Zagros and Makran during the Mesozoic–Cenozoic.* Geological Society, London, Special Publications, **330**, 121–138.
DOI: 10.1144/SP330.7 0305-8719/10/$15.00 © The Geological Society of London 2010.

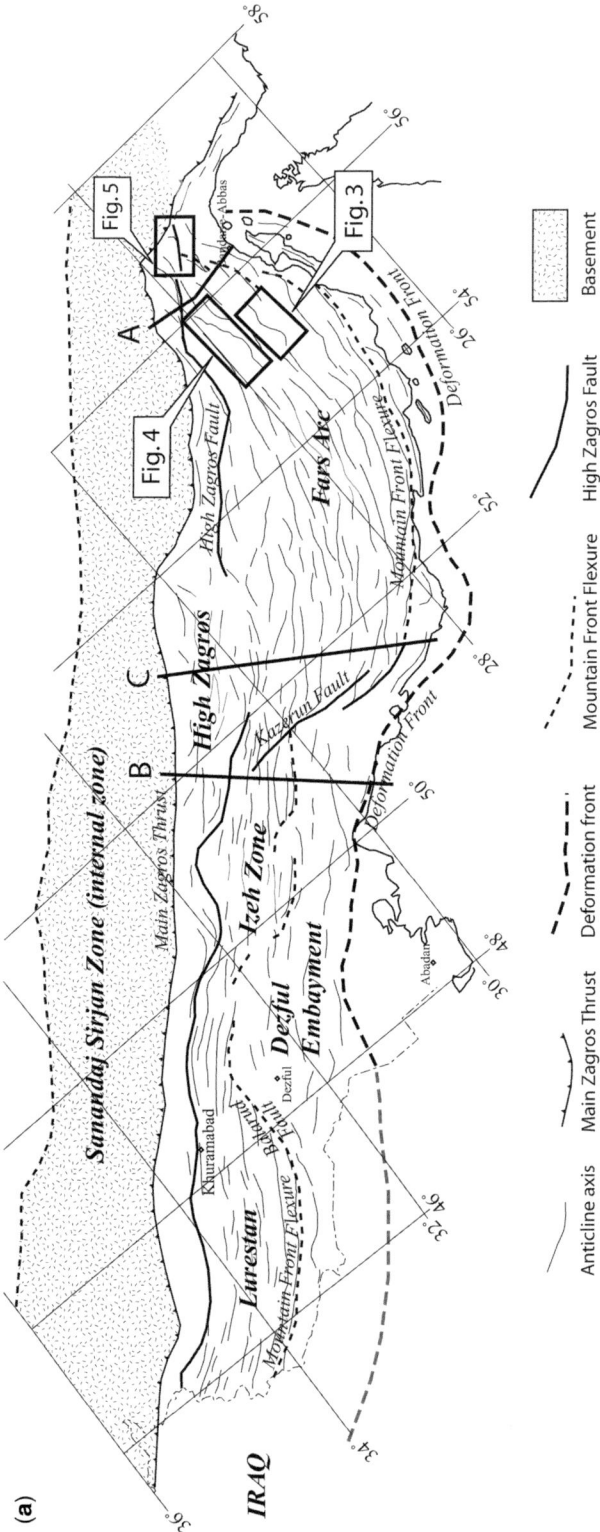

(a)

Fig. 1. General map of the Zagros fold-and-thrust belt. (**a**) Map showing the structural pattern. Lines A and B refer to sections in Figures 2 and 12; line C refers to section in Figure 12b. (**b**) Topographic map with the main tectonic subdivisions, main fault lines, seismicity and a selection of focal mechanisms.

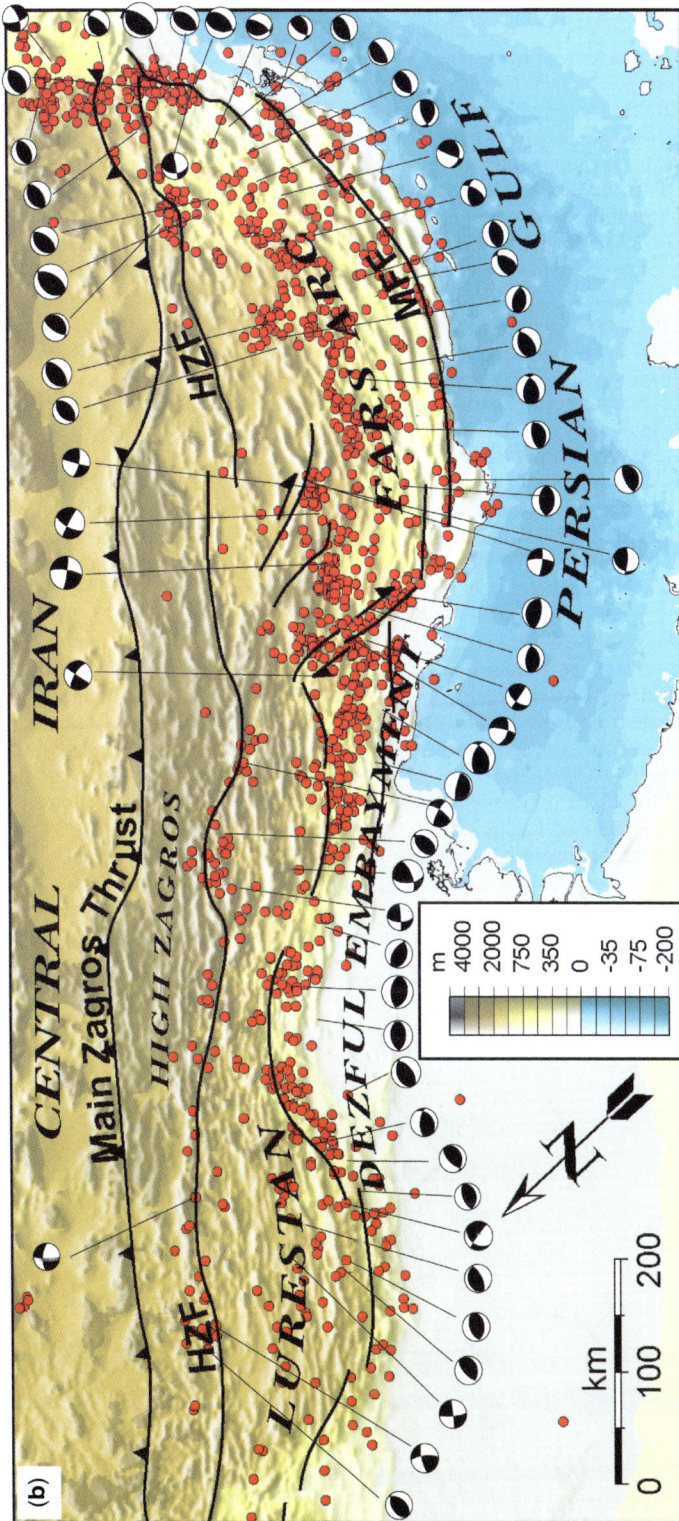

Fig. 1. (*Continued*)

the Hormuz Formation (Sharland 2001; Ziegler 2001). In particular, the present-day segmented geometry of the ZFTB, characterized by salients (Fars and Luristan arcs) and re-entrants (Dezful Embayment and Hormuz Strait), is interpreted as reflecting the influence of inherited north–south fault trends within the Pan-African basement (Fig. 1) (Falcon 1974; Kent 1979; Fürst 1990; McQuillan 1991; Berberian 1995; Moitei 1993; Hessami *et al.* 2001; Sherkati & Letouzey 2004; Sherkati *et al.* 2006). Among these fault zones, the Oman line (Gansser 1955; O'Brien 1957; Kadinski-Cade & Barazangi 1982; Molinaro *et al.* 2004; Talebian & Jackson 2004) and the Kazerun fault zone (Sepehr & Cosgrove 2004; Authemayou *et al.* 2005, 2006) play a major role (Fig. 1). According to Sherkati *et al.* (2006), the Kazerun fault zone has been the site of quasi-constant movements throughout Mesozoic and Cenozoic times. The question of the timing of basement involvement during the Zagros orogeny is more controversial. Molinaro *et al.* (2005) and Sherkati *et al.* (2005, 2006) have defended a general scenario in which the present thick-skinned style is recent (post-Pliocene) following a previous thin-skinned style during the Miocene. In contrast, Bahroudi & Talbot (2003) and Mouthereau *et al.* (2007) have proposed a more classical view, in which the main basement faults have been accommodating reverse movement since the beginning of the collision, particularly along the Main Frontal Flexure (MFF), which is considered to be the deformation front of the belt.

The main purpose of this paper is to present a general scenario of basement involvement during Cenozoic deformations in the Fars Arc of the ZFTB. Data used for this work were high-resolution topographic data (Shuttle Radar Topography Mission; SRTM) combined with satellite images and field observations. We start by describing in detail the geometry and morphology of a selection of basement-involved structures, located in the eastern side of the Fars Arc where basement faults strike obliquely to folds in the cover. This allows us to address the problem of basement v. cover deformation and their relative chronology. We then analyse the river network, knickpoints on river profiles and their relations with structures. Finally, we discuss the timing and segmentation of basement faults in the Fars Arc.

Geology of the Zagros Fold–Thrust Belt and current evidence for basement faulting

In the general context of the Alpine–Himalayan orogenic system, the Zagros Fold–Thrust Belt (ZFTB) is a young and active collisional belt developed at the frontier between the converging Arabian and Eurasian plates (Fig. 1). The present convergence is characterized by a N13°E direction and a rate of 22 ± 2 mm a^{-1} increasing to 26 ± 2 mm a^{-1} in eastern Oman, according to recent estimations based upon GPS measurements (Vernant *et al.* 2004). In the ZFTB, the Arabian passive margin sequence has been decoupled from its basement and deformed by large-scale folding and thrusting. Recent geological cross-sections across the ZFTB (Blanc *et al.* 2003; McQuarrie 2004; Molinaro *et al.* 2005; Sherkati *et al.* 2006) show that the Mesozoic cover is detached along a thick evaporitic horizon (the Hormuz Salt) and folded regularly into large detachment anticlines (Fig. 2) (Colmann-Sadd

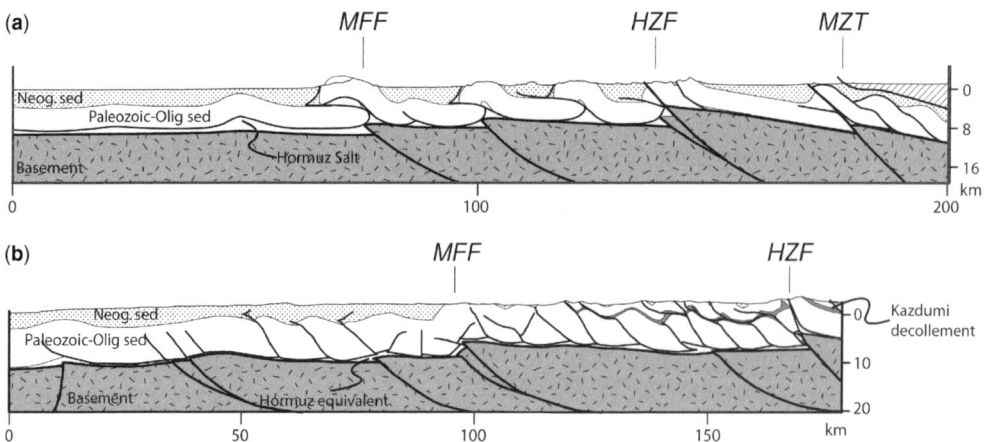

Fig. 2. Cross-sections across the Zagros fold-and-thrust belt (see location in Fig. 1). (**a**) Balanced cross-section in the southeastern Fars modified after Molinaro *et al.* (2005). (**b**) Balanced cross-section in the central part of the Fars modified from Sherkati *et al.* (2006).

1978; for recent reviews see Sherkati *et al.* 2005; Sepehr *et al.* 2006). The style of deformation in the overlying Cenozoic units is more complex and appears to be controlled by changing mechanical stratigraphy across the ZFTB (Sherkati *et al.* 2005). In the Eastern Fars pre-existing salt diapirs may have play an important role in the localization and geometry of folds (Callot *et al.* 2007; Jahani *et al.* 2007, 2009). These sections also depict major reverse faults cutting through the basement. The strongest line of evidence for basement involvement in the Zagros comes from the relatively intense mid-crustal seismic activity that characterizes the ZFTB (Jackson & Fitch 1981; Berberian 1995; Talebian & Jackson 2004; Hatzfeld *et al.* 2009). The focal mechanisms determined for seismic events are predominantly of reverse type (Fig. 1b) and occur at depths generally confined between 7 and 20 km. In the ZFTB, the thickness of the cover overlying Hormuz Salt is between 8 and 12 km, therefore most of the seismic events occur in the brittle upper crust (Talebian & Jackson 2004). Strike-slip solutions are much less frequent and are mainly associated with right lateral displacement along the Kazerun fault zone in the Central Zagros (Baker *et al.* 1993) and to a lesser degree with strike-slip movement along the Main Recent Fault (MRF) (Tchalenko & Braud 1974; Talebian & Jackson 2004; Fig. 1b).

Another argument for basement faulting that is invoked is the existence of irregular chains of strong topographic and structural uplift (of the order of several kilometres) visible throughout the Zagros. These chains of uplift were identified by early workers in the Zagros. Falcon (1961, 1974), analysing serial structural sections throughout the Zagros, identified a 'major geo-flexure' running along the entire length of the orogen and speculated that it originated from vertical isostatic adjustment within the collision zone. More recently Berberian (1995), on the basis of the identification of structural and topographic uplifts and seismicity, proposed a map of the main active basement thrust faults in the ZFTB (Fig. 1).

Remote sensing data (essentially satellite images) have also been used to demonstrate the existence of basement strike-slip faults beneath the Zagros cover (Fürst 1990; Hessami *et al.* 2001; Authemayou *et al.* 2005, 2006). These approaches essentially rely on the identification of deflections in the trends of fold axes, offset markers and alignments of salt diapirs (Kazerun fault zone).

Detailed examples of the geometry and morphological expression of basement faults

It has been proposed (Molinaro *et al.* 2005; Sherkati *et al.* 2005) that in the Eastern Zagros Mountains basement control on surface structures occurred only at a late stage of the tectonic evolution. In other words, the current thick-skinned style of Zagros deformation succeeded a more general thin-skinned phase of orogeny. This chronology is particularly well illustrated by spectacular interference patterns, in which early detachment folds are cut by late oblique basement faults. Such patterns are observed recurrently throughout the Eastern Zagros. Elsewhere in the Zagros faults and folds strike parallel to each other so the distinction between separate tectonic phases is less obvious. In the Eastern Zagros most of the folds have a general east–west trend; hereafter we focus on three structures showing anomalous orientations compared with this general trend and we propose a conceptual model of development explaining their different morphologies.

Shab anticline

The majority of folds visible on the satellite photograph in Figure 3a, located in the eastern limb of the Fars Arc, display east–west trends. However, one anticline (Shab anticline) presents a marked deflection in trend to a NE–SW orientation in its central part. Such deflections and sinuosities are frequently observed in folds in the eastern limb of the Fars Arc and have been classically interpreted as the effect on the cover of strike-slip movement along underlying basement trends (Ricou 1974; Hessami *et al.* 2001). More recently, Ramsey *et al.* (2008) have interpreted the sinuous trend of the frontal fold of the Fars Arc as the result of the coalescence of several growing anticlines. According to this study, a diagnostic feature of this process is a saddle-like geometry of the topographic crest and the underlying layers in the short deflected parts of the fold. With sandbox experiments and field examples in the Fars Arc, Callot *et al.* (2007) and Jahani *et al.* (2009) have proposed that sigmoid folds may be related to the presence of salt plugs, their weakness favouring the lateral propagation of folds toward them. They have also demonstrated that pre-existing salt walls localize deformation whatever their direction and may generate oblique structures.

A topographic profile extracted from a relief map of the area along the entire length of the Shab anticline shows that the deflection is associated with a distinct step in topography (Fig. 3c). This step becomes more apparent when considering the position of the top of the Asmari formation, reconstructed from geological maps. Such a step, of *c.* 1500 m, cannot be produced by strike-slip movement alone but is compatible with reverse movement along an underlying NE–SW-trending basement fault. This is in agreement with the observation, directly south of the Shab anticline and in line with

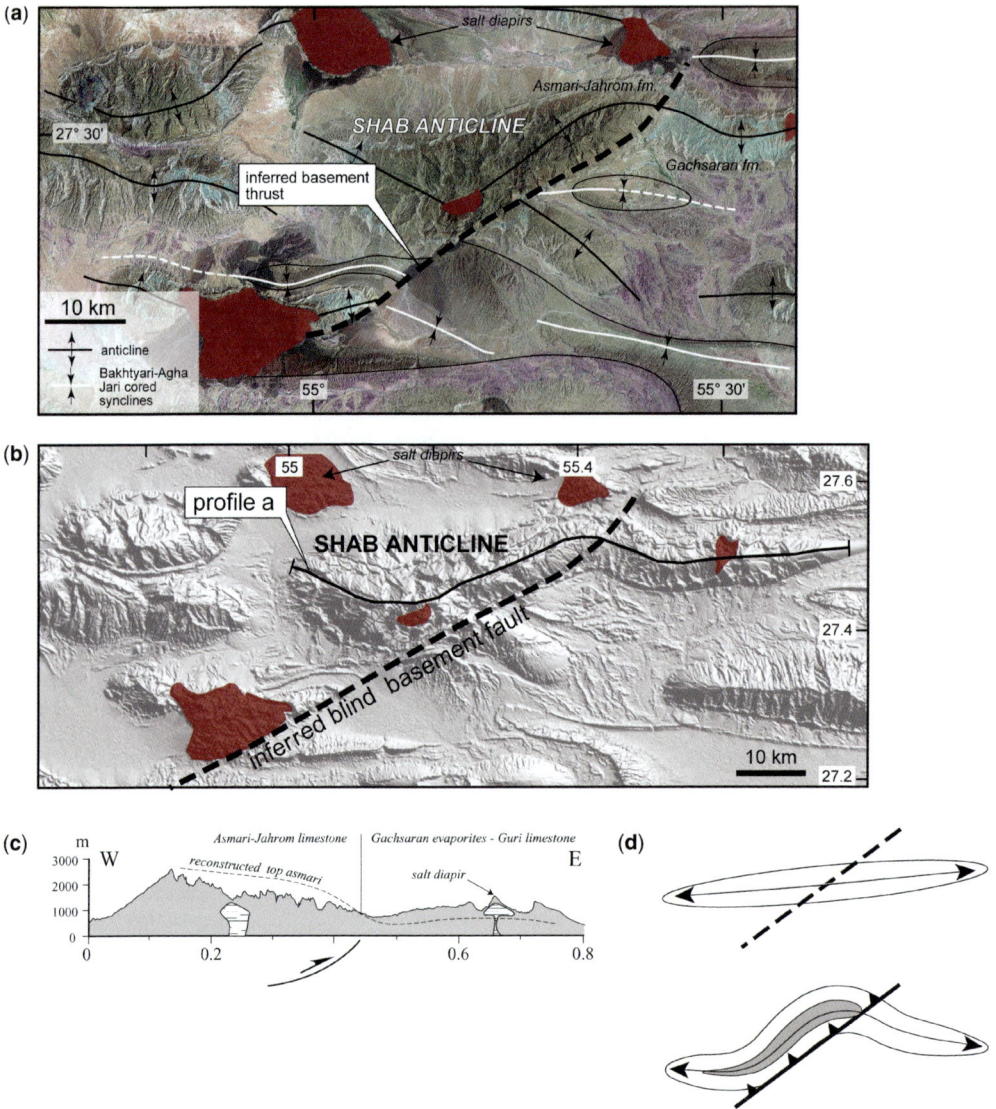

Fig. 3. (**a**) Structural (Landsat 7 photograph) and (**b**) shaded relief (SRTM data) maps of folds in the eastern limb of the Fars Arc. (**c**) Topographic profile across the Shab anticline [location shown in (a)] with the geometry of the Asmari Fm. (length of profile in degrees). (**d**) Conceptual model for the evolution of the Shab anticline.

the NE–SW trend, of another anticline thrusting eastward over a syncline (Fig. 3a). Furthermore, the NE–SW direction of the reverse basement fault inferred from the morphological study is consistent with focal mechanisms recorded in Eastern Fars (Fig. 1b). Finally, because the deflected segment of the fold is long and is not associated with a saddle-like geometry, we exclude the coalescence of anticlines as a possible mechanism explaining the sinuosity of the Shab anticline even if salt domes are observed in the core of the structure (Fig. 3).

Kuh-e-Muran structure

The Kuh-e-Muran structure (Fig. 4) is located in the Eastern Fars Arc just south of the HZF. This 80 km long structure has a main NE–SW trend parallel to the trend of the deflected Shab anticline. However, careful analysis of its geometry (Fig. 4a) reveals folds of short lateral extent superimposed on the structure. These small folds have a general east-west orientation, which changes to NE–SW when approaching the southeastern limb of the

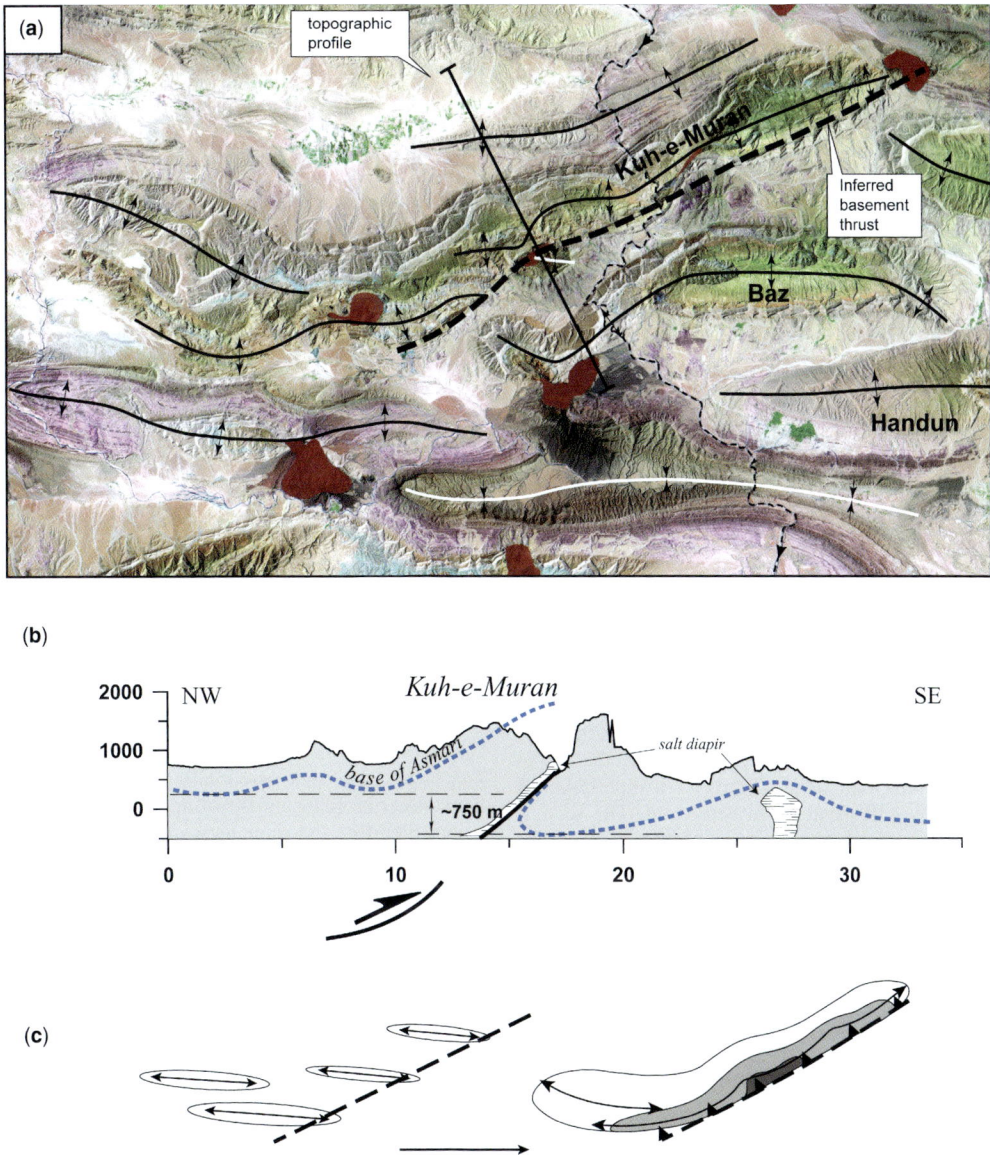

Fig. 4. (a) Structural (Landsat 7 photograph) map of structures in the eastern Fars (white lines, synclines; black lines, anticlines; bold dashed line, basement fault; fine dashed line, main river course (Rud-e-Shur); red areas, salt diapirs). (b) Topographic profile with the geometry of the Asmari Fm. [location shown in (a)] showing a 750 m step between the backlimb and the forelimb synclines. (c) Conceptual model for the evolution of the Kuh-e-Muran structure.

Kuh-e-Muran structure. Like many structures in the Eastern Fars, the Kuh-e-Muran structure is surrounded by salt plugs. This led Jahani et al. (2009) to propose that this structure is linked to an initial NE–SW salt wall that controls the position of Kuh-e-Muran. In this interpretation the obliquity of the salt wall to the shortening direction necessitates a strike-slip component that generates the

bending of the small folds and explains the geometry of the structure. Below we propose an alternative interpretation based on morphological arguments. A topographic profile through this structure shows a 350 m topographic step between the two synclines located on both sides of the structure. This step increases to 750 m when considering the position of the top of the Asmari formation reconstructed from geological maps (Fig. 4b). We interpret such a pattern as the result of the activation of a NE–SW reverse basement fault parallel to the Kuh-e-Muran structure and uplifting the previous folds located on its hanging wall. Deflection of the east–west fold axes along its trace reflects the passive draping of the cover accommodating the vertical uplift along the fault. Further amplification of the initial folds may have partly contributed to their coalescence in a process similar to that described by Ramsey *et al.* (2008). A two-step evolution with a late NE–SW basement fault cutting obliquely through an already formed train of east–west detachment folds (with east–west axes) is consistent with the geological and morphological data (Fig. 4c). The NE–SW direction of the reverse basement fault inferred from the morphological study is also consistent with focal mechanisms recorded in this zone (Fig. 1b).

Kuh-e-Khush structure

Culminating at 2400 m, the Kuh-e-Khush anticline is the easternmost structure linked to the High Zagros Fault. In map view, the Kuh-e-Khush anticline exhibits a 'butterfly' shape (Fig. 5a) and has already been described by Molinaro *et al.* (2005), Sherkati *et al.* (2005) and Jahani *et al.* (2009). In map view it is easy to recognize an ENE–WSW trend linked to the High Zagros Fault superimposed on a NW–SE trend of an early fold. The section (Fig. 5b) shows the position of the base of the Mishan marls in the synclines on both sides of the structure (redrawn from Sherkati *et al.* 2005). A step of *c.* 1500 m is observed between the bases of the synclines and is related to a basement fault. Molinaro *et al.* (2005) proposed that this structural pattern is the result of a two-step evolution with a late basement fault cutting obliquely through an already formed detachment fold (Fig. 5c). For this structure Jahani *et al.* (2009) proposed an interpretation similar to that for the Kuh-e-Muran structure (i.e. with a salt wall controlling the geometry of the structure) but they did not reject the existence of a basement fault (the HZF) reaching the surface.

Summary

Three kinds of interference patterns between detachment folds and basement faults are observed in the Fars Arc and summarized in Figure 6. Although the resulting geometries are different in map view, each of them can be explained by a two-step evolution with a late basement fault cutting obliquely through already existing detachment folds. We propose that the final resulting geometry depends both on the development stage of the initial fold and the amount of displacement along the underlying basement fault. The Shab anticline (Fig. 6a) is the result of a large and long fold cut by a basement fault with a moderate vertical throw. In this case the displacement along the basement fault has not been sufficient to completely overprint the previously formed fold, which shows only a deflection parallel to the basement fault. The Kuh-e-Muran structure (Fig. 6c) is the result of the interference between 'small' folds (i.e. with a small shortening ratio and small lateral extension) and a late basement fault with a large vertical throw that obliterates almost completely the previous folds. Finally, between these two extreme cases is the Kuh-e-Kush example (Fig. 6b), where the ratio between shortening by folding and shortening by faulting remains moderate.

Structures, rivers and incision

Since the work of Oberlander (1965), very few studies have analysed the drainage network of the Zagros fold-and-thrust belt. However, the geometry of the drainage system can provide valuable new information on large-scale tectonics (Brookfield 1998; Snyder *et al.* 2000; d'Agostino *et al.* 2001; Leturmy *et al.* 2003) or on kinematics of structures (Jackson *et al.* 1996; van der Beek *et al.* 2002; Champel *et al.* 2002; Ramsey *et al.* 2008). The Zagros domain was essentially marine until the final onset of folding and thrusting related to the collision, therefore the river network developed on a newly formed topographic domain. Topography is controlled only by tectonic activity and syntectonic sedimentation, and no antecedent rivers interact with the new drainage network. The Zagros fold–thrust belt is therefore well suited to study the interactions between a growing topography and the development of a drainage network.

Below we analyse the drainage system in the Fars Arc, paying specific attention to its geometry in the vicinity of basement-involved structures, to understand the kinematics of these structures and the possible feedback on the river system.

River network

We used SRTM topographic data with a 3″ (*c.* 90 m) horizontal resolution (USGS) to extract the river network presented in Figure 7 and analysed the longitudinal profiles of three rivers.

The drainage divide between streams with outlets in the Persian Gulf and rivers flowing northward coincides everywhere with the Main Zagros Thrust except in the eastern part of the Fars Arc, where streams flowing southward have sources located more to the north in the Sanandaj Sirjan Zone, which belongs to the internal zone of the

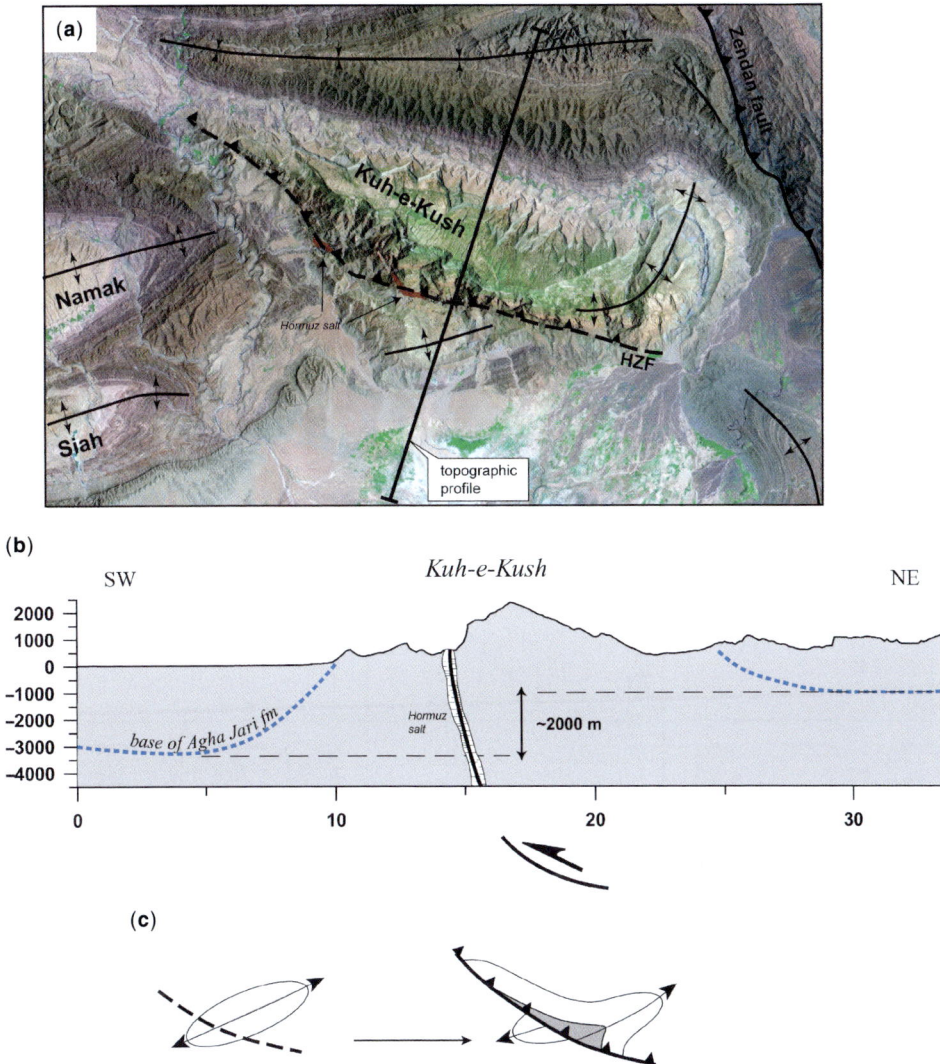

Fig. 5. (**a**) Structural (Landsat 7 photograph) map of structures along the eastern limb of the HZF (white lines, synclines; black lines, anticlines; bold dashed line, basement fault). (**b**) Topographic profile [location shown in (a)] with the geometry of the Agha Jari Fm. (constrained from Sherkati *et al.* 2005) showing 2000 m of vertical displacement north of the HZF. (**c**) Conceptual model for the evolution of the Kuh-e-Kush structure.

Zagros. The coincidence between the position of the MZT and the drainage divide indicates that this major structure has been a topographic barrier since the beginning of the topographic building of the Fars Arc. In the Eastern Fars the northward displacement of the drainage divide could be linked to reactivation of deep basement thrusts hidden by the MZT (Fig. 2a). This is also suggested by seismic events (Fig. 1b) recorded in this zone, which is the only place along the Zagros where seismicity occurs in the upper crust north of the MZT.

The Fars Arc is divided in two large sub-basins with outlets on each side of the arc; the two main rivers are Rud-e-Shur flowing to the SE and Rud-e-Mand flowing to the west. Flow directions in those two sub-basins are dominated by a N90 to N20 (Figs 7 & 8a) direction, which corresponds to the main direction of fold axes in the Fars arc (Fig. 8b). A secondary flow direction is around N50. It could be linked to the second N50–70 structural direction corresponding to the sigmoidal part of coalescent folds and to the NW–SE direction of

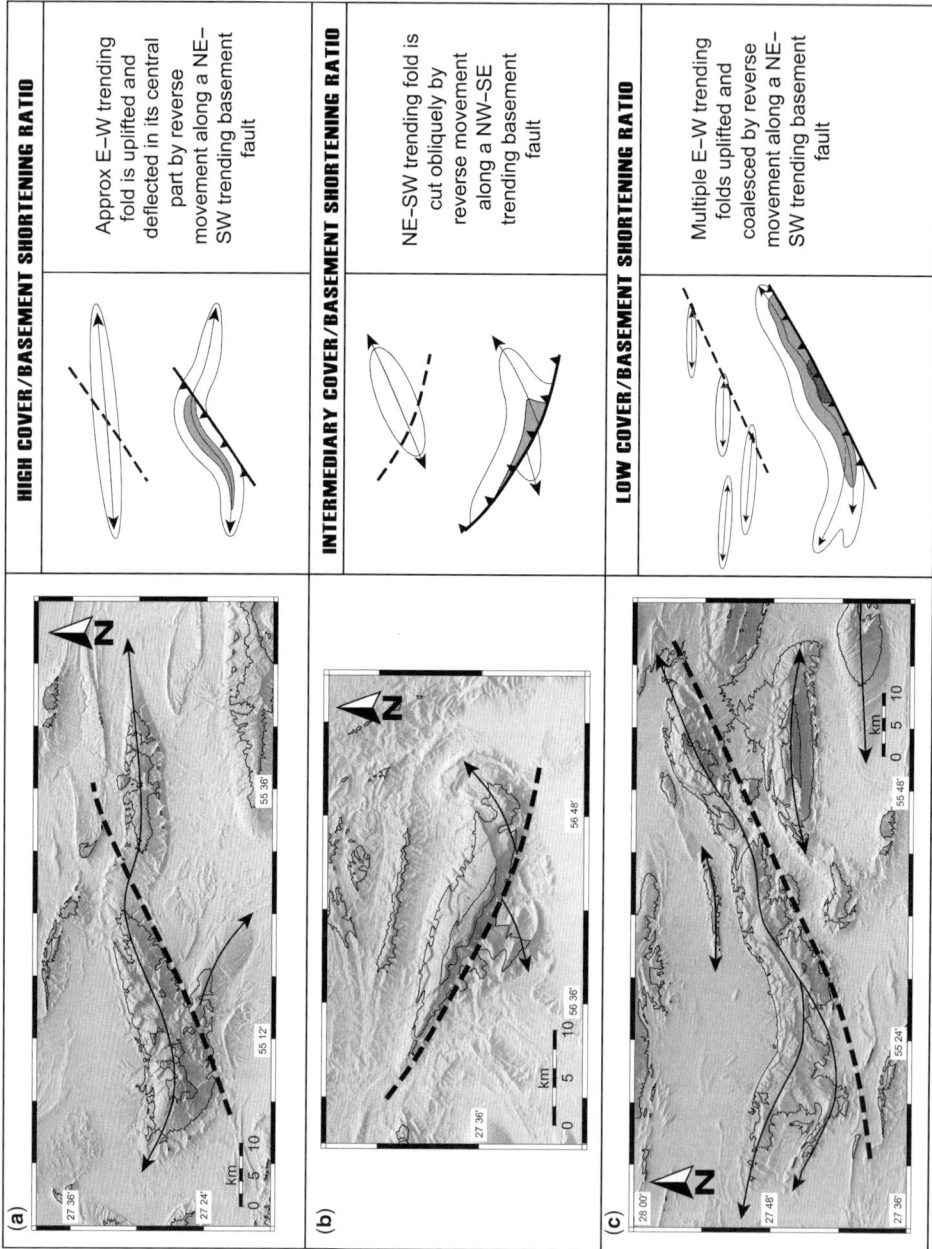

Fig. 6. Synthesis of observed interference patterns in the Fars Arc showing shaded relief maps (contoured at 1000 and 2000 m) and kinematic models for (**a**) the Shab anticline, (**b**) the Kuh-e-Kush structure along the HZF and (**c**) the Kuh-e-Muran structure.

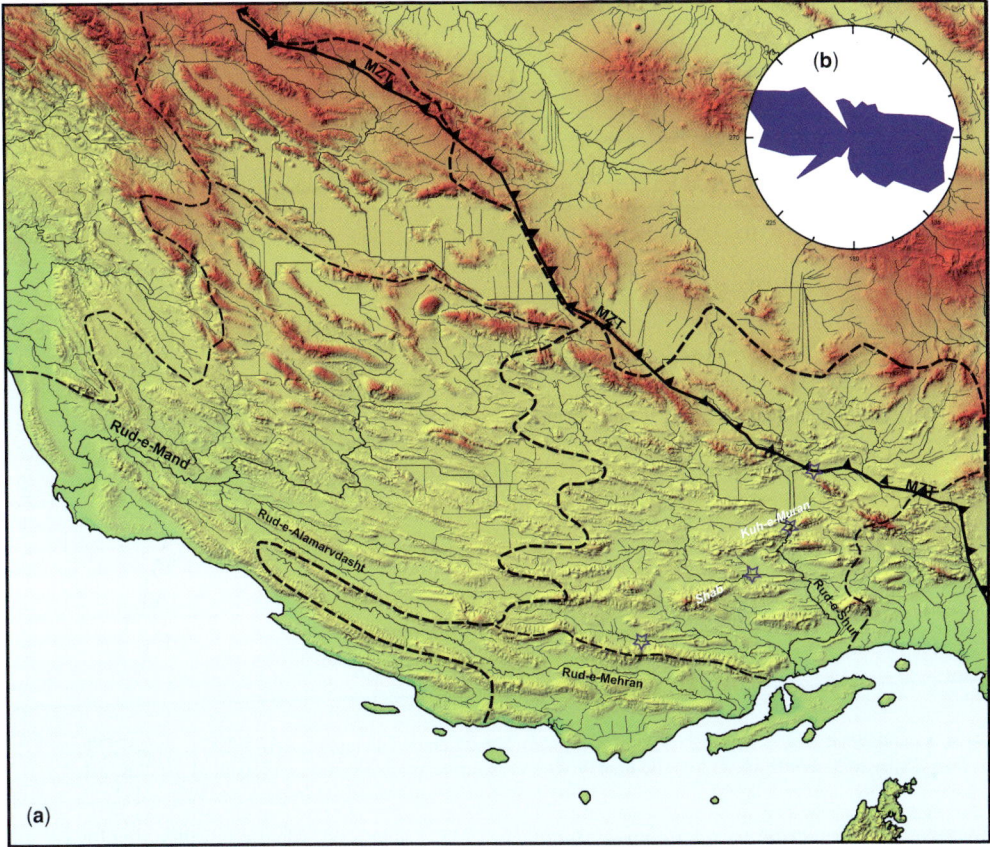

Fig. 7. (**a**) River network of the Fars Arc extracted from SRTM data. Dashed lines, drainage divides; stars, position of knickpoints on river profiles (Fig. 9). (**b**) Rose diagram of river flow directions.

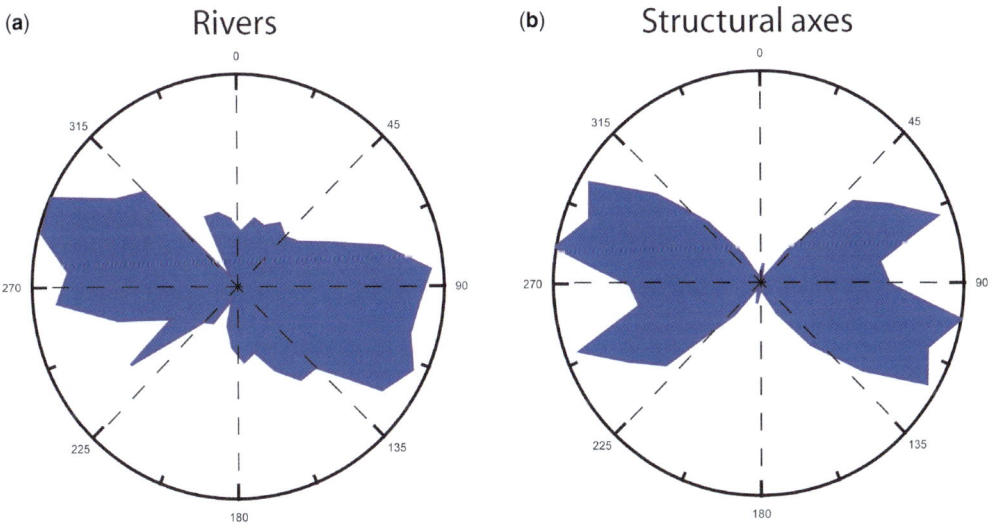

Fig. 8. (**a**) Rose diagram of river flow directions and (**b**) rose diagram of structural trends.

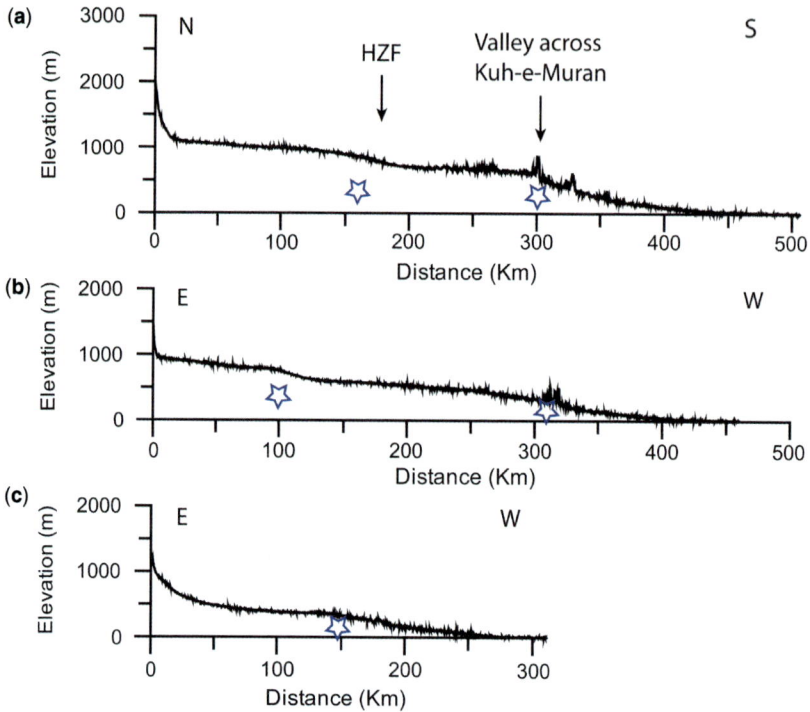

Fig. 9. Longitudinal profiles of three rivers of the eastern Fars. Stars at knickpoints correspond to those in Figure 7. (a) Rud-e-Shur river profile. (b, c) Longitudinal profiles of two rivers flowing north and south of the Shab anticline.

basement faults indicated by interference patterns. It is important to notice that this structural trend has only slight influence on the flow directions of the hydrographic network. The drainage divide between rivers flowing eastward and those flowing westward is highly sinuous and mimics the periclinal terminations of folds. As proposed by Ramsey *et al.* (2008), we think that the flow direction of these tributaries was controlled by the lateral growth and coalescence of anticlines making rivers 'prisoners' of synclines. The sinuosity of the drainage divide and flow directions therefore reflects a structural control on rivers.

River profiles

River profiles in a declining relief have a characteristic concave shape with a slope declining tangentially toward the outlet. Local convexity or knickpoints with growing slopes toward the outlet can be attributed to a tributary junction, a lithological change or a local tectonic displacement. Rivers are generally very sensitive to vertical tectonic displacements, and they adjust surface warping by deflection, avulsion, erosion and aggradation along the profile or changes in channel pattern (Goodrich 1898; Howard 1967; Ouchi 1985; Merrits *et al.*

1994; Sklar & Dietrich 1998; Holbrook & Schumm 1999).

Three river profiles crossing the eastern Fars are presented in Figure 9. Rud-e-Shur (Fig. 9a) is 500 km long and is the principal river of the eastern drainage basin. This river crosses the HZF and the Kuh-e-Muran structure (Fig. 4), but its flow direction does not seem to be controlled by the structural pattern in the basement. It longitudinal profile (Fig. 9a) shows two significant knickpoints with a local convex pattern typical of retrogressive erosion in response to tectonic activity. The northern one is located close to the Main Zagros Thrust and the southern one is linked to the crossing of the Kuh-e-Muran structure. The time scale of the change in river slope is difficult to estimate as it depends on rock strength, stream power and magnitude of tectonic movements (Whipple & Tucker 1999). Nevertheless, the observation of knickpoints suggests that these two structures have been active recently.

In a scenario where the main course of rivers initially controlled by the position of fold axes (Fig. 10a) and where deformation along basement faults occurs in a late stage of deformation, rivers would be constrained to cross-cut basement structures (Fig. 10b) as no escape solution exists. The river profile would show a local convexity linked

(a)

(b)

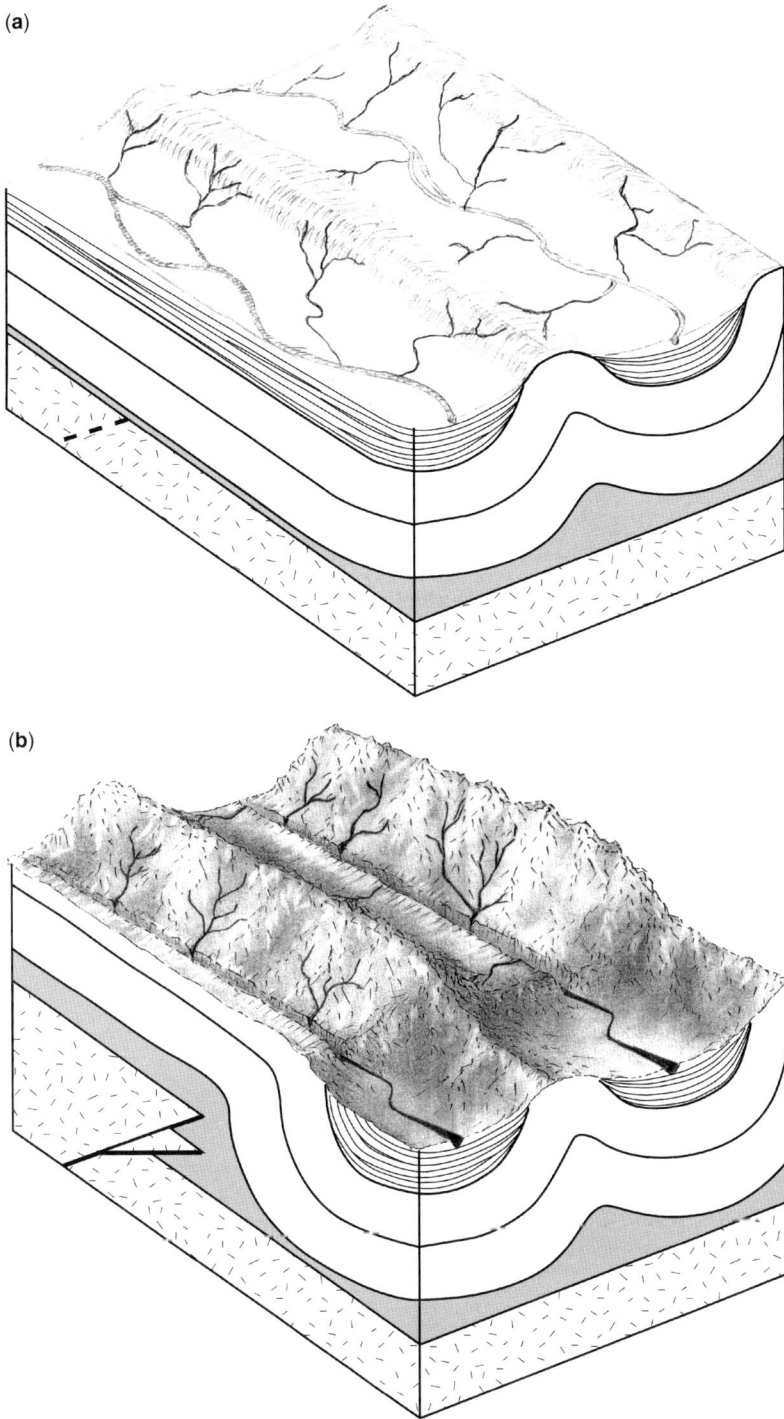

Fig. 10. A 3D view of the two-step kinematic evolution showing the relationships between folding, basement faulting, and the geometry of the river network. During the first step folds develop and main rivers flow in the synclines. When basement faults are activated obliquely to folds and rivers, the latter are constrained to cross-cut the uplifted zone and their longitudinal profiles show a convex shape in the vicinity of the basement fault.

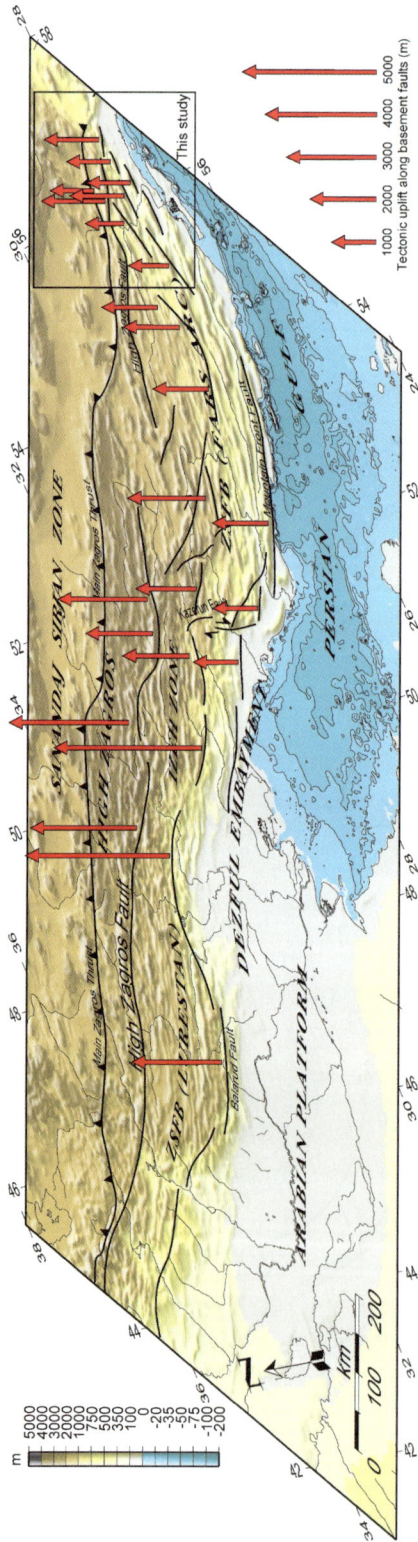

Fig. 11. Synthesis of tectonic uplift linked with the main basement faults in the Zagros. Data are from published sections (Letouzey *et al.* 2002; Molinaro *et al.* 2004, 2005; Sherkati & Letouzey 2004; Mouthereau *et al.* 2007; Verges *et al.* 2007; Jahani *et al.* 2009) and this study.

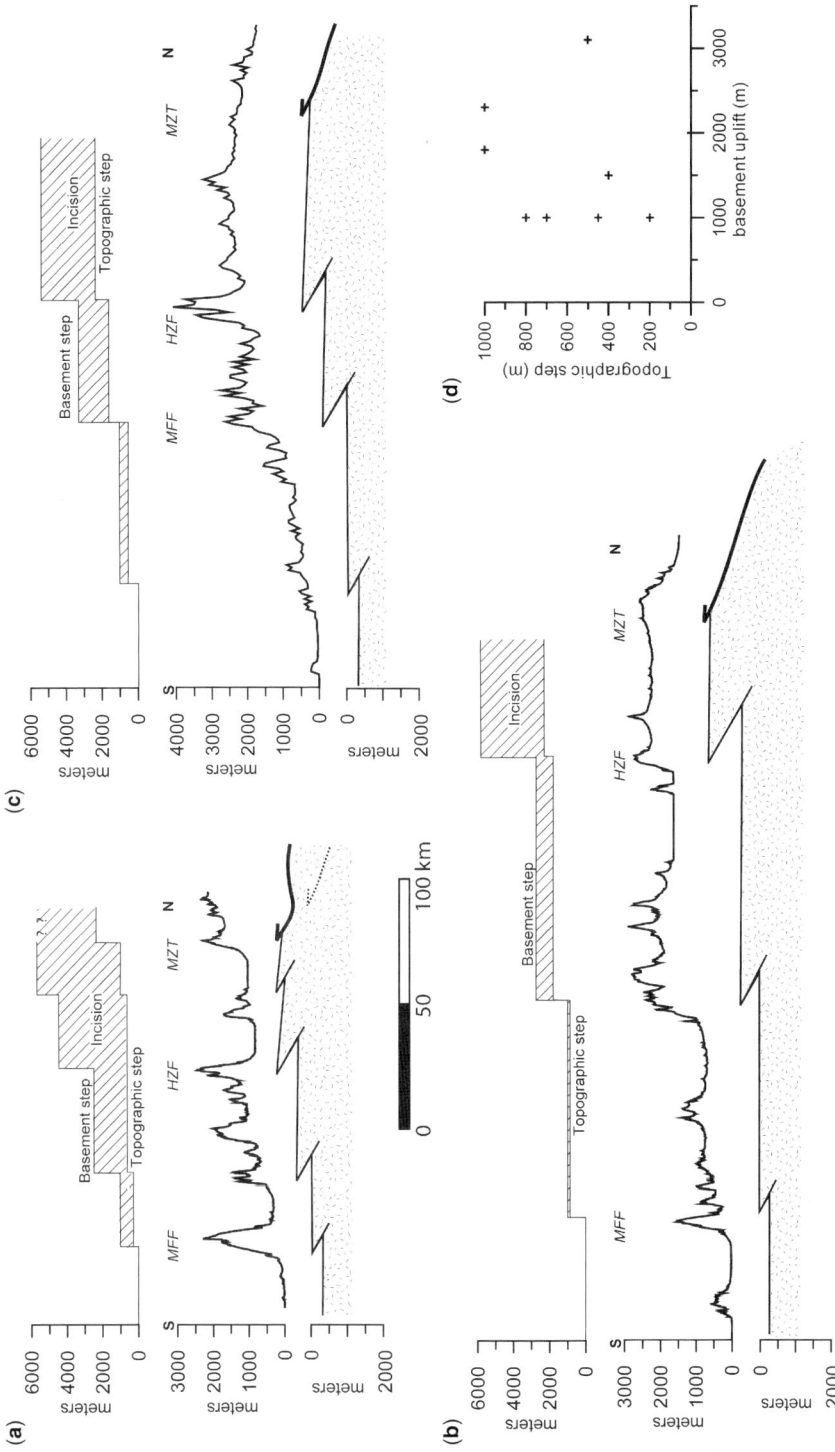

Fig. 12. Incision in the Fars Arc and Central Zagros deduced from the present-day topographic step link with tectonic uplift along basement faults. (**a**) Eastern Fars; tectonic uplift along basement faults is calculated from the Molinaro *et al.* (2005) balanced section (see also section in Fig. 2a). (**b**) Central Zagros; tectonic uplift along basement faults is calculated from the Sherkati *et al.* (2006) balanced cross-section (see also section in Fig. 2b) (**c**) Western Fars; tectonic uplift along basement faults is calculated from the Sherkati & Letouzey (2004) balanced cross-section. (**d**) Diagram showing that there is no relationship between the amount of tectonic uplift and the topographic step produced by basement faulting.

to the uplift of the hanging wall along the basement fault (Fig. 11c).

The two other longitudinal profiles flow north and south of the Shab anticline. Each of them shows a knickpoint located on both sides of the deflected part of the Shab anticline and confirms the presence of an active basement fault associated with it. The northern knick point on Profile B corresponds to the location of the MFF defined by Berberian (1995).

Discussion

Based on published sections we can now have an overview of basement faulting throughout the Zagros (Fig. 11). In the Central Zagros, the basement faults limit the main palaeogeographical domains and their long history can be reconstructed (Sherkati & Letouzey 2004; Sherkati *et al.* 2006). At present, the basement deformation is concentrated along two fault zones, marked by strong tectonic uplift (Fig. 11). In the Fars Arc, in contrast, the tectonic uplift is distributed between several basement faults, which are not restricted to the boundary between structural or palaeogeographical zones. As a consequence, the amount of uplift on each fault in this area ranges between 1000 and 2000 m compared with the 5000 m or more observed along the basement faults in the Central Zagros (Fig. 11). Fault segmentation appears to increase southeastward when approaching the Zagros–Makran transform zone. This change in fault geometry is also reflected by a marked decrease from west to east in the aspect ratio (half wavelength/axial length ratio) of folds (Molinaro *et al.* 2004). This change in geometry may be related to the influence of inherited transverse structures within the Arabian basement and is probably related to the 'Oman line' transform fault system, which developed during the Permo-Triassic Neotethys rifting (Stampfli 2000).

According to Sherkati *et al.* (2005), it is assumed that in the ZFTB folding occurred in discrete steps rather than as a continuous process. Good seismic data and accurate timing have been given by these workers for the Dezful Embayement. The first discrete episode of folding occurred during the deposition of the Gachsaran Formation (i.e. early to middle Miocene) and was followed by a period of tectonic quiescence during the deposition of the Mishan marls and lower Agha Jari sandstones. The main episode of folding occurred during the deposition of the upper Agha Jari Fm. This event is well expressed by syntectonic sedimentation and growth strata described everywhere in the Zagros (Hessami *et al.* 2001; Sherkati *et al.* 2005; Mouthereau *et al.* 2007; Homke *et al.* 2004). During this stage uplift linked to folding was largely compensated by subsidence; progressive unconformities with onlaps and overlaps on folded strata (Fig. 10a) are commonly encountered in the upper

Agha Jari Fm in the field as well as on seismic lines. These sedimentary features indicate that no high topography developed before the end of deposition of the upper Agha Jari Fm. A question that arises is when the major topographic steps (Fig. 1) developed. It has been noticed by several researchers (Berberian 1995) that the steps are associated with alignments of seismic activity in the basement (events deeper than 10 km). In Figure 12 we compare the topographic steps observed on SRTM data and the basement uplift calculated from three balanced cross-sections (Sherkati & Letouzey 2004; Molinaro *et al.* 2005; Sherkati *et al.* 2006) constrained by field and subsurface data. The topographic step corresponds to the difference between mean elevations of valleys on both sides of the basement fault. The difference between the topographic step and the basement uplift corresponds to the mean incision in valleys in response to basement uplift. The graph in Figure 12 shows that there is no correlation between the value of basement uplift and the topographic step. However, on the three sections we can observe that incision increases systematically towards the hinterland, from south to north. In a context where the climate does not change drastically from south to north and where rivers are larger from north to south in response to progressive advection of tributaries, this could be the result of there being a longer time for rivers to respond to the basement uplift; that is, activation of basement faults may have occurred earlier in the inner part of the belt. The progressive increase of incision from south to north should therefore reflect the forward propagation of basement faulting from north to south. Such an observation contradicts the kinematic scenario proposed by Mouthereau *et al.* (2007) in which, in the ZFTB, the outermost basement faults are activated before the inner ones. On the contrary, it confirms the kinematic model proposed by Molinaro *et al.* (2005) showing that the post-folding basement faults are activated following a general 'piggy-back' sequence.

This work has been financially supported by the MEBE Programme sponsored by AGIP-ENI, BP, CNRS, Petronas, Shell, Total and UPMC, and managed by E. Barrier and M. F Brunet (CNRS-Université Pierre et Marie Curie, Paris 6). We gratefully acknowledge the support in the field by the Geological Survey of Iran. This work benefited greatly from discussions with J. Letouzey, S. Sherkati, S. Jahani, J.-P. Callot, J. Vergès and J. C. Ringenbach.

References

AUTHEMAYOU, C., BELLIER, O., CHARDON, D., MALEKZADE, Z. & ABASSI, M. 2005. Role of the Kazerun Fault system in active deformation of the Zagros fold-and-thrust belt (Iran). *Comptes Rendus Géosciences*, **337**, 539–545.

AUTHEMAYOU, C., CHARDON, D., BELLIER, O., MALEKZADEH, Z., SHABANIAN, E. & ABBASSI, M. R. 2006. Late Cenozoic partitioning of oblique plate convergence in the Zagros fold-and-thrust belt (Iran). *Tectonics*, **25**, article TC3002.

BAHROUDI, A. & TALBOT, C. J. 2003. The configuration of the basement beneath the Zagros basin. *Journal of Petroleum Geology*, **26**, 257–282.

BAKER, C., JACKSON, J. & PRIESTLEY, K. 1993. Earthquakes on the Kazerun Line in the Zagros Mountains of Iran: strike-slip faulting within a fold–thrust belt. *Geophysical Journal International*, **115**, 41–61.

BERBERIAN, M. 1995. Master 'blind' thrust faults hidden under the Zagros folds: active basement tectonics and surface morphotectonics. *Tectonophysics*, **241**, 193–224.

BLANC, E. J.-P., ALLEN, M. B., INGER, S. & HASSANI, H. 2003. Structural styles in the Zagros Simply Folded Zone, Iran. *Journal of the Geological Society, London*, **160**, 401–412.

BROOKFIELD, M. E. 1998. The evolution of the great river systems of southern Asia during the Cenozoic India–Asia collision: rivers draining southwards. *Geomorphology*, **22**, 285–312.

CALLOT, J. P., JAHANI, S. & LETOUZEY, J. 2007. The role of pre-existing diapirs in fold and thrust belt development. *In*: LACOMBE, O., LAVÉ, J., ROURE, F. & VERGÉS, J. (eds) *Thrust Belts and Foreland Basins*. Springer, Berlin, 307–323.

CHAMPEL, B., VAN DER BEEK, P. A., MUGNIER, J. L. & LETURMY, P. 2002. Growth and lateral propagation of fault-related folds in the Siwaliks of western Nepal: rates, mechanisms and geomorphic signature. *Journal of Geophysical Research*, **107**, doi: 10.1029/2001JB000578.

COLMAN-SADD, S. P. 1978. Fold development in Zagros Simply Folded Belt, southwest Iran. *AAPG Bulletin*, **62**, 984–1003.

D'AGOSTINO, N., JACKSON, J. A., DRAMIS, F. & FUNICIELLO, R. 2001. Interactions between mantle upwelling, drainage evolution and active normal faulting: an example from the central Apennines (Italy). *Geophysical Journal International*, **147**, 475–497.

FALCON, N. L. 1961. Major earth-flexuring in the Zagros Mountains of southwest Iran. *Quarterly Journal of the Geological Society of London*, **117**, 367–376.

FALCON, N. L. 1974. Southern Iran: Zagros Mountains. *In*: SPENCER, A. (ed.) *Mesozoic–Cenozoic Orogenic Belts*. Geological Society, London, Special Publications, **4**, 199–211

FÜRST, M. 1990. Strike slip faults and diapirism of the southeastern Zagros Range. *In*: *Proceedings of International Symposium on Diapirism with Special Reference to Iran 2*, 150–182.

GANSSER, A. 1955. New aspects of the geology of Central Iran. *In*: *Proceedings of the 4th World Petroleum Congress, Rome, Section 1*, 297.

GOODRICH, H. B. 1898. *Geology of the Yukon Gold District, Alaska, United States*. Annual Report, Geological Society, Washington, DC.

HATZFELD, D., AUTHEMAYOU, C. *ET AL.* 2009. The kinematics of the Zagros Mountains (Iran). *In*: LETURMY, P. & ROBIN, C. (eds) *Tectonic and Stratigraphic Evolution of Zagros and Makran during the* *Mesozoic–Cenozoic*. Geological Society, London, Special Publications, **330**, 19–42.

HESSAMI, K., KOYI, H. A. & TALBOT, C. J. 2001. The significance of strike-slip faulting in the basement of the Zagros fold and thrust belt. *Journal of Petroleum Geology*, **24**, 5–28.

HOLBROOK, J. & SCHUMM, S. A. 1999. Geomorphic and sedimentary response of rivers to tectonic deformation: a brief review and critique of a tool for recognizing subtle epeirogenic deformation in modern and ancient settings. *Tectonophysics*, **305**, 287–306.

HOMKE, S., VERGÉS, J., GARCÉS, M., EMAMI, H. & KARPUZ, R. 2004. Magnetostratigraphy of Miocene–Pliocene Zagros foreland deposits in the front of the Push-e Kush Arc (Lurestan Province, Iran). *Earth and Planetary Science Letters*, **225**, 397–410.

HOWARD, A. D. 1967. Drainage analysis in geological interpretation: a summary. *AAPG Bulletin*, **51**, 2246–2259.

JACKSON, J. 1980. Errors in focal depth determination and depth of seismicity in Iran and Turkey, *Geophysical Journal of the Royal Astronomical Society*, **61**, 285–301.

JACKSON, J. A. & FITCH, T. 1981. Basement faulting and the focal depths of the larger earthquakes in the Zagros mountains (Iran). *Geophysical Journal of the Royal Astronomical Society*, **64**, 561–586.

JACKSON, J., NORRIS, R. & YOUNGSON, J. 1996. The structural evolution of active fault and fold systems in central Otago, New Zealand: evidence revealed by drainage patterns. *Journal of Structural Geology*, **18**, 217–234.

JAHANI, S., CALLOT, J. P., FRIZON DE LAMOTTE, D., LETOUZEY, J. & LETURMY, P. 2007. The salt diapers of the eastern Fars province (Zagros, Iran): a brief outline of their past and present. *In*: LACOMBE, O., LAVÉ, J., ROURE, F. & VERGÉS, J. (eds) *Thrust Belts and Foreland Basins*, Springer, Berlin, 289–308.

JAHANI, S., LETOUZEY, J., CALLOT, J. P. & FRIZON DE LAMOTTE, D. 2009. The eastern termination of the Zagros fold–thrust Belt (Iran): relationships between salt plugs, folding and faulting. *Tectonics*, doi: 10.1029/2008TC002426.

KADINSKI-CADE, K. & BARAZANGI, M. 1982. Seismotectonics of southern Iran: the Oman line. *Tectonics*, **1**, 389–412.

KENT, P. E. 1979. The emergent Hormuz salt plugs of (Zagros mountains) Southern Iran. *Journal of Petroleum Geology*, **2**, 117–144.

LETOUZEY, J., SHERKATI, S., MENGUS, J. M., MOTIEI, H., EHSANI, M., AHMADRIA, A. & RUDKIEWICZ, J. L. 2002. A regional structural interpretation of the Zagros Mountain Belt in Northern Fars and High Zagros (SW Iran). *AAPG Annual Meeting 2002* (Abstract).

LETURMY, P., LUCAZEAU, F. & BARAZANGI, F. 2003. Dynamic interactions between the Gulf of Guinea passive margin and the Congo river drainage basin. Part I: morphology and mass balance. *Journal of Geophysical Research*, **108**, 2383, doi: 10.1029/2002JB001927.

MCQUARRIE, N. 2004. Crustal scale geometry of the Zagros fold–thrust belt, Iran. *Journal of Structural Geology*, **26**, 519–535.

MCQUILLAN, H. 1991. The role of basement tectonics in the control of sedimentary facies, structural patterns and salt plug emplacements in the Zagros fold belt of southwest Iran. *Journal of SE Asian Earth Sciences*, **5**, 453–463.

MERRITS, D. J., VINCENT, K. R. & WOHL, E. E. 1994. Long river profiles, tectonism, and eustasy: a guide to interpreting fluvial terraces. *Journal of Geophysical Research*, **99**, 14031–14050.

MITRA, S. & MOUNT, S. V. 1998. Foreland basement-involved structures. *AAPG Bulletin*, **82**, 70–109.

MOITEI, H. 1993. *Stratigraphy of Zagros*. Publications of the Geological Survey of Iran [In Farsi].

MOLINARO, M., GUEZOU, J. C., LETURMY, P., FRIZON DE LAMOTTE, D. & ESHRAGHI, S. A. 2004. The origin of changes in structural style across the Bandar Abbas syntaxis, SE Zagros (Iran). *Marine and Petroleum Geology*, **21**, 735–752.

MOLINARO, M., LETURMY, P., GUEZOU, J. C. & FRIZON DE LAMOTTE, D. 2005. The structure and kinematics of the south-eastern Zagros fold–thrust belt, Iran: from thin-skinned to thick-skinned tectonics. *Tectonics*, **24**, TC3007, doi: 10.1029/2004TC001633.

MOLNAR, P. & CHEN, W. P. 1983. Focal depths of intracontinental and intraplate earthquakes and their implications for the thermal and mechanical properties of the lithosphere. *Journal of Geophysical Research*, **88**, 4183–4214.

MOUTHEREAU, F., LACOMBE, O., TENSI, J., BELLAHSEN, N., KARGAR, S. & AMROUCH, K. 2007. Mechanical constraints on the development of the Zagros folded belt (Fars). *In*: LACOMBE, O., LAVÉ, J., ROURE, F. & VERGÉS, J. (eds) *Thrust Belts and Foreland Basins from Fold Kinematics to Hydrocarbon Systems*. Frontiers in Earth Sciences. Springer, Berlin, 247–266.

NARR, W. & SUPPE, J. 1994. Kinematics of basement-involved compressive structures. *American Journal of Science*, **294**, 802–860.

OBERLANDER, T. M. 1965. *The Zagros Streams: A New Interpretation of Transverse Drainage in an Orogenic Zone*. Syracuse University, Geographical Series, **1**.

O'BRIEN, C. A. E. 1957. Salt diapirism in south Persia. *Geologie en Mijnbouw*, **19**, 357–376.

OUCHI, S. 1985. Response of alluvial rivers to slow active tectonic movement. *Geological Society of America Bulletin*, **96**, 504–515.

RAMSEY, L. A., WALKER, R. T. & JACKSON, J. 2008. Fold evolution and drainage development in the Zagros mountains of Fars province, SE Iran. *Basin Research*, **20**, 23–48.

RICOU, L. E. 1974. *L'etude géologique de la région de Neyriz (Zagros iranien) et l'évolution structurale des Zagrides*. PhD thesis, Université Paris-Sud, Orsay.

SEPEHR, M. & COSGROVE, J. W. 2004. Structural framework of the Zagros Fold–Thrust Belt, Iran. *Marine and Petroleum Geology*, **21**, 829–843.

SEPEHR, M., COSGROVE, J. & MOIENI, M. 2006. The impact of cover rock rheology on the style of folding in the Zagros fold–thrust belt. *Tectonophysics*, **427**, 265–281.

SHARLAND, P. R., ARCHER, R. *ET AL.* 2001. Arabian plate sequence stratigraphy. *In*: GeoArabia Special Publication, **2**, 371.

SHERKATI, S. & LETOUZEY, J. 2004. Variation of structural style and basin evolution in the central Zagros

(Izeh zone and Dezful Embayment), Iran. *Marine and Petroleum Geology*, **21**, 535–554.

SHERKATI, S., MOLINARO, M., FRIZON DE LAMOTTE, D. & LETOUZEY, J. 2005. Detachment folding in the Central and Eastern Zagros fold-belt (Iran): salt mobility, multiple detachments and late basement control. *Journal of Structural Geology*, **27**, 1680–1696.

SHERKATI, S., LETOUZEY, J. & FRIZON DE LAMOTTE, D. 2006. Central Zagros fold–thrust belt (Iran): new insights from seismic data, field observation and sandbox modelling. *Tectonics*, **25**, doi: 10.1029/2004TC001766.

SKLAR, L. & DIETRICH, W. E. 1988. River longitudinal profiles and bedrock incision models: Stream power and the influence of sediment supply. *In*: TINKLER, K. J. & WOHL, E. E. (eds) *Rivers over Rock: Fluvial Processes in Bedrock Channels*. American Geophysical Union, Geophysical Monograph, **107**, 237–260.

SNYDER, N. P., WHIPPLE, K. X., TUCKER, G. E. & MERRITTS, D. J. 2000. Landscape response to tectonic forcing: Digital elevation model analysis of stream profiles in the Mendocino triple junction region, Northern California. *Geological Society of America Bulletin*, **112**, 1250–1263.

STAMPFLI, G. 2000. Tethyan oceans. *In*: BOZKURT, E., WINCHESTER, J. A. & PIPER, J. D. A. (eds) *Tectonics and Magmatism in Turkey and the Surrounding Area*. Geological Society, London, Special Publications, **173**, 1–23.

TALEBIAN, M. & JACKSON, J. 2004. A reappraisal of earthquake focal mechanisms and active shortening in the Zagros mountains of Iran. *Geophysical Journal International*, **156**, 506–526.

TCHALENKO, J. S. & BRAUD, J. 1974. Seismicity and structure of the Zagros (Iran): the Main Recent Fault between 33° and 35° N. *Philosophical Transactions of the Royal Society London*, **1262**, 1–25.

VAN DER BEEK, P., CHAMPEL, B. & MUGNIER, J.-L. 2002. Control of detachment dip on drainage development in regions of active fault-propagation folding. *Geology*, **30**, 471–474.

VERGES, J., CASCIELLO, E. *ET AL.* 2007. Cover and basement structure across the Iranian Zagros Fold Belt from Fars to Lurestan Provinces. International Symposium on Middle East Basin Evolution, Paris, December 2007.

VERNANT, P., NILFOROUSHAN, F. *ET AL.* 2004. Contemporary crustal deformation and plate kinematics in the Middle East constrained by GPS measurements in Iran and Northern Oman. *Geophysical Journal International*, **157**, 381–398.

WHIPPLE, K. X. & TUCKER, G. E. 1999. Dynamics of the stream-power river incision for height limits of mountain ranges, landscape response timescales, and research model. *Journal of Geophysical Research*, **104**, 17661–17674.

WOODWARD, N. B. 1988. Primary and secondary basement controls on thrust sheet geometries. *In*: SCHMIDT, C. J. & PERRY, W. J., JR. (eds) *Interaction of the Rocky Mountain Foreland and the Cordilleran Thrust Belt*. Geological Society of America, Memoris, **171**, 353–366.

ZIEGLER, M. A. 2001. Late Permian to Holocene paleofacies evolution of the Arabian plate and its hydrocarbon occurrences. *GeoArabia*, **6**, 445–504.

A study of fold characteristics and deformation style using the evolution of the land surface: Zagros Simply Folded Belt, Iran

C. M. BURBERRY[1,2]*, J. W. COSGROVE[2] & J.-G. LIU[2]

[1]*Energy & Geoscience Institute, University of Utah, 423 Wakara Way,*
Salt Lake City, UT 84102, USA

[2]*Department of Earth Science and Engineering, Imperial College London,*
London, SW7 2AZ, UK

*Corresponding author (e-mail: cburberry@egi.utah.edu)

Abstract: Deformation styles within a fold–thrust belt can be understood in terms of the spatial organization and geometry of the fold structures. In young fold–thrust belts such as the Zagros, this geometry is reflected topographically by concordant landform morphology. Thus, the distribution of deformation structures can be characterized using satellite image analysis, digital elevation models, the drainage network and geomorphological indicators. The two distinct fold types considered in this study (fault-bend folds and detachment folds) both trending NW–SE, interact with streams flowing NE–SW from the High Zagros Mountains into the Persian Gulf. Multiple abandoned stream channels cross fault-bend folds related to deep-seated thrust faults. In contrast, detachment folds, which propagate laterally relatively rapidly, are characterized by diverted major stream channels and dendritic minor channels at the fold tips. Thus these two fold types can be differentiated on the basis of their geometry (fault-bend folds, being long, linear and asymmetrical, can be distinguished from detachment folds, which tend to be shorter and symmetrical) and on their associated geomorphological structures. The spatial organization of these structures in the Zagros Simply Folded Belt indicates that deformation is the result of the interaction of footwall collapse and the associated formation of long, linear fault-bend folds, and serial folding characterized by relatively short periclinal folds. Footwall collapse occurs first, followed by serial folding to the NE (i.e. in the hanging wall of the fault-bend folds), often on higher detachments within the sediment pile.

Fold–thrust belts form in the frontal regions of major contractional orogens, and are frequently important hydrocarbon provinces. Deformation within a fold–thrust belt varies both along and across its strike, on scales of tens to hundreds of kilometres. Variation in deformation style is related to the distribution of major basement faults, the rheological profile of the cover rocks and the depth of the detachment horizon, usually evaporites or over-pressured shales. (Sepehr & Cosgrove 2004, 2007; Sepehr *et al.* 2006; Sherkati *et al.* 2006). Different deformation styles may also be seen in units separated by a detachment horizon, or with different rheologies (Wiltschko & Chapple 1977; Nickelsen 1988; Sepehr & Cosgrove 2007).

In this paper, the geometry and geomorphological attributes of the surface folds and the variation in drainage patterns have been used to investigate the distribution of deformation styles and the sequence of deformation within the Zagros Simply Folded Belt. Fold–thrust belts were previously considered to be dominated by detachment folds (Sattarzadeh 1997; Vita-Finzi 2001, 2005). However, seismic studies, neotectonic and geomorphological investigations of the Zagros Belt reveal the presence of numerous fault-bend folds (Berberian & Qorashi 1994; Talebian & Jackson 2004). In addition, published cross-sections of the Zagros (McQuarrie 2004; Sepehr & Cosgrove 2004; Sherkati & Letouzey 2004; Sherkati *et al.* 2006; Carruba *et al.* 2007; Mouthereau *et al.* 2007; Stephenson *et al.* 2007) and theoretical studies (Suppe & Connors 1999) indicate that fold types other than detachment and fault-bend folds can also develop in compressional orogens (Fig. 1). Two such fold types are asymmetric detachment folds with thrusts cutting the forelimb, also known as faulted detachment folds, and fault-propagation folds, strongly asymmetric folds developed above a propagating blind fault tip (Rodgers & Rizer 1981; Mitra 2002; McQuarrie 2004; Sherkati & Letouzey 2004; Sherkati *et al.* 2005; Sepehr *et al.* 2006). Fault-bend folds may also develop from fault-propagation folds (Cosgrove & Ameen 2000).

As noted above, in fold–thrust belts the fold style is found to be a function of the stratigraphy, depth and the regional tectonic setting. Competing buckling or faulting instabilities determine which

From: LETURMY, P. & ROBIN, C. (eds) *Tectonic and Stratigraphic Evolution of Zagros and Makran during the Mesozoic–Cenozoic*. Geological Society, London, Special Publications, **330**, 139–154.
DOI: 10.1144/SP330.8   0305-8719/10/$15.00 © The Geological Society of London 2010.

(a)

(b)   **(c)**

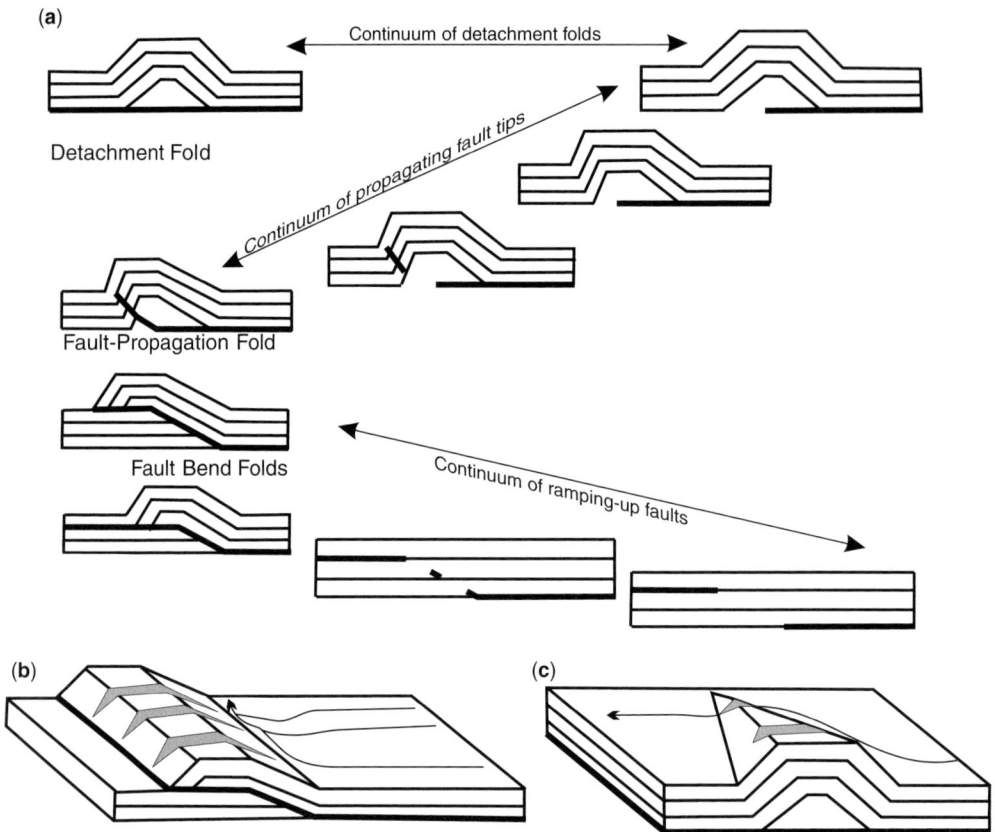

Fig. 1. (**a**) Continua of possible fold geometries in cross-section, including detailed models of the distinct geometries considered in this study, showing characteristic aspect ratio, asymmetry and drainage diversion; (**b**) fault-bend fold with a high aspect ratio and long hinge length showing wind gaps and defeated streams diverted parallel to the fold hinge line; (**c**) detachment fold with low aspect ratio and short hinge length showing a wind gap in the centre of the structure and a water gap at the end.

fold style develops, thus distinct fault-bend, fault-propagation and detachment folds may all occur in the same stratigraphy (Jamison 1992). The two distinct fold types considered in this study (fault-bend folds and detachment folds) have characteristic 3D fold geometries and amplification histories. Detachment folds have a low aspect ratio (length:width ratio) and near-perfect symmetry, whereas fault-bend folds have high aspect ratios and a more asymmetric profile (Sattarzadeh 1997; Cosgrove & Ameen 2000; Blanc *et al.* 2003). Exceptions to this do occur. For example, where detachment folds interact and link, the aspect ratio is increased (Sattarzadeh *et al.* 2000; Mitra 2002). In these instances, it is not easy to differentiate between detachment folds and fault-propagation folds (Jamison 1987). Fault-bend folds, however, can be identified from the characteristic broad, asymmetric hinges and long hinge lengths (Suppe 1983).

Periclinal detachment folds have a characteristic amplification pattern predicted from the theoretical analyses of folding, which varies in time and space and is reflected in the fold geometry. As these folds amplify, lateral propagation of the fold tips occurs (Mercier *et al.* 2007), thus at a specific point on the fold the uplift rate will vary with time and at a particular time the uplift rate will vary along the fold hinge (Cobbold 1976; Summers 1979). As the fold amplifies and begins to lock up, continued shortening may be accommodated by the development of thrust faults in the forelimb, creating asymmetric folds. In contrast, fault-bend folds are formed by uplift above a ramp in a thrust fault and the lateral extent of the fault determines the length of the fold hinge line (Butler 1992). It has been demonstrated that lateral propagation of fault tips and the associated folds can occur (Medwedeff 1992; Delcaillau *et al.* 1998;

Champel *et al.* 2002), but in regions where fault tips interact, this lateral propagation is restricted and faults are seen to maintain near-constant lengths throughout their evolution (Walsh *et al.* 2002). Significant lateral propagation is not noted in the study area, thus the amplification histories of the two fold-types can be distinguished by their interaction with drainage patterns. In models of fault-bend folding where the fold asymmetry does not increase with time, also assumed in this study, the uplift rate of the fold crest is relatively uniform along the length of the fold hinge line (Wilkerson *et al.* 1991) but decreases above the fault tips as displacement on the fault dies out (Azor *et al.* 2002).

Landscapes are often considered to be linear systems with characteristic response times (Beaumont *et al.* 2000). In fact, a complex feedback system operates, which modifies this simple interaction. Elevation changes related to rifting or to folding drive the evolution of the landscape (Cowie *et al.* 2006; Whittaker *et al.* 2007). Drainage systems adapt to changes in the surface slope, recording fold growth and evolution (Jackson *et al.* 1996). Therefore, the river network is a good indicator of the landform organization, as the points of lowest altitude in the system are embedded into this network. Structural patterns can be seen from statistical analyses of the network (Cudennec & Fouad 2006). Long-term folding produces topographic features that can affect drainage. Streams are often diverted, frequently with associated ponding of sediment. In regions of active folding, asymmetric patterns in river incision and abandoned alluvial fans also occur (Walker 2006). The relative ages of folds and their associated thrusts within a fold–thrust belt can be gained from an examination of the sinuosity of their foreland edge when viewed from above using aerial or satellite images. The amount of incision from stream channels or other subaerial erosion processes produces a fold or mountain front which, when viewed from above, becomes more sinuous with increasing exposure age (Burbank & Anderson 2001).

Wind and water gaps are produced as streams interact with growing folds (Jackson *et al.* 1996; Burbank *et al.* 1999). Water gaps occur where fold uplift rate is slow relative to the stream downcutting rate; for example, at the end of detachment folds. Wind gaps are caused by abandonment of the stream channel where the uplift rate is higher; that is, nearer the central point of a detachment fold. In contrast, the uplift rate along the hinge of a fault-bend fold is relatively uniform and depends on the uplift rate of the underlying block. Thus, streams will either keep pace with uplift, in which case water gaps will occur along the length of the fold, or the stream will be deflected and entrained by the orogen-parallel drainage fabric, resulting in wind gaps along the fold (Burbank & Pinter 1999; Burbank *et al.* 1999; Hovius 2000). This current growth of folds in the Zagros Simply Folded Belt is apparent from the deflected and upwarped Roman canal that crosses the Shaur anticline (Lees 1955).

Based on the above differences in geometry and amplification rate recorded in the drainage patterns, this study aims to demonstrate the value of remotely sensed images in the identification of different fold types and deformation styles in inaccessible areas such as the Zagros of Iran. In addition, the use of the fold-front sinuosity index provides a measure of the relative age of these structures.

Regional geology of the Zagros orogenic belt

The Zagros orogenic belt formed as a result of the collision between the Arabian craton and the Iranian plate (Fig. 2). The belt can be divided into the Urumieh–Dokhtar Magmatic arc, Sanandaj–Sirjan zone and the Simply Folded Belt (Alavi 1994). The Simply Folded Belt, which is situated between the High Zagros Fault and the Zagros Frontal Fault, consists of periclinal, *en echelon* flexural slip folds, which change from gentle and open in the SW to closed and locally overturned in the NE (Alavi 1994). Marked changes in the type and distribution of deformation occur along the belt and these changes coincide with important basement faults such as the north–south-trending Kazerun Fault (Sattarzadeh *et al.* 2000; Bahroudi & Talbot 2003; Walpersdorf 2006; Sepehr & Cosgrove 2007). Since the onset of collision in the Late Cretaceous, the deformation front has migrated from the suture to the current position at present just offshore of the Zagros Frontal Fault, as demonstrated by migrating depocentres deduced from the study of the sedimentary cover sequence and the location of recent earthquake epicentres (Alavi 1994; Talebian & Jackson 2004).

The 13 km thick sediment pile that rests on the Precambrian basement can be divided into the lower mobile group (Hormuz Salt), the competent group, the upper mobile group (Miocene evaporites) and the incompetent group (Plio-Pleistocene clastic deposits). The Precambrian Hormuz formation forms the lower detachment, and other important detachment horizons are the Triassic Dashtak Formation and the Miocene Gachsaran Formation (Alavi 2004; Sepehr & Cosgrove 2007). In the Zagros belt, the resistant Oligo-Miocene Asmari limestone and the Cretaceous Bangestan group form the whaleback anticlines currently seen at the surface (Blanc *et al.* 2003; McQuarrie 2004).

Active deformation in the Zagros is dominated by folding above the various subsurface

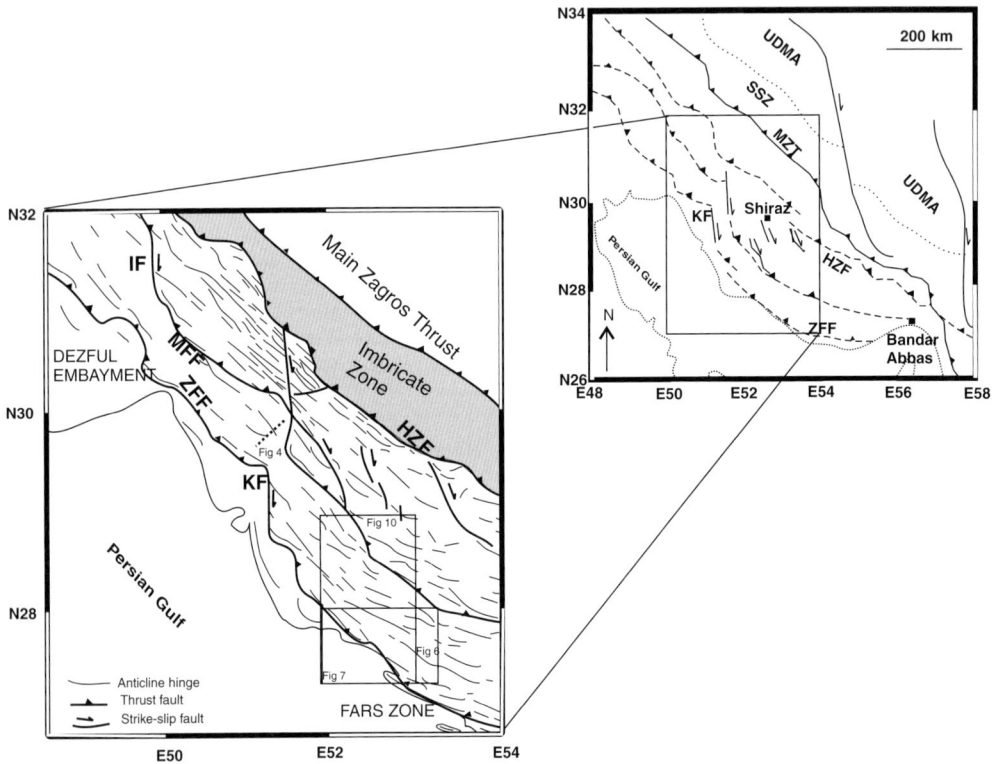

Fig. 2. The setting of the Zagros orogenic belt (Molinaro et al. 2005). The High Zagros Fault (HZF) forms the NE limit of the Simply Folded Belt, and the Zagros Frontal Fault (ZFF) marks the edge of the surface expression of deformation. UDMA, Urumieh–Dohktar Magmatic Arc; SSZ, Sanandaj–Sirjan Zone; KF, Kazerun Fault. The locations of the cross-section in Figure 4, as well as boxed areas corresponding to Figures 6, 7 and 10, are marked.

detachments, and the variable aspect ratio of the anticlines implies that some are fault bend folds and some are detachment folds (Blanc et al. 2003). The Zagros anticlines show variable vergence, implying the presence of a thick, low-friction décollement beneath the belt. In the Fars zone, this is inferred to be the Hormuz Salt. No outcropping Hormuz Salt is seen in the Dezful Embayment, but the low-angle belt taper, rounded folds and variable vergence imply that there is a thick, low-friction ductile detachment below these folds (Carruba et al. 2007). The Gachsaran Salt acts as a second ductile décollement in many parts of the Fars zone.

Periclinal detachment folds are predominantly found in the Zagros Simply Folded Belt. However, deformation is also accommodated on blind thrusts on the Gachsaran and Hormuz evaporite layers (Berberian 1995) that core asymmetric anticlines. The geometry of many long linear anticlines with a high aspect ratio along the Mountain Front Fault and the Zagros Frontal Fault implies thrust involvement (Sepehr & Cosgrove 2005). Some thrust-cored folds are inferred to form above reactivated normal

faults, which fold the weak, ductile Hormuz Salt. Deformation translates to the competent group above as a brittle thrust (Sattarzadeh et al. 2000). In contrast, linear, en echelon arrays of folds with aspect ratios similar to the detachments folds are linked to basement strike-slip faults (Cosgrove & Ameen 2000). Finally, the influence of salt diapirs sourced from the basal Hormuz Salt may produce domal features. Faulting and folding are at least in part coseismic, but it is probable that much of the seismic deformation is not accommodated on major emergent faults, but rather by episodic fold growth (Berberian & Qorashi 1994).

Fold geometry in the Zagros varies significantly, both horizontally and vertically, as a function of the mechanics of the detachment zone and the presence of intermediate décollements (Sherkati et al. 2006). Duplexes and disharmonic folding form where the detachment surface is deep and whaleback anticlines form where the décollement is shallow; for example, on the secondary Triassic Dasktak evaporite décollement in the Fars zone. Variation in folding styles can be seen in different segments of the belt, attributed to variable depocentres as major north–south-trending faults dissect the belt (Sepehr & Cosgrove 2007). Shortening rates are also found to vary across these faults (Walpersdorf 2006). Across-strike variation in the amount of shortening implies that most deformation is located at the foreland of the belt (Oveisi et al. 2007) in contrast to the more widely scattered seismicity, with folds such as the Mand Anticline near the Kazerun Fault accommodating 20–35% of the shortening across the belt at this point.

Derivation of the drainage network and recognition of landform organization

The Zagros Simply Folded Belt is eminently suitable for satellite image analysis because of low vegetation coverage and well-exposed outcrops in an arid climate. For this study, a section of the Fars region, to the east of the Kazerun fault zone (Fig. 2) has been chosen. In this region, the folds are capped by the Oligocene Asmari limestone and the Hormuz Salt, Dashtak Formation and Gachsaran Formation are the major décollement horizons (Bahroudi & Koyi 2003; Alavi 2004). All folds investigated in this study have been taken from the Fars region to minimize any variation in deformation style caused by variations in the rheological profile of the cover succession along the strike of the belt. Within the study area, immediately adjacent to the Kazerun fault zone, the shorter wavelength anticlines detach on the Triassic Dashtak evaporites (Sepehr & Cosgrove 2007).

The two distinct fold types can be recognized based on characteristic drainage pattern diversions, aspect ratio, hinge length and symmetry. Landsat-7 ETM+ images were used to delimit fold shapes and to mark locations and diversions of the stream network. Colour composites were created by combining Landsat-7 ETM+ bands 5, 3, 1 as red, green and blue, which highlights variation in lithology and suppresses vegetation (Drury 2001). Carbonate layers appear in pink, here picking out the folds that are capped by the Asmari limestone. Fold axial traces were identified from the closures of the top Asmari marker. Major stream channels have also been identified. In areas of confusion, the concentration of vegetation around streams visible on a 432-RGB image (Sabins 1996) was used to separate stream beds from other features. The visualization of colour composites was optimized via contrast enhancement techniques. Interactive contrast enhancement relies on manually adjusting the distribution of the data so that the full value range of the image is used. Another technique removes colour bias by balancing the colour range between different bands using an algorithm known as balanced contrast enhancement technique (Liu 1991). To clarify the relationship between lithology and topography and to identify wind and water gaps in the folds, the enhanced satellite images were draped over a digital elevation model (DEM) covering the study area N27–30°, E51–54°. Lastly, a synthetic river network was generated from the DEM tiles. The algorithm used picks out relative topographic lows and constructs a downslope flow network from these points.

Mio-Pliocene deformation in the Zagros Simply Folded Belt has altered the drainage network by defeating and diverting streams (Molinaro et al. 2005). Water gaps where the stream has cut through the end of the growing fold, and wind gaps where the stream system has been defeated and diverted around the growing fold, were identified by comparing satellite images and the artificial stream network. The artificially generated stream network was compared with the manually picked network. Errors occurred in the region of namakirs (salt glaciers) where the software interpreted the relative topographic low as a stream channel. Once this comparison was completed, a map of the locations of stream diversions and abandoned channels was generated.

Folds identified from satellite image analysis were categorized using geomorphological indices aspect ratio and fold symmetry index (Fig. 3) calculated from fold width, hinge length and forelimb width measured off a contour map, generated from the DEM tiles. Both aspect ratio (hinge length : fold width) and fold symmetry index (width of forelimb : half-width) can be used to separate fold types and were calculated for each fold. A perfectly symmetric fold will have a symmetry index of unity, whereas an asymmetric fold will have a lower

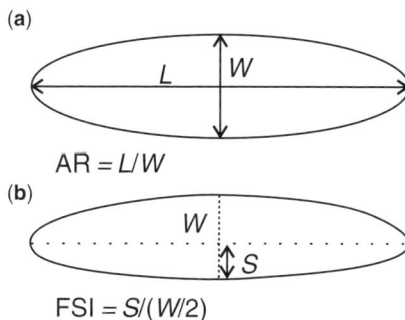

Fig. 3. Geomorphological indices used to differentiate fold types: (**a**) aspect ratio (L/W); (**b**) fold symmetry index [$S/(W/2)$]. S is defined as the width of the shorter limb, and for a symmetrical detachment fold the fold symmetry index value will be equal to unity.

(a)

| Mish | Khami | Razi | Anneh | Eshger | Zardrud | Sepidar | Darishk | Mokhtar |
| Anticline | Anticline | Anticline | Anticline | Anticline | Anticline | Anticline | Anticline | Anticline |

2 km

(b)

AR vs Hinge length (km)

(c)

SI vs Length (km)

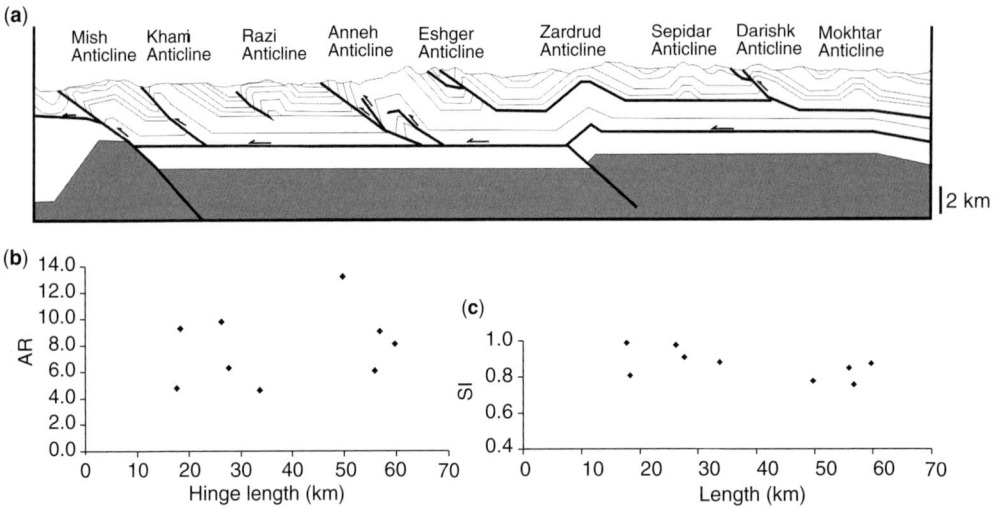

Fig. 4. Cross-section (**a**) of a region where the structure of the folds has been determined by fieldwork (Sepehr & Cosgrove 2007) and associated geomorphological indices (**b**; AR, **c**; SI) measured from remote data. The folds in the section can be clearly separated into those likely to be fault-bend folds and those likely to be detachment folds. It should be noted that the major basement-involved thrust is the Mountain Front Fault. Detachment folds form behind the thrust, on multiple detachments within the sediment pile. Smaller folds detach on shallower horizons.

symmetry and therefore a lower value of the fold symmetry index will be calculated.

The method was tested on a region in the Dezful Embayment, where a structural cross-section has been constructed from field evidence (Sepehr *et al.* 2006). Four folds in this section (the Mish, Khami, Anneh and Eshger anticlines) have hinge lines longer than 50 km and corresponding aspect ratios indicate that these folds are fault-bend folds as in the cross-section (Fig. 4). Consideration of the symmetry index allows the folds to be clearly divided into two end-member groups, fault-bend folds and detachment folds. Erosion of the land surface will increase the symmetry of the fold structures, as steep slopes are eroded more rapidly than gentle slopes. Over time, therefore, the slopes of the landforms will diverge from the original concordant morphology and tend to a symmetric structure. For folds where there is an independent field dataset of structural dips, the symmetry is shown to be consistently overestimated by about 5%; that is, landform symmetry is higher (fold symmetry index closer to unity) than the symmetry indicated by the dipping beds. This implies that, despite the expected influence of surface processes, the observed variation in fold symmetry index can still be used as an indicator of fold type.

In addition, the relative age of the structures was derived, using the sinuosity of the front of the fold when viewed from above as a proxy. This index is the ratio of the actual mountain front length to the fold hinge length (Fig. 5). A higher sinuosity

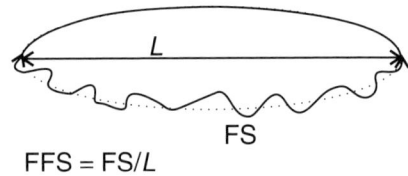

$$FFS = FS/L$$

Fig. 5. A third geomorphological index, fold front sinuosity (FFS), used as a proxy for the age of the fold structure. FFS is calculated as the actual length of the fold front/fold length. Older structures, having been exposed to subaerial weathering processes for longer, will have a higher FFS.

measurement implies a heavily incised fold front and thus an older fold structure (Azor *et al.* 2002; Silva *et al.* 2003).

Development of the present-day drainage network

The pre-folding drainage system is likely to have been a dendritic system (Summerfield 1991) with the dominant direction of flow approximately from NE to SW (i.e. towards the Persian Gulf). The development of NW–SE-trending folds has caused the present drainage pattern to deviate significantly from the original pattern, producing drainage basins that mimic the fold trends. Present-day major streams display a trellis morphology (Fig. 6a),

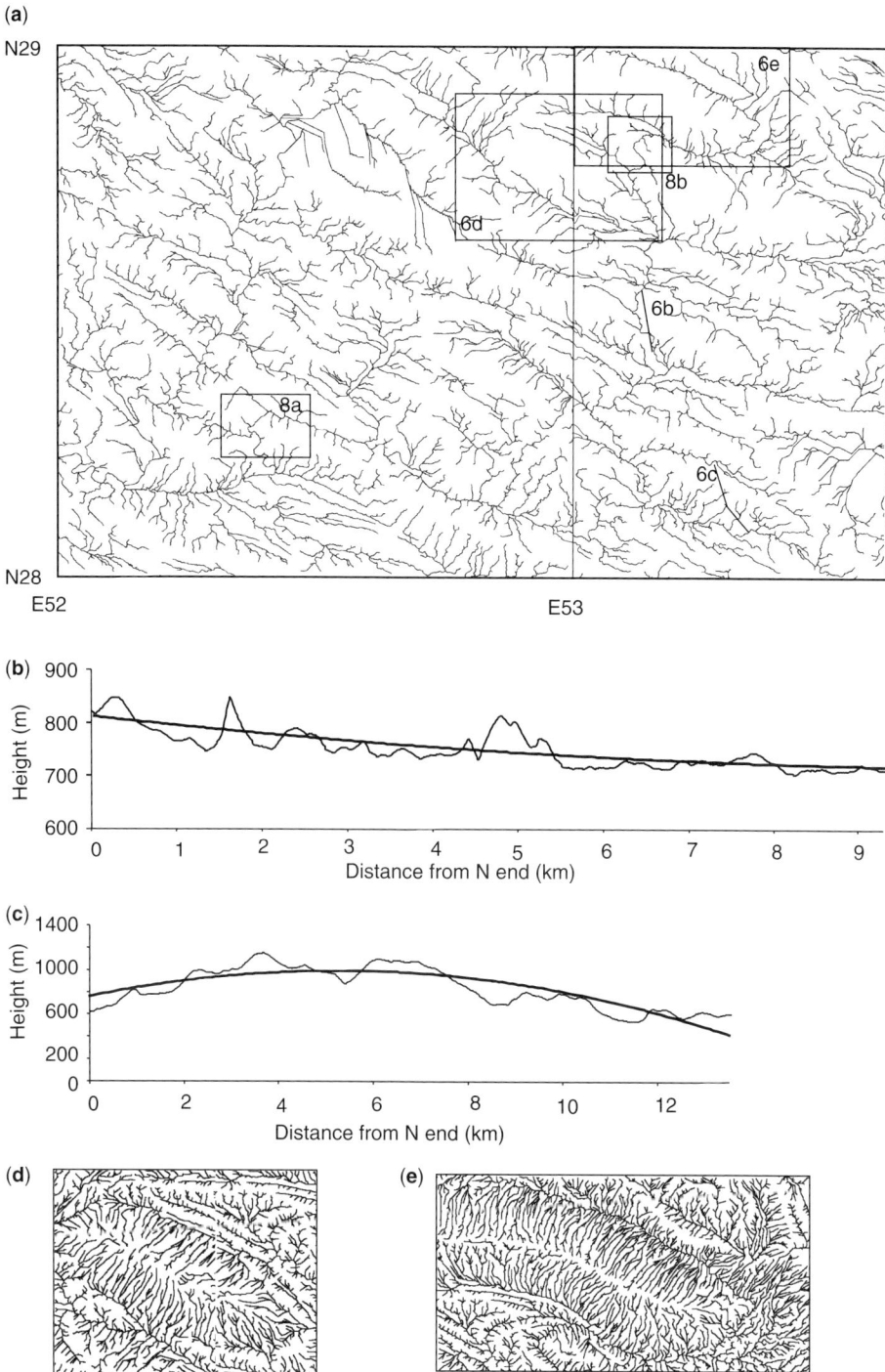

Fig. 6. (**a**) Drainage patterns in the Zagros Simply Folded Belt showing high-order streams displaying a trellis pattern with the prominent direction paralleling the fold hinge lines, with detail of (**b**) the typical long profile of a water gap, (**c**) the typical long profile of a wind gap, (**d**) low-order streams showing a symmetrical pattern around a detachment fold and (**e**) low-order streams showing an asymmetric pattern around a fault-bend fold. The location of Figure 8a, b is also marked.

with the prominent stream direction parallel to the fold hinges. This drainage morphology modifies the area and shape of the drainage basins. In some instances, stream channels can be seen to bend around the nose of a fold structure. In many of these cases, the downcutting, erosive power of the stream has been greater than the uplift rate of the fold and a water gap has been incised at the nose of the fold. Water gaps still display the concave-up profile of an equilibrium stream channel (Fig. 6b). In contrast, wind gaps may be located on the fold crest, adjacent to the point of significant stream diversion. Wind gaps display a characteristic convex-up profile, as the abandoned stream channel has been uplifted with continuing fold amplification (Fig. 6c). Wind gaps are found at the crests of detachment folds because of the high uplift rate in these areas of the fold, and along the full length of fault-bend folds, as a result of these folds having a relatively uniform rate of uplift.

Single fold shapes are delimited by minor streams and some asymmetry in the stream network around these landforms can also be recognized. Symmetric stream network patterns are taken as representative of detachment folds (Fig. 6d) and asymmetric stream network patterns are found across fault-bend folds (Fig. 6e). The watershed of the minor streams is the crest of the landform, in this study taken to be a proxy for the position of the fold crest. Each fold has three distinct morphometric domains within the drainage pattern: a dendritic region towards each fold tip, and a central section where streams flow parallel to dip (Fig. 6d, e). The small dendritic tip regions imply limited incorporation of undeformed material into the fold structure, and therefore limited lateral propagation, even for fault-related folds (Walsh *et al.* 2002).

Spatially, wind and water gaps are not uniformly distributed across the Simply Folded Belt. In the hinterland of the belt, and in the Dezful Embayment,

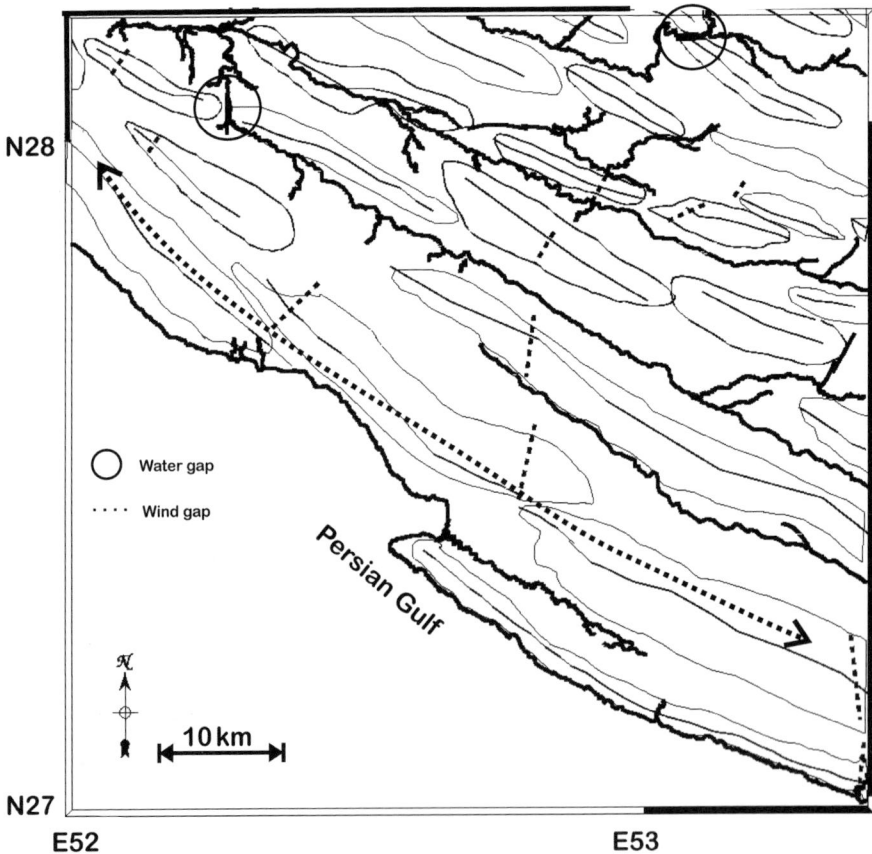

Fig. 7. Spatial arrangement of wind and water gaps in the Zagros Simply Folded Belt. Long structures crossed by multiple wind gaps (dotted lines) are considered to be fault-bend folds (e.g. the main fold structures along the coast of the Persian Gulf, the present-day deformation front). In contrast, folds with shorter hinge lines and water gaps at the tips (circled) are considered to be detachment folds.

streams are generally observed to be transverse to the structures (Oberlander 1985). In the foreland, in the Fars region, distinct fold structures are crossed by multiple wind gaps (Fig. 7) and many folds have water gaps at the tips. Multiple wind gaps crossing a long fold crest imply uniform and rapid uplift along this fold structure, causing abandonment and diversion of multiple stream channels. This in turn implies that these folds are fault-bend folds. Fault-bend folds can therefore be characterized by the presence of multiple cross-cutting wind gaps along the hinge-line of the structure.

Drainage can be further distorted from a general dendritic pattern by folds linking with neighbouring structures. In this case, the river system appears to cut across the axial trace of a fold but closer inspection reveals that there is a topographic low at the point where the two fold noses interact (Fig. 8a).

The same process has been documented in other actively deforming regions (Jackson *et al.* 1996). A river that has been diverted around a growing fold is unlikely to be further diverted given the low dips and correspondingly low amplification rates of the folds in the linking region, and the morphology is markedly different from that of a typical water gap (Fig. 8b).

Geomorphological analysis of the fold structures

Fold axial traces are seen to trend consistently NW–SE in this section of the Simply Folded Belt. In some areas, the fold geometry is altered by the proximity of major thrust faults. A large variation can be found in fold hinge length, with the shortest folds having a hinge length around 10 km and the longest around 100 km.

Each fold was classified using the geomorphological indices aspect ratio and fold symmetry index, together with the hinge length and the locations of wind and water gaps. The fold population in the Zagros can be broadly separated into two groups based on the relationship between length and aspect ratio. When these indices are compared with the length of the hinge line, folds with a hinge length greater than about 60 km show a linear relationship between aspect ratio and the hinge length (Fig. 9a) Many of these folds are also distinctly asymmetric (Fig. 9b). Detachment folds are seen to cluster with aspect ratios less than 10, and hinge lengths less than 60 km (Fig. 9a) and can also be seen to have a symmetry close to unity (Fig. 9b). These folds do not display the linear relationship between aspect ratio and fold length. These two datasets are interpreted as representative of two distinct folding processes operating in the region, forming the two distinct fold types. Fault-bend folds, having no lateral propagation of the fault tips, as described above, show a linear relationship between the aspect ratio and the hinge length. A distinct cluster of detachment folds can be clearly marked, but the spread in the dataset implies that this cluster may include the other fold geometries as discussed above. The linear relationship given for the fault-bend fold dataset could potentially be extended into the clustered region, raising questions about the minimum and maximum geometrically and kinematically viable hinge lengths for the two distinct fold types that have yet to be satisfactorily answered.

Asymmetric detachment folds and fault-propagation folds form another significant class of folds, which can be identified by the combination of low symmetry and a short hinge length. The dominant vergence of the asymmetric detachment

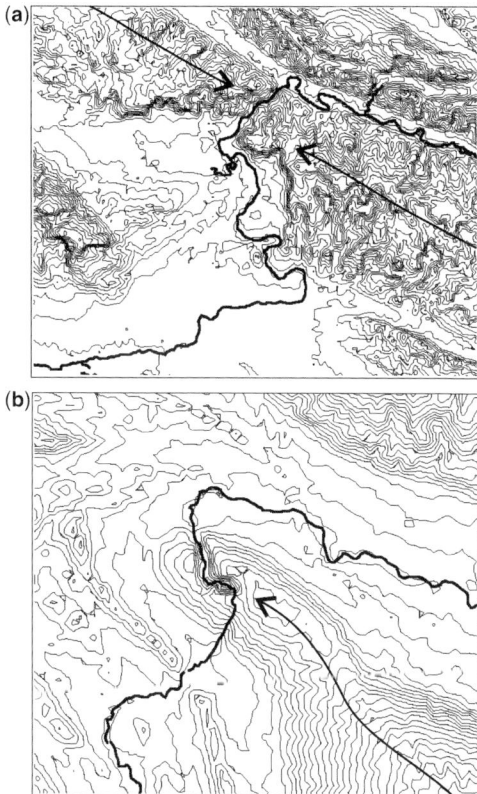

Fig. 8. (**a**) Contour map of linked fold structures in the Zagros Simply Folded Belt, showing interaction of fold tips and a river channel incised at the topographic low, in contrast to (**b**) the narrow canyon of a water gap created by river incision at the tip of a growing fold. The locations of these figures are shown in Figure 6.

(a)

(b)

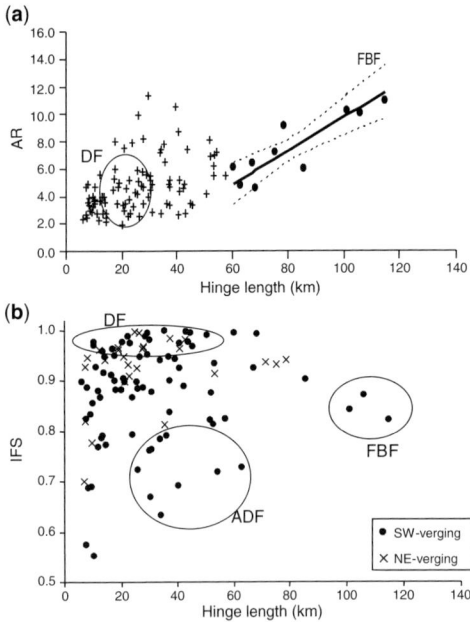

Fig. 9. The separation of detachment and fault-bend folds by comparison between hinge length and (**a**) aspect ratio and (**b**) fold symmetry index. The large cluster of data marked at low hinge lengths corresponds to detachment folds (DF). The data points along the trendline (with 95% confidence interval marked) are identified as fault-bend folds (FBF). A third cluster marked by short hinge lengths but low symmetry is likely to be asymmetric detachment folds (ADF). Folds generally verge SW but those structures with a NE vergence are marked by crosses in (b).

folds is SW, although a minority verge to the NW. These folds often have a lower symmetry than the fault-bend folds and can be identified as a third, less well-defined cluster on a graph of symmetry v. hinge length (Fig. 9b). Structures with an aspect ratio close to unity are folds related to salt domes rather than true detachment folds.

Spatial arrangement and relative age of fold structures

Folds with short hinge lengths and low aspect ratios were found to be spatially coincident with the locations of diverted streams and water gaps at the fold tips and were inferred to be detachment folds. In turn, anomalously long, high aspect ratio, asymmetric folds identified from geomorphological indices are also the fold structures cross-cut by multiple wind gaps, confirming that these are fault-bend folds. The majority of long fault-bend folds are

found to lie directly above the outcropping major thrusts within the Zagros Simply Folded Belt, with the largest, prominent fault-bend folds along the Zagros Frontal Fault, the terrestrial deformation front (Figs 1 & 10). Detachment folds cluster between the major thrusts, and more asymmetric detachment folds are often found close to the fault-bend folds.

The spatial variation in the locations of fault-bend folds and detachment folds is also spatially coincident with the location of the major topographic steps in the landscape (McQuillan 1973; Sattarzadeh 1997; McQuarrie 2004; Mouthereau *et al.* 2006). In the longwave topographic signal, the major steps can be seen corresponding to the Mountain Front Fault and the Zagros Frontal Fault. Fault-bend folds are seen to lie at the locations of the major steps, suggesting that these folds are related to the major thrust faults.

Folds in the eastern section of the study area, away from the influence of lateral ramps in the thrust faults and major north–south-trending faults such as the Kazerun Fault, show a broadly linear decrease in relative age from the High Zagros Fault to the deformation front of the belt (Fig. 11). Superimposed on this trend are a number of fold structures with older relative ages than expected. These lie close to the positions of the thrust faults and are interpreted to indicate the structures formed by footwall collapse as the deformation front migrates to the foreland. A peak in the fold front sinuosity (FFS) between the major thrust faults corresponds to a branch line thrust of the rear thrust, the High Zagros Fault.

The wavelength of a detachment fold structure is determined by the depth to the detachment, with lower wavelength folds detaching on higher horizons within the sedimentary succession. In the Zagros Simply Folded Belt, the detachment fold amplitude structures are consistently an order of magnitude smaller than the fault bend fold amplitudes, indicating that these folds are detaching on an intermediate décollement level rather than the major basal décollement. Variation in shortening occurs above and below intermediate detachment horizons and detachment folds decrease in amplitude towards the SW with each discrete block. A forward-developing sequence of fold initiation is thus inferred, with a corresponding decrease in the percentage shortening and a general drop in elevation. In addition, the FFS index indicates that serial detachment folding is occurring in discrete blocks between thrusts that have formed sequentially by footwall collapse.

Movement up a thrust ramp creates a fault-bend fold. Compressive stresses build up, leading to serial folding in the cover behind the fault-bend fold. Voids are filled by the ductile detachment material.

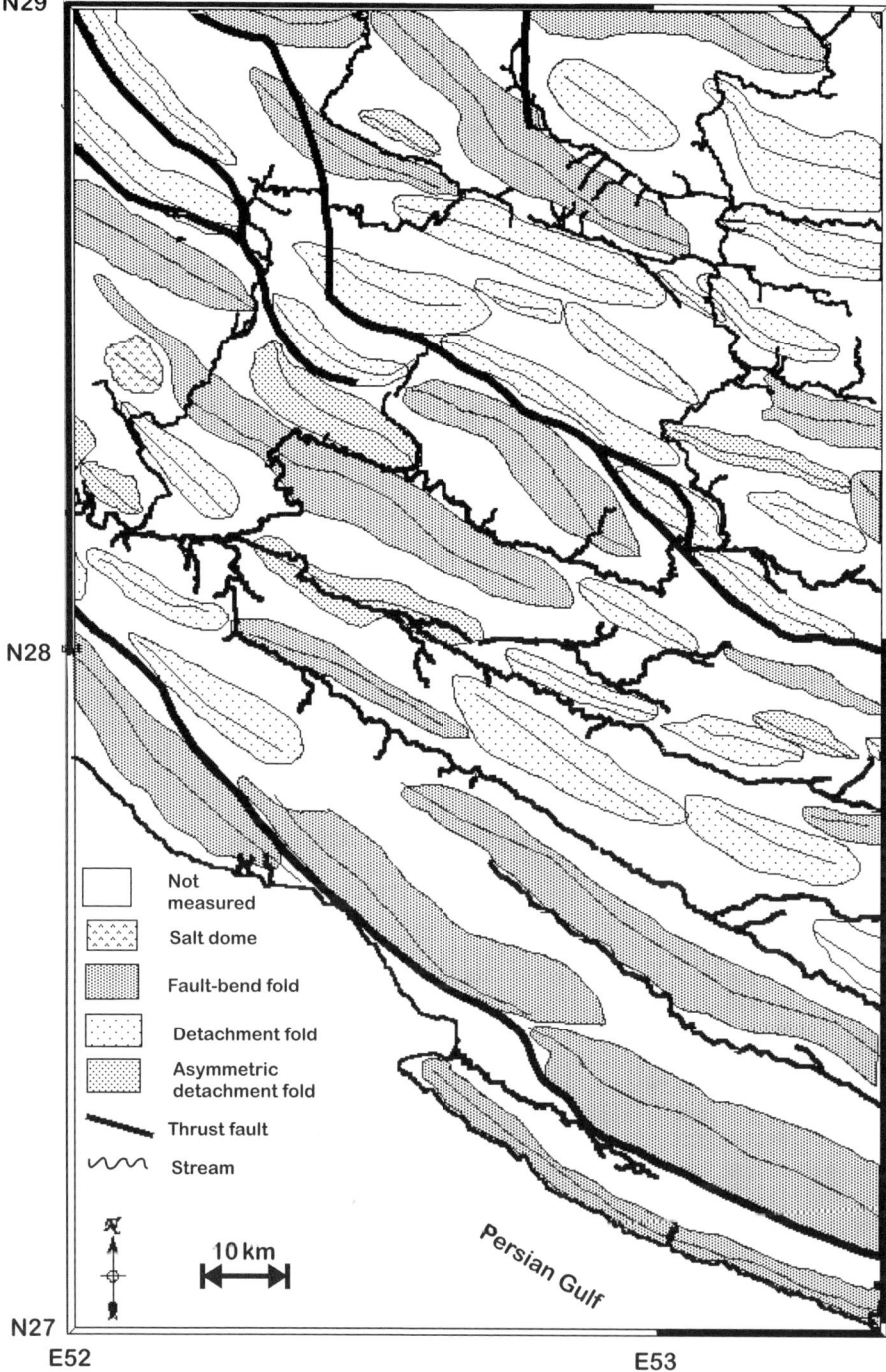

Fig. 10. Spatial arrangement of the end-member fold types in the Zagros Simply Folded Belt: (**a**) shaded by fold type and (**b**) showing both an actual view of the surface and a 3D section of the belt. Long fault-bend folds are found associated with the major thrust faults and detachment folds are found in discrete blocks between these thrust slices. The base image is a Landsat-7 ETM+ false colour composite, draped over an SRTM DEM. The full shaded map has been published by Burberry *et al.* (2007).

(b)

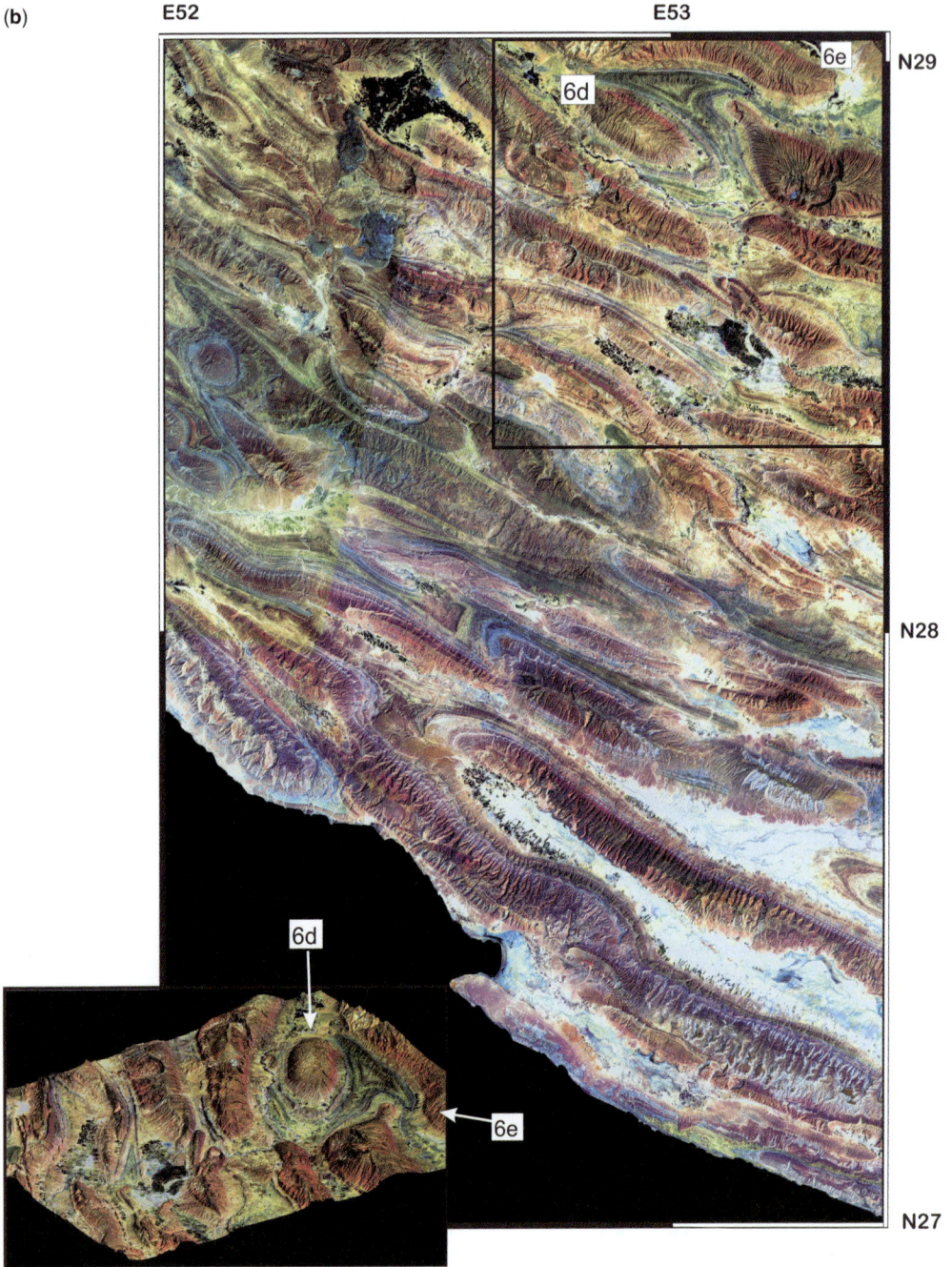

Fig. 10.

The presence of asymmetric, detachment folds implies that additional shortening may have been accommodated by thrust development in the core of detachment folds. Eventually, deformation of the block requires stresses in excess of those required to create a new footwall thrust. The original thrust is abandoned, the footwall collapses and the process repeats (Fig. 12).

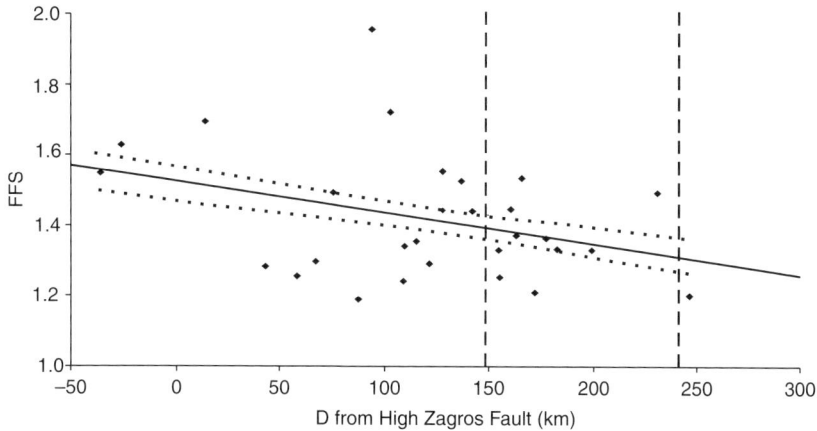

Fig. 11. Relative age of fold structures within the Zagros Simply Folded Belt, decreasing from the hinterland towards the foreland. Anomalously old folds are located close to the thrust faults (marked by vertical dashed lines) implying the interaction of footwall collapse and subsequent serial folding as the deformation front migrates to the SW.

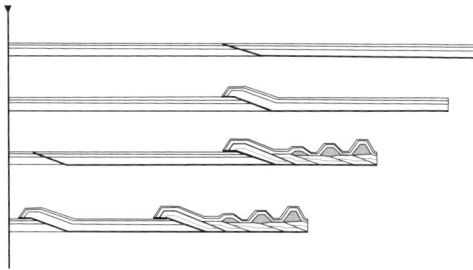

Fig. 12. Model of combined brittle and ductile deformation by both footwall collapse and serial folding in the Zagros Simply Folded Belt. Detachment folds form on higher décollement horizons within the sediment pile, whereas major thrust faults detach on the basal Hormuz Salt formation. Voids created by the formation of detachment folds are filled by flow of material from the décollement horizon, and the lower portion of the sediment pile must also undergo shortening, potentially on imbricate thrust faults, so that the section balances.

Conclusions

It has been proposed that the different fold types that develop in a fold–thrust belt (primarily detachment folds and fault-bend folds) can be identified on satellite images using their geometry (specifically aspect ratio and symmetry) and interaction with drainage patterns. Fault-bend folds are characteristically asymmetric with long hinge lines and high aspect ratios, and detachment folds are characteristically symmetric and have shorter hinge lines and lower aspect ratios. Remote sensing and the analysis of satellite images have been used to distinguish between fold types in the Zagros fold–thrust

belt and help elucidate the sequence of their development, using the geomorphological indices described above. This study demonstrates that the two distinct fold types described can be differentiated with some statistical confidence.

In addition, the spatial organization of the two distinct fold types (three ridges of fault-bend folds, separated by two regions of detachment folds) reveals the process by which the deformation front has migrated to the SW, away from the collision zone. The topographic ridges representing the fault-bend folds, which are spatially associated with the major thrust faults (the High Zagros Fault, the Mountain Front Fault and the Zagros Frontal Fault), formed sequentially by footwall collapse as the deformation front migrated towards the foreland basin.

It has previously been suggested that the detachment folds within a fold–thrust belt form in a non-synchronous serial fashion linked to the propagation of deformation away from the collision zone towards the foreland (Blay *et al.* 1977; Vita-Finzi 2005). This study reveals that, within the Zagros Simply Folded Belt, the migration of deformation involves the interaction of both footwall collapse and the simplified serial folding model. As the fault-bend folds amplified, propagation of the thrusts became progressively more difficult and the compressive stress rose. This resulted in the development of detachment folds by serial folding on higher detachment horizons in the cover behind the fault-bend folds, which acted as buttress structures.

This research was supported by a Janet Watson Scholarship from the Department of Earth Science and Engineering, Imperial College London. We thank J. Carter

(Imperial College) for assistance with the statistical analyses of the data. Satellite data used in this study were
sourced from the Global Landcover Facility, www.
landcover.org. The DEM tiles used in this study were
sourced from the NASA Land Processes Distributed
Active Archive Centre (ftp://e0srp01u.ecs.nasa.gov/
srtm/). We gratefully acknowledge the comments of
P. Van der Beek and E. A. Keller, whose contributions
significantly improved the original manuscript.

References

ALAVI, M. 1994. Tectonics of the Zagros orogenic belt of
Iran: new data and interpretations. *Tectonophysics*,
229, 211–238.

ALAVI, M. 2004. Regional stratigraphy of the Zagros
fold–thrust belt of Iran and its proforeland evolution.
American Journal of Science, **304**, 1–20.

AZOR, A., KELLER, E. A. & YEATS, R. S. 2002. Geomorphic indicators of active fold growth: South
Mountain–Oak Ridge anticline, Ventura basin,
southern California. *Geological Society of America
Bulletin*, **114**, 745–753.

BAHROUDI, A. & KOYI, H. A. 2003. Effect of spatial distribution of Hormuz salt on deformation style in the
Zagros fold and thrust belt: an analogue modelling
approach. *Journal of the Geological Society, London*,
160, 719–733.

BAHROUDI, A. & TALBOT, C. J. 2003. The configuration
of the basement beneath the Zagros Basin. *Journal of
Petroleum Geology*, **26**, 257–282.

BEAUMONT, C., KOOI, H. & WILLETT, S. D. 2000.
Coupled tectonic–surface process models with appli-
cations to rifted margins and collisional orogens. *In*:
SUMMERFIELD, M. A. (ed.) *Geomorphology and
Global Tectonics*. Wiley, New York, 29–55.

BERBERIAN, M. 1995. Master blind thrust faults hidden
under the Zagros Folds—active basement tectonics
and surface morphotectonics. *Tectonophysics*, **241**,
193–224.

BERBERIAN, M. & QORASHI, M. 1994. Coseismic
fault-related folding during the South Golbaf earthquake of November 20, 1989, in southeast Iran.
Geology, **22**, 531–534.

BLANC, E. J. P., ALLEN, M. B., INGER, S. & HASSANI, H.
2003. Structural styles in the Zagros Simple Folded
Zone, Iran. *Journal of the Geological Society,
London*, **160**, 401–412.

BLAY, P., COSGROVE, J. W. & SUMMERS, J. M. 1977.
An experimental investigation of the development of
structures in multilayers under the influence of
gravity. *Journal of the Geological Society, London*,
133, 329–342.

BURBANK, D. W. & ANDERSON, R. S. 2001. *Tectonic
Geomorphology*. Blackwell Science, Oxford.

BURBANK, D. W. & PINTER, N. 1999. Landscape evolution: the interactions of tectonics and surface processes. *Basin Research*, **11**, 1–6.

BURBANK, D. W., MCLEAN, J. K., BULLEN, M.,
ABDRAKHMATOV, K. Y. & MILLER, M. M. 1999.
Partitioning of intermontane basins by thrust-related
folding, Tien Shan, Kyrgyzstan. *Basin Research*, **11**,
75–92.

BURBERRY, C. M., COSGROVE, J. W. & LIU, J. G. 2007.
Stream network characteristics used to infer the
distribution of fold types in the Zagros Simply
Folded Belt, Iran. *Journal of Maps, Student Edition*,
2007, 32–45.

BUTLER, R. W. H. 1992. Structural evolution of the
western Chartreuse fold and thrust system, NW
French Subalpine chains. *In*: MCCLAY, K. R.
(ed.) *Thrust Tectonics*. Chapman & Hall, London,
287–298.

CARRUBA, S., PEROTTI, C. R., BUONAGURO, R.,
CALABRO, R., CARPI, R. & NAINI, M. 2007. Structural pattern of the Zagros fold and thrust belt in the
Dezful Embayment (SW Iran). *In*: MAZZOLI, S. &
BUTLER, R. W. H. (eds) *Styles of Continental Contraction*. Geological Society of America, Special Papers,
414, 11–32.

CHAMPEL, B., VAN DER BEEK, P., MUGNIER, J. L. &
LETURMY, P. 2002. Growth and lateral propagation
of fault-related folds in the Siwaliks of western
Nepal: rates, mechanisms, and geomorphic signature.
Journal of Geophysical Research—Solid Earth, **107**,
doi: 10.1029/2001JB000578.

COBBOLD, P. R. 1976. Fold shapes as functions of progressive strain. *Philosophical Transactions of the
Royal Society of London, Series A*, **283**, 129–138.

COSGROVE, J. & AMEEN, M. 2000. A comparison of the
geometry, spatial organisation and fracture patterns
associated with forced folds and buckle folds. *In*:
COSGROVE, J. W. & AMEEN, M. S. (eds) *Forced
Folds and Fractures*. Geological Society, London,
Special Publications, **169**, 7–21.

COWIE, P. A., ATTAL, M., TUCKER, G. E., WHITTAKER,
A. C., NAYLOR, M., GANAS, A. & ROBERTS, G. P.
2006. Investigating the surface process response to
fault interaction and linkage using a numerical modelling approach. *Basin Research*, **18**, 231–266.

CUDENNEC, C. & FOUAD, Y. 2006. Structural patterns
in river network organization at both infra- and supra-
basin levels: the case of a granitic relief. *Earth Surface
Processes and Landforms*, **31**, 369–381.

DELCAILLAU, B., DEFFONTAINES, B. ET AL. 1998. Morphotectonic evidence from lateral propagation of an
active frontal fold; Pakuashan anticline, foothills of
Taiwan. *Geomorphology*, **24**, 263–290.

DRURY, S. 2001. *Image Interpretation in Geology*. Nelson
Thornes, Cheltenham.

HOVIUS, N. 2000. Macroscale process systems of mountain belt erosion. *In*: SUMMERFIELD, M. A. (ed.) *Geomorphology and Global Tectonics*. Wiley, New York,
77–105.

JACKSON, J., NORRIS, R. & YOUNGSON, J. 1996. The
structural evolution of active fault and fold systems
in central Otago, New Zealand: evidence revealed by
drainage patterns. *Journal of Structural Geology*, **18**,
217–234.

JAMISON, W. R. 1987. Geometric analysis of fold development in overthrust terranes. *Journal of Structural
Geology*, **9**, 207–219.

JAMISON, W. R. 1992. Stress controls on fold–thrust style.
In: MCCLAY, K. R. (ed.) *Thrust Tectonics*. Chapman
& Hall, London, 155–164.

LEES, G. M. 1955. Recent earth movements in the Middle
East. *Geologische Rundschau*, **43**, 221–226.

LIU, J. G. 1991. *Digital image processing for automatic lithological mapping using Landsat TM imagery*. Imperial College, London.

MCQUARRIE, N. 2004. Crustal scale geometry of the Zagros fold–thrust belt, Iran. *Journal of Structural Geology*, **26**, 519–535.

MCQUILLAN, H. 1973. Small-scale fracture density in Asmari-Formation of Southwest Iran and its relation to bed thickness and structural setting. *AAPG Bulletin*, **57**, 2367–2385.

MEDWEDEFF, D. A. 1992. Geometry and kinematics of an active laterally propagating wedge thrust, Wheeler Ridge, California. *In*: MITRA, S. & FISHER, G. W. (eds) *Structural Geology of Fold–Thrust Belts*, John Hopkins University Press, Baltimore, 3–28.

MERCIER, E., RAFANI, S. & AHMADI, R. 2007. Fold kinematics in fold–thrust belts: the hinge migration question. *In*: LACOMBE, O., LAVE, J., ROURE, F. & VERGES, J. (eds) *Thrust Belts and Foreland Basins*. Springer, Berlin, 135–147.

MITRA, S. 2002. Structural models of faulted detachment folds. *AAPG Bulletin*, **86**, 1673–1694.

MOLINARO, M., LETURMY, P., GUEZOU, J. C., FRIZON DE LAMOTTE, D. & ESHRAGHI, S. A. 2005. The structure and kinematics of the southeastern Zagros fold–thrust belt, Iran: from thin-skinned to thick-skinned tectonics. *Tectonics*, **24**, doi: 10.1029/2004TC001633.

MOUTHEREAU, F., LACOMBE, O. & MEYER, B. 2006. The Zagros folded belt (Fars, Iran): constraints from topography and critical wedge modelling. *Geophysical Journal International*, **165**, 336–356.

MOUTHEREAU, F., LACOMBE, O., TENSI, J., BELLAH-SEN, N., KARGAR, S. & AMROUCH, K. 2007. Mechanical constraints on the development of the Zagros Folded Belt (Fars). *In*: LACOMBE, O., LAVÉ, J., ROURE, F. & VERGÉS, J. (eds) *Thrust Belts and Foreland Basins from Fold Kinematics to Hydrocarbon Systems*. Frontiers in Earth Sciences. Springer, Berlin, 247–266.

NICKELSEN, R. P. 1988. Structural evolution of folded thrusts and duplexes on a first-order anticlinorium in the Valley and Ridge Province of Pennsylvania. *In*: MITRA, G. & WOJTAL, S. (eds) *Geometries and Mechanism of Thrusting, with Special Reference to the Appalachians*. Geological Society of America, Special Papers, **222**, 89.

OBERLANDER, T. M. 1985. Origin of drainage transverse to structures in orogens. Tectonic geomorphology. *Binghamton Symposia in Geomorphology*, **15**, 155.

OVEISI, B., LAVE, J. & VAN DER BEEK, P. A. 2007. Rates and processes of active folding evidenced by Pleistocene terraces at the central Zagros front, Iran. *In*: LACOMBE, O., LAVÉ, J., ROURE, F. & VERGÉS, J. (eds) *Thrust Belts and Foreland Basins*. Springer, Berlin, 267–287.

RODGERS, D. A. & RIZER, W. D. 1981. Deformation and secondary faulting near the leading edge of a thrust fault. *In*: MCCLAY, K. R. & PRICE, N. J. (eds) *Thrust and Nappe Tectonics*. Geological Society, London, Special Publications, **9**, 65–77.

SABINS, F. F. 1996. *Remote Sensing Principles and Interpretation*. W. H. Freeman, New York.

SATTARZADEH, Y. 1997. *Active tectonics in the Zagros Mountains, Iran*. PhD thesis, Imperial College, London.

SATTARZADEH, Y., COSGROVE, J. & VITA-FINZI, C. 2000. The Interplay of faulting and folding during the evolution of the Zagros deformation belt. *In*: COSGROVE, J. W. & AMEEN, M. S. (eds) *Forced Folds and Fractures*. Geological Society, London, Special Publications, **169**, 187–196.

SEPEHR, M. & COSGROVE, J. W. 2004. Structural framework of the Zagros Fold–Thrust Belt, Iran. *Marine and Petroleum Geology*, **21**, 829–843.

SEPEHR, M. & COSGROVE, J. W. 2005. Role of the Kazerun Fault Zone in the formation and deformation of the Zagros Fold–Thrust Belt, Iran. *Tectonics*, **24**, doi: 10.1029/2004TC001725.

SEPEHR, M. & COSGROVE, J. W. 2007. The role of major fault zones in controlling the geometry and spatial organisation of structures in the Zagros Fold–Thrust Belt. *In*: RIES, A. C., BUTLER, R. W. H. & GRAHAM, R. H. (eds) *Deformation of the Continental Crust: The Legacy of Mike Coward*. Geological Society, London, Special Publications, **272**, 419–436.

SEPEHR, M., COSGROVE, J. W. & MOIENI, M. 2006. The impact of cover rock rheology on the style of folding in the Zagros fold–thrust belt. *Tectonophysics*, **427**, 265–281.

SHERKATI, S. & LETOUZEY, J. 2004. Variation of structural style and basin evolution in the central Zagros (Izeh zone and Dezful Embayment), Iran. *Marine and Petroleum Geology*, **21**, 535–554.

SHERKATI, S., MOLINARO, M., FRIZON DE LAMOTTE, D. & LETOUZEY, J. 2005. Detachment folding in the Central and Eastern Zagros fold–belt (Iran): salt mobility, multiple detachments and late basement control. *Journal of Structural Geology*, **27**, 1680–1696.

SHERKATI, S., LETOUZEY, J. & FRIZON DE LAMOTTE, D. 2006. Central Zagros fold–thrust belt (Iran): new insights from seismic data, field observation, and sandbox modeling. *Tectonics*, **25**, doi: 10.1029/2004TC001766.

SILVA, P. G., GOY, J. L., ZAZO, C. & BARDAJI, T. 2003. Fault-generated mountain fronts in southeast Spain: geomorphologic assessment of tectonic and seismic activity. *Geomorphology*, **50**, 203–225.

STEPHENSON, B., KOOPMAN, A., HILLGARTNER, H., MCQUILLAN, H., BOURNE, S., NOAD, J. & RAWNSLEY, K. 2007. Structural and stratigraphic controls on fold-related fracturing in the Zagros Mountains, Iran: implications for reservoir development. *In*: LONERGAN, L., JOLLY, R. J. H., RAWNSLEY, K. & SANDERSON, D. J. (eds) *Fractured Reservoirs*, Geological Society, London, Special Publications, **270**, 1–21.

SUMMERFIELD, M. A. 1991. *Global Geomorphology*. Prentice Hall, Harlow.

SUMMERS, J. M. 1979. *An experimental and theoretical investigation of multilayer fold development*. PhD thesis, Imperial College, London.

SUPPE, J. 1983. Geometry and kinematics of fault-bend folding. *American Journal of Science*, **283**, 684–721.

SUPPE, J. & CONNORS, C. 1999. Shear fault-bend and fault-propagation folding; new theory and examples. Geological Society of America, 1999 annual meeting.

Geological Society of America, Abstracts with Programs, **31**, 237.

TALEBIAN, M. & JACKSON, J. 2004. A reappraisal of earthquake focal mechanisms and active shortening in the Zagros Mountains of Iran. *Geological Journal International*, **156**, 506–526.

VITA-FINZI, C. 2001. Neotectonics at the Arabian plate margins. *Journal of Structural Geology*, **23**, 521–530.

VITA-FINZI, C. 2005. Serial deformation. *Proceedings of the Geologists' Association*, **116**, 293–300.

WALKER, R. T. 2006. A remote sensing study of active folding and faulting in southern Kerman province, S.E. Iran. *Journal of Structural Geology*, **28**, 654–668.

WALPERSDORF, A. 2006. Difference in the GPS deformation pattern of north and central Zagros (Iran). *Geophysical Journal International*, **167**, 1077–1088.

WALSH, J. J., NICOL, A. & CHILDS, C. 2002. An alternative model for the growth of faults. *Journal of Structural Geology*, **24**, 1669–1675.

WHITTAKER, A. C., COWIE, P. A., ATTAL, M., TUCKER, G. E. & ROBERTS, G. P. 2007. Bedrock channel adjustment to tectonic forcing: implications for predicting river incision rates. *Geology*, **35**, 103–106.

WILKERSON, M. S., MEDWEDEFF, D. A. & MARSHAK, S. 1991. Geometrical modeling of fault-related folds—a pseudo-3-dimensional approach. *Journal of Structural Geology*, **13**, 801–812.

WILTSCHKO, D. V. & CHAPPLE, W. M. 1977. Flow of weak rocks in Appalachian plateau folds. *AAPG Bulletin*, **61**, 653–670.

Structure of the Mountain Front Flexure along the Anaran anticline in the Pusht-e Kuh Arc (NW Zagros, Iran): insights from sand box models

H. EMAMI[1,2], J. VERGÉS[1]*, T. NALPAS[3], P. GILLESPIE[4], I. SHARP[5], R. KARPUZ[6], E. P. BLANC[7] & M. G. H. GOODARZI[8]

[1]*Group of Dynamic of Lithosphere (GDL), Institute of Earth Sciences 'Jaume Almera', CSIC, Lluís Solé i Sabarís s/n, Barcelona, 08028, Spain*

[2]*Statoil ASA, 15th Floor, 30 Bokharest Blg., 9th St., Bokharest Ave., Tehran 115137 Iran*

[3]*Géosciences Rennes (UMR 6118), Université de Rennes 1, Campus de Beaulieu, 35042 Rennes cedex, France*

[4]*Statoil, Tectonics and Structural Geology, Forushagen, Stavanger, Norway*

[5]*Statoil ASA, Technology and New Energy, Research Centre, Sandsliveien 90, Sandsli, NO-5020 Bergen, Norway*

[6]*OMV Exploration and Production, Vienna, Austria*

[7]*Statoil, Vækerø Bygg B311B, Drammensveien 261, N0246 Oslo, Norway*

[8]*National Iranian Oil Company, Exploration Directorate, Surface Geology, 1st Seoul St., Sheaikh Bahaie Ave., Tehran 19948-14695, Iran*

*Corresponding author (e-mail: jverges@ija.csic.es)

Abstract: The Mountain Frontal Flexure shows a single step along the front of the Pusht-e Kuh Arc with about 3 km of structural relief. This front has been interpreted as being formed by a basement monocline above a blind crustal-scale and low-angle thrust with a ramp–flat geometry (the ramp dips 12–15° towards the inner part of the orogen and cuts the entire crust). The Anaran anticline on top of the Mountain Frontal Flexure shows an irregular geometry in map view and consists of four segments with diverse directions of which the SE Anaran, the Central Anaran and the NW Dome are culminations. The North–South Anaran segment may form a linking zone developed during the rise and amplification of single culminations, the NW Dome and the Central Anaran, above the Mountain Frontal Flexure. The asymmetric Anaran anticline is characterized by the existence of multiple normal faults, some of them with significant dip-slip displacements of up to 1000 m. These faults limit grabens located along the crests of the anticline segments. Cross-cutting relationships show that the normal faults along the Central Anaran are older than along the North–South Anaran, reinforcing the temporal constraints on the later growth of this segment of the anticline. The geometry of the Anaran anticline is asymmetric with the subvertical forelimb very little exposed. This forelimb is cut above and below by a thrust system that seems to develop along the fold hinges. The lower thrust, with a ramp–flat geometry, carries the entire anticline towards the foreland on top of slightly deformed rocks in the footwall. The thrust flattens in the Gachsaran evaporitic level forming a typical triangular zone filled with evaporites, which produce a strong fold disharmony between the overburden (Passive Group) and the underlying rocks (Competent Group). The growth of the Anaran anticline lasted for about 6 Ma and was the consequence of detachment folding that was subsequently thrust, rotated and uplifted above the Mountain Frontal Flexure with coeval reactivation of earlier crestal layer-parallel extension normal faults to accommodate the large increase of structural relief between the foreland and the tectonic arc. Three main results from analogue modelling have been combined with field data to resolve the geometry of the Anaran anticline as well as its evolution: (1) a thickening of intermediate evaporites (Gachsaran Formation) is produced above the flat segment of the thrust carrying the anticline on top of foreland strata; (2) growth strata deposited in the adjacent syncline modify the geometry of the anticline by increasing the dip and the length of its forelimb; (3) coeval erosion to anticline growth, as well as thick growth strata deposition, increases fold amplification rather than foreland propagation of deformation. The proposed fold model may be applied to other anticlines on top of this major basement-related thrust, such as the Siah Kuh and Khaviz anticlines in the Pusht-e Kuh Arc and Dezful Embayment domains.

From: LETURMY, P. & ROBIN, C. (eds) *Tectonic and Stratigraphic Evolution of Zagros and Makran during the Mesozoic–Cenozoic.* Geological Society, London, Special Publications, **330**, 155–178.
DOI: 10.1144/SP330.9   0305-8719/10/$15.00 © The Geological Society of London 2010.

Most foreland fold-and-thrust belts show basement-involved structures, as has been classically recognized in the Rocky Mountains (e.g. Narr & Suppe 1994). These structures are commonly characterized by significant structural and topographic relief and usually by folding at the cover levels (Mitra & Mount 1998). Although there is an extensive literature on these compressive basement structures it is not straightforward to determine their geometry at depth or the amount of propagation of faulting in the cover. The typical profile of basement-involved structures is characterized by a long gently dipping backlimb (0–20°) that mimics the basement–cover contact and a shorter gently dipping to steep forelimb (30° to overturned) (Mitra & Mount 1998). The backlimbs of these structures are generally well imaged on seismic profiles whereas the forelimbs are generally poorly imaged, especially the upright ones. Basement-related thrusts have been given various names such as drape or forced folds

(Stearns 1978), upthrust folds (Foose *et al.* 1961; Berg 1962; Prucha *et al.* 1965), or fold thrusts (Berg 1962; Brown 1988; Stone 1993). The basement-involved structures can also be associated with a component of the strike-slip faulting. Mitra & Mount (1998) developed balanced geometric and kinematic models to interpret a wide variety of foreland basement structures.

The Zagros Fold Belt (also named the Simply Folded Belt) is characterized by a folded 12–14 km thick sedimentary cover deposited on the northeastern continental border of the Arabian plate (e.g. Falcon 1974; Colman-Sadd 1978). Its maximum width is about 280 km in the centre of the Fars Arc and 230 km in the Pusht-e Kuh Arc (Fig. 1). These two arcs constitute belts of anticlines showing a regular NW–SE trend and are limited by a major geo-flexure (Falcon 1961), variously named the Main Front Fault (Berberian 1995), the Mountain Front Flexure (McQuarrie 2004), and the Zagros

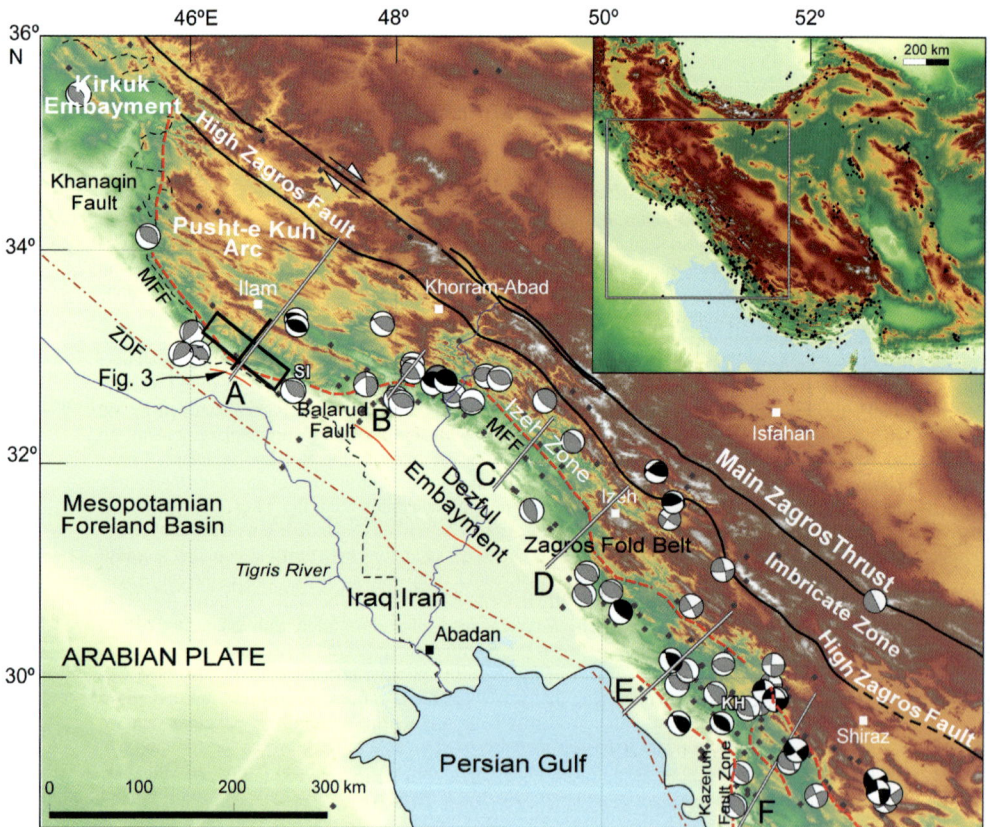

Fig. 1. Tectonic map of the Zagros Fold Belt showing the position and geometry of the Mountain Front Flexure (MFF). Earthquakes of M ≥ 5 are indicated by small black diamonds. Focal mechanisms from Talebian & Jackson (2004) are also shown, in black (M$_w$ ≥ 5.3) and grey (M$_w$ ≤ 5.3). Bold lines show the position of the morphotectonic transects of Figure 4 (A–F). Box shows the extent of the geological map depicting the Anaran anticline (Fig. 5). KH, Khavir anticline; SI, Siah Kuh anticline; ZDF, Zagros Deformation Front.

Frontal Fault (Sepehr & Cosgrove 2004). This structural and topographic front, which we will define as the Mountain Front Flexure (MFF in Fig. 1), has an irregular geometry that defines tectonic salients or arcs and re-entrants or embayments; these are, from SE to NW, the Fars Arc (Fars stratigraphic province), the Dezful Embayment (Khuzestan stratigraphic province), the Pusht-e Kuh Arc (Lurestan stratigraphic province), and the Kirkuk Embayment (Kurdistan in Iraq) (Fig. 1). The Mountain Front Flexure bounding the Pusht-e Kuh Arc has an east–west-trending segment along the Balarud Fault, a NW–SE frontal segment in which the Anaran anticline is located and an north–south segment along the Khanaqin Fault. The frontal segment of the Mountain Front Flexure is subparallel to the folds in the Pusht-e Kuh Arc whereas the Balarud Fault and the Khanaqin Fault represent deep oblique structures above which the NW–SE-trending anticlines of the arc are bent into the Dezful Embayment and the Kirkuk Embayment, respectively (Fig. 1).

The Anaran anticline is located along the frontal segment of the Mountain Front Flexure in the Pusht-e Kuh Arc and shows excellent and continuous outcrops. Nevertheless, the geometry of the Anaran anticline at depth as well as its relation to the Mountain Front Flexure is little known at present, although new seismic lines have been acquired to image the prolongation of the foreland to the NE and its connection with the Anaran anticline.

The objectives of this work are to determine the potential geometry of the Anaran anticline at depth and its relation to the Mountain Front Flexure using a combination of field geology and analogue modelling to help to understand the interaction of basement and cover structures, the geometry of normal faults in the cover, and the role of erosion and syntectonic sedimentation during the evolution of the anticline. The geometry at depth and the evolution of the anticline are important for oil exploration and could be applied to other anticlines located above the Mountain Front Flexure across the entire Zagros Fold Belt.

The stratigraphy of the Pusht-e Kuh Arc

The stratigraphy of the Lurestan Province consists of a 10–12 km thick succession that encompasses the Palaeozoic and Mesozoic Arabian passive margin deposits followed by the sediments corresponding to the long-lived Cenozoic Zagros foreland infilling (Fig. 2). This thick pile of sediments was probably deposited on top of the Proterozoic–Early Cambrian Hormuz evaporites (not salt), although this is not directly verified in the Lurestan Province.

Most of the stratigraphy described in this section is based on the work of James & Wynd (1965) and Colman-Sadd (1978).

The mechanical behaviour of the 10–12 km thick sedimentary pile in response to folding was discussed first by O'Brien (1950) and Dunnington (1968), and has been the focus of several recent papers (e.g. Sattarzadeh et al. 2000; Molinaro et al. 2004; Sherkati et al. 2005; Sepehr et al. 2006) with specific studies for Lurestan (Casciello et al. 2009; Farzipour-Saein et al. 2009). O'Brien (1950) and Dunnington (1968) used regional stratigraphy and divided the succession of the Zagros Fold Belt into five major structural–mechanical units, which, from bottom to top, are (Fig. 2) the Basement Group, the Lower Mobile Group, the Competent Group, the Upper Mobile Group and the Passive Group. At the bottom of this succession the Precambrian metamorphic basement forms the Basement Group. Above it, the Lower Mobile Group is formed by Late Proterozoic–Early Cambrian Hormuz evaporites. This stratigraphic layer is not confirmed below the Pusht-e Kuh Arc domain but the similar shape, trend and distribution of folding and shortening across the Zagros seems to support a weak layer at the basement–cover limit (see Vergés et al. 2009). The Competent Group has greatest stratigraphic thickness (it is almost 6 km thick), and consists of Palaeozoic deposits at the base followed by limestones, marls and evaporites from the Triassic to the Cretaceous Khami and Bangestan groups and up to the Miocene Asmari Formation. The Upper Mobile Group is mostly formed by the c. 800 m thick Gachsaran evaporites in the study region. The Passive Group consists of a 3–4 km thickness of foreland clastic deposits made up of the Agha Jari and Bakhtyari formations in the study area.

The Mountain Front Flexure

The Mountain Front Flexure is a major bend in the cover succession characterized by strong variations of topography and structural relief across it (Falcon 1961) that must correspond to a large basement blind thrust (Fig. 3). In addition, major seismic events are active in the basement, which may be related to the Mountain Front Flexure as observed in map view (Fig. 1) (Berberian 1995). The Mountain Front Flexure is responsible for the main tectonic divisions in the Zagros Fold Belt and their morphotectonic characteristics: low topography in the Kirkuk and Dezful embayments, high topography in the Pusht-e Kuh Arc and intermediate topography in the Fars Arc.

Falcon (1961) presented an integrated structural and topographic dataset from anticlines and

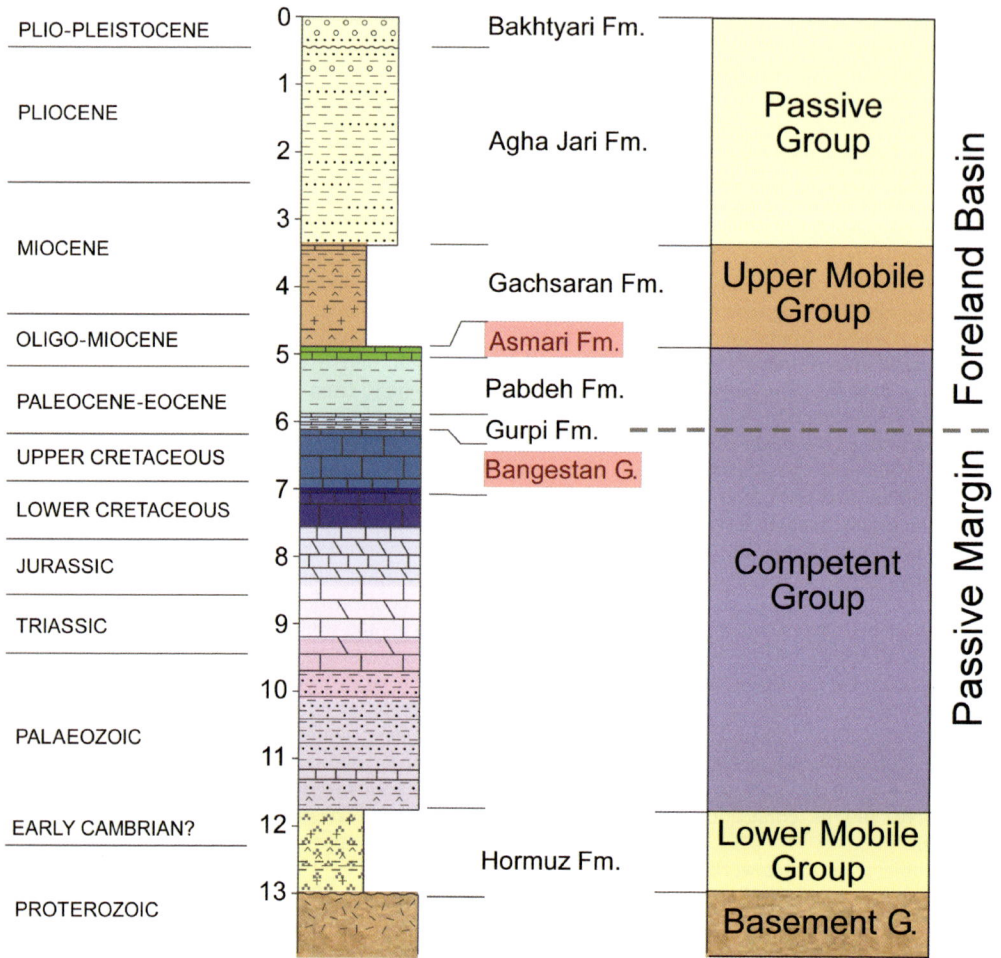

Fig. 2. Simplified stratigraphic column of the Lurestan Province, which forms the Pusht-e Kuh Arc, combined with mechanical strength based on O'Brien (1950) and Dunnington (1968) (vertical scale in km).

synclines across the Mountain Front Flexure, covering the entire Dezful Embayment, to quantify the amount of uplift of this basement-involved structure. More recently, Sherkati *et al.* (2006) presented five regional structural cross-sections that show the geometry of the Mountain Front Flexure in the Dezful Embayment and NW Fars Arc (their location is shown in Fig. 1). Here we present six transects from cross-sections in Figure 3 (transect A in Fig. 4), from Falcon (1961) (transect B in Fig. 4) and Sherkati *et al.* (2006) (transects C to F in Fig. 4) to better define the topographic and structural relief as well as the morphological variations along the Mountain Front Flexure (Fig. 4). In the transects three lines are marked: (1) the bold black line corresponds to the topographic profile; (2) the green line shows the folding at the level of the Asmari

Formation (based on structural cross-sections); (3) the red line connects the hinges of the synclines to show the amount of structural relief across the different steps that form the Mountain Front Flexure along strike.

Along transect A crossing the Pusht-e Kuh Arc the topography increases about 900 m in less than 10 km of horizontal distance and the structural relief increases about 3 km between the foreland in Iraq and the Pusht-e Kuh tectonic arc in Iran as determined by the position of the hinges of the synclines in both regions (Fig. 4). Along transect B crossing the Balarud Fault Zone the increase of topography is of about 700 m whereas the structural variation is 3.8 km (Fig. 4). Transect C crossing the boundary between the Dezful Embayment and the Izeh Zone shows an increase of topography of

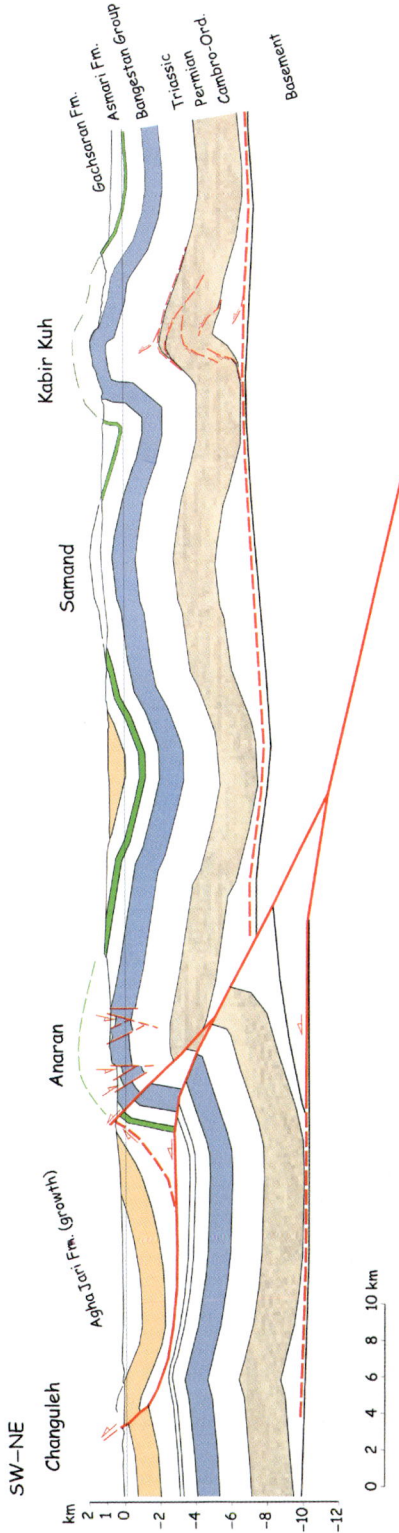

Fig. 3. Simplified cross-section of the front of the Pusht-e Kuh Arc showing the structure of the cover and basement. The basement thrust system has been constructed to fit the variations of topography and structural relief across the Mountain Front Flexure. The structure of the Anaran anticline and Mountain Front Flexure integrate the results obtained in this paper.

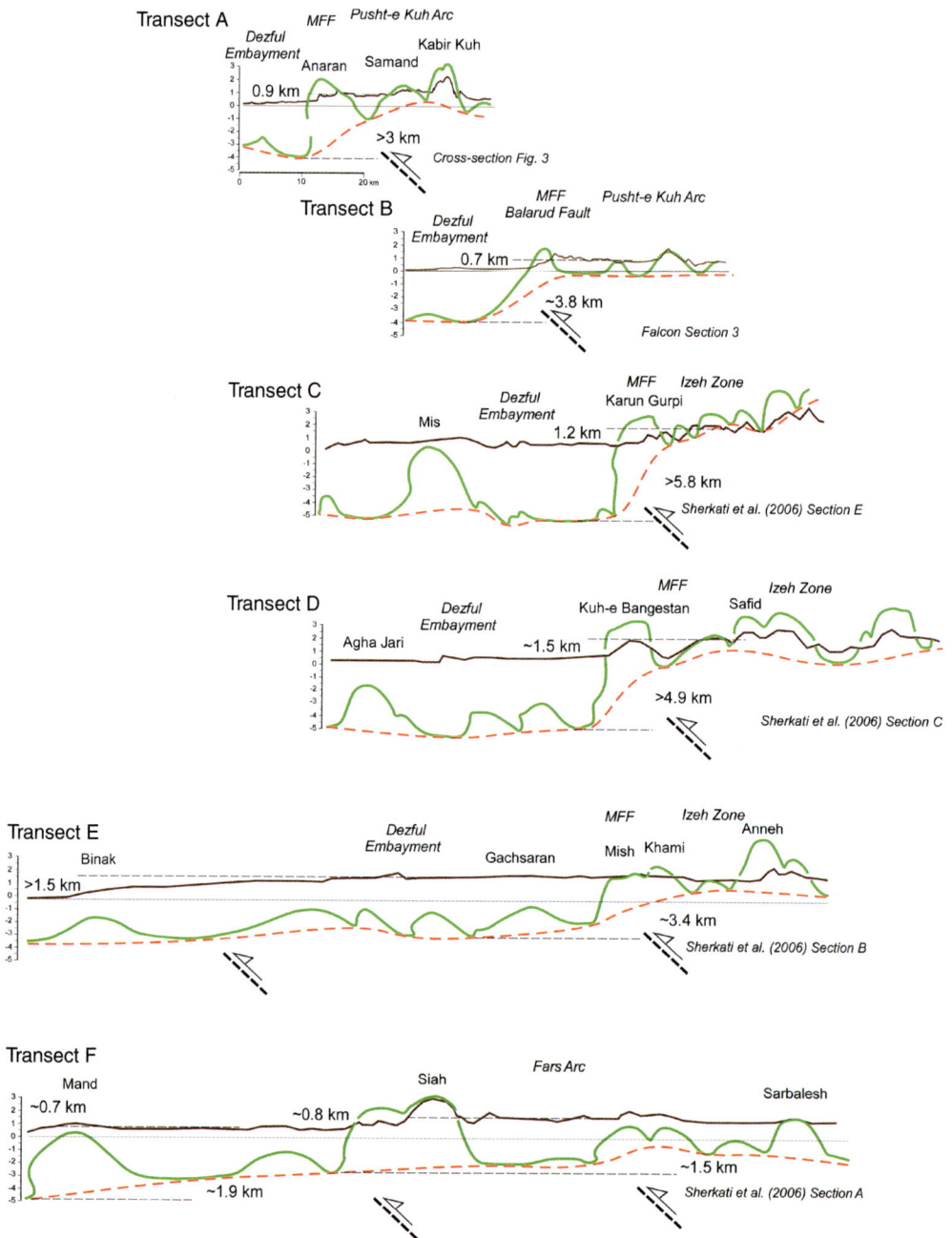

Fig. 4. Six morphotectonic transects across the Pusht-e Kuh Arc (Transects A and B), Dezful Embayment (Transects C–E) and Fars Arc (Transect F) domains. Transect A is from the cross-section in Figure 3, transect B is from transect 3 of Falcon (1961), and transects C–F correspond to regional cross-sections from Sherkati *et al.* (2006). In each transect the green line shows the trace of Asmari limestones defining the geometry of folding and the red line connects the synclinal hinges across the Mountain Front Flexure to allow calculation of the structural relief along the geoflexure (2× vertical exaggeration).

1.2 km and a structural variation of more than 5 km across the Mountain Front Flexure (Fig. 4). Transect D to the SE shows similar variations to transect C in both topography and structural relief across the Mountain Front Flexure (Fig. 4). Transect E to the SE shows a different pattern with a structural relief variation of 3.4 km with little topographic change across the Mountain Front Flexure. The main increase in topography along this transect occurs along the coastline of the Persian Gulf, where there is about 1.5 km of topography increase (Fig. 4). Transect F in the Fars Arc shows two clear steps in the basement that produce changes in both topography and structural elevation (c. 0.8 and 1.9 km for the SW step and c. 0.8 and 1.5 km for the NE step) (Fig. 4). This double step marks the configuration of the Fars Arc characterized by several splays of the Mountain Front Flexure, especially in its NW limit along the Kazerun Fault Zone (Fig. 1).

The present seismicity of the Zagros fold and thrust belt is largely concentrated along the Mountain Front Fault (Talebian & Jackson 2004; Fig. 1). The seismic activity along various segments of the Mountain Front Fault indicates that moderate to large magnitude earthquakes occur in the basement (Berberian 1995). The fault plane solution of earthquakes show a nearly pure thrust faulting with nodal planes striking parallel to the trend of the regional structures. The north to NE c. 60° dipping planes of focal mechanisms suggest dominant reverse movements for these faults.

The lack of both surficial thrusting and good quality seismic data crossing the Mountain Front Flexure makes its interpretation difficult. Consequently, geological cross-sections across this regional structure show different interpretations for it at depth. Berberian (1995) referred to a segmented blind thrust with 15–115 km long segments with a total length of 1350 km along the entire Zagros Fold Belt. Sherkati et al. (2006) interpreted the Mountain Front Flexure as a thrust fault at depth showing typical ramp–flat geometry in the Fars Arc and Dezful Embayment domains. McQuarrie (2004), contrarily, attributed the Mountain Front Flexure to the tectonic accumulation of Hormuz Salt beneath the Pusht-e Kuh and Fars arcs, whereas she proposed a thrust fault with ramp and flat geometry in the sedimentary cover beneath the Dezful Embayment.

We also propose a crustal low-angle blind thrust fault below the Mountain Front Flexure that represents the frontal tip line of the basement thrust system of the Zagros as constructed in Figure 3. The geometry of this thrust assumes that the topographic step across the Mountain Front Flexure as well as the high elevation of the Pusht-e Kuh Arc are related to a single thrust at depth because in this region there is only a single step, in contrast to the regions to the SE (Sherkati et al. 2006; Fig. 4). The regional uplift of the Pusht-e Kuh Arc maintains a wide constant elevation through the Pusht-e Kuh Arc for more than 150 km of width with no apparent additional steps. The best geometrical fit to reproduce this wide structural uplift consists of a low-angle NE-dipping thrust (12–15°) affecting the entire crust down to its base (Fig. 3). In its upper part the blind thrust flattens at the base of the 12–14 km thick cover sequence along the Lower Mobile Group, producing the gentle regional blind monocline along the Mountain Front Flexure with a low-dipping forelimb at depth. The displacement of the low-angle thrust to produce the observed regional uplift in the Pusht-e Kuh Arc is about 11 km. Although we are proposing a single low-angle thrust to account for most of the deeper deformation beneath the Pusht-e Kuh Arc, internal distributed deformation along steeper NE-dipping thrusts with small displacements is consistent with the clustering of seismic events at depths between 11 and 15 km within the crystalline basement, as has been determined in the central Fars Arc (e.g. Tatar et al. 2004).

Most of the published papers showing regional cross-sections across the Zagros Fold Belt have discussed the involvement of the basement rocks below the folded cover sequence, which has been solved using various approaches, none of them geometrically satisfactory. Blanc et al. (2003), Molinaro et al. (2005) and Sherkati et al. (2005) proposed basement thrusts with variable geometry but little displacement that branch along the seismogenic layer at 20 km of depth. Sepehr et al. (2006) showed high-angle reverse faults with small displacement beneath some of the anticlines, although the continuation of these faults at depth was not represented in their cross-sections. Mouthereau et al. (2007) showed listric thrust faults that cut the entire crust down to the Moho, again showing little displacement. In all these interpretations the listric geometry of single thrusts produces a hinge in the ramp–flat passage at depth, which does not deform the apparently undisturbed structure and topography of the folded cover rocks at surface, and only distributed internal deformation would fit with the proposed geometries. Our proposed interpretation fits with the cover structure as well as with the variations in topography between the foreland and the Pusht-e Kuh Arc. In addition, the basement blind thrust shows about 11 km of shortening that accounts for a large part of the Neogene shortening calculated for the cover folding.

The Anaran anticline

The Anaran anticline on top of the Mountain Front Flexure is the most external fold of the SE region of the Pusht-e Kuh Arc (Fig. 5). It is connected to the open Samand anticline to the NE and to the Changuleh anticline to the SW, which is partially buried beneath the Mesopotamian plains, forming the frontal group of anticlines belonging to the SE region of the Pusht-e Kuh Arc (Figs 3 & 5). In previous studies the Anaran anticline has been interpreted together with the Samand and Kabir Kuh anticlines as formed above thrust ramps directly above basement uplifts (Blanc *et al.* 2003) or as buckling folds developed above thick Hormuz Salt (McQuarrie 2004). The Anaran, Samand and Kabir Kuh anticlines have been recently interpreted as forming part of the multi-detachment fold system of the Pusht-e Kuh Arc governed by the varying mechanical stratigraphy (Vergés *et al.* 2009).

The Anaran anticline is 85 km long and 5.5 km wide with maximum topography of about 1.6 km in its central part. Its geometry in map view is very irregular (having a zigzag shape) and formed by four segments, which are either independent like the SE Anaran or linked together but showing different orientation (Central Anaran, north–south Anaran and NW Dome) (Figs 5 & 6). The SE Anaran, Central Anaran and NW Dome form clear culminations whereas the north–south Anaran seems to form the connection between the Central Anaran and the NW Dome culminations, which may have developed as independent anticlines before their linkage. This interpretation of the anticline as small culminations linked together rather than having its growth controlled by an inherited irregular geometry seems to fit better the cross-cutting relationships of the various structural elements of the anticline as discussed below.

The shape of the anticline shows a relatively regular and long backlimb dipping at about 30° and a steep forelimb masked by large dip-slip normal faults predominantly along the Central and north–south Anaran segments that produce a significant change in structural relief across the anticline (Figs 5 & 6). Along the front, the Upper Cretaceous Sarvak Formation is in contact with the Miocene Gachsaran thick evaporites at a high-angle fault that runs beneath the Quaternary deposits for 20 km along the front of the Central Anaran anticline (Fig. 5). Towards the foreland, the thick Agha Jari deposits form a rounded syncline with growth patterns in Changuleh (Homke *et al.* 2004). At depth, seismic lines show a significant divergence between the base of the Agha Jari Formation, reproducing the syncline, and the Asmari Formation, which displays subhorizontal reflectors limiting a thick wedge mostly filled by evaporites of the Gachsaran Formation (Vergés *et al.* 2009; Fig. 5).

The most characteristic feature of the Anaran anticline, however, is the large number of roughly planar dip-slip normal faults that cut the anticline,

especially along its Central and north–south segments. On the map the normal faults display a trend subparallel to the strike of the anticline for the Central Anaran and a slightly oblique trend for the north–south Anaran segment (Figs 5 & 6). These normal faults affect the forelimb of the Central Anaran and both flanks of the north–south segment although most of the faults terminate northwards at the softer Gurpi-Pabdeh marls delineating the anticline backlimb (Figs 5 & 6). When observed in cross-section, well-stratified marls of the Gurpi-Pabdeh formations are folded above the propagating tip of the normal faults as documented in extensional settings in other geological contexts (e.g. Sharp *et al.* 2000, 2009).

Normal faults of the Central Anaran segment form conjugate pairs, dipping at *c.* 60°, which define grabens located in the crestal domain of the anticline (Fig. 7a). A second set of dip-slip normal faults truncates the Central Anaran anticline towards the foreland of the previous one but showing a 30°-rotated (about a horizontal axis) geometry limited by a subvertical forelimb-facing fault and by 30°-dipping backlimb-facing faults as depicted in Figure 7a. These normal faults along the Central Anaran anticline are cut by the normal faults of the north–south Anaran anticline segment, which also define a 30°-rotated graben structure towards the foreland (Figs 5, 6 & 7b–d).

Along the forelimb of the Central Anaran anticline the Upper Cretaceous Sarvak Formation is tectonically juxtaposed to the Miocene Gachsaran Formation for 20 km along the strike (Figs 5 & 6). This contact shows a smoothly curved trace and a reverse displacement of more than 1 km, given that the Asmari, Pabdeh and Gurpi formations are missing along the inferred thrust fault. An important characteristic of this frontal thrust is that it limits the entire north–south Anaran anticline normal fault system in its hanging wall and thus is younger than it.

According to cross-cutting relationships the present-day geometry of the Anaran anticline seems to have developed by the linkage of two anticlines (Central Anaran and NW Dome) through the north–south Anaran segment as proposed for folds in the Fars Arc (Ramsey *et al.* 2008). This sequence seems to be corroborated by the fact that the north–south Anaran segment and associated oblique normal faults cut the east–west-trending grabens along the Central Anaran anticline. The high-angle thrust along the front of the Central Anaran anticline is the latest structure cutting the north–south Anaran normal faults (Fig. 6). Reactivation of SW-dipping normal faults showing about 1000 m of dip-slip displacement occurred during the growth of the Mountain Front Flexure to accommodate the large increase in structural relief across the anticline.

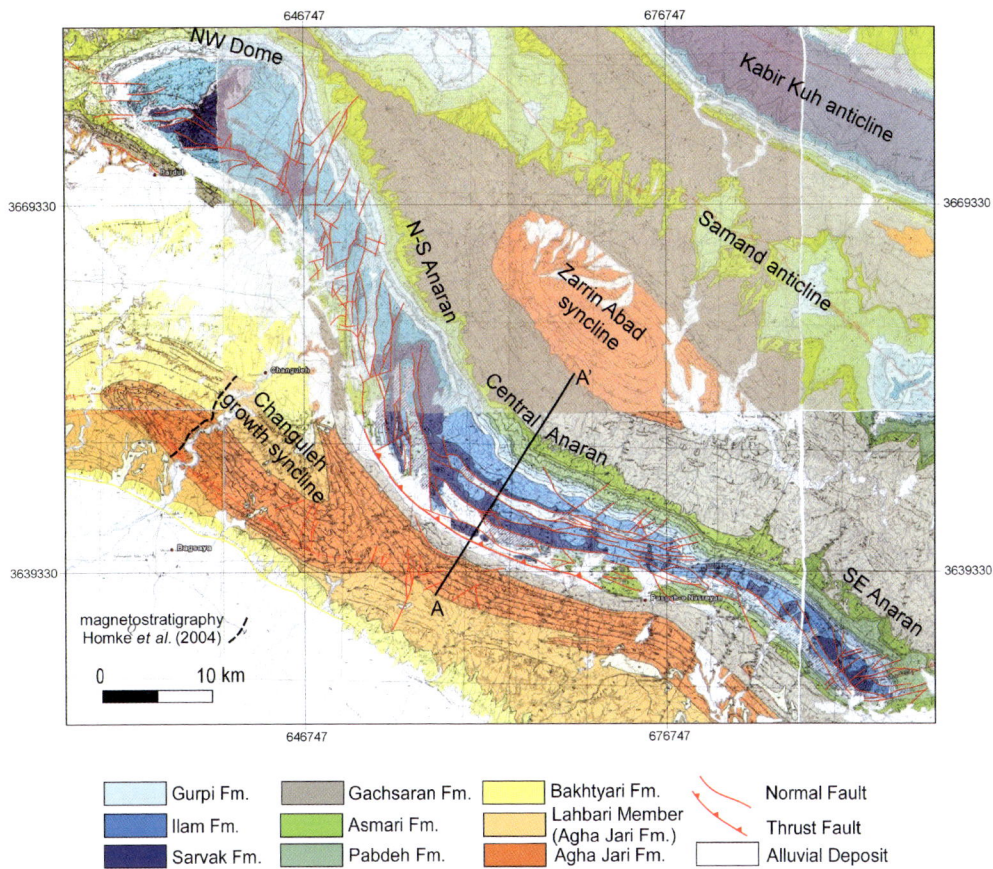

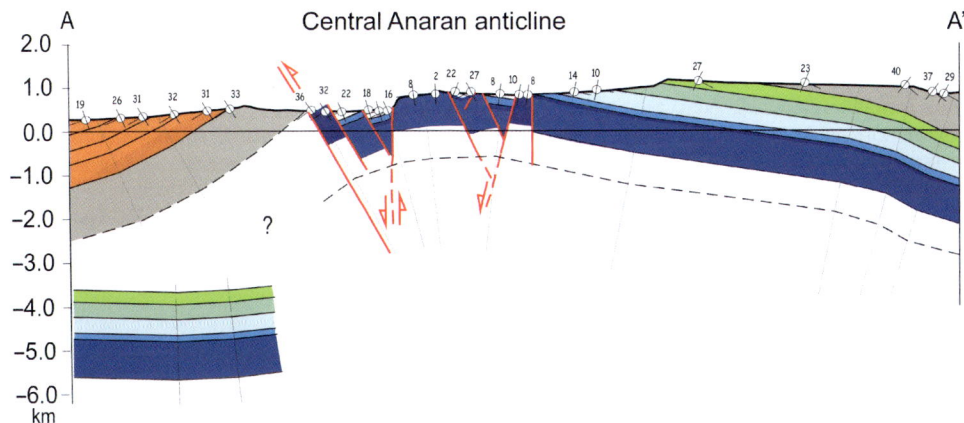

Fig. 5. Map view of the Anaran anticline using map sheets of Kuh-e Varzarin (MacLeod & Roohi 1970) and Kuh-e Anaran (Setudehnia 1967) in which normal faults have been highlighted. The trace of the thrust along the SW limb of the Central Anaran anticline has been added to the map. Black line across the Central Anaran anticline shows the position of cross-section A–A′ based on field and seismic data (same V and H scales).

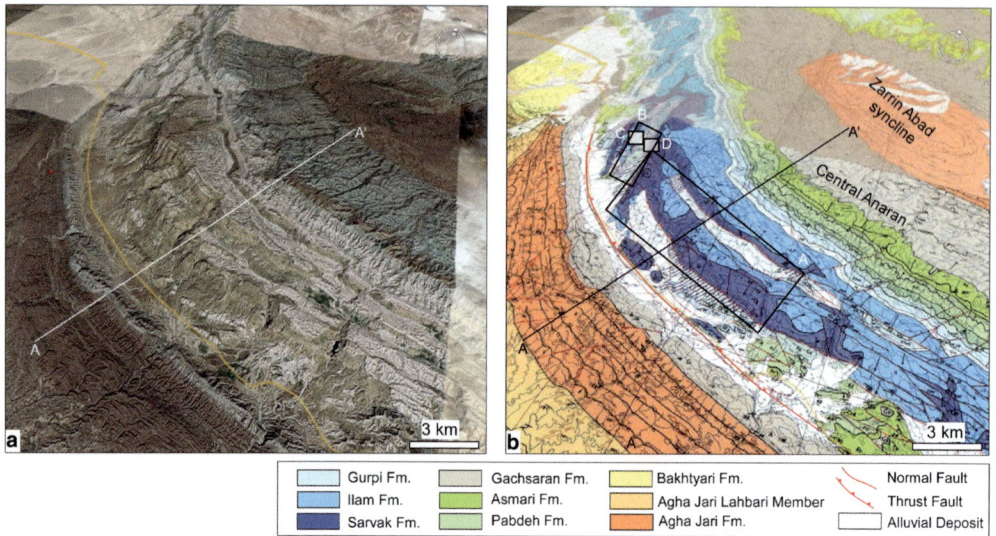

	Gurpi Fm.		Gachsaran Fm.		Bakhtyari Fm.		Normal Fault
	Ilam Fm.		Asmari Fm.		Agha Jari Lahbari Member		Thrust Fault
	Sarvak Fm.		Pabdeh Fm.		Agha Jari Fm.		Alluvial Deposit

Fig. 6. Two views of the Anaran anticline from the south (shaded relief from Google Earth with 1.5 × vertical exaggeration) to show the topography of the Anaran anticline, which is mostly controlled by the displacements of normal faults (**a**) and adding the geological maps Kuh-e Varzarin (MacLeod & Roohi 1970) and Kuh-e Anaran (Setudehnia 1967) (**b**). Boxes A–D show the position of photographs in Figure 7.

The cross-cutting relationships between structures related to the Anaran anticline show an evolution marked by its initial growth between the Samand and the Changuleh anticlines. Crestal grabens that developed along the Central Anaran anticline were later rotated and reactivated during fold amplification. The north–south Anaran segment formed later to link two separated anticlines (the Central and NW Dome culminations) during the growth of the Mountain Front Flexure. The significant increase of tectonic relief beneath the Anaran anticline produced both the continuous slip of well-oriented crestal faults resulting in displacement of about 1000 m and the final emergence of a high-dipping thrust along the front of the Anaran anticline. The combined downthrown of the foreland-facing normal fault along the Central Anaran anticline and the high-angle thrust in its front is more than 2 km. The unambiguous relationship between the large displacement normal faults along the Central and the north–south segments of the Anaran anticline, and the position and extent of the high-angle frontal thrust and the Mountain Front Flexure indicate their causal relationship.

Nevertheless, the Anaran anticline initial growth and the Mountain Front Flexure are not genetically linked but converge in this segment of the long Mountain Front Flexure. This statement is confirmed by the detailed regional cross-sections across the entire Dezful Embayment by Sherkati *et al.* (2006). These sections show that the Mountain

Front Flexure coincides along its trace with multiple anticlines and synclines that are at least slightly rotated towards the foreland above the frontal slope of the basement monocline as observed in the Anaran anticline (Fig. 4).

Experimental procedure

Two sets of analogue models were created to test different options for the development of the Anaran anticline. The first set was designed to reproduce potential geometries of folding and thrusting using two different detachments: one at the base of the basement–cover contact (Hormuz evaporites or equivalent) and a second one at the Gachsaran level (Models 1 & 2). The second set of models was intended to test the proposed sequence of deformation starting with folding and subsequent basement thrusting together with folding assuming a single detachment at the basement–cover contact (Model 3). This second set of models shows in addition syntectonic sedimentation (part of the Gachsaran, Agha Jari and Bakhtyari formations) as potential modifiers of the anticline evolution (Model 4).

The modelling techniques are similar to those used for experiments on brittle–ductile systems at the Laboratory of Experimental Tectonics of Géosciences Rennes (Rennes 1 University, France), as described in previous studies (e.g. Faugère & Brun

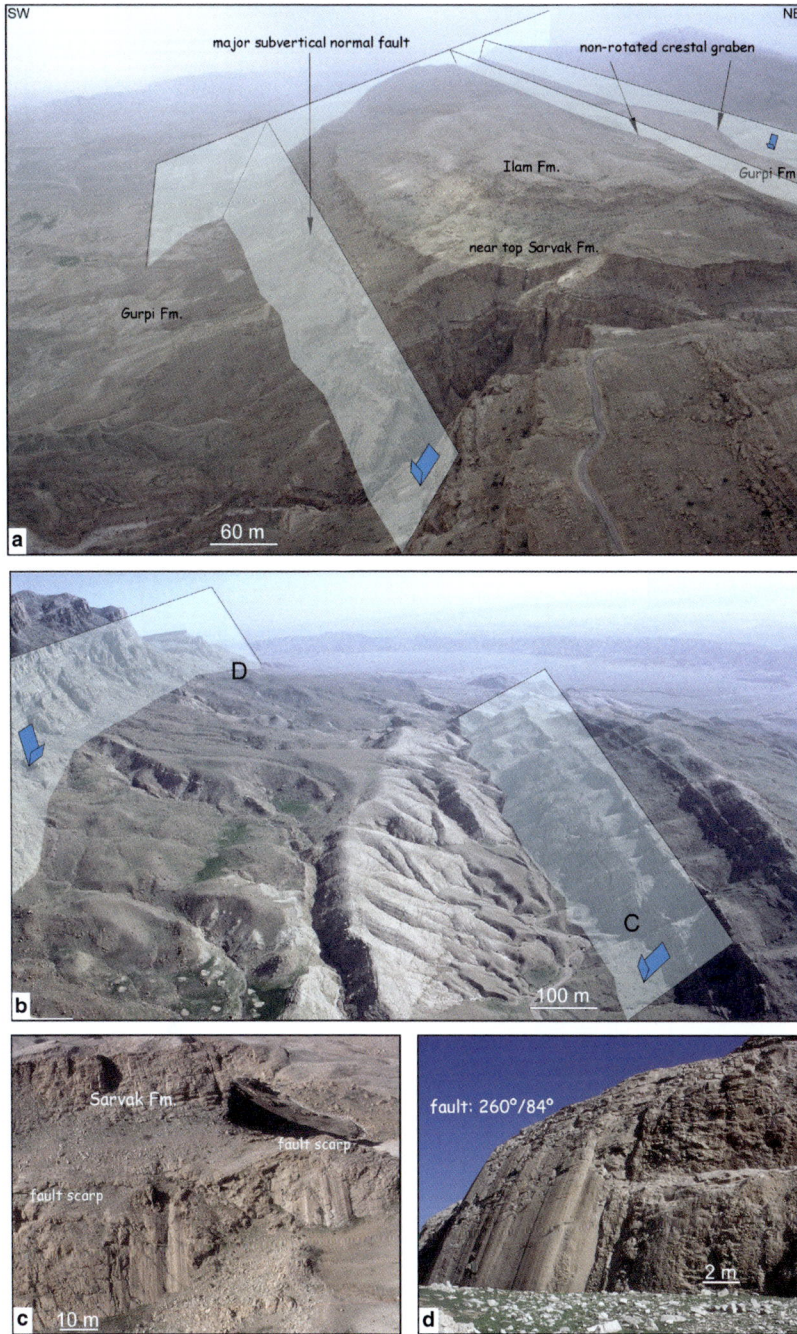

Fig. 7. Field and helicopter photographs of the Anaran anticline. (**a**) Helicopter photograph across the Central Anaran anticline along the line of the cross-section A–A′ (Fig. 5), showing the non-rotated crestal graben to the right filled with Gurpi marls, and the vertical cliff (subvertical normal fault) corresponding to the NE side of the rotated graben. This very steep fault shows almost 1000 m of displacement, based on the juxtaposition of stratigraphic units (**b**) Helicopter view to the south of the normal faults that limit the graben along the north–south segment of the Anaran anticline. (**c**) Helicopter view of the fault plane dipping at 30°, which corresponds to the western boundary of the rotated graben. (**d**) View of the fault plane dipping at 85°, corresponding to the eastern side of the rotated graben (conjugate system of faults in (c) Location of these photographs is indicated in Figure 6.

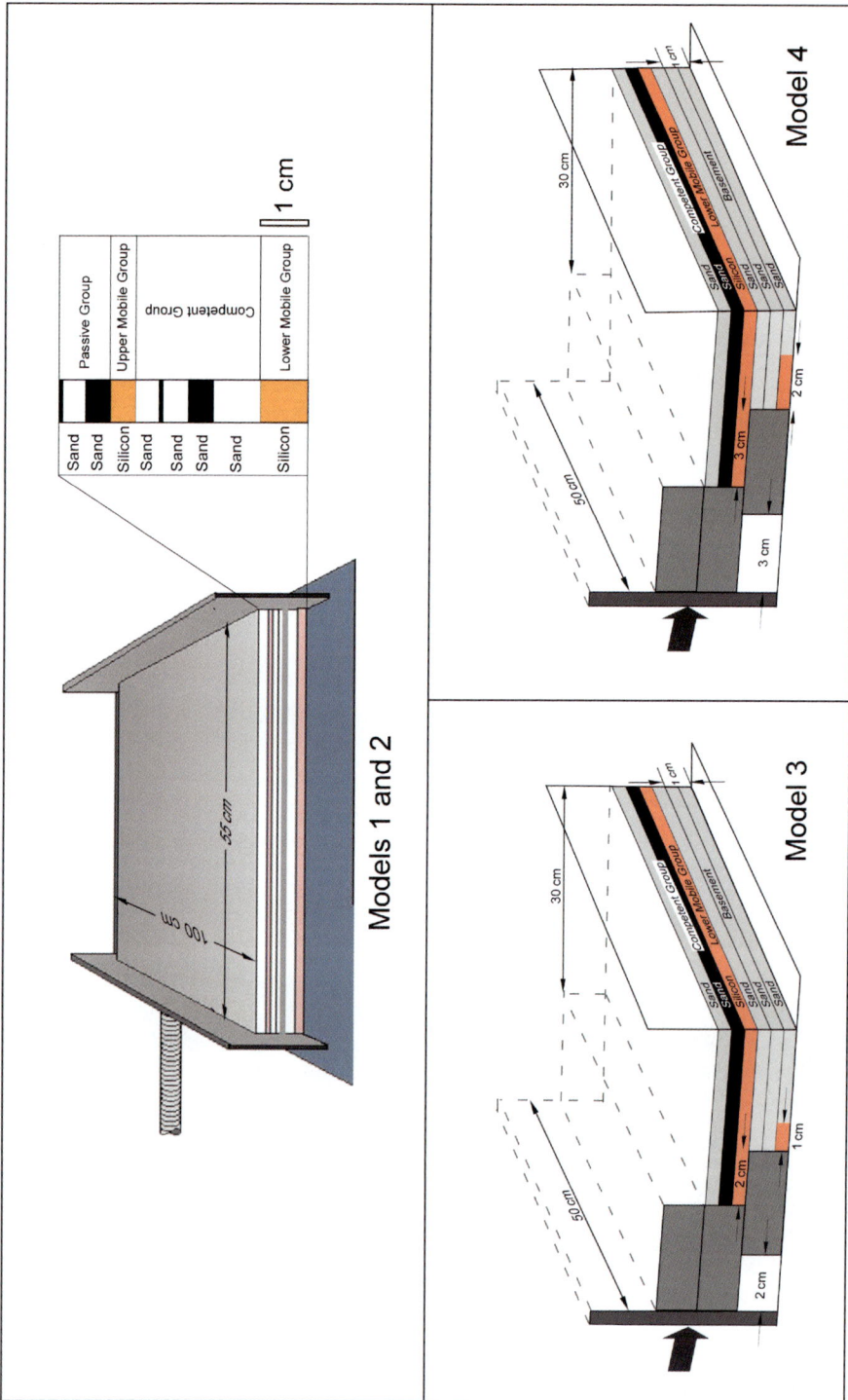

Fig. 8. Diagrams showing the initial configuration of analogue models presented in this work. Models 1 and 2 reproduce the complete mechanical stratigraphy of the Pusht-e Kuh Arc simulating the Lower Mobile, Competent, Upper Mobile and Passive groups. Model 3 shows only the Lower Mobile and the Competent groups, whereas syntectonic deposits in Model 4 represent both Upper Mobile and Passive groups. The thin layers of silicone at the base of Model 3 and Model 4 are needed to locate the basement fault below the cover fold during the second step of compression.

1984; Vendeville *et al.* 1987; Davy & Cobbold 1991). Basement and brittle sediments (pre- and syntectonic) are represented by sand, with an angle of internal friction close to 30° (Krantz 1991) and a density of $\rho = 1400$ kg m^{-3}. Weak ductile sediments such as shale, clay, marl or salt are represented by silicone putty with a viscosity of $\mu = 10^4$ Pa s at 30 °C and a density of $\rho = 1400$ kg m^{-3}. The experimental apparatus is composed of a rigid mobile wall pushed at a constant rate with a compression velocity ($V = 1$ cm h^{-1}) (Fig. 8).

Models 1 and 2 are similar in size and initial configuration. The models are 100 cm × 55 cm, and are limited by fixed and mobile walls on two sides and free on the other two sides. The model is wide enough to achieve a relatively large amount of shortening without edge effects. Compression is simulated by a mobile wall moved at a constant velocity by an electric motor. The models are built with four alternating layers presenting brittle–ductile behaviour to account for the different mechanical units in the sedimentary pile of the Pusht-e Kuh Arc (Vergés *et al.* 2009). The 1 cm thick basal silicone layer represents the basal detachment of the Hormuz Formation or equivalent (evaporitic deposits or overpressured shales of the Lower Mobile Group) (Nalpas & Brun 1993; Weijermars *et al.* 1993; Letouzey *et al.* 1995). This basal unit is overlain by a 2.5 cm thick sand layer representing the Competent Group (Palaeozoic strata, Khami Group and Bangestan Group and up to the Asmari Formation). The intermediate silicone layer is 0.5 cm thick and is overlaid by a 1.0 cm thick sand layer. This second silicone layer represents the Upper Mobile Group (Gachsaran Formation)

(Sherkati *et al.* 2005, 2006; Vergés *et al.* 2009) whereas the sand unit on top corresponds to the Passive Group (Agha Jari and Bakhtyari formations) (Fig. 8).

Models 3 and 4 are smaller than Models 1 and 2 (50 cm × 30 cm) and also have a different initial configuration. They are limited on two opposite sides by one fixed and one mobile wall whereas the other two sides are free, to eliminate edge effects during shortening (Fig. 8). The general configuration of these experiments corresponds to three layers of brittle–ductile alternation. The basal layer, used to represent basement, is formed by a 1.5–cm thick layer of sand overlain by a cover composed of a 0.5 cm thick silicone layer representing the basal detachment (Lower Mobile Group). The Competent Group (from Palaeozoic to Miocene Asmari Formation) is represented by the upper 1 cm thick sand unit. Syntectonic sedimentation was added in Model 4 (see also Nalpas *et al.* 1999, 2003; Barrier *et al.* 2002). Blue and white thin alternations represent the Gachsaran Formation (pre-growth strata in the study area) and Agha Jari and Bakhtyari formations (pre-growth and growth units in the study region) (Homke *et al.* 2004). Because the second silicone layer is not included in Models 3 and 4 the Gachsaran Formation is considered as part of the Passive Group in this model (Fig. 8).

For the second set of models, 14 experiments were carried out to calibrate the velocity of the compression as well as the position of the basement fault below the early formed anticline. This was achieved by using a different model setup with a thin layer of silicone in front of the mobile wall to initiate

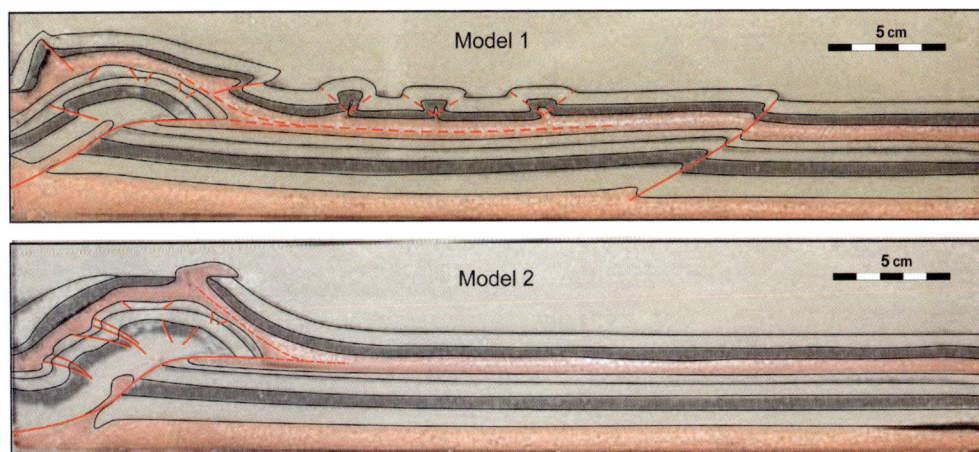

Fig. 9. Models 1 and 2 show the final stages of shortening corresponding to 16.6% (7.5 cm). Model 1 was produced with no erosion during shortening whereas coeval erosion was applied to Model 2 during deformation. Erosion has a remarkable impact on the way shortening propagates, as well as on the geometry of the entire cover succession.

thrusting at the base of the model (Fig. 8). In addition, three plastic blocks were used in front of the mobile wall to divide the shortening between the sedimentary cover and the basement. In Model 3, 2 cm of shortening was applied first to the cover (basal silicone and sand) and then 2 cm more was applied to both the basement and the cover. In Model 4, an initial 3 cm of shortening was applied to the cover (basal silicone and sand) and then 2 cm more to both the basement and the cover (Fig. 8).

Photographs of the surface of each model were taken at regular time intervals to observe their evolution in map view. After deformation the models were covered with sand to avoid further deformation during subsequent operations. Finally, the models were saturated with water to increase their internal cohesion before cutting them into vertical slices parallel to the shortening direction (Figs 9–11).

Analogue Model Results

As described above, Models 1 and 2 simulate the entire stratigraphy of the Pusht-e Kuh Arc with Lower Mobile, Competent, Upper Mobile and Passive groups (Fig. 9). Models 1 and 2 share constant velocity of $1 \, \text{cm h}^{-1}$ and final shortening of 7.5 cm, corresponding to 13.63%. Models 3 and 4 were planned to observe a sequence of deformation in which first we fold the cover above a basal detachment and then we develop a basement-involved thrust below one of the already formed anticlines (numbered 2 in Figs 10 & 11). These models show a slightly different stratigraphy from previous ones given that only Basement, Lower Mobile and Competent groups are displayed in Model 3. Model 4 incorporates growth strata units that might correspond to the Upper Mobile and Passive groups in nature.

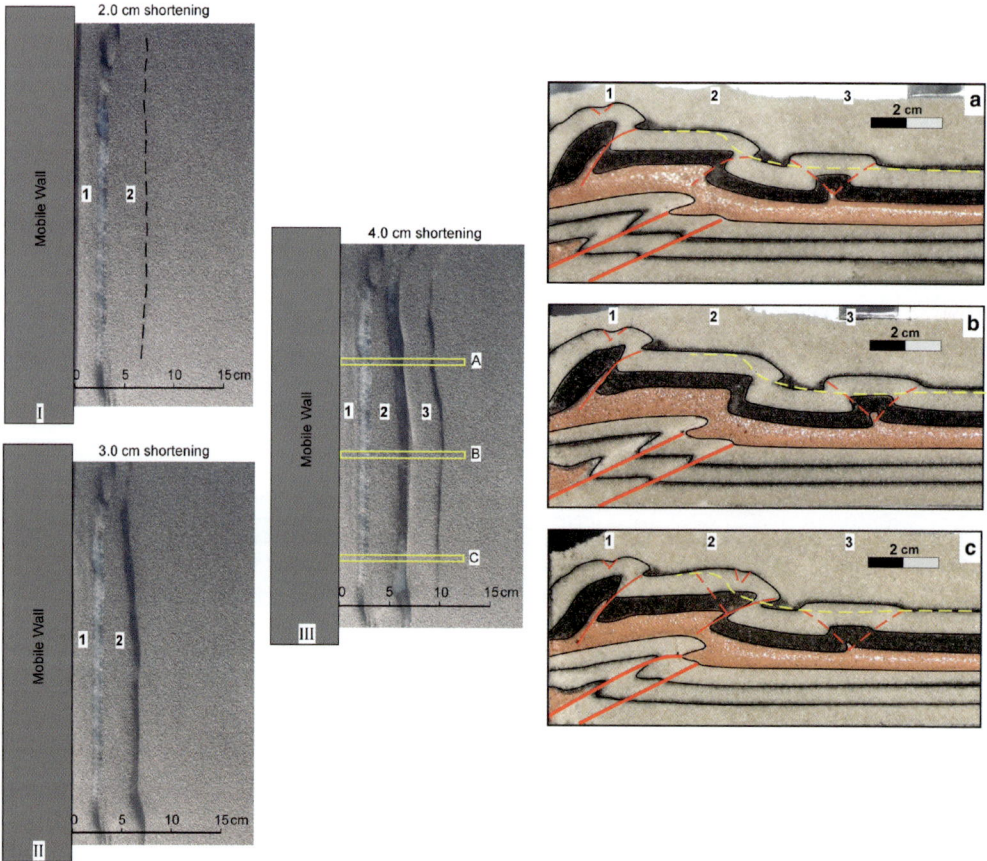

Fig. 10. Map and cross-section composition for Model 3 to show the evolution (map view) as well as the final geometry of the simulated Anaran anticline, which is shown by anticline 2. This model shows only the Lower Mobile and the Competent groups. In cross-sections **a–c**, the topmost thin black layer may represent the Asmari Formation. The dotted yellow line shows the connection of the hinges of the synclines across the simulated Mountain Front Flexure.

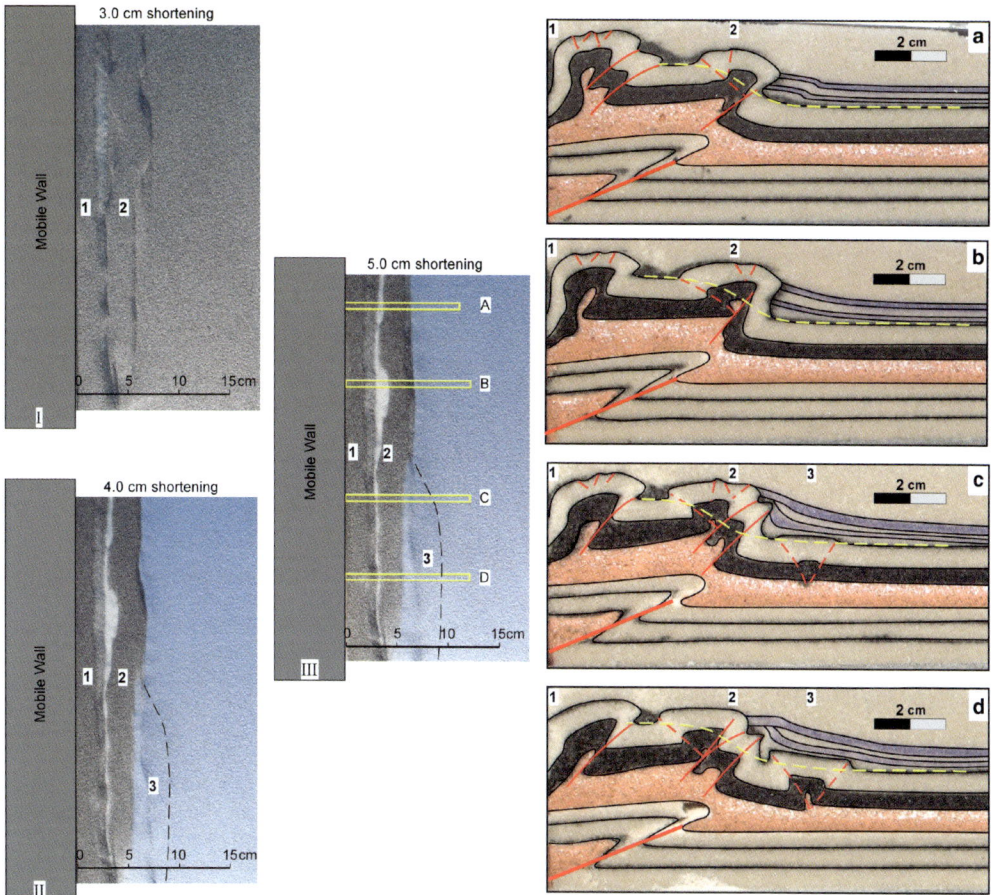

Fig. 11. Map and cross-section composition for Model 4 to show the evolution (map view) as well as the final geometry of the simulated Anaran anticline, which is shown by anticline 2. This model shows only the Lower Mobile and the Competent groups. In cross-sections **a–d**, the topmost thin black layer may represent the Asmari Formation. The dotted yellow line shows the connection of the hinges of the synclines across the simulated Mountain Front Flexure. Blue layers in both maps and cross-sections correspond to growth strata units deposited during shortening (they may correspond to the Gachsaran, Agha Jari and Bakhtyari formations in nature). These syntectonic units have a profound impact on deformation propagation as well as on the evolution of the Anaran anticline geometry.

Model 1

As soon as compression started a major thrust initiated close to the base of the mobile wall cutting through the basal detachment (Lower Mobile Group) and the sedimentary cover (Competent Group). This thrust flattened at the intermediate silicone layer (Upper Mobile Group), thus defining a large anticline in its hanging wall that shows a significant layer-parallel extension along the outer arc of the fold (Fig. 9). The upper brittle unit (Passive Group) deformed differently from the lower brittle unit; in particular, because of the migration of silicone material from the growing anticline to the adjacent synclines but especially towards the anticline in the foreland. Folding and thrusting in this upper brittle unit (Passive Group) are not directly linked to the structures below and thus show tectonic decoupling. Further shortening is then transferred to the intermediate detachment layer (Upper Mobile Group) producing several short-wavelength detachment folds on top of this intermediate detachment (corresponding to the Passive Group in nature) (Fig. 9). These anticlines show symmetric box folding geometries with very steep and short flanks. The first anticline is localized just in front of the main anticline, and the evolution of the deformation is in sequence: the shortening is more important in the first anticline, with more local uplift, and the deformation decreases progressively towards the foreland. Coeval with the folding of the upper brittle unit, a second major thrust cutting the lower stiff unit developed far from the mobile wall, leading to formation of a monocline in the sedimentary cover. Above the major anticline, the upper brittle unit shows thinning in the crest of the anticline that could partly balance the amount of shortening observed towards the foreland, thus indicating that gravity sliding may be an additional factor above the intermediate ductile layer, as probably occurred in some of the Zagros folds.

Model 2

This model behaves very like Model 1 during initial shortening, forming a major thrust and hanging-wall anticline for the lower brittle layer (Competent Group) (Fig. 9). The coeval removal of material from the top of the growing anticline to simulate concomitant erosion controls the further development of the fold system. The first significant result is the simplicity of the anticline geometry of the upper brittle unit (Passive Group), with no satellite folds and no thrusting, compared with that produced in Model 1. The second major difference is that when erosion reaches the intermediate ductile unit it flows out of the anticline by a process similar to that for salt extrusions in nature. The geometry of the fold system also varies at the scale of the entire sand box model because coeval erosion inhibits the propagation of deformation into the foreland. The amount of foreland shortening in Model 1 needs to be taken up by existing structures in Model 2. The geometry of the hanging-wall anticline in the lower brittle unit is more complex than in Model 1 especially in the backlimb, showing two groups of backthrusts, and shows higher structural relief. In the upper brittle unit the forelimb of the anticline shows an increase in dip as well as the typical rounded concave-upward shape of detachment folds, although backthrusting parallel to the ductile intermediate unit is needed to transfer the displacement of the upper brittle unit towards the back of the model, where it is possibly partly eroded.

Model 3

Initial shortening in the cover produced two anticlines (numbered 1 and 2 in Fig. 10). Anticline 2 represents the Anaran anticline just above the future basement-involved thrust forming the Mountain Front Flexure. Anticline 1 corresponds to an edge effect related to the mobile wall. Subsequent shortening involved both basement and cover units, creating thrusting in the basement and continuous folding in the cover (the position of the basement thrust has been forced with a short strip of silicone at the base of the model) (Fig. 8). The

generation of the basement-involved thrust that bends into the main detachment (Lower Ductile Group) allowed the migration of shortening along this plane, producing a new symmetric box fold (number 3) in the brittle unit (Competent Group) towards the foreland. In cross-section C (Fig. 10), anticline 2 shows a final asymmetric geometry and topography with a high gentle backlimb (showing a small displacement backthrust as along the backlimb of the SE Anaran segment) and a low and subvertical forelimb showing a marked forethrust. The entire structure mimics the topography of the basement and shows a regional rotated position of the anticline and adjacent syncline to the foreland as occurs across the Mountain Front Flexure in nature. Basement and cover structures are decoupled in Section C (Fig. 10).

Model 4

The main difference from Model 3 consists of the addition of syntectonic sedimentation in the frontal areas of the model after initial shortening (Fig. 11). Initially, Model 4 evolved similarly to Model 3 by forming two anticlines of which anticline 2 simulates the Anaran fold. Syntectonic sands (alternation of white and blue layers) were sprinkled manually every 15 min, to try to keep the rate of sedimentation comparable with the rate of the frontal fold growth (e.g. Barrier *et al.* 2002; Nalpas *et al.* 2003). These sediments correspond in nature to the growth Agha Jari and Bakhtyari formations in addition to the pre-growth Gachsaran Formation (included in the growth sediments). Development of anticline 2 along strike is variable in this experiment. In cross-sections A and B deformation did not migrate into the foreland, the length of the straight forelimb of anticline 2 increased with time and the geometry of the growth strata is simple, showing an onlap relationship against the forelimb of anticline 2. The dip of the forelimb in cross-sections A and B is slightly overturned. In cross-sections C and D, however, some shortening is transferred to the foreland, initiating a new detached box fold, but only at the beginning of the shortening (Fig. 11). This fold modified the geometry of the growth strata, showing internal major unconformities, and complicated the forelimb geometry of anticline 2 but not its subvertical shape. One significant result of this experiment with respect to the previous one is the increase in length of the forelimb of the anticline against which the growth strata impinged, which produces an increase of the uplift of the anticline in relation to the foreland. Also important is the low deformation in the foreland during syntectonic sedimentation. This increase in forelimb length and its subvertical to overturned dip is clearly sustained by the continuous deposition of growth strata. As in Model 3, the structures of base-ment and cover are decoupled across the detachment level, although the cover structure shows rotation to accommodate the differential tectonic relief created by the basement thrusting.

The potential removal by erosion of the growth strata units adjacent to the subvertical and long anti-cline forelimbs of the Zagros Fold Belt probably triggered the gravitational collapse of many of these unstable forelimbs, especially during major seismic events, which still affect this folded region.

Discussion

In this section the geometry of the normal faults and the role of both syntectonic sedimentation and erosion during Anaran fold growth are discussed to propose a geometric model for the Anaran anti-cline and its evolution.

The geometry of the normal faulting

The Anaran anticline shows numerous well-preserved normal faults that are mostly subparallel to the fold axis for different segments of the anti-cline (Figs 5 & 6). These normal faults are an impor-tant characteristic of other anticlines located above the Mountain Front Flexure, such as the Siah Kuh anticline along the SE continuation of the Anaran anticline, and the Khaviz anticline in the Central Zagros (KH in Fig. 1) (Wennberg et al. 2007). As discussed above, the normal faults in the Anaran anticline are divided in two sets: one set dipping c. 60° and limiting crestal grabens and another set that shows rotation on a horizontal axis of about 30° towards the forelimb, which is coeval with the amplification of the anticline.

Some of the results from analogue and numerical models may apply to the Anaran anticline structure. Models 1 and 2 show well-developed normal faults in the crestal region of the anticlines (Fig. 9). Similar normal faults also grew in Models 3 and 4, although they were not as well developed as in previous models (Figs 10 & 11). Already published numerical and analogue models reproducing basement-involved thrusts propagating through the cover show interesting results on normal fault formation in the anticline crests (e.g. Friedman et al. 1976; Barrier et al. 2002; Finch et al. 2003; Hardy & Finch 2006). Hardy & Finch (2006) showed that uniform strong layered cover bends above the propagating basement thrust and produces a narrow anticline with a subvertical forelimb with an extensional graben slightly rotated in the tran-sition between crestal and forelimb domains. Their models also show that crestal grabens develop better when the propagation of the thrust fault is

efficient through the cover and flattens upwards on top of the footwall, developing maximum layer-parallel extension. Similar conclusions are also pro-vided from analogue models by Friedman et al. (1976). Those workers showed that crestal grabens developed after 2–5% extension along the outer arc and that displacement along these normal faults is relatively small, terminating downwards near the neutral surface of the fold.

These results have been integrated in the cross-section across the Anaran anticline in which two successive crestal grabens developed during folding. The faults delimiting the present crestal region are thus affecting only the shallow domains of the anticline. However, the subvertical fault of the rotated graben shows about 1000 m of vertical displacement and is related in this study to the reac-tivation of previous normal faults to accommodate the large structural and topographic variations between the footwall and the hanging wall across the blind propagating basement-involved thrust at depth (Fig. 5).

The role of the syntectonic sedimentation and erosion

Other important factors acting during the growth of an anticline are syntectonic sedimentation and erosion, which have been proved important for con-straining the fold geometry, as may occur in the Anaran anticline, in which thick units of pre-growth and growth strata deposits have been identified in front of the anticline (Homke et al. 2004). More than 800 m of Gachsaran evaporites and about 800 m of Agha Jari fluvial sediments constitute the pre-growth units. Growth units are composed of a total of about 1200 m of fluvial Agha Jari and allu-vial Bakhtyari deposits. By means of magnetostrati-graphic studies it has been possible to constrain the timing of folding from 7.6 Ma to at least 2.5 Ma or younger (Homke et al. 2004).

According to sand box models, syntectonic sedi-mentation and erosion act in a similar way during the growth of a fold, inhibiting propagation of defor-mation into the undeformed foreland (e.g. Storti & McClay 1995; Barrier et al. 2002; Bonini 2003; Nalpas et al. 2003). This can be seen in Model 4, in which the syntectonic sedimentation keeps defor-mation along the same structure. This prolonged deformation acting along the same anticline pro-duces the lengthening of the forelimb against which the growth strata onlap. Coeval erosion with folding produces a similar effect to syntectonic sedimentation by keeping the deformation at the already active structures rather than propagating it towards the foreland. However, its effect on the forelimb geometry is negligible (Model 2 in

Fig. 9). If intermediate weak units become get exposed they may flow and create gravitational salt extrusions. An example of this occurs at the NW termination of the Central Anaran forelimb, where the Gachsaran evaporites occur (Figs 5 & 6).

The geometry of the Anaran anticline

A cross-section across the Anaran anticline shows that the geometry of the crestal and backlimb domains can be constructed by using surface data, but not the forelimb geometry at depth, which can either be parallel to the base of the Agha Jari Formation or steeper (Fig. 5). The correct geometry of this forelimb is important for the imaging of the seismic surveys across this area and their potential conversion to depth, a key point for current hydrocarbon exploration. Below, we use the analogue modelling results to constrain the geometry of the Anaran anticline at depth.

Models 1 and 2, displaying a similar stratigraphic configuration to the Anaran anticline (Lower Mobile, Competent, Upper Mobile and Passive groups), show a complete decoupling between Competent and Passive groups above the intermediate detachment level that develops a thick wedge of evaporites. This wedge develops above the flat portion of the thrust by the push of the hanging-wall anticline with the concomitant migration of sandwiched evaporites between the Competent and Passive groups (Fig. 9). This evaporitic tectonic thickening forms a typical triangular zone enhancing the disharmony of the overburden folds (Fig. 9). Model 4 shows, in addition, that the existence of thick growth strata filling the adjacent growth syncline ahead of the anticline increases both the length of the anticline forelimb and its dip, which in the model is subvertical (Fig. 12).

Numerical models based on a basement-involved thrust propagating into the cover sequence also result in similar final geometries for the folded cover, as described, for example, by Finch *et al.* (2003) and Hardy & Finch (2006). These workers indicated that a homogeneous weak cover sequence above a propagating basement blind thrust forms a broad anticline with a wide extensional zone whereas a more rigid layered cover succession (which gives a closer fit with the mechanical stratigraphy of the upper part of the Competent Group in the study area) shows a much narrower anticline with a steeper forelimb and minimum changes of thickness across the anticline (Hardy & Finch 2006; fig. 8). Numerical models also show that the propagation of the basement-involved thrust in the stiff cover sequence produces a band of deformation linked to the fault propagation as documented by Hardy & Finch (2006) (Fig. 12). In agreement with this deformed triangular band, analogue models of

Friedman *et al.* (1976) indicate an out-of-sequence propagation of thrusts ending with the development of a steeper thrust in the hanging wall of the previous ones (Fig. 12).

Combining field data with model results we construct a cross-section that consists of an asymmetric anticline with a high, long and gently dipping backlimb and a short subvertical forelimb, which is transported above a thrust that branches downwards into the main crustal low-angle thrust and flattens upwards into the Gachsaran intermediate detachment level (Fig. 12). The footwall of the Anaran anticline shows a subhorizontal geometry gently dipping either towards the foreland or to the hinterland close to the thrust. The thrust carrying the Anaran anticline transfers shortening towards the foreland to the Changuleh thrust with less than 1 km of displacement. A smaller thrust developed along the footwall beneath the Asmari Formation along the local Kalhur Member evaporites as described by Vergés *et al.* (2009) and a larger thrust developed in its hanging wall, which crops out for about 20 km (Fig. 5). A passive backthrust developed along the backlimb of the Changuleh syncline, which is observed along the front of the SE Anaran anticline (Fig. 5). The geometry of the forelimb of the Anaran anticline is complex in detail, showing a significant compartmentalization of the structure, as has been observed in front of the Mountain Front Flexure as in the Kuh-e Bangestan anticline (Sherkati *et al.* 2006). An alternative model for the Anaran anticline would be to root the cover thrust system at the basement–cover contact along the Lower Mobile Group. This thrust system with a listric geometry would not be locally linked to the basement thrust system (Fig. 12b).

Evolution of the Anaran anticline and Mountain Front Flexure

In this section a sequential restoration is presented to show the potential evolution of the Anaran anticline, assuming initial detachment folding and then basement-involved thrusting uplifting the arc from the Mesopotamian foreland as two separate tectonic processes that converge in the study area. Although an early Tertiary phase of folding for the Anaran anticline cannot be disregarded (Homke *et al.* 2009), we refer in this section to the Neogene evolution of the fold in which the timing of the deformation has been determined as starting at 7.65 Ma and possibly ending at about 2.5–1.5 Ma and thus lasting for about 6 Ma (Homke *et al.* 2004). This period encompasses both the growth of the Anaran anticline and its amplification above the basement-involved monocline that defines the Mountain Front Flexure. The original shape of the Anaran

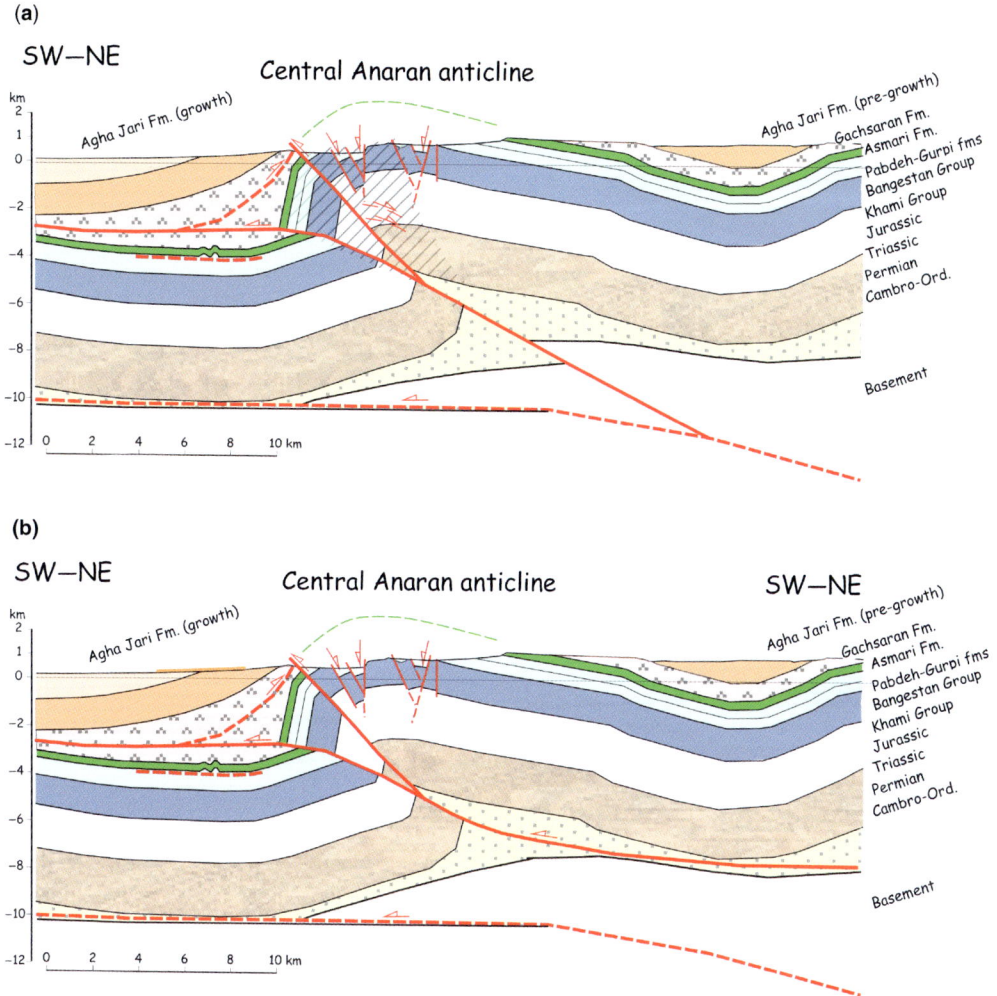

Fig. 12. (**a**) Proposed geometrical model for the Anaran anticline and Mountain Front Flexure using field data as well as analogue modelling results. A combination of thrusting and normal faulting could coexist along the anticline, which possibly shows a large region of intense faulting outlined by dashed oblique lines (Hardy & Finch 2006). (**b**) Alternative model in which the thrust system affecting the Anaran anticline is detached from the basement–cover contact.

anticline is conjectural, from nearly symmetric like the Samand anticline or the recently analysed Mand anticline near the Persian Gulf shoreline (Oveisi *et al.* 2007) to foreland verging. In this work, however, the initial configuration of the Anaran anticline is constructed at the onset of the Agha Jari growth deposition, after 7.65 Ma, by unfolding the present anticline geometry before the inception of the basement-involved thrusting (Fig. 13a). A first set of normal faults limiting the crestal graben possibly formed at that time to account for layer-parallel fold extension along the Central Anaran anticline. Growth strata indicating the inception of folding started at 7.65 Ma after the deposition of

800 m thick Gachsaran and 800 m thick pre-growth Agha Jari formations (Homke *et al.* 2004; Fig. 13a).

As mentioned above, there are no direct indications for the age of the uplift of the Pusht-e Kuh Arc in this region. The deposition of the *c.* 800 m thick fine-grained deposits of the Lahbari Member starting at 5.5 Ma along the footwall of the Mountain Front Flexure could tentatively be associated with this major event (Homke *et al.* 2004). This interpretation is based on the fact that these fine-grained sediments, deposited about 2.1 Ma after fold growth initiation, could be sourced from the softer Gachsaran and Agha Jari deposits regionally uplifted above the proposed low-angle blind

(a) After 7.65 Ma

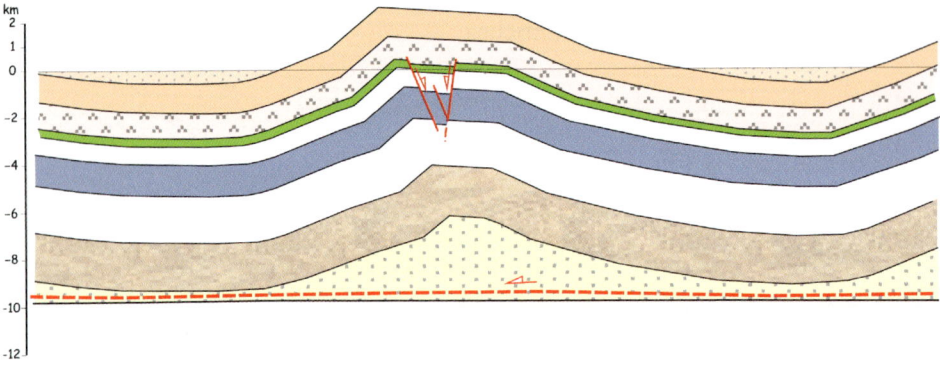

(b) After 5.5 Ma

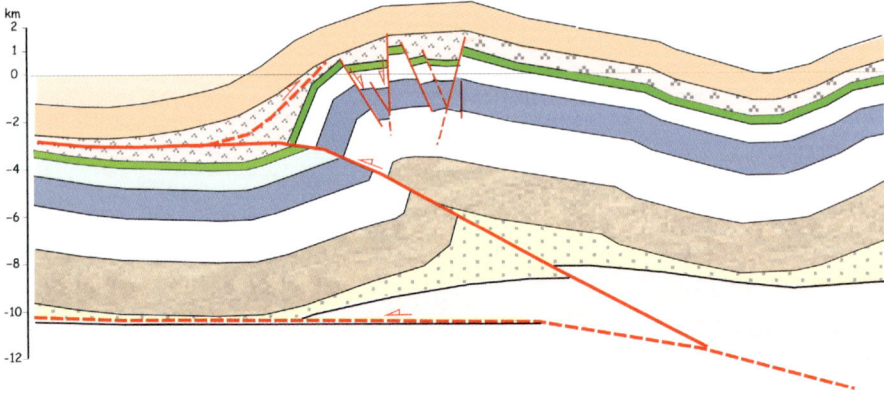

(c) Before 1.5 Ma   Central Anaran anticline   SW–NE

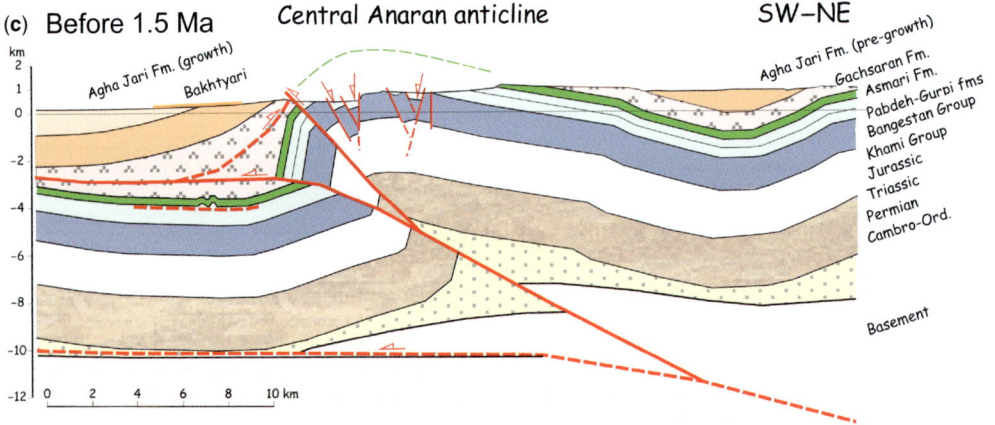

Fig. 13. Proposed evolution for the Central Anaran anticline above the Mountain Front Flexure. (**a**) Cover folding above a detachment layer at basement–cover contact after 7.6 Ma. (**b**) Regional uplift of the entire Pusht-e Kuh Arc above a low-angle crustal thrust with amplification of the anticline and reactivation of normal faults to account for the 3 km of structural relief increase. This regional uplift may have started at about 5.5 Ma during deposition of the fine-grained Lahbari Member. (**c**) Final stages of anticline growth with steep thrusting. Bakhtyari conglomerates cover the already folded Agha Jari fluvial strata at 1.5 Ma.

thrust, which generated the Pusht-e Kuh Arc rise (Vergés 2007; Fig. 13b). This phase is characterized by the formation of the basement monocline and the tectonic transport of the Anaran anticline above the propagating thrust that flattens into the Gachsaran evaporites. The Anaran anticline rotated and previous well-oriented normal faults were reactivated to accommodate the different structural relief between the anticline and the foreland. The foreland-oriented normal fault shows about 1000 m of dip-slip displacement (Fig. 13b). The movement of the Anaran anticline in the hanging wall of this thrust produced the push and thickening of the Gachsaran evaporites in its front as well as the development of backthrusting near the Gachsaran–Agha Jari contact. The position of the thrusts separating the forelimb of the Anaran anticline mostly reactivated the fold hinges to propagate as observed in many other examples in the study region (Vergés *et al.* 2009).

These tectonic processes forming the Anaran anticline occurred during deposition of the *c.* 800 m thick Lahbari Member, which shows growth strata patterns. Partially influenced by this thick syntectonic deposition and by fold tightening (Sans & Vergés 1995) the foreland advance of the Anaran anticline forelimb become inactive and a new thrust developed in the hanging wall of the previous one, uplifting the fold crest about 1000 m more (Fig. 13c). The only time control for this period indicates that the northeastern limb of the Changuleh syncline was already formed during the deposition of the youngest Bakhtyari deposits, which may have an age of about 1.5 Ma (Homke *et al.* 2004; Fig. 13c). However, these conglomerates were folded in the Changuleh anticline to the SE, indicating that the forelimb of the Anaran anticline was already very steep during its translation to the SW above the flat thrust in the Gachsaran evaporites.

Large undeformed erosive straths located at about 300, 350 and 400 m above sea level, some of them covered by thin aggradational fluvial or alluvial fill terraces, constitute the most typical landscape along the frontal region of the Mountain Front Flexure in the Anaran anticline. Taking into account that the youngest extrapolated age for the top of the Bakhtyari Formation is *c.* 1.5 Ma (Homke *et al.* 2004), post-deposition river incision is calculated by the difference between the top of the Bakhtyari Formation in the Changuleh growth syncline (610 m) and the 250 m level of the present river course crossing those conglomerates. These values indicate a mean rate of fluvial incision since the end of the Bakhtyari deposition of about 0.24 mm a^{-1}.

Present deep tectonic activity along the Mountain Front Flexure is indicated by the occurrence of numerous deep basement earthquakes that define a narrow band along the Mountain Front Flexure (Berberian 1995; Talebian & Jackson 2004; Fig. 1). This recent seismic activity could easily trigger large mass movements that contribute to the rapid denudation of the anticlines, as occurred in historical times along the Kabir Kuh anticline by the complete failure of a 15 km long segment of its NE flank, which produced the most catastrophic and gigantic landslide in the area ($>$160 km^3), the Seymareh landslide (Harrison & Falcon 1934).

Conclusions

Conclusions from this study are related to the regional significance of the Mountain Front Flexure, to the structure of the Anaran anticline as representative of a fold located above this regional basement blind thrust-related monocline, and to the analogue models constructed to understand the deep structure along the front of the Pusht-e Kuh Arc in the NW Zagros Fold Belt.

The Mountain Front Flexure shows a single step along the front of the Pusht-e Kuh Arc, along the Balarud Fault (SE termination of the Pusht-e Kuh Arc) and through the Izeh Zone. In these segments between 3 and 5 km of structural relief is observed across the Mountain Front Flexure. In the NW Fars Arc, the Mountain Front Flexure shows more than one step in the basement as well as through the Kazerun Fault Zone. In this paper, the Mountain Front Flexure is interpreted as a major basement-involved thrust flattening upwards at the basement–cover contact to form a monocline. This low-angle thrust (dips of 12–15°) cuts through the entire crust and forms the tip line of the Zagros basement thrust system. A minimum of 11 km of shortening is needed to account for the structural and topographic variations between the Pusht-e Kuh Arc and the Mesopotamian foreland basin.

The Anaran anticline is formed of four segments, of which the SE Anaran, the Central Anaran and the NW Dome form culminations. This anticline developed as a detachment fold between the Samand and Kabir Kuh anticlines in its back and the Changuleh anticline in its front. These anticlines formed by multiple detachment levels with different shapes depending on the local mechanical stratigraphy. Normal faults developed during the growth of the Anaran anticline by layer-parallel extension during folding and were subsequently rotated and reactivated during the rise of the anticline above the Mountain Front Flexure, showing dip-slip displacements of about 1 km (foreland downthrow).

The proposed geometry for the Anaran anticline shows an asymmetric fold with a subvertical forelimb cut by two thrusts: the higher and steeper one crops out through the front of the Central Anaran segment for about 20 km and the lower one is a major blind thrust with a ramp–flat geometry. This lower thrust flattens at the Gachsaran evaporites and carries the Anaran anticline on top of the foreland syncline. The Anaran anticline at depth shows a compartmentalized structure, which is totally decoupled from the overlying Passive Group (Agha Jari and Bakhtyari formations). Additional shortening may occur in the footwall of the main thrust, either by folding or by duplicating the top of the Competent Group depending on the regional mechanical stratigraphy. In the study area the Miocene Kalhur evaporites act as a detachment, producing disharmonic folding of the Asmari Formation.

Analogue models with two detachment levels show that a triangular zone filled with evaporites (Gachsaran Formation in the study area) formed above the flat segment of the main thrust as a result of the push of the hanging-wall anticline towards the foreland. Models with syntectonic erosion concentrate deformation in the already active folds and inhibit shortening propagation towards the foreland. Models with syntectonic deposition show similar results, focusing deformation in already active folds. In addition, syntectonic deposition has a strong impact in controlling the geometry of the forelimb against which growth strata impinge. The anticline forelimb increases both its dip and its length through growth deposition in the adjacent growth syncline. All these analogue model results are compatible with observed field data in the study region and have been integrated in the proposed geometric model of the Anaran anticline.

We therefore propose the evolution of the Anaran anticline as follows: (1) an initial cover folding above the basement–cover contact linked to the suite of folds cropping out in the present Pusht-e Kuh Arc starting at 7.65 Ma with coeval crestal normal faulting growth; (2) a regional uplift of the entire Pusht-e Kuh Arc above a low-angle crustal thrust with amplification of the Anaran anticline and reactivation of normal faults to account for the 3 km of structural relief increase between the foreland and the tectonic arc, probably starting at 5.5 Ma during fine-grained Lahbari Member deposition; (3) a final tightening of the Anaran anticline above the Mountain Front Flexure producing the high-angle thrust elevating the crest and backlimb of the fold by almost 1 km. At about 1.5 Ma the front of the Anaran anticline was already folded, as demonstrated by the subhorizontal unconformity of Bakhtyari conglomerates above it. Since the end of Bakhtyari coarse-grained deposition, uplift has been recorded by river incision, anticline gravitational collapses and recent basement seismic activity.

The characteristic geometry of the Anaran anticline is closely related to its singular position on top of the Mountain Front Flexure and thus has little applicability to other anticlines forming part of the Pusht-e Kuh Arc. However, a large number of anticlines are located on top of the Mountain Front Flexure along the Pusht-e Kuh Arc, Dezful Embayment and Fars Arc, such as the Siah Kuh and Khaviz anticlines, and thus the Anaran anticline geometry and evolution presented here can provide clues for their interpretation.

This is a contribution of the Group of Dynamics of the Lithosphere (GDL), financed by a collaborative project between the Institute of Earth Sciences 'Jaume Almera', CSIC of Barcelona (Spain) and StatoilHydro Tehran and StatoilHydro Research Centre, with the partial support of project Team Consolider-Ingenio 2010 No. CSD 2006-00041. We also acknowledge support in the field by StatoilHydro and NPA staff, and thank the National Iranian Oil Company (NIOC) for its collaboration during this project. We are grateful to S. Grelaud for providing us with results for Models 1 and 2. We also thank S. Sherkati and many others for the fruitful discussions during the first trips to this wonderful region of the Zagros Mountains, as well as J. Cosgrove and D. Frizon de Lamotte, who provided excellent reviews that helped to improve this study.

References

BARRIER, L., NALPAS, T., GAPAIS, D., PROUST, J. N., CASAS, A. & BOURQUIN, S. 2002. Influence of syntectonic sedimentation on thrust geometry. Field examples from the Iberian Chain (Spain) and analogue modelling. *Sedimentary Geology*, **146**, 91–104.

BERBERIAN, M. 1995. Master 'blind' thrust faults hidden under the Zagros folds: active basement tectonics and surface morphotectonics. *Tectonophysics*, **241**, 193–224.

BERG, R. R. 1962. Mountain flank thrusting in Rocky Mountain foreland, Wyoming and Colorado. *AAPG Bulletin*, **46**, 2019–2032.

BLANC, E. J.-P., ALLEN, M. B., INGER, S. & HASSANI, H. 2003. Structural styles in the Zagros Simple Folded Zone, Iran. *Journal of the Geological Society, London*, **160**, 401–412.

BONINI, M. 2003. Detachment folding, fold amplification, and diapirism in thrust wedge experiments. *Tectonics*, **22**, 1065, doi: 10.1029/2002TC001458.

BROWN, W. G. 1988. Deformational styles of Laramide uplifts in the Wyoming foreland. *In*: SCHMIDT, C. J. & PERRY, W. J. (eds) *Interaction of the Rocky Mountain Foreland and the Cordilleran Thrust Belt*. Geological Society of America, Memoirs, **171**, 1–25.

CASCIELLO, E., VERGÉS, J., SAURA, E., CASINI, G., FERNÁNDEZ, N., BLANC, E., HOMKE, S. & HUNT, D. 2009. Fold patterns and multilayer rheology of the Lurestan Province, Zagros Simply Folded Belt (Iran). *Journal of the Geological Society, London*, **166**, 1–13. doi: 10.1144/0016-7649.2008.138.

COLMAN-SADD, S. P. 1978. Fold development in Zagros simply folded belt, southwest Iran. *AAPG Bulletin*, **62**, 984–1003.

DAVY, P. & COBBOLD, P. R. 1991. Experiments on shortening of 4-layers model of continental lithosphere. *Tectonophysics*, **188**, 1–25.

DUNNINGTON, H. V. 1968. Salt-tectonic features of northern Iraq. *In*: MATTOX, R. B. (ed.) *A symposium based on paper from the International Conference on Saline Deposits*. Geological Society of America, Special Papers, **88**, 183–227.

FALCON, N. L. 1961. Major earth-flexuring in the Zagros Mountains of southwest Iran. *Quarterly Journal of the Geological Society of London*, **117**, 367–376.

FALCON, N. L. 1974. Southern Iran: Zagros Mountains. *In*: SPENCER, A. M. (ed.) *Mesozoic–Cenozoic Orogenic Belts. Data for Orogenic Studies*. Geological Society, London, Special Publications, **4**, 199–211.

FARZIPOUR-SAEIN, A., YASSAGHI, A., SHERKATI, S. & KOYI, H. 2009. Basin evolution of the Lurestan region in Zagros fold-and-thrust belt, Iran. *Journal of Petroleum Geology*, **32**, 5–20.

FAUGÈRE, E. & BRUN, J. P. 1984. Modélisation expérimentale de la distension continentale. *Comptes Rendus de l'Académie des Sciences*, **299**, 365–370.

FINCH, E., HARDY, S. & GAWTHORPE, R. 2003. Discrete element modelling of contractional fault-propagation folding above rigid basement fault blocks. *Journal of Structural Geology*, **25**, 515–528.

FOOSE, R. M., WISE, D. U. & GARBARINI, G. S. 1961. Structural geology of the Beartooth Mountains, Muntana and Wyoming. *Geological Society of America Bulletin*, **72**, 1143–1172.

FRIEDMAN, M., HANDIN, J., LOGAN, J. M., MIN, K. D. & STERNS, D. W. 1976. Experimental folding of rocks under confining pressure: Part III. Faulted drape folds in multilithologic layered specimens. *Geological Society of America Bulletin*, **87**, 1049–1066.

HARDY, S. & FINCH, E. 2006. Discrete element modelling of the influence of cover strength on basement-involved fault-propagation folding. *Tectonophysics*, **415**, 225–238.

HARRISON, J. V. & FALCON, N. 1934. Collapse structures. *Geological Magazine*, **71**, 529–539.

HOMKE, S., VERGÉS, J., GARCÉS, M., EMAMI, H. & KARPUZ, R. 2004. Magnetostratigraphy of Miocene–Pliocene Zagros foreland deposits in the front of the Pusht-e Kuh Arc (Lurestan Province, Iran). *Earth and Planetary Science Letters*, **225**, 397–410.

HOMKE, S., VERGÉS, J. *ET AL.* 2009. Late Cretaceous–Paleocene formation of the early Zagros foreland basin: biostratigraphy and magnetostratigraphy of the Amiran, Taleh Zang and Kashkan sequence in Lurestan Province, SW Iran. *Geological Society of America Bulletin*, **121**, 963–978. doi: 10.1130/B26035.1.

JAMES, G. A. & WYND, J. G. 1965. Stratigraphic nomenclature of Iranian oil consortium agreement area. *AAPG Bulletin*, **49**, 2182–2245.

KRANTZ, R. W. 1991. Measurements of friction coefficients and cohesion for faulting and fault reactivation in laboratory models using sand and sand mixtures. *Tectonophysics*, **188**, 203–207.

LETOUZEY, J., COLLETTA, B., VIALLY, R. & CHERMETTE, J. C. 1995. Evolution of Salt-Related Structures in Compressional Settings. *In*: JACKSON, M. P. A., ROBERTS, D. G. & SNELSON, S. (eds) *Salt Tectonics: a Global Perspective*. American Association of Petroleum Geologists, Memoirs, **65**, 41–60.

MACLEOD, C. J. & ROOHI, M. 1970. *Kuh-e Varzarin Geological Compilation Map, 1:100 000 scale (Sheet 29236 E)*. Iranian Oil Operating Companies (IOOC), Tehran.

MCQUARRIE, N. 2004. Crustal scale geometry of the Zagros fold–thrust belt, Iran. *Journal of Structural Geology*, **26**, 519–535.

MITRA, S. & MOUNT, V. S. 1998. Foreland basement-involved structures. *AAPG Bulletin*, **82**, 70–109.

MOLINARO, M., GUEZOU, J. C., LETURMY, P., ESHRAGHI, S. A. & FRIZON DE LAMOTTE, D. 2004. The origin of changes in structural style across the Bandar Abbas syntaxis, SE Zagros (Iran). *Marine and Petroleum Geology*, **21**, 735–752.

MOLINARO, M., LETURMY, P., GUEZOU, J.-C., FRIZON DE LAMOTTE, D. & ESHRAGHI, S. A. 2005. The structure and kinematics of the south-eastern Zagros fold thrust belt; Iran: from thin-skinned to thick-skinned tectonics. *Tectonics*, **24**, TC3007, doi: 10.1029/2004TC001633.

MOUTHEREAU, F., LACOMBE, O., TENSI, J., BELLAHSEN, N., KARGAR, S. & AMROUCH, K. 2007. Mechanical constraints on the development of the Zagros folded belt (Fars). *In*: LACOMBE, O., LAVÉ, J., ROURE, F. & VERGÉS, J. (eds) *Thrust Belts and Foreland Basins from Fold Kinematics to Hydrocarbon Systems*. Frontiers in Earth Sciences, Springer, Berlin, 247–266.

NALPAS, T. & BRUN, J. P. 1993. Salt flow and diapirism related to extension at crustal scale. *Tectonophysics*, **228**, 349–362.

NALPAS, T., GYÖRFI, I., GUILLOCHEAU, F., LAFONT, F. & HOMEWOOD, P. 1999. Influence de la charge sédimentaire sur le développement d'anticlinaux synsédimentaires. Modélisation analogique et exemple de terrain (bordure sud du bassin de Jaca). *Bulletin de la Société Géologique de France*, **170**, 733–740.

NALPAS, T., GAPAIS, D., VERGÉS, J., BARRIER, L., GESTAIN, V., LEROUX, G. & ROUBY, D. 2003. Effects of rate and nature of synkinematic sedimentation on the growth of compressive structures constrained by analogue models and field examples. *In*: MCCANN, T. & SAINTOT, A. (eds) *Tracing Tectonic Deformation Using the Sedimentary Record*. Geological Society, London, Special Publications, **208**, 307–319.

NARR, W. & SUPPE, J. 1994. Kinematics of basement-involved compressive structures. *American Journal of Science*, **294**, 802–860.

O'BRIEN, C. A. E. 1950. Tectonic problems of the oil field belt of southwest Iran. *In*: BUITER, J. (ed.) *Proceedings of 18th International Geological Congress, Great Britain, Part 6*, 45–58.

OVEISI, B., LAVÉ, J. & VAN DER BEEK, P. 2007. Rates and processes of active folding evidenced by Pleistocene terraces at the central Zagros front (Iran). *In*: LACOMBE, O., LAVÉ, J., ROURE, F. & VERGÉS, J. (eds) *Thrust Belts and Foreland Basins from Fold Kinematics to Hydrocarbon Systems*. Frontiers in Earth Sciences, Springer, Berlin, 267–288.

PRUCHA, J. J., GRAHAM, J. A. & NICKELSONE, R. P.
1965. Basement controlled deformation in Wyoming
province of Rocky Mountain foreland. *AAPG Bulletin*,
49, 966–992.

RAMSEY, L. A., WALKER, R. T. & JACKSON, J. 2008.
Fold evolution and drainage development in the
Zagros mountains of Fars province, SE Iran. *Basin
Research*, **20**, 23–48.

SANS, M. & VERGÉS, J. 1995. Fold development related
to contractional salt tectonics: Southestern Pyrenean
Thrust Front, Spain. *In*: JACKSON, P. A., ROBERTS,
D. G. & SNELSON, S. (eds) *Salt Tectonics: a Global
Perspective*. American Association Petroleum Geol-
ogists, Memoirs, **65**, 369–378.

SATTARZADEH, Y., COSGROVE, J. W. & VITA-FINZI, C.
2000. The interplay of faulting and folding during the
evolution of the Zagros deformation belt. *In*: COS-
GROVE, J. W. & AMEEN, M. S. (eds) *Forced Folds
and Fractures*. Geological Society, London, Special
Publications, **169**, 187–196.

SEPEHR, M. & COSGROVE, J. W. 2004. Structural framework of the Zagros Fold–Thrust Belt, Iran. *Marine and
Petroleum Geology*, **21**, 829–843.

SEPEHR, M., COSGROVE, J. W. & MOIENI, M. 2006. The
impact of cover rock rheology on the style of folding in
the Zagros Fold–Thrust Belt. *Tectonophysics*, **427**,
265–281.

SETUDEHNIA, A. 1967. *Kuh-e Anaran Geological Compilation Map 1:100 000 scale (Sheet 25466 E)*. Iranian
Oil Operating Companies (IOOC), Tehran.

SHARP, I. R., GAWTHORPE, R. L., UNDERHILL, J. R. &
GUPTA, S. 2000. Fault-propagation folding in extensional settings: examples of structural style and
synrift sedimentary response from the Suez rift,
Sinai, Egypt. *Geological Society of America Bulletin*,
112, 1877–1899.

SHARP, I., GILLESPIE, P. *ET AL.* 2010. Stratigraphic
architecture and fracture controlled dolomitization
of the Cretaceous Khami and Bangestan groups: an
outcrop case study, Zagros Mountains, Iran. *In*: VAN
BUCHEM, F., GERDES, K. & ESTEBAN, M. (eds)
*Mesozoic and Cenozoic Carbonate Systems of the
Mediterranean and the Middle East: Stratigraphic
and Diagenetic Reference Models*. Geological
Society, London, Special Publications, **329**, 341–394.

SHERKATI, S., MOLINARO, M., FRIZON DE LAMOTTE,
D. & LETOUZEY, J. 2005. Detachment folding in the
Central and Eastern Zagros fold-belt (Iran). *Journal
of Structural Geology*, **27**, 1680–1696.

SHERKATI, S., LETOUZEY, J. & FRIZON DE LAMOTTE, D.
2006. The Central Zagros fold–thrust belt (Iran): new
insights from seismic data, field observation and sandbox modeling. *Tectonics*, **25**, article number TC4007.

STEARNS, D. W. 1978. Faulting and forced folding in the
Rocky Mountain foreland. *In*: MATTHEWS, V., III
(ed.) *Laramide Folding Associated with Basement
Block Faulting in the Western United States*. Geological Society of America, Memoirs, **151**, 1–37.

STONE, D. S. 1993. Basement-involved thrust-generated
folds as seismically imaged in the subsurface of
the central Rocky Mountain foreland. *In*: SCHMIDT,
C. J., CHASE, R. B. & ERSLEV, E. A. (eds)
*Laramide Basement Deformation in the Rocky
Mountain Foreland of the western United States*.
Geological Society of America, Special Papers, **280**,
271–318.

STORTI, F. & MCCLAY, K. 1995. Influence of syntectonic
sedimentation on thrust wedges in analogue models.
Geology, **23**, 999–1002.

TALEBIAN, M. & JACKSON, J. 2004. A reappraisal of
earthquake focal mechanisms and shortening in the
Zagros mountains of Iran. *Geophysical Journal International*, **156**, 506–526.

TATAR, M., HATZFELD, D. & GHAFORI-ASHTIANY, M.
2004. Tectonics of the Central Zagros (Iran) deduced
from microearthquake seismicity. *Geophysical
Journal International*, **156**, 255–266.

VENDEVILLE, B., COBBOLD, P. R., DAVY, P., BRUN,
J. P. & CHOUKROUNE, P. 1987. Physical models
of extensional tectonics at various scales. *In*:
COWARD, M. P., DEWEY, J. F. & HANCKOCK,
P. L. (eds) *Continental Extensional Tectonics*. Geological Society, London, Special Publications, **28**,
95–107.

VERGÉS, J. 2007. Drainage responses to oblique and
lateral thrust ramps: a review. *In*: NICHOLS, G.,
PAOLA, C. & WILLIAMS, E. (eds) *Sedimentary Pro-
cesses, Environments and Basins: a Tribute to Peter
Friend*. International Association of Sedimentologists
Special Publications, **38**, 29–47.

VERGÉS, J., GOODARZI, M. G. H., EMAMI, H., KARPUZ,
R., EFSTATIOU, J. & GILLESPIE, P. 2009. Multiple
detachment folding in Pusht-e Kuh Arc, Zagros. Role
of mechanical stratigraphy. *In*: MCCLAY, K., SHAW,
J. & SUPPE, J. (eds) *Thrust Fault Related Folding*.
AAPG Memoirs, **94**, 1–26.

WEIJERMARS, R., JACKSON, M. P. A. & VENDEVILLE,
B. C. 1993. Rheological and tectonic modeling of
salt provinces. *Tectonophysics*, **217**, 143–174.

WENNBERG, O. P., AZIZZADEH, M. *ET AL.* 2007. The
Khaviz Anticline: an outcrop analogue to giant fractured Asmari Formation reservoirs in SW Iran. *In*:
LONERGAN, L., JOLLY, R. J. H., RAWNSLEY, K. &
SANDERSON, D. J. (eds) *Fractured Reservoirs*. Geological Society, London, Special Publications, **270**,
23–42.

Mesozoic deep-water carbonate deposits from the southern Tethyan passive margin in Iran (Pichakun nappes, Neyriz area): biostratigraphy, facies sedimentology and sequence stratigraphy

CÉCILE ROBIN[1]*, SPELA GORICAN[2], FRANÇOIS GUILLOCHEAU[1], PHILIPPE RAZIN[3], GILLES DROMART[4] & HAMID MOSAFFA[5]

[1]*Géosciences-Rennes, UMR 6118 Université de Rennes 1–CNRS, 35042 Rennes cedex, France*

[2]*Institute of Paleontology, ZRC SAZU, Gosposka 13, Sl-1000 Ljubljana, Slovenia*

[3]*EGID, Université de Bordeaux III, BP 06, 33401 Talence Cedex, France*

[4]*Laboratoire Sciences Terre, UMR 5570 Ecole Normale Supérieure de Lyon–Université de Lyon 1–CNRS, 69364 Lyon Cedex, France*

[5]*Geological Survey of Iran, PO Box 13185-1494, Tehran, Iran*

Corresponding author (e-mail: cecile.robin@univ-rennes1.fr)

Abstract: The objective of this work is to study the Mesozoic turbiditic sediments from the southern Tethys margin in Iran. These sediments are exposed as nappes in the Pichakun Mountains (i.e. the Zagros Mountains in the Neyriz area), which inverted during latest Cretaceous time. Radiolarians are used to both define and date four main lithostratigraphic formations: (1) the Bar Er Formation (undated, probably Late Triassic to Early Jurassic); (2) the Darreh Juve Formation (Aalenian–early Bajocian to middle Callovian–early Oxfordian); (3) the Imamzadeh Formation (middle Callovian–early Oxfordian to Aptian); (4) the Neghareh Khaneh Formation (late Aptian to Turonian–Coniacian). Most of the sediments are deep-sea gravity-flow lobe deposits. Channel deposits occurred during the Bajocian (i.e. the Darreh Juve Fm) and deeply incised channels (canyons?) occurred during the Albian (i.e. the Neghareh Khaneh Fm). Twenty-seven facies, grouped into eight facies associations, are defined. Based on a sequence stratigraphic study (i.e. the stacking pattern), five second-order cycles (10–30 Ma duration), defined between two successive distal facies time-intervals, are characterized: (1) the J2 (Toarcian?–middle Oxfordian, unconformity: Late Toarcian–Aalenian); (2) the J3 (middle Oxfordian–Berriasian, unconformity: middle? Tithonian); (3) the K1.1 (Berriasian–undated top); (4) the K1.2 (undated base–early Aptian, unconformity: late Hauterivian); (5) the K1.3 (early Aptian–at least Turonian–Coniacian, unconformity: Aptian–Albian boundary). The most important tectonic event recorded occurred at the Aptian–Albian boundary (a deposition of olistoliths, from a few metres to 100 m thick, in debris flows; related to Austrian deformations). The Arabian-scale late Toarcian and early Tithonian deformations have been recorded as unconformities. It is expected that another tectonic event occurred during the late Hauterivian. The unconformity of cycle K1.1 could be a late Valanginian eustatic fall of climatic origin.

During Jurassic and Cretaceous times, the Tethyan southern margin was a broad carbonate platform, stretching from Oman to Tunisia (Dercourt *et al.* 2000; Philip 2003). Because of their economic interest, these carbonate platform sediments have been extensively studied and several sequence stratigraphic and palaeogeographical publications are available (Murris 1980; Le Nindre *et al.* 1990; Grabowski & Norton 1994; Al-Husseini 1997; Sharland *et al.* 2001; Ziegler 2001; Haq & Al-Qahtani 2005). Conversely, only a limited number of available studies are dedicated to the deep-sea and the outer parts of the margin. These studies are mainly concerned with the nappes of deep-sea sediments

of the Oman Mountains (Bernouilli & Weissert 1987; Béchennec *et al.* 1990; Le Métour *et al.* 1995; Pillevuit *et al.* 1997; Blechschmidt *et al.* 2004). Furthermore, most of these studies deal with palaeogeographical reconstructions, with an emphasis on the location, on a shelf–abyssal plain profile, of shallow isolated carbonate platforms (interpreted as seamount caps) and their relationships with different turbiditic systems.

A similar pattern seems to have developed along the southern Tethyan margin in Iran, with the occurrence of the Kermanshah and Neyriz exotic deposits (Ricou 1976; Hallam 1976; Braud 1989). Unfortunately, no sedimentological or

From: LETURMY, P. & ROBIN, C. (eds) *Tectonic and Stratigraphic Evolution of Zagros and Makran during the Mesozoic–Cenozoic.* Geological Society, London, Special Publications, **330**, 179–210.
DOI: 10.1144/SP330.10   0305-8719/10/$15.00 © The Geological Society of London 2010.

sequence stratigraphic studies have been performed on these carbonate platforms or the associated turbiditic facies.

This study focuses on the turbiditic facies of the Iranian palaeomargin in the Neyriz area (i.e. the Zagros Mountains). The primary objective of the present study is to use radiolarians to date deepwater sediments from the Iranian Mesozoic Tethyan southern margin and to characterize their depositional environments, the associated stratigraphic cycles, and their tectonic or eustatic causes.

Geological setting

The studied outcrops are located SW of Neyriz (Fig. 1), in the Zagros Mountains along the 'Crush' Zone (i.e. the Zagros Thrust Zone or High Zagros Belt; Stocklin 1968; Berberian & King 1981; Alavi 1994), which is the suture of the Tethys Ocean closure. All of the sediments from the outer margin and the oceanic crust crop out as nappes, which have been thrust over the autochthonous shallow carbonate deposits of the Arabian platform. Four tectonic units were defined in the study area (Ricou 1976; Hallam 1976): (1) the autochthonous Arabian platform (Sarvak Fm, Cenomanian–Turonian); (2) the Pichakun nappes (Pichakun Series, Late Triassic to Cenomanian); (3) the 'Mélange' or Bakhtegan Beds of Hallam (1976), with carbonate blocks of Late Triassic age (Ricou 1976); (4) the ophiolite suite (Neyriz Ophiolites). The three former formations are stacked, thrust units, which are unconformably overlain by shallow-water reefal limestones of Maastrichtian age (James & Wynd 1965). These latest Cretaceous shallow-water sediments were thrust during the Pliocene (Ricou 1976) by the Sanadaj–Sirjan metamorphic units along the Main Zagros Thrust. A margin inversion took place during the Late Cretaceous, between the Cenomanian (the youngest sediments of the Pichakun nappes) and the Maastrichtian (unconformably overlapping sediments).

The 'Mélange' or Bakhtegan Beds and equivalent rocks along the Arabian platform (i.e. the Bisitoun or Bisotun Limestones in the Kermanshah area (Braud 1989), and the Kawr Group in Oman (Bernouilli & Weissert 1987; Pillevuit *et al.* 1997), are interpreted as deposits on top of an isolated carbonate platform located on top of a seamount (Searle & Graham 1982; Stampfli *et al.* 1991; Dercourt *et al.* 2000). The Pichakun unit is made up of turbiditic deposits ('flysch') located between the Arabian carbonate platform and the isolated carbonate platform (or possible seamount) named the 'Mélange' by Ricou (1976) (i.e. the Neyriz Seamount). The nature of the crust below these turbiditic sediments and the significance of these isolated carbonate platforms are a matter of discussion: is the crust oceanic (Searle & Graham 1982) or thinned continental (Braud 1989; Lapierre *et al.* 2004)? Do carbonates represent real top seamount sediments (Searle & Graham 1982) or deposits of subsiding tilted blocks (Braud 1989)?

The Pichakun nappes are exposed in the Pichakun Mountains, between Lake Bakhtegan and Lake Tashk (Fig. 1b), and consist of a stack of at least 10 southward thrust units (Ricou 1976). Because of refolding of the stacked nappes, the geometrical relationship of the thrust units between the southern and northern sides of these nappes remains unclear (Ricou 1976).

Methods

Since the studies by Hallam (1976) and Ricou (1976), no study has applied the modern principles of sedimentology and stratigraphy to deep-sea marine sediments from the southern Tethyan margin in Iran.

The Pichakun nappes (Fig. 1b), which are composed of different types of gravity-flow deposits, do not represent the whole margin. Most of the tectonic units either have been eroded or do not crop out. Thus, this study does not aim to reconstruct the complete geometry of the margin. Only correlations between measured sections, based on biostratigraphic data (radiolarians) and the stacking pattern (sequence stratigraphy) of the gravity-flow deposits, were carried out.

Three main tectonic units (Ricou 1976) were studied (Fig. 1b): the Imamzadeh unit (three measured sections); the Darreh Juve unit (one measured section); the Bar Er unit (one measured section).

Lithostratigraphy and biostratigraphy

The definition of the lithostratigraphic units is based on the recommendations of the International Stratigraphy Guide of the International Commission of Stratigraphy (ICS; http://www.stratigraphy.org).

The biostratigraphy is based on the study of radiolarians in cherts and siliceous limestones. Siliceous limestones were processed with acetic acid (10%) and then with hydrofluoric (5%) acid, whereas the chert samples were processed with hydrofluoric acid only. Dating is primarily based on the following zonations: Baumgartner *et al.* (1995) for the Middle and Late Jurassic; Jud (1994) for the late Tithonian to Barremian; O'Dogherty (1994) for the Aptian to Turonian interval.

Facies sedimentology

Several sedimentary facies (lithology, sedimentary structures, rare trace fossils) were defined and

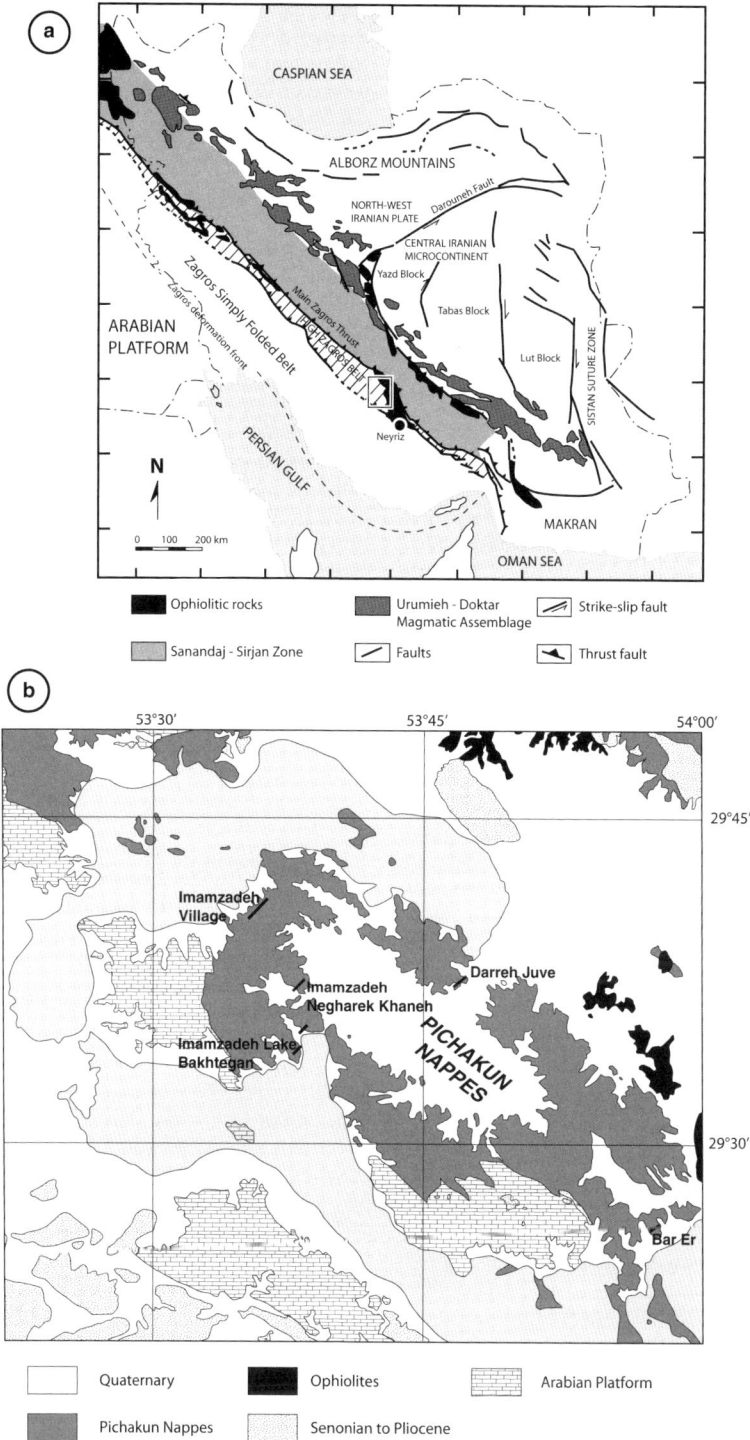

Fig. 1. Location map of the study area. (**a**) Structural map of Iran (modified from Stocklin 1968). (**b**) Simplified geological map of the Pichakun Nappes and location of the sections (from the National Iranian Oil Company, geological map of Shiraz at 1:250 000, Tehran 1979, 2nd ed.).

grouped into facies associations. Facies associations are more characteristic of depositional environments, rather than, as in the case of deep-sea gravity-flow deposits, facies that record both elementary hydrodynamic processes and the nature of upstream sediment sources. A palaeocurrent study was performed to determine the location of the source areas and the hydrodynamic processes. All of the studied sections are now preserved as thrust units. Palaeomagnetic analyses to quantify possible tectonic rotations occurring at the time of nappe emplacements were not carried out on these outcrops. This implies that palaeocurrent data must be used with caution.

Sequence stratigraphy: correlations

The application of sequence stratigraphy, and mainly the stacking pattern technique (i.e. the correlation of wells or outcropping sections, Van Wagoner *et al.* 1988, 1990; Homewood *et al.* 1999), to deep-sea gravity-flow deposits has been mainly carried out in siliciclastic systems (e.g. Mutti 1985; Posamentier *et al.* 1988). The interpretation of stratigraphic sequences in carbonate gravity-flow deposits is more complex than for terrigenous sediments (Eberli 1991). Carbonate production on the shelf and thus the nature and volume of sediments available for resedimentation along the slope are highly dependent on the response of the carbonate factory to relative sea-level fluctuations (Eberli 1991; Handford & Loucks 1993; Pomar 2001). No suitable predictive models are available, in contrast to those that are refined for siliciclastic gravity-flow deposits (e.g. Mutti 1985; Mutti *et al.* 1999; Posamentier *et al.* 1988). Since the publication of the classical works by Eberli (1991) and Schlager *et al.* (1994), most researchers agree that carbonate turbiditic deposition occurs during high relative sea level (i.e. highstand shedding) and that almost no deposition occurs during low relative sea level (i.e. emersion and no carbonate sediment production). In contrast, most of the siliciclastic turbidite deposits occur at the time of the maximum rate of relative sea-level fall (Posamentier *et al.* 1988).

The 'stacking pattern' technique is based on the migration of the depositional profile along a proximal (landward) to distal (seaward) trend (Van Wagoner *et al.* 1988, 1990; Homewood *et al.* 1999). This implies the building of an 'ideal' depositional profile that will be the basis for defining the distal-up and proximal-up trends throughout the different sections. However, this general facies model remains an approximation of the real conditions. First, facies may change laterally along the depositional profile; not only along the channels but also on the lobes. The lateral substitution of some facies needs to be taken into account. Second, depositional profiles may change according to the volume and the grain size of the sediments. At a time of low carbonate supply from the platform, condensation or radiolarite deposition has to be expected and the use of the complete depositional model becomes meaningless. Probably, most of the distal–proximal cycles defined in this paper do not record migrations of the same profile, but rather a change in the carbonate supply from the platform.

The relationship between the three depositional sequence surface types on the platform (i.e. unconformity, flooding or transgressive surface and maximum flooding surface) for sedimentary cycles in deep-sea environments is subject to discussion (e.g. Posamentier *et al.* 1988; Mutti *et al.* 1999). The unconformity surface is the only one that is fully accepted by all researchers. It is interpreted as recording a sharp facies change from distal to proximal facies (i.e. a downward shift of facies). In this paper, the most distal facies time interval of a cycle is the equivalent of the maximum flooding surface on the platform. For the most proximal facies time period of the cycle, we have defined what is here named the proximal turn-around surface. This surface either may coincide with the unconformity or may occur later. An *a priori* relationship with the flooding surface on the shelf is not expected.

Lithostratigraphy

Four main lithostratigraphic units, corresponding to the four lithological units recognized by Ricou (1976), are presented in Figure 2. They are mappable units that range from several tens of metres to a few hundreds of metres thick and that correspond to lithostratigraphic formations. All of these new formations (Fm) are named from type-section localities in the Neyriz area; they are, from base to top, the Bar Er Fm, the Darreh Juve Fm, the Imamzadeh Fm and the Neghareh-Khaneh Fm (Fig. 2).

The Bar Er Formation

The Bar Er Fm corresponds to the black marls ('Marnes noires') previously described by Ricou (1976). This formation was defined near the village of Bar Er (29°26.254′N, 53°55.584′E). It is mainly composed of black shales, more or less calcareous, alternating with thick conglomerates (1–25 m). Most of the conglomerates are matrix-supported (clays to silts) with reef clasts a few centimetres to metres in size. Some bioclastic limestones, a few decimetres thick, are interbedded within these alternations. The conglomeratic beds range from a

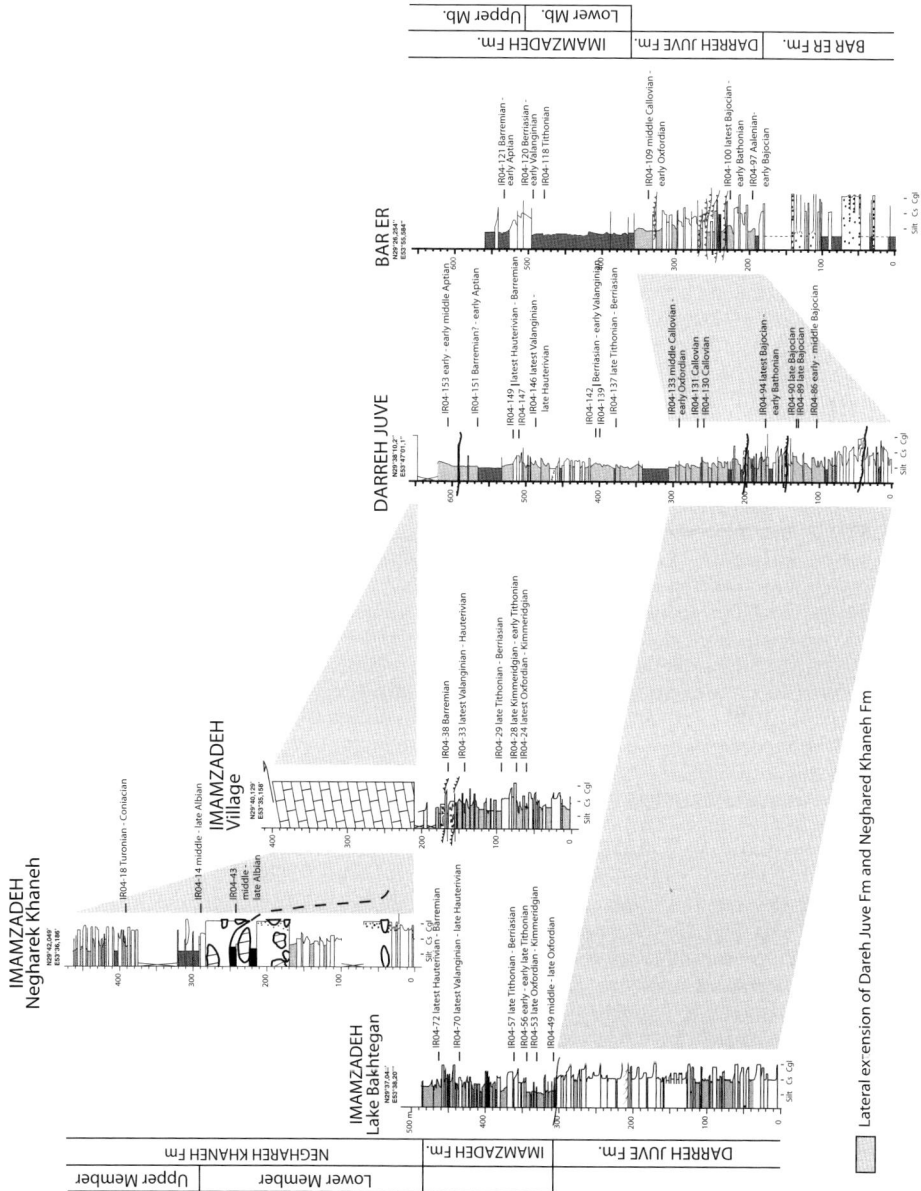

Fig. 2. Lithostratigraphy and biostratigraphy of the five studied sections from the Pichakun area.

few metres to 15 m thick. The base of this formation is always thrust. Its minimum thickness is about 180 m in the Bar Er type-section.

As suggested by Ricou (1976), the large clasts from the reef are resedimented from the carbonate platforms. The reworked foraminifers (M. Lys, cited by Ricou 1976) indicate a minimum Late Triassic age. Therefore, this formation could be Late Triassic to Early Jurassic in age.

The Darreh Juve Formation

The Darreh Juve Formation corresponds to the siliceous limestones ('Calcaires siliceux') described by Ricou (1976). The type-section of this formation is located near the village of Darreh Juve (29°38′10.2″N, 53°47′01.1″E). It is mainly composed of massive limestone rock units (the grain sizes range from sands to conglomerates) with some heterolithic levels (i.e. alternations of siliciclastic claystones or muddy limestones with sand-grained clastic limestones). The thickness of this formation (170 m) can be estimated in the Bar Er section, the only place where the boundary with the underlying formation was observed (Fig. 2). A thickness of 300 m has been preserved (because of the thrust nature of the lower boundary in this unit) in the type-section and in the Imamzadeh–Lake Bakhtegan section. The top of the Darreh Juve Formation could be either sharp, with an erosional surface (i.e. the Imamzadeh–Lake Bakhtegan section) or progressive, with a decrease in the bed thickness. This formation passes upward into the green and/or reddish cherts and shales of the Imamzadeh Formation.

The lithology of this unit changes throughout the various thrust units, from more massive homolithic coarse-grained limestones with matrix-supported conglomerates (i.e. the Imamzadeh–Lake Bakhtegan section) to more heterolithic fine-grained limestones (i.e. the Bar Er section). The massive homolithic facies of this unit are dominated by ooid packstones to grainstones, with some echinoid fragments, or other bioclasts, in a micritic matrix. These coarse-grained limestones alternate with heterolithic levels composed of green and reddish shales and silicified limestones. The ratio between the homolithic and heterolithic levels varies laterally. The more heterolithic facies are dominated by shaly and silicified fine-grained limestone alternations.

The Imamzadeh Formation

The Imamzadeh Fm corresponds to the radiolarites described by Ricou (1976). The type-section of this formation is located along two sections in the Imamzadeh area: near Lake Bakhtegan

(29°37.048′N, 53°38.207′E, the base of the formation) and near the village (29°40.129′N, 53°35.158′E, the top of the formation). It is mainly composed of an alternation of shales to true cherts with silicified limestones. Some coarser-grained bioclastic levels (medium- to coarse-grained sands with rare conglomerates) can occur. The thickness of this formation (which is composed of two members) was estimated to be about 240 m in the Darreh Juve section, the only place where the base and top boundaries are observed (Fig. 2).

The lower member is exposed in the southern part of the Bar Er village. The type-section crops out along the valley, between two hills (29°26.254′N, 53°55.584′E). It is primarily composed of red to green shales, cherts (radiolarites) and silicified limestones (an alternation of muddy limestones and fine sand-grained bioclastic limestones). Its thickness varies from 40 m (the Imamzadeh–Lake Bakhtegan section) to 140 m (the Bar Er section). The lithology of this member changes throughout the various thrust units, from more heterolithic (the Imamzadeh–Lake Bakhtegan section) to more homolithic shaly facies (the Bar Er section).

The upper member has been defined near Lake Bakhtegan, on the northern border of the Bakhtegan Lake and north of the Kuhe Na Anhir hills (29°37.048′N, 53°38.207′E). This member records a more or less sharp transition from the underlying member and is made of an alternation of fine-grained (almost completely silicified) limestones with locally silicified medium- to very coarse sand-grained bioclastic limestones (10 cm to 1 m thick) with some clast-supported conglomerates. The characteristic of this upper member, compared with the lower member, is the lack of shales and primary cherts. Generally, the top contact of this member is thrust.

The Neghareh-Khaneh Formation

The Neghareh-Khaneh Fm corresponds to the conglomeratic limestones ('Calcaires conglomératiques') described by Ricou (1976). The type-section of this formation is located on the outcrops overhanging the sanctuary of Imazadeh, near the hamlet of Neghareh-Khaneh (29°42.049′N, 53°36.186′E). It is made up of massive conglomerates interbedded with alternations of shales and limestones that are a few decimetres thick. The upper part of the formation is composed only of shale and limestone alternations. The base of this formation is sharp and erosional, whereas its top is always thrust. In the type-section, the formation, which is composed of two members, is 450 m thick.

The lower member is made up of thick conglomerates (5–250 m) infilling palaeocanyons with very large blocks (10–200 m). These blocks (true olistoliths) are composed of various types of carbonate platform deposits (mainly inner shelf: mud, stromatolites, oncoids, etc.). Either sand-grained bioclastic limestones or alternations of claystones and cherts may be interbedded with these conglomerates.

The upper member is an alternation of claystones, cherts and clastic limestones. Their grain sizes range from fine-grained limy sands to conglomerates.

Biostratigraphy

Radiolarians were analysed from samples of radiolarian cherts and siliceous limestones from the

Darreh Juve, Imamzadeh and Neghareh Khaneh Formations. The preservation of radiolarians ranges from poor to moderate. A list of stratigraphically important species is given in Table 1. Sample locations are shown in Figure 2. The characteristic species are illustrated in Figures 3–5.

The oldest radiolarians are Aalenian to Bajocian in age (Fig. 3 and Table 1) and were found in the lower part of the Darreh Juve Formation in an interval several tens of metres thick, composed almost entirely of cherts with only rare resedimented-limestone beds. In sample IR04-89 from Darreh Juve, the co-occurrence of *Williriedellum marcucciae* Cortese (FAD in UA Zone 4 of Baumgartner *et al.* 1995) with *Hexasaturnalis suboblongus* (Yao) (the entire range in the Bajocian; see

Table 1. *Radiolarian inventory and age of samples*

Sample	Radiolarians	Age
Darreh Juve section		
IR04-86	*Acaeniotylopsis variatus* (Ozvoldova) *Hsuum* cf. *matsuokai* Isozaki & Matsuda *Thurstonia* sp. *Trillus* sp. *Tympaneides* sp. *Unuma* cf. *typicus* Ichikawa & Yao	Early–middle Bajocian
IR04-89	*Dictyomitrella*? *kamoensis* Mizutani & Kido *Hexasaturnalis suboblongus* (Yao) *Protunuma fusiformis* Ichikawa & Yao *Stichomitra*? *takanoensis* Aita *Tethysetta dhimenaensis* (Baumgartner) *Williriedellum marcucciae* Cortese	Late Bajocian
IR04-90	*Dictyomitrella*? *kamoensis* Mizutani & Kido *Stichomitra*? *takanoensis* Aita *Striatojaponocapsa plicarum* (Yao) *Unuma darnoensis* Kozur	Late Bajocian
IR04-94	*Williriedellum tetragonum* (Matsuoka) *Williriedellum yaoi* (Kozur)	Latest Bajocian–early Bathonian
IR04-130	*Eucyrtidiellum nodosum* Wakita *Eucyrtidiellum ptyctum* (Riedel & Sanfilippo) *Dictyomitrella*? *kamoensis* Mizutani & Kido *Plicaforacapsa catenarum* (Matsuoka) *Praewilliriedellum robustum* (Matsuoka) *Pseudoeucyrtis firma* Hull *Pseudoristola tsunoensis* (Aita) *Striatojaponocapsa conexa* (Matsuoka) *Williriedellum yaoi* (Kozur)	Callovian
IR04-131	*Higumastra imbricata* (Ozvoldova) *Guexella nudata* (Kocher) *Pseudoeucyrtis firma* Hull *Ristola procera* (Pessagno) *Striatojaponocapsa conexa* (Matsuoka)	Callovian
IR04-133	*Angulobracchia purisimaensis* (Pessagno) *Cinguloturris carpatica* Dumitrica *Eucyrtidiellum nodosum* Wakita *Mirifisus guadalupensis* Pessagno	Middle Callovian–early Oxfordian

(Continued)

Table 1. *Continued*

Sample	Radiolarians	Age
IR04-137	*Protunuma japonicus* Matsuoka & Yao *Tritrabs casmaliaensis* (Pessagno) *Williriedellum carpathicum* Dumitrica *Angulobracchia portmanni* Baumgartner *Cinguloturris cylindra* Kemkin & Rudenko *Dicerosaturnalis dicranacanthos* (Squinabol) *Emiluvia chica* (Foreman) *Hsuum mclaughlini* Pessagno & Blome *Neorelumbra buwaydahensis* Kiessling *Pseudodictyomitra carpatica* (Lozyniak) *Tethysetta boesii* (Parona)	Late Tithonian–Berriasian
IR04-139	*Angulobracchia portmanni* Baumgartner *Cinguloturris cylindra* Kemkin & Rudenko *Dicerosaturnalis dicranacanthos* (Squinabol) *Emiluvia chica* (Foreman) *Hiscocapsa kaminogoensis* (Aita) *Hsuum* sp. *Neorelumbra buwaydahensis* Kiessling *Pantanellium squinaboli* (Tan) *Pseudodictyomitra carpatica* (Lozyniak) *Tethysetta boesii* (Parona)	Berriasian–early Valanginian
IR04-142	*Angulobracchia portmanni* Baumgartner *Dicerosaturnalis dicranacanthos* (Squinabol) *Emiluvia chica* (Foreman) *Hiscocapsa kaminogoensis* (Aita) *Hsuum* sp. *Neorelumbra buwaydahensis* Kiessling *Pantanellium squinaboli* (Tan) *Tethysetta boesii* (Parona)	Berriasian–early Valanginian
IR04-146	*Aurisaturnalis variabilis* (Squinabol) *Cecrops septemporatus* (Parona) *Crolanium* sp. *Crucella?* *inflexa* (Rüst) *Dicerosaturnalis dicranacanthos* (Squinabol) *Dicerosaturnalis trizonalis* (Rüst) *Hemicryptocapsa capita* (Tan) *Tethysetta usotanensis* (Tumanda)	Latest Valanginian–late Hauterivian
IR04-147	*Aurisaturnalis carinatus* (Foreman) *Cryptamphorella conara* (Foreman) *Dactyliodiscus lenticulatus* (Jud) *Dictyomitra pseudoscalaris* Tan *Hiscocapsa uterculus* (Parona) *Tethysetta boesii* (Parona)	Latest Hauterivian–Barremian
IR04-149	*Archaeodictyomitra lacrimula* (Foreman) *Dibolachras tytthopora* Foreman *Hiscocapsa uterculus* (Parona) *Pseudodictyomitra lanceloti* Schaaf *Suna hybum* (Foreman) *Tethysetta boesii* (Parona) *Tethysetta usotanensis* (Tumanda) *Thanarla pacifica* Nakaseko & Nishimura	Latest Hauterivian–Barremian
IR04-151	*Archaeodictyomitra lacrimula* (Foreman) *Aurisaturnalis carinatus perforatus* Dumitrica & Dumitrica-Jud *Dictyomitra pseudoscalaris* Tan *Mictyoditra pseudodecora* (Tan) *Pantanellium* spp. *Pseudodictyomitra* cf. *carpatica* (Lozyniak) *Pseudodictyomitra lanceloti* Schaaf	Barremian?–early Aptian

(*Continued*)

Table 1. *Continued*

Sample	Radiolarians	Age
IR04-153	*Suna hybum* (Foreman) *Tethysetta boesii* (Parona) *Tethysetta usotanensis* (Tumanda) *Thanarla pacifica* Nakaseko & Nishimura *Archaeodictyomitra lacrimula* (Foreman) *Dactyliodiscus lenticulatus* (Jud) *Pseudodocrolanium puga* (Schaaf)	Early–early middle Aptian
Barer Section		
IR04-97	*Elodium pessagnoi* Yeh & Cheng *Hsuum altile* Hori & Otsuka *Hsuum exiguum* Yeh & Cheng *Parahsuum?* *natorense* (El Kadiri) *Praeparvicingula nanoconica* (Hori & Otsuka) *Trillus* sp.	Aalenian–?early Bajocian
IR03-03	*Stichomitra?* *takanoensis* Aita *Williriedellum marcucciae* Cortese *Williriedellum tetragonum* (Matsuoka)	Latest Bajocian–early Bathonian
IR04-109	*Cinguloturris carpatica* Dumitrica *Emiluvia premyogii* Baumgartner *Eucyrtidiellum ptyctum* Riedel & Sanfilippo *Protunuma japonicus* Matsuoka & Yao *Spongocapsula palmerae* Pessagno *Transhsuum brevicostatum* (Ozvoldova) *Tritrabs casmaliaensis* (Pessagno)	Middle Callovian–early Oxfordian
IR04-118	*Archaeodictyomitra apiarium* (Rüst) *Dicerosaturnalis dicranacanthos* (Squinabol) *Emiluvia chica* Foreman *Eucyrtidiellum pyramis* (Aita) *Mirifisus dianae minor* (Baumgartner) *Paronaella?* *tubulata* Steiger *Podocapsa amphitreptera* Foreman *Protunuma japonicus* Matsuoka & Yao *Syringocapsa longituba* Steiger & Steiger	Tithonian
IR04-120	*Archaeodictyomitra apiarium* (Rüst) *Cinguloturris cylindra* Kemkin & Rudenko *Dicerosaturnalis dicranacanthos* (Squinabol) *Fultacapsa tricornis* (Jud) *Mirifisus dianae minor* (Baumgartner) *Obesacapsula polyedra* (Steiger) *Pseudodictyomitra carpatica* (Lozyniak) *Pantanellium squinaboli* (Tan) *Ristola cretacea* (Baumgartner) *Tethysetta boesii* (Parona)	Berriasian–early Valanginian
IR04-121	*Aurisaturnalis carinatus perforatus* Dumitrica & Dumitrica-Jud *Archaeodictyomitra lacrimula* (Foreman) *Cryptamphorella conara* (Foreman) *Dicerosaturnalis trizonalis* (Rüst) *Podobursa* sp. *Tethysetta usotanensis* (Tumanda)	Barremian–early Aptian
Imamzadeh–Lake Bakhtegan section		
IR04-49	*Acastea diaphorogona* (Foreman) *Cinguloturris carpatica* Dumitrica *Emiluvia ordinaria* Ozvoldova *Emiluvia orea* Baumgartner *Protunuma japonicus* Matsuoka & Yao	Middle–late Oxfordian

(Continued)

Table 1. *Continued*

Sample	Radiolarians	Age
	Sethocapsa aff. *horokanaiensis* Kawabata	
	Tethysetta mashitaensis (Mizutani)	
	Transhsuum brevicostatum (Ozvoldova)	
	Williriedellum crystallinum Dumitrica	
	Zhamoidellum ovum Dumitrica	
IR04-53	*Cinguloturris carpatica* Dumitrica	Latest Oxfordian–Kimmeridgian
	Mirifusus dianae (Karrer)	
	Orbiculiforma? *lowreyensis* Pessagno	
	Parahsuum sp.	
	Paronaella cava (Ozvoldova)	
	Podocapsa amphitreptera Foreman	
	Protunuma japonicus Matsuoka & Yao	
	Pseudoeucyrtis reticularis Matsuoka & Yao	
	Ristola altissima altissima (Rüst)	
	Spongocapsula perampla (Rüst)	
	Syringocapsa longituba Steiger & Steiger	
	Tethysetta mashitaensis (Mizutani)	
	Williriedellum crystallinum Dumitrica	
	Zhamoidellum ovum Dumitrica	
IR04-56	*Archaeodictyomitra apiarium* (Rüst)	Early–early late Tithonian
	Cinguloturris fusiforma Hori	
	Cinguloturris cylindra Kemkin & Rudenko	
	Dicerosaturnalis trizonalis (Rüst)	
	Eucyrtidiellum pyramis (Aita)	
	Mirifusus dianae (Karrer)	
	Podocapsa amphitreptera Foreman	
	Protunuma japonicus Matsuoka & Yao	
	Pseudodictyomitra carpatica (Lozyniak)	
	Ristola cretacea (Baumgartner)	
	Sethocapsa horokanaiensis Kawabata	
	Syringocapsa longituba Steiger & Steiger	
IR04-57	*Cinguloturris cylindra* Kemkin & Rudenko	Late Tithonian–Berriasian
	Deviatus diamphidius (Foreman)	
	Dicerosaturnalis dicranacanthos (Squinabol)	
	Emiluvia chica (Foreman)	
	Emiluvia hopsoni Pessagno	
	Eucyrtidiellum pyramis (Aita)	
	Hiscocapsa kaminogoensis (Aita)	
	Hsuum raricostatum Jud	
	Mirifisus dianae minor (Baumgartner)	
	Neorelumbra tippitae Kiessling	
	Pantanellium squinaboli (Tan)	
	Paronaella? *tubulata* Steiger	
	Praeparvicingula cosmoconica (Foreman)	
	Pseudodictyomitra carpatica (Lozyniak)	
	Fultacapsa tricornis (Jud)	
	Svinitzium depressum (Baumgartner)	
	Tethysetta boesii (Parona)	
	Tricolocapsa? *campana* Kiessling	
	Williriedellum aff. *crystallinum* Dumitrica	
IR04-70	*Acaeniotyle umbilicata* (Rüst)	Latest Valanginian–late Hauterivian
	Aurisaturnalis variabilis variabilis (Squinabol)	
	Cecrops septemporatus (Parona)	
	Crolanium bipodium (Parona)	
	Crucella bossoensis Jud	
	Dicerosaturnalis dicranacanthos (Squinabol)	
	Dictyomitra pseudoscalaris Tan	
	Hemicryptocapsa capita (Tan)	

(*Continued*)

Table 1. *Continued*

Sample	Radiolarians	Age
IR04-72	*Hiscocapsa kaminogoensis* (Aita) *Suna hybum* (Foreman) *Syringocapsa limatum* Foreman *Tethysetta usotanensis* (Tumanda) *Archaeodictyomitra lacrimula* (Foreman) *Crolanium* sp. *Dibolachras tytthopora* Foreman *Dicerosaturnalis dicranacanthos* (Squinabol) *Dicerosaturnalis trizonalis* (Rüst) *Hiscocapsa uterculus* (Parona) *Pantanellium* sp. *Pseudodictyomitra lanceloti* Schaaf *Suna hybum* (Foreman) *Tethysetta boesii* (Parona) *Thanarla pacifica* Nakaseko & Nishimura	Latest Hauterivian–Barremian

Imamzadeh Village Section

Sample	Radiolarians	Age
IR04-24	*Cinguloturris fusiforma* Hori *Emiluvia ordinaria* Ozvoldova *Eucyrtidiellum ptyctum* (Riedel & Sanfilippo) *Orbiculiforma? lowreyensis* Pessagno *Pantanellium oligoporum* (Vinassa) *Podocapsa amphitreptera* Foreman *Protunuma japonicus* Matsuoka & Yao *Pseudoeucyrtis reticularis* Matsuoka & Yao *Ristola altissima altissima* (Rüst) *Syringocapsa longituba* Steiger & Steiger *Tetratrabs bulbosa* (Baumgartner) *Williriedellum crystallinum* Dumitrica *Zhamoidellum ovum* Dumitrica	Latest Oxfordian–Kimmeridgian
IR04-28	*Archaeodictyomitra apiarium* (Rüst) *Emiluvia ordinaria* Ozvoldova *Eucyrtidiellum ptyctum* (Riedel & Sanfilippo) *Hsuum mclaughlini* Pessagno & Blome *Mirifusus dianae dianae* (Karrer) *Mirifusus dianae minor* (Baumgartner) *Pantanellium oligoporum* (Vinassa) *Podocapsa amphitreptera* Foreman *Protunuma japonicus* Matsuoka & Yao *Pseudoeucyrtis reticularis* Matsuoka & Yao *Syringocapsa longituba* Steiger & Steiger *Williriedellum crystallinum* Dumitrica *Zhamoidellum ovum* Dumitrica	Late Kimmeridgian–early Tithonian
IR04-29	*Angulobracchia portmanni* Baumgartner *Archaeodictyomitra apiarium* (Rüst) *Cinguloturris cylindra* Kemkin & Rudenko *Emiluvia chica* (Foreman) *Hıscocapsa kaminogoensis* (Aita) *Pantanellium squinaboli* (Tan) *Praeparvicingula cosmoconica* (Foreman) *Svinitzium depressum* (Baumgartner) *Tethysetta boesii* (Parona)	Late Tithonian–Berriasian
IR04-33	*Acaeniotyle umbilicata* (Rüst) *Cecrops septemporatus* (Parona) *Crolanium* sp. *Crucella bossoensis* Jud *Dicerosaturnalis dicranacanthos* (Squinabol)	Latest Valanginian–Hauterivian

(Continued)

Table 1. *Continued*

Sample	Radiolarians	Age
IR04-38	*Dicerosaturnalis trizonalis* (Rüst) *Hiscocapsa uterculus* (Parona) *Mictyoditra pseudodecora* (Tan) *Suna hybum* (Foreman) *Archaeodictyomitra lacrimula* (Foreman) *Aurisaturnalis carinatus perforatus* Dumitrica & Dumitrica-Jud *Dicerosaturnalis dicranacanthos* (Squinabol) *Dicerosaturnalis trizonalis* (Rüst) *Cecrops septemporatus* (Parona) *Hemicryptocapsa capita* (Tan) *Mictyoditra* sp. *Pantanellium* sp. *Tethysetta boesii* (Parona)	Barremian

Imamzadeh Negharek Khaneh Section

IR04-43	*Archaeocenosphaera? mellifera* O'Dogherty *Holocryptocanium barbui* Dumitrica *Mita gracilis* (Squinabol) *Rhopalosyringium mosquense* (Smirnova & Aliev)	Middle–late Albian
IR04-14	*Diacanthocapsa* sp. *Dictyomitra montisserei* (Squinabol) *Holocryptocanium barbui* Dumitrica *Mita gracilis* (Squinabol) *Mita obesa* (Squinabol) *Mita spoletoensis* (O'Dogherty) *Pseudodictyomitra lodogaensis* Pessagno *Rhopalosyringium adriaticum* O'Dogherty *Rhopalosyringium mosquense* (Smirnova & Aliev)	Middle–late Albian
IR04-18	*Alievium* sp. *Annikaella omanensis* De Wever, Bourdillon-de Grissac & Beurrier *Archaeodictyomitra squinaboli* (Pessagno) *Dictyomitra formosa* Squinabol *Hemicryptocapsa polyhedra* Dumitrica *Patellula verteroensis* (Pessagno) *Theocampe ascaglia* Foreman *Theocampe urna* (Foreman)	Turonian–Coniacian

Dumitrica & Dumitrica-Jud 2005) indicates a late Bajocian age for the top of this cherty interval. *Unuma* cf. *typicus* Ichikawa & Yao in sample IR04-86, below, suggests that this sample is probably not older than the early Bajocian (FAD of *U. typicus* in UA Zone 3; i.e. the early–middle Bajocian). Sample IR04-97 from Bar Er contains typical Aalenian species such as *Elodium pessagnoi* Yeh & Cheng, *Hsuum altile* Hori & Otsuka, and *H. exiguum* Yeh & Cheng (see Gorican *et al.* 2006), together with *Parahsuum? natorense* (El Kadiri) (LAD in UA Zone 3).

In the overlying succession, thin chert is interbedded within calcareous turbidites and only rarely contains determinable radiolarians (Fig. 3, 16–18). A latest Bajocian–early Bathonian age (UA Zone 5) was determined on the basis of *Williriedellum*

tetragonum (Matsuoka) in samples IR04-94 (Darreh Juve) and IR03-03 (Bar Er).

Callovian to early Oxfordian (UA Zones 7 and 8) radiolarians were obtained from the top of the Darreh Juve Formation in the Darreh Juve and Bar Er sections. The most characteristic species, which last occurs in UA Zone 7, is *Striatojaponocapsa conexa* (Matsuoka), which is abundant in sample IR04-130 (Fig. 3, 19–28) and is also present in sample IR04-131. In sample IR04-133 and in the correlative sample IR04-109 from Bar Er, only long-ranging species occur. However, the absence of *Striatojaponocapsa conexa* suggests that these two samples are younger (i.e. probably assignable to UA Zone 8; middle Callovian–early Oxfordian).

Middle–late Oxfordian to mid-Tithonian radiolarians (UA Zones 9–12; Fig. 4, 1–13) were found in the proximal successions of the Imamzadeh Formation in the Imamzadeh sections. The oldest sample, IR04-49, already contains *Emiluvia ordinaria* Ozvoldova, which first appears in UA Zone 9 (mid–late Oxfordian). *Williriedellum crystallinum* Dumitrica was also identified. *Podocapsa amphitreptera* Foreman was the most characteristic species in the following samples: IR04-53 to 56 and IR04-24 to 28. Other species, not extending above UA Zone 11 (late Kimmeridgian–early Tithonian), are found together in sample IR04-28; for example *Emiluvia ordinaria* Ozvoldova, *Pseudoeucyrtis reticularis* Matsuoka & Yao and *Eucyrtidiellum ptyctum* (Riedel & Sanfilippo). Sample IR04-56 contains *Cinguloturris cylindra* Kemkin & Rudenko, *Eucyrtidiellum pyramis* (Aita) and *Ristola cretacea* (Baumgartner), together with *Protunuma japonicus* Matsuoka & Yao and *Syringocapsa longituba* Steiger & Steiger (determined as *S. spinellifera* Baumgartner by Baumgartner *et al.* 1995), and is thus assigned to UA Zone 12 (early–early late Tithonian). *Podocapsa amphitreptera* starts in UA Zone 9 of Baumgartner *et al.* (1995), but is found somewhat later in other zonations (e.g. Gorican 1994). Sample IR04-49 lacks *P. amphitreptera* and is therefore assigned to middle–late Oxfordian. The overlying samples are from the latest Oxfordian to mid-Tithonian. Time-equivalent deposits of the distal successions from Darreh Juve and Bar Er are shales devoid of determinable radiolarians. The youngest Jurassic cherts of the proximal successions (sample IR04-56) are approximately correlative with the transition from shale to pure radiolarian chert in distal successions (see sample IR04-118, below).

Late Tithonian to Aptian radiolarian assemblages were obtained from both the proximal and distal successions. In both settings, the radiolarians are relatively well preserved throughout the sections. In this time interval, three major radiolarian turnovers occur around the following horizons: early–late Valanginian, Hauterivian–Barremian and early–middle Aptian (Jud 1994; O'Dogherty 1994; O'Dogherty & Guex 2002). It is therefore relatively easy to divide this interval into three main assemblages.

The lowest assemblage (i.e. late Tithonian to early Valanginian; Fig. 4, 14–26) is characterized by the first appearance of *Angulobracchia portmanni* Baumgartner, *Paronaella? tubulata* Steiger, *Praeparvicingula cosmoconica* (Foreman) and *Svinitzium depressum* (Baumgartner) in UA Zone 13 of Baumgartner *et al.* (1995), and corresponds to zones D1 to base D2 of Jud (1994). Therefore, samples IR04-29, IR04-57 and IR04-137 are certainly at least late Tithonian in age. For sample IR04-118, the distinction between early and late Tithonian is not possible because *Paronaella?*

tubulata Steiger (FAD in UA Zone 13) coexists with *Protunuma japonicus* Matsuoka & Yao and *Syringocapsa longituba* Steiger & Steiger (both LAD in UA Zone 12). Late Tithonian to early Valanginian radiolarians were recovered from pure radiolarian cherts in the Darreh Juve and Bar Er sections (samples IR04-137 to 142 and IR04-120). The common species in this earliest Cretaceous assemblage is also *Cingulotarris cylindra* Kemkin & Rudenko, which does not extend above the Valanginian.

The most distinctive species in the second early Cretaceous assemblage (Fig. 5, 1–15) is *Cecrops septemporatus* (Parona), which ranges from the late Valanginian to the Barremian (zones E2–G1 of Jud 1994). It is also important to note the simultaneous first appearance of *Crolanium* spp. (zone E2) and somewhat later appearance of *Suna hybum* (Foreman) (zone F1). This assemblage was found in the Imamzadeh sections (samples IR04-70 and IR04-33) and at Darreh Juve (IR04-146). *Aurisaturnalis variabilis* (Squinabol), ranging from the latest Valanginian to the late Hauterivian (zones F1–F3) is common in samples IR04-70 and IR04-146.

The latest Hauterivian to early Aptian (Fig. 5, 6–11) is characterized by *Aurisaturnalis carinatus* (Foreman), which rapidly evolved from *Aurisaturnalis variabilis* (Squinabol) across the Hauterivian–Barremian boundary and became extinct in the early Aptian (Dumitrica & Dumitrica-Jud 1995). The transitional subspecies are very rare around the boundary level; however, the advanced subspecies *A. carinatus perforatus* Dumitrica & Dumitrica-Jud is more common, easily recognizable and very useful stratigraphically. This subspecies first appears near the top of the Barremian G1 zone (Dumitrica & Dumitrica-Jud 1995). The assemblage further contains many species that became extinct either during the early Aptian [e.g. *Tethysetta boesii* (Parona), *Tethysetta usotanensis* (Tumanda), *Mictyoditra pseudodecora* (Tan)] or the earliest middle Aptian (e.g. *Archaeodictyomitra lacrimula* (Foreman), *Suna hybum* (Foreman), *Thanarla pacifica* Nakaseko & Nishimura, *Pantanellium* spp.) (see O'Dogherty 1994). The following samples are assigned to this interval: IR04-72 and IR04-38 from the Imamzadeh sections, IR04-121 from Bar Er, and IR04-147 to IR04-151 from Darreh Juve. In sample IR04-38, the coexistence of *Cecrops septemporatus* (Parona) and *Aurisaturnalis carinatus perforatus* Dumitrica & Dumitrica-Jud allows a more precise age assignment to the Barremian (i.e. zone G1 of Jud 1994; see Dumitrica & Dumitrica-Jud 1995). In the Darreh Juve samples, the latest Hauterivian to early Aptian interval is well documented across a succession more than 50 m thick. Thus, it seems likely that the upper sample

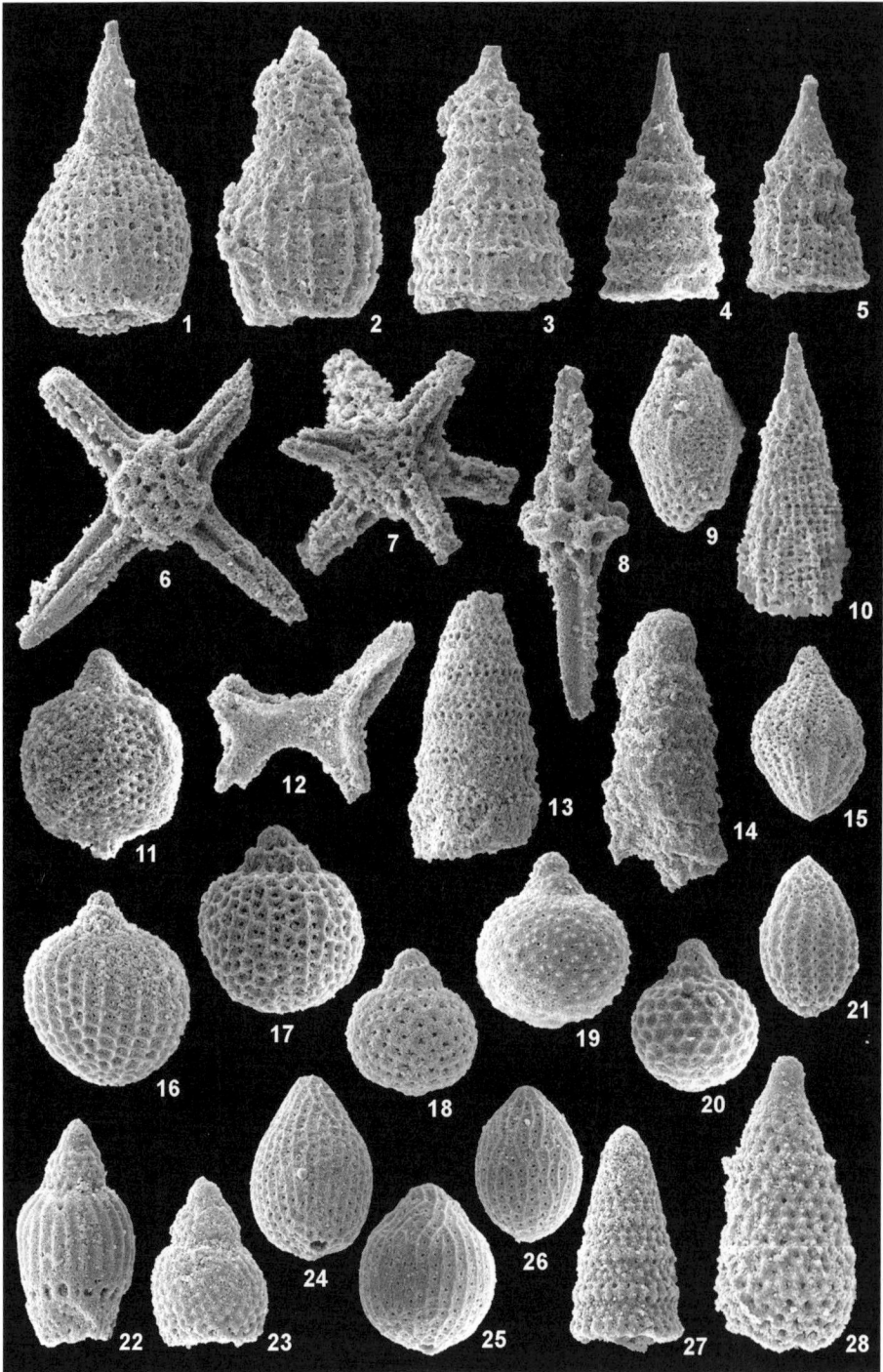

Fig. 3. Middle Jurassic radiolaria from the Bar Er and Dareh Juve sections. The sample number, SEM number and magnification are indicated for each illustration. 1, *Parahsuum? natorense* (El Kadiri), IR04-97, 071702, 150×. 2, *Hsuum altile* Hori & Otsuka, IR04-97, 071711, 200×. 3, *Elodium pessagnoi* Yeh & Cheng, IR04-97, 071708, 200×. 4, *Praeparvicingula nanoconica* (Hori & Otsuka), IR04-97, 071706, 200×. 5, *Hsuum exiguum* Yeh & Cheng, IR04-97,

(IR04-151) is early Aptian in age. The highest productive sample of this section (IR04-153) contains a poorly preserved radiolarian assemblage. The only stratigraphically important species is *Archaeodictyomitra lacrimula* (Foreman), whose range does not extend above the base of the middle Aptian (base of Costata subzone according to O'Dogherty 1994).

Late Aptian and early Albian radiolarians were not recovered from the studied sections. Sediments of this age are apparently missing, either because the upper boundary of the Imamzadeh Formation is erosional or the succession is truncated by a fault on the top. Middle–late Albian radiolarians (Fig. 5, 19–26) were found in samples IR04-43 and IR04-14, both below and above a thick conglomerate with boulders at Negareh-Khaneh (i.e. the lower member of the Neghareh Khaneh Formation). The age is constrained by *Archaeocenosphaera*? *mellifera* O'Dogherty, *Mita gracilis* (Squinabol), *Rhopalosyringium mosquense* (Smirnova & Aliev), which first occur in the middle Albian, and by *Mita spoletoensis* (O'Dogherty), which is the index taxon of the Spoletoensis zone (middle Albian to the base of the Cenomanian; O'Dogherty 1994). The highest sample of this section (IR04-18, the upper member from the Neghareh Khaneh Formation) (Fig. 5, 19–26) contains *Hemicryptocapsa polyhedra* Dumitrica, which is a typical Turonian species (O'Dogherty 1994). On the other hand, it also contains *Theocampe urna* (Foreman) and the genus *Annikaella*, both known to occur from the Coniacian onward (Sanfilippo & Riedel 1985; De Wever *et al.* 1988; see also Dumitrica *et al.* 1997, p. 15). Because O'Dogherty's zonation extends only to the early Turonian and because an accurate radiolarian zonation is not available for the successive stages, the age of sample IR04-18 is broadly determined as Turonian–Coniacian.

Not a single radiolarian, either *in situ* or reworked, was found in the shales of the Bar Er Fm.

The oldest age obtained for the Darreh Juve Fm is Aalenian–early Bajocian? (in the Bar Er section).

The youngest age is middle Callovian–early Oxfordian (in the Darreh Juve and Bar Er sections).

The base of the Imamzadeh Fm is middle–late Oxfordian (Imamzadeh–Lake Bakhtegan section). The boundary between the two members (i.e. the lower and upper) is diachronous, from early–early late Tithonian (Imamzadeh–Lake Bakhtegan) to late Tithonian–early Valanginian (Bar Er section). The top of the Imamzadeh Fm is at least early Aptian in age.

The lower member of the Neghareh-Khaneh Fm, primarily made of coarse-grained sediments, is poor in radiolarians, with a middle–late Albian age on top. It might be late Aptian?–Albian in age. Its upper member is of Albian (late?) to Turonian–Coniacian age, although it could be younger.

Sedimentology: facies description

In this study, 27 facies are described throughout the measured sections. These are grouped into eight major facies associations (see Fig. 6 for description and Fig. 7 for photographs). All of these facies are typical gravity-flow deposits, according to the classification of Mutti (1992). However, one group of facies, here named 'slaty' medium-grained sands, is specific to this study (not included in the study by Mutti 1992). The eight main facies associations are defined from the grain size of the sediments, their homolithic or heterolithic characters, and, in two cases, their sedimentary structures.

The two first associations (C.ms and C.cs, Figs 6 & 7a, b) are coarse-grained, conglomeratic sediments, either matrix- or clast-supported. The petrology and shape of both the matrix particles and the clasts are highly variable (either rounded or angular). If rounded, the pebbles are part of clast-supported conglomerates and vice versa. The matrix of the bimodal matrix-supported conglomerates ranges from small angular to subangular pebbles to carbonate silts. In the case of the conglomerates,

Fig. 3. (*Continued*) 071704, 200×. 6, *Tympaneides* sp., IR04-86, 071604, 200×. 7, *Thurstonia* sp., IR04-86, 071606, 200×. 8, *Trillus* sp., IR04-86, 071607, 200×. 9, *Unuma* cf. *typicus* Ichikawa & Yao, IR04-86, 071613, 200×. 10, *Hsuum* cf. *matsuokai* Isozaki & Matsuda, IR04-86, 071611, 200×. 11, *Williriedellum marcucciae* Cortese, IR04-89, 071508, 250×. 12, *Hexasaturnalis suboblongus* (Yao), IR04-89, 071502, 250×. 13, *Dictyomitrella*? *kamoensis* Mizutani & Kido, IR04-89, 071513, 250. 14, *Stichomitra*? *takanoensis* Aita, IR04-89, 071514, 200×. 15, *Protunuma fusiformis* Ichikawa & Yao, IR04-89, 071510, 200×. 16, *Williriedellum tetragonum* (Matsuoka), IR04-94, 071313, 250×. 17, *Williriedellum* aff. *tetragonum* (Matsuoka), IR04-94, 071306, 250×. (Note that some frames include more than one pore.) 18, *Williriedellum yaoi* (Kozur), IR04-94, 071308, 250×. 19, *Williriedellum* cf. *yaoi* (Kozur) (corroded specimen), IR04-130, 070416, 250×. 20, *Williriedellum yaoi* (Kozur), IR04-130, 070415, 250×. 21, *Plicaforacapsa catenarum* (Matsuoka), IR04-130, 070406, 250×. 22, *Eucyrtidiellum ptyctum* (Riedel & Sanfilippo), IR04-130, 070408, 250×. 23, *Eucyrtidiellum nodosum* Wakita, IR04-130, 070417, 250×. 24–26, *Striatojaponocapsa conexa* (Matsuoka), IR04-130 (24, 070427; 25, 070430; 26, 070435) 250×. 27, *Dictyomitrella*? *kamoensis* Mizutani & Kido, IR04-130, 070420, 250×. 28, *Pseudoristola tsunoensis* (Aita), IR04-130, 070407, 250×.

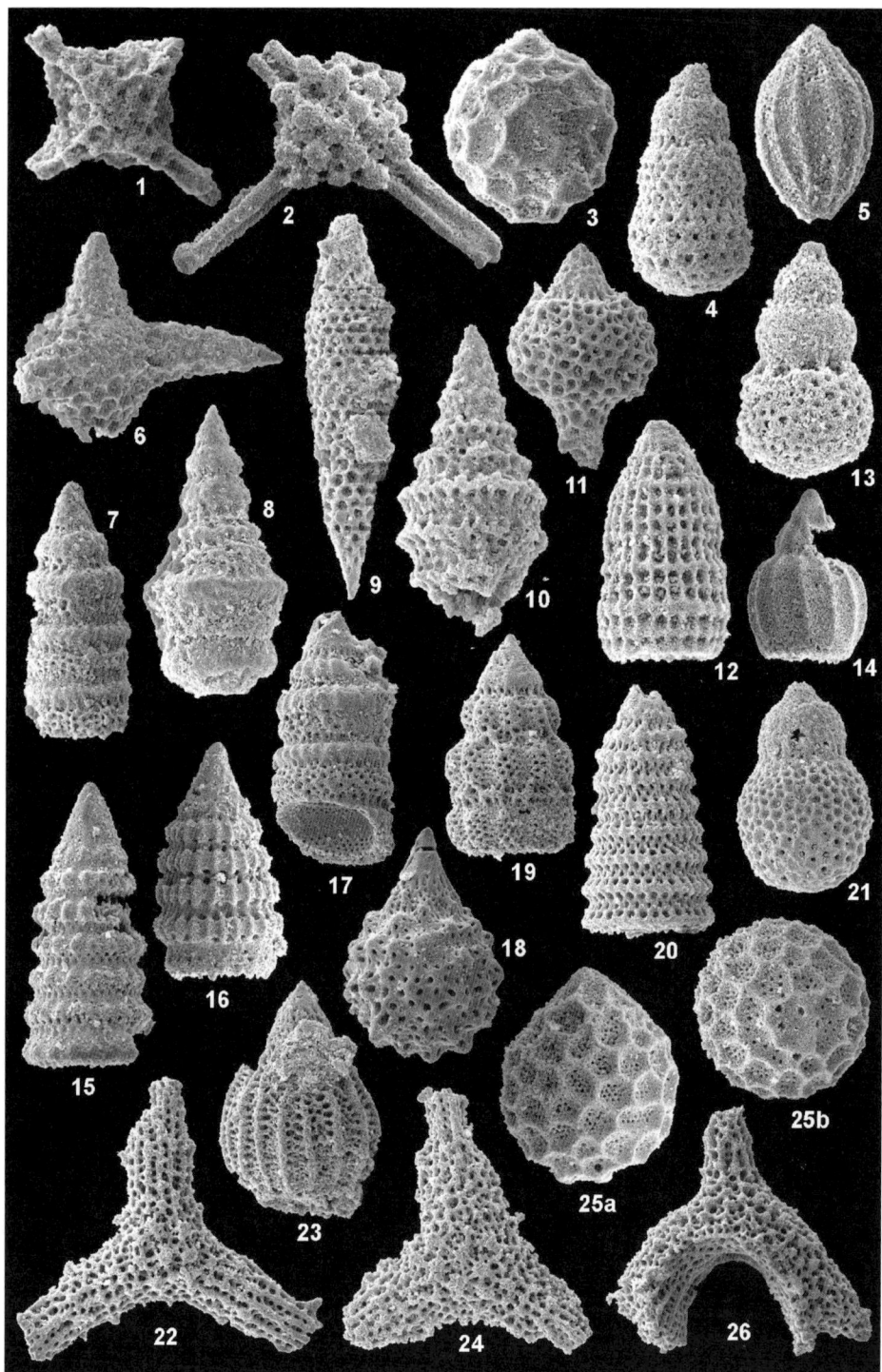

Fig. 4. Late Jurassic and earliest Cretaceous radiolarians from the Imamzadeh II section. The sample number, SEM number and magnification are indicated for each illustration. 1, *Emiluvia ordinaria* Ozvoldova, IR04-49, 071001, 150×. 2, *Emiluvia orea* Baumgartner, IR04-49, 071003, 150×. 3, *Williriedellum crystallinum* Dumitrica, IR04-49, 071005, 200×. 4, *Sethocapsa* aff. *horokanaiensis* Kawabata, IR04-49, 071018, 250×. 5, *Protunuma japonicus* Matsuoka & Yao,

the nature of the source area seems to be of primary importance, more so than the process itself (i.e. debris flows v. high-density turbidity current; Shanmugam 1996, 2000). For the rounded clast-supported conglomerates, the source seems to be pebbles rounded by waves on the shore or in rivers in an upstream catchment, and all of the fine-grained particles have been winnowed. For the subangular to angular matrix-supported conglomerates, the pebbles come from a local relief: they mixed with the overlying soft sediments and were transported toward the basin during a single episode from this source by gravity flows.

The third association (Sc.meg, Fig. 6) is made up of coarse-grained sands with possible granules and megaripples. It corresponds to the classical Mutti's by-pass facies (e.g. F6, Mutti, 1979, 1992).

The fourth association (Sc, Figs 6 & 7c), which is the most common of the homolithic facies, is characterized by poorly sorted coarse-grained sands. All of the intermediates between bimodal and unimodal sands exist from granule-rich sands (bimodal) to poorly sorted sands (unimodal). Both are mostly structureless. Some planar laminations can occur in the unimodal sands. The exact nature of the related transport process is difficult to determine (debris flows v. high-density turbidity current).

The fifth association (Sm.sl, Figs 6 & 7d) is specific to the southern Tethys margin (these facies also develop along the Oman palaeomargin). These facies, which are either homolithic or heterolithic, are composed of stacked subplanar lamina-sets with a thickness of a few centimetres. These lamina-sets may be planar, but generally they are slightly undulating (subplanar laminations). In some cases, the undulation amplitude is high enough to generate structures similar to growing hummocky cross-stratification (HCS; Bourgeois 1983; Fig. 7e). They differ from a true HCS by their grain size (which is coarser than the very well-sorted fine to medium sands from the storm-induced HCS) and the lack of an association with true wave ripples. The significance of the lamina-sets is unclear: they may represent flow velocity variations or amalgamated deposits of distinct gravity flows. They can be bundled into bedsets that could record different stacked depositional events. The process of transport and deposition is difficult to determine: the subplanar lamination and the growing HCS suggest a deposition from a suspension (i.e. the draping of vertically growing structures), but the occurrence of parting lineation on the planar laminations indicates a dominant bedload component (Harms 1975; Allen 1982).

The sixth and the seventh associations (Sm.he1, Sm.he2, Fig. 6) represent classical turbiditic facies (i.e. medium-density turbidity flows) that can be described using the classical Bouma model (Bouma 1962). The planar laminations include two distinct subfacies: the first is characterized by true graded laminations and the second by crude planar (subplanar) laminations. This latter subfacies corresponds to the distal equivalent of facies association Sm.sl. Two types of current ripple bedding are recognized: strictly bedload and a mixed bedload-suspended load (type A and B climbing ripples; Allen 1973). The subfacies with bedload current ripples is rare and seems to be a local record of facies association Sm.sl.

The eighth association (Sf.he, Figs 6 & 7f) is composed of fine-grained thin layers (a few centimetres thick), with either planar to subplanar laminations or undulating laminations (from HCS-like structures to in-phase current climbing ripples; type S of Allen 1973). They correspond to low-density turbidity currents. No true Stow sequences (Shanmugam 2000) were encountered. Based on petrology, two sub-associations can be defined: carbonate and siliceous (radiolarians). In the study area, radiolarians behave as particles and are reworked, transported and deposited under the action of gravity turbulent flows. They show evidence of grading and laminations, with the existence of ripple undulations on top of the strata.

Fig. 4. (*Continued*) IR04 56, 071222, 200×. 6, *Podocapsa amphitreptera* Foreman, IR04-53, 071103, 150×. 7, *Cinguloturris cylindra* Kemkin & Rudenko, IR04-56, 071201, 200×. 8, *Cinguloturris fusiforma* Hori, IR04-56, 071241, 200×. 9, *Pseudoeucyrtis reticularis* Matsuoka & Yao, IR04-53, 071128, 150×. 10, *Tethysetta mashitaensis* (Mizutani), IR04-53, 071108, 150×. 11, *Syringocapsa longituba* Steiger & Steiger, IR04-56, 071221, 150×. 12, *Archaeodictyomitra apiarium* (Rüst), IR04-56, 071211, 200×. 13, *Sethocapsa horokanaiensis* Kawabata, IR04-56, 071233, 250×. 14, *Eucyrtidiellum pyramis* (Aita), IR04-57, 070227, 200×. 15, *Pseudodictyomitra carpatica* (Lozyniak), IR04-57, 070324, 200×. 16, *Svinitzium depressum* (Baumgartner), IR04-57, 070207, 200×. 17, *Cinguloturris cylindra* Kemkin & Rudenko, IR04-57, 070206, 150×. 18, *Fultacapsa tricornis* (Jud), IR04-57, 070224, 150×. 19, *Neorelumbra tippitae* Kiessling, IR04-57, 070215, 150×. 20, *Praeparvicingula cosmoconica* (Foreman), IR04-57, 070213, 150×. 21, *Tricolocapsa? campana* Kiessling, IR04-57, 070241, 200×. 22, 24, *Paronaella? tubulata* Steiger, IR04-57 (22, 070312; 24, 070317), 150×. (Note tritrabid structure on distal extension of arms.) 23, *Hsuum raricostatum* Jud, IR04-57, 070221, 150×. 25a and b, *Williriedellum* aff. *crystallinum* Dumitrica, IR04-57 (25a, 070229; 25b, 070230, antapical view showing closed aperture with three pores), 200×. 26, *Deviatus diamphidius* (Foreman), IR04-57, 070304, 150×.

Fig. 5. Cretaceous radiolarians from the Imamzadeh, Dareh Juve and Negareh-Khaneh sections. The sample number, SEM number and magnification are indicated for each illustration. 1, *Cecrops septemporatus* (Parona), IR04-70, 070901, 200×. 2, *Suna hybum* (Foreman), IR04-70, 070911, 150×. 3, *Aurisaturnalis variabilis variabilis* (Squinabol), IR04-70, 070905, 200×. 4, *Crolanium bipodium* (Parona), IR04-70, 070923, 200×. 5, *Syringocapsa limatum* Foreman,

In the field, along cliffs, sometimes with an exposure a few kilometres long, most of the deposits are laterally continuous with minor changes in facies and thicknesses, which correspond to lobe deposits. Channel structures have been recognized in the Darreh Juve Formation. They correspond to low erosional structures (maximum depth 1 m) with a width ranging from tens to hundreds of metres. A deeply incised channel (canyon?), filled by conglomerates and blocks, was identified in the Lower Member of the Neghareh Khaneh Formation. The few occurring slumps are *in situ* slumps with a low horizontal displacement (a few metres) on a local décollement level.

Discussion: facies model

Facies zonation (Fig. 8)

As mentioned above, the main objective of this sedimentological study was to establish a proximal–distal facies zonation to define the main discontinuities in the deep-sea sedimentary record of the southern Tethyan passive margin in Iran. However, field data are not abundant enough to construct precise 3D geometries of these gravity-flow systems. The main difficulty of this exercise is to take into account the lateral substitution of two or more facies that, with time, can be deposited at the same place along the depositional profile. The model proposed here was based on hydrodynamical considerations (i.e. the same type of processes and flow velocity) and on field data (i.e. rare lateral variations on outcrops and the close vertical association of two facies in a given area).

Coarse-grained sands with a megaripple facies association (the sc.meg or F6 by-pass facies of Mutti 1992) are closely associated (Fig. 8) with the coarse-grained sand facies association (Cs). They record the transition between the two types of gravity-flow systems (Mutti 1992; Mutti et al. 1999). Conglomeratic facies associations (C.ms,

C.cs), with evidence of channels, are located upstream of this by-pass facies association (Cs.meg). The conglomeratic facies associations (C.ms, C.cs) were not ranked along the profile: they record changes in the lithology of the source area, rather than any proximal–distal trends (see discussion, below).

Downstream, lobe deposits start with thick homolithic coarse-grained sands (Sc). They can laterally pass into two types of facies successions (Fig. 8): (1) the 'classical' Bouma sequence evolution (Sm.he) or (2) the 'slaty' facies evolution (Sm.sl). Discerning the reason for the two types of evolution is not an aim of this study. However, a possible explanation could be a change in carbonate production on the platform and then a change in the sediment grain sizes (i.e. poorly sorted coarse-grained sands v. better sorted medium-grained sands). The Bouma sequence evolution (Sm.he) displays a classical lateral evolution with Ta to Tab sequences upstream and Tc downstream. The 'slaty' facies evolution (Sm.sl) shows a more complex pattern with a more 'planar to subplanar lamination' trend (the most common) and an 'HCS-like' trend (Fig. 8).

The distal transition toward fine-grained sand facies (Sf.he) is unclear: this does not correspond to the classical evolution through space of the Bouma sequence. The strata, made of Tc sequences (Sm.he2), are laterally continuous and are interbedded with Sf.he strata. The grain size and the sedimentary structures of the Sf.he strata are more compatible with a lateral evolution from the 'slaty' facies (Sm.sl) trend. This is supported by the same high-frequency rate of gravity-flow event preservation for both the Sm.sl and the Sf.he facies associations.

Two laterally equivalent trends have been defined. The Bouma sequence evolution passes from massive structureless beds to planar (subplanar) laminated strata and then to in-phase climbing current ripple beddings. The 'slaty' facies evolution evolves from Sm.sl facies associations to

Fig. 5. (*Continued*) IR04-70, 070921, 150×. 6, *Aurisaturnalis carinatus perforatus* Dumitrica & Dumitrica-Jud, IR04-38, 070703, 200×. 7, *Hemicryptocapsa capita* (Tan), IR04-38, 070711, 200×. 8, 9, *Archaeodictyomitra lacrimula* (Foreman), IR04-151 (8, 070814; 9, 070809), 200×. 10, *Pseudodictyomitra* cf. *carpatica* (Lozyniak), IR04-151, 070822, 200×. 11, *Thanarla pacifica* Nakaseko & Nishimura, IR04-151, 070831, 200×. 12, *Diacanthocapsa* sp., IR04-14, 070625, 200×. 13, *Pseudodictyomitra lodogaensis* Pessagno, IR04-14, 070637, 200×. 14, *Mita gracilis* (Squinabol), IR04-14, 070614, 200×. 15, *Mita obesa* (Squinabol), IR04-14, 070606, 150×. 16, *Mita spoletoensis* (O'Dogherty), IR04-14, 070602, 150×. 17, *Rhopalosyringium adriaticum* O'Dogherty, IR04-14, 070619, 300×. 18, *Rhopalosyringium mosquense* (Smirnova & Aliev), IR04-14, 070618, 200×. 19, *Annikaella omanensis* De Wever, Bourdillon-de Grissac & Beurrier, IR04-14, 070503, 300×. 20, *Dictyomitra formosa* Squinabol, IR04-18, 070502, 200×. 21, *Theocampe* sp., IR04-18, 070505, 300×. 22, *Theocampe ascalia* Foreman, IR04-18, 070508, 300×. 23, *Theocampe urna* (Foreman), IR04-18, 070504, 300×. 24a and b, *Hemicryptocapsa polyhedra* Dumitrica, IR04-18 (24a, 070510; 24b, 070511, antapical view), 300×. 25, *Patellula verteroensis* (Pessagno), IR04-18, 070522, 150×. 26, *Alievium* sp., IR04-18, 070519, 150×.

(a)

FACIES ASSOCIATIONS

1 - [C.ms] MATRIX-SUPPORTED CONGLOMERATES

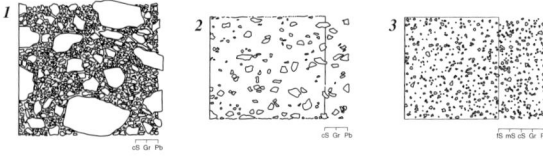

Description:
- Homolithic matrix-supported conglomerates
- Angular to subangular pebbles
- High variability of both matrix (fine sand to small pebbles) and clast (granules to large pebbles) size

Processes:
Debris flows
Mutti's F1-F2-F5? facies

2 - [C.cs] CLAST-SUPPORTED CONGLOMERATES

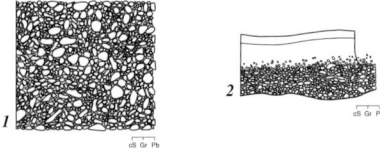

Description:
- Homolithic clast-supported conglomerates
- Subangular to rounded pebbles (0.5 to 30 cm)
- Structureless, sometimes graded

Processes:
High-density turbidity currents
(hyperconcentrated flow)
Mutti's F3 facies

3 - [Sc.meg] COARSE-GRAINED SANDS WITH GRANULES AND CURRENT MEGARIPPLES

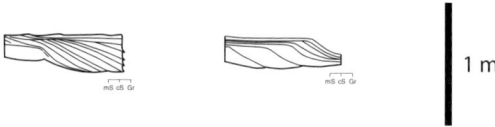

1 m

Description:
- Poorly sorted coarse-grained sands with possible granules
- Current megaripple bedding:
    (1) filling incised furrows or
    (2) climbing (type B)

Processes:
By-pass facies
Mutti's F6 facies

4 - [Sc] COARSE-GRAINED SANDS WITH GRANULES

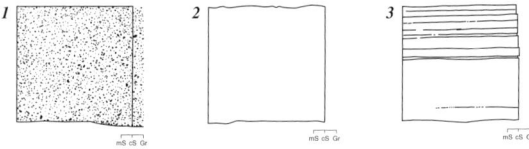

Description:
- Homolithic (0.4-2m thick beds)
- Poorly sorted coarse-grained sands with possible granules and small pebbles
- Three main subfacies with all the intermediate ones:
    (1) structureless matrix-supported "conglomerates" (small pebbles and granules in a sandy matrix, F4-F7) rare crude "planar laminations"
    (2) massive -structureless coarse-grained sands (F8)
    (3) more or less graded coarse to medium-grained sands with planar laminations and rare water escape structures (F8,F9a)

Processes:
Sandy debris flows to high-density turbidity flows
(hyperconcentrated to concentrated flows)
Mutti's F4, F5, F7 and F8 facies

Fig. 6. Facies and facies associations: description and interpretation.

most of the Sf.he facies, including carbonate strata upstream and siliceous (radiolaritic) strata downstream. The radiolaritic facies may laterally replace the distal carbonate silt layers. The significance of the radiolarite facies evolution, including the type of the source area, is not understood.

Palaeocurrent data and palaeotopographic implications

Palaeocurrent measurements were taken from both the Darreh Juve and Imamzadeh Formations (Fig. 9). For the Darreh Juve Formation (Aalenian–Callovian), all three sections (i.e. Imamzadeh–Lake Bakhtegan, Darreh Juve and Bar Er) show the same dominant pattern toward the NW quarter: Imamzadeh–Lake Bakhtegan is NW dominant, with some WSW and ENE directions; Darreh Juve is west dominant, with some east and north directions; and Bar Er is WNW dominant, with NNW and north directions. However, other directions can exist, sometimes at 180° to each other. This can imply either a multisource system, indicating a complex topography of the platform, or a reflection of turbidity currents on the opposite slopes of a relatively narrow basin (Pickering & Hiscott 1985; Kneller *et al.* 1991).

The palaeocurrent pattern is more complex for the Imamzadeh Formation (Oxfordian–Aptian): Imamzadeh–Lake Bakhtegan is NNW dominant, with an ESE direction; Imamzadeh village is SE

(b)

5 - [Sm.sl] "SLATY" MEDIUM-GRAINED SANDS

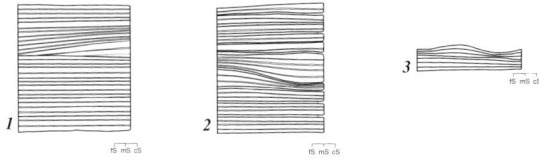

1   *2*   *3*

fS mS cS   fS mS cS   fS mS cS

Description:
- Homolithic (0,3 to 5 m thick)
- Stacking of few centimetres-thick laminasets of subplanar laminations
- Possible metric-scale undulations, similar to HCS-like structures (growing-up HCS)
- Possible bundling of the laminasets into bedsets (amalgamation?)

Processes:
Unknown: medium-density turbidity currents ?

6 - [Sm.he1] MEDIUM-GRAINED SANDS - HETEROLITHICS

1   *2*   *3*

fS mS cS   fS mS cS   fS mS cS

Description:
- Heterolithic : alternation of marly shales with medium-grained sands (0,2 to 0,5 m thick)
- (1) structureless or slightly graded (Bouma Ta)
- (2) planar laminations graded (Bouma Tb)
- (3) crude planar laminations (Bouma Tb)

Processes:
Medium-density turbidity currents
Mutti's F9a facies

7 - [Sm.he2] MEDIUM-GRAINED SANDS - HETEROLITHICS

1   *2*   *3*

fS mS cS   fS mS cS   fS mS cS

Description:
- Heterolithic : alternation of marly shales with medium-grained sands (0,1 to 0,4 m thick)
- (1) crude planar laminations and bedload current ripple bedding (Bouma Tbc)
- (2) subplanar lamination and climbing (type A and B) current ripple bedding (mixed bedload-suspended load) (Bouma Tbc)
- (3) climbing (type A and B) current ripple bedding (mixed bedload-suspended load) (Bouma Tc)

Processes:
Medium-density turbidity currents
Mutti's F9a facies

8 - [Sf.he] FINE-GRAINED SANDS TO SILTS - HETEROLITHICS

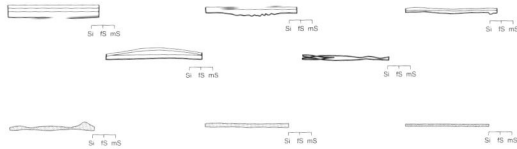

Si fS mS   Si fS mS   Si fS mS

Si fS mS   Si fS mS

Si fS mS   Si fS mS   Si fS mS

Description:
- Heterolithic: alternation marly shales with carbonate silts- fine sands or cherts (1-10 cm thick)
- Cherts can be a diagenetic silicification of the carbonate silts or true clastic radiolarites
- All the beds have a sharp base and top - the top is sometimes undulated (ripple current shape)
- Fine- grained sands: crude grading, subplanar laminations and HCS-like structures (growing-up)
- Carbonate silts and radiolarites: structureless or subplanar

Processes:
Medium- to low-density turbidity current with a deposition from mainly suspended load
Mutti's F9a facies

Fig. 6. (*Continued*)

dominant, with north and south directions; Darreh Juve is WNW dominant, with an east direction; and Bar Er is WNW dominant. The SE directions of the Imamzadeh village section could be explained by tectonic rotation, but because the SE directions are at 180° to the other NW dominant directions, the best explanation for this SE-dominant direction is again either multisource systems on opposite platforms or reflected turbidity currents.

These palaeocurrent measurements suggest a complex topographic pattern in the basin with a narrow (100 km large) deep-sea system. This is in good agreement with the palaeogeographical data provided by Murris (1980) and Ziegler (2001), and the idea of isolated carbonate platforms (so-called seamounts) close to the transition between continental and oceanic crust.

Sequence stratigraphy: correlations

Five major stratigraphic cycles were identified and correlated along the five measured sections (Fig. 10). Two cycles have been defined for the Jurassic (labelled J2 and J3) and three for the lower Cretaceous (labelled K1.1, K1.2 and K1.3). The undated base of the Bar Er Formation records an overall distal-up trend, with a maximum of distal facies (Fig. 10) on top of the formation.

Based on biochronostratigraphic data, the order of magnitude of duration of these five major cycles is around 10 Ma, which corresponds to second-order cycles. At a lower time-scale duration, cycles lasting a few million years (i.e. third-order cycles) can be defined. The resolution of the biostratigraphic data is not sufficient to validate

Fig. 7. Gravity-flow facies. (**a**) Matrix-supported conglomerate (C.ms) with a pebbly matrix (the Neghareh Khaneh Fm, lower member, Imamzadeh–Neghareh Khaneh section). (**b**) Matrix-supported conglomerate (C.ms) with a matrix of fine-grained carbonate sands (the Darreh Juve Fm, Bar Er section). (**c**) Massive coarse-grained carbonate sands with granules and/or small pebbles (Sc) (the Imamzadeh Fm, Imamzadeh Village). (**d**) 'Slaty' (Sm.sl) medium-grained carbonate sands (the Imamzadeh Fm, Imamzadeh–Lake Bakhtegan). (**e**) Growing HCS-like structure (Sm.sl to Sf.he) in heterolithic fine- to medium-grained carbonate sands (the Imamzadeh Fm, Imamzadeh Village). (**f**) Radiolarites (Sf.he) (the Imamzadeh Fm, Imamzadeh Village).

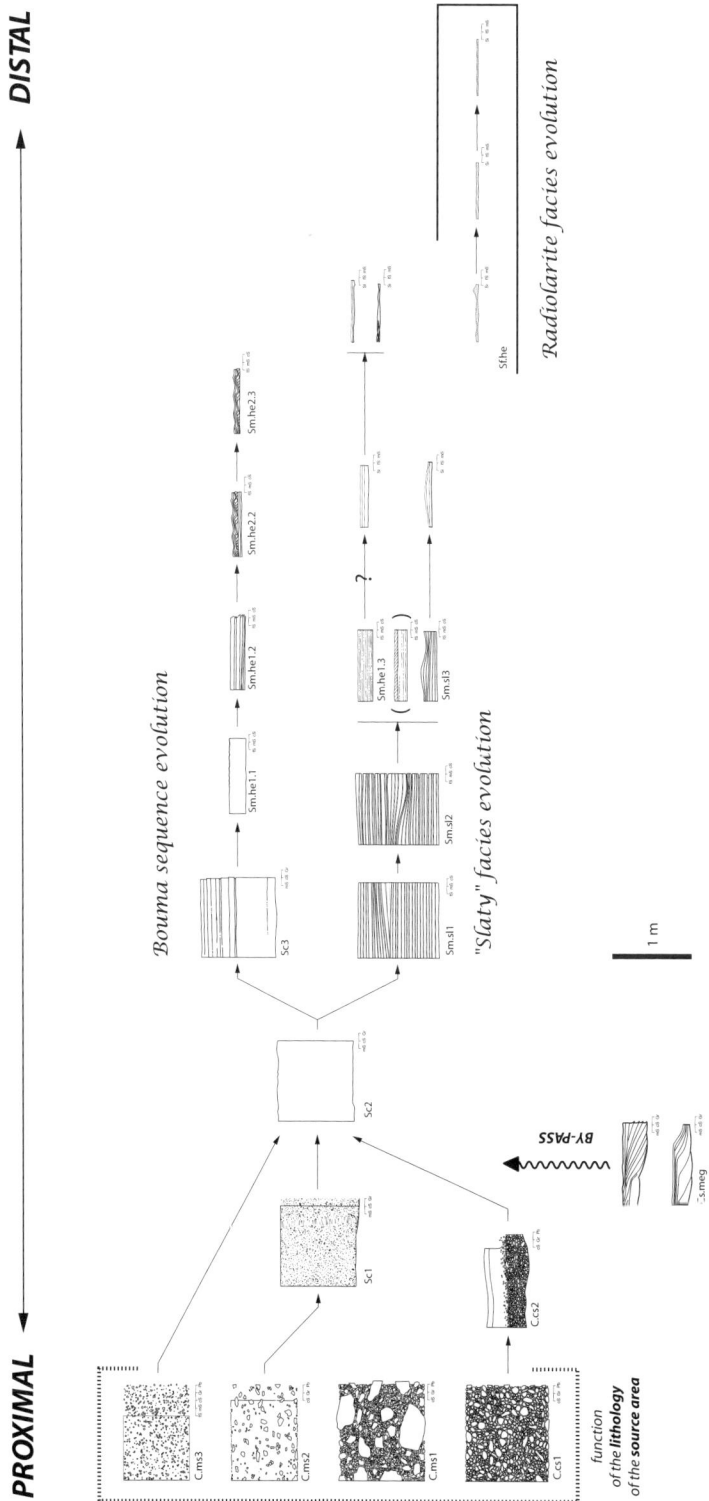

Fig. 8. Facies model: facies zonation and lateral substitution along a proximal–distal trend.

BAR ER

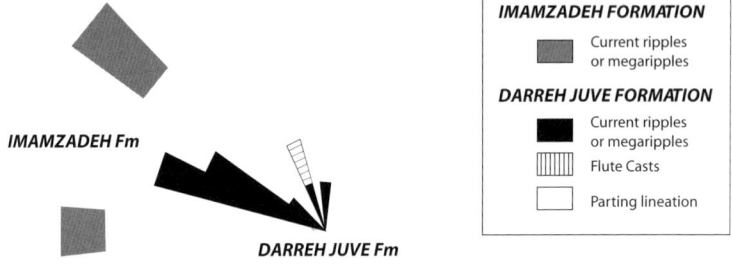

DARREH JUVE

IMAMZADEH VILLAGE

**IMAMZADEH
LAKE BAKHTEGAN**

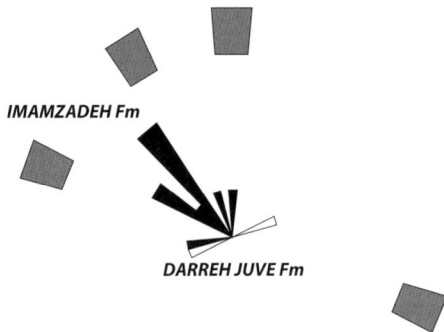

Fig. 9. Palaeocurrent measurements.

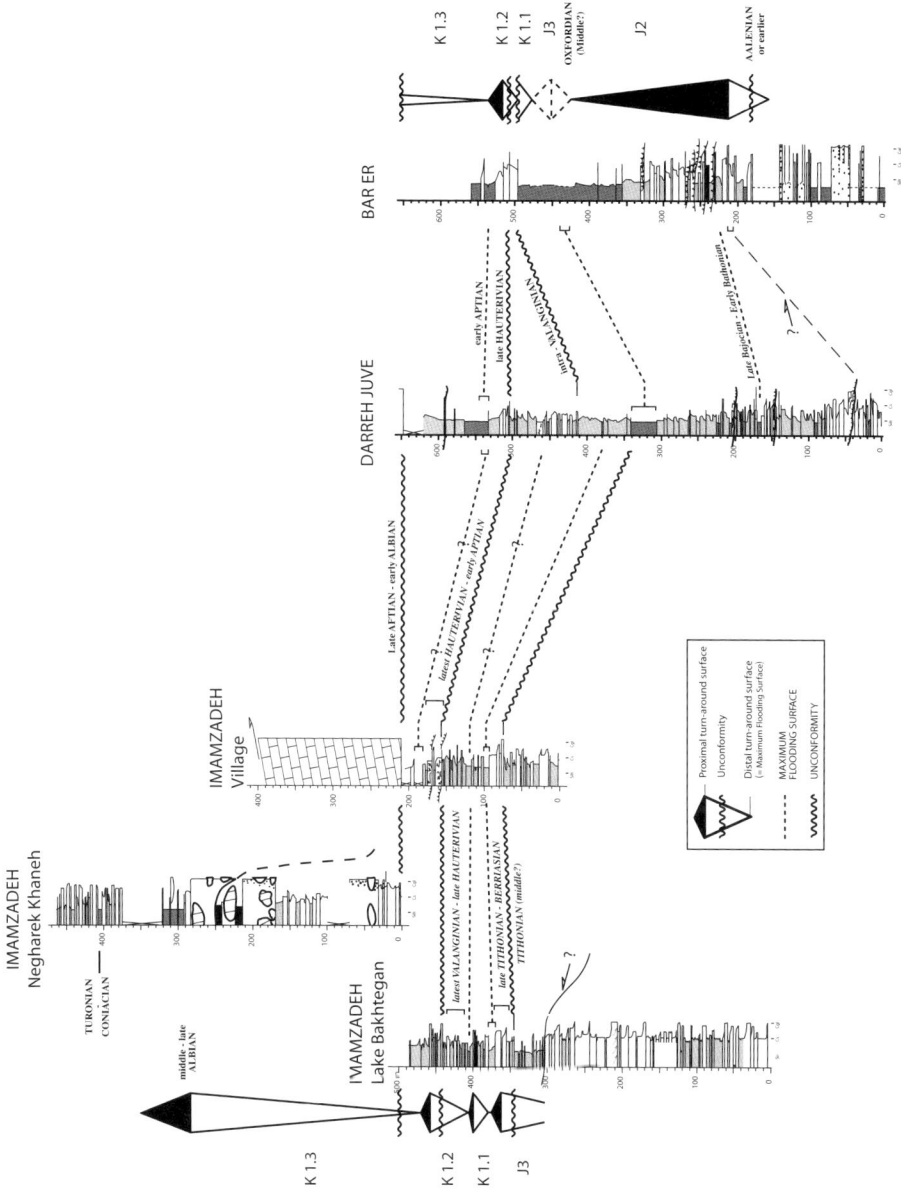

Fig. 10. Biostratigraphic and sequence stratigraphic correlations of the five studied sections: sequence definitions. The proximal turn-around surface corresponds to the occurrence of the most proximal facies along the depositional profile (Fig. 8). This surface either may coincide with the unconformity or may occur later. An *a priori* relationship with the flooding surface on the shelf is not expected.

correlations at this time scale. These third-order cycles are well recorded in the Darreh Juve Fm. At a higher time-scale duration (around 100 Ma), the two periods recording the most proximal facies (large debris flows, in this case) are the base of the Bar Er Fm (Triassic) and the base of the Neghareh-Khaneh Fm (latest Aptian to middle–late Albian).

Cycle J2 (Fig. 10) corresponds to the Darreh Juve Formation, which displays different stratigraphic patterns: the Imamzadeh–Lake Bakhtegan section is on one side and the Darreh Juve–Bar Er sections are on the other. In the Darreh Juve–Bar Er sections, cycle J2 is dominated by a distal-up trend. A sharp unconformity is recorded during the proximal-up trend. The age of the unconformity is poorly constrained: it occurred during the Aalenian or earlier. The proximal turn-around surface is Early Bajocian, possibly latest Aalenian. In the Imamzadeh–Lake Bakhtegan section, an aggradational to proximal-up trend is recorded. No age is available. The most proximal facies are recorded in Imamzadeh–Lake Bakhtegan (debris flows, hyperconcentrated to concentrated flows, with some low erosional channels: proximal lobe deposits) and the most distal one is in Bar Er (rare concentrated flows and by-pass deposits, numerous medium-density turbidity currents, some slumps: middle to distal lobes). In the Imamzadeh–Lake Bakhtegan section, a sharp erosional transition occurred at the top of the formation, directly overlain by medium- to low-density turbidity current deposits of middle to late Oxfordian age. This top erosional discontinuity could be a hiatus, a time-equivalent of the retrogradational trend recorded in the Darreh Juve–Bar Er sections.

The age constraint (i.e. the third-order maximum flooding surface of Late Bajocian–Early Bathonian age) between the Darreh Juve and the Bar Er sections indicate a pinching out of the sediments from Darreh Juve to Bar Er. It may record a truncation (onlap) of the base of the formation. This truncation suggests a tilting of this domain after the basal sharp unconformity of Aalenian age or earlier, characterized by intraformational conglomerates in the Bar Er section. The tilting can explain the occurrence of numerous *in situ* slumps in the middle to distal lobe deposits.

Third-order cycles (several tens of metres thick) are bounded by reddish to greenish claystones to siltstones. They probably record a break in the carbonate supply from the platform (i.e. a condensed interval and maximum flooding surface).

Cycle J3 (Fig. 10) corresponds to the base of the Imamzadeh Formation. Two stratigraphic patterns were recognized. In the Imamzadeh area, the cycle is very well recorded with a sharp unconformity (a downward shift of facies). In the Darreh Juve and Bar Er sections, the time-equivalent sediments have been identified only in biostratigraphic correlations. The unconformity is recorded only in the Darreh Juve section. The base maximum flooding surface (i.e. the most distal facies) overlaid sediments of middle Callovian to early Oxfordian age (i.e. the Darreh Juve Formation) and is capped by late Oxfordian to Kimmeridgian sediments (i.e. the Imamzadeh village sequence) and probably middle to late Oxfordian deposits (the age of the sediments overlying the hiatus of the Imamzadeh–Lake Bakhtegan section, which may contain the maximum flooding surface). An Oxfordian age, probably middle Oxfordian, can be proposed for this maximum flooding surface. The unconformity occurs approximately during the middle part of the Tithonian (Imamzadeh–Lake Bakhtegan).

Cycle K1.1 (Fig. 10) corresponds to the middle part of the Imamzadeh Formation. This cycle does not show a well-recorded unconformity, except in the Imamzadeh–Lake Bakhtegan section. This cycle is more condensed (less than 10 m thick) in the Bar Er section. It is stacked with an overlying sequence at the base of the upper member of the Imamzadeh Formation. The maximum time span of the entire cycle is late Tithonian to Valanginian; a more precise dating within this interval is not possible with radiolarian biostratigraphy. Based on the position within the cycle, the base maximum flooding surface is probably of Berriasianage. The unconformity probably occurred during Valanginian time. No age is available for the proximal turn-around surface.

Cycle K1.2 (Fig. 10) corresponds to the top of the Imamzadeh Formation. This cycle is characterized by a sharp downward shift of facies (unconformity) in all four sections where this cycle is exposed. No age is available for the maximum flooding surface at the base. The unconformity is bounded below by sediments of late Valanginian to Hauterivian age and just above by sediments of late Hauterivian to Barremian age. This suggests a Late Hauterivian age for this unconformity, with an error bar starting earlier in the Hauterivian and ending during the early Barremian. The proximal turn-around surface is of Barremian age.

Cycle K1.3 (Fig. 10) corresponds to the Neghareh Khaneh Formation. This cycle is characterized by a sharp unconformity. This is the most important downward shift of facies in all of the studied unconformities. The top maximum flooding surface is difficult to localize. Only one section (Imamzadeh–Neghareh Khaneh) shows this transition (Fig. 10). Unfortunately, this interval of around 50 m is badly exposed. Scarce outcrops of fine-grained sediments suggest distal environments for these deposits. The relationship with the overlying sediments (i.e. the top of the Neghareh

Khaneh Formation), which is mainly aggradational, is unknown.

The maximum flooding surface at the base is of early Aptian age (i.e. the Bar Er section). The unconformity is bounded below by early to middle Aptian age sediments and overlain by middle to late Albian sediments. A late Aptian to early Albian age is suggested for this unconformity. The proximal turn-around surface occurred during the middle to late Albian. No age is available for the maximum flooding surface at the top.

Discussion: tectonic v. eustatic control of the stratigraphic cycles (Fig. 11)

The J2 cycle unconformity (Aalenian or earlier, Fig. 11) is the time-equivalent of a major tectonic unconformity of late Toarcian age (176 ± 2 Ma, GTS 2004), occurring at the Arabian platform scale (Sharland et al. 2001; the base of the AP7 tectonostratigraphic megasequence). Palaeogeographical changes (Ziegler 2001) suggest a reactivation of the more or less northward-trending Precambrian structures, dividing the Arabian plate into two domains: the Central Arabian Arch and the Ghawar–Safanyia fault system. Eastward of this axis (i.e. the present-day Arabian Gulf to Oman), the consequence of this deformation is a southward extension of the subsiding domain (i.e. the Dhruma Fm carbonate platform). The continental siliciclastic input sharply decreases.

This intraplate deformation event could have recorded the first stage of the continental rifting of the Somali Ocean, which is well known on its southern Madagascar margin (the Toarcian Morondova Basin, Geiger et al. 2004).

The J3 cycle maximum flooding surface (MSF; Oxfordian, probably middle Oxfordian, Fig. 11) occurred at the time of major marine flooding across the Arabian platform in the Jurassic period. This flooding event ranged from the Middle Callovian (MFS J40 of Sharland et al. 2001; Haq & Al-Qahtani 2005) to the Late Kimmeridgian (MFS J100, Sharland et al. 2001). Six minor flooding events have been defined (Sharland et al. 2001). The only one occurring in the Oxfordian is of Middle Oxfordian age (MFS J50, 158 Ma, GTS 2004). It corresponds to the most important MFS of the Late Jurassic (Haq & Al-Qahtani 2005); condensed organic-rich limestones of the Hanifa Fm).

This Late Jurassic (Middle Oxfordian to Kimmeridgian) flooding event occurred across the entire Arabian platform and East African domain (i.e. from southern Tanzania to western Ethiopia). It is limited to the Somali Ocean margin. It does not correspond to the major eustatic sea-level high event that occurred later (Tithonian, Haq et al. 1987).

The J3 cycle unconformity (Tithonian, probably middle Tithonian, Fig. 11) is also a time-equivalent to a major tectonic unconformity. It has been dated from the Early Tithonian (149.5 Ma, Gradstein 2004) by Sharland et al. (2001); the base of the AP8 tectonostratigraphic megasequence of Sharland. Again, palaeogeographical changes (Ziegler 2001) suggest a reactivation of the more or less northward-trending Precambrian structures (i.e. the Central Arabian Arch and Ghawar–Safanyia fault system), and of the NNE–SSW Dibba fault, bounding the present-day Oman Mountains westward. This unconformity records a long-wavelength tilting of the Arabian plate westward (Murris 1980) with superimposed reactivated Precambrian structures. In Oman, this tilting is coeval with a shoreline retreat of at least 200 km (Droste & van Steenwinkel 2004).

This unconformity probably records an intraplate deformation, related to the end of the Somali Ocean rift and the beginning of the oceanic accretion between Africa–Arabia on one side and Madagascar–India on the other (i.e. the first preserved magnetic anomaly on the oceanic crust: chron M22, the base Tithonian, Cochran 1988; Coffin & Rabinowitz 1988).

The K1.1 cycle maximum flooding surface (probably Berriasian, Fig. 11) on the platform could be the time-equivalent of the base Berriasian maximum flooding surface (MFS K10; marine shales from the Sulaiy Fm (Qatar and Abu Dhabi) and Rayda Fm (Oman), Sharland et al. 2001). According to Haq & Al-Qahtani, (2005), it corresponds to the major MFS of the cycle, ranging from the base Tithonian unconformity to the Late Valanginian unconformity.

The K1.1 cycle unconformity (Fig. 11), undated but ranging from the Berriasian to the base Hauterivian, could correspond either to the late Valanginian unconformity of Sharland et al. (2001) or to the Early Valanginian major eustatic fall of Haq et al. (1987). The relationship of Sharland's Late Valanginian unconformity with Haq's eustatic fall is unclear, but the major eustatic sea-level fall, of probable climatic origin, is clearly recorded on the Oman carbonate platform (Le Bec 2003).

The K1.2 cycle unconformity (late Hauterivian, Fig. 11) was not recognized by Sharland et al. (2001) or Haq & Al-Qahtani (2005) as a major unconformity on the Arabian platform. It does not correspond to any eustatic sea-level fall (Haq et al. 1987). The Hauterivian carbonate deposits of the Oman platform (Le Bec 2003) are tectonically controlled. This instability seems to be limited to the outer platform at the transition to the slope.

The K1.3 cycle maximum flooding surface (early Aptian, Fig. 11) on the platform could be

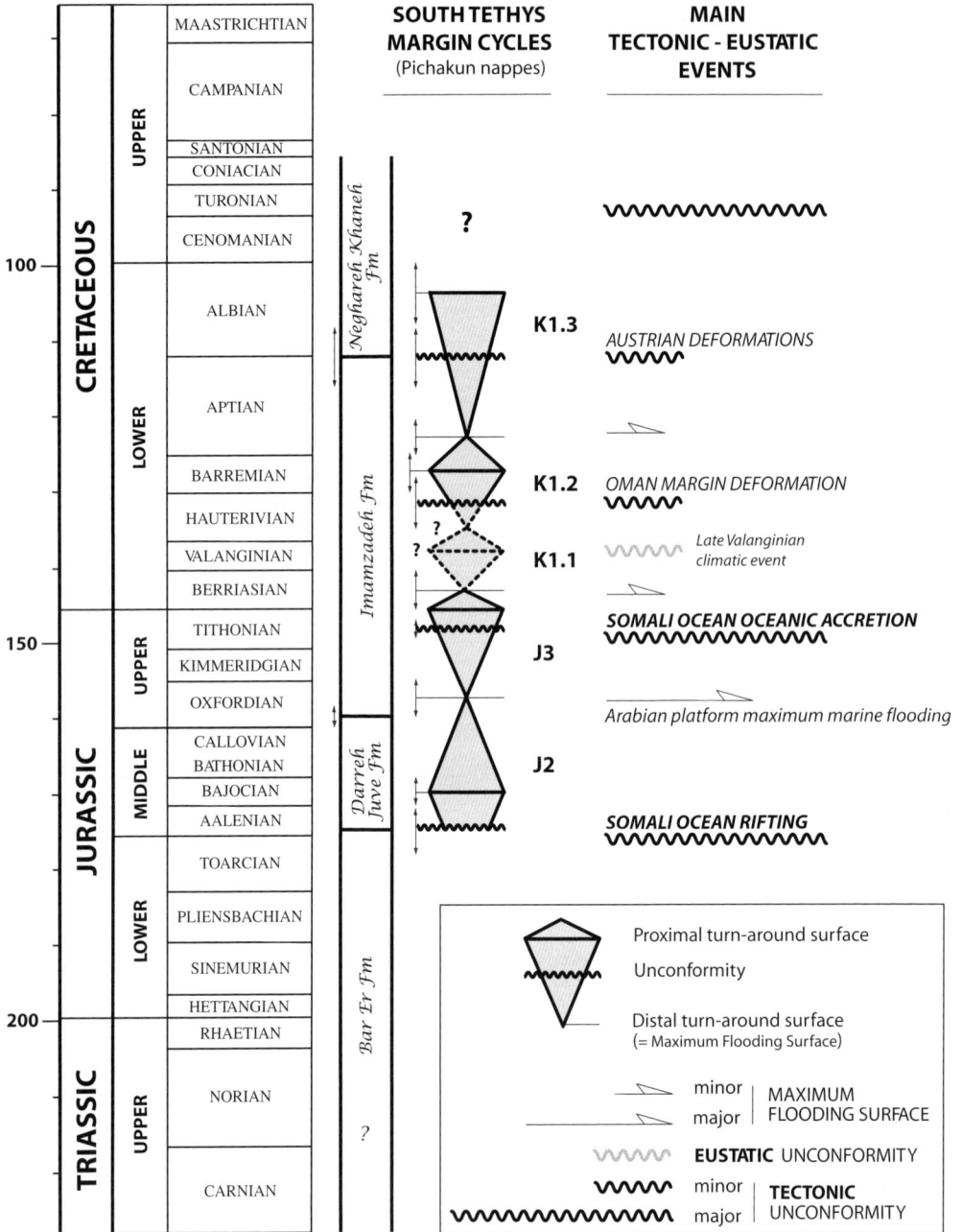

Fig. 11. Evolution of the southern Tethys margin based on the Pichakun Nappes data: biochronostratigraphy, sequence stratigraphy, tectonic and eustatic controls.

the time-equivalent of the base Aptian maximum flooding surface (K70, Sharland *et al.* 2001); that is, the base Shu'aiba Fm or the Hawar Fm. According to Haq & Al-Qahtani (2005), it corresponds to the major MFS of a cycle, ranging from the Late

Valanginian unconformity to the Late Aptian unconformity.

This MFS is time-equivalent to a minor eustatic sea-level high (Haq *et al.* 1987), but records a major marine flooding along the southern margin of

Tethys and the African margin of the Somali and Mozambique oceans (Harris *et al.* 1984; Dercourt *et al.* 2000).

The K1.3 cycle unconformity (late Aptian–early Albian, Fig. 11) corresponds to a major tectonic unconformity (Harris *et al.* 1984; Christian 1997; Droste & van Steenwinkel 2004), dated as Late Aptian by Sharland *et al.* (2001). It corresponds to the boundary between the carbonate Dariyan Fm and the shaly Kazhdumi Fm (Iran, James & Wynd 1965) and between the Shu'aiba Fm and the Nahr Umr Fm (Bahrain to Oman, Droste & van Steenwinkel 2004). According to Sharland *et al.* (2001), this unconformity is not as important as the Late Toarcian and Early Tithonian ones. This tectonic unconformity is coeval with a widespread siliciclastic influx over the Arabian plate (clays to sands). This unconformity probably records the intraplate deformation related to the beginning of the convergence between Africa–Arabia and Eurasia (Austrian deformations *sensu lato*).

Conclusion

Four main lithostratigraphic formations have been defined and dated in the Pichakun Nappes, an inversion of the southern Tethyan margin during the latest Cretaceous: (1) the Bar Er Formation (probably Late Triassic to Early Jurassic); (2) the Darreh Juve Formation (Aalenian–early Bajocian to middle Callovian–early Oxfordian); (3) the Imamzadeh Formation (middle Callovian–early Oxfordian to Aptian); (4) the Neghareh Khaneh Formation (late Aptian to Turonian–Coniacian).

Most of the sediments are deep-sea gravity-flow lobe deposits. Few channel deposits occur (i.e. the Darreh Juve Formation in the Imanzadeh–Lake Bakhtegan section). Some deeply incised channels (canyons?) were observed in the lower member of the Neghareh Khaneh Formation (Albian). Twenty-seven facies, grouped into eight facies associations, were defined. The texture of the proximal conglomeratic facies (i.e. clast-supported v. matrix-supported) is highly dependent of the nature (i.e. petrology, grain size) of the source area sediments. The boundary between debris flows and high-density turbidity currents is difficult to characterize in these conglomeratic facies. A specific facies association, in comparison with the 'classical' turbiditic facies model of Mutti (1992) (i.e. the 'slaty' facies), has been defined here. It corresponds to a stacking of planar to subplanar lamina-sets that are a few centimetres thick. Growing HCS-like structures can develop. In this study, radiolarians are reworked by low-density currents. Pelagic radiolarites were not encountered.

Two main palaeogeographical units have been characterized in the Pichakun nappes: proximal deep-sea deposits (i.e. nappes from the Imanzadeh area; lobes and channels) and distal ones (i.e. nappes from Bar Er and Darreh Juve; only lobes). Palaeocurrent measurements suggest a complex topographic pattern. The deposition occurred in a narrow deep-sea plain (a few hundred kilometres in size), located between the Arabian platform and seaward isolated carbonate platforms.

The most important discontinuity is located at the boundary between the Imamzadeh and Neghareh Khaneh Formations (i.e. the Aptian–Albian boundary, a deposit of olistoliths that are few metres to hundreds of metres thick in debris flows). Five second-order cycles (10–30 Ma duration), defined between two successive distal facies time-intervals, are proposed: (1) J2 (Toarcian?–middle Oxfordian, unconformity: Late Toarcian–Aalenian); (2) J3 (middle Oxfordian–Berriasian, unconformity: middle? Tithonian); (3) K1.1 (Berriasian–undated top); (4) K1.2 (undated base–early Aptian, unconformity: late Hauterivian); (5) K1.3 (early Aptian to at least Turonian–Coniacian, unconformity: Aptian–Albian boundary).

All of the main tectonic events recorded by the Arabian platform during this time interval (Sharland *et al.* 2001) have been identified as major unconformities in the Iran deep-sea record: the late Toarcian (plate deformation as a result of the beginning of the rifting of the Somali Ocean), early Tithonian (oceanic accretion in the Somali Ocean) and Aptian–Albian (Austrian deformations) events. In the study area, the most important event corresponds to the Aptian–Albian deformations. It is expected that another tectonic event occurred during the late Hauterivian. The late Valanginian eustatic fall, of climatic origin, seems to be recorded as an unconformity in the Iran southern Tethys margin sediments.

This study has been carried out and funded by the MEBE (Middle East Basins Evolution) programme, and was sponsored by AGIP-ENI, BP, the CNRS, Petronas, Shell, Total and UPMC. We thank E. Barrier and M. F. Brunet (CNRS-Université Pierre et Marie Curie, Paris 6), the managers of this programme and O. Saidi, the director of the Iran Geological Survey, for their help. A review by F. H. Nader (Institut Français du Pétrole) greatly improved the quality of the paper.

References

ALAVI, M. 1994. Tectonics of the Zagros orogenic belt of Iran: new data and interpretations. *Tectonophysics*, **229**, 211–238.

AL-HUSSEINI, M. I. 1997. Jurassic sequence stratigraphy of the western and southern Arabian Gulf. *GeoArabia*, **2**, 361–382.

ALLEN, J. R. L. 1973. A classification of climbing-ripple cross-lamination. *Journal of the geological Society, London*, **129**, 537–541.

ALLEN, J. R. L. 1982. *Sedimentary Structures: their Character and Physical Basis*. Developments in Sedimentology, **30**, Elsevier, Amsterdam.

BAUMGARTNER, P. O., BARTOLINI, A. *ET AL.* 1995. Middle Jurassic to Early Cretaceous radiolarian biochronology of Tethys based on Unitary Associations. *In*: BAUMGARTNER, P. O., O'DOGHERTY, L., GORICAN, S., URQUHART, E., PILLEVUIT, A. & DE WEVER, P. (eds) *Middle Jurassic to Lower Cretaceous Radiolaria of Tethys: Occurrences, Systematics, Biochronology*. Mémoires de Géologie (Lausanne), **23**, 1013–1038.

BÉCHENNEC, F., LE MÉTOUR, J., RABU, D., BOURDILLON-DE-GRISSAC, C., DE WEVER, P., BEURRIER, M. & VILLEY, M. 1990. The Hawasina Nappes: stratigraphy, palaeogeography and structural evolution of a fragment of the south-Tethyan passive continental margin. *In*: ROBERTSON, A. H. F., SEARLE, M. P. & RIES, A. C. (eds) *The Geology and Tectonics of the Oman Region*. Geological Society, London, Special Publications, **49**, 213–223.

BERBERIAN, M. & KING, G. C. P. 1981. Towards a paleogeography and tectonic evolution of Iran. *Canadian Journal of Earth Sciences*, **18**, 210–265.

BERNOUILLI, D. & WEISSERT, D. 1987. The Upper Hawasina nappes in the central Oman mountains: stratigraphy, palinspastics and sequence of nappe emplacement. *Geodinamica Acta*, **1**, 47–58.

BLECHSCHMIDT, I., DUMITRICA, P., MATTER, A., KRYSTYN, L. & PETERS, T. 2004. Stratigraphic architecture of the northern Oman continental margin; Mesozoic Hamrat Duru Group, Hawasina Complex, Oman. *GeoArabia*, **9**, 81–132.

BOUMA, A. H. 1962. *Sedimentology of some Flysch Deposits: A Graphic Approach to Facies Interpretation*. Elsevier, Amsterdam.

BOURGEOIS, J. 1983. Hummocks – do they grow? *AAPG Bulletin*, **67**, 428.

BRAUD, J. 1989. *La suture du Zagros au niveau de Kermanshah (Kurdistan iranien): reconstitution paléogéographique, évolution géodynamique, magmatique et structurale*. Thèse d'Etat, Université de Paris-Sud (Paris XI).

CHRISTIAN, L. 1997. Cretaceous subsurface geology of the Middle East region. *GeoArabia*, **2**, 239–256.

COCHRAN, J. R. 1988. Somali Basin, Chain Ridge, and origin of the northern Somali Basin gravity and geoid low. *Journal of Geophysical Research*, **93**, 11985–12008.

COFFIN, M. F. & RABINOWITZ, P. D. 1988. Evolution of the conjugate East African–Madagascan margins and the western Somali Basin. Geological Society of America, Special Paper, **226**, 1–78.

DERCOURT, J., GAETANI, M. *ET AL.* 2000. *Peri-Tethys Atlas; Palaeogeographical Maps; Explanatory Notes*. Commission for the Geologic Map of the World, Paris.

DE WEVER, P., BOURDILLON-DE GRISSAC, C. & BEURRIER, M. 1988. Radiolaires sénoniens de la Nappe de Samail (Oman). *Revue de Micropaléontologie*, **31**, 166–179.

DROSTE, H. & VAN STEENWINKEL, M. 2004. Stratal geometries and patterns of platform carbonates: the Cretaceous of Oman. *In*: EBERLI, G. P., MASAFFERO, J. L. & SARG, J. F. R. (eds) *Seismic Imaging of Carbonate Reservoirs and Systems*. American Association of Petroleum Geologists, Memoirs, **81**, 185–206.

DUMITRICA, P. & DUMITRICA-JUD, J. 1995. *Aurisaturnalis carinatus* (Foreman), an example of phyletic gradualism among saturnalid-type radiolarians. *Revue de Micropaléontologie*, **38**, 195–216.

DUMITRICA, P. & DUMITRICA-JUD, R. 2005. *Hexasaturnalis nakasekoi* nov. sp., a Jurassic saturnalid radiolarian species frequently confounded with *Hexasaturnalis suboblongus* (Yao). *Revue de Micropaléontologie*, **48**, 159–168.

DUMITRICA, P., IMMENHAUSER, A. & DUMITRICA-JUD, R. 1997. Mesozoic radiolarian biostratigraphy from Masirah Ophiolite, Sultanate of Oman. Part I: Middle Triassic, uppermost Jurassic and Lower Cretaceous spumellarians and multisegmented nassellarians. *Bulletin of National Museum of Natural Science*, **9**, 1–106.

EBERLI, G. P. 1991. Calcareous turbidites and their relationship to sea-level fluctuations and tectonism. *In*: EINSELE, G., RICKEN, W. & SEILACHER, A. (eds) *Cycles and Events in Stratigraphy*. Springer, Berlin.

GEIGER, M., CLARK, D. N. & METTE, W. 2004. Reappraisal of the timing of the breakup of Gondwana based on sedimentological and seismic evidence from the Morondava Basin, Madagascar. *Journal of African Earth Sciences*, **38**, 363–381.

GORICAN, S. 1994. *Jurassic and Cretaceous radiolarian biostratigraphy and sedimentary evolution of the Budva Zone (Dinarides, Montenegro)*. Mémoires de Géologie (Lausanne), **18**, 176.

GORICAN, S., CARTER, E. S. *ET AL.* 2006. *Catalogue and systematics of Pliensbachian, Toarcian and Aalenian radiolarian genera and species*. Zalozba ZRC/ZRC Publishing, ZRC SAZU, Ljubljana.

GRABOWSKI, G. J., JR. & NORTON, I. O. 1994. Tectonic controls on the stratigraphic architecture and hydrocarbon systems of the Arabian Plate. *In*: AL-HUSSEINI, M. I. (ed.) *Geo'94. The Middle East Petroleum Geosciences, I*. GulfPetroLink, Bahrain, 413–436.

GRADSTEIN, F. M., OGG, J. G. & SMITH, A. G. (eds) 2004. *A Geological Time Scale*. Cambridge University Press, Cambridge.

HALLAM, A. 1976. Geology and plate tectonics interpretation of the sediments of the Mesozoic radiolarite–ophiolite complex in the Neyriz region, southern Iran. *Geological Society of America Bulletin*, **87**, 47–52.

HANDFORD, C. R. & LOUCKS, R. G. 1993. Carbonate depositional sequences and systems tracts—responses of carbonate platforms to relative sea-level changes. *In*: LOUCKS, B. & SARG, R. J. (eds) *Carbonate Sequence Stratigraphy: Recent Developments and Applications*. American Association of Petroleum Geologists, Memoirs, **57**, 3–41.

HAQ, B. U. & AL-QAHTANI, A. M. 2005. Phanerozoic cycles of sea-level change on the Arabian Platform. *GeoArabia*, **10**, 127–160.

HAQ, B. U., HARDENBOL, J. & VAIL, P. R. 1987. Chronology of fluctuating sea levels since the Triassic. *Science*, **235**, 1156–1167.

HARMS, J. C. 1975. Stratification and sequences in prograding shoreline deposits. *In*: HARMS, J. C., SOUTHARD, J. B., SPEARING, D. R. & WALKER, R. G. (eds) *Depositional Environments as Interpreted from Primary Sedimentary Structures and Stratification Sequences*. Society of Economic Paleontologists and Mineralogists, Short Courses, **2**, 81–102.

HARRIS, P. M., FROST, S. H., SEIGLIE, G. A. & SCHNEIDERMANN, N. 1984. Regional unconformities and depositional cycles, Cretaceous of the Arabian Peninsula. *In*: SCHLEE, J. S. (ed.) *Interregional Unconformities and Hydrocarbon Accumulation*. American Association of Petroleum Geologists, Memoirs, **36**, 67–80.

HOMEWOOD, P. W., MAURAIAUD, P. & LAFONT, F. 1999. *Best practices in sequence stratigraphy for explorationists and reservoir engineers*. Centre de Recherche Elf Exploration–Production, Mémoire, **25** [in French].

JAMES, G. A. & WYND, J. G. 1965. Stratigraphic nomenclature of Iranian oil consortium agreement area. *AAPG Bulletin*, **49**, 2182–2245.

JUD, R. 1994. *Biochronology and Systematics of Early Cretaceous Radiolaria of the Western Tethys*. Mémoires de Géologie (Lausanne), **19**, 147.

KNELLER, B., EDWARDS, D., McCAFFREY, W. & MOORE, R. 1991. Oblique reflection of turbidity currents. *Geology*, **14**, 250–252.

LAPIERRE, H., SAMPER, A. ET AL. 2004. The Tethyan plume: geochemical diversity of Middle Permian basalts from the Oman rifted margin. *Lithos*, **74**, 167–198.

LE BEC, A. 2003. *Distribution et dynamique des systèmes carbonatés de la plate-forme crétacé inférieur du Sultanat d'Oman*. Thèse de Doctorat, Université Michel de Montaigne (Bordeaux 3), Bordeaux.

LE MÉTOUR, J., MICHEL, J. C., BÉCHENNEC, F., PLATEL, J. P. & ROGER, J. 1995. *Geology and mineral wealth of the Sultanate of Oman*. MPM Geological Documents. Directorate General of Minerals, Ministry of Petroleum and Minerals, Sultanate of Oman, Muscat.

LE NINDRE, Y. M., MANIVIT, J., MANIVIT, H. & VASLET, D. 1990. Stratigraphie séquentielle du Jurassique et du Crétacé en Arabie Saoudite. *Bulletin de la Société Géologique de France*, **VI**, 1025–1034.

MURRIS, R. J. 1980. Middle East: Stratigraphic evolution and oil habitat. *AAPG Bulletin*, **64**, 597–618.

MUTTI, E. 1979. Turbidites et cônes sous-marins profonds. *In*; HOMEWOOD, P. (ed.) *Sédimentation détritique (fluviatile, littorale et marine)*. Institut de Géologie, Université de Fribourg, Fribourg, 353–419.

MUTTI, E. 1985. Turbidite systems and their relations to depositional sequences. *In*: ZUFFA, G. G. (ed.) *Provenance of Arenites*. NATO Advanced Studies Institutes Series. Serie C: Mathematical and Physical Sciences, 148. D. Reidel Publishing, Dordrecht, 65–93.

MUTTI, E. 1992. *Turbidite Sandstones*. Agip, Milan.

MUTTI, E., TINTERRI, R., REMACHA, E., MAVILLA, N., ANGELLA, S. & FAVA, L. 1999. An introduction to the analysis of ancient turbidite basins from an outcrop perspective. *American Association of Petroleum Geologists, Continuing Education Course Note Series*, **39**, 96.

O'DOGHERTY, L. 1994. *Biochronology and Paleontology of Mid-Cretaceous Radiolarians from Northern Apennines (Italy) and Betic Cordillera (Spain)*. Mémoires de Géologie (Lausanne), **21**, 413.

O'DOGHERTY, L. & GUEX, J. 2002. Rates and pattern of evolution among Cretaceous radiolarians: relation with global paleoceanographic events. *Micropaleontology*, **48**, Supplement 1, 1–22.

PHILIP, J. 2003. Peri-Tethyan neritic carbonate areas: distribution through time and driving factors. *Palaeogeography, Palaeoclimatology, Palaeoecology*, **196**, 19–37.

PICKERING, K. T. & HISCOTT, R. N. 1985. Contained (reflected) turbidity currents from the Middle Ordovician Cloridorme Formation, Quebec, Canada: an alternative to the antidune hypothesis. *Sedimentology*, **32**, 373–394.

PILLEVUIT, A., MARCOUX, J., STAMPFLI, G. & BAUD, A. 1997. The Oman Exotics: a key to the understanding of the Neotethyan geodynamic evolution. *Geodinamica Acta*, **10**, 209–238.

POMAR, L. 2001. Types of carbonate platforms: a genetic approach. *Basin Research*, **13**, 313–334.

POSAMENTIER, H. W., JERVEY, M. T. & VAIL, P. R. 1988. Eustatic controls on clastic deposition I—Conceptual framework. *In*: WILGUS, C. K., HASTINGS, B. S., KENDALL, C. G. S. C., POSAMENTIER, H., ROSS, C. A. & VAN WAGONER, J. (eds) *Sea-level Changes: an Integrated Approach*. Society of Economic Paleontologists and Mineralogists, Special Publications, **42**, 109–124.

RICOU, L. E. 1976. *Evolution structurale des Zagrides: la région-clé de Neyriz (Zagros iranien)*. Mémoire de la Société Géologique de France, Nouvelle Série, **125**, 140.

SANFILIPPO, A. & RIEDEL, W. R. 1985. Cetaceous Radiolaria. *In*: BOLLI, H. M., SAUNDERS, J. B. & PERCH-NIELSEN, K. (eds) *Plankton Stratigraphy*. Cambridge University Press, Cambridge, 573–630.

SCHLAGER, W., REIJMER, J. J. G. & DROXLER, A. 1994. Highstand shedding of carbonate platforms. *Journal of Sedimentary Petrology*, **B64**, 270–281.

SEARLE, M. P. & GRAHAM, G. M. 1982. 'Oman exotics': oceanic carbonate build-ups associated with the early stages of continental rifting. *Geology*, **10**, 43–49.

SHANMUGAM, G. 1996. High-density turbidity currents: are they sandy debris flows? *Journal of Sedimentary Research*, **66**, 2–10.

SHANMUGAM, G. 2000. 50 years of the turbidite paradigm (1950s–1990s): deep-water processes and facies models—a critical perspective. *Marine and Petroleum Geology*, **17**, 285–342.

SHARLAND, P. R., ARCHER, R. ET AL. 2001. *Arabian Plate Sequences Stratigraphy*. GeoArabia, Special Publication, **2**, 371.

STAMPFLI, G., MARCOUX, J. & BAUD, A. 1991. Tethys margins in space and time. *Palaeogeography, Palaeoclimatology, Palaeoecology*, **87**, 373–409.

STOCKLIN, J. 1968. Structural history and tectonics of Iran—a review. *AAPG Bulletin*, **52**, 1229–1258.

VAN WAGONER, J. C., POSAMENTIER, H. W., MITCHUM, R. M., VAIL, P. R., SARG, J. F., LOUTIT, T. S. & HARDENBOL, J. 1988. An overview of the fundamentals of sequence stratigraphy and key definitions. *In*: WILGUS, C. K., HASTINGS, B. S., KENDALL, C. G. S. C., POSAMENTIER, H., ROSS, C. A. & VAN WAGONER, J. (eds) *Sea-level Changes: an Integrated Approach.* Society of Economic Paleontologists and Mineralogists, Special Publications, **42**, 39–45.

VAN WAGONER, J. C., MITCHUM, R. M. J., CAMPION, K. M. & RAHMANIAN, V. D. 1990. *Siliciclastic sequence stratigraphy in well logs, cores, and outcrops: concepts for high resolution correlation of time and facies.* American Association of Petroleum Geologists, Methods in Exploration Series, **7**, 55.

ZIEGLER, M. 2001. Late Permian to Holocene paleofacies evolution of the Arabian Plate and its hydrocarbon occurences. *GeoArabia*, **6**, 445–504.

The influence of Late Cretaceous tectonic processes on sedimentation patterns along the northeastern Arabian plate margin (Fars Province, SW Iran)

ALIREZA PIRYAEI[1,2,3]*, JOHN J. G. REIJMER[2,4], FRANS S. P. VAN BUCHEM[1,5], MOHSEN YAZDI-MOGHADAM[6], JALIL SADOUNI[6] & TANIEL DANELIAN[7,8]

[1]*IFP, 1 & 4, avenue de Bois-Préau, 92852 Rueil-Malmaison Cedex, France*

[2]*Université de Provence (Aix-Marseille I), Laboratoire de Géologie des Systèmes et des Réservoirs Carbonatés EA 4234–LGSRC, 3, place Victor Hugo, Case 67, F-13331 Marseille Cédex 3, France*

[3]*Present address: NIOC Exploration Directorate, 1st Dead-end, Seoul St., NE Sheikh Bahaei Sq., PO Box 19395-6669 Tehran, IRAN*

[4]*Present address: VU University Amsterdam, Faculty of Earth and Life Sciences, Department of Sedimentology and Marine Geology, De Boelelaan 1085, 1081 HV Amsterdam, Netherlands*

[5]*Present address: Maersk Oil Qatar AS, PO Box 22.050, Doha, Qatar*

[6]*NIOC Exploration Directorate, 1st Dead-end, Seoul St., NE Sheikh Bahaei Sq., PO Box 19395-6669 Tehran, Iran*

[7]*Université Pierre et Marie Curie-Paris 6, UMR 5143 'Paléobiodiversité et Paléoenvironnements', C. 104, 4 place Jussieu, 75252 Paris Cedex 05, France*

[8]*Present address: Université Lille 1, Laboratoire Géosystèmes (UMR 8157 CNRS), UFR des Sciences de la Terre—bâtiment SN5, 59655 Villeneuve d'Ascq cedex, France*

*Corresponding author (e-mail: a.piryaei@niocexp.ir)

Abstract: During the Late Cretaceous the northeastern margin of the Arabian plate (Zagros–Fars Area) was characterized by significant variations in sedimentary facies, sedimentation patterns and accommodation space, and by shifting depocentres. A succession of events recording the evolution of the region from a passive to an active margin is documented by the study of eight outcrop sections and one well. This new study uses new age dating (benthic and planktonic foraminifers, nannoplankton and radiolarian biozonations and strontium isotope stratigraphy). The new observations provide a detailed overview of the response of the sedimentary system to changes in the tectonic regime related to obduction processes. These changes are very well shown in regional cross-sections and palaeogeographic maps. Three tectono-sedimentary phases are recognized indicating the evolution from a passive to an active margin: Phase I (Late Albian to Cenomanian, before obduction) comprises three depositional third-order sequences comparable with those of the other parts of the Zagros and Arabian plate. This interval is composed of shallow-water platform carbonates and intra-shelf basins. The platform facies consists of rudist and benthic foraminifer-dominated assemblages, whereas the intra-shelf basins contain an '*Oligostegina*' facies. Eustatic sea-level variations and local differential subsidence controlled sediment deposition during this phase. Phase II (Turonian to Late Campanian, obduction phase) is characterized by major changes in depositional environments and sedimentary facies, as a result of obduction and foreland basin creation. It consists of pelagic and platform carbonates in the south, and a foreland basin with obducted radiolarites, ophiolitic and olistoliths or thrust slices in the north. During this phase, large volumes of turbidites and gravity flows with olistoliths were shed from both the SW and NE into the foreland basin. The age of the tectonic slices increases upward through the section, from Early Cretaceous at the base to Permian at the top. Based on various dating methods used on the far-travelled sediments, the depositional age of the radiolarites can be attributed to the Albian–Cenomanian, whereas the planktonic foraminifers are of Santonian to Campanian age. Phase III (Late Campanian to Maastrichtian, after obduction) shows the development of rudist-dominated carbonates in the NE prograding onto the deep basinal facies in the centre of study area. In the extreme NE no sediments of this age have been recorded, suggesting uplift at that time.

From: LETURMY, P. & ROBIN, C. (eds) *Tectonic and Stratigraphic Evolution of Zagros and Makran during the Mesozoic–Cenozoic*. Geological Society, London, Special Publications, **330**, 211–251.
DOI: 10.1144/SP330.11   0305-8719/10/$15.00 © The Geological Society of London 2010.

The transition from a tectonically passive to a tectonically active regime along a plate margin is one of the most dramatic events that may take place within a given palaeogeographical area. It marks the complete shift of accommodation space creation and destruction related to eustatic processes by tectonic forces, associated with the development of a flexural basin, tectonic accretion, uplift and erosion. These events are accompanied by the emplacement of far-travelled sediments, depocentre shifts, and deposition of sediments from different origins (e.g. reworked crystalline and sedimentary rocks). Thus, the influence of the compressional tectonic regime on the sedimentation pattern of an active margin is twofold: a foreland basin is created that migrates through time, and sedimentary and crystalline rocks are mobilized, through uplift and erosion (Allen & Homewood 1986; Cross 1986; Robertson 1987; DeCelles & Giles 1996).

In this study we document the influence of the Late Cretaceous active margin tectonics on the sedimentation patterns of the northeastern margin of the Arabian plate in SW Iran (interior Fars province; Fig. 1). Previous workers have identified the presence of Late Cretaceous thrust tectonics and the emplacement of ophiolites on the autochthonous series along the northeastern margin of the Arabian plate (Iran: Gray 1950; Ricou 1968*a*, *b*, 1970, 1971*a*, *b*; Wells 1969; Haynes & McQuillan 1974; Stoneley 1975, 1981; Hallam 1976; Oman: Glennie & Hughes Clarke 1973; Glennie *et al.* 1973). For Iran, however, differences in opinion exist, about the age and allochthonous or autochthonous nature of the radiolarian cherts, as well as the nature and mechanism of emplacement of the ophiolites (Ricou 1968*a*, *b*, 1970, 1971*a*, *b*; Stoneley 1974, 1975). Central to this debate are the outcrops in the Dalneshin Mountain region between Shiraz and Neyriz, which were mapped and provisionally dated by Stoneley (1981; Fig. 2). The complexity of the area, and the difficulty in dating the radiolarite beds, led Stoneley (1981) to propose two alternative models for the origin and emplacement of the allochthonous rocks in the Late Cretaceous, one involved gliding from an aborted spreading ridge along the continental margin, and the other based on subduction processes.

The focus of the present study is on the transition period from a passive to an active plate margin, and the infill history of the foreland basin during the Late Cretaceous. To document this transition a 160 km south–north transect has been investigated that covers tectonically undisturbed Cretaceous strata from the central Fars province, to the actual thrust belt present in the Kuh-e Dalneshin area (Transect 1 in Fig. 3). A second transect was constructed to document the lateral variability in the thrust belt

zone through time (Transect 2 in Fig. 3). The investigated strata include shallow-water carbonate shelf deposits of Late Albian to Cenomanian age (representative of the passive margin setting), and tectonically controlled foreland basin deposits of Turonian to Maastrichtian age. Particular points of interest are the change in basin configuration, the shifts in depocentre and the associated sedimentary features during the transition in tectonic regimes.

The methods used include structural analysis and sequence stratigraphy to subdivide the rock record into tectono-sedimentary (genetically) related packages. A variety of dating methods were used to understand the nature of the far-travelled units. Compared with the work of Stoneley (1981), this study is based on a more detailed sedimentological analysis, combined with a microfacies analysis, and covers a larger study area including stratigraphic sections located on the Arabian plate.

Following an introduction to the stratigraphic and tectonic background of the study area, the paper provides a detailed description of the sedimentary facies and environments. Findings from eight outcrop sections and one well will be presented, to define three successive tectono-sedimentary phases. Finally we discuss our results and evaluate the three tectonic models that have been proposed for the Late Cretaceous of the eastern margin of the Arabian plate.

Geological background

The most comprehensive work on the Mesozoic and Cenozoic lithostratigraphic units of the Zagros was carried out by James & Wynd (1965). That study was linked to a study addressing biofacies nomenclature of Wynd (1965). Other more recent, stratigraphic studies include those of Player (1969), Kheradpir (1975), Khalili (1976), Setudehnia (1978), Szabo & Kheradpir (1978) and Shakib (1994). A sequence stratigraphic study of the Cretaceous in the Zagros area was recently carried out by van Buchem *et al.* (2001, 2006).

The Late Cretaceous interval exposed in the study area comprises five lithostratigraphic units (Fig. 4): (1) the Cenomanian to Turonian Sarvak Formation, which developed throughout the area and contains sediments that were deposited in a shallow continental to intra-shelf basin environment; (2) the Late Turonian to Santonian Surgah and Ilam formations, which developed at a few locations and comprise sediments deposited on isolated platforms that interfinger with basinal environments; (3) the Santonian to Maastrichtian Gurpi Formation consists of sediments deposited in pelagic to hemipelagic settings; (4) the

(a)

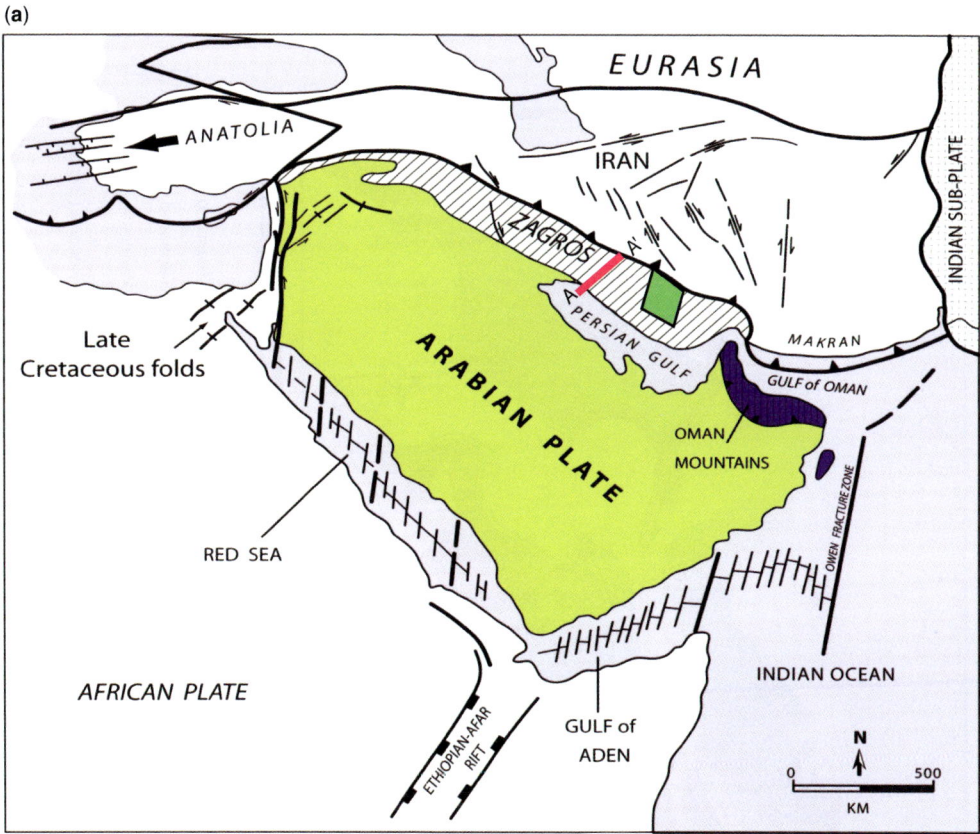

(b)

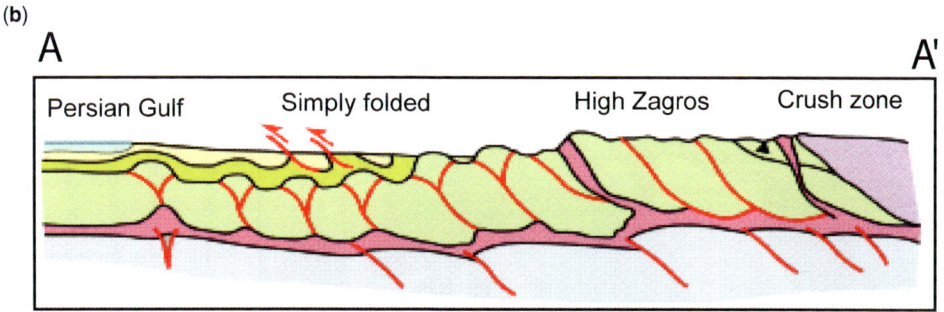

Fig. 1. (a) Regional map of the Arabian plate showing Zagros Fold–Thrust Belt and study area (modified after Sharland *et al.* 2001). (b) Schematic SW–NE cross-section of the Zagros Foreland Fold–Thrust Belt showing its structural zones (after Sherkati & Letouzey 2004).

Maastrichtian Tarbur Formation, with reefal sediments deposited mainly in marginal settings; (5) the Late Maastrichtian Sachun Formation, with evaporites deposited in restricted platform sub-environments (James & Wynd 1965; Kheradpir 1975; Khalili 1976; Bolz 1985; Motiei 1993; van Buchem *et al.* 2001, 2006). The majority of the

petroleum source and reservoir rocks in the region, as well as seal rocks, are concentrated in this interval (Bordenave & Burwood 1990; Beydoun *et al.* 1992; Bordenave 2000; Bordenave & Herge 2005).

The major tectonic events in the geological history of the Zagros Foreland Fold–Thrust Belt (FFTB) took place in the Late Cretaceous to Late

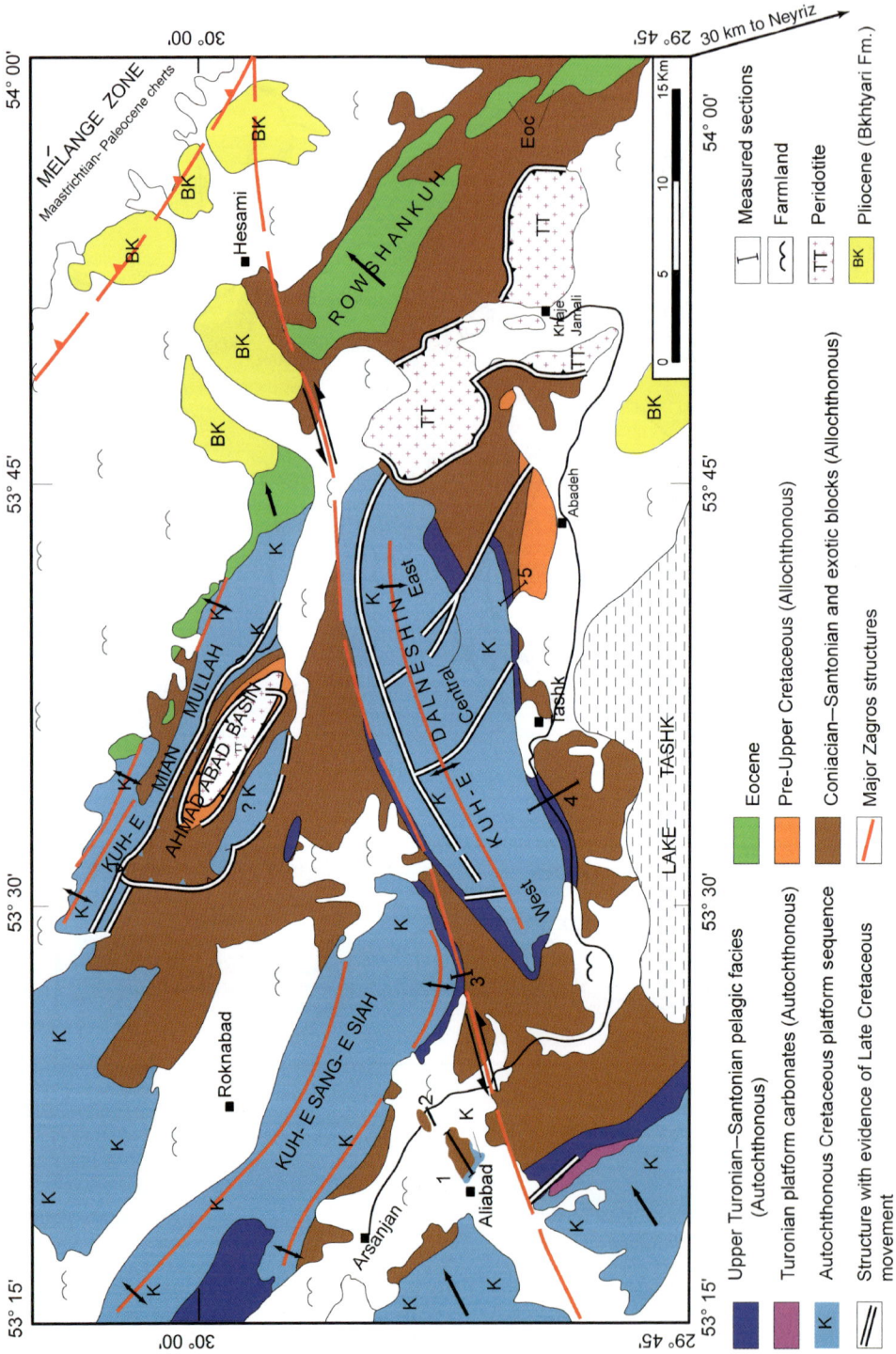

Fig. 2. Geological map of Kuh-e Dalneshin and Arsanjan area (modified from figs 3 & 4 of Stoneley 1981), showing the relative distribution of autochthonous and allochthonous facies associated with major structural features related to late Cretaceous and Neogene tectonic processes in the Zagros area.

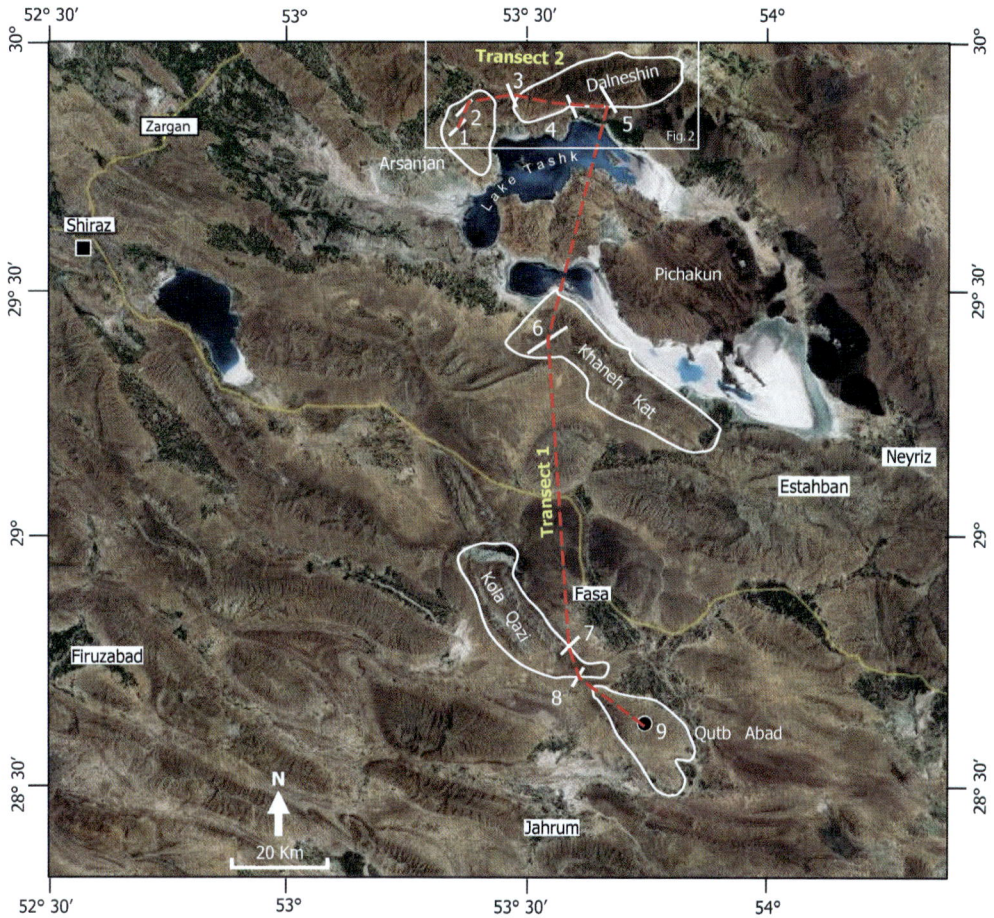

Fig. 3. Satellite photograph of the study area, indicating the studied outcrop sections and the well location. The outcrop sections are located in the Kuh-e Dalneshin and Arsanjan area, Khaneh-Kat monocline and Kola-Qazi anticline. The Qutb Abad well is located in the southeastern part of the map within the Qutb Abad anticline.

Pliocene time interval (Stocklin 1968, 1974, 1977; Takin 1972; Falcon 1974; Ricou 1976; Berberian & King 1981; Koop & Stoneley 1982; Alavi 2004). This mountain belt is the result of at least two main tectonic events, the first being the beginning of the Neo-Tethys closure, which lead to thrusting, obduction and the creation of a foreland basin in the Late Cretaceous (Gray 1950; Ricou 1968a, b, 1970, 1971a; Wells 1969; Haynes & McQuillan 1974; Stoneley 1975, 1981; Hallam 1976; Alavi 1994, 2004), and the second being the final collision, which caused the closure of the Neo-Tethys in the Miocene–Pliocene (Glennie et al. 1973, 1974; Ricou et al. 1977; Stoneley 1981; Glennie 2000, 2001). The tectonic event that is of interest for this study is the Late Cretaceous obduction.

From SW to NE the study area can be subdivided into three tectono-sedimentary zones.

(1) The simply folded area situated south of the Bakhtegan and Tashk lakes shows the classic formations of the Zagros, with large anticlines and synclines (Fig. 3), in which the sedimentary record is composed of autochthonous rocks with some reworked units in the Santonian to Campanian time interval.

(2) The ophiolitic zone exposed around the Bakhtegan and Tashk lakes that is terminated by the main thrust fault in the NE includes the same formations as the simply folded area, but also deep facies of the Neo-Tethys ocean basin and ophiolites, radiolarites and highly tectonized units. This succession of sediments (Pichakun Series of Ricou 1968b) also includes thrust slices of Late Triassic limestone to Middle Jurassic oolitic and micro-brecciated limestone and Late Cretaceous conglomeratic limestone. The Pichakun Series is wedged between two

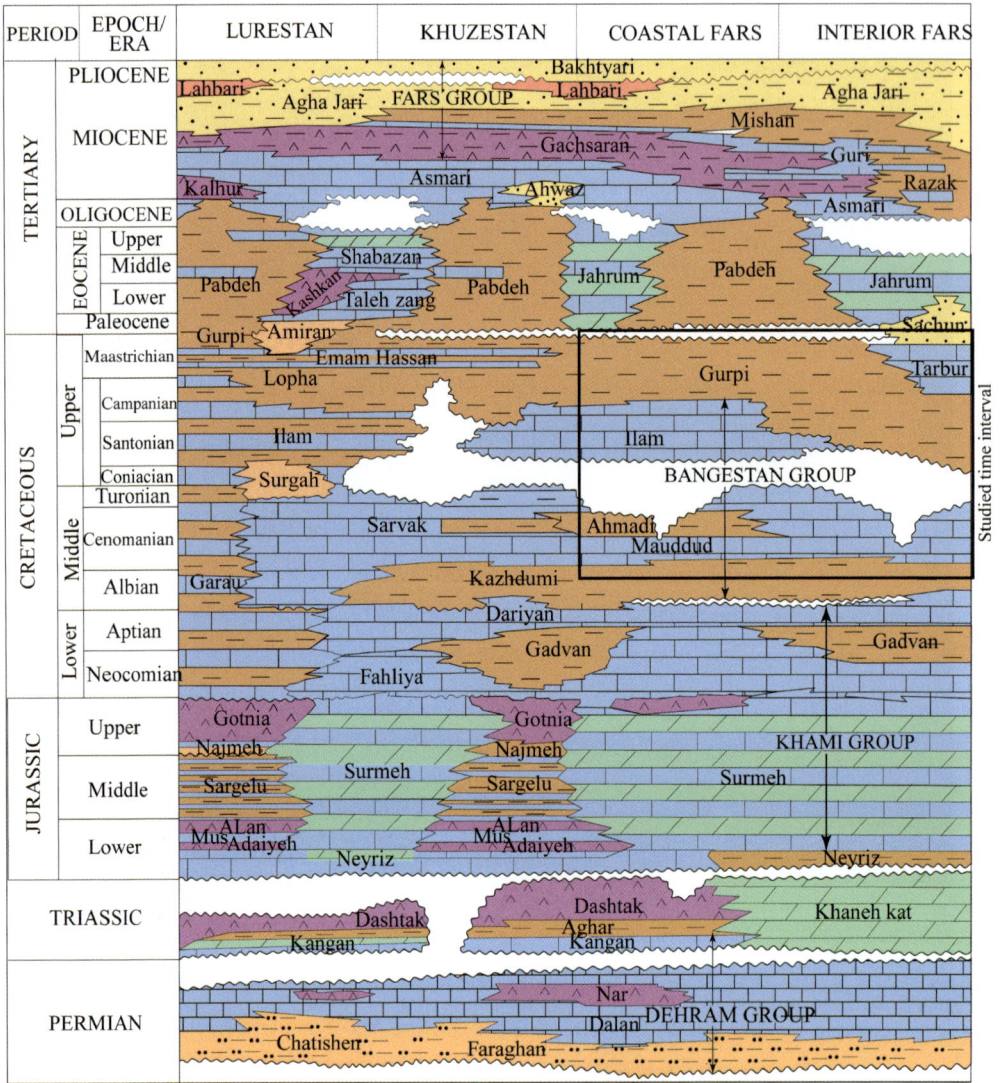

Fig. 4. Lithostratigraphic units of the Zagros Fold–Thrust Belt (modified from Gulf Petrolink 1998).

thrust sheets with a sheared sedimentary mélange (Haynes & McQuillan 1974; Pamic & Adib 1982).

(3) In the northeastern part of the study area the lithological series is more complicated and shows ophiolite thrust sheets that override Late Cretaceous shallow- and deep-water carbonates. These deposits are unconformably overlain by Campanian to Maastrichtian pelagic and platform carbonate deposits and evaporites (James & Wynd 1965; Alavi 1994). According to Ricou (1974), this series of rocks is of Turonian to end-Cretaceous age, and was refolded in the Pliocene. These rocks represent remnants of lithosphere plates that were moved to this region by thrusting during obduction. This assemblage occurs associated with layered gabbros, sheeted dykes and pillow lavas, and also contains chaotic and regularly bedded radiolarian cherts intervals (of Jurassic to Late Cretaceous age) and Santonian to Campanian pelagic marls. The initial emplacement as slivers of Neo-Tethyan oceanic crust over the Afro-Arabian continental shelf must have been a Cenomanian–Maastrichtian event (Alavi 1994).

Methods and materials

A high-resolution sedimentary sequence analysis is used for the pre-obduction carbonate units to reconstruct the stratigraphic architecture of the sedimentary system. This method follows the sequence stratigraphic approach used for shallow-water carbonate systems (van Wagoner et al. 1988; Goldhammer et al. 1990; Pomar 1991; Loucks & Sarg 1993; Read et al. 1995; Kerans & Tinker 1997; Homewood & Eberli 2000; van Buchem et al. 2002). A time-based classification of depositional sequences has been used following Haq et al. (1988) and Vail et al. (1991). The allochthonous (far-travelled) sediments have been dated by different methods to find an age range and to unravel the relationship between the various constituents.

The method used for this study consists of four steps. The first step comprises the detailed description of sedimentary features such as bedding geometry, textures, lithological compositions, faunal content, ichnofacies and special surfaces. The second step deals with 1D sequence analysis. Fluctuations of the accommodation space/sediment supply ratio are dependent on the palaeobathymetric interpretations of the depositional environments, the occurrence of distinct surfaces, and the preservation of sedimentary structures (sets and co-sets). The third step is a time-based correlation, constrained by (1) the bedding geometries, (2) the hierarchy of the depositional sequences as defined in each section, and (3) biostratigraphy and chemostratigraphy. The fourth step shows the high-resolution sequence stratigraphic model, which is a 3D dynamic depositional model constrained by the chronostratigraphic framework, and the third order sedimentary sequence stacking.

The study area is located in the southeastern Zagros (Fars area) (Fig. 3). It lies between the main Zagros thrust zone, the Kazerun Fault, the Razak Fault and the northern part of the Persian Gulf. Eight outcrop sections were measured during this study (1–8; Table 1), which comprise 4660 m of sediments in which 1100 samples were taken for thin-section analysis. The outcome of this analysis is presented in transects 1 and 2. In addition, one well (9; Table 1) was analysed. One of the transects connects the Outer Zagros (simply folded zone) to the Inner Zagros (highly tectonized unit) whereas the other runs along the Late Cretaceous nappes to make a comparison between proximal and distal parts of the far-travelled sediments. The palaeogeographical and associated isopach maps presented in this study are, however, based on more than 40 outcrop sections and wells that are confidential NIOC property.

Where possible, biostratigraphic analysis of foraminifers, ammonites, nannofossils and radiolaria

Table 1. Studied sections and their characteristics

No.	Section name	Coordinates (GPS)	Nearby town	Measured thickness (m)	Time range
1	Arsanjan (Hossein Abad)	053°19'41"E 29°50'28"N	Arsanjan	590	Cenomanian–Santonian
2	Ali Abad	053°20'33"E 29°52'06"N	Arsanjan	220	Cenomanian–Santonian
3	Tang-e Gomban	053°25'24"E 29°52'51"N	Arsanjan	50	Cenomanian–post-Turonian
4	Gardaneh-ye Abadeh-Tashk	053°32'08"E 29°49'47"N	Arsanjan	650	Albian–post-Cenomanian
5	Tang-e Jazin	053°36'47"E 29°55'30"N	Abadeh-Tashk	820	Aptian–post-Cenomanian
6	Khaneh Kat	053°33'38"E 29°33'45"N	Kharameh	1210	Albian–Maastrichtian
7	Kola Qazi north	053°39'23"E 28°47'11"N	Fasa	480	Santonian–Maastrichtian
8	Kola Qazi south	053°33'19"E 28°48'09"N	Fasa	643	Santonian–Maastrichtian
9	Qutb Abad well	053°44'01"E 29°41'57"N	Fasa, Jahrum	1000	Albian–Maastrichtian

was carried out on thin sections or washed or HF acid-etched samples. The palaeontological analysis of foraminifers was carried out in the Palaeontological Department of the NIOC, using the biozonation model proposed by Wynd (1965).

Sedimentary facies and depositional environments

In the study area a wide variety of lithologies and sedimentary facies can be found. Based on lithology, sedimentary features and palaeontological analysis, four main lithofacies groups are distinguished: carbonate-dominated facies, resedimented carbonate and siliciclastic facies, radiolarites and igneous–ophiolitic facies. They are presented below with their environmental interpretation. The stratigraphic relationships between these lithofacies are discussed in the sequence stratigraphic section.

Carbonate-dominated facies

The Late Cretaceous carbonate sediments examined in this study represent an evolution from shallow-water (inner–mid-ramp, outer ramp–slope) to deep-water (intra-shelf and foreland basin) environments. Because of the changing fauna and geometrical organization of the Late Cretaceous carbonate systems, they are presented in chronological order. A summary diagram of these systems is provided in Figure 5, and illustrations of the most characteristic facies are provided in Figures 6–10. Detailed information on the faunal composition is presented in the fossil range charts of the sedimentary logs as key sections (Figs 11, 12, 13 & 14). These sedimentary systems in each time interval can be related to a particular lithostratigraphic unit (Fig. 4).

Muddy ramp system dominated by large benthic foraminifers (Late Albian, Mauddud Mb.)

The Late Albian carbonate facies observed in the study area consists of wackestone with a purely benthic faunal association dominated by large foraminifers, in particular discoidal *Orbitolina* and *Trocholina*, and to a lesser extent *Hemicyclammina*. The sediment is organized in decimetre- to metre-scale beds, locally with iron-stained horizons, and separated by marl layers with varying clay content, which have *Thalassinoides* burrows.

No time-equivalent deep-water deposits were recorded in the study area. However, elsewhere in the Zagros, they correspond to the Kazhdumi intra-shelf basin deposits (e.g. James & Wynd

1965; van Buchem *et al.* 2006). In lithostratigraphic nomenclature the Late Albian ramp system is called Mauddud Member of the Sarvak Formation (Fig. 4).

Rimmed carbonate platform and intra-shelf basin system (Cenomanian–Turonian, Sarvak Fm., Ahmadi Mb.)

In this system four main environments are distinguished: the low-energy platform top facies, the high-energy platform top to margin facies, the slope facies and intra-shelf basinal facies.

(1) The low-energy platform top facies consists of bioturbated wackestone to packstone containing varying amounts of benthic foraminifers (i.e. *Nezzazata*, orbitolinids, alveolinids), coral, bivalves and algal debris. The bedding pattern differs from decimetre to metre scale. The small- to medium-scale bedding normally has a wavy to nodular appearance with a considerable amount of bioturbation. The facies are mostly related to either lagoonal or relatively deeper sub-environments (below fair-weather wave base) (Fig. 6a, b).

(2) The high-energy platform top to margin facies consists of fine- to coarse-grained rudist grainstone to rudstone, associated with benthic foraminifers (orbitolinids, trocholinids, alveolinids), peloids and intraclasts. This facies is organized in metre- to decimetre-scale beds, displaying planar lamination. Within the upper part of Cenomanian platform deposits, thickening upward cycles are present. The thick and massive bedded facies end with rudist floatstone to rudstone, which can be very porous, that locally are topped by a silicified surface (Fig. 6c, d). The high-energy facies occurs associated with intraformational conglomerates that occasionally fill intraplatform channels, which are tens of metres wide and several metres deep (Fig. 6c, Arsanjan area). In these channels the grain-supported sediments eroded the underlying muddier facies and also show loading.

(3) The slope environment is clearly observed in Gardaneh-e Abadeh-Tashk (Section 3) where the platform carbonate facies in the middle part of the Kuh-e Dalneshin laterally changes to the basinal facies, as can be observed in the eastern part of the Kuh-e Dalneshin (Section 5; Fig. 14). In this section lateral changes in bedding pattern took place from thick platform carbonate to thin basinal oligosteginid facies. The sediment, mainly wackestone, contains a mix of platform-derived fauna (such as alveolinids, *Nezzazata*, and fine-grained rudist debris) and intra-shelf basinal fauna (oligosteginids, sponge spicules).

(4) The intra-shelf basin facies consists of oligosteginid wackestone to packstone (Fig. 6e), organized in decimetre-scale beds with wavy contacts.

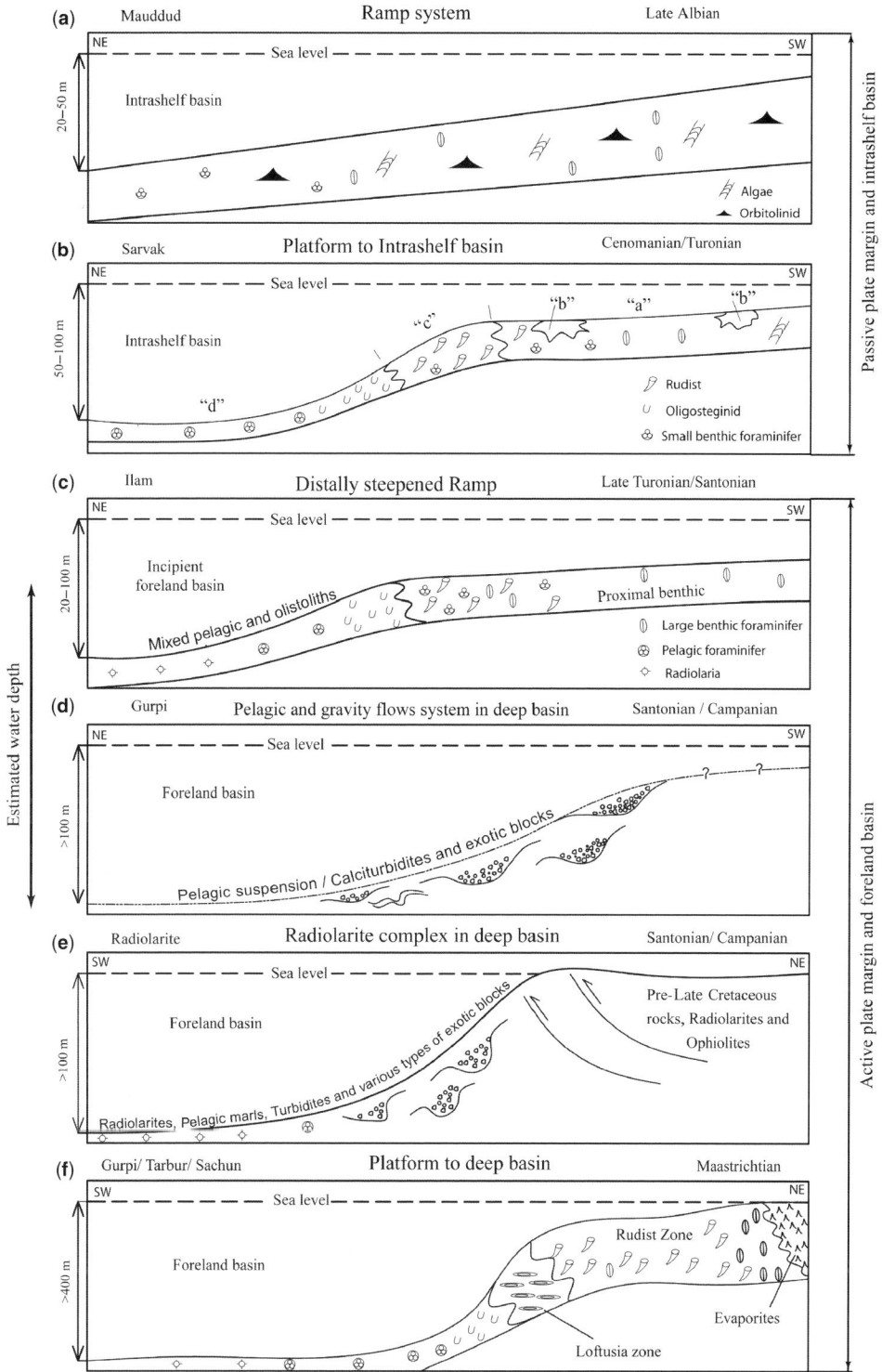

Fig. 5. Summary of carbonate depositional environments within the Fars area based on sedimentary facies and depositional environments. The Albian ramp system is variously called the Mauddud Member or Sarvak Formation.

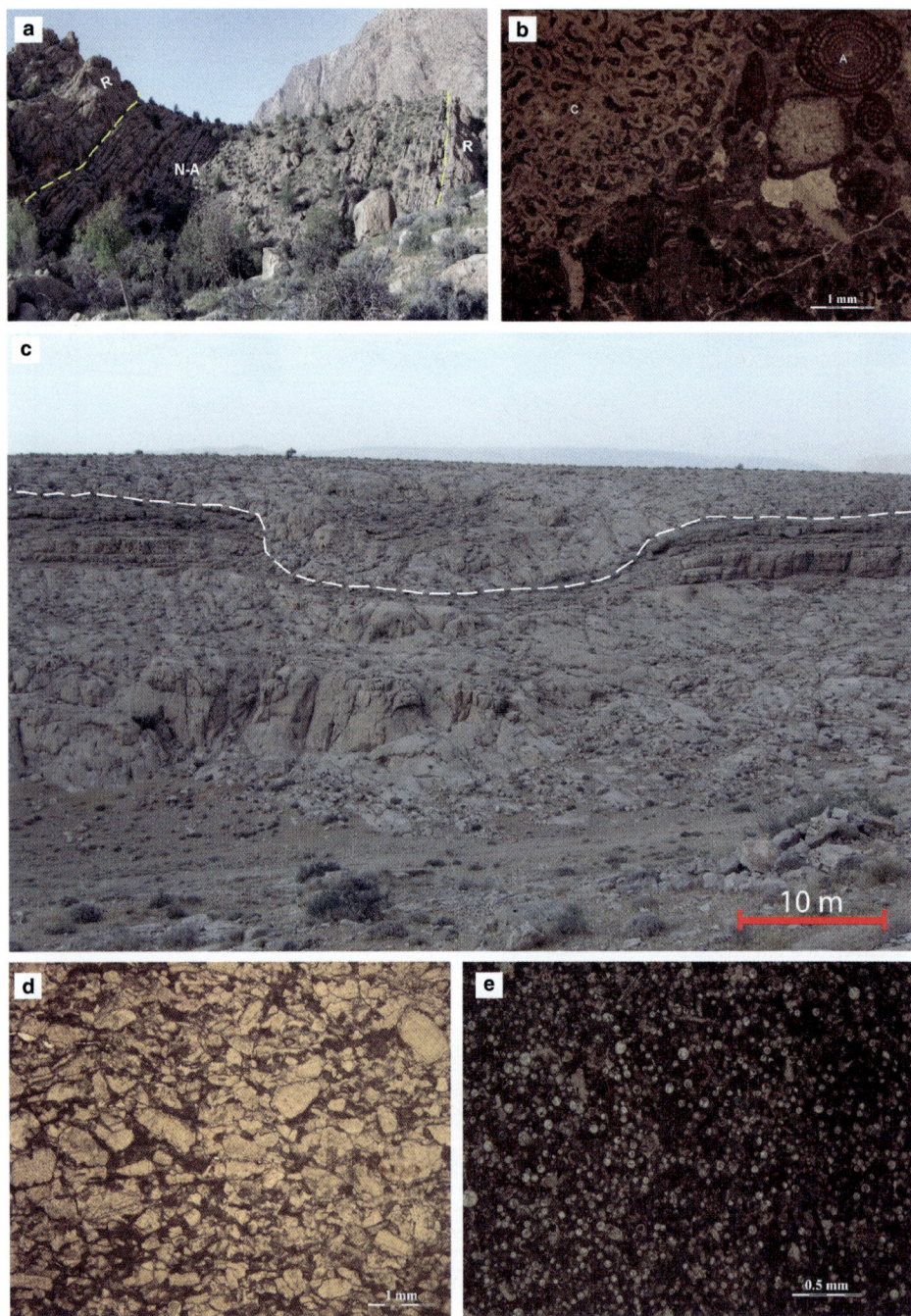

Fig. 6. Facies variations within the Cenomanian–Turonian sedimentary system (Sarvak Fm.). (**a**) Outcrop showing medium to massive beds of shallow-water carbonate facies. Yellow dashed lines mark the boundaries between rudist debris unit (R) and *Nezzazata*–alveolinid (NA) units in the Sarvak Formation. The trees for scale are 3–5 m in height. (**b**) Thin-section microphotograph of shallow-water facies with alveolinid (A) and coral (C) in a bioclast packstone texture. (**c**) Intra-platform channel filled with grainy sediments of the Sarvak Formation. Width of the channel is *c.* 10 m. (**d, e**) Thin-section microphotographs: (d) rudist debris facies of the platform edge; (e) oligosteginid facies of the Cenomanian intra-shelf basin.

Fig. 7. Facies illustrations of the Campanian to Maastrichtian sedimentary system (Gurpi and Tarbur fms.).
(**a**) Decimetre-scale bedding of pelagic marls, Gurpi Formation (Kola-Qazi south, Section 8; Fig. 11). (**b**) Thin-section microphotograph of pelagic bioclast packstone of the same marly facies with various species of globigerinid foraminifers and oligosteginids. (**c**) Interfingering and progradation of rudist-dominated facies (Maastrichtian Tarbur Formation) and pelagic marls (Gurpi Formation), Khaneh Kat anticline (Section 6); the trees for scale are 3–5 m in height. (**d**, **e**) Thin-section microphotographs: (d) floatstone sedimentary texture of rudist-dominated facies, Tarbur Formation in Section 6 (Khaneh Kat); (e) *Loftusia*-bearing facies, Tarbur Formation.

Fig. 8. Facies illustrations of early Santonian to Maastrichtian Gurpi Formation (Kola Qazi, Section 8; Fig. 11).
(**a**) Olistostrome unit within Campanian pelagic marls (Gurpi Formation). (**b**) Channel fill deposits with a lensoid form.
(Note three fining upward sedimentary cycles.) (**c**) Photomicrograph of clast-dominated gravity deposits shown in
(**b**). Litho-facies of slumped neritic deposits show close similarity to the underlying Santonian platform facies.
(**d**) Depositional cycles in gravity deposits within the uppermost part of the Gurpi Formation. (**e**) Sediments consisting
of clast-dominated conglomerate and coarse-grained carbonate sandstone.

Locally this facies is black and has a distinct smell,
suggesting that organic matter is still preserved in
the rocks. The sediments have a massive appearance
in outcrop, but at close inspection show bedding on

a decimetre scale (Figs 13b & 14, Section 5). The
interfingering of the intra-shelf basin and slope
deposits suggests a water depth of the order of
tens of metres (*c.* 50 m).

Fig. 9. Facies illustrations of the Santonian–Campanian carbonate and siliciclastic systems (Gurpi Formation and radiolarite complex). (**a**) Planar lamination and X-bedding in the uppermost beds of the Maastrichtian Gurpi Formation. (**b**) Microphotograph of the texture, showing peloidal packstone to grainstone. (**c**) Field photograph of Bouma cycle of turbiditic facies in radiolarite complex, with microphotographs of: (**d**) the clastic turbiditic facies in Section 1, with peloids, ooid and skeletal grains in a radiolarite matrix; (**e**) erosional surface at the base of sandstone facies; (**f**) pebbly sandstone containing igneous and carbonate rock fragments.

Large benthic foraminifer- and rudist-dominated ramp system (Santonian, Ilam and Gurpi fms.)

Shallow-water sediments consist of packstone to floatstone with rudist and echinoderm debris, algae and benthic foraminifers (*Rotalia*, *Dicyclina*, *Cuneolina*). These sediments were deposited in an open–mid-ramp setting. Locally interbedding with facies bearing a pelagic fauna (*Heterohelix*, oligosteginids) can be observed, which is interpreted as an outer ramp environment. The lithostratigraphic

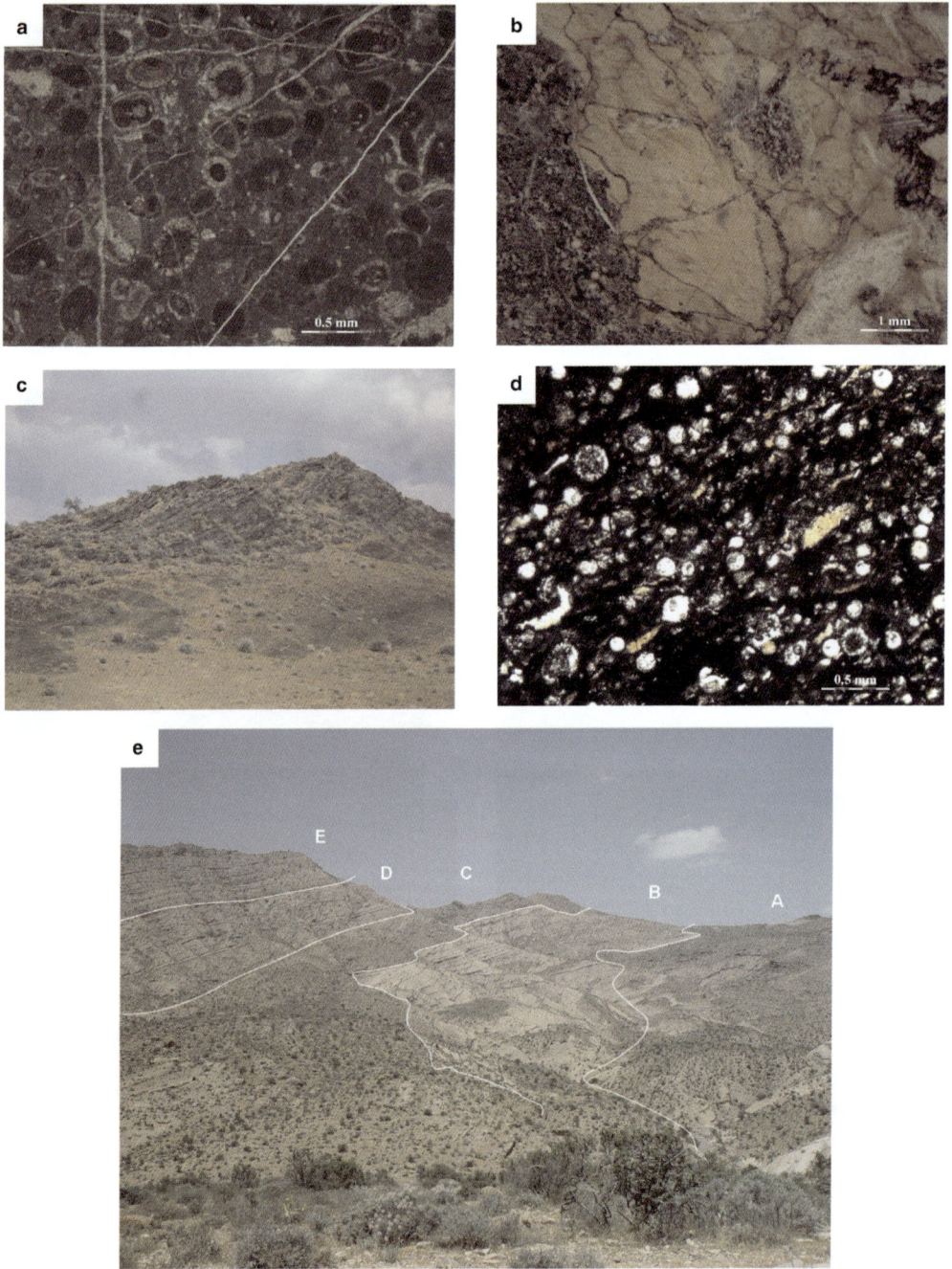

Fig. 10. Facies illustrations of the radiolarite–ophiolite deposits in the Dalneshin area. Microphotographs of (**a**) oolitic and peloidal carbonate sandstone among turbiditic facies (Section 1); (**b**) igneous peridotite part of the ophiolite series (Section 5). (**c**) Decimetre-scale bedding with varicoloured radiolarite deposits, Pichakun Mountain range south of Dalneshin. (**d**) Microphotograph of poorly laminated radiolarian facies. (**e**) Overview of the Gurpi Formation in the Kola Qazi anticline (Section 8; Fig. 11). Unit shows gravity-flow deposits (Unit A), regular bedded pelagic marls (Unit B), slumped platform carbonate and host pelagic chaotic facies (Unit C), alternation of thin- to medium-bedded lime mudstone to wackestone with marls (Unit D) and calciturbiditic beds in the Maastrichtian part of the Gurpi Formation (Unit E). The trees are <3 m in height.

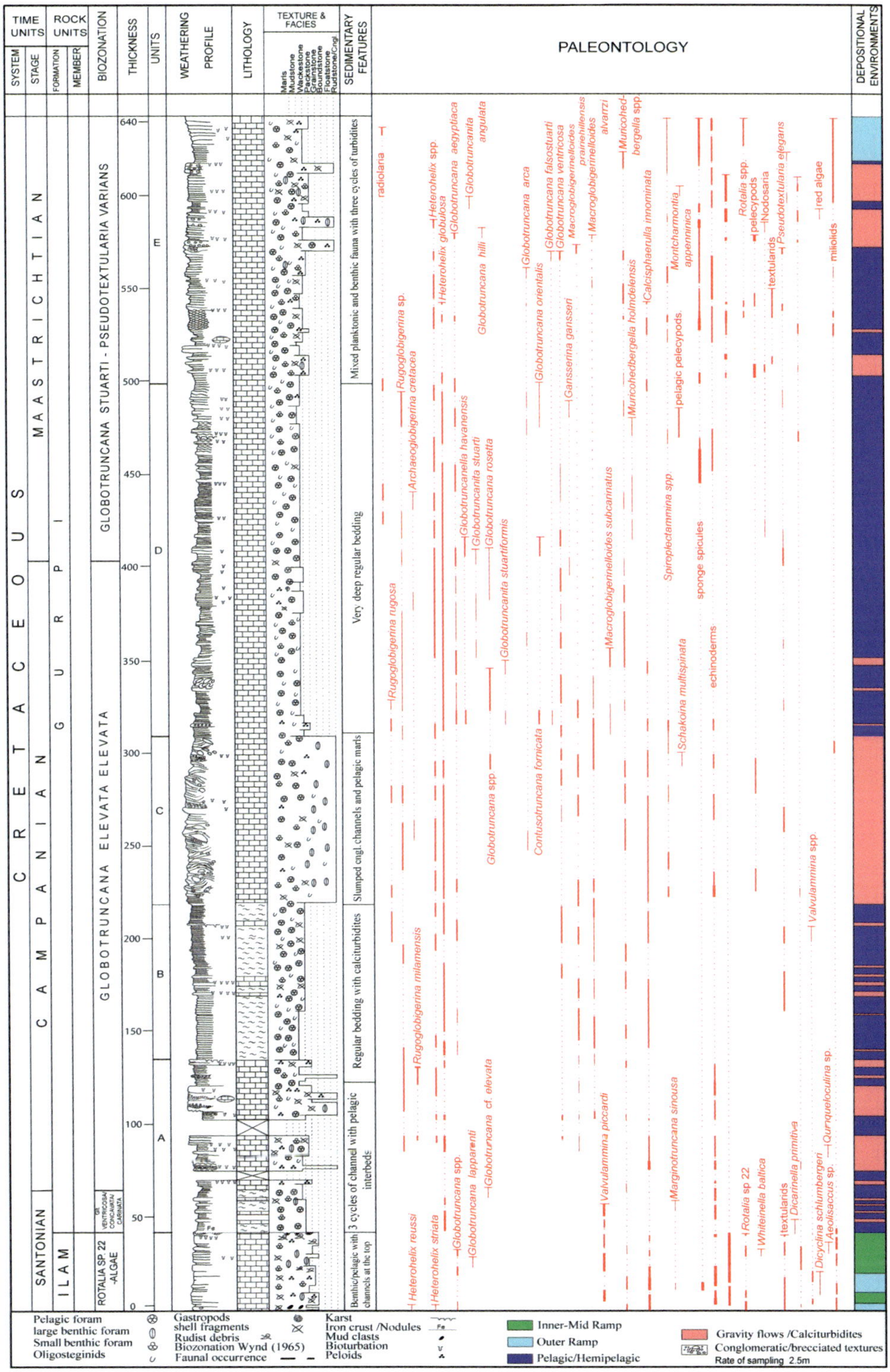

Fig. 11. Kola Qazi section (Section 8). Sedimentary and palaeontological log based on field data and microscopic observations.

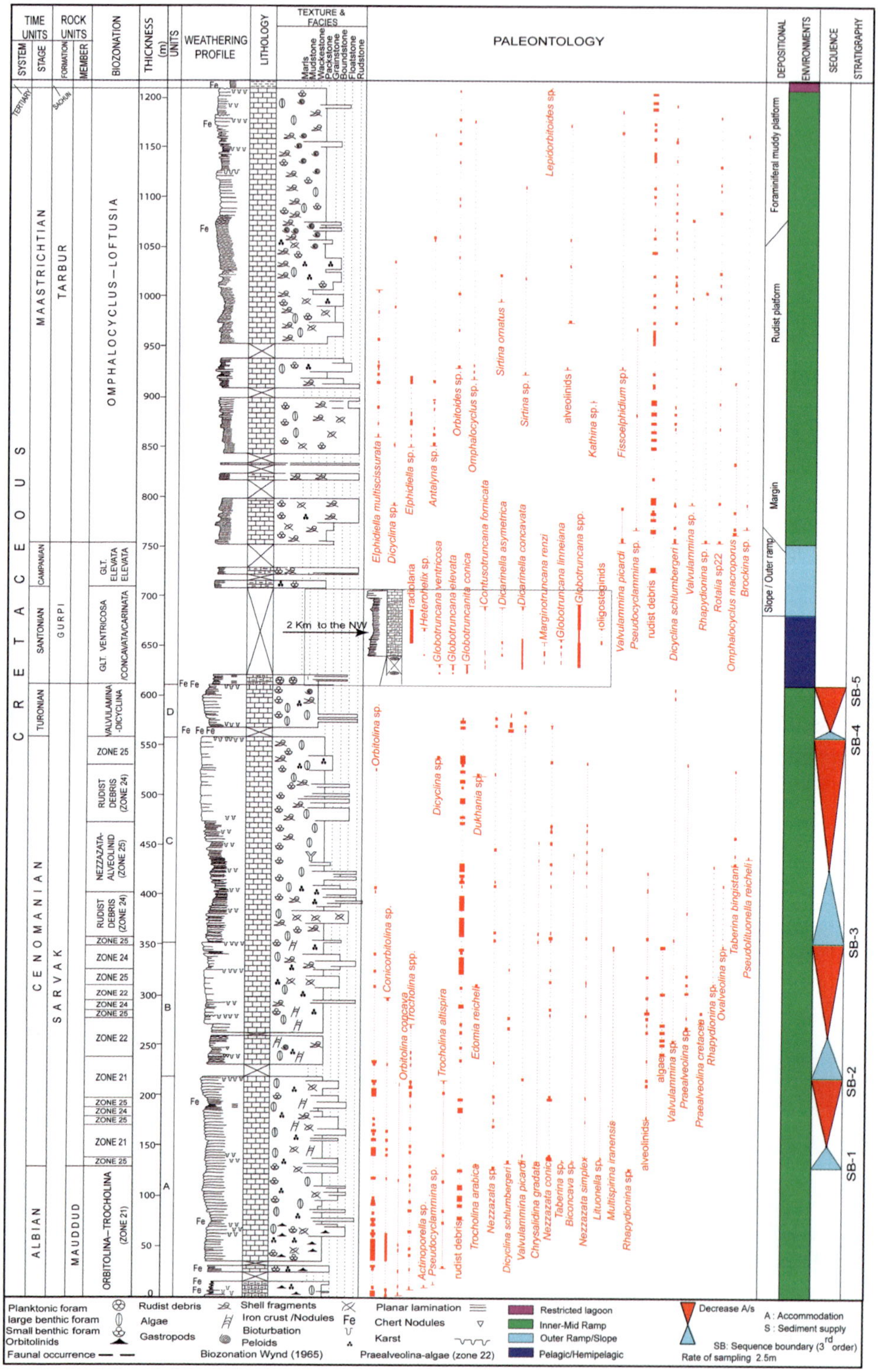

Fig. 12. Khaneh Kat section (Section 6). Sedimentary and palaeontological log based on field data and microscopic observations, showing the lateral variation in the Santonian sequences, from pure pelagic marls to detrital facies.

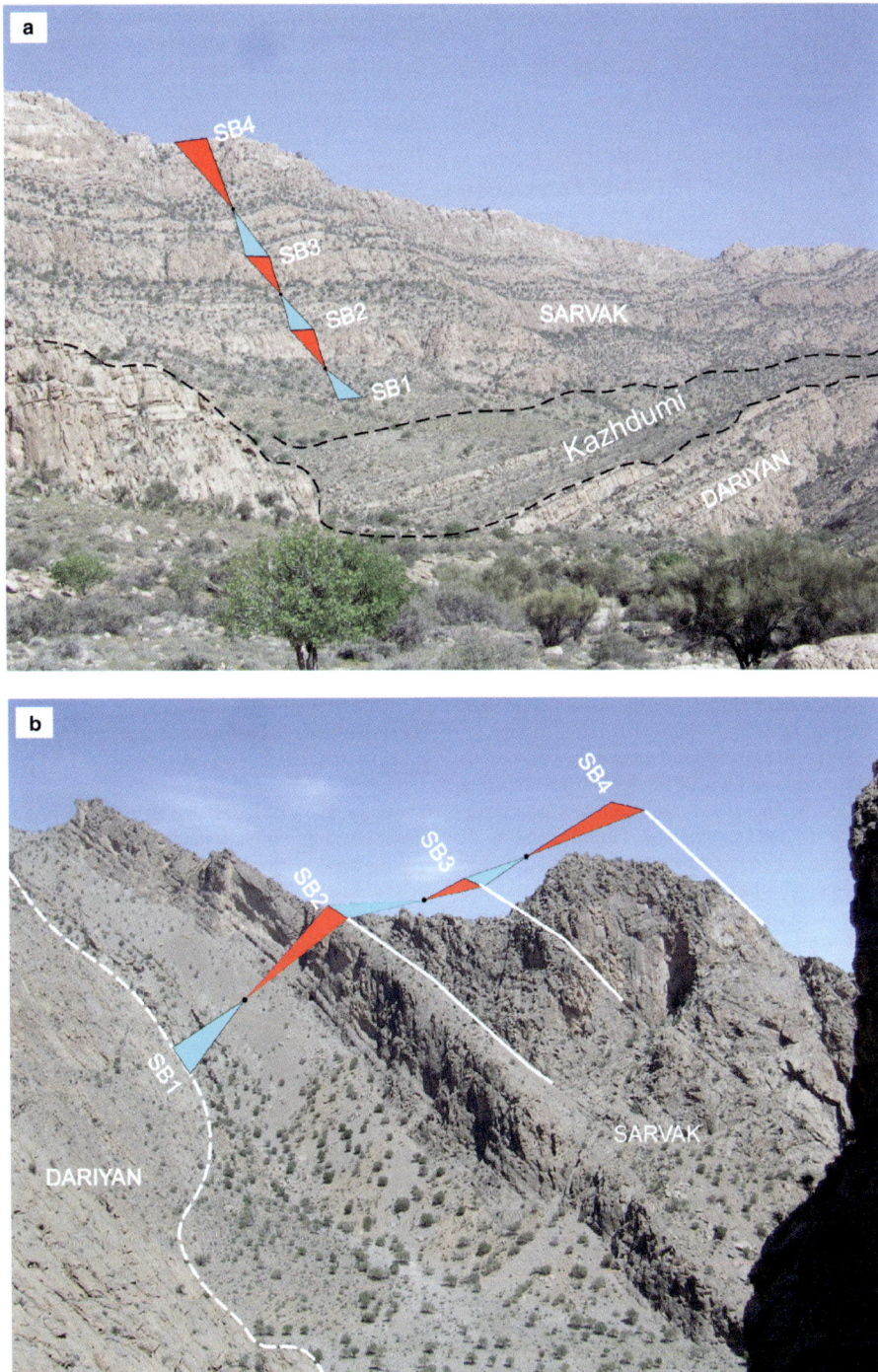

Fig. 13. (**a**) Overview of Aptian (Dariyan Formation) to Albian Kazhdumi and Cenomanian–Turonian Sarvak formations, Khaneh Kat anticline (Section 6). The trees for scale are 3–5 m in height. (**b**) Sarvak Formation with three third-order sequences and *Oligostegina* facies unconformably underlain by *Orbitolina*-bearing facies of the Aptian Dariyan Formation.

nomenclature for this facies is the Ilam Formation; locally, where the argillaceous content of the basinal facies is high, it is called the Gurpi Formation.

Deep basin pelagic facies (Late Turonian– Maastrichtian, Surgah and Gurpi Formations)

The pelagic basinal facies were deposited in the deepest part of the basin during the Late Turonian– Early Coniacian (Surgah Formation) and Late Santonian–Maastrichtian (Gurpi Formation). The mudstones to packstones contain numerous types of planktonic foraminifers, such as globigerinids and globotruncanas, oligosteginids and various ammonite species. The bedding pattern varies from millimetre to decimetre scale (Fig. 7a, b).

In the northern flank of the Kola Qazi (Section 7) and West Dalneshin (Section 3), which do not show any slumped deposits, the amount of black deposits (organic matter?) increases considerably. In these locations the very thin-bedded marly facies is sometimes interbedded with medium wavy bedded limestone intervals. These mud-supported carbonate facies contain iron nodules, tar sands and a considerable amount of ammmonite and echinoderm fragments, which sometimes occur associated with condensed surfaces and *Zoophycus* ichnofossils.

The Maastrichtian basinal facies is well developed in the southern flank of the Kola Qazi anticline (see Section 8, Fig. 10e), and contains bioclastic wackestone to packstone. The faunal content of the basinal facies consists of a variety of planktonic foraminifers, radiolaria, oligosteginids and pelagic macrofossils in association with benthic elements reworked from the platform–ramp top. Compared with the underlying Campanian pelagic facies, the sediments show different decimetre- to metre-scale bedding patterns that alternate with marly intervals (Figs 10e & 11, units D and E).

Large benthic foraminifers- and rudist-dominated platform carbonate (Maastrichtian, Tarbur Fm.)

Several depositional environments can be distinguished in this depositional system, as follows.

(1) The inner platform facies is characterized by mudstones to packstones, with *Orbitoides*, miliolids, *Dicyclina* and non-skeletal grains such as peloids.

(2) The open platform facies consists of echinoid and rudist debris wackestone–packstone.

(3) The thick packages of coarse- to very coarse-grained rudist-bearing limestone are interpreted as a high-energy barrier that occurred at the platform top. This facies is thick to massive bedded (metre scale), and shows considerable lateral and vertical

variation in grain size and bedding pattern (Fig. 7c). This unit is predominantly composed of rudist debris floatstone, packstone, rudstone and grainstone (Fig. 7d). It is associated with benthic foraminifers (*Orbitoides*, *Dicyclina*) and also echinoid debris.

(4) A *Loftusia*-dominated facies is developed in the northern flank of Kola Qazi anticline in the upper part of Section 7. The *Loftusia* facies contains both *Loftusia persica* and *Loftusia coxi*. Salinity and temperature are two important factors that control their distribution (Meriç *et al.* 2001). These large foraminifers are characteristic of an outer shelf environment (Zambetakis-Lekkas & Kemeridou 2006). They occur associated with rudist debris, *Orbitoides*, *Siderolites calcitrapoides* and *Omphalocyclus macroporus* (Fig. 7e). The various species of *Loftusia* occur especially during the Maastrichtian in the Middle East.

(5) The Maastrichtian evaporitic facies (Sachun Formation) is poorly developed in the study area. It consists mainly of gypsiferous limestone, dolomite and red marls deposited in a sabkha environment. Sedimentary structures are bird's eyes and salt clasts. These facies are well exposed at the top of Sections 6 and 7.

Resedimented carbonate and siliciclastic facies

These facies are developed in both carbonate and siliciclastic domains, which within the proximal to distal parts of the basin include gravity deposits, debris flows and turbidites.

Gravity deposits

The carbonate gravity deposits are mainly developed in the southern part of the study area (Kola Qazi anticline, Section 8), whereas the siliciclastic deposits originate from the radiolarite and ophiolitic nappes and tectonic slices that occur in the NE parts of the study area (Dalneshin Mountain, Sections 1, 2, 4 and 5).

(1) Olistostromes. This chaotic slumped facies has dominantly been observed in the Kola Qazi anticline (Fig. 8a; Section 8). It contains metre-sized olistoliths, with a facies that resembles the underlying Santonian platform carbonates (Ilam Fm.), embedded in a fine-grained matrix of Campanian marls (Gurpi Fm.). This facies also contain various-sized, well-rounded carbonate clasts with the same composition and origin. Extensive and thick olistostrome intervals usually do not result from normal slope redeposition processes. They may indicate the presence of tectonic activity in the area at the time of their deposition.

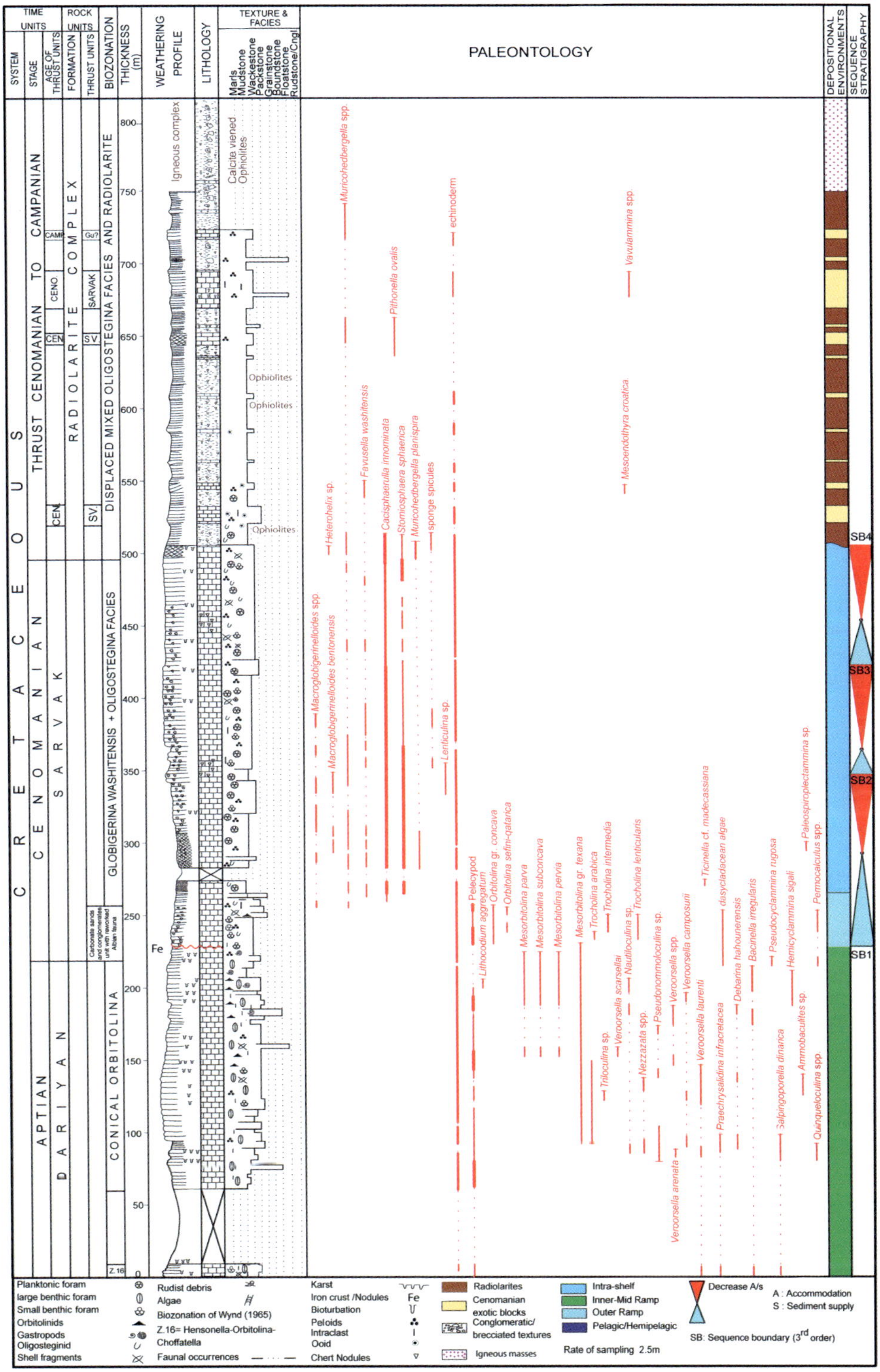

Fig. 14. Tang-e Jazin section (Section 5). Sedimentary and palaeontological log based on field data and microscopic observations.

(2) Slump folding. This process has been observed in the southern flank of Kola Qazi anticline (Section 8; Fig. 11) and laterally wedges out. The folded strata are Campanian in age and consist of pelagic marly facies. Slump folding is generally created as a result of slope steepening and downslope movement, which may be caused by tectonic instability, rapid basin subsidence or high sedimentary input (Helwig 1970; Hall 1973; Bradley & Hanson 1998).

Debris flows

Debris flows have been observed in Sections 1, 2, 5 and 8 (Table 1) in the Santonian to Campanian intervals. Debris-flow facies in Sections 1, 2 and 5 were deposited associated with the radiolarite complexes. In Section 8, which is dominated by pelagic facies, debris flows occur frequently in the lower and upper parts of the section. Approximately four discrete depositional cycles are present in the lower part (Fig. 8b) and three in the upper part of this section (Fig. 8d), which all show normal grading. Each cycle consists of grain-supported conglomerate at the base passing into muddier and carbonate-rich facies at the top. The conglomerates are poorly sorted and contain subangular to rounded clasts with a wide range of grain sizes (up to tens of centimetres in diameter). The clasts are monomict and mainly of Santonian shallow-water origin (Fig. 8c). In the upper part of Section 8 the cycles show a progradational stacking pattern (Fig. 8d, e) and in the topmost cycles the texture is packstone to grainstone with planar bedding (Fig. 9a, b). The debris-flow facies is commonly associated with gravity-flow deposits, in which the larger clasts are matrix-supported, indicative of plastic behaviour of the matrix, with a finite yield-strength that was able to support blocks while moving (Major & Iverson 1999; Young 2000; Elverhøi *et al.* 2005). The alternation between debris-flow cycles and interbedded pelagic sediment may relate to a series of rises and falls in relative sea level (Karner & Watts 1983; Weaver & Kuijpers 1983; Hilbrecht 1989; Andresen *et al.* 2003; Lantzsch *et al.* 2007).

The debris-flow facies shows strong lateral thickness changes and is organized in thinning upward sedimentary cycles. Each cycle represents a channel-fill deposit that commences with an erosional surface and grain-supported conglomerate at the base, followed by mud-supported conglomerate in the middle, and bioclast wackestone to packstone carbonate facies at the top.

Turbidites

Calciturbidites are well developed in most of the outcrops through the Santonian to Maastrichtian interval. The calciturbidite beds contain both shallow-water and hemipelagic fragments such as pelagic fauna, small benthic foraminifers and peloids. They usually show an erosional base and mud clasts derived from the underlying pelagic units at the base, as well as trough and planar cross-lamination. They are interpreted to have been deposited by stream flows or concentrated density flows (Mulder & Alexander 2001). The calciturbidite facies has a wide distribution and is well exposed in the southern flank of Kola Qazi (Section 8; Fig. 11), east Kuh-e Dalneshin (Section 5; Fig. 14), Arsanjan (Section 1) and Khaneh Kat (Section 6). In Section 8, the amount of calciturbidites increases upwards and shows sets of medium- to thick-bedded packstone to grainstone associated with planar bedding (Fig. 9a, b).

The turbiditic system in the northern part of the study area contains more clastic and igneous materials. The siliciclastic sediments mainly occur in Sections 1 and 5 associated with radiolarite and other exotic materials derived from allochthonous units tectonically emplaced during the obduction process. The sedimentary texture of the facies varies between conglomeratic, sandstone and siltstone. The siliciclastic deposits are composed of carbonate and igneous rock fragments and quartz, which are probably derived from obducted radiolarites, ophiolites and carbonate blocks. The siliciclastic deposits interfinger with marls and shales of Santonian to Campanian age.

In the upper part of Sections 1 and 5 (radiolarite complex) siliciclastic sediments are very abundant. They comprise fine- to very coarse-grained pebbly sandstone, which locally is underlain by a fine- to medium-grained conglomerate. The sandstones predominantly form medium- to coarse-grained units varying from a few centimetres to several metres in thickness, with considerable lateral variations (Fig. 9c, d). These sediments are interbedded with autochthonous and allochthonous radiolarian and carbonate pelagic facies. However, when conglomerates are present, the amount of marly pelagic facies is much less. The base of many beds is erosional and associated with clast-bearing beds (Fig. 9e). Planar lamination, trough bedding, normal and reverse graded bedding, Bouma sequences, flute casts and convolute lamination are common features. Locally (Arsanjan section) the sandstones are very coarse-grained, showing prominent erosional bases, and usually are highly chertified.

Whereas the dominant components are mostly reworked chert and radiolarite particles, bivalve fragments, ooids, peloids, intraclasts, and coarse carbonate and igneous rock fragments are also observed (Figs 9f & 10a). The fragments are subangular, and poorly sorted trace fossils are also present.

The siliciclastic deposits are interpreted to have been deposited from concentrated density flows and high-density turbidity flows (Mulder & Alexander 2001). Convolute lamination and pebbly sandstone masses reflect slumping caused by channel wall failure (Elliott & Williams 1988).

Radiolarite complex

The siliceous sediments include radiolarite, siliceous shale–mudstone and chert. These facies are composed of radiolaria and other siliceous elements that at times occur associated with non-siliceous allochems in a fine-grained matrix. The radiolarite complex was emplaced from other areas and subsequently mixed with pelagic facies in the frontal part of the nappes. Therefore, based on their composition, sedimentary features and associated facies, the radiolarite intervals can be divided into three types.

Type 1: chaotic and poorly bedded units

These units are characterized by interfingering with calcareous pelagic deposits. These sediments show some displacements and therefore are relatively chaotic and structureless, with no special bedding pattern. The calcareous intercalations are either deposited *in situ* or as exotic blocks that frequently are interbedded with clastic turbidites. This radiolarite facies and associated sediments represent mainly mixed red to brownish radiolarian cherts and siliceous shales.

The sedimentary textures show a mixture of various species of radiolarian fossils with oligosteginids and planktonic foraminifers. This type of facies is strongly influenced by tectonic processes that displace radiolarian and other basinal deposits onto the platform setting.

Type 2: well-bedded less siliceous units

This type of facies is composed of decimetre-scale bedded radiolarian facies with no considerable carbonate interbeds and pelagic fauna. The facies shows sparse to packed textures with red to brown and green colours. The faunal assemblage is mainly composed of radiolaria, which are poorly preserved.

These types of sediments illustrate a repetition between grain-supported radiolarian facies with no lamination, graded lamination that often shows an erosional base and parallel-laminated sediments, and brown to green matrix-supported interbeds. The laminations show a normal grading from radiolaria and clastic particles near the base to the matrix-dominated sediments with a concentration of smaller radiolaria at the top. This

grading normally continues into siliceous mudstone interbeds.

The radiolarian fossils are mostly replaced by calcite and occasionally associated with calcareous fossils such as globigerinids, oligosteginids and sponge spicules. The clastic components such as quartz grains, ferruginous mud clasts and ophiolitic fragments are common and usually show a particular orientation (rough lamination). The bedding pattern rarely shows slump folding.

Type 3: well-bedded highly siliceous units

These thin- to medium-bedded radiolarite units are mainly black to dark grey and green in colour. The components show sparse to packed textures with very rare carbonate elements. The siliceous radiolarites are predominantly fine-grained, densely vitreous to slightly granular in texture and mostly show parallel bedding. These facies do not show a large variation in their components or any orientation. Locally, the dissolution of the radiolarian facies resulted in the formation of chert, hematite and clay-rich horizons. The radiolarite facies is well developed in the North and South Dalneshin areas and Kuh-e Pichakun (Fig. 10c, d, Sections 1 and 4).

In the middle part of Dalneshin, the radiolarites are well bedded in the lower part, whereas upwards they are interrupted by carbonate thrust slices. In the lower unit the radiolarian facies does not contain any calcareous fauna. This interval consists of an alternation between red shale–radiolarites and light to dark grey marls, which were dated as Coniacian to Santonian (Stoneley 1981). In this study the radiolarites are dated as Albian to Early Cenomanian. The upper unit shows a mix of radiolarites and turbidite facies, which occasionally is interbedded with olistoliths of platform-derived carbonate material.

Southward between Kuh-e Dalneshin and Khaneh Kat, the radiolarites are very well bedded, more siliceous and have different colours (Fig. 10c).

Igneous rocks–ophiolites

In the study area igneous–ophiolite rocks are found either as part of the obducted ophiolite complex (Stoneley 1981; Fig. 2), or as constituents of turbidites that occur interbedded in the radiolarite complex (Fig. 10b). The ophiolitic series of the Dalneshin–Neyriz region are associated with thrust sheets of former Tethyan oceanic sediments. These series are analogous to a part of the Oman ophiolite, but they differ from the central and eastern Iranian ophiolite mélange (Ricou 1971; Stocklin 1974; Stoneley 1981). The full ophiolitic sequence is not exposed in the study area and the series exposed represents isolated masses of pillow lava and numerous sheets of other igneous related rocks. The ophiolitic complex sometimes occurs intercalated with marly pelagic facies and turbiditic deposits.

Peridotite is one of the main components of the igneous rocks present in the study area. It occurs in the lower igneous portion of the ocean floor underlying the radiolarites and other pelagic sediments of oceanic crust (Desmons & Radelli 1989). Peridotites are mainly exposed in the eastern part of Kuh-e Dalneshin (around Section 5) and were reported by Stoneley (1981) in the northern of Dalneshin area (Ahmad Abad, Fig. 2). Stoneley (1981) also mentioned that peridotite locally changed to chromite, lherzolite and serpentinite. In this region it also occurs associated with other ultramafic and mafic rocks such as harzburgite, dunite, websterite, gabbro, dolerite dykes and basaltic pillow lavas. The amount of these types of rocks increases eastward in the Dalneshin Mountain range. In the eastern part of the study area (Section 5; Fig. 14) they are more abundant and occur in the upper part of the section. In the central and western part of the Dalneshin area the presence of igneous–ophiolite components as large blocks is limited, but they can be found frequently as igneous rock fragments in the turbidites (Fig. 10b).

Correlations and sedimentary–tectonic evolution

The regional stratigraphic variation is illustrated through the description of three outcrops that are used as reference sections representing the southern, central and northern part of the study area along Transect 1 (Fig. 15).

Southern area: Kola Qazi south flank (Section 8, Figs 10e & 11)

The Kola Qazi mountain range is a well-developed anticline in which the uppermost Ilam and Gurpi formations are exposed. Sediments found within the southern and northern flank show major facies differences over a relatively short distance. The Kola Qazi sections used in this study are positioned on the SE pericline of the anticline near the town of Fasa (Fig. 3). This section is representative of the Late Cretaceous slope to central foreland basin setting.

The section starts with the upper part of the Ilam Formation (40 m), containing thick-bedded to massive limestones, predominantly bioclastic packstone which locally grades to grainstone. They contain small benthic foraminifers, rudist debris, shell fragments and some planktonic foraminifers. Non-skeletal grains such as peloids and intraclasts

are also common. At the top of the formation some channels are present; these are filled with coarse-grained sediments. Large chertified rudists are present just below the interval where the Gurpi Formation conformably overlies the Ilam Formation. Based on the faunal assemblage a Santonian age can be proposed for the Ilam sediments in this section (Fig. 11).

The Gurpi Formation reaches a thickness of 585 m and includes autochthonous and allochthonous gravity-flow facies. It can be subdivided into five units based on bedding pattern and sedimentary features (Figs 10e & 11).

Unit A (42–135 m) consists of 93 m of marl and marly limestone with a pelagic facies interbedded with massive monomict conglomerates including Santonian clasts. The conglomerates appear to be laterally discontinuous (Fig. 8b). The interval contains three cycles; each cycle starts with medium- to coarse-grained, monomict conglomerates and ends with marl and marly limestone showing a pelagic facies. Highly bioturbated surfaces with iron staining are common in the conglomerates. Faunal components include planktonic and benthic foraminifers, and suggest Santonian and Campanian ages (Fig. 11).

Unit B (135–220 m) consists of 85 m of pelagic facies with globigerinid and oligosteginid bioclast wackestone–packstone and argillaceous lime mudstone interbedded with thin- to medium-bedded calciturbidites containing reworked Santonian elements (Fig. 7a, b). The faunal assemblage comprises a diverse association of planktonic and benthic foraminifers as well as other shell debris. Non-skeletal grains include peloids, which mainly occur in the calciturbidites. The faunal assemblage (i.e. *Rugoglobigerina rugosa*) suggests a middle Campanian age for this unit.

Unit C (220–310 m) is 90 m thick, of Campanian age, and characterized by large-scale slumping into the deep-water pelagic facies realm of bioclast wackestone containing large amounts of clasts to boulders of probably Santonian age (Fig. 8a). The reworked boulders may reach sizes exceeding 1 m in diameter. This unit occurs only in the SW flank of the Kola Qazi anticline and is traceable over long distances, despite the absence of any continuous bedding. No fossils predating the Ilam and Gurpi formations (Santonian) were found in the exotic constituents. The fauna assemblage includes the above-mentioned pelagic fauna (Unit B) and benthic foraminifers such as miliolids and textularids, as well as pelecypods and shell fragments.

Unit D (310–500 m) is 190 m thick, and consists of thin- to thick-bedded lime-mudstone to wackestone. The fossil content largely overlaps with the fauna of the underlying unit, but shows more varieties of *Globotruncana*, *Globigerina* and oligosteginid

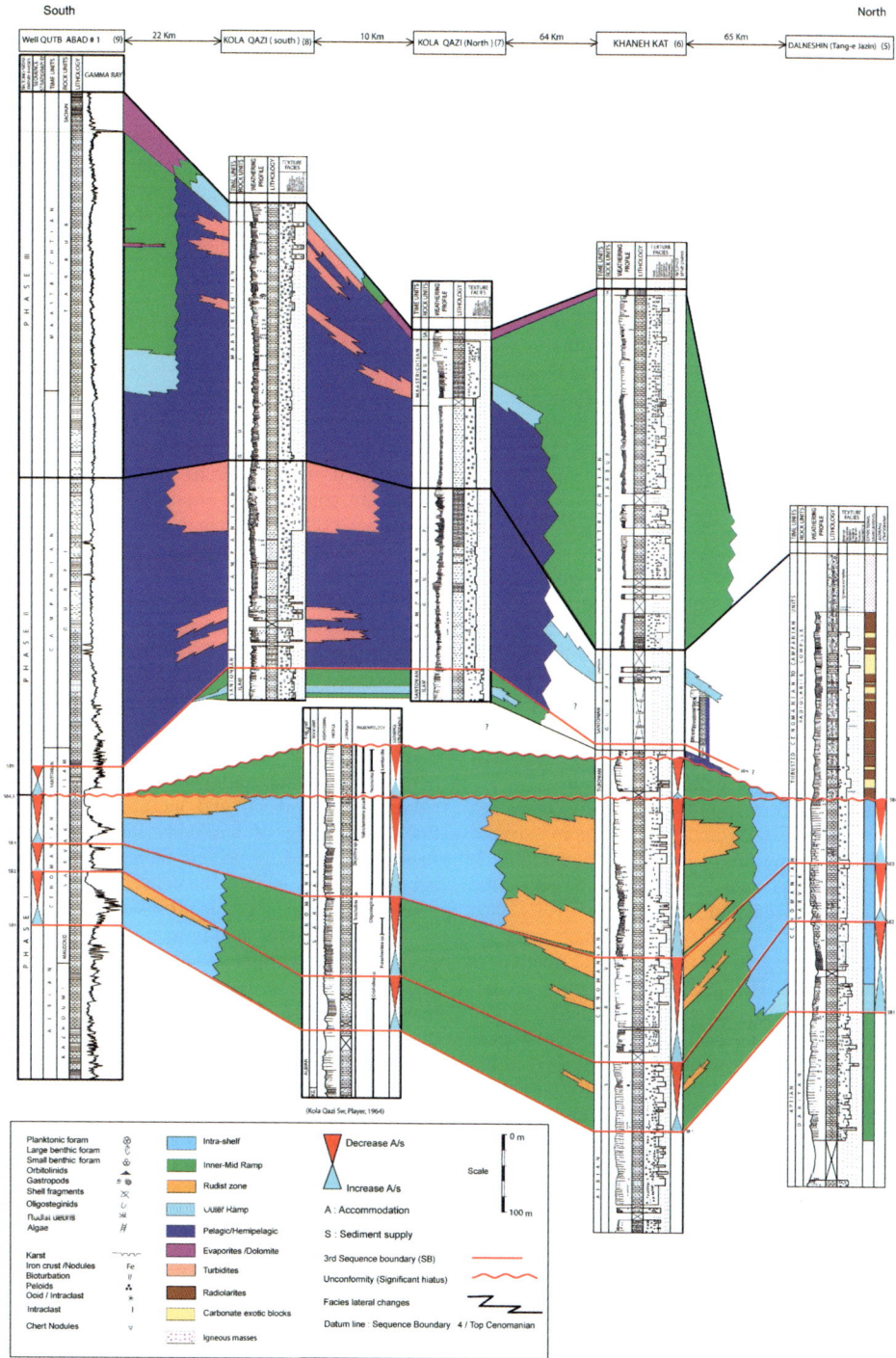

Fig. 15. Transect 1 (south–north). Cross-section with high-resolution sequence stratigraphy through the Sarvak, Ilam, Gurpi and Tarbur formations and the radiolarite complex. The important lateral change between carbonate platform deposits containing rudist debris and intra-shelf basinal facies during the Cenomanian–Turonian should be noted. The lateral change during Santonian to Campanian time took place between platform, basin and radiolarite–ophiolite facies, which were overlain by marginal facies in the Maastrichtian.

species. Some ammonites, only casts, were also present.

Unit E has a generally different appearance. It consists of 120 m of coarse-grained, bioclastic limestone and micro-conglomerates, forming massive to thick beds with planar bedding (Fig. 9a, b). In addition, the unit shows a thickening and coarsening upward trend. The components consist of rudist debris and other shell fragments, benthic foraminifers, echinoderm fragments, peloids and intraclasts. Iron-stained bioturbated surfaces are also frequent. This part of the Gurpi Formation is Late Campanian to Maastrichtian in age. Unit E at the top is capped by a sequence boundary that is overlain by marly facies, probably the Sachun Formation (the section ends with a flat and not exposed zone).

Middle area: Khaneh Kat section
(Section 6, Fig. 12)

The Khaneh Kat anticline lies between Kuh-e Dalneshin and Kola-Qazi with a north–south orientation (Fig. 3). In these outcrops a continuous succession from upper Triassic to lower Tertiary units is exposed. The Sarvak, Gurpi and Tarbur formations have been studied in great detail in this outcrop. The section is best characterized by its well-developed Cenomanian and Maastrichtian shallow-water platform deposits, whereas the deeper-water deposits Gurpi facies are very thin.

The Sarvak Formation is fully exposed, reaches a thickness of 630 m, and consists of massive to thick-bedded limestones with a benthic fauna. It can be subdivided into four subunits based on the faunal composition and sedimentary cycles (see Figs 12 & 13a).

In the main section the Gurpi Formation is covered (60 m). The section starts with several decimetre-scale beds of carbonate sandstone and thin- to medium-bedded pelagic mudstone to wackestone. A few kilometres further to the NW, at the periclinal termination of the Khaneh Kat anticline, a comparable section was measured and dated as Santonian. This section, however, lacks the calciturbidites and carbonate sandstone facies. Based on the fauna of the *Globotruncana ventricosa concavata* assemblage zone (Wynd 1965) a Santonian to Campanian age is attributed to the sediments.

Within the Gurpi Formation several clast-dominated decimetre- to metre-scale beds occur that contain radiolarite elements such as chert and clastic sandstone rock fragments.

A transition zone can be defined (720–840 m, Figs 7c & 12) where interfingering is observed between the pelagic marls of the Gurpi Fm. and the rudist- and benthic foraminifer-rich sediments derived from the nearby Tarbur platform.

The sediments of the Tarbur Formation consist at this location of 470 m of rudist debris packstone–floatstone–rudstone, and massive grainstones to packstones with bivalve debris, miliolids, Orbitoidae, *Dicyclina* and other skeletal and non-skeletal grains. Based on the fauna of the *Omphalocyclus–Loftusia* assemblage zone (Wynd 1965) a Maastrichtian age is assigned to this formation. This formation is capped by tens of metres of evaporites of the Sachun Formation.

Northern area: Tang-e Jazin (Kuh-e Dalneshin) (Figs 13b & 14)

The Tang-e Jazin section is located in a valley, situated 5 km north of the Jahan Abad village, in the central part of the southern flank of the Dalneshin mountain front. Jurassic to Late Cretaceous sediments are continuously exposed in this area. The Albian interval is eroded in this section and redeposited as carbonate conglomerate at the base of the Cenomanian interval. The Late Cretaceous part of this succession consists of three main units: a lower carbonate pelagic unit (280 m), the middle radiolarite nappe (200 m), and the upper ophiolitic series (more than 100 m).

The lower unit, which is attributed to the Sarvak Formation, can be subdivided into a lower part, 37 m thick, that consists mostly of muddy, large benthic foraminifer-dominated, shallow-water facies (Mauddud Member), and an upper part, 240 m thick, consisting of a muddy intra-shelf basinal facies dominated by pelagic fauna (Ahmadi member). This part is very brecciated probably as a result of Cenozoic tectonic processes.

The middle unit of this section is composed of decimetre to centimetre alternations of red to green radiolarite with calcareous shale and pelagic marls. Quartz sandstone turbidites alternate with the radiolarites in the upper part of this unit. In addition, exotic blocks of the Cenomanian oligosteginid facies are common and their number increases upward. These exotic carbonate blocks mostly consist of wackestone to packstone that contain oligosteginids (*Pithonella ovalis*), *Praealveolina*, *Muricohedbergella* and echinoderm fauna. As a result of synsedimentary tectonics these sediments sometimes are brecciated or conglomeratic in appearance and locally contain abundant calcite veins. In this section the dimensions of the exotic blocks do not exceed tens of metres. A few kilometres to the west of the Dalneshin area the exotic blocks reach vertical dimensions of hundreds of metres and their horizontal dimensions exceed a few kilometres.

The lower contact of radiolarite complex with the Sarvak Formation is not exposed. Based on foraminifer and nannofossil age determination the radiolarite interval can be attributed to the Santonian to Campanian.

The uppermost unit comprises igneous–ophiolite obducted blocks, in which different types of ultramafic and mafic igneous rocks of peridotite and basalt types are common. The igneous–ophiolite components were not analysed in detail in this study. These rocks occur associated with turbiditic sandstone and conglomerate and metamorphic fragments at the base, and are overlain by crystalline green igneous blocks. Some deformation structures can be observed in the slightly lithified cherty and carbonate sediments.

The succession ends with a few hundred metres of igneous basaltic rocks. In total 320 m of the radiolarite–ophiolite interval were logged (see Fig. 14). It is very difficult to determine whether the marly intervals in the upper parts of the succession are *in situ* or were displaced in a later stage.

Depositional sequences

To document the stepwise evolution of the depositional history in this area, we have applied sequence stratigraphic principles to identify depositional sequences. It is clear that in a setting with a strong tectonic influence, rapid lateral facies and thickness changes can be expected. In this type of setting, independent age dating is thus of prime importance. Well-established biostratigraphic schemes have been used (James & Wynd 1965; Wynd 1965), as well as those from the more recent literature (Wells 1969; Gollestanneh 1974; Kheradpir 1975; Setudehnia 1978; Murris 1980; Davoudzadeh & Schmidt 1984; Bordenave & Huc 1995). The radiolarites are a notoriously difficult facies for age dating, as they are not very well preserved, and are often recrystallized and deformed.

The results of the sequence stratigraphic correlations are presented in three ways, as follows.

Cross-sections. Transect 1 (Fig. 15) has north–south orientation, is 160 km long, and is perpendicular to the deformation front, and thus is ideally suited to illustrate the main facies and thickness changes. Transect 1 runs from Kuh-e Dalneshin in the north, passing through the Khaneh Kat anticline to the Kola-Qazi and Qutb Abad anticlines. Transect 2 (Fig. 16) has an east–west orientation, is 66 km long, and is located in the Kuh-e Dalneshin area, parallel to the deformation front. It illustrates lateral facies changes both in the Cenomanian shelf carbonates and in the radiolarite complex. The correlation datum for these transects is the top of the Cenomanian Sarvak platform.

Summary cross-sections. These have a different datum for each main phase, illustrating the geometrical and facies evolution of the system (Fig. 19).

Combined palaeogeographical and isopach maps. These maps show the main sedimentary environments and thickness variations for the Cenomanian, Campanian and Maastrichtian intervals (Fig. 17).

The depositional history of the study area can be subdivided into three main phases: Phase I, covering the Late Albian to Cenomanian, is characterized by shallow-water platform carbonate sedimentation, with the presence of intra-shelf basinal areas. Phase II covers the Turonian to Late Campanian, during which the platform carbonate system changes to a foreland basin followed by obduction processes. Phase III covers the Late Campanian to Maastrichtian, a period of renewed carbonate sedimentation and foreland basin infill.

The stratigraphic surface that forms the base of the studied interval is a sequence boundary that cuts across the lithostratigraphic boundaries of the Mauddud Mb. and Sarvak Fm. This choice is supported by recent work carried out in Fars and Khuzestan, where a sequence boundary has been placed at the top of the Late Albian–earliest Cenomanian shallow-water carbonates, rich in oribitolinid and trocholinid benthic foraminifers, which is overlain by deeper water sediments, containing earliest Cenomanian ammonites. The orbitolinid–trocholinid assemblage at this location has thus been shown to be of latest Albian age.

Phase I: Late Albian to Cenomanian (Figs 17a & 19a)

Throughout the study area Phase I is characterized by shallow-water carbonate deposition. It contains a rudist–benthic foraminifer-dominated assemblage that locally changes to intra-shelf basinal carbonate mudstone sedimentation (Sarvak Formation, including Ahmadi Member). Three third-order sequences are distinguished. The age dating of this interval is based on benthic and planktonic foraminiferal assemblages.

Sequence 1 is positioned on top of a shallowing-up trend of *Orbitolina–Trocholina*-dominated shallow-water carbonates (Late Albian Mauddud Mb.). The succeeding transgression locally introduces intra-shelf basinal conditions. The lower sequence boundary (SB1) is well defined because of a distinct deepening of the depositional environment (bedding pattern and fauna). In the east Dalneshin, the Albian sequence is not present; therefore, this surface is positioned at the top of the Aptian sequence. At this location SB1 is succeeded by a few metres of

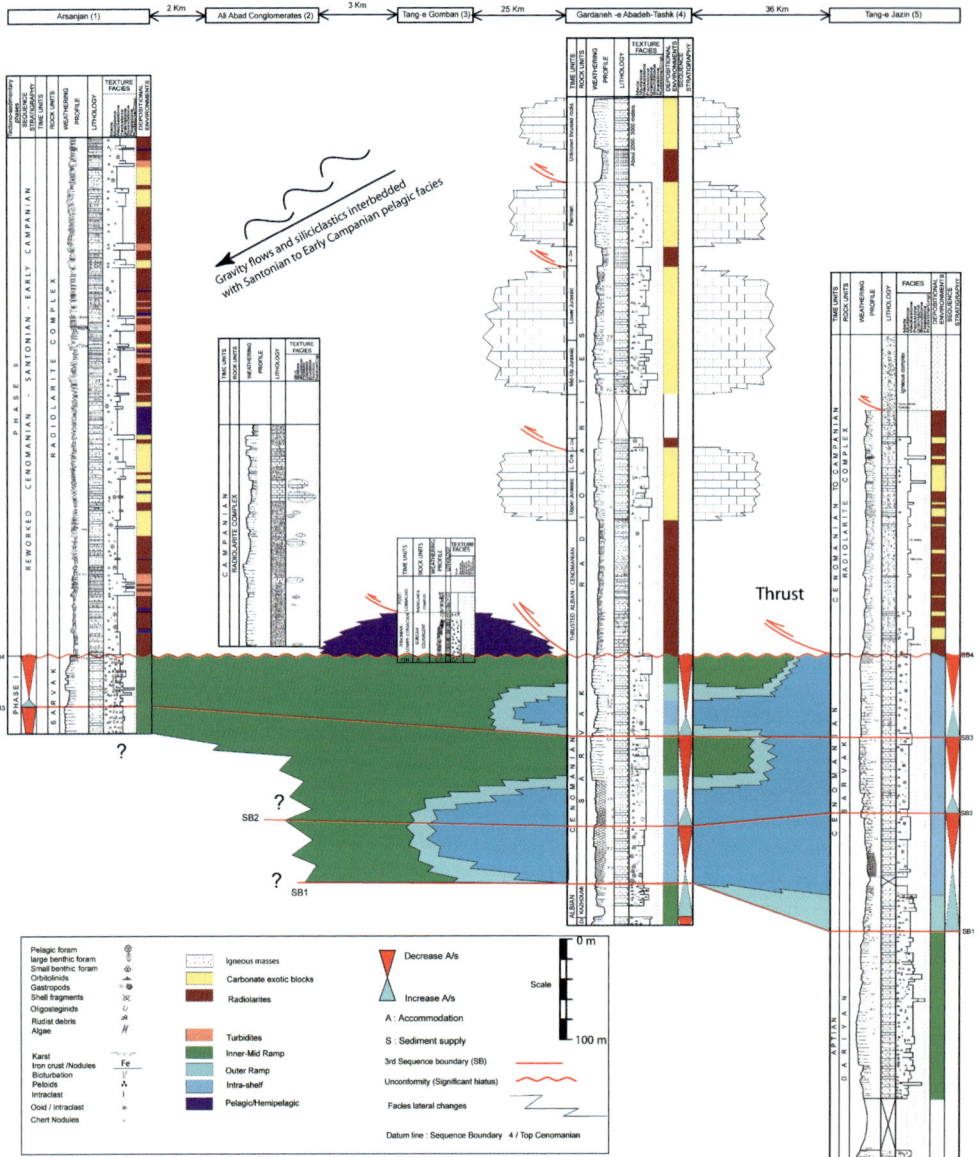

Fig. 16. Transect 2 (east–west). Cross-section with high-resolution sequence stratigraphy through the Sarvak Formation and the radiolarite complex, illustrating the displacement of the nappe system over the Cenomanian–Turonian carbonate platform. The amount of igneous deposits increases towards the east, whereas the amount of turbiditic facies decreases in the same direction. The cross-section also shows the well-bedded radiolarite facies in the centre.

carbonate conglomerate containing Albian fauna (Figs 14 & 16).

In Khaneh Kat, where the entire Cenomanian–Turonian succession consists of shallow-water deposits, the *Orbitolinia–Trocholina* zone (Wynd

1965) occurs several times (Fig. 12). At this location the first *Orbitolina–Trocholina* zone is attributed to the Late Albian whereas the other deposits are considered as Cenomanian. SB1 is placed at the top of the first *Orbitolinia–Trocholina* interval, which is

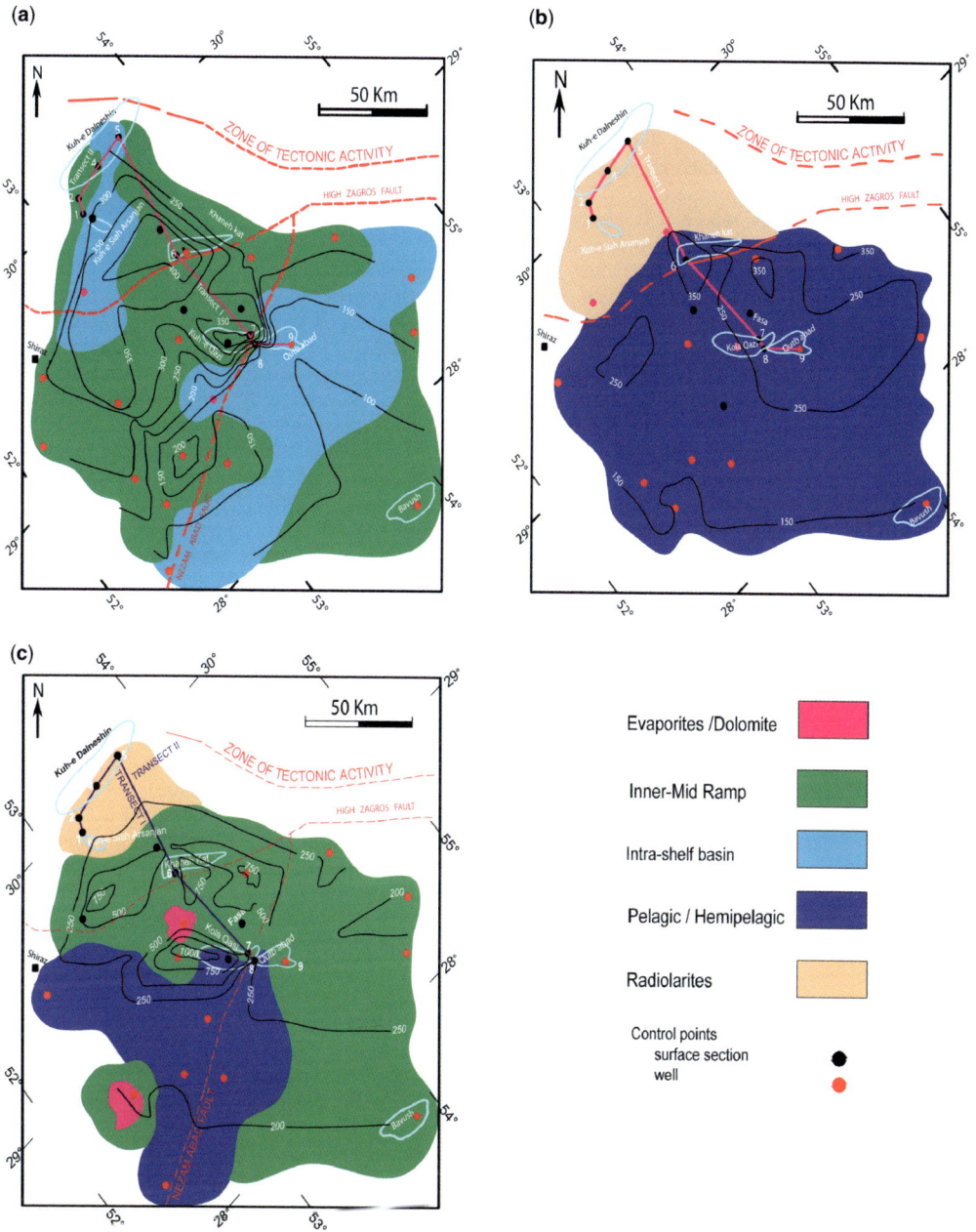

Fig. 17. Palaeogeographical and isopach maps of the study area during: (**a**) Cenomanian, (**b**) Campanian and (**c**) Maastrichtian times.

succeeded by an interval with a higher faunal content. Further to the south in the Kola Qazi anticline and Qutb Abad well, SB1 is overlain by a marly unit and intra-shelf basin oligosteginid facies, as shown in Figures 11 and 15.

Sequence 2 is characterized by subtle changes in the bedding pattern and depositional facies, both in the shallow-water platform carbonate (Khaneh Kat; Fig. 12), and the intra-shelf basin (Kuh-e Dalneshin). In the Qutb Abad well Sequences 1, 2 and 3 are marked by a large shift in the gamma-ray logs because of intra-shelf basin flooding in the lower part of the sequence, which sometimes grades to rudist-dominated sediments in the upper part of the sequence (Fig. 15).

Sequence 3 is well expressed along Transects 1 and 2 (Figs 15 & 16). Sequence boundary 3 (SB3) is best marked in the area of the Central Dalneshin and Kola Qazi sections, where it is placed at the top of a shallowing upward trend that is overlain by a transgressive intra-shelf basin facies. In the Khaneh Kat (Fig. 12) this sequence consists entirely of platform carbonate but it shows a sharp shift in bedding. SB4 corresponds to the top of the Cenomanian, and is overlain by Turonian shallow-water carbonates of the Sarvak Fm. (Khaneh Kat), Late Turonian pelagic marls of the Surgah Formation, Santonian pelagic marls of the Gurpi Formation, shallow-water carbonates of the Ilam Formation (e.g. Kola Qazi; Fig. 11) or radiolarites, as shown in the Kuh-e Dalneshin area. In the Zagros area exposure features characterize this widespread stratigraphic surface. In the Arsanjan section incised channels filled with slumped sediments can be observed at the topmost part of the Cenomanian platform carbonate deposits (Fig. 6c). The channel dimensions vary a few metres in depth and width. In this area the Cenomanian top shows a distinct conglomerate, which resulted from karstification processes. Other sedimentary features that occur are chertification, bioturbation and iron staining.

The Cenomanian rocks mainly represent a platform carbonate–intra-shelf basin system. The dataset from our study area is not large enough to produce a detailed palaeogeographical map, but the data do show that during the Cenomanian in the northeastern part intra-shelf basinal conditions prevailed, and that towards the west (see Transect 2; Fig. 16) and the south (see Transect 1; Fig. 15) lateral facies changes occur to shallow-water platform sediments. Based on the overall geometry of the Cenomanian platform, and observations made in the High Zagros, the depth of the intra-shelf basin is estimated to have been of the order of 50–70 m.

In the northern transect, the Cenomanian succession has a comparable thickness, with 235 m in the Tang-e Jazin section (Fig. 14) and 230 m in central Dalneshin. In the north–south transect, however, significant lateral thickness changes occur in the Cenomanian sequences: the Khaneh Kat section reaches a thickness of 560 m, all in shallow-water facies, whereas the intra-shelf basinal Tang-e Jazin section is 235 m thick. In the Kola Qazi section (Player 1969), the thickness of the Cenomanian deposits is reduced to 370 m, which includes 200 m of platform carbonate

sediments and 170 m of deeper intra-shelf deposits, whereas in the Qutb Abad well, the Cenomanian is further reduced to *c.* 200 m, with a clear expression of the three sequences. Constrained by the correlations, the thickness differences only can be explained by regional differences in subsidence.

Phase II: Turonian to Late Campanian (Figs 17b, 18 & 19b, c)

The definition of depositional sequences in this interval is difficult because tectonic processes have had a strong impact on the sedimentation system. In the lower part two sequences can still be defined. The surfaces marking the top of the Turonian and Santonian are characterized by numerous erosional sedimentary features and can be considered as SB5 and SB6, respectively, introducing Sequences 4 and 5.

Sequence 4, of Turonian age, is laterally very discontinuous. In the Khaneh Kat anticline a covered interval a few metres thick, consisting of a marly facies, probably represents the Turonian (Fig. 15). This sequence is succeeded by a series of bivalve-dominated boundstones and by massive to thick-bedded bioclastic wackestone to packstone of the Turonian *Valvulammina–Dicyclina* assemblage zone (James & Wynd 1965). In the Dalneshin area the Turonian is represented by several tens of metres of dark grey pelagic facies attributed to *Marginotruncana sigali* and *Dicarinella primitiva* assemblage zones (Wynd 1965). SB5 is well marked in the Khaneh Kat section (Fig. 12), with intensive bioturbation, chertification and iron staining. Turonian radiolarian deposits, reported by Stoneley (1981) in the Dalneshin and Arsanjan area, were not reconfirmed by our study.

In this study no Middle to Late Coniacian aged rocks have been recorded; even the presence of Early Coniacian sediments seemed very doubtful, although they were reported by Stoneley (1981). In the sedimentary succession of the Dalneshin Mountain range Late Turonian–Early Coniacian(?) sediments are immediately covered by far-travelled sediments. Therefore it is difficult to assess the depositional history during the Coniacian interval; whether sediments were deposited and then eroded or a non-depositional interval prevailed during this period.

The lower sequence boundary (SB5) is the uppermost Sarvak surface presented above. The upper boundary, SB6, marks the transition from the platform carbonates of the Ilam Fm. to the deep-water pelagic marls of the Gurpi Fm. It is locally incised by channels that are several decimetres deep and several metres wide, and filled with intraformational carbonate clasts. However, in areas where no shallow-water Ilam facies was deposited, this sequence boundary is positioned in the

Santonian part of the Gurpi marls. Sequence 5 shows significant lateral facies changes. In the Kola Qazi sections it is represented by shallow-water carbonates of the Ilam Formation, to the south it changes to pelagic marls, and in Khaneh Kat it changes to marls that contain carbonate sandstone and conglomerate beds interbedded with pelagic facies (Figs 11 & 12). In the Khaneh Kat section the first trace of radiolarite fragments is found.

No depositional sequence can be distinguished in the Late Santonian to Late Campanian interval because of the absence of lithology contrasts or a traceable surface, and probably reflect the reduction in the intensity of tectonic activity. In Section 8 (Kola Qazi south) the facies evolution shows a fining-up and deepening trend in the lower part, gradually losing the conglomeratic lenses and grading to thin-bedded marls. From 170 m onwards calciturbidites occur more often, containing pelagic and benthic foraminifers, echinoderm fragments, oligosteginids, and non-skeletal grains such as peloids and mud clasts. At 220 m in Section 8 this pattern is abruptly interrupted by a 90 m thick chaotic slumped unit (unit C in Figs 10e & 11) with eroded sediments of underlying platform carbonate. Within the Kola Qazi anticline this unit laterally wedges out towards the northern flank. At a larger scale, towards the south in the Qutb Abad, this unit grades to basinal facies with no apparent change in faunal content and log signature. The Late Santonian to Late Campanian interval contains stacked rudstone units in the Khaneh Kat section (Fig. 12) interbedded with non-exposed intervals (probably marly facies). The rudstone units comprise large rudist and other bivalve debris as well as echinoderm debris, intraclasts and mud clasts in a packstone to rudstone texture. This alternation is deposited in a slope facies (Fig. 7c).

In the Dalneshin and Arsanjan areas (along Transect 2) the Santonian to Campanian interval is represented by the radiolarite complex. The sedimentological organization of the radiolarite complex is complicated, and it is not possible to define sedimentary sequences in this succession. An order of tectonically driven events can, however, be deduced (see Discussion). In the Arsanjan area the radiolarite is less siliceous compared with the Pichakun area. In Section 1, it comprises significant amounts of fine- to very coarse-grained siliciclastic turbidites (reworked radiolarites) and calciturbidites, containing ooids, peloids, benthic and planktonic foraminifers and chert rock fragments (see Figs 9 & 10a) embedded in marls. Based on planktonic foraminifers, nannofossils and Sr isotopes, a Santonian to Campanian age has been confirmed for the marly intervals (at least in the first 200 m of the radiolarite part). About 500 m of the radiolarite succession was measured in this

hilly area in a syncline between Kuh-e Dalneshin, Kuh-e Sang Siah and Kuh-e Siah. It is very likely that the section displays the maximum thickness of this interval and that only several tens of metres of the topmost part were not studied.

In Section 2, a 200 m thick package of carbonate conglomerates is found, with locally some interbedded pelagic marly sediments of Campanian age. The clasts are all of Cenomanian and Santonian age, and represent both platform carbonates and pelagic marls. The carbonate conglomerate unit of Section 2 occurs isolated in the field, has an iron-stained chertified upper surface, and is overlain by chaotic red to brown radiolarites. It is surrounded by radiolarites but radiolarite clasts are sparse amongst the components (a few small blocks, tens of centimetres in diameter). The low content of radiolarites in the conglomerate unit can be interpreted as the result of a localized, fast sedimentation of the conglomerates, such as in a submarine fan or an incised canyon.

In the middle part of Dalneshin, the lower part of the interval (270–400 m) consists of centimetre-scale bedded radiolarites. This is followed by a 2000–3000 m thick succession with numerous exotic blocks up to 100 m in size. The age of these blocks becomes systematically older upsection, from Early Cretaceous to Permian. These thrust tectonic slices most probably were embedded within the radiolarite during the Santonian and Campanian.

In the northeastern part of the study area the radiolarite consists of a poorly bedded and coloured facies. This unit is characterized by an alternation of sand-sized reworked radiolarites, and igneous–ophiolitic materials. Section 5 covers only the first 100 m of this series whereas laterally further to the east the overlying ophiolite mélange is thousands of metres thick. The latter mélange is basically dominated by chromite (the Khajeh Djamali Chromite Mine can be found in this area).

The above-described observations suggest that major tectonic activity took place during the Campanian, with the interbedding of clasts that vary from a centimetre to hundreds of metres in size in the radiolarite facies. The virtual absence of the reworked carbonate facies in the Pichakun area seems to suggest that the source of this material must have been in the NE, and the increasing age of the olistoliths–tectonic slices indicates a gradual exhumation–thrusting of the older platform carbonates (see below for further discussion).

Phase III: Late Campanian to Maastrichtian (Figs 17c & 19d, e)

Phase III is characterized by an overall shallowing-upward trend, ranging from the Gurpi marls to the shallow-water platform carbonates of the Tarbur Formation, and finally the evaporites of the Sachun Formation. This phase was recorded only in the carbonate domain in the middle and southern parts of the area. No sediments of Maastrichtian age have been found in the Dalneshin area.

The boundary between Phase II and Phase III is placed at the top of the slumped and resedimented carbonates in the Kola Qazi and Khaneh Kat anticlines. In the Khaneh Kat and Qutb Abad anticlines the sedimentary system grades into platform carbonate facies (Figs 11, 12 & 15). The upper contact is located at the top of the Cretaceous, where the carbonate system passes vertically into an evaporitic sedimentary system.

In Transect 1 a clear lateral facies change can be observed from north to south (see also the facies model in Fig. 15): In the Khaneh Kat anticline over 450 m of shallow-water carbonate facies has been recorded, with massive to thick-bedded rudist debris rudstone and floatstone of the Tarbur Formation. The platform facies thins laterally to 80 m of mainly packstone facies with large benthic foraminifers such as *Loftusia* and *Omphalocyclus*, in the Kola Qazi north flank, and is absent in the southern flank.

In the Kola Qazi sections a mix of Maastrichtian pelagic and platform slope deposits can be observed. The pelagic succession gradually becomes dominated by turbidites and debris-flow deposits. In the upper part of this section at least three fining upward, debris-flow cycles are present. The debris flows consist of channelled monomict conglomerates that vertically grade to planar-bedded grainstone–packstone facies. In the southern flank the topmost 135 m of Phase III shows a mix of benthic and pelagic fauna (505–640 m). In the upper 50 m the benthic fauna prevails. The debris-flow–turbidite fining upward cycles show a range of very coarse-grained, monomict conglomerate at the base passing into highly mature carbonate grainstone at the top. The cycles represent the progradation of the platform carbonate to the SW (Fig. 8a, b). In the Qutb Abad well a shallowing-up trend, from pelagic marls to the *Loftusia–Omphalocylus*-bearing outer ramp–upper platform slope facies, and to the Sachun evaporites, has been observed.

The evaporitic facies of the Sachun Fm. have been found in the Khaneh Kat, Kola Qazi and Qutb Abad anticlines, whereas they are absent in the Dalneshin area (Figs 11, 12 & 15).

The palaeogeographical reconstruction suggests an uplifted area in the north (no sedimentation during the Maastrichtian), a Tarbur platform bordering this high, a pelagic basin in the centre (south of Kola Qazi), and another margin further south (at the Qutb Abad well).

Discussion

In the previous section it has been shown that the studied succession can be subdivided in three tectono-sedimentary phases: Phase I (Late Albian to Cenomanian), characterized by relatively little tectonic activity, and a passive margin setting of the Arabian plate; Phase II (Turonian to Late Campanian), which was a time of strong tectonic activity, corresponding to an active margin setting with obduction, and foreland basin formation; Phase III (Late Campanian to Maastrichtian), which again was a tectonically calmer period. During activation of the plate margin, the study area was split into two regions: (1) the interior of the tectonic plate, where gradual tectonic activity controlled the depocentre evolution of a carbonate-dominated system; (2) the plate margin, where the sedimentation was completely controlled by the obduction tectonics, with rapid subsidence caused by loading of the plate margin by the obducted ocean-floor deposits (radiolarites, ophiolites) and active destruction of part of the sediments (from Cretaceous to Permian) during thrusting. To better understand the relationship between these two geographically different areas, we will look in more detail at the radiolarite complex and at the carbonate gravity deposits.

Dating of the radiolarite complex

The radiolarite complexes along the Zagros Fold–Thrust Belt (Kuh-e Dalenshin, Arsanjan area, Pichakun Mountain and Neyriz region; Fig. 3) show variations in their origin (autochthonous and/or allochthonous deposits), the timing of their deposition and their subsequent emplacement. To understand the timing of their deposition reliable age dates are of great importance. The radiolarite complex is a series of rock units that are mainly composed of radiolarite associated with exotic shallow- or deep-water carbonate blocks, resedimented facies and occasionally autochthonous pelagic marl interbeds. Wells (1969) proposed that the pre-Cretaceous fossils in the radiolarite complexes were derived from the exotic blocks. He suggested, based on more reliable *in situ* microfossils, a Senonian age for the radiolarites throughout southern and western Iran. This interpretation was supported by James & Wynd (1965). Haynes & McQuillan (1974), however, proposed a pre-Cretaceous age for radiolarites found west and south of Dalneshin and around the Pichakun Mountain region. In the eastern part of the Pichakun Mountain region a varicoloured and highly siliceous regular bedded radiolarite sequence occurs (Fig. 10c) that in places is interlayered with calciturbidites. This setting suggests deep-water deposition with gravity-induced input. Ricou (1968a) proposed a Cretaceous to Triassic age for these deposits, whereas Hallam (1976) suggested a Cretaceous to Jurassic age.

The only detailed work in the Dalneshin and Arsanjan areas has been carried out by Stoneley (1981), who dated the radiolarian facies, and inter-bedded marly pelagic facies, containing oligostegi-nids and planktonic foraminifers, as mid-Turonian to Santonian (Fig. 2). Unfortunately, no specimen names and no sample locations in measured sections were given. In the present study, five sections were measured in the same area (Figs 3 & 16), and a similar lithological succession to that reported by Stoneley (1981) was found. The results of our study agree with the findings of Stoneley (1981) with respect to the type of sediment, but the age ranges are different, varying from Albian to Campa-nian for carbonate and radiolarite sediments (Tables 2 & 3). None of the aforementioned workers used radiolaria for age dating. It is impor-tant to mention that the radiolarite complexes rep-resent two different ages: the age of deposition and the age of their emplacement. Based on our work, at least in the Dalneshin and Arsanjan areas the age of deposition of the radiolarites ranges between Albian to Cenomanian. The age of empla-cement is considered to be post-Cenomanian–Turonian.

In his work, Stoneley (1981) considered all the radiolarite facies and accompanying pelagic marls as radiolarite sequences and dated them, based on their foraminifersl contents, to Turonian–? Early Coniacian and Mid–Late Coniacian–Santonian intervals. In our study the carbonates and radio-larites, which are underlain by a Cenomanian–Turonian platform carbonate, are separated by pelagic marl, mixed radiolarites and pelagic marls, and radiolarite intervals.

Four methods have been used in this study to date the radiolarite complex. It should be noted that the obtained age dates are the age of sedimen-tation rather than the age of emplacement. As mentioned above, the time of emplacement mainly falls within the Santonian–Campanian time interval.

(1) Planktonic foraminifers based on the biozo-nation of Wynd (1965). This method is used for the planktonic foraminifer-bearing units deposited as either autochthonous or allochthonous sediments. The results obtained for the various sections show Cenomanian to Late Turonian and probably Early Coniacian ages for these deposits. These findings partly agree with the age dates of the lower sequence of Stoneley (1981). The preservation/erosional ratio in the Turonian–? Early Coniacian pelagic interval, which is overlain by radiolarite facies and allochthonous material is very variable along the Dalneshin Mountain range. Therefore this auto-chthonous unit at some sites might contain sedi-ments of middle Coniacian to Santonian age as proposed by Stoneley (1981).

Table 2. *Nannofossil determinations of four spot samples in Dalneshin*

Section name	Sample no. and location	Nannofossils	Age
Arsanjan (1)	053°20′30″E 29°49′20″N (two samples from lower 50 m in lateral sampling)	*Eiffellithus eximius, Eiffellithus turriseiffeli, Eprolithus floridanus, Lithastrinus grillii, Lucianorhabdus cayeuxi, Gartnerago obliquum, Manivitella pemmatoidea, Marthasterites furcatus, Micula staurophora, Prediscosphaera cretacea, Prediscosphaera cretacea, Parhabdolithus embergeri, Reinhardtites anthophorus, Watznaueria barnesae*	Santonian
Arsanjan (1)	ARP-2719 (175 m)	*Arkhangelskiella cymbiformis, Broinsonia parca, Cribrosphaerella ehrenbergii, Eiffellithus eximius, Eiffellithus turriseiffeli, Microrhabdolus decoratus, Reinhardtites anthophorus, Tetralithus aculeus, Watznaueria barnesae*	Early Campanian
Tang-e Jazin (5)	ARP-2563 (710 m)	*Micula staurophora, Tetralithus aculeus?, Watznaueria barnesae*	Campanian?

Table 3. *Sr isotopic composition and calculated ages of the Dalneshin sections and surrounding area*

No.	(Section), location or sample no., metres in section	$^{87}Sr/^{86}Sr$	2 SEM (%)	Age* (Ma) Mc GTS04	Curve error (Ma)	Analytical age uncertainty[†] (Ma)
1	(1) 053°19′00″E 29°50′00″N	0.707552	0.0015	77.2	>77.0 and <77.4	76.5–77.7
2	(1) 053°19′30″E 29°49′30″N	0.707488	0.0015	80.7	>80.2 and <81.1	79.4–81.6
3	(1) 053°20′00″E 29°50′30″N	0.707455	0.0013	82.6	>82.4 and <82.9	81.8–83.5
4	(1) 053°20′00″E 29°50′00″N	0.707525	0.0011	78.2	>78.0 and <78.5	77.6–79.0
5	(4) 2541 (490 m)	0.707553	0.0013	77.1		
6	(4) 2542 (494 m)	0.707440	0.0014	83.5		
7	(4) 2543 (497 m)	0.707449	0.0014	83.0		
8	(4) 2544 (500 m)	0.707457	0.0013	82.5		
9	(4) 2549 (537 m)	0.707464	0.0011	82.1		
10	(4) 2556 (635 m)	0.707537	0.0015	77.7		
11	(4) 2557 (645 m)	0.707525	0.0011	78.2		
12	(4) 2561 (690 m)	0.707436	0.0013	83.8		
13	(4) 2563 (710 m)	0.707592	0.0013	75.8		
14	(3) 2620 (258 m)	0.707481	0.0012	81.2		
15	(3) 2621 (262 m)	0.707428	0.0013	84.8 or >99		
16	(3) 2622 (265 m)	0.707566	0.0011	76.5		
17	(1) 2695 (70 m)	0.707762	0.0013	69.1		
18	(1) 2696 (73 m)	0.707493	0.0012	80.3		
19	(1) 2697 (76 m)	0.707627	0.0014	74.1		
20	(1) 2698 (79 m)	0.707994	0.0015	29.3		
21	(1) 2705 (115 m)	0.707721	0.0014	70.9		
22	(1) 2709 (140 m)	0.707722	0.0013	70.8		
23	(1) 2716 (165 m)	0.707566	0.0015	76.5		
24	(1) 2736 (305 m)	0.707680	0.0012	72.2		
25	(1) 2743 (370 m)	0.707675	0.0012	72.4		
26	(1) 2763 (580 m)	0.707561	0.0013	76.9		

The values have been based on whole-rock sample analysis. The spot samples are shown by their coordinates and numbered samples are positioned in the outcrop sections. $^{87}Sr/^{86}Sr$ ratios normalized to $^{86}Sr/^{88}Sr = 0.1194$, $^{87}Sr/^{86}Sr$ ratios normalized to NBS987 $^{87}Sr/^{86}Sr = 0.710235$.
*Look-up table, version 4: 08/03, McArthur *et al.* (2000).
[†]Look-up table age range based on ± 0.000015.

(2) Nannoplankton. Only a few samples contained nannofossils; two samples are from the Arsanjan area and one sample from east Dalneshin. The list of nannofossils is given in Table 2. The age dating varies from Santonian to Early Campanian for the pelagic marls intercalated with the radiolarites. This age range matches the strontium isotope results, but no foraminifers were found to back this up.

(3) Radiolaria. A number of samples were examined through the radiolarite intervals of Sections 1 and 4. These samples were taken from radiolarian and mixed radiolarian–marly facies. Some of the processed samples yielded relatively well-preserved radiolaria, namely *Archaeodictyomitra montisserei* (Squinabol) and *Pseudodictyomitra paronai* (Aliev), allowing us to assign them to the *Spoletoensis* zone of O'Dogherty (1994) (Unitary Associations 10–15) and thus correlate them with the Albian to Early Cenomanian time interval.

(4) Strontium isotopes (Table 3). In total 26 samples were dated using strontium isotope variations. All samples were taken from the Dalneshin

area (Sections 1, 4 and 5). The samples were taken from the Sarvak platform carbonate to carbonate pelagic facies present within the radiolarite complex. They all give ratios suggesting a Late Cretaceous age, c. 77–83 Ma (Early Campanian). These results are at odds with the well-known Cenomanian age of the Sarvak platform carbonates.

Although for the marly units above the platforms a younger age is likely, a Campanian age seems to be too young. Nannofossils appear to back up this Campanian age locally. There is a possibility that diagenetic processes around the time of the uppermost Sarvak deposition influenced the Sr isotopes. Faulting could be another possibility to explain some of the variation, but through the Cenomanian interval no considerable faulting is seen. None of the samples studied have any obvious dolomite or other detrital contaminants, or signs of unusual diagenesis. Hence the whole-rock isotopic composition may accurately reflect a seawater signature. Some of the samples appear to have had more alteration of primary fabric during diagenesis, but there are insufficient data to evaluate any possible effects on the Sr ratio. These results remain for the moment enigmatic, and are not used in the age dating of the studied sediments.

Nature of the contact

The character of the contact between the radiolarite complex and underlying platform deposits of the Sarvak Formation is another point of discussion. Gray (1950) and Ricou (1970) interpreted the contact as a fault contact in which the deposits are not in stratigraphic concordance. The Sarvak Formation is generally speaking of Late Albian to Cenomanian age, but locally ranges to Turonian age (James & Wynd 1965). According to Stoneley (1981) the Sarvak platform is overlain abruptly, conformably or with a slight discordance, by both radiolarian facies and pelagic marls, and he thus favoured an *in situ* location and deposition of these sediments.

In this study a regional trend in the amount of interbedded marly pelagic facies is observed, which decreases from west to east in the Dalneshin area. The bedding pattern of the radiolarite is well ordered in the central part of Dalneshin (Section 4), whereas in the eastern and western parts of the Dalneshin and Arsanjan area (Sections 1, 2 and 5) a more irregular pattern is observed. No radiolarite *sensu stricto* was deposited in the studied section, but radiolarites were found as rock fragments in the clastic sedimentary system that alternated with pelagic limestone beds. The depositional environment of radiolarite during the sedimentation is proposed as a transitional setting between the planktonic foraminifer-dominated basin of the plate interior and the radiolarite-dominated deep ocean setting (possibly of tens to several hundreds of metres water depth).

It is now well established that not all radiolarites were formed in deep oceanic settings. There are well-known examples of radiolarian-rich sediments that accumulated on submarine highs of continental margins (Danelian *et al.* 1997) or even in shallow-water environments of continental shelve (Ellis & Baumgartner 1995). The accumulation of radiolarian-rich sediments is not only dependent on the water depth but is also related to areas with upwelling ocean currents that bring cold nutrient-rich waters to environments just below the storm wave base.

The significance of the Santonian–Campanian gravity deposits

During the Santonian–Campanian a maximum volume of gravity deposits is observed, varying from fine-grained calciturbidites to large olistostromes.

In the Kola Qazi anticline (south flank), Santonian debris flows show more textural maturity and have better developed sedimentary cycles than in the Campanian (Fig. 11). In the Santonian units amalgamated channels with a lensoid shape and erosional surfaces are present, whereas in the Campanian units the slumped deposits do not show any small-scale sedimentary features. Most of the reworked clasts are of shallow-water platform carbonate origin with probably Santonian age. The clasts or blocks increase in size in the Campanian. The direction of these flows could not be determined.

In the Khaneh Kat anticline (Section 6, middle part of Transect 2; Fig. 12) the Santonian sequence consists of carbonate sandstone with planar lamination, which gradually passes into grainier or coarser carbonate conglomerate–rudstone. In the Khaneh Kat anticline the detritus at times contains radiolarian cherts. The conglomerates–rudstone units, however, show minor variation in their components. Although most of the Santonian–Campanian interval is not exposed no direct evidence confirms the existence of *in situ* radiolarites in this area.

In the Arsanjan area, in Section 1, the debris flows are of Campanian age, and consist of multiple debris flows and lithoclastic calciturbidites that are interbedded with thin- to medium-bedded oligosteginid–marly pelagic and radiolarian facies. The lensoid channelized conglomeratic units contain clasts of pre-late Cretaceous age associated with ophiolitic and radiolarite constituents. In Section 2, a dense carbonate conglomerate unit consisting of Cenomanian–Santonian platform and intra-shelf basinal facies is observed, locally interbedded with Campanian marls. The density of the conglomerates suggests deposition very close to the sediment source.

In the central part of the Dalneshin area the age dating is problematic, because of the scarcity of the autochthonous pelagic interbeds. The age is tentatively interpreted as possibly Coniacian–Santonian at the base (after Stoneley 1981, but not confirmed by our work), to Campanian for the upper part, possibly passing into Maastrichtian. In this area a 2000–3000 m thick succession of detached mountain-sized blocks with a shelf-edge carbonate facies occurs. They range in age from early Cretaceous to Permian, and occur embedded in radiolarian facies. The blocks are highly fractured and chertified. One of the Jurassic blocks is in an overturned position, as is shown by the sedimentary structures and age dating. This setting suggests the presence of a sharp-edged escarpment close to a steep, rapidly subsiding(?) basin margin. The systematic increase in age of the olistoliths–tectonic slices is proof of the gradual exhumation–thrusting and destruction of the sediments.

The eastern part of Dalneshin area (Section 5; Fig. 14) is dominated by ophiolite–igneous rocks, which occur in the lower part of the succession as reworked debris flows and towards the top as an ophiolitic mass.

Based on the above observations, it is proposed that two main directions of gravity deposit sedimentation occurred.

(1) From SW to NE. During the Santonian the Arabian plate started to dip towards the north at a small angle, as a result of tectonic loading at the beginning of the early obduction phase, thus creating the foreland basin. At the same time, local uplift occurred in the more internal parts of the Arabian plate as a result of the development of a foreland basin flexural bulge (Murris 1980; Robertson 1987; J. Letouzey, pers. comm.). These tectonic events caused the gradual deepening of the environment in the northern part of the study area (mixed pelagic carbonates and radiolarites), and the deposition of the erosional products (Santonian platform sediments) in the newly created foreland basin (Kola Qazi area; Sections 7 and 8). As the uplift of the plate increased, the size and volume of the eroded sediments increased significantly (Campanian chaotic units in the Kola Qazi area, Section 8; Fig. 11).

(2) From NE to SW. In a later stage of the obduction process the sedimentation process reversed in the northern part of the study area, when sediments were shed from the advancing nappes that were thrusted from the NE and distributed their erosional products to the SW. Initially in the study area, these sediments consist of fine-grained turbidites that occur interbedded with oligosteginids, calcareous pelagic facies and radiolarite. With the advancing obduction front, and continued thrusting, the basin deepened and steep, nappe-bounded escarpments developed in which the exhumed–thrust rocks were deposited as exotic blocks (Central Dalneshin, Section 3; Hallam 1976). In the final phase ophiolitic nappes were overthrusted and emplaced (Figs 14 & 16).

Sedimentation and tectonic models

In this study a subdivision of three tectono-sedimentary phases has been used; these phases are briefly summarized here and then compared with models from the literature (Glennie *et al.* 1973; Stoneley 1981; Alavi 1994, 2004).

Phase I. During the Late Albian to Cenomanian the margins of the Arabian plate were still passive. Sedimentation occurred on a vast, regional carbonate shelf with intra-shelf basinal depressions, but accommodation space varied substantially because

of a variable tectonic regime on the Arabian plate, locally influenced by salt-tectonics (Sherkati *et al.* 2005). In our study area this is reflected in the thick succession observed at the Khaneh Kat section (Fig. 12). There is, however, a possibility that major tectonic deformation started earlier in the Dalneshin area, where uncertainty exists about the presence of Turonian radiolarites, which were reported by Stoneley (1981) but were not confirmed by our work.

Intra-shelf basin formation is a common phenomenon during the Cenomanian of the Middle East and has been observed elsewhere in the Zagros area (Taati Qorayem *et al.* 2003), the united Arab Emirates (Burchette 1993) and Oman (van Buchem *et al.* 1996, 2002). The depositional system consisted of rudist-rimmed platform carbonates surrounding intra-shelf basinal depressions characterized by an oligosteginid-dominated deeper water fauna.

The Cenomanian third-order depositional sequences correspond well to the sequences described in the Zagros (Taati Qorayem *et al.* 2003), in Oman (van Buchem *et al.* 1996, 2002) and in the Arabian plate in general (Sharland *et al.* 2001).

Summarizing, in this phase sedimentation was mostly eustatically driven, with local evidence for differential sedimentation rates. At the end of this stage the continental plate started to tilt towards the NE in the northern part of the study area (Fig. 19).

Phase II. The major change in the sedimentation patterns on the Arabian plate took place during the Turonian to Campanian, when the sedimentary system showed a series of major tectono-sedimentary events affecting both continental and margin settings.

During the Santonian, the continental plate was tilted more towards the NE (Murris 1980; Robertson 1987), thus creating a foreland basin between the Arabian and Iranian plates. In our study area this phenomenon was accompanied by the channelling and slumping of Santonian platform carbonate deposits from the SW. The Santonian deposits display a northward change from platform carbonate with gravity-flow deposits of the same age at the top (Kola Qzai), to marly pelagic facies in the Khaneh Kat and Dalneshin areas that occasionally are interbedded with radiolarite, debris flows and turbidites with different composition and age.

A comparison of the results of our study with the palaeogeographical map of Murris (1980) shows that during the Albian to Cenomanian the Arabian plate represents more differentiation between shallow- and deep-water carbonate systems. Despite considerable subaerial exposure at the end of Cenomanian–Turonian, even in most parts of the Fars area, deep-water sedimentation persisted in the northeastern sector of Fars through the Turonian and Santonian.

During the Campanian, sedimentation changed dramatically in the northern part, with the subsequent emplacement along a SW–NE transect of (1) conglomerates of Cenomanian–Santonian platform deposits, (2) olistoliths of exhumed autochthonous platforms, and (3) ophiolite masses. This sequence is interpreted as a tectono-sedimentary succession related to obduction-related thrust belt tectonics.

Phase III. During the Maastrichtian, platform carbonates developed again, bordering the foreland basin. The presence of these well-developed platform carbonates just south of the Dalnshin area is interpreted as evidence for a strongly reduced shedding of erosional products from the obduction zone, and a halt in obduction. No sediments of Maastrichtian age were observed in the Dalneshin area, which is interpreted as evidence for exposure of this area at that time.

The depocentre that was located in the Dalneshin area during the Campanian shifted towards the SW in the Maastrichtian, possibly to the area around Kaneh Kat. The Maastrichtian palaeogeographical map of Murris (1980) shows a shallow-water carbonate system covering the allochthonous sediments towards the southwestern end of the Zagros thrust belt. In the northern part of the Zagros, where ultrabasic rocks are absent and continental crust is present, the clastic material was shed southwestward as a result of isostatic rise of the orogenic zone. Our observations illustrate that in the northern part of the Fars area no sedimentation took place whereas in the central part the shallow carbonate deposits unconformably onlapped the radiolarite and ophiolite nappes. In the southwestward region the Campanian deep-water system persisted until the end of the Maastrichtian. For the Late Maastrichtian to Palaeocene–Eocene, Murris (1980) suggested the presence of an open-marine basin and a wide shallow-water evaporitic platform in the SW. Our data show that the carbonate depositional system became shallower and reached evaporitic conditions in the late Maastrichtian to Palaeocene–early Eocene.

Other contemporaneous platform carbonates in the region are the Qahlah and Simsima formations in the United Arab Emirates and in the Northern Oman Mountains, respectively. These formations show interfingering with each other and are underlain by the obducted Semail Ophiolite, of Santonian–Campanian age, in the Northern Oman Mountains. The sediments of these formations were deposited in an open-marine environment that laterally passed into shallower marine conditions with larger foraminifersl species such as *Loftusia*

morgani, Orbitoides media and *Omphalocyclus* sp. (Abdelghany 2003).

In the literature, several tectonic models have been proposed to account for the subduction process in the Zagros area. Of the three most cited models (Stoneley 1981; Alavi 1994, 2004; Glennie 2005) the second and third consider that the subduction of the Neo-Tethyan oceanic plate beneath the Iranian lithospheric plate took place during the Late Cretaceous. This agrees with the model presented here (Fig. 18), assuming that the first obduction in the region took place in the Turonian and Coniacian.

The model of Glennie (2005) shows a large-scale evolution of the Zagros from mid–late Permian to Plio-Pleistocene, and proposes that the end of the subduction occurred in the Campanian, which agrees with our findings in the studied sections.

The model of Alavi (1994, 2004) agrees the best with the model proposed in this study (Fig. 15). The model is general and based on the tectono-sedimentary evolution of the Zagros Fold–Thrust Belt evidenced by litho-stratigraphic units. However, in the Zagros area the onset and the end of the subduction phase vary for the different localities. The model has the following stages.

(1) In the latest Turonian to middle Maastrichtian obduction occurred, resulting in the deposition of a turbiditic facies, which is in agreement with our observations in the Inner Fars.

(2) In the middle Maastrichtian the foreland basin became more restricted, resulting in the deposition of the evaporitic facies of the Sachun Formation. In the studied successions, however, the evaporitic facies were deposited late in the Maastrichtian and Palaeocene.

(3) In the a latest Cretaceous to Middle Eocene time interval terrigeneous material was shed from the uplifted area.

Most relevant for this study is the model of Stoneley (1981). His west–east cross-section along the Kuh-e Dalneshin corresponds to our Transect 2 (Fig. 16). According to Stoneley (1981) a shallow-water platform carbonate developed in the area during the Cenomanian. The basin deepened towards the east and it contained *Oligostegina* facies deposited within an intra-shelf basin. Our results account for a similar distribution of facies along the transect, but the timing of their deposition shows some differences. Stoneley (1981) proposed maximum uplift and associated slumping in the Santonian. The results of our study, however, strongly suggest that these events took place in the Late Santonian to Early Campanian (Figs 15, 16 & 19).

Stoneley (1981) mentioned that the ophiolites could not be the adjacent crust of the Tethys, in view of their Cenomanian–Coniacian age (Oman) as opposed to the Triassic age of rifting. The

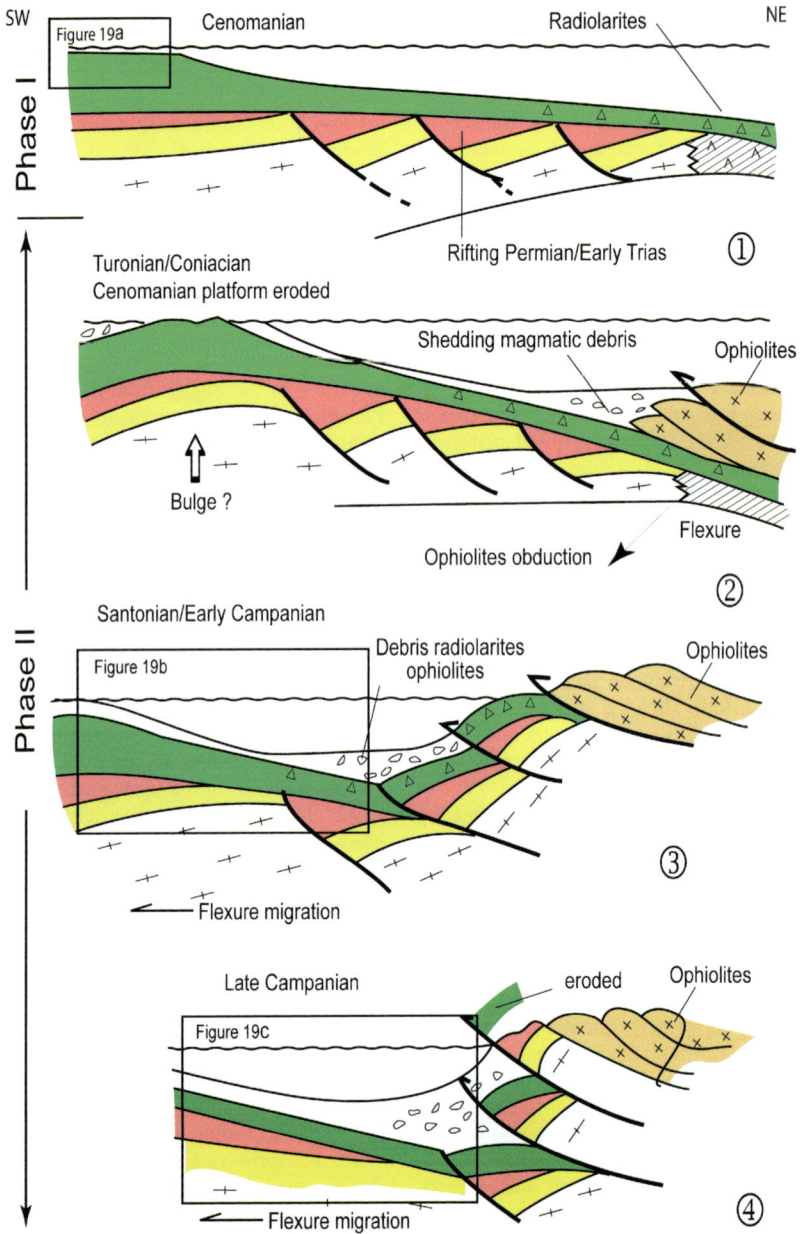

Fig. 18. Structural profiles of the Arabian continental margin evolution from Cenomanian to Late Campanian time.

ophiolite formed either in a Late Cretaceous oceanic realm, which could only have been located NE of the contemporaneous subduction zone, perhaps as a neo-formed arc spreading centre or at the site of the still active mid-oceanic ridge. In the case of a back-arc setting the ophiolite obduction on the top of the Arabian margin would have required a second subduction zone to allow for its SW-directed transport over the former arc system. Associated with the first subduction zone is the submarine volcanism interleaved with the Coniacian cherts, which now occurs incorporated with slices of shallow-water sediments from the down-going Arabian margin. Therefore it should display a calc-alkaline geochemical signature. The aborted spreading ridge model sounds simpler and seems better

(e)

Phase III b (Late Maastrichtian-Paleocene Sachun and Jahrum formations)

Section 9   Section 8   Section 7   Section 6   Section 5

Eocene Jahrum platform

Evaporites Maastrichtian/
Paleocene platform

100 m

161 Km

Section No. 9   8   7   6   1   2   4   5
Distance (Km)   22   10   64   64   5   25   36

(d)

Subaerial exposure

P   Ophiolite

T

C   J   P

C   J   P

J   P

Phase III a - Maastrichtian   150 m

C- Cretaceous   T- Triassic
J- Jurassic   P- Permian

(c)

9   8   7   6   1   2   4   5

Santonian clasts

ophiolite

±2000–3000 m

J

P

Santonian / Cenomanian
clasts

C   J

P

scale 100 m

scale ±250 m   Note
change
in scale

Phase II (b) - Late Campanian

(b)

Camp.

radiolaria

SB6
SB5   5
SB4   4

100 m

Subsidence due to loading
(regional tilting)

Phase II (a) - Santonian / Early Campanian

(a)

Phase I + Sequences 5 and 6 (Sarvak, Ilam and Radiolarite)   N 46° W

SB6   SB 5   Santonian
SB4   Turonian   ?   SB4
SB3   SB3
SB2   Cenomanian   Rudist Zone   SB2
SB1   SB1

Cenomanian

Late Albian

200 m   oligosteginids
    32 Km   64 Km   65 Km
9   7-8   6   5

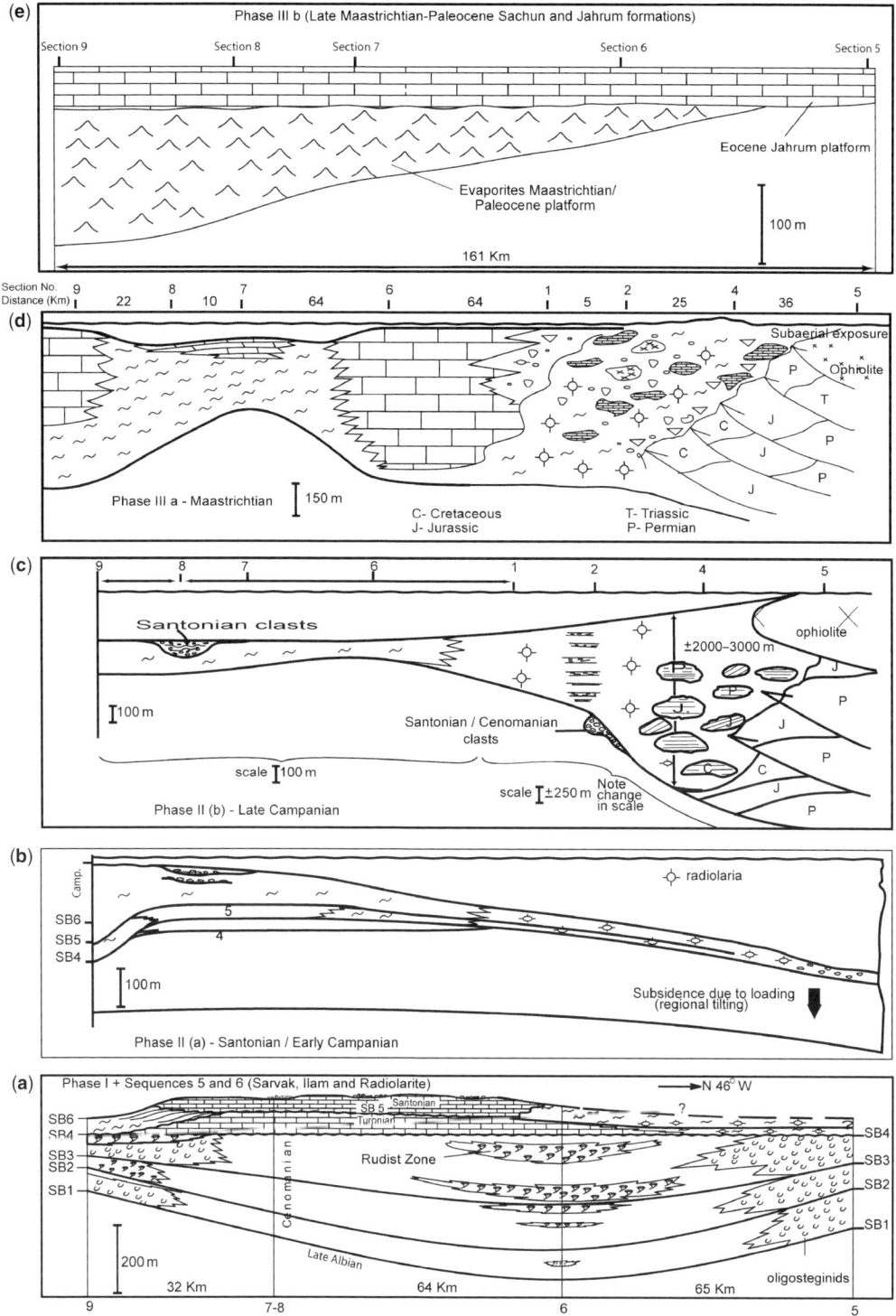

Fig. 19. Cross-sections showing the three tectono-sedimentary phases: (**a**) Phase I, Cenomanian; (**b**) Phase II(a), Santonian–early Campanian; (**c**) Phase II(b), late Campanian; (**d**) Phase III(a), Maastrichtian; (**e**) Phase III(b), late Maastrichtian–Paleocene.

supported by the data. Stoneley (1981) did not present any experimental results to support his models. In the case of palaeo-subduction, this needs to be recognized by the existence of HP–LT metamorphism (blueschists) under the tectonic mélange. In the case of a palaeo-arc–back arc, it requires geochemical analysis of the pillow lava samples.

Our observations provide support for the obduction model based on the following findings: (1) the precise dating on the different types of sediments to document various tectonosedimentary stages; (2) the upward older age of the olistoliths (from Early Cretaceous to Permian) that represent different thrusting stages of the tectonic mélange; (3) the overall facies succession (Transect 2), which shows a turbiditic system interfingering with obducted radiolarites and pelagic marls in the WSW (distal part) to ophiolite masses, Phanerozoic platform olistoliths and radiolarites (proximal part, ENE) which all are underlain by the Cenomanian to Turonian platform carbonate.

Conclusions

The following conclusions can be drawn from this study. In this study new stratigraphic data were obtained from outcrops located close to the margin of the Arabian plate in the interior Fars area in Iran. Based on these field observations in combination with a regional set of well logs, three tectonosedimentary phases could be distinguished. These phases reflect distinct changes in the depositional environment, the palaeogeography, the dominant controlling process steering sedimentation patterns (eustatics v. tectonics) and the position of the depocentres. The following phases were recognized.

Phase I (Late Albian–Cenomanian)

During this time interval the Arabian plate still had a passive margin, and sedimentation occurred on a vast carbonate shelf with intra-shelf basins. Accommodation space varied locally as a result of tectonic control on the Arabian plate that was partly influenced by salt-tectonics. The depositional system consisted of rudist-rimmed platform carbonates surrounding intra-shelf basinal depressions that were characterized by an *Oligostegina*-dominated deeperwater fauna. In this phase sedimentation was mostly eustatically driven, with local evidence for differential sedimentation rates.

Phase II (Turonian–Late Campanian)

During this phase a major change in the sedimentation patterns on the Arabian plate took place. A series of major tectono-sedimentary events affected both continental and margin settings. A foreland basin developed between the Arabian and Iranian plates, which was infilled from the SW with channelled and slumped Santonian platform carbonate deposits, during the flexure of the plate. In the northern part sedimentation changed most dramatically, with the subsequent emplacement of radiolarites and debris of obducted slice of ophiolites, within which a succession of olistoliths of increasingly older age is found. This succession of facies testifies of the different stages of thrusting, with the gradual erosion of tectonic slices of exhumed deposits, eventually covered by the main ophiolite masses.

Phase III (Late Campanian–Maastrichtian)

During this phase platform carbonates developed again bordering the foreland basin in the north and in the south. In the north these were deposited in an onlap position against the obducted units described in Phase II. The strongly reduced shedding of erosional products from the obduction zone probably suggests a halt in obduction during this time interval.

In comparison with the existing tectonic models our observations have provided biostratigraphic data and sedimentological observations that allow for a more precise reconstruction of the deformation steps, by the dating of the radiolarites and interbedded marls, the dating of the olistoliths–tectonic slices, and the reconstruction of a regional cross-section based on measured sections.

We would like to thank the staff of the National Iranian Oil Company (NIOC) for their permission to perform this study and for their help in organizing the field campaigns. We are grateful to the following NIOC staff members, who assisted during the field campaigns: A. Amirkhani, B. Ariyfar, A. Bakhshi, R. Charoosaei, F. Farzaneh, M. Masibeigi and M. Mousaviyan. In addition, NIOC provided help with the palaeontological analysis and supplied all thin sections. We also thank F. Roure and J. Letouzey for their constructive reviews of a first draft of the manuscript. C. Muller is thanked for the nannoplankton determinations, and T. Allen for the strontium datings. TOTAL is thanked for its financial support of this project. IFP, the Université de Provence and the VU University Amsterdam provided research facilities. This paper forms part of the first author's PhD thesis.

References

ABDELGHANY, O. 2003. Late Campanian–Maastrichtian foraminifers from the Simsima Formation on the western side of the Northern Oman Mountains. *Cretaceous Research*, **24**, 391–405.
ALAVI, M. 2004. Regional stratigraphy of the Zagros fold–thrust belt of Iran and its proforeland evolution. *American Journal of Science*, **304**, 1–20.

ALAVI, M. 1994. Sedimentary and structural characteristics of the paleo-Tethys remanent in NE Iran. *Geological Society of America Bulletin*, **103**, 983–992.

ALLEN, P. A. & HOMEWOOD, P. (eds) 1986. *Foreland Basins*. International Association of Sedimentologists, Special Publications, **8**, 453.

ANDRESEN, N., REIJMER, J. J. G. & DROXLER, A. 2003. Timing and distribution of calciturbidites around a deeply submerged platform carbonate in a seismically active setting (Pedro Bank, northern Nicaragua Rise, Caribbean Sea). *International Journal of Earth Sciences*, **92**, 573–592.

BERBERIAN, M. & KING, G. C. P. 1981. Towards a paleogeography and tectonic evolution of Iran. *Canadian Journal of Earth Science*, **18**, 210–265.

BEYDOUN, Z. R., CLARKE, M. W. H. & STONELEY, R. 1992. Petroleum in the Zagros basin: a Late Tertiary foreland basin overprinted onto outer edge of a vast hydrocarbon-rich Paleozoic–Mesozoic passive-margin shelf. *In*: MACQUEEN, R. W. & LECKIE, D. A. (eds) *Foreland Basins and Fold Belts*. AAPG, Memoirs, **55**, 309–339.

BOLZ, H. 1985. *The paleogeographic evolution during the Cretaceous in the operating area with special emphasis on the Bangestan Group*. NIOC, Tehran.

BORDENAVE, M. L. & BURWOOD, R. 1990. Source rock distribution and maturation in the Zagros orogenic belt: provenance of the Asmari and Bangestan reservoir oil accumulations. *Organic Geochemistry*, **16**, 369–387.

BORDENAVE, M. L. & HUC, A. Y. 1995. The Cretaceous source rocks in Zagros foothills of Iran. *Reviews of Institut Français du Pétrole*, **50**, 727–753.

BORDENAVE, M. L. 2000. Zagros domain of Iran holds exploration, EOR opportunities. *Oil and Gas Journal*, **8**, 36–38.

BORDENAVE, M. L. & HEGRE, J. A. 2005. The influence of tectonics on the entrapment of oil in the Dezful embayment, Zagros fold belt, Iran. *Journal of Petroleum Geology*, **28**, 339–368.

BRADLEY, D. & HANSON, L. 1998. Paleoslope analysis of slump folds in the Devonian flysch of Maine. *Journal of Geology*, **106**, 305–318.

BURCHETTE, T. P. 1993. Mishrif Formation (Cenomanian–Turonian), southern Persian Gulf: Platform carbonate growth along a cratonic basin margin. *In*: SIMO, J. A. T., SCOTT, R. W. & MASSE, J. P. (eds) *Cretaceous Platform Carbonates*. AAPG Memoirs, **56**, 185–200.

CROSS, T. A. 1986. Tectonic controls of foreland basin subsidence Laramide style deformation, western United States. *In*: ALLEN, P. A. & HOMEWOOD, P. (eds) *Foreland Basins*, International Association of Sedimentologists, Special Publications, **8**, 15–40.

DANELIAN, T., DE WEVER, P. & AZÉMA, J. 1997. Palaeoceanographic significance of new and revised palaeontological datings for the onset of Vigla Limestone sedimentation in the Ionian zone of Greece. *Geological Magazine*, **134**, 869–872.

DAVOUDZADEH, M. & SCHMIDT, K. 1984. A review of the Mesozoic paleogeography and paleotectonic evolution of Iran. *Neues Jahrbuch für Geologie und Palaeontologie, Abhandlungen*, **168**, 182–207.

DECELLES, P. G. & GILES, K. A. 1996. Foreland basin system. *Basin Research*, **8**, 105–123.

DESMONS, J. & RADELLI, L. 1989. Comment and reply on 'Ophiolite generation and emplacement by rapid subduction hinge retreat on a continent-bearing plate'. *Geology*, **17**, 480–481.

ELLIOTT, C. G. & WILLIAMS, P. F. 1988. Sediment slump structures: A review of diagnostic criteria and application to an example from Newfoundland. *Journal of Structural Geology*, **10**, 171–182.

ELLIS, G. & BAUMGARTNER, P. O. 1995. Austral shallow-water radiolarites—implications for the global cycling of silicon and carbon in the Mid-Cretaceous. European Union of Geosciences 8. *Terra Abstracts*, **7**, Abstracts Supplement 1, 225.

ELVERHØI, A., ISSLER, D., DE BLASIO, F. V., ILSTAD, T., HARBITZ, C. B. & GAUER, P. 2005. Emerging insights into the dynamics of submarine debris flows. *Natural Hazards and Earth System Science*, **5**, 633–648.

FALCON, N. L. 1974. Southern Iran–Zagros Mountains. *In* SPENCER, A. M. (ed.) *Time and Place in Orogenic Belts*. Geological Society, London, Special Publications, **3**, 9–22.

GLENNIE, K. W., BOEUF, M. G. A., HUGHES CLARKE, M. W., MOODY STUART, M., PILAAR, W. H. F. & REINHARDT, B. M. 1973. Late Cretaceous nappes in Oman Mountains and their geologic evolution. *AAPG Bulletin*, **57**, 5–27.

GLENNIE, K. W. & HUGHES CLARKE, M. W. 1973. Late Cretaceous nappes in Oman Mountain and their geologic evolution. Reply. *AAPG Bulletin*, **57**, 2287–2290.

GLENNIE, K. W., HUGHES CLARKE, M. W., MOODY-STUART, M., POLAAR, W. F. H. & REINHARDT, B. M. 1974. *Geology of the Oman mountains*. Verhandelingen Koninklijke Nederlands Geologisch Mijnbouwkundig Genootschaap, **31**.

GLENNIE, K. W. 2000. Cretaceous tectonic evolution of Arabia's eastern plate margin: a tale of two oceans. *In*: ALSHARHAN, A. S. & SCOTT, R. W. (eds) *Middle East Models of Jurassic/Cretaceous Carbonate Systems*. SEPM Special Publications, **69**, 9–20.

GLENNIE, K. W. 2001. The age of the Hawasina and other problems of Oman Mountains Geology: Discussion. *Journal of Petroleum Geology*, **24**, 477–484.

GLENNIE, K. W. 2005. *The Geology of the Oman Mountains: An Outline of their Origin*, 2nd edn. Scientific Press Ltd, UK.

GOLDHAMMER, R. K., DUNN, P. A. & HARDIE, L. A. 1990. Depositional cycles, composite sea-level changes, cycle stacking patterns, and the hierarchy of stratigraphic forcing: examples from Alpine Triassic platform carbonates. *Geological Society of America Bulletin*, **102**, 535–562.

GOLLESTANEH, A. 1974. The biostratigraphy of the Khami Group and the Jurassic–Cretaceous boundary in Fars Province (Southern Iran). *Bulletin du BRGM (Deuxième Série) section IV*, **3-1974**, 165–179.

GRAY, K. W. 1950. A tectonic window in southwestern Iran. *Journal of the Geological Society, London*, **106**, 461–464.

Gulf Petrolink 1998. Exploration and production features United Arab Emirates and Iran. *Geo Arabia*, **3**, 427–455.

HALL, B. A. 1973. Slump folds and the determination of paleoslope. *Geological Society of America, Abstracts with Programs,* **5**, 648.

HALLAM, A. 1976. Geology and plate tectonics interpretation of the sediments of the Mesozoic radiolarite–ophiolite complex in the Neyriz region, southern Iran. *Geological Society of America Bulletin,* **87**, 47–52.

HAQ, B. U., HARDENBOL, J. & VAIL, P. R. 1988. Mesozoic and Cenozoic chronostratigraphy and eustatic cycles. *In*: WILGUS, C. K., HASTINGS, B. S., KENDALL, C. G. ST. C., POSAMENTIER, H. W., ROSS, C. A. & VAN WAGONER, J. C. (eds) *Sea-level Changes: an Integrated Approach.* Society of Economic Paleontologists and Mineralogists, Special Publications, **42**, 71–108.

HAYNES, S. J. & MCQUILLAN, H. 1974. Evolution of the Zagros Suture Zone, Southern Iran. *Geological Society of America Bulletin,* **85**, 739–744.

HELWIG, J. 1970. Slump folds and early structures, northeastern Newfoundland Appalachians. *Journal of Geology,* **78**, 172–187.

HILBRECHT, H. 1989. Redeposition of Late Cretaceous pelagic sediments controlled by sea-level fluctuations. *Geology,* **17**, 1072–1075.

HOMEWOOD, P. W. & EBERLI, G. P. (eds) 2000. *Genetic stratigraphy on the exploration and the production scales.* Centre de Recherches Elf Exploration–Production Elf-Aquitaine, Mémoires, **24**.

JAMES, G. A. & WYND, J. G. 1965. Stratigraphic nomenclature of Iranian Oil Consortium agreement area. *AAPG Bulletin,* **49**, 2182–2245.

KARNER, G. D. & WATTS, A. B. 1983. Gravity anomalies and flexure of the lithosphere at mountain ranges. *Journal of Geophysical Research,* **88**, 10449–10477.

KERANS, C. & TINKER, S. (eds) 1997. *Sequence Stratigraphy and Characterisation of Carbonate Reservoirs.* Society of Economic Paleontologists and Mineralogists, Short Course Notes, **40**.

KHALILI, M. 1976. *The biostratigraphic synthesis of the Bangestan Group in southwest Iran.* NIOC Report, **1219**.

KHERADPIR, A. 1975. *Stratigraphy of the Khami Group in southwest Iran.* NIOC Report, **1235**.

KOOP, W. J. & STONELEY, R. 1982. Subsidence history of the Middle East Zagros Basin, Permian to Recent. *Philosophical Transactions of the Royal Society of London, Series A,* **305**, 149–168.

LANTZSCH, H., ROTH, S., REIJMER, J. J. G. & KINKEL, H. 2007. Sea-level related resedimentation processes on the northern slope of Little Bahama Bank (Middle Pleistocene to Holocene). *Sedimentology,* **54**, 1307–1322.

LOUCKS, R. G. & SARG, J. F. (eds) 1993. *Carbonate Sequence Stratigraphy, Recent Developments and Applications.* AAPG Memoirs, **57**.

MCARTHUR, J. M., CRAME, J. A. & THIRLWALL, M. F, 2000. Definition of Late Cretaceous stage boundaries in Antarctica using strontium isotope stratigraphy. *Journal of Geology,* **108**, 623–640.

MAJOR, J. J. & IVERSON, R. M. 1999. Debris flow deposition: Effects of pore-fluid pressure and friction concentrated at flow margins. *Geological Society of America Bulletin,* **111**, 1424–1432.

MERIÇ, E., ERSOY, S. & GORMU, M. 2001. Palaeogeographical distribution of the species of *Loftusia* (Foraminiferida) in the Tethyan Ocean during the Maastrichtian (Late Cretaceous). *Cretaceous Research,* **22**, 353–364.

MOTIEI, H. 1993. *Stratigraphy of Zagros.* Geological Survey of Iran, Tehran [in Farsi].

MULDER, T. & ALEXANDER, J. 2001. The physical character of subaqueous sedimentary density flows and their deposits. *Sedimentology,* **48**, 269–299.

MURRIS, R. J. 1980. Middle East: Stratigraphic evolution and oil habitat. *AAPG Bulletin,* **64**, 597–618.

O'DOGHERTY, L. 1994. *Biochronology and palaeontology of Mid-Cretaceous radiolarians from Northern Apennines (Italy) and Betic Cordillera (Spain).* Mémoires de Géologie (Lausanne), **21**.

PAMIC, J. & ADIB, D. 1982. High-grade amphibolites and granulites at the base of the Neyriz peridotities in southeastern Iran. *Neues Jahrbush für Mineralogie, Abhandlungen,* **143**, 113–121.

PLAYER, R. A. 1969. *The Hormuz salt plugs of southern Iran.* PhD thesis, Reading University.

POMAR, L. 1991. Reef geometries, erosion surfaces and high frequency sea level changes, Upper Miocene reef complex, Mallorca, Spain. *Sedimentology,* **38**, 243–269.

READ, J. F., KERANS, C. & WEBER, L. J. (eds) 1995. *Milankovitch Sea Level Changes, Cycles and Reservoirs on Platform Carbonates in Green-house and Icehouse Worlds.* Society of Economic Paleontologists and Mineralogists, Short Course Notes, **35**.

RICOU, L. E. 1968*a*. Sur la mise en place au Crétacé supérieur d'importantes nappes a radiolarites et ophiolites dans les monts Zagros (Iran). *Comptes Rendus de l'Académie des Sciences, Série D,* **267**, 2272–2275.

RICOU, L. E. 1968*b*. Une coupe à travers les séries à radiolarites des Monts Pichakun (Zagros, Iran). *Bulletin de la Société Géologique de France, 7 Série,* **10**, 478–85.

RICOU, L. E. 1970. Comments on radiolarite and ophiolite nappes in the Iranian Zagros Mountains. *Geological Magazine,* **107**, 479–80.

RICOU, L. E. 1971*a*. Le métamorphisme au contact des péridotites de Neyriz (Zagros interne, Iran): développement de skarns à pyroxène. *Bulletin de la Société Géologique de France,* **1–2**, 146–155.

RICOU, L. E. 1971*b*. Le croissant ophiolitique peri-arabe; Une ceinture de nappes mises en place au Cretacé superieur. *Revue Géographie et de Géologie Dynamique,* **13**, 327–349.

RICOU, L. E. 1974. *L'évolution géologique de la région de Neyriz (Zagros iranien) et l'évolution structurale de Zagrides.* Thesis, University of Orsay, No. 1269.

RICOU, L. E. 1976. *Evolution structurale des Zagrides: la region clef de Neyriz (Zagros Iranien).* Mémoires de la Société Géologique de France, Nouvelle Série 40, **125**.

RICOU, L. E., BRAUD, J. & BRUNN, J. H. 1977. *Le Zagros.* Mémoires de la Société Géologique de France Série, **8**, 33–52.

ROBERTSON, A. H. F. 1987. Upper Cretaceous Muti Formation transition of a Mesozoic platform carbonate to a foreland basin in the Oman Mountains. *Sedimentology,* **4**, 1123–1142.

SETUDEHNIA, A. 1978. The Mesozoic sequence in southwest Iran and adjacent areas. *Journal of Petroleum Geology,* **1**, 3–42.

SHAKIB, S. S. 1994. Paleoenvironmental and biostratigraphic significance of foraminiferal associations from the Early Cretaceous sediments of southwest Iran. *In*: SIMMONS, M. D. (ed.) *Micropaleontology and Hydrocarbon Exploration in the Middle East.* Chapman & Hall, London, 127–150.

SHARLAND, P. R., ARCHER, R. *ET AL*. 2001. Arabian plate sequence stratigraphy. GeoArabia, Special Publications, **2**, 371.

SHERKATI, S. & LETOUZEY, J. 2004. Variation of structural style and basin evolution in the central Zagros (Izeh zone and Dezful Embayment), Iran. *Marine Petroleum Geology*, **21**, 535–554.

SHERKATI, S., MOLINARO, M. & FRIZON DE LAMOTTE, D. 2005. Detachment folding in the central and eastern Zagros fold trust belt (Iran): Salt mobility, multiple detachments and late basement control. *Journal of Structural Geology*, **27**, 1680–1696.

STOCKLIN, J. 1968. Structural history and tectonics of Iran: a review. *AAPG Bulletin*, **52**, 1229–1258.

STOCKLIN, J. 1974. Possible ancient continental margins in Iran. *In*: BURK, C. A. & DRAKE, C. L. (eds) *The Geology of Continental Margins.* Springer, New York, 873–887.

STOCKLIN, J. 1977. *Structural correlation of the Alpine ranges between Iran and central Asia.* Société Géologique de France, Mémoire Hors-Série, **8**, 333–353.

STONELEY, R. 1974. Evolution of the continental margin bounding a former southern Tethys. *In*: BURK, C. A. & DRAKE, C. L. (eds) *The Geology of Continental Margins.* Springer, New York, 889–903.

STONELEY, R. 1975. On the origin of ophiolite complexes in the southern Tethys region. *Tectonophysics*, **25**, 303–322.

STONELEY, R. 1981. The geology of Kuh-e-Dalneshin area of southern Iran and its bearing on the evolution of southern Tethys. *Journal of the Geological Society, London*, **138**, 509–526.

SZABO, F. & KHERADPIR, A. 1978. Permian and Triassic stratigraphy, Zagros basin, southwest Iran. *Journal of Petroleum Geology*, **1**, 57–82.

TAATI QORAYEM, F., VAN BUCHEM, F. & RAZIN, P. 2003. High resolution sequence stratigraphy of the Bangestan Group in a tectonically active setting (Dezful–Izeh) Zagros–Iran. AAPG International Conference, Barcelona, Spain, Abstract.

TAKIN, M. 1972. Iranian geology and continental drift in the Middle East. *Nature*, **235**, 147–150.

VAIL, P. R., AUDEMARD, F., BOWMAN, S. A., EISNER, P. N. & PEREZ-CRUZ, C. 1991. The stratigraphic signatures of tectonics, eustacy and sedimentology— an overview. *In*: EINSELE, G., RICKEN, W. &

SEILACHER, A. (eds) *Cycles and Events in Stratigraphy.* Springer, Berlin, 617–659.

VAN BUCHEM, F. S. P., RAZN, P. *ET AL*. 1996. High resolution sequence stratigraphy of the Natih Formation (Cenomanian/Turonian) in Northern Oman: distribution of source rocks and reservoir facies. *GeoArabia*, **1**, 65–91.

VAN BUCHEM, F. S. P., LETOUZEY, J. *ET AL*. 2001. The petroleum systems of the Dezful Embayment and Northern Fars (S. W. IRAN). NIOC–IFP Joint Study Research Project 2000. Internal Report of Institut Français du Pétrole (IFP).

VAN BUCHEM, F. S. P., RAZIN, P., HOMEWOOD, P. W., OTERDOOM, H. & PHILIP, J. 2002a. Stratigraphic organization of carbonate ramps and organic-rich intrashelf basins: Natih Formation (middle Cretaceous) of northern Oman. *AAPG Bulletin*, **86**, 21–54.

VAN BUCHEM, F. S. P., BAGHBANI, D. *ET AL*. 2006. Aptian organic-rich intra-shelf basin creation in the Dezful Embayment (Kazhdumi and Dariyan formations, SW Iran). AAPG Annual Meeting, Houston, TX. Abstract.

VAN WAGONER, J. C., POSAMENTIER, H. W., MITCHUM, R. M., VAIL, P. R., SARG, J. F., LOUTIT, T. S. & HARDENBOL, J. 1988. An overview of the fundamentals of sequence stratigraphy and key definitions. *In*: WILGUS, C. K., POSAMENTIER, H., HASTINGS, B. S., VAN WAGONER, J., ROSS, C. A. & KENDALL, C. G. St. C. (eds) *Sea-level Change: an Integrated Approach*. Society of Economic Paleontologists and Mineralogists, Special Publications, **42**, 39–45.

WEAVER, P. P. E. & KUIJPERS, A. 1983. Climatic control of turbidite deposition on the Madeira abyssal plain. *Nature*, **306**, 360–363.

WELLS, A. 1969. The crush zone of the Iranian Zagros Mountains and its implications. *Geological Magazine*, **106**, 385–394.

WYND, J. G. 1965. *Biofacies of the Iranian Oil Consortium Agreement area*. Iranian Oil Operating Companies, Geological and Exploration Division, Tehran, Report, **1082**.

YOUNG, K. 2000. Depositional processes of submarine debris flows in the Miocene fan deltas, Pohang Basin, SE Korea with special reference to flow transformation. *Journal of Sedimentary Research*, **70**, 491–503.

ZAMBETAKIS-LEKKAS, A. & KEMERIDOU, A. 2006. New data on the palaeobiogeography of *Loftusia* genus (Foraminiferida). An *in situ* presence of the genus in eastern Greece (Boeotia). *Comptes Rendus Géosciences*, **338**, 632–640.

Subaerial exposure and meteoric diagenesis of the Cenomanian–Turonian Upper Sarvak Formation, southwestern Iran

E. HAJIKAZEMI[1]*, I. S. AL-AASM[1] & M. CONIGLIO[2]

[1]*Department of Earth and Environmental Sciences, University of Windsor, Windsor, Ontario N9B 3P4 Canada*

[2]*Department of Earth and Environmental Sciences, University of Waterloo, Waterloo, Ontario N2L 3G1, Canada*

*Corresponding author (e-mail: hajikaz@uwindsor.ca)

Abstract: The Sarvak Formation (Cenomanian–Turonian) forms one of the main reservoir rocks in many oilfields in southern Iran. Extensive lateral and vertical facies variations as well as effects caused mainly by the subaerial exposure associated with the regional Turonian unconformity have resulted in variable porosity and permeability. Dissolution affected the entire upper part of the Sarvak Formation and is the most important process related to subaerial exposure. Brecciation, development of palaeosol and formation of bauxite deposits are also limited to the upper few metres of the top of the Sarvak Formation and indicate warm and humid climatic conditions. Subaerial exposure had varying effects on the diagenesis depending on its duration, palaeotopography and the availability of meteoric water. The $\delta^{18}O$ and $\delta^{13}C$ values obtained from calcitic matrix, various generations of calcite cements and calcitic rudist shells in the Upper Sarvak overlap to a large extent, indicating their equilibration with fluids of similar isotopic composition. Negative $\delta^{18}O$ values (i.e. $-6.6‰$ to $-1.7‰$) suggest a significant meteoric component. More ^{18}O-depleted values (e.g. $-12.3‰$) obtained from late calcite cements indicate their precipitatation from warm fluids. Positive $\delta^{13}C$ values (i.e. $0.00‰$ to $3.4‰$) in the various carbonate phases reflect values of seawater coeval with an Oceanic Anoxic Event and later modified by meteoric waters.

The Middle Cretaceous Sarvak Formation in southern Iran contains more than 20% oil-in-place of Iranian oil reservoirs, forming the second most important reservoir rocks after the Asmari Formation of the Oligo-Miocene. In addition to its petroleum significance, the Sarvak Formation forms productive groundwater aquifers feeding karstic springs in the Zagros region (Raeisi & Karami 1997; Ahmadipour 2002).

Sarvak carbonates were deposited in a ramp environmental setting with extensive lateral and vertical facies variations (Setudehnia 1978). These carbonates were affected by a major unconformity (Turonian) that involved extensive erosion and karstification. Therefore, the variability and extent of diagenesis and the distribution of diagenetic phases could be due to the presence of the unconformity and associated karstification. This unconformity has been reported in several localities in southern Iran as well as elsewhere in the Persian Gulf region (James & Wynd 1965; Harris *et al.* 1984; Setudehnia 1978).

Karstification of carbonate rocks by meteoric water during subaerial exposure is an important diagenetic process that contributes to the development of petroleum reservoirs by enhancing their porosity and permeability (e.g. Wright 1991; Lousk 1999; Wang & Al-Aasm 2002; Fu *et al.* 2006). Karstified systems are commonly affected by later diagenesis including cementation.

Karstic features have great implications for sequence stratigraphy, palaeoenvironmental interpretation and, more importantly, the development of hydrocarbon reservoirs. However, identification of these features may be problematical in some instances in drilled cores. The high porosity and permeability of the karstified carbonates and the importance of palaeokarst processes in developing porosity have been widely documented in the literature (Mylroie & Carew 1995; Lousk 1999; among many others). Karstification could result in modification and redistribution of porosity and permeability (Choquette & James 1988; Hopkins 1999), often significantly complicating exploration for oil and gas, as well as delineation of reservoir characteristics.

The aim of this investigation is to identify diagenetic processes in the Upper Sarvak Formation, including extensive karstification, that were mainly controlled by subaerial exposure. These processes also greatly influenced reservoir quality. To obtain

From: LETURMY, P. & ROBIN, C. (eds) *Tectonic and Stratigraphic Evolution of Zagros and Makran during the Mesozoic–Cenozoic*. Geological Society, London, Special Publications, **330**, 253–272.
DOI: 10.1144/SP330.12   0305-8719/10/$15.00 © The Geological Society of London 2010.

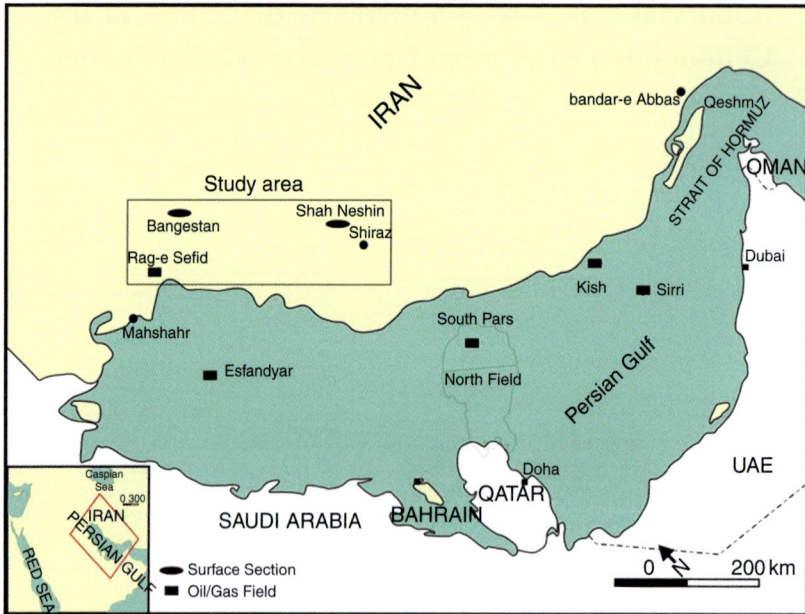

Fig. 1. Map of SW Iran and Persian Gulf showing the location of the study area. Inset map at bottom left shows regional location.

a better understanding of the diagenetic processes, petrographic and stable carbon and oxygen isotope results of the Upper Sarvak carbonates from surface sections and subsurface cores from southwestern Iran have been integrated (see Fig. 1 for location of the study area). Data obtained from these investigations have been used to characterize the diagenetic fluid(s) involved. In addition, these data have been utilized to substantiate the greenhouse conditions and amplification of karst formation and the evolution of porosity in the Upper Sarvak Formation.

Previous study of the Sarvak Formation

The stratigraphy and sedimentology of the main lithostratigraphic units of the Zagros region in SW Iran are now well established (Alavi 2004). However, understanding the various petroleum systems in the region requires palaeogeographical reconstructions of these units, focusing on their regional and local lithofacies variations.

The Middle Cretaceous Sarvak Formation of the Bangestan Group (i.e. Albian–Santonian) in the Zagros Mountains was deposited in a ramp setting during Cenomanian to Early Turonian time (Setudehnia 1978; Taghavi *et al.* 2006). Stratigraphically, this formation is equivalent to the Mauddud, Ahmadi and Mishrif formations in the Persian Gulf (Ghazban 2007), the Wara, Ahmadi

and Mishrif formations in Iraq and Saudi Arabia (Fig. 2) and the Natih Formation in Oman (van Buchem *et al.* 2002).

James & Wynd (1965) conducted the first systematic stratigraphic investigation of the Zagros Mountains including the Cenomanian–Turonian Sarvak Formation. Their work was followed by that of Setudehnia (1978), who reviewed the geology of the formation and examined depositional facies and their cyclicity in southwestern Iran.

Boltz (1978) recognized the deposition of the Sarvak carbonates within a restricted intrashelf basin in Dezful Embayment and suggested a significant tectonic influence on general sedimentation patterns.

Hajikazemi *et al.* (2002) demonstrated that much of the variability in diagenetic features and their spatial distribution in the Upper Sarvak Formation in the Sirri oilfields in the northeastern Persian Gulf could be related to the presence of a major Turonian unconformity and associated karstification.

Taghavi *et al.* (2006) investigated the Upper Sarvak Formation in the Dehluran Field (*c.* 120 km NW of Ahwaz) and identified three major transgressive–regressive sequences. According to those workers the diagenetic processes, especially the regressive parts of each major sequence, were controlled by sea-level fluctuations.

Beiranvand *et al.* (2007) recognized five facies association in the Sarvak Formation and identified

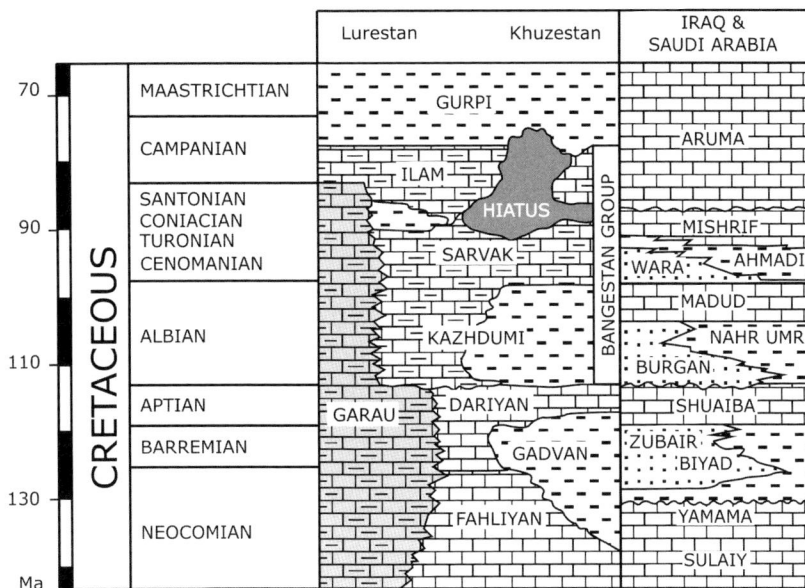

Fig. 2. Simplified table of Cretaceous rock units from southwestern Iran (after Bordenave 2002).

different reservoir rock types with significant vertical and lateral heterogeneities in a giant oilfield near the northern head of the Persian Gulf.

Farzadi & Hesthmer (2007) integrated the 3D geophysical and well-log data from the Sirri oilfields and introduced the volume-based interpretation technique as a suitable method for detailed stratigraphic interpretation of the Turonian palaeokarst in the Sarvak carbonates.

Based on these investigations it is apparent that lateral and vertical facies variations are important stratigraphic characteristics of the Upper Sarvak Formation. Furthermore, the major Turonian unconformity, which represents deep erosion and diagenesis, greatly influenced the development of porosity and determined the reservoir characteristics in various areas. Despite some important findings, the diagenetic processes and evolution of these heterogeneous carbonate reservoir rocks in southern Iran are not well understood and warrant further investigation.

Regional geological setting

The Permian to Cretaceous evolution of southern Iran and the Persian Gulf region ended with the establishment of an extensive carbonate shelf as the region became a passive margin of the Neo-Tethys Ocean (e.g. Murris 1980; Harris et al. 1984; Van Buchem et al. 2002). The Early Cretaceous was a period of maximum rifting of the Neo-Tethys Ocean, whereas the major phase

of tectonics occurred in the Late Cretaceous, resulting in the closure of the Neo-Tethys Ocean accompanied by major continental collision.

The Middle Cretaceous sedimentary succession that includes the Sarvak Formation was dominated by widespread carbonate accumulation with an extensive development of carbonate platforms and rudist–foraminifer bioherms occurring widely throughout the Neo-Tethys Ocean (Burchette & Wright 1992; Droste & Van Steenwinkel 2004). In the early Cenomanian, benthic foraminifers and rudists were widely distributed throughout the region and the latter formed rudist-debris rimmed shelves (Burchette & Britton 1985).

During the Cenomanian–Turonian, sedimentation rates varied considerably, from low deposition (e.g. NE Persian Gulf) to accumulation of thick successions (e.g. Sarvak type section in the study area). This variability in sedimentation rate is evident when considering the total thickness of the formations and the absence or very low thickness of some microfacies, especially the open marine microfacies of the lower Sarvak, in the southern Iran and the Persian Gulf region. The variable thicknesses are evident from the isopach map of the middle Cretaceous for the sedimentary basin (Fig. 3). The thickness variation suggests a period of instability, characterized in some areas by active salt tectonism (Burchette & Britton 1985). Thickness variations within the Sarvak Formation between some of the drilled wells in the Persian Gulf area could also be due to erosion at the

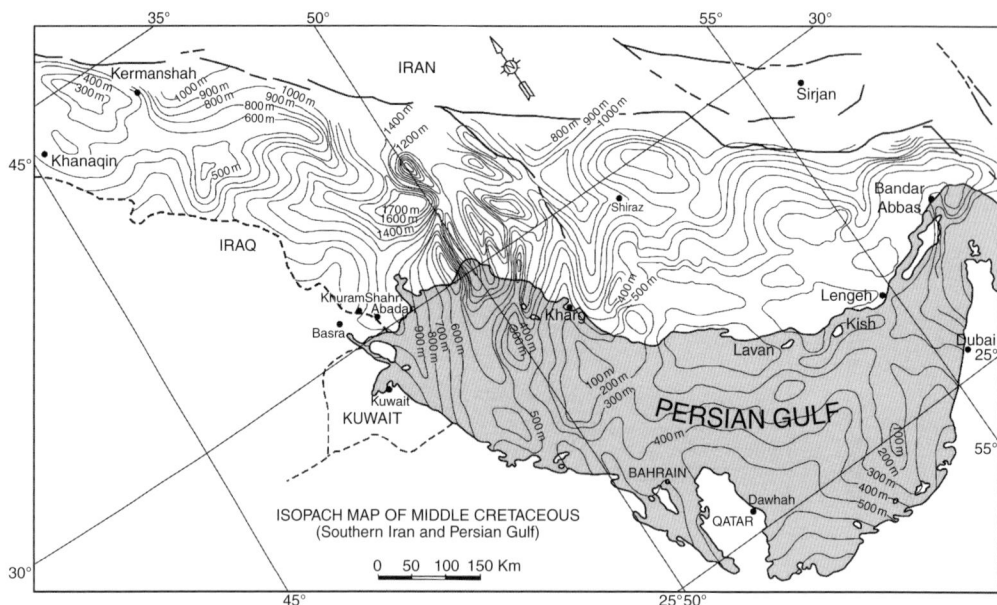

Fig. 3. Isopach map of Middle Cretaceous carbonate sedimentary rocks in southern Iran and Persian Gulf region (modified from Motiei 1993).

Turonian unconformity or variable subsidence rates in different parts of the basin (Ghazban 2007).

The Middle to Late Cretaceous was also a period of major environmental changes with global sea levels steadily rising (Sharland *et al.* 2001). This is indicated by the early Turonian global maximum flooding surface (Haq *et al.* 1988). The Sarvak carbonates were deposited when the global sea level was at a highstand. Following the deposition of the Upper Sarvak units, a major sea-level fall in mid-Turonian time exposed these carbonates and resulted in a regional Turonian unconformity. This widespread unconformity is recorded by rubbly breccia, karstic topography and hematite nodules capping Cenomanian platform carbonates (Motiei 1993) in southern Iran. It resulted from both a combination of localized uplift, following the initiation of ophiolite obduction, and global eustatic sea-level fall (James & Wynd 1965; Setudehnia 1978; Motiei 1993; Ghazban 2007). In addition, there is evidence for the presence of a local erosional unconformity at the Cenomanian–Turonian boundary, observed in drilled wells and outcrops of the Sarvak Formation examined in southwestern Iran (e.g. Beiranvand *et al.* 2007).

Sea-level fluctuation left a recognizable imprint on sedimentation patterns; however, local tectonic controls, notably salt tectonics and diapirism, were also of significant importance (Burchette 1993).

Stratigraphy of the Sarvak Formation

The Sarvak Formation is named after Tang-e-Sarvak (Sarvak gorge) in the central part of the southern flank of Bangestan Mountain in southwestern Iran. At its type section, the Sarvak Formation consists of three limestone units reaching a maximum thickness of 821 m (Motiei 1993). The lower boundary of the Sarvak Formation conformably overlies the Kazhdumi Formation with a transitional contact, whereas the nature of the upper contact, which is the Turonian unconformity, is variable. For example, in the type locality marls of the Gurpi Formation (Maestrichtian) overlie the Sarvak Formation, whereas in the Dezful Embayment the Sarvak carbonates and the overlying Ilam Formation (Santonian–Campanian) form an extensive continuous limestone unit, which appears to have been deposited in a shallow-marine environment.

The lower part of the Sarvak Formation (254.5 m thick) consists of argillaceous micritic limestones with lenticular bedding and thin-layered marl interbeds. The middle part of the succession (524.5 m thick) includes massive chalky limestones with iron-rich siliceous carbonate concretions and rudist fragments. Cross-bedding is evident in a number of horizons. The top 42 m of the formation includes massive limestones with the topmost strata consisting of a weathered brecciated ferruginous limestone (Fig. 4).

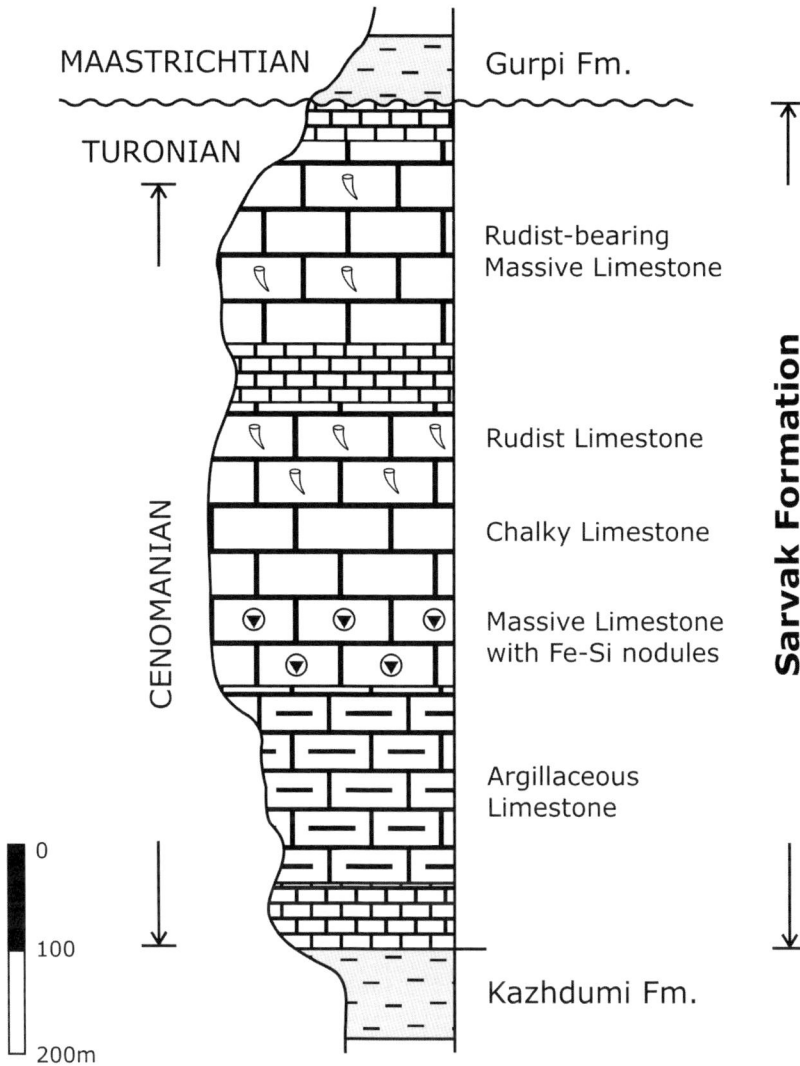

Fig. 4. General stratigraphic column of the Sarvak Formation at its type section in the Bangestan Mountain in southwestern Iran (modified from Motiei 1993).

The formation is interpreted to have been deposited in a shallow-marine setting during the Cenomanian–Turonian. Three third-order sequences have been distinguished within the Upper Sarvak Formation. Van Buchem *et al.* (2002) interpreted these sequences in the Natih Formation (Upper Sarvak equivalent in Oman) as an alternation of two types of depositional systems, a mixed carbonate–clay ramp with benthic foraminifers as the dominant fauna at the base and a carbonate-dominated ramp bordering an intrashelf basin with rudists in a mid-ramp environment and organic-rich carbonates representing the basinal facies

forming the top of each sequence. Van Buchem *et al.* invoked eustatic control as the main mechanism for the third-order sequences.

Material and methods

This study is based on petrographic investigation of 360 representative samples derived from detailed field study of the Sarvak Formation type section, the Upper Sarvak Formation from another surface section at the Shahneshin Mountain area 90 km NW of Shiraz and samples from two cored wells in the Rag-e Sefid oilfield in southwestern Iran (see Fig. 1 for study area). The type section and cores were logged and sampled based on facies and/or diagenetic variations.

Petrographic thin sections were stained with Alizarin red-S and potassium ferricyanide and carefully analysed by standard optical transmitted light, fluorescence and cathodoluminescence (CL) microscopy, as well as by scanning electron microscopy (SEM). The CL microscopy of 98 representative carbonate samples was carried out using a Technosyn 8200 MKII model cold CL stage with a 12–15 KV beam and a current intensity of 420–430 μA on the unstained halves of the uncovered thin sections.

Carbonate rocks were classified according to Dunham's modified carbonates classification (Wright 1992). In addition, 48 core samples from mainly the upper most part of the formation were impregnated with an epoxy resin mixed with a blue dye for porosity identification.

Eighty samples of the least altered carbonate matrix (i.e. micrite), various generations of calcite cements and rudist shells were micro-sampled using a miscoscope-mounted dental drill assembly. Extracted powders were reacted with 100% pure phosphoric acid at 25 °C for 4 h (Al-Aasm *et al.* 1990). The evolved CO_2 gas was analysed for oxygen and carbon isotopic ratios using a Finnigan Mat Delta Plus mass spectrometre. All analyses for oxygen and carbon isotopes are reported in per mil (‰) notation relative to the Pee Dee Belemnite (PDB) standard. Precision for both isotopes was better than 0.05‰.

Depositional environment

Based on field and detailed petrographic investigations four main depositional environments were identified in the ramp setting in the Upper Sarvak Formation using the Burchette & Wright (1992) classification, comprising the inner-ramp, mid-ramp, outer-ramp, and open marine or basinal environments. Each environment is characterized by several microfacies.

Inner ramp

This environment can be subdivided into lagoonal and shoal or barrier environments in the lagoonal environment; benthic foraminiferal wackestone and packstones are the dominant microfacies. Miliolids, *Textularia*, *Praealveolina*, *Nezzazata* and *Orbitolina* are the main skeletal grains. Echinoderms and gastropods are of lesser importance. Rudist biostromes are seen in surface sections and can be identified as rudist floatstone or rudstone in core samples. Sedimentary features such as burrowing and geopetal fabric are abundant in this environment.

The shoal or barrier environment is a high-energy environment with grainstones as the dominant microfacies. Benthic foraminifer grainstone, rudist–echinoderm grainstones, rudist rudstones and peloidal–bioclastic grainstones or packstones are the main microfacies representing this environment.

Mid-ramp

The mid-ramp environment is mainly represented by bioclastic wackestone to packstones. Bioclasts include benthic foraminifers (*Praealveolina*, *Dicyclina*, *Ovalveolina* and miliolids) and fragments of rudists and gastropods. Burrows, usually filled with calcite cement, are the most common sedimentary structure in this facies. At the Sarvak type locality this facies includes cross-bedding, which is a good indicator of a higher energy environment. This depositional environment appears to be restricted mainly to the Upper Sarvak Formation.

Outer ramp

The outer-ramp environment is composed of metre-scale-bedded mudstone to wackestone with less variety of foraminifers (*Praealveolina* and *Orbitolina*), some gastropod and echinoderm fragments, sponge spicules and rarely pelagic foraminifers. This environment is more dominant in the lower Sarvak Formation but also occurs in the lower part of the Upper Sarvak Formation.

Basinal environment

The basinal environment is represented by organic-rich wackestones and packstones in the lowermost part of the Upper Sarvak Formation. *Hedbergella washitensis* and various oligosteginids are the main skeletal grains in this microfacies, indicating a low-energy open-marine depositional environment.

Based on wells and outcrops, intervals representing each sedimentary environment with their corresponding microfacies have variable thicknesses ranging from tens to hundreds of metres obtained from different wells and surface sections. Local thickness variations and the presence or absence of each microfacies largely depend on topography and palaeogeography of the basin at the time of deposition. Several factors including sea-level change and salt movements (diapirism) affected basinal configurations and topography of the Cenomanian–Turonian basin in the area (Videtich *et al.* 1988).

Diagenesis

The Upper Sarvak Formation has been subjected to extensive and variable diagenetic modifications

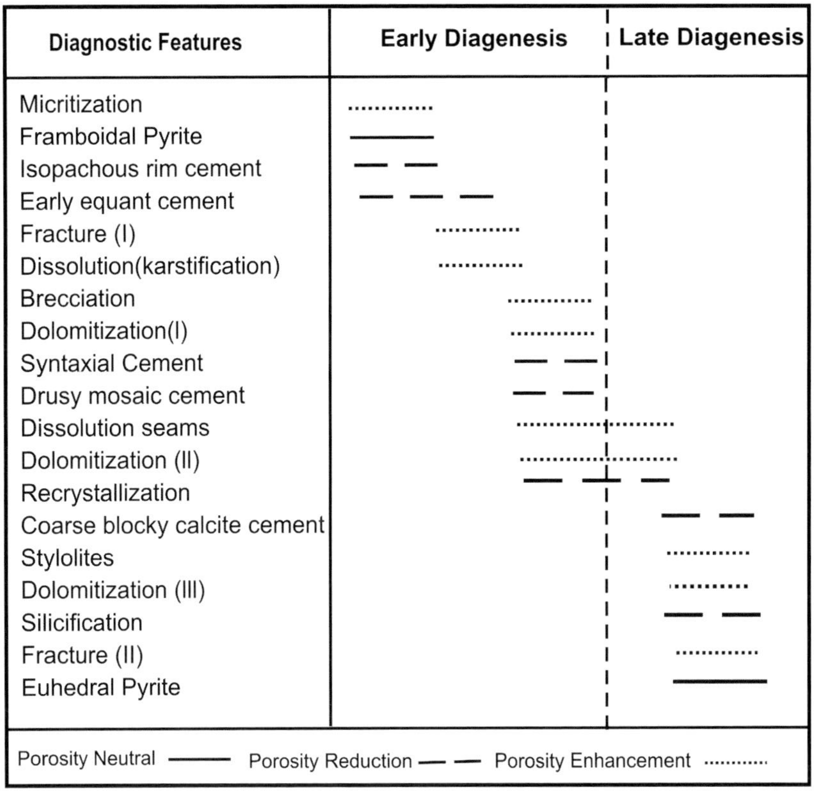

Diagnostic Features	Early Diagenesis	Late Diagenesis
Micritization	············	
Framboidal Pyrite	──	
Isopachous rim cement	─ ─	
Early equant cement	─ ─ ─	
Fracture (I)	············	
Dissolution(karstification)	············	
Brecciation	············	
Dolomitization(I)	············	
Syntaxial Cement	─ ─	
Drusy mosaic cement	─ ─	
Dissolution seams	············|············	
Dolomitization (II)	············|············	
Recrystallization	─ ─ ─ ─	
Coarse blocky calcite cement		─ ─
Stylolites		············
Dolomitization (III)		············
Silicification		─ ─
Fracture (II)		············
Euhedral Pyrite		──

Porosity Neutral ──── Porosity Reduction ── ── Porosity Enhancement ············

Fig. 5. Paragenetic sequence of the diagenetic processes affecting the Upper Sarvak Formation.

after its deposition in a shallow-marine environment, during subaerial exposure and subsequent burial. Major diagenetic processes affecting the Upper Sarvak carbonates include micritization, dissolution, karstification, compaction, dolomitization, calcite cementation, recrystallization, silicification and pyrite formation (Fig. 5).

Micritization

Micritization identified here is an early diagenetic process. Micritic envelopes surround some rudist and echinoderm fragments, and wholesale micritization of some benthic foraminifers in bioclastic packstone microfacies makes the recognition of skeletal grains difficult in some intervals.

Dissolution

Dissolution and associated karstification was the most important diagenetic process to affect the porosity and permeability in the Upper Sarvak carbonates, especially in grain-supported intervals. According to well data, porosity in the Upper Sarvak Formation varies from 0.5 to more

than 14%. However, the lower part of the Sarvak Formation has less than 1% porosity. Porosity is mainly secondary in nature and dominated by vuggy, mouldic and intercrystalline types. Vugs are sub-spherical or irregular in shape, and are up to several millimetres in size in core samples and a few centimetres in surface sections. The vugs are partially or completely filled by ferroan calcite cement (Fig. 6a). Cavernous porosity is also observed in upper part of the Sarvak Formation below the unconformity surface (Fig. 6b). Intercrystalline porosity occurs between calcite cement or dolomite crystals (Fig. 6c, d). Moulds of benthic foraminifers and gastropods are totally or partially filled with calcite cement, whereas bitumen fills some of the vuggy and intercrystalline porosities (Fig. 6e, f). Primary porosity is of lesser importance and mostly includes intraparticle porosity. The body cavity of the rudists forms the most important intraparticle porosity (Fig. 6g, h). The microfacies representing the shoal or barrier deposits, especially rudist grainstone or rudstone, are the most porous intervals in the Upper Sarvak Formation, and they contain both primary and secondary porosity.

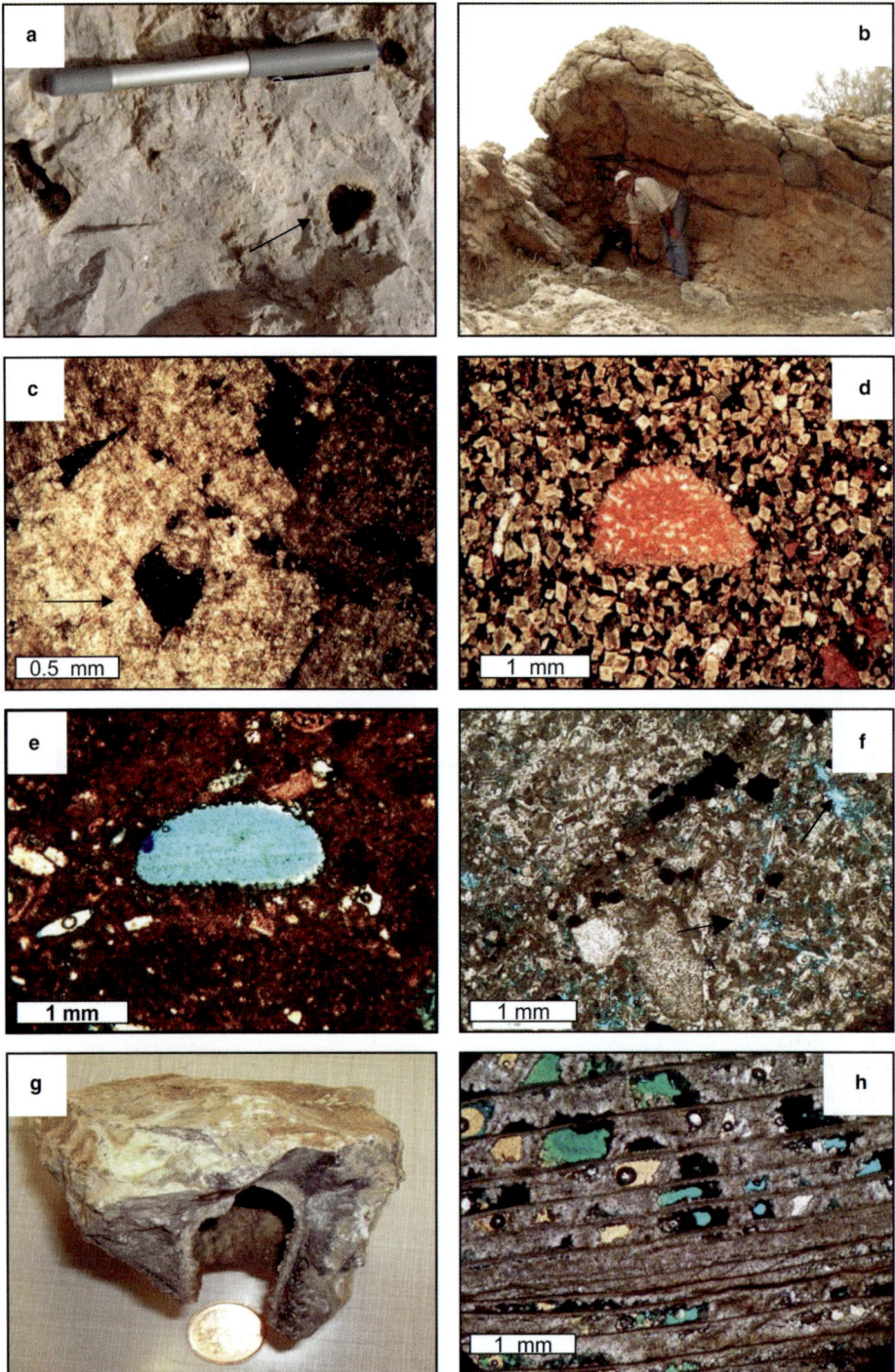

Fig. 6. Field photograph and thin-section photomicrographs of various porosity types in the Upper Sarvak Formation. (**a**) Vuggy porosity partially filled with rusty-weathering, ferroan calcite cement. (**b**) Cavernous porosity in the Upper Sarvak Formation, Shahneshin Mountain. (**c**) Intercrystalline porosity in calcite cement. (**d**) Intercrystalline porosity in a matrix selective dolomite in *Orbitolina* wackestone or packstone filled with organic matter. (**e**) Mouldic porosity in bioclastic wackestone filled with blue dye. (**f**) Interparticle porosity filled with blue dye. (**g, h**) Preserved primary intraparticle porosity in rudist shells.

Karst formation

Karstification can occur in various diagenetic realms, including meteoric (via dissolved CO_2 releasing carbonic acid), deep burial (thermal regime associated with deep CO_2 or H_2S production) and in mixing zones (mixing of fresh water and seawater in coastal areas) (Ford 1988; Wright 1991; Ford and Williams 2007). The effects discussed here are mainly related to meteoric dissolution. This is based on the presence of the prominent unconformity surfaces in the Upper Sarvak Formation, resulted from regional exposure associated with the Turonian unconformity and local exposure of some areas at the Cenomanian–Turonian boundary. Karst development in the Upper Sarvak Formation occurred continuously on emergent portions of the ramp, especially in the inner ramp environment.

Subaerial exposure-related features in the Upper Sarvak Formation seen in the Shahneshin Mountain surface section include: brecciation of carbonates; presence of cavernous porosity; boxwork texture showing extensive carbonate dissolution; erosional surfaces and incised channels filled with pisoids and clays; development of palaeosol and bauxite.

Breccia related to dissolution of carbonate rock is widespread in this area. Breccia fragments are angular to sub-angular, varying in size from 2 mm to tens of centimetres. The breccia is either matrix-supported with calcite matrix being stained with Fe-oxides or grain-supported with bituminous calcite matrix (Fig. 7a, b). This breccia is interpreted to have resulted from dissolution widening associated with karstification and collapse of the overlying beds rather than being related to faults or major fractures. The brecciation observed in surface sections and core samples from Rag-e Sefid oilfield provides strong support for our interpretation of subaerial exposure and karst development.

Boxwork texture (also known as honeycomb texture; Sweeting 1973, p. 142) is a palaeokarst feature consisting of a network of thin blades of calcium carbonate projecting out from the rock surface. The blades enclose irregularly shaped and variably sized spaces, and the overall appearance is that of sets of open, connected spaces (Fig. 7c). Boxwork texture forms in meteoric diagenetic environments.

A scoured surface occurs on the top of the Sarvak Formation. In a 2D view the erosional feature appears as flat-topped U-shaped depressions, typically 2–3 m wide and with maximum depths ranging from 1 to 2 m (Fig. 7d). However, based on their lateral extension in 3D views, the scoured surfaces have channel morphologies extending at least for 30 m on the outcrop scale. The eroded surface appeared to be the highly karstified limestones, later covered by fluvial deposits. Material filling the channels is composed mainly of black pisolites within a matrix composed of brown clay and silt-size calcitic sediments (Fig. 7e) and contained within the channel. Similar erosional surfaces have been reported within the Middle Cretaceous Natih Formation of Oman. Emersion features and erosional surfaces have been observed and interpreted to have developed during Turonian sea-level fall and filled during the subsequent transgression (Van Buchem et al. 1996, 2002).

Palaeosols in the form of bauxite and laterite developed in the top 10–50 cm of the Sarvak Formation at Shahneshin Mountain (Fig. 7f). They overlie pisolites along the exposure surface. Although only one body of bauxite has been mined, many smaller masses, up to 3 m thick, are found in the region. The mined bauxite horizon, in the Firoozabad area near Shiraz, is c. 30 m thick covering an area of about 1 km², and is developed along the contact between Sarvak and Ilam formations (Liaghat & Hosseini Zarasvandi 2003). The relatively small thickness of most bauxite layers may indicate a short duration of exposure, considering the humid climate at the time of their development.

Bauxite and laterite found in the study area could be classified as karst bauxite (Bardossy 1982). They are perhaps the most persuasive and clear evidence of emergence, subaerial exposure and karstification of the Upper Sarvak Formation.

Clearly, the breccia, the occurrence of bauxite and laterites, palaeosol and extensive erosional truncation and karstification must have formed prior to the deposition of the overlying formation. The existence of warm and humid climatic conditions at this time would have intensified the hydrological cycle. Under such conditions carbonate dissolution and karstification could be increased below the unconformity surface, enhancing the overall porosity. This could be the main reason for porosity enhancement up to 15% in some intervals obtained from well data and surface section samples.

Compaction

Both mechanical and chemical compaction are observed in the Upper Sarvak Formation. Mechanical compaction resulted in porosity reduction by breakage of the grains in packstones. Dissolution seams and stylolites are among the most common chemical compaction features in most of the rock units within the Sarvak succession (Fig. 7a).

Recystallization

Partial recrystallization of the calcitic matrix occurs in some intervals (Fig. 8a). Recrystallization has also been locally important in modifying the internal

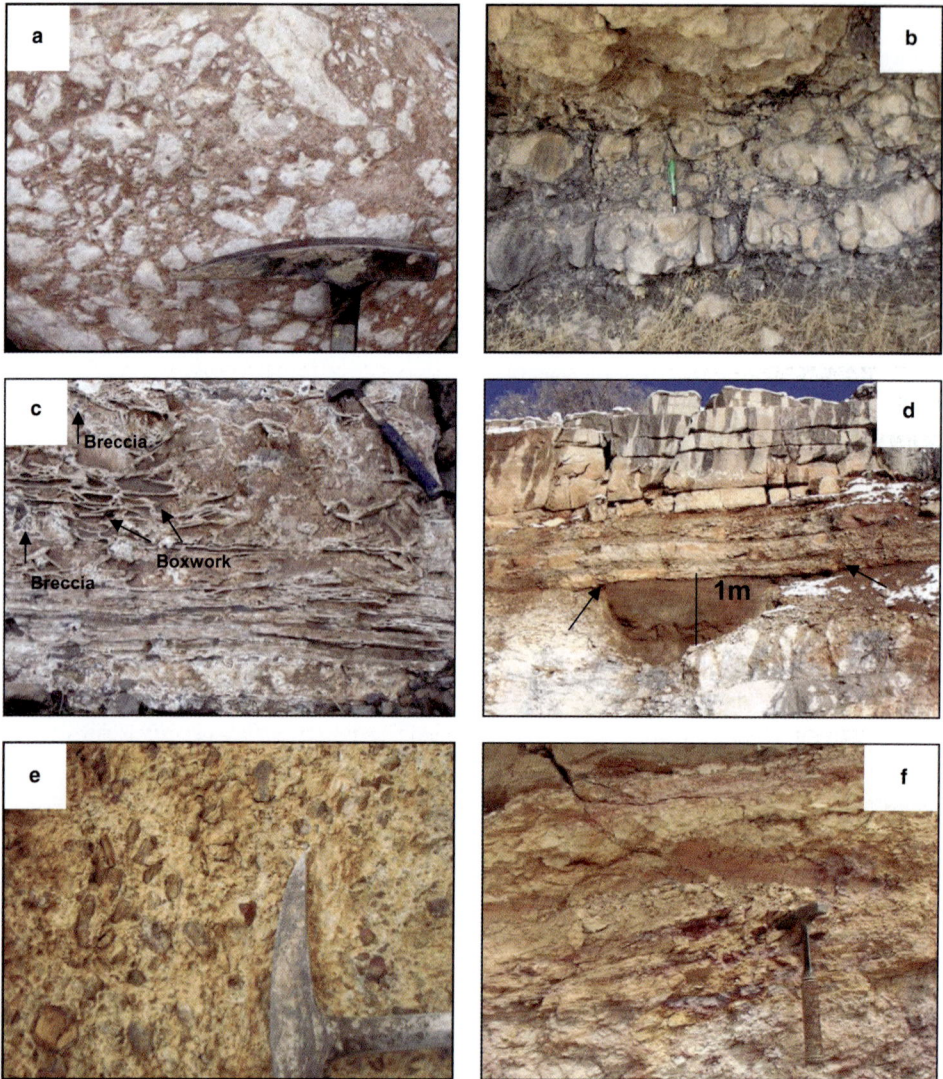

Fig. 7. Field photographs of karst-controlled diagenetic features in the Upper Sarvak Formation (Shahneshin Mountain). (**a**) Chaotic, solution-collapse breccia within the Sarvak Formation cemented by Fe-stained carbonate matrix. (**b**) Solution-collapse breccia with bituminous carbonate cement on top of the Sarvak Formation. (**c**) Boxwork texture with fragments of Sarvak Formation cemented by fringing calcite. Calcite blades form a honeycomb or network of open space. (**d**) Subaerial exposure surface (arrows) and channel erosion of the Upper Sarvak Formation. (**e**) Ferroan pisolites filling the channel shown in (d). (**f**) Limonitic and reddish soil crust overlying pisolites along the exposure surface.

structure in some of bioclasts such as benthic foraminifers.

Fracturing

Several sets of fractures with variable orientations occur in the Upper Sarvak Formation, but two sets are more common: vertical fractures and sub-horizontal fractures. Vertical fractures appear to have a wider opening (more than 2 mm in some cases) compared with the second set and are partially or completely sealed with calcite cement (Fig. 8b). The second set of fractures cross-cuts the earlier set and is approximately horizontal. Apertures do not exceed a few micrometres and are devoid of cement. This second set also postdates formation

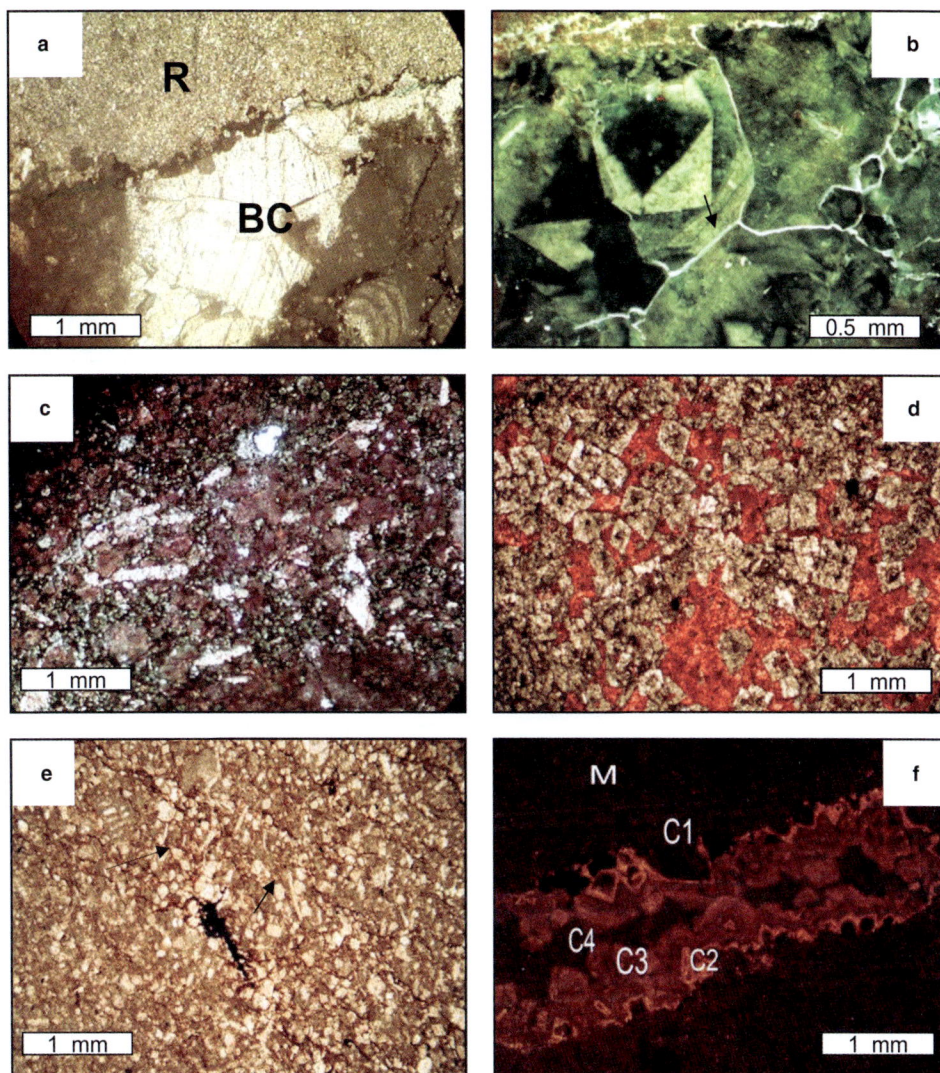

Fig. 8. Photomicrographs of diagenetic features in the Upper Sarvak Formation. (**a**) Partial recrystallization (R) of matrix and blocky calcite cement (BC) in bioclastic wackestone. (**b**) Fluorescence photomicrograph of equant calcite cement filling vertical fractures retaining the hairline fractures. (**c**) Dolomitic cement in dissolution breccia. (**d**) Dolomite with cloudy centre and clear rim (dolomite III) partially replacing matrix. (**e**) Clear dolomite partially replacing the calcitic matrix (dolomite II). (**f**) A CL view of calcite cement microstratigraphy in a fracture (dolomite I). M, carbonate matrix with limited pore space filled with luminescent orange calcite; C1, C2, C3, drusy mosaic calcite cement; C4, coarse blocky calcite cement.

of stylolites and dissolution seams. The origin(s) of these small-scale fractures is not completely understood. They could be related to minor tectonic events (e.g. salt tectonics) in the area. Large-scale fractures are seen in the uppermost part of the Sarvak Formation in surface sections, and have various trends. They are devoid of cement and seem to be the product of the dissolution caused by the influence of meteoric water during the subaerial exposure.

Dolomitization

Three types of dolomite are recognized in the Upper Sarvak Formation; (1) fine crystalline (<20 μm) dolomite cement (dolomite I) filling the interparticle porosity in brecciated intervals underlying the unconformity; (2) compaction-associated subhedral–euhedral fine- to medium-crystalline (40–60 μm) clear dolomite replacing carbonate matrix in association with dissolution seams (dolomite II); (3) compaction-associated euhedral medium-crystalline (>70 µm) dolomite with a cloudy centre and clear rim (dolomite III), replacing the carbonate matrix and mainly associated with stylolites (Fig. 8c–e). Dolomite II and III are more common in mud-supported microfacies such as bioclastic wackstones.

Calcite cementation

Five main types of calcite cements are recognized in the Upper Sarvak Formation; from earliest to latest these are: (1) fine-crystalline isopachous rim cement; (2) fine- to medium-grained equant calcite cement; (3) syntaxial overgrowth calcite; (4) drusy mosaic calcite; (5) blocky sparry calcite.

The fine-crystalline isopachous rim cement occurs around some of the skeletal grains growing into the primary intergranular pores in grainstones, preventing the compaction of grains. It appears non-luminescent under CL. Equant calcite cements partially fill some of the rudist and foraminifer chambers and primary pores in packstones or grainstones, reducing the porosity in some intervals. They consist of dark to non-luminescent crystals with thin bright luminescent rims surrounding them and projecting towards the centre of the pores. Syntaxial overgrowth calcite commonly surrounds echinoderm fragments in grainstone shoal deposits. Under CL, this type of cement appears to be non-luminescent. Drusy mosaic and blocky sparry calcite cement are the most abundant cements and fill the first generation of fractures, and partially or completely fill some moulds and primary intraparticle pores in some of the rudists.

A consistent CL zonal succession is recognized in the fracture-filling drusy mosaic calcite cements. Based on CL microstratigraphy this cement is divided into three subzones, which are referred to, from earliest to latest, as C1 to C3 (Fig. 8f).

The earlier generation of fracture-filling calcite cement (C1) is dull to non-luminescent with a small portion showing concentric zoning consisting of alternating bright, dull and non-luminescent bands.

The bright-rimmed luminescent portion (C2) reflects a notable change in composition and chemistry of fluids involved in its precipitation. The following cement generation is a zoned cement (C3) and shows dull red to orange concentric luminescence zoning.

Coarse blocky sparry calcite cement is the latest cement (C4) and it partially fills the remaining voids within the earlier sets of fractures and some rudist body cavities, and partially or completely occludes some vugs, hence reducing the porosity and reservoir quality. This calcite cement is commonly ferroan and characterized by dull to non-luminescent crystals. Two-phase fluid inclusions are visible in some of the larger crystals.

Discussion on cement stratigraphy

The non-luminescent characteristic of the isopachous rim cement, equant calcite cements and syntaxial cement indicates their precipitation under oxidizing conditions in either marine or meteoric vadose environments. The non-luminescent equant calcite crystals could be considered as both vadose meteoric or marine cement, and the bright rims surrounding the crystals could indicate the influence of reducing conditions of a phreatic meteoric environment (Moldovany & Lohmann 1984).

The CL characteristics observed in the drusy mosaic calcite cement (C1–C3), which fills fractures (I) throughout the Upper Sarvak Formation, could also signify meteoric water influence. The succession of non-luminescent to bright yellow rimmed luminescent cement (C1–C3) could be interpreted as meteoric or shallow burial in origin (Meyers 1974; Moldovany & Lohmann 1984; Carpenter & Lohmann 1989). However, the meteoric origin of these cements is further supported by the stable carbon and oxygen isotope results obtained from them (see below). Therefore, the cement succession is interpreted to reflect the effect of meteoric water on the entire Upper Sarvak Formation during subaerial exposure. The repeating of dark and bright banding of this cement might be due to sea-level fluctuation and periodic emplacement of the Upper Sarvak Formation in phreatic and vadose meteoric environments and changes in the available oxygen level. The dull luminescent blocky spar (C4) could be interpreted as burial cement. The two-phase fluid inclusions observed in this type of cement confirm its precipitation in an environment with ambient temperature greater than 50 °C.

Pyrite formation

Two types of pyrite morphology occur in the Upper Sarvak Formation: (1) fine-crystalline framboidal pyrite (pyrite I); (2) coarse-crystalline authigenic pyrite (pyrite II), which replaces either matrix or calcite cement in some intervals.

The framboidal morphology shows spheroidal clusters consisting of smaller, discrete, equant single crystals or microcrysts. Framboidal pyrite is interpreted to have formed during the early stages of diagenesis and under reducing conditions (Butler & Rickard 2000). It is important to note that some of the Fe-staining of the carbonates in the uppermost part of the Sarvak Formation might be the product of alteration of this pyrite. Authigenic pyrite (i.e. pyrite II) occurs as octahedral crystals, as well as cubic grains. Partial replacement of fracture-filling calcite cement with pyrite II in some intervals could indicate its occurrence during the last stages of diagenesis.

Silicification

Minor silica precipitation partially replaces some of the calcitic rudist shells and the latest fracture-filling blocky calcite cements, in the latter case retaining the shape of the original calcite crystals. The silica was probably provided by the dissolution of sponge spicules, which are common in the intervals representing the outer ramp environment of the Upper Sarvak Formation.

Stable isotope geochemistry

On the basis of petrography, single generations of cement, matrix and rudist shells were micro-sampled for their carbon and oxygen isotopic composition to further determine the diagenetic environments and their significance (Table 1 and Fig. 9).

Carbonate matrix

The $\delta^{13}C$ and $\delta^{18}O$ values of the carbonate matrix range from $-0.2‰$ to $2.8‰$ and -6.2 to $-0.8‰$, respectively. Compared with the postulated $\delta^{13}C$ and $\delta^{18}O$ values of Middle Cretaceous marine carbonates (Veizer et al. 1999; see Fig. 5 for values), the $\delta^{13}C$ values of the Upper Sarvak Formation carbonates reported here fall well within the range of marine carbonates, whereas the $\delta^{18}O$ values reflect a 1–2‰ depletion.

Calcite cements

Drusy mosaic calcite cements in the fractures (referred to as early cement in Table 1 and Fig. 9), display $\delta^{13}C$ and $\delta^{18}O$ values ranging from 0.00‰ to 3.4‰ and $-6.6‰$ to $-1.7‰$, respectively. The blocky calcite cement (referred to as late cement in Table 1 and Fig. 9) also displays extreme depletion in $\delta^{18}O$ compared with matrix and early cements, and $\delta^{13}C$ values show a significant overlap. The $\delta^{13}C$ and $\delta^{18}O$ values for the blocky calcite cement range from 0.4‰ to +2.4‰ and from $-12.3‰$ to $-4.4‰$, respectively. Compared with the matrix the $\delta^{13}C$ and $\delta^{18}O$ values obtained from calcite cements show a wider range.

Rudist shells

The $\delta^{13}C$ and $\delta^{18}O$ values of the rudist shells range from 0.9‰ to 2.1‰ and $-5.2‰$ to $-3.0‰$, respectively. The range of isotopic compositions of the rudist shells is similar to the range of carbonate matrix.

Palaeosol

Four samples of palaeosol were analysed and yielded $\delta^{13}C$ and $\delta^{18}O$ values of -5.7 to $-2.9‰$ and -6.1 to $-4.2‰$, respectively.

Discussion of stable isotope geochemistry

Oxygen isotopes

The $\delta^{18}O$ values of diagenetic carbonate phases are mainly controlled by fluid composition, temperature and water/rock ratios (Brand & Veizer 1981), and consequently the $\delta^{18}O$ values are expected to be reset by diagenetic alteration. The similarity of stable isotopic compositions of the drusy mosaic calcite cements (early cement), carbonate matrix and rudist shells reflects their equilibration with pore fluids of similar isotopic composition. In contrast, the blocky calcite cements (late cement) have $\delta^{18}O$ values that are relatively depleted.

In most Cretaceous carbonates studied by Scholle & Arthur (1980), the $\delta^{18}O$ data show significant depletion in $\delta^{18}O$ values at or near the Cenomanian–Turonian boundary. Such depletion corresponds closely to the major $\delta^{13}C$ maxima (see Scholle & Arthur 1980, Fig. 3). The majority of the cements and matrix analysed here displayed similar depletion in $\delta^{18}O$ values. Such depleted $\delta^{18}O$ values in conjunction with CL and petrographic observations, in association with extensive karstification, suggest an important meteoric water influence.

Meteoric diagenesis of marine carbonates typically results in a shift toward more negative values of both $\delta^{13}C$ and $\delta^{18}O$ (e.g. Al-Aasm & Veizer 1986). The $\delta^{18}O$ values for some of the Sarvak matrix and cements are more negative than Middle Cretaceous marine values (Fig. 9), suggesting their modification as a result of either influence by meteoric water with variable composition or precipitation from warm fluids, or a combination of both (see Given & Lohmann 1985; Carpenter et al. 1991). The interaction with isotopically light waters would have depleted the ^{18}O signature of the carbonates. This could have occurred while meteoric water percolated through the Sarvak carbonates during the exposure of these rocks, a hypothesis that is supported by our field and petrographic observations.

Table 1. *Isotopic composition of various carbonate components in the Upper Sarvak Formation*

Sample no.	Mineralogy	$^{13}C_{VPDB}(‰)$	$^{18}O_{VPDB}$ (‰)	Notes
8952.8 shell	Calcite	2.0	−3.6	Rudist shell
91-shell	Calcite	1.9	−3.6	Rudist shell
92-shell	Calcite	1.9	−3.3	Rudist shell
98-shell	Calcite	1.8	−3.0	Rudist shell
BH-101	Calcite	0.9	−5.2	Rudist shell
BH104-3	Calcite	1.9	−3.7	Rudist shell
BH104-4	Calcite	2.0	−3.2	Rudist shell
BH-4 shell	Calcite	2.1	−4.1	Rudist shell
SE-21	Calcitic soil	−3.0	−4.2	Palaeosol
SE-22	Calcitic soil	−1.6	−4.5	Palaeosol
SK-1	Calcitic soil	−3.5	−6.1	Palaeosol
SK-2	Calcitic soil	−3.0	−5.8	Palaeosol
R2525-M	Calcite	2.8	−3.4	Matrix
R2517-M	Calcite	2.4	−5.2	Matrix
BH-6-M	Calcite	1.5	−4.2	Matrix
BH-3-M	Calcite	2.6	−5.2	Matrix
BH-2-M	Calcite	2.4	−4.8	Matrix
BH-12-M	Calcite	2.5	−4.6	Matrix
BH-107	Calcite	0.8	−6.2	Matrix
BH-106-M	Calcite	2.9	−5.3	Matrix
BH-104-M	Calcite	2.2	−5.2	Matrix
BH-0-M	Calcite	2.5	−4.4	Matrix
9211.5-M	Calcite	1.2	−4.2	Matrix
8961-M	Calcite	1.9	−3.6	Matrix
8943-M	Calcite	1.4	−3.1	Matrix
8700-M	Calcite	1.0	−3.6	Matrix
8688-M	Calcite	1.6	−3.1	Matrix
8686-M	Calcite	1.8	−2.7	Matrix
166-M	Calcite	−0.2	−3.3	Matrix
165-M	Calcite	0.1	−2.0	Matrix
164-M	Calcite	1.5	−2.2	Matrix
162-M	Calcite	1.6	−3.8	Matrix
161-M	Calcite	2.2	−2.3	Matrix
159-M	Calcite	1.7	−2.6	Matrix
151-M	Calcite	0.0	−3.1	Matrix
150-M	Calcite	1.9	−1.0	Matrix
143-M	Calcite	2.1	−0.8	Matrix
139-M	Calcite	2.4	−1.6	Matrix
129-M	Calcite	2.4	−3.5	Matrix
128-M	Calcite	2.6	−3.8	Matrix
120-M	Calcite	2.4	−1.1	Matrix
R2517-C	Calcite	2.4	−4.8	Early cement
BH-4-C1 R	Calcite	1.8	−6.2	Early cement
BH-2-C1	Calcite	1.5	−5.9	Early cement
BH-2-C	Calcite	2.5	−4.9	Early cement
BH-12-2	Calcite	2.0	−6.3	Early cement
BH104-C1	Calcite	0.7	−5.6	Early cement
BH104-5	Calcite	2.0	−5.7	Early cement
BH101-1	Calcite	1.2	−4.1	Early cement
BH-0-1	Calcite	2.2	−6.6	Early cement
9211.5-C	Calcite	2.2	−2.1	Early cement
8961-C	Calcite	2.1	−4.9	Early cement
8943-C1	Calcite	0.4	−3.9	Early cement
8943.5-C1	Calcite	0.2	−3.4	Early cement
8688-C	Calcite	0.3	−3.4	Early cement
166-C	Calcite	0.1	−6.2	Early cement
165-C	Calcite	0.0	−4.4	Early cement
160-C	Calcite	1.4	−1.7	Early cement

(Continued)

Table 1. *Continued*

Sample no.	Mineralogy	$^{13}C_{VPDB}$(‰)	$^{18}O_{VPDB}$ (‰)	Notes
159-C	Calcite	3.1	−3.9	Early cement
158-C	Calcite	3.3	−3.7	Early cement
155-C	Calcite	0.6	−2.2	Early cement
153-C	Calcite	0.6	−2.1	Early cement
147-C	Calcite	1.7	−3.1	Early cement
143-C	Calcite	2.8	−3.2	Early cement
140-C	Calcite	2.9	−2.8	Early cement
129-C	Calcite	2.8	−3.3	Early cement
128-C	Calcite	3.4	−5.2	Early cement
120-C	Calcite	2.4	−3.1	Early cement
8688-C3	Calcite	1.6	−4.4	Late cement
8951-C2	Calcite	1.6	−5.7	Late cement
8952.8 C2	Calcite	1.3	−5.4	Late cement
BH-104-C3	Calcite	1.6	−5.9	Late cement
BH-2-C2	Calcite	2.4	−5.6	Late cement
BH-4-C2 R	Calcite	2.0	−6.6	Late cement
165-C2	Calcite	0.4	−8.0	Late cement
BH-106-C2	Calcite	2.4	−10.2	Late cement
BH106-CR	Calcite	0.8	−9.2	Late cement
BH106-C	Calcite	1.3	−12.3	Late cement
BH104-C2	Calcite	1.7	−5.2	Late cement

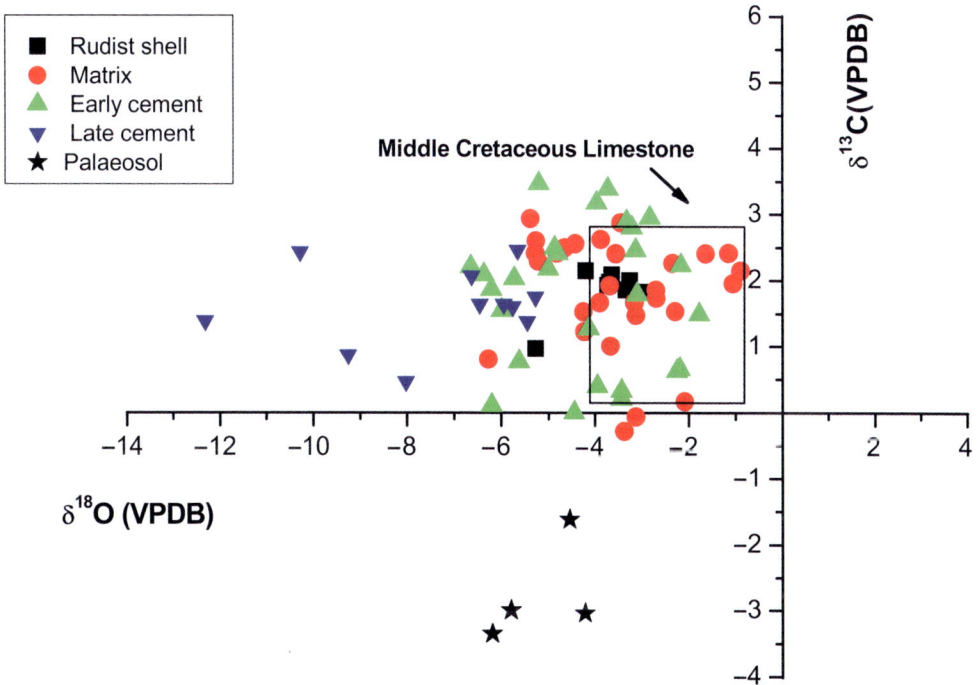

Fig. 9. Carbon and oxygen isotopic composition of carbonate components of the Upper Sarvak Formation in the study area.

The $\delta^{18}O$ values of the blocky calcite cement (late cement) are lower than those of the matrix or early cement, suggesting their precipitation either from meteoric water or under higher ambient temperatures as a result of increase of burial depth, or a combination of both.

Carbon isotopes

The magnitude of $\delta^{13}C$ exchange between carbonates and meteoric water is controlled by the duration of meteoric diagenesis and extent of the water–rock interaction (James & Choquette 1990), where the $\delta^{13}C$ values of diagenetic phases are mainly buffered by the isotopic composition of the precursor rock (Brand & Veizer 1981). Stable-isotope patterns in peritidal carbonates also suggest that micrites can retain the original $\delta^{13}C$ if the duration of subaerial exposure is brief (Joachimski 1994).

The analysed Upper Sarvak carbonates at the type section lack the negative $\delta^{13}C$ values commonly seen in meteoric diagenetic environments below subaerial exposure surfaces and unconformities (see Allan & Matthews 1982). One of the most important processes that could account for the absence of negative $\delta^{13}C$ excursions at the top of the Sarvak Formation is a missing sedimentary section or removal of the rocks (see Allan & Matthews 1982) because of extensive erosion of the uppermost layer during the palaeoexposure. However, depleted $\delta^{13}C$ (i.e. -1.6 to $-5.7‰$) has been observed in palaeosols and recrystallized matrix sampled at the unconformity surfaces at the top of the Sarvak Formation in the Shahneshin area.

The Cenomanian–Turonian transition is marked by a major positive $\delta^{13}C$ excursion (i.e. 3–5‰, Arthur *et al.* 1988; Voigt 2000). The widespread accumulation of organic carbon-rich sediment accompanied by a positive $\delta^{13}C$ excursion in marine carbonates across the Cenomanian–Turonian boundary is one of the most prominent episodes of the Mesozoic, described as an Oceanic Anoxic Event (OAE) (Jenkyns 1980; Arthur *et al.* 1988). In addition, carbonates immediately above and below the Cenomanian–Turonian boundary level have exhibited $\delta^{13}C$ values of $+2.0$ to $+3.0‰$ (Schlanger *et al.* 1983).

Some of the depleted $\delta^{13}C$ values measured for Sarvak carbonate matrix (i.e. $<+1.0‰$) underlie the Cenomanian–Turonian boundary and display a departure from the $\delta^{13}C$ positive excursion. Such depleted $\delta^{13}C$ values are due to subaerial exposure and influence of the unconformity surface.

The highly enriched $\delta^{13}C$ values characteristic of the Cenomanian–Turonian boundary are missing in the Sarvak carbonates as a result of a fall in sea level and non-deposition.

Palaeotemperature and climatic considerations

The Middle Cretaceous was a time of unusually warm climate and is one of the best examples of 'greenhouse' conditions in the geological record (Barron 1983; Bice & Norris 2002). Diverse sets of biotic and marine geological data provide convincing evidence that Middle Cretaceous oceans were extremely warm by today's standards. Global average surface temperatures were more than 10 °C higher than today (Bice & Norris 2002). The cause of the high temperatures is unclear but it has been widely attributed to high levels of atmospheric greenhouse gases (Barron 1987). Atmospheric CO_2 concentrations up to four times the present-day concentrations increased average precipitation rates by 25% (Barron 1987). Global climate changes at the Cenomanian–Turonian boundary have been linked to changes of oceanic water temperature (Arthur *et al.* 1988) and to changes in atmospheric CO_2 (Bice & Norris 2002).

Existing marine records based on $\delta^{18}O$ thermometry suggest that mean sea surface temperatures (SSTs) were highest during the Cretaceous thermal maximum, mostly in Turonian time (Huber *et al.* 2002; Wilson *et al.* 2002). To be more specific, SSTs exceeding 34 °C are evident from the late Cenomanian to late Turonian (Forster *et al.* 2007). During the greenhouse period, meteoric water with increased P_{CO2} would be more aggressive (James & Choquette 1990). In such warm climates the supply of meteoric water was important in determining the rate, intensity and direction of diagenetic alteration. Thus, the Cenomanian–Turonian interval constitutes an ideal candidate for investigations of the changing climatic conditions. The dominance of warm climate and high amounts of rainfall during exposure of the Sarvak carbonates are important factors in determining the ultimate effect of subaerial exposure.

Because the $\delta^{18}O$ of calcite is dependent on the temperature of the water from which it is precipitated, in this study we have used this variable to provide a thermometer for the Middle Cretaceous sea by analysing the shells of rudists collected from the Upper Sarvak Formation. Calcite palaeotemperatures have been calculated using the equation modified by Anderson & Arthur (1983) that is specifically based on bivalve calcite. This expresses the $\delta^{18}O$ of the water ($\delta^{18}O_{water}$), directly relative to the SMOW standard:

$$T(°C) = 16.0 - 4.14(\delta^{18}O_{calcite} - \delta^{18}O_{water})$$
$$+ 0.13(\delta^{18}O_{calcite} - \delta^{18}O_{water})^2.$$

To obtain the best palaeotemperature results, we analysed the $\delta^{18}O$ of the rudist shells for their potential to preserve the original stable isotope signatures because of their outer shell layer mineralogy, which originally consisted of low-Mg calcite (Al-Aasm & Veizer 1986). To produce quantitative temperature estimates it was essential that the original shell $\delta^{18}O$ did not change over time. The $\delta^{18}O$ values obtained from the rudist shells (-3.0 to $-5.2‰$) are similar to those from the well-preserved rudist shells derived from Campanian carbonates obtained by Immenhauser *et al.* (2005).

The $\delta^{18}O$ values obtained from the Sarvak rudist shells would provide a temperature minimum of $25\,°C$ (using $\delta^{18}O_{rudist\ shell} = -3.0‰$) and a maximum of $36\,°C$ (with $\delta^{18}O_{rudist\ shell} = -5.0‰$), assuming seawater compositions for the Middle Cretaceous (no ice build-up) of $-1.2‰$ PDB (White *et al.* 2001). Such calculated temperatures are in accordance with the proposed Middle Cretaceous SST values ranging between 28 and 36 °C for tropical open-marine settings (Steuber 1996; Huber *et al.* 2002; Wilson *et al.* 2002; Immenhauser *et al.* 2005; Forster *et al.* 2007). Therefore, the high oceanic temperatures that characterize the Middle Cretaceous greenhouse conditions appear to have been preserved in the rudist shells found in the Upper Sarvak Formation.

A possible objection to using $\delta^{18}O$ values here is that they may have been affected by meteoric water, which would bias the $\delta^{18}O$ values towards more negative values and therefore higher palaeotemperatures. We do not consider this to be a significant problem for the samples analysed here for several reasons. The setting is in a shallow-marine environment and naturally represents a very warm environment compared with the deep oceans. We might also expect that meteoric water in equatorial environments generally has a $\delta^{18}O$ closer to that of seawater than in higher latitudes, making it difficult to greatly affect the $\delta^{18}O$ even with a very large meteoric water influence. Finally, under careful examination we have found no evidence of recrystallization within the rudist shells.

The temperature of cement precipitation can be also estimated by assuming a value for the isotopic composition of local meteoric water and then calculating a temperature range over which the calcite cements precipitated in equilibrium with this water. Estimation of the ancient meteoric water composition depends on palaeolatitude and palaeotopography at the time of recharge, as well as the degree of water–rock interaction in the infiltration zone. During the Middle Cretaceous, the study area was at an equatorial position, with a humid tropical climate. Thus, the $\delta^{18}O$ of meteoric water could not have been less than $-6‰$ (White *et al.* 2001). Precipitation of early cements with $\delta^{18}O$ of $-4.1 \pm 1.4‰$ and formation of palaeosol with $\delta^{18}O$ of $-4.2‰$ with such meteoric water gives

palaeotemperature estimates of about 25 °C. Such calculated high palaeotemperatures are consistent with a warm climate in the Middle Cretaceous.

That Middle Cretaceous precipitation rates were probably elevated relative to present-day values is supported by the globally widespread distribution of coeval laterites (Sigleo & Reinhardt 1988). Bauxite development was also widespread during the Middle Cretaceous, a consequence of the periodic emergence of the carbonate shelves (Mameli *et al.* 2007). Starting in the Middle Cretaceous a large-scale expansion of lateritic bauxite formation took place on the surfaces of several continents. Bauxites occurred with periodic emergences of the carbonate shelf, under favourable climatic conditions (e.g. warm and humid) that allowed lateritic weathering.

The Upper Sarvak lateritic horizons are sound geological evidence and in good accordance with the global warming and increased precipitation of the Middle Cretaceous period. Limonitic and reddish soil covers on the exposure surface and bauxite deposits are an indication of humid climate at the time of their development (Wopfner & Schwarzbach 1976). The palaeosols observed in the study area could be the result of the cumulative weathering effects of long-term exposure largely associated with sea-level lowstands.

Conclusions

By combining detailed field and core examination with petrographic and geochemical analyses, several conclusions can be reached regarding the diagenesis and evolution of reservoir porosity of the Sarvak Formation, as follows:

(1) Emergence of the Sarvak carbonate ramp as a result of a relative sea-level fall, local salt tectonics and uplift caused large areas of the carbonate succession to be subaerially exposed and subjected to warm and humid conditions. This resulted in profound changes such as the development of vuggy and cavernous porosity within the system, which contributed to elevated rates of additional karst formation.

(2) Petrographic, sedimentological and stable isotope evidence indicates that the subaerial exposure and meteoric diagenesis were the most important diagenetic process affecting the Upper Sarvak Formation.

(3) Meteoric dissolution resulted in enhancement of the porosity and permeability especially in rudist-bearing grainstone or rudstone intervals of the Upper Sarvak Formation.

(4) The positive $\delta^{13}C$ values of the matrix carbonates, rudist shells and calcite cements are within the range of marine Cretaceous carbonates whereas the

$\delta^{18}O$ values show depletions caused by precipitation or equilibration with meteoric water. The $\delta^{18}O$ values of the late cement are distinctly lighter than those of the early cement, indicating some exchange reaction or contribution from isotopically light waters or precipitation in a burial environment under higher ambient temperatures.

(5) Recognition of the Cenomanian–Turonian and mid-Turonian unconformities is a key factor in locating high-porosity and -permeability zones in the succession and is based on the presence of (a) sediment filling a scoured surface, (b) preserved palaeosols and bauxite-rich horizons, (c) iron oxide staining of underlying horizons, and (d) solution-collapse breccia. These features are compatible with warm tropical and greenhouse conditions dominating the region during Cenomanian–Turonian time.

(6) Depleted $\delta^{13}C$ values measured within the Sarvak succession are indicative of subaerial exposure associated with porosity enhancement.

E. H. wishes to thank the National Iranian South Oil Company for access to samples and making the fieldwork possible. The reviewers are thanked for critically reading an earlier version of this manuscript and their constructive comments. The co-operation of A. M. Zadeh Mohammadi, M. A. Kavoosi and F. Pakar is greatly appreciated. F. Ghazban and F. Keyvani are thanked not only for leading us to excellent outcrop locations but also for their valuable comments and discussions. Funding for this project was provided by the Natural Science and Engineering Research Council of Canada (NSERC) to I. S. A. and M. C. Additional funding was provided by StatoilHydro.

References

AHMADIPOUR, M. R. 2002. The role of Sarvak Formation in supplying Pol-e Dokhtar town (Iran) with drinking water. *Acta Carsologica*, **31**, 93–103.

AL-AASM, I. S. & VEIZER, J. 1986. Diagenetic stabilization of aragonite and low-Mg calcite, II. Stable isotopes in rudists. *Journal of Sedimentary Petrology*, **56**, 763–770.

AL-AASM, I. S., TAYLOR, B. E. & SOUTH, B. 1990. Stable isotope analysis of multiple carbonate samples using selective acid extraction. *Chemical Geology*, **80**, 119–125.

ALAVI, M. 2004. Regional stratigraphy of the Zagros fold–thrust belt of Iran and its proforeland evolution. *American Journal of Science*, **304**, 1–20.

ALLAN, J. R. & MATTHEWS, R. K. 1982. Isotope signature associated with early meteoric diagenesis. *Sedimentology*, **29**, 797–897.

ANDERSON, T. F. & ARTHUR, M. A. 1983. Stable isotopes of oxygen and carbon and their application to sedimentologic and palaeoenvironmental problems. *In*: ARTHUR, M. A., ANDERSON, T. F., KAPLAN, I. R., VEIZER, J. & LAND, L. S. (eds) *Stable Isotopes in Sedimentary Geology*. Society of Economic Paleontologists and Mineralogists Short Course Series, **10**, 1–15.

ARTHUR, M. A., DEAN, W. E. & PRATT, L. M. 1988. Geochemical and climatic effects of increased marine organic carbon burial at the Cenomanian–Turonian boundary. *Nature*, **335**, 714–717.

BARDOSSY, G. 1982. *Karst Bauxites*. Elsevier Scientific, Amsterdam.

BARRON, E. J. 1983. A warm, equable Cretaceous: the nature of the problem. *Earth-Science Reviews*, **19**, 305–338.

BARRON, E. J. 1987. Cretaceous plate tectonic reconstructions. *Palaeogeography, Palaeoclimatology, Palaeoecology*, **59**, 3–29.

BEIRANVAND, B., AHMADI, A. & SHARAFODIN, M. 2007. Mapping and classifying flow units in the upper part of the Middle Cretaceous Sarvak formation (western Dezful embayment, SW Iran) based on a determination of the reservoir types. *Journal of Petroleum Geology*, **30**, 357–373.

BICE, K. L. & NORRIS, R. D. 2002. Possible atmospheric CO_2 extremes of the Middle Cretaceous (late Albian–Turonian). *Palaeoceanography*, **17**, 1070.

BOLTZ, H. 1978. *The palaeogeographical evolution during the Cretaceous in the Operating Area with special emphasis on the Bangestan Group*. Oil Service Company (OSCO) Internal Report, **1274**.

BORDENAVE, M. L. 2002. The Middle Cretaceous to Early Miocene Petroleum System in the Zagros Domain of Iran and its Prospect Evaluation. AAPG Annual Meeting, 10–13 March 2002, Houston, TX.

BRAND, U. & VEIZER, J. 1981. Chemical diagenesis of a multicomponent carbonate system—2. Stable isotopes. *Journal of Sedimentary Petrology*, **51**, 987–997.

BURCHETTE, T. P. 1993. Mishrif Formation (Cenomanian–Turonian), southern Arabian Gulf: carbonate platform growth along a cratonic margin basin. *In*: SIMO, J. A. T., SCOTT, R. W. & MASSE, J.-P. (eds) *Cretaceous Carbonate Platforms*. American Association of Petroleum Geologists, Memoirs, **56**, 185–199.

BURCHETTE, T. P. & BRITTON, S. R. 1985. Carbonate facies analysis in the exploration for hydrocarbons: a case study from the Cretaceous of the Middle East. *In*: BRENCHLEY, P. J. & WILLIAMS, B. P. J. (eds) *Sedimentology. Recent Developments and Applied Aspects*. Blackwell, Oxford, 311–338.

BURCHETTE, T. P. & WRIGHT, V. P. 1992. Carbonate ramp depositional systems. *Sedimentary Geology*, **79**, 3–57.

BUTLER, I. B. & RICKARD, D. 2000. Framboidal pyrite formation via the oxidation of iron (II) monosulfide by hydrogen sulphide. *Geochimica et Cosmochimica Acta*, **64**, 2665–2672.

CARPENTER, S. J. & LOHMANN, K. C. 1989. $\delta^{18}O$ and $\delta^{13}C$ variations in late Devonian marine cements from the Golden Spike and Nevis reefs, Alberta, Canada. *Journal of Sedimentary Petrology*, **59**, 792–814.

CARPENTER, S. J., LOHMANN, K. C., HOLDEN, P., WALTER, L. M., HUSTON, T. J. & HALLIDAY, A. N. 1991. $\delta^{18}O$ values, $^{87}Sr/^{86}Sr$ and Sr/Mg ratios of Late Devonian abiotic marine calcite: implications for the composition of ancient seawater. *Geochimica et Cosmochimica Acta*, **55**, 1991–2010.

CHOQUETTE, P. W. & JAMES, N. P. 1988. *Palaeokarst*. Springer, New York.

DROSTE, H. & VAN STEENWINKEL, M. 2004. Stratal geometries and patterns of platform carbonates: The Cretaceous of Oman. *In*: EBERLI, G., MASAFERRO, J. L. & SARG, J. F. R. (eds) *Seismic Imaging of Carbonate Reservoirs and Systems*. American Association of Petroleum Geologists, Memoirs, **81**, 185–206.

FARZADI, P. & HESTHMER, J. 2007. Diagnosis of the Upper Cretaceous palaeokarst and turbidite systems from the Iranian Persian Gulf using volume-based multiple seismic attribute analysis and pattern recognition. *Petroleum Geoscience*, **13**, 227–240.

FORD, D. 1988. Characteristics of dissolutional cave systems in carbonate rocks. *In*: JAMES, N. P. & CHOQUETTE, P. W. (eds) *Paleokarst*. Springer, New York, 25–57.

FORD, D. C. & WILLIAMS, P. 2007. *Karst Hydrogeology and Geomorphology*. Wiley, New York.

FORSTER, A., SCHOUTEN, S., BASS, M. & SINNINGHE DAMSTE, J. 2007. Middle Cretaceous (Albian–Santonian) sea surface temperature record of the tropical Atlantic Ocean. *Geology*, **35**, 919–922.

FU, Q., QING, H. & BERGMAN, C. M. 2006. Paleokarst in Middle Devonian Winnipegosis mud mounds, subsurface of south–central Saskatchewan, Canada. *Bulletin of Canadian Petroleum Geology*, **54**, 22–36.

GHAZBAN, F. 2007. *Petroleum geology of the Persian Gulf*. Joint publication, Tehran University Press and National Iranian Oil Company, Tehran.

GIVEN, R. K. & LOHMANN, K. C. 1985. Derivation of the original isotopic composition of Permian marine cements. *Journal of Sedimentary Petrology*, **55**, 430–439.

HAJIKAZEMI, E., GHAZBAN, F. & YOUSEFPOUR, M. R. 2002. Depositional and diagenetic history of the Middle Cretaceous sedimentary sequence in the Sirri Oil fields in the Persian Gulf, Iran. Middle East and North Africa Oil and Gas Conference, Imperial College, London.

HAQ, B. U., HARDENBOL, J. & VAIL, P. R. 1988. Mesozoic and Cenozoic chronostratigraphy and cycles of sea-level change. *In*: WILGUS, C. K., HASTINGS, B. S., POSAMENTIER, H., VAN WAGONER, J., ROSS, C. A. & RENDAL, C. G. S. (eds) *Sea Level Changes: An Integrated Approach*. Society of Economic Paleontologists and Mineralogists, Special Publications, **42**, 71–108.

HARRIS, P. M., FROST, S. H., SEIGLIE, G. A. & SCHNEIDERMANN, N. 1984. Regional unconformities and depositional cycles, Cretaceous of the Arabian peninsula. *In*: SCHLEE, J. S. (ed.) *Interregional Unconformities and Hydrocarbon Accumulations*. American Association of Petroleum Geologists, Memoirs, **36**, 67–80.

HOPKINS, J. C. 1999. Characterization of reservoir lithologies within subunconformity pools: Pekisko Formation, Medicine River Field, Alberta, Canada. *AAPG Bulletin*, **83**, 1855–1870.

HUBER, B. T., NORRIS, R. D. & MACLEOD, K. G. 2002. Deep-sea paleotemperature record of extreme warmth during the Cretaceous. *Geology*, **30**, 123–126.

IMMENHAUSER, A., NGLERB, T. F., STEUBER, T. & HIPPLER, D. 2005. A critical assessment of mollusk $^{18}O/^{16}O$, Mg/Ca, and $^{44}Ca/^{40}Ca$ ratios as proxies for Cretaceous seawater temperature seasonality. *Palaeogeography, Palaeoclimatology, Palaeoecology*, **215**, 221–237.

JAMES, N. P. & CHOQUETTE, P. W. 1990. Limestone diagenesis, the meteoric environment. *In*: MCILREATH, I. & MORROW, D. (eds) *Sediment Diagenesis*. Geological Association Canada, Reprint Series, **4**, 36–74.

JAMES, G. A. & WYND, J. G. 1965. Stratigraphic nomenclature of Iranian oil consortium Agreement area. *AAPG Bulletin*, **49**, 2182–2245.

JENKYNS, H. C. 1980. Cretaceous anoxic events: from continent to ocean. *Journal of the Geological Society, London*, **137**, 171–188.

JOACHIMSKI, M. 1994. Subaerial exposure and deposition of shallowing upward sequences: evidence from stable isotopes of Purbeckian peritidal carbonates (basal Cretaceous), Swiss and French Jura Mountains. *Sedimentology*, **41**, 805–824.

LIAGHAT, S. & HOSSEINI ZARASVANDI, A. 2003. Determination of the origin and mass change geochemistry during bauxitization process at the Hangam deposit, SW Iran. *Geochemical Journal*, **37**, 627–637.

LOUSK, R. 1999. Paleocave carbonate reservoirs; origins, burial-depth modifications, spatial complexity, and reservoir implications. *AAPG Bulletin*, **83**, 1795–1834.

MAMELI, P., MONGELLI, G., OGGIANO, G. & DINELLI, E. 2007. Geological, geochemical and mineralogical features of some bauxite deposits from Nurra (Western Sardinia, Italy): insights on conditions of formation and parental affinity. *Geologische Rundschau*, **96**, 887–902.

MEYERS, W. J. 1974. Carbonate cement stratigraphy of the Lake Valley (Mississippian) Sacramento Mountains, New Mexico. *Journal of Sedimentary Petrology*, **44**, 837–861.

MOLDOVANY, E. P. & LOHMANN, K. C. 1984. Isotopic and petrographic record of phreatic diagenesis: Lower Cretaceous Sligo and Cupido Formations. *Journal of Sedimentary Petrology*, **54**, 927–958.

MOTIEI, H. 1993. *Geology of Iran. The stratigraphy of Zagros*. Geological Survey of Iran, Tehran [in Farsi].

MURRIS, R. J. 1980. Middle East: stratigraphic evolution and oil habitat. *AAPG Bulletin*, **64**, 597–618.

MYLROIE, J. E. & CAREW, J. L. 1995. Karst development on carbonate islands. *In*: BUDD, D. A., SALLER, A. H. & HARRIS, P. M. (eds) *Unconformities and Porosity in Carbonate Strata*. American Association of Petroleum Geologists, Memoirs, **63**, 55–76.

RAEISI, E. & KARAMI, G. 1997. Hydrochemographs of Berghan karst spring as indicators of aquifer characteristics. *Journal of Cave and Karst Studies*, **59**, 112–118.

SCHLANGER, S. O., ARTHUR, M. A., JENKYNS, H. C. & SCHOLLE, P. A. 1983. Stratigraphic and paleoceanographic setting of organic carbon-rich strata deposited during the Cenomanian–Turonian oceanic anoxic event. *AAPG Bulletin*, **67**, 545.

SCHOLLE, P. A. & ARTHUR, M. A. 1980. Carbon isotope fluctuations in Cretaceous pelagic limestones: potential stratigraphic and petroleum exploration tool. *AAPG Bulletin*, **64**, 67–87.

SETUDEHNIA, A. 1978. The Mesozoic succession in S.W. Iran and adjacent areas. *Journal of Petroleum Geology*, **1**, 3–42.

SHARLAND, P. R., ARCHER, R. *ET AL.* 2001. *Arabian Plate Sequence Stratigraphy*. GeoArabia Special Publication, **2**.

SIGLEO, W. & REINHARDT, J. 1988. Palaeosols from some Cretaceous environments in the southeastern United States. *In*: REINHARDT, J. & SIGLEO, W. R. (eds) *Palaeosols and Weathering through Geologic Time*. Geological Society of America, Special Paper, **216**, 123–142.

STEUBER, T. 1996. Stable isotope sclerochronology of rudist bivalves: growth rates and Late Cretaceous seasonality. *Geology*, **24**, 315–318.

SWEETING, M. M. 1973. *Karst Landforms*. Macmillan, London.

TAGHAVI, A. A., MORK, A. & EMADI, M. A. 2006. Sequence stratigraphically controlled diagenesis governs reservoir quality in the carbonate Dehluran Field, southwest Iran. *Petroleum Geoscience*, **12**, 115–126.

VAN BUCHEM, F. S. P., RAZIN, P. *ET AL.* 1996. High resolution sequence stratigraphy of the Natih Formation (Cenomanian/Turonian) in Northern Oman: distribution of source rocks and reservoir facies. *GeoArabia*, **1**, 65–91.

VAN BUCHEM, F. S. P., RAZIN, P., HOMEWOOD, P. W., OTERDOOM, H. & PHILIP, J. 2002. Stratigraphic organization of carbonate ramps and organic-rich intrashelf basins: Natih Formation (middle Cretaceous) of northern Oman. *AAPG Bulletin*, **86**, 21–54.

VEIZER, J., ALA, D. *ET AL.* 1999. $^{87}Sr/^{86}Sr$, $\delta^{13}C$ and $\delta^{18}O$ evolution of Phanerozoic seawater. *Chemical Geology*, **161**, 59–88.

VIDETICH, P. E., MCLIMANS, R. K., WATSON, H. K. & NAGY, R. M. 1988. Depositional, diagenetic, thermal and maturation histories of Cretaceous Mishrif Formation, Fateh Field, Dubai. *AAPG Bulletin*, **72**, 1143–1159.

VOIGT, S. 2000. Cenomanian–Turonian composite $\delta^{13}C$ curve for Western and Central Europe: the role of organic and inorganic carbon fluxes. *Palaeogeography, Palaeoclimatology, Palaeoecology*, **160**, 91–104.

WANG, B. & AL-AASM, I. S. 2002. Karst-controlled diagenesis and reservoir development: example from the Ordovician main reservoir carbonate rocks on the eastern margin of the Ordos basin, China. *AAPG Bulletin*, **86**, 1639–1658.

WHITE, T., GONZALES, L., LUDVIGSON, G. & POULSON, C. 2001. Middle Cretaceous greenhouse hydrologic cycle of North America. *Journal of Geology*, **29**, 363–366.

WILSON, P. A., NORRIS, R. D. & COOPER, M. J. 2002. Testing the Cretaceous greenhouse hypothesis using glassy foraminiferal calcite from the core of the Turonian tropics on the Demerara Rise. *Geology*, **30**, 607–610.

WOPFNER, H. & SCHWARZBACH, M. 1976. Ore deposits in the light of paleoclimatology. *In*: WOLF, K. H. (ed.) *Handbook of Strata-bound and Stratiform Ore Deposits*. Elsevier, Amsterdam, Volume 3, 43–92.

WRIGHT, V. P. 1991. Palaeokarst: types, recognition, controls and associations. *In*: WRIGHT, V. P., ESTEBAN, M. & SMART, P. L. (eds) *Paleokarsts and Paleokarstic Reservoirs*. Postgraduate Research Institute for Sedimentology, University of Reading Contribution, **152**, 56–88.

WRIGHT, V. P. 1992. A revised classification of limestones. *Journal of Sedimentary Geology*, **76**, 177–185.

Spatial and temporal diagenetic evolution of syntectonic sediments in a pulsatory uplifted coastal escarpment, evidenced from the Plio-Pleistocene, Makran subduction zone, Iran

MAHBOUBEH HOSSEINI-BARZI

Department of Geology, Faculty of Earth Sciences, Shahid Beheshti University, Tehran, Iran
(e-mail: m_hosseini@sbu.ac.ir)

Abstract: Plio-Pleistocene sediments in coastal escarpments of Makran in SE Iran shows different patterns of diagenetic features in diverse depositional facies and therefore implies a facies-controlled diagenetic evolution mainly related to different porosity patterns. Although geochemical characteristics of various cements in this sediment are masked by pervasive meteoric diagenesis, remnants of primary geochemical characteristics (Mg, Mn, Fe and Sr concentrations, and $\delta^{18}O$ and $\delta^{13}C$ values), cathodoluminescence of these cements and paragenetic sequences of diagenetic features help us to unravel the diagenetic history and to reconstruct the history of exposures. The correlation of diagenetic features between rock units in each sedimentary column, inferred by using two distinct dolomite cementation phases, suggests pulsatory changes in the fluid chemistry related to eight episodes of seaward and landward migration of fluids as a result of relative sea-level fluctuation during eight episodes of uplift and relaxation and/or eustatic sea-level fluctuation. Moreover, lateral correlation of paragenetic sequences (eastward increase in the number of phases of cementation and dissolution) implies that pulsation in fluid migration has developed toward the eastern part of each fault-limited block as uplift and erosion rate increased. In the last stage, a net seaward migration of fluids seems to have occurred as relative sea level fell (as a result of regional uplift and/or eustacy). In the last two stages of uplift, reverse fault activities enhanced methane seepage, as indicated by the geochemistry of dolomite cements ($\delta^{13}C - 16.15‰$ to $20.16‰$).

The diagenetic evolution of mixed siliciclastic–carbonate sediments and pore-water chemistry (marine, mixed, and meteoric) are postulated to be strongly controlled by climate, tectonics, and eustacy (Allan & Matthews 1977, 1982; Strasser & Davaud 1989; Budd & Harris 1990; Harwood & Sullivan 1991; Tucker 1993; Moss & Tucker 1995; Vecsei & Hoppie 1996; Morad 1998). Therefore, in active tectonic settings, evidence such as diagenetic history (e.g. Lincoln & Schlanger 1991) in conjunction with stratigraphic relationships should be useful for reconstructing burial and exposure history related to the sea-level fluctuations and the local tectonic movements.

Along the Makran coastline, the formation of the coastal escarpments and fluid migration has been attributed to repeated pulses of vertical movement (e.g. Page *et al.* 1979; Fowler *et al.* 1985; Platt *et al.* 1985; Vita-Finzi 1987; Fruehn *et al.* 1997; Reyss *et al.* 1998; Hosseini-Barzi & Talbot 2003). Although tectonic events interrupted the formation of complete sedimentary cycles in the shelf and shoreline exposed in these escarpments, the effects of these tectonic movements did not uniformly influence the sedimentary records along the coastline (Vita-Finzi 1987; Hosseini-Barzi & Talbot 2003). Despite its importance for sea-level

reconstructions, only very few studies have addressed diagenetic processes along the Makran coast (e.g. von Rad *et al.* 2000).

The main objectives of this study include presenting the diagenetic history of sediments that have been exposed in coastal escarpments of Makran SE of Chabahar, Iran. Based on the diagenetic history, the controlling factors of the relative sea-level changes (i.e. global sea-level changes and/or vertical tectonic movements) are constrained by examining the diagenetic features of mixed sediments occuring along part of the coast. Also, this paper documents the repeated pumping of diagenetic fluids and methane seepage during episodic relative sea-level falls, which were probably due to tectonic activity in the Iranian Makran setting.

Geological setting

The Makran accretionary prism results from active subduction of the oceanic lithosphere of the Oman Sea beneath the Lut and Afghan continental blocks at a rate of about 2.5 cm a^{-1} (Stoneley 1974; Farhoudi & Karig 1977; White 1982; Kidd & McCall 1985; Platt *et al.* 1985; Vernant *et al.* 2004). As subduction proceeded, from Paleocene time, the occanic depo-trough shifted episodically

From: LETURMY, P. & ROBIN, C. (eds) *Tectonic and Stratigraphic Evolution of Zagros and Makran during the Mesozoic–Cenozoic*. Geological Society, London, Special Publications, **330**, 273–289.
DOI: 10.1144/SP330.13   0305-8719/10/$15.00 © The Geological Society of London 2010.

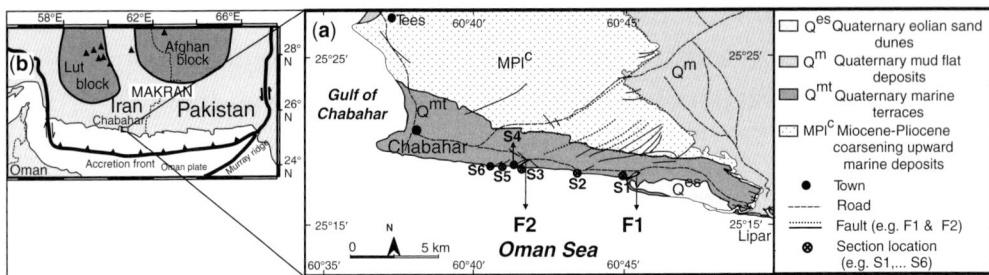

Fig. 1. (**a**) Location map of Makran, indicating location of study area (after Kukowski *et al.* 2001, fig. 1); ▲, volcanoes. (**b**) Geological map of study area (modified after Geological Survey of Iran 1996) showing location of stratigraphic sections at Sites 1, 2, 3, 4, 5 and 6 (S1–S6) and block-bounding faults (F1 and F2). Thrust parallel to the coastline from seismotectonic data (Byrne *et al.* 1992).

southward; with each shift a thrust slice of the sedimentary prism accreted onto the northern continental blocks (e.g. White & Ross 1979; McCall 1993). The differing rate of emergence along the coastline resulted in the formation of marine terraces of different heights (Page *et al.* 1979; Vita-Finzi 1987; Hosseini-Barzi & Talbot 2003).

The studied sediments are exposed east of Gulf of the Chabahar, along 9 km of the Makran coast of the Oman Sea (Fig. 1). This coastline defines the southern boundary of the Coastal Makran Zone of Farhoudi & Karig (1977).

The Pliocene–Pleistocene sediment (Geological Survey of Iran 1996) of the Coastal Makran Zone consists of mixed siliciclastic–carbonate deposits, which are known to be shallow-marine molasse deposits and comprise mostly material recycled from Makran flyschs and older molasses exposed in the northern highlands (White & Klitgord 1976; Farhoudi & Karig 1977; Kidd & McCall 1985; Critelli *et al.* 1990; McCall 1993). The exposed sedimentary rocks in the study area show four depositional facies (Hosseini-Barzi & Talbot 2003) (Fig. 2): (1) backshore to upper shoreface; (2) middle to lower shoreface; (3) lower shoreface; (4) distal storm-generated beds (upper offshore). Such a facies association suggests a high-energy shallow shelf environment (Hosseini-Barzi & Talbot 2003). The relations between these depositional facies and unconformities define four rock units (Fig. 2).

In the study area, two NW-dipping faults (F1 and F2) divide single terraces into two blocks (I and II), and the terraces have different heights from west to east on each block (see Fig. 2). These terraces have a slight back-tilt (5–10°) toward the NNW. The sedimentary record (spatial and temporal change of depositional facies, thickness of rock units and unconformities) shows four episodes of pulsed reverse slip on these faults, with each reverse movement being followed by a relaxation

phase and normal slip along these listric faults (Hosseini-Barzi & Talbot 2003).

Methods

Representative samples were collected from six well-exposed cliffs (7.5–36 m height) (Figs 1 & 2) along a 20 km length of the Makran coast. Because of the friability of collected samples ($n = 130$), they were vacuum-impregnated with blue resin prior to thin-section preparation. Half of each thin section was stained with an alizarin Red-S and ferricyanide solution (Dickson 1965) to facilitate differentiation of the carbonate minerals.

The prepared thin sections were examined using a standard petrographic microscope, and with a Technosyn cold-cathodoluminescence (CL) unit (operating at 20 kV and 400 mA) and an epifluorescence attachment with a blue-extinction filter block (480 nm main wavelength).

Twenty-seven representative samples were thin polished and carbon coated for electron microprobe analysis. Eight samples were chosen from S1 (54 analyses: rock unit D, 18; C, 1; B_u, 26; A, 9), six samples from S2 (51 analyses: rock unit D, 16; B_u, 16; B_l, 7; A, 12), two samples from S3 (10 analyses: rock unit D, 3; A, 7), five samples from S4 (39 analyses: rock unit D, 10; A, 29), three samples from S5 (33 analyses: rock unit D, 22; B, 11) and three samples from S6 (14 analyses: rock unit D, 1; B, 13). Examinations used a Cameca Camebax BX50™ electron microprobe equipped with three spectrometers, and a back-scattered electron (BSE) detector. The operating conditions during analysis included an acceleration voltage of 15 kV, a measured beam current of 8 nA for carbonates and a beam diameter of 1 μm. The standards and count times used were wollastonite (Ca; 10 s), MgO (Mg; 10 s), strontianite (Sr; 10 s), MnTiO$_3$ (Mn; 20 s), hematite (Fe; 10 s), and albite (Na and Si; 5 and 10 s, respectively). The precision of

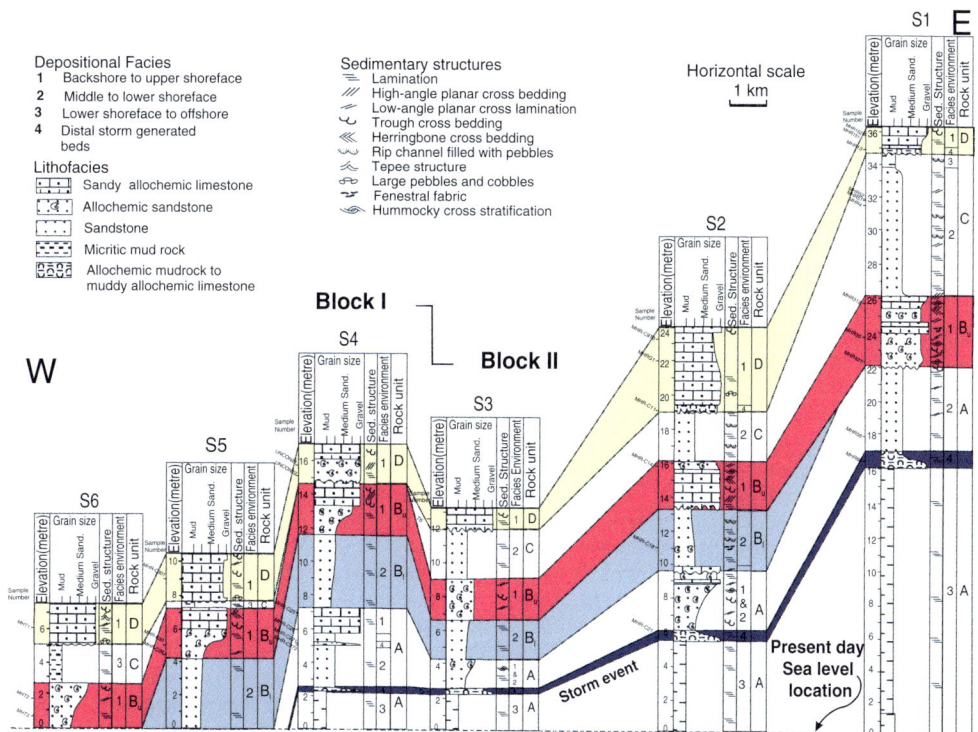

Fig. 2. Stratigraphic sections and interpreted correlations of rock units and related cycles in Blocks I and II.

analyses was better than 0.1 mol%. The detection limit was 50 ppm.

Twenty representative samples were coated with a thin layer of gold and examined with a JEOL™ JSM-T330 and a Zeiss DSM 960A scanning electron microscope (SEM) at an accelerating voltage of 20 kV.

Carbon and oxygen isotope analyses of 19 selected samples (which are representative and have distinct cement types large enough to be drilled for sample powder) were performed using a SIRA-12 mass spectrometer. Samples containing both calcite and dolomite were subjected to chemical separation, using techniques described by Al-Aasm *et al.* (1990). The phosphoric acid fractionation factors used were 1.01025 for calcite at 25 °C, and 1.01065 for dolomite. Data are presented in the normal δ notation (‰) relative to PDB. Precision (1σ) was better than 0.05‰ for both $\delta^{13}C$ and $\delta^{18}O$.

According to the petrography of stained thin sections, CL and BSE images, and electron microprobe data, the paragenetic sequences of diagenetic features were reconstructed for each depositional facies and rock unit. Then, rock units in each sedimentary column were correlated based on diagenetic features, and finally the diagenetic histories of the sedimentary columns were compared.

Details of this procedure are discussed in the section 'Paragenetic sequence of diagenetic features and diagenetic evolution'.

Diagenetic features

As described below, various microscopic diagenetic features such as calcite and dolomite cementation, neomorphism and secondary porosity show different patterns in different lithofacies.

Diagenetic features are used to infer the diagenetic environment based on some index features. For example, micritic rims may reflect a marine cement, or dissolution phases may produce secondary porosity through meteoric interaction (e.g. Bathurst 1975; Meyers 1989; Caron & Nelson 2003). These dissolution phases are present as dissolved (corroded) outer margin of successive generations of cement (Kaufman *et al.* 1988; Braithwaite 1993; Caron & Nelson 2003). Moreover, stable isotope geochemistry and CL can provide supporting data for interpretations of diagenetic history. For example, the isotopic content of dolomite cement ($\delta^{18}O$ 0.21– 1.93 and $\delta^{13}C$ −20.16 to −16.15) and its bright luminescence imply a mixed meteoric–marine condition (e.g. Warren 2003). Also, CL images in conjunction with trace element data from electron

microprobe analysis are used to distinguish diagenetic environments (e.g. non-luminescence and
Fe-, Sr- and Mg-rich calcite fringe imply a marine
diagenetic environment whereas bright luminescence and Mn- and Fe–Mn-rich calcite cements
imply meteoric or mixed diagenetic environments)
(e.g. Brand & Veizer 1980; Veizer 1983; Mucci
1988; Pingitorre *et al.* 1988; Meyers 1989, 1991;
Rao & Adabi 1992; Caron & Nelson 2003).

Calcite cement

Calcite cements show various petrographical
characteristics in different facies defined by
Hosseini-Barzi & Talbot (2003) (Fig. 2). In fact,
the cementation pattern follows the type and the
amount of porosity, which allows fluid circulation
and chemical precipitation of cement. As different
facies have different porosity types and amounts,
different cement types are recognized in each;
such as (Fig. 3) early marine fine cements and micritic cements, blocky mosaic and drusy, fringes,
scalenohedral and rhombohedral blades, void-filling
discrete rhombs and their overgrowth.

Although the electron microprobe data show a
low-magnesium calcite composition for all kinds
of calcite cements (Table 1 and Fig. 4), the stable
isotope compositions of various calcite cements
are not similar (Table 2 and Fig. 5).

A Marine micrite rim, which coated the
bioclasts, indicates the first marine diagenetic
phases. In some cases, micrite rim cement developed even after dissolution of bioclasts and inside
the mouldic porosity (Figs 3a & 6a, b), which indicates repetition of marine and meteoric diagenetic
conditions. These micrite cements are common in
the backshore to upper shoreface facies (with
porous and, therefore, open diagenetic system),
but in storm-generated beds they are rare (with a
closed diagenetic system).

Micritic cements have corroded the feldspars,
calcified silicates, and dispersed grains in a carbonate matrix, and seem to have formed as replacive–
displacive cements (e.g. Braithwaite 1993;
Figs 3a, c, d & 6a). The formation of this kind of
cement could begin during early diagenesis (e.g.
Bathurst 1975; Harris *et al.* 1985) in marine environments or in late diagenesis in mixed or meteoric
environments (e.g. Hanor 1978; Esteban & Klappa
1983). Micrite cements are common among the
grains in the backshore to upper shoreface, lower
shoreface and upper offshore facies. These
cements have euhedral rhombic crystal shapes in
SEM images (Fig. 6c), which implies a non-clastic
origin. Many of these micritic cements changed to
microspars during neomorphism.

Micritic and fine cements show higher Mg
contents (average 1550 ppm) than other calcite

cements (Fig. 4). Stable isotope analysis of micrite
rims was carried out on two samples: the youngest
micrite rim-bearing sample (MHRG1 in the top of
rock unit D) and the oldest micrite rim-bearing
sample (MHR-c 27-2; the storm-generated bed in
rock unit A). The $\delta^{18}O$ value of micrite rims from
the younger sample is $-1.48\%o$, and its $\delta^{13}C$ value
is $-0.44\%o$, which is close to marine cement
isotopic composition. However, these values are
different in the older sample of rock unit A
($\delta^{18}O = -5.6\%o$ and $\delta^{13}C = -3.15\%o$) (sample
MHR-C272 in Table 2 and Fig. 5). These variable
isotopic compositions could be ascribed to a
meteoric diagenetic overprint on these marine
origin cements.

Fringes developed as stubby, elongate and
needle-like crystals (Fig. 3). They either covered
the micrite rims or grew directly over the substrate
as first-generation marine cement. This type of
cement is common in the backshore to upper shoreface facies (Fig. 3a). These cements are relatively
enriched in Sr (average 3850 ppm) (aragonite in
origin) and Mg (average 2650 ppm) and depleted
in Fe and Mn contents, which may imply a marine
diagenetic condition (Fig. 4). The measured $\delta^{18}O$
value for needle fringes is $-3.33\%o$ and $\delta^{13}C$ is
$-1.78\%o$, indicating a meteoric diagenetic overprint.

Scalenohedral and rhombohedral blades appear
as void-filling cement in the large intergranular
(backshore to upper shoreface facies; Fig. 3a) and
mouldic porosity (backshore to upper shoreface
facies and storm-generated beds; Fig. 3a, d). They
commonly formed as epitaxial overgrowth cement.
In some cases, dissolution either smoothed sharp
angles of the crystal boundaries or corroded outer
margins of successive generations of cement
(Fig. 3a, d), which suggests a meteoric diagenetic
phase. Zoning in the stained thin sections and
CL images reflects compositional variations in
the diagenetic fluids during cement precipitation
(Fig. 6d). Accordingly, microprobe analysis and
BSE images revealed zoning by the presence of
variable trace element content as indicated by CL;
that is, relatively Sr-rich zones (e.g. 3850 ppm)
with dark luminescence (which may be marine aragonite in origin) separate bright luminescence zones
relatively rich in Fe and Mn (which might imply
meteoric condition).

Void-filling discrete rhombs and their overgrowth are common in the middle to lower shoreface and backshore to upper shoreface (Fig. 3a, b).
They commonly precipitated as replacive cements
(penetrating into feldspar grains) and as displacive
cements (which dispersed grains during their
growth) (e.g. Braithwaite 1993). Occasionally,
some generations of these calcite overgrowths dissolved before or after precipitation of next zones
(Figs 3b & 6e). As dissolution of calcite is

(a)

Diagenetic feature in backshore to upper shoreface (Facies1)	Early Time Late
Calcite mosaic	
Dolomicrite rim, discrete rhombs and overgrowth of dolomite in 2 phases	
Dissolution of calcifed grains, neomorphosed crystals, cements and fossils	
Neomorphism	
Overgrowth of calcite	
Fringes and epitaxial growth of scalenohedral blades in 4 phases	
Micrite rim	
Replacive and displacive microspar (calcite & dolomite)	

(b)

Diagenetic feature in lower to middle shoreface (Facies 2)	Early Time Late
Discrete rhombs and overgrowth of dolomite in 2 phases	
Discrete rhombs and overgrowth of calcite in 2 phases	
Dissolution of calcified grains and calcite rhombs in 2 phases	

(c)

Diagenetic feature in lower shoreface to offshore lithofacies (Facies 3)	Time — Early ... Late
Replacive and displacive microspar (calcite & dolomite)	

(d)

Diagenetic feature in storm generated beds (Facies 4)	Early Time Late
Dissolution of fossils and cements	
Fringes and epitaxial growth of scalenohedral and rhombohedral blades in 8 phases	
Micrite and microspar rim in 2 phases	
Replacive and displacive microspar (calcite & dolomite)	

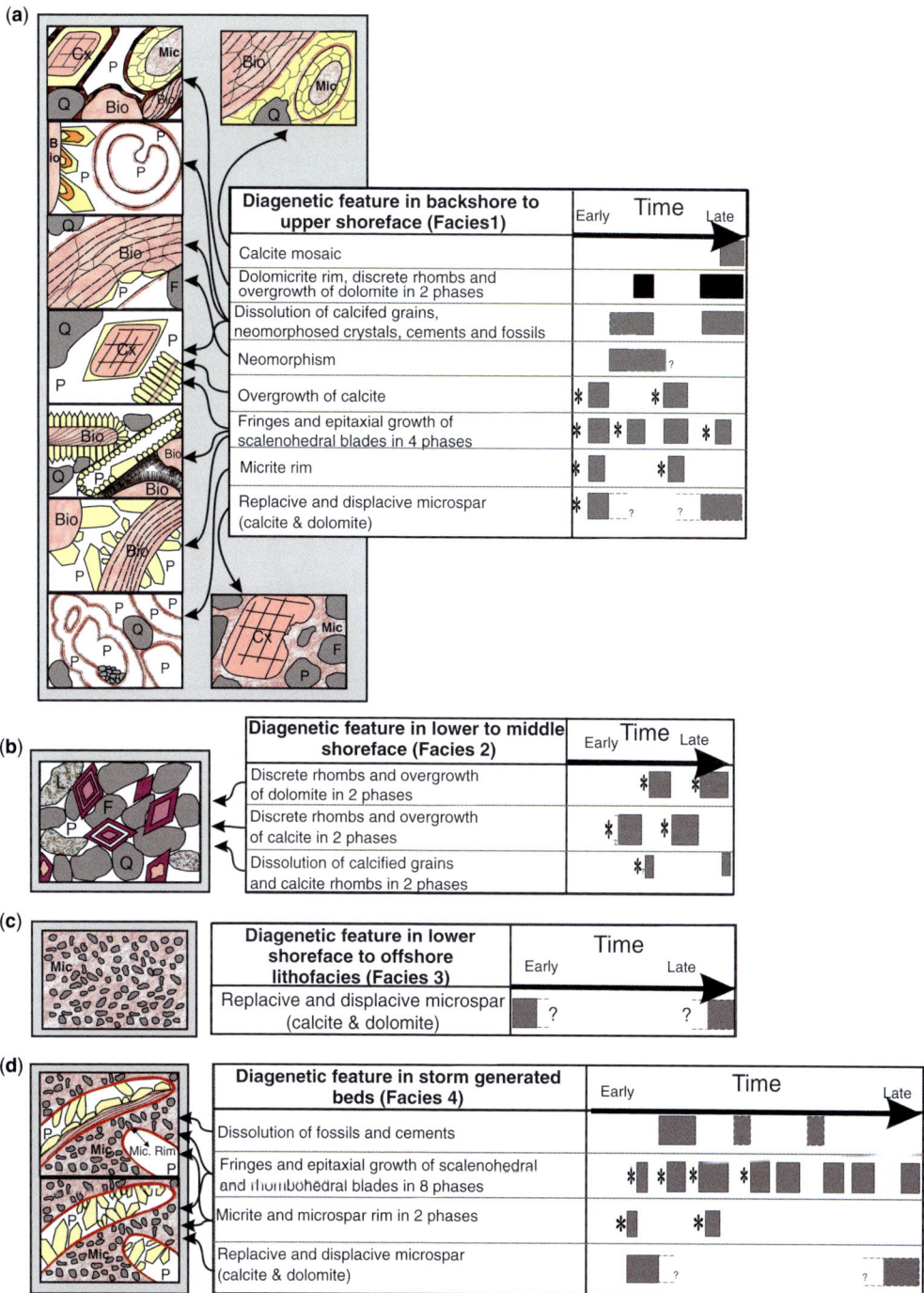

Fig. 3. Variations of diagenetic processes and paragenetic sequences of diagenetic features in different depositional facies (**a–d**). *, Not present everywhere; dash-line box indicates probable occurrence; ?, time limitation is not certain. Time axis is not scaled.

Table 1. *Average trace element contents (in ppm) in various cements and carbonate particles from almost 200 electron microprobe analyses on chosen samples*

Constituents	Mn (ppm)	Fe (ppm)	Sr (ppm)	Mg (ppm)
Micrite	1400	950	850	1550
Fringe	50	50	3850	2650
Blade	1600	0	800	1450
Overgrowth	750	1150	700	400
Mosaic and drusy	1100	1000	0	550
Dolomite	2150	600	600	106 500
Neomorphism	100	250	2000	750

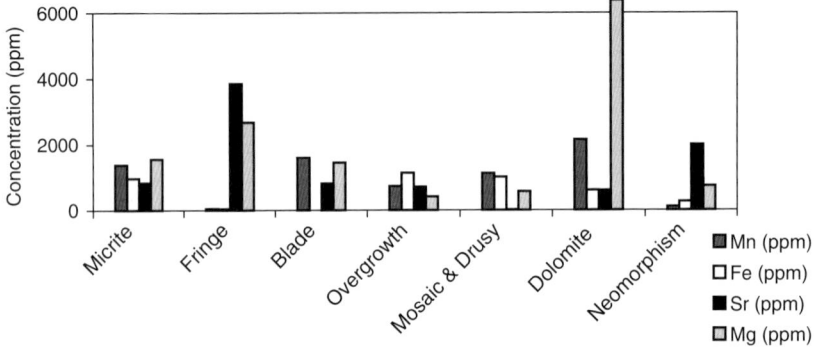

Fig. 4. Average trace elements contents (in ppm) in various cements and carbonate particles from almost 200 electron microprobe analyses on chosen samples (see Table 1).

Table 2. $\delta^{18}O$ v. $\delta^{13}C$ of various carbonate cements and particles from representative samples

Sample no.	Section	Rock unit	Mineralogy/description	$\delta^{18}O$(VPDB)	$\delta^{13}C$
MHR39	S1	Bu	Dol	0.68	−20.16
MHR421	S1	Bu	Dol/rim	0.21	−21.23
MHR56	S1	A	Dol	1.93	−16.15
MHR39	S1	Bu	Cal	−5.55	−7.3
MHR421	S1	Bu	Cal	−3.14	−10.86
MHR56	S1	A	Cal	−5.4	−8.06
UNCON2-1	S4	D	Cal	−5.91	−4.71
UNCON2-2	S4	D	Cal	−4.28	−3.02
MHR-C14	S2	BU	Cal	−3.82	−5.86
MHT-2	S6	B	Cal	−3.97	−4.42
MHR15-2	S1	D	Cal/fossil	−1.48	−1.05
MHR15-1	S1	D	Cal/bladed	−0.95	−0.65
MHR-C25	S4	A	Cal/drusy cement	−3.65	−2.47
MHR5-1	S1	C	Cal/mosaic	−5.94	−2.31
MHR5-2	S1	C	Cal/neomorphosed fossil	−3.8	−1.22
MHR-C-27-1	S4	A	Cal/scalenohedral cement	−2.04	−2.13
MHRG2	S5	D	Cal/rim cement	−3.33	−1.78
MHR-C-27-2	S4	A	Cal/micrite rim	−5.56	−3.15
MHRG1	S2	D	Cal/micrite rim	−1.48	−0.44

VPDB, Vienna Pee Dee Belemnite.

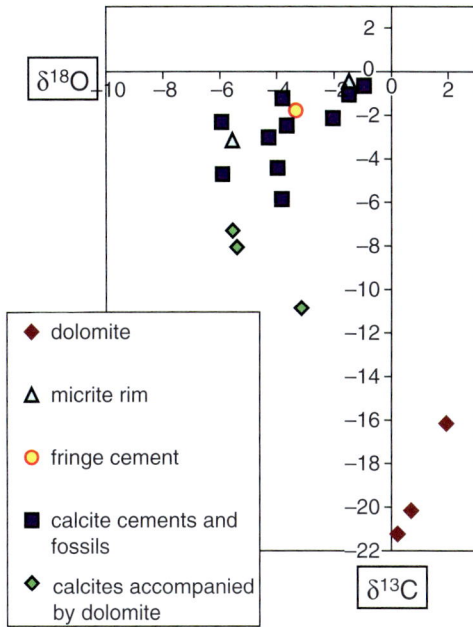

Fig. 5. $\delta^{18}O$ (VPDB, Vienna Pee Dee Belemnite) v. $\delta^{13}C$ of various carbonate cements and particles from representative samples (see Table 2 for sample numbers).

common in meteoric conditions, the precipitation of calcite cement and its dissolution implies a change of diagenetic fluids from marine to meteoric and again to marine conditions during the growth of cement. BSE and CL images (Fig. 6f, g) and staining of carbonates show distinct zoning in overgrowths, which commonly were exposed to dissolution before precipitation of the next generation of cement. Although there are no significant concentrations of measured trace elements for calcite overgrowth (Fig. 4) the electron microprobe results in rock units D and B_1 show relative enrichment in Sr (average 700 ppm, with a range of 1550–4400 ppm) and Mg (average 400 ppm, with a range of 1900–3200 ppm) and depletion in Mn and Fe (below detection limit), probably indicating their primary high-magnesium calcite and aragonite mineralogy in a marine diagenetic environment.

Blocky mosaic and drusy cements usually occur as void-filling cements in the intragranular (inside bioclasts) and intergranular porosity, and as the last generation they cover the micrite coatings, especially in the backshore to upper shoreface facies (Fig. 3a). Despite slight Mn (average 1100 ppm) and Fe (average 1000 ppm) enrichments (Fig. 4), the general concentrations of trace elements in these cements are not significant and most probably imply a meteoric diagenetic condition.

Dolomite cement

Dolomites in the studied sediments display various textures in different lithofacies, including discrete rhombs and syntaxial overgrowth and microcrystalline dolomite.

Discrete rhombs and syntaxial overgrowth of dolomite are common in sandstones of lower to middle shoreface facies, and are followed by syntaxial overgrowth of dolomite and/or calcite (Figs 3b & 6e, h, i). Also, in backshore to upper shoreface and lower to middle shoreface facies, overgrowths of dolomite cover other cements and single-crystal carbonate particles.

Microcrystalline dolomite occasionally forms as rims or scattered microcrystals over the pores in backshore to upper shoreface facies (Fig. 3a). They cover calcite overgrowth and micrite rims, and some have rhombic crystal shape (Fig. 6j, k). Moreover, in all facies, microcrystalline dolomite accompanies calcite as replacive–displacive cement, which fills intergranular pores and corrodes and/or disperses grains.

The geochemical data for dolomite are completely different from those for calcite in this sediment. The Mg concentration measured by microprobe analysis in various dolomites ranges from 30.8 to 41.2 mol% with an average of 36.2 mol%, implying a non-stoichiometric composition of these dolomites. The Mn content of dolomite is higher than in calcite cements (average of 21 500 ppm) (Fig. 4). Their $\delta^{18}O$ values range from 0.21 to 1.93‰, and their $\delta^{13}C$ values from $-16.15‰$ to $-21.23‰$ (Fig. 5), indicating an isotopically light carbon input.

Neomorphism

The backshore to upper shoreface facies exhibits common neomorphosed particles with relics of precursor microstructures, which could be due to their high porosity. Recrystallization of micrite to microspar occurs in all facies. The existence of impurities (e.g. organic matter) in molluscan shells is shown by linear arrangements of inclusions that cut across the calcite mosaic (Fig. 3a, d). Remnants of high-magnesium calcite as paramorphic calcitization (Sandberg 1985) are found, especially in crinoids. The pseudospars have a brownish colour and irregular shapes. Aragonite stabilization of partially dissolved fossils explains dissolution–precipitation (Sandberg, 1985) events in the backshore to upper shoreface facies, which is the most porous facies. These pseudospars have been covered by epitaxial overgrowth cements with rhombohedral habit (Fig. 3a, d). Neomorphosed constituents with their low-magnesium calcite mineralogy have relatively high Sr and low Mn contents (Fig. 4). This diagenetic feature can occur during the flushing

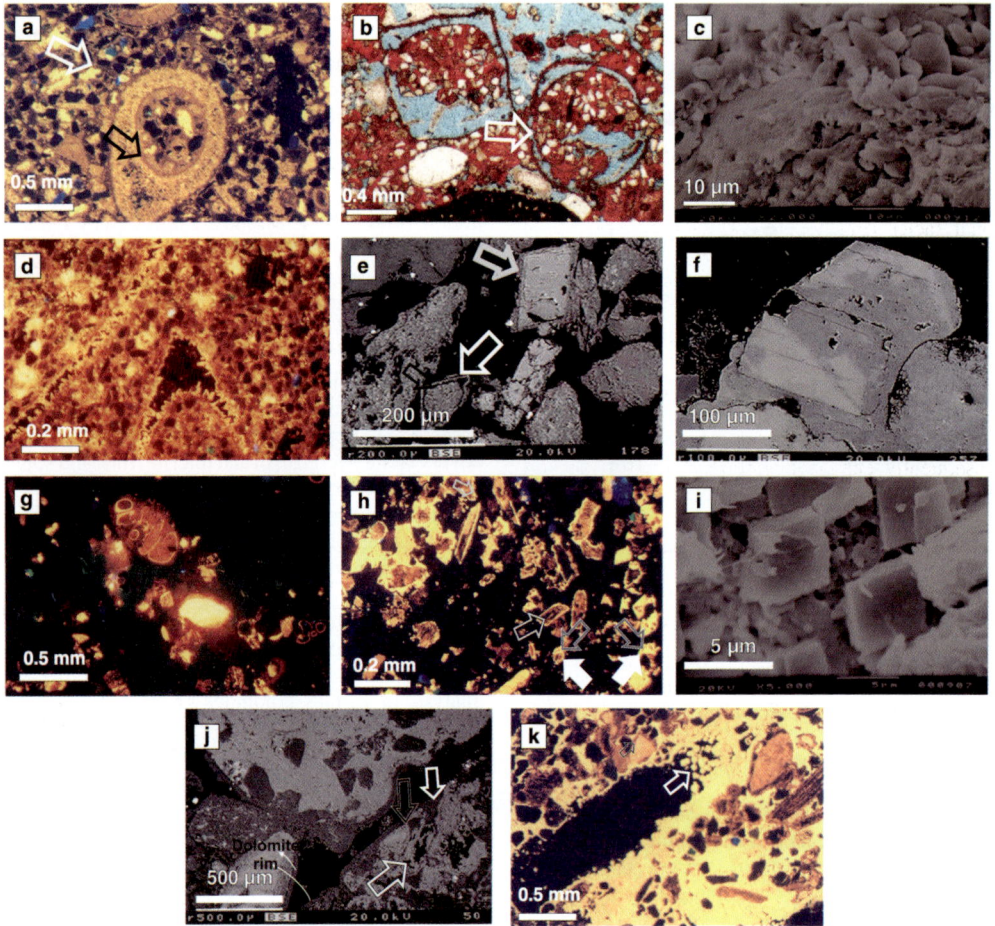

Fig. 6. (**a**) Growth of micrite rim before dissolution of bioclast (black-line arrow) and as replacive–displacive cement among the grains (white-line arrow) (MHRG1 from rock unit D). (**b**) Growth of micrite rim before and after dissolution of bioclast (MHR-C18 from rock unit B_l). (**c**) SEM image showing micrite cement rhombs (MHR56 from rock unit A). (**d**) CL zoning of scalenohedral calcite blades (MHR-C21 from rock unit A). (**e**) Dissolution of calcite overgrowth cement (black-line arrow) and between dolomite cements (white-line arrow). Grey arrow shows calcified silicate (MHR4 from rock unit C). (**f**) BSE image of zoning inside the overgrowth cements (MHR-C18 from rock unit B_l). (**g**) CL image of zoning inside the overgrowth cements (MHR15-1 from rock unit D). (**h**) CL image of succession of calcite (darker zones, shown by white-line arrow) and dolomite (brighter luminescence, shown by white-filled arrow). Grey-line arrows point to calcified silicates (MHR39 from rock unit B_u). (**i**) SEM image of euhedral dolomite rhombs (MHR56 from rock unit A). (**j**) Microcrystalline dolomite rims (dark rims) in BSE image, which cover the calcite (white line arrow) and dolomite (black arrow) overgrowth cements. Grey-line arrow shows dissolution of calcified silicates accompanied by iron oxide precipitation (MHR421 from rock unit B_u). (**k**) Bright luminescence of scattered rhombs and rim of microcrystalline dolomite (white-line arrow), which cover the micrite rims (black-line arrow) in a CL image (MHR421 from rock unit B_u).

of meteoric fluids into the marine sediments (Jorgensen 1976; Hanor 1978; Strasser *et al.* 1989).

Secondary Porosity

Secondary porosities caused by dissolution of fossils are common in the backshore to upper shoreface and upper offshore facies. Dissolution of

cements in all facies causes rounded edges in the cement crystals (Fig. 3a, d) and removes or corrodes margins of the successive generations of cement zones (Figs 3b & 6e). Episodic attacks by meteoric fluids in marine sediments produce mouldic porosities and dissolve the edges of cement crystals that precipitated in the mouldic porosities and in inter-particle porosities. The porosity could be effective in facies with large intergranular porosities such as the backshore to upper shoreface facies.

Discussion

Facies-controlled diagenesis pattern

The diagenetic features vary in different depositional facies (Fig. 3). In the backshore to upper shoreface sediment, large intergranular primary porosity (effective porosity) in sandy allochemic limestone enhanced the circulation of fluids inside the sediment body. This, in turn, induced formation of secondary mouldic porosity and neomorphism of bioclasts, as well as precipitation of a variety of cements and significant lithification (Fig. 3).

Micritic cements and a fine matrix, which subsequently enhanced moderate lithification and preserved mouldic porosity, reduced the intergranular porosity in the allochemic mudrock and muddy allochem limestone of storm-generated beds. Nevertheless, water circulation was strong enough for congruent dissolution of large bivalves and production of mouldic porosity. This implies an active meteoric water circulation (open system), which could be due to a rapid relative sea-level fall and consequence meteoric fluid flux into the marine diagenetic environment. This well-developed dissolution of bioclasts increased the concentration of ions and provided supersaturation of pore fluids. Therefore, coarse crystalline cement (scalenohedral blades) precipitated in large mouldic pores, whereas microcrystalline cement precipitated in the small intergranular pores (Fig. 3). These different precipitating cements explain the exceptionally complete induration of this sediment in the stratigraphic column.

The lower to middle shoreface friable sandstones show high intergranular porosity. This could be due to the lack of bioclasts that could behave as an internal source for carbonate cements. Consequently, cements would be limited to small rhombs and delicate overgrowths.

Commonly, sandstones are limited in the amount, size and variability of cement morphology but limestones have a high frequency of various cement types and size, which is due to the difference in accessibility of ions and fluid circulation (porosity types).

Paragenetic sequence of diagenetic features and diagenetic evolution

An important key to interpreting carbonate diagenesis and cementation is the recognition of the timing and frequency of diagenetic events, and their link to such factors as sea-level fluctuations, tectonic events and burial (Harris *et al.* 1985).

The procedure for reconstruction of diagenetic evolution was performed in three steps, as follows.

(1) Reconstruction of paragenetic sequences in each depositional facies. On the basis of the petrography of stained thin sections, CL and BSE images, and electron microprobe data, paragenetic sequences of diagenetic features were reconstructed for each depositional facies and rock unit. To reconstruct paragenetic sequences of the most recent sedimentary cycle, the diagenetic features of recycled particles (such as rounded quartz overgrowths, or diagenetic features of sedimentary lithic fragments, which are recognizable especially in CL images) were omitted from the results to remove the effect of diagenesis of older cycles. Recognition of diagenetic phases and their relative timing was performed simply: the younger diagenetic features cover or affect the older ones. The back-stripping of diagenetic history from the present day to the time of deposition in each facies provides different paragenetic sequences (Fig. 3), which imply that the diagenetic patterns were facies controlled.

(2) Correlation of diagenetic features in different rock units for each sedimentary column. The rock units in each sedimentary column were correlated based on diagenetic features. For each rock unit, the paragenetic sequences from different facies were compiled and the paragenetic sequence for each rock unit was constructed (Fig. 7). However, as the rock units are limited by unconformity surfaces, during pulses of relative sea-level fall (mostly caused by local uplift) along this coastline (Hosseini-Barzi & Talbot 2003), the diagenetic suites are bounded by dissolution phases as discontinuities during pulses of meteoric fluid flux on a thin-section scale.

Also, as mentioned above, dolomite that precipitated in two distinct phases is well developed in rock units B_u, B_l, A, and the bottom of C. In contrast, rock unit D has only scarce development of the later phase of dolomicrite in the matrix. The two dolomitic phases (which can be distinguished from each other by the phase of calcite cementation between them; Figs 3a, b & 6e, h) appear as distinct diagenetic events. Therefore, the two phases of dolomite cementation as well as dissolution events have been utilized as base-lines for correlation of diagenetic features. Moreover, zoning in CL images is used to constrain the relative timing of diagenetic events and to correlate them (Fig. 8). This method was introduced by Meyers (1978) as 'cement stratigraphy,' by Cander *et al.* (1988) as 'cathodoluminescence stratigraphy', and by Caron & Nelson (2003) as 'high resolution diagenetic stratigraphy'.

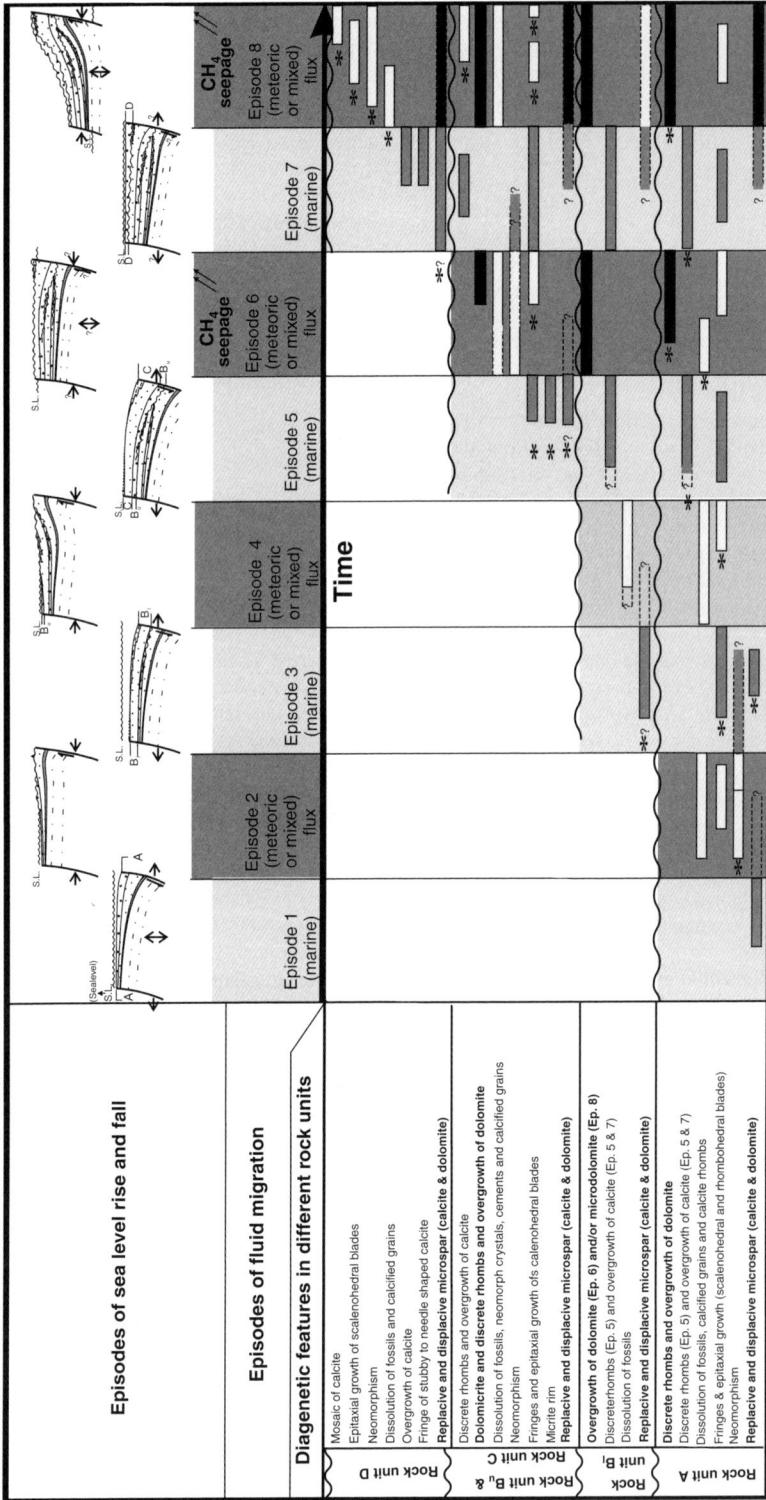

Fig. 7. Paragenetic sequence of diagenetic features in the various rock units obtained by compiling the data from sections S2 and S1. ?, Time limitation is not certain; *, not present everywhere; dash-line boxes indicate probable occurrence; black boxes are related to dolomite cements; wavy lines indicate unconformity; Ep., episode. Time axis is not scaled. Episodes of relative sea-level rise and fall (sketches at top of figure) and erosion surfaces are after Hosseini-Barzi & Talbot (2003).

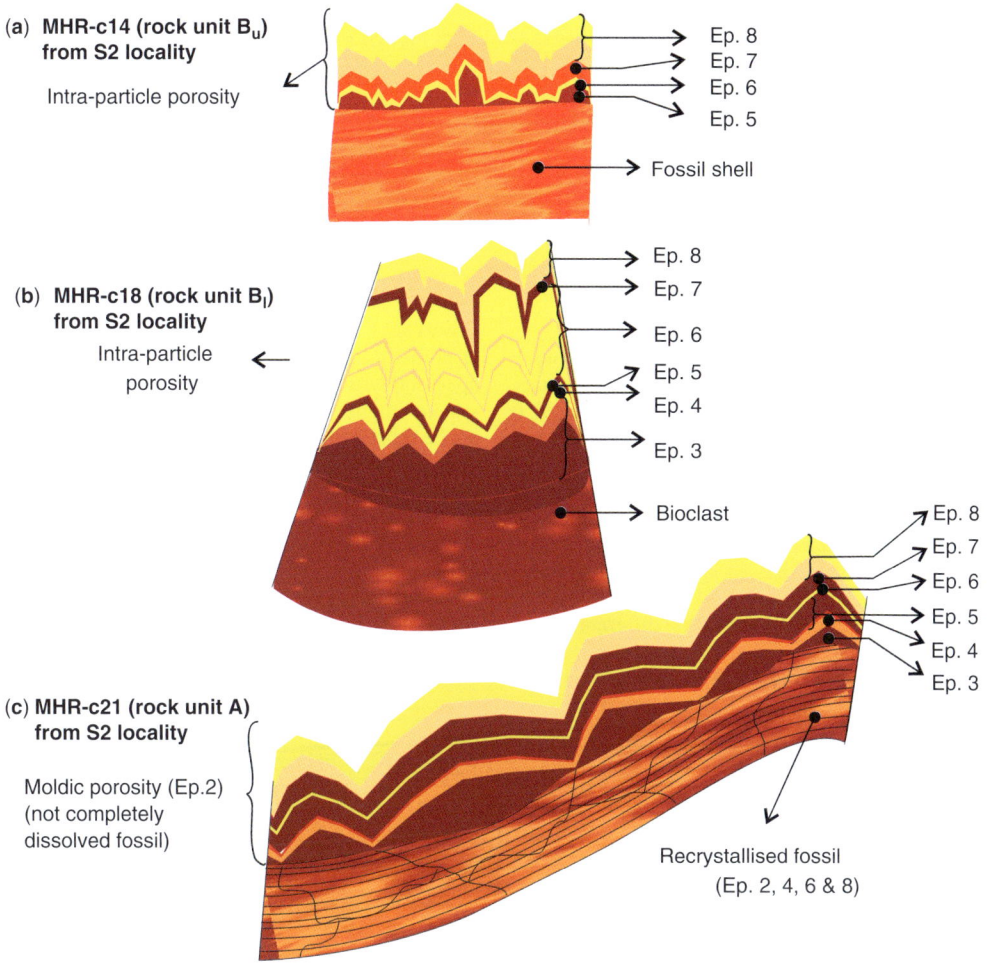

Fig. 8. Schematic illustration showing how correlation between the zoning in CL images from different samples is possible. Ep. 2–8, episodes of fluid migration. Samples were collected from S2.

Precise reconstruction of diagenetic environments (as marine or mixed–meteoric) requires detailed geochemical studies (e.g. stable isotopes) in each zoning band of cements (e.g. Brand & Veizer 1981; Machel & Burton 1991). Moreover, the chemical characteristics of previous diagenetic features are overprinted by low-magnesium calcite that resulted from pervasive meteoric diagenesis in an open system in the studied sediments. Despite all these limitations, petrography of stained thin sections, CL and BSE images, and electron microprobe data were used in a provisional approach to identify the pulses of changes in fluid chemistry, which are shown by episodes 1–8 (Fig. 7). The correlation of diagenetic histories shows similarities among rock units in S1 and S2, except for rock unit B_l, which was eroded in S1. Therefore, in constructing

a complete sedimentary column, diagenetic data from S1 and S2 were combined. The correlation of rock units according to the relative timing of diagenetic features in the combined column is shown in Figure 7.

The episodes of marine environments were distinguished using micrite rims, early fringe cement, dark colour of zoning bands in CL images, or violet colour in stained thin sections. Also, these bands have relatively high contents of Fe (usually from 1250 to 2950 ppm), Sr (usually from 1050 to 4400 ppm), and Mg (often below the detection limit and occasionally from 2050 to 5350 ppm). The Mn content is always below the detection limit. The high Fe content can be explained by the replacement of Mg in high-Mg calcite cements by Fe during diagenetic overprinting by Fe- and

Mn-rich meteoric fluids. The high Sr content could be a remnant of this element from precursor aragonite. The Mg content shows the remnants of this mobile element during meteoric diagenesis. Diagenetic episodes of mixed or meteoric environments are characterized by dissolution phases, dolomite cementation, bright colour of zoning bands in CL images, and trace element contents that are usually below the detection limits, along with occasional high Mn content (1000–3150 ppm, with an average of 1900 ppm) compared with sub-oxic conditions, especially from mixed environments.

(3) Comparisons of diagenetic features between columns. The diagenetic histories of sedimentary columns were compared, as shown in Figure 9. It is known that the diagenetic features are not necessarily spatially synchronous (Meyers 1978; Caron & Nelson, 2003). Therefore, it is expected that the paragenetic sequences of rock units in a sedimentary column are occasionally different from their equivalents in other columns. In the studied sections, despite differences in details of diagenetic patterns between the same rock units in different stratigraphic columns, the repetition in fluid chemistry is almost the same (Fig. 9). This implies that fluid migration had similar patterns in different columns, even though it could have been diachronous along the studied segments of the coastline, although, where the number of diagenetic episodes is not similar along the coastline, there is an increasing trend in their number toward the east of each block (Fig. 9).

In closed systems the precipitation of cements affects the composition of the fluid remaining in the pore space, in terms of trace element contents. Sequentially, the cement will precipitate with different trace element contents and provide zoning (e.g. Brand & Veizer 1980; Machel & Burton 1991). However, in the studied sediments high amounts of porosity and active fluid circulation provided an open diagenetic system. The existence of this open system implies that the zoning in cements could not be related to rock–water interactions in isolated microenvironments and might be due to extrinsic factors such as seismic pumping (Sibson *et al.* 1975), and/or rapid relative sea-level changes, and consequent diagenetic fluid migration and changes in fluid chemistry (Tucker 1993). This is supported by common, similarly repeated patterns along the coast (Fig. 9), which show more regional controls on fluid flow than a cause related to microenvironments. Moreover, where there are changes in the

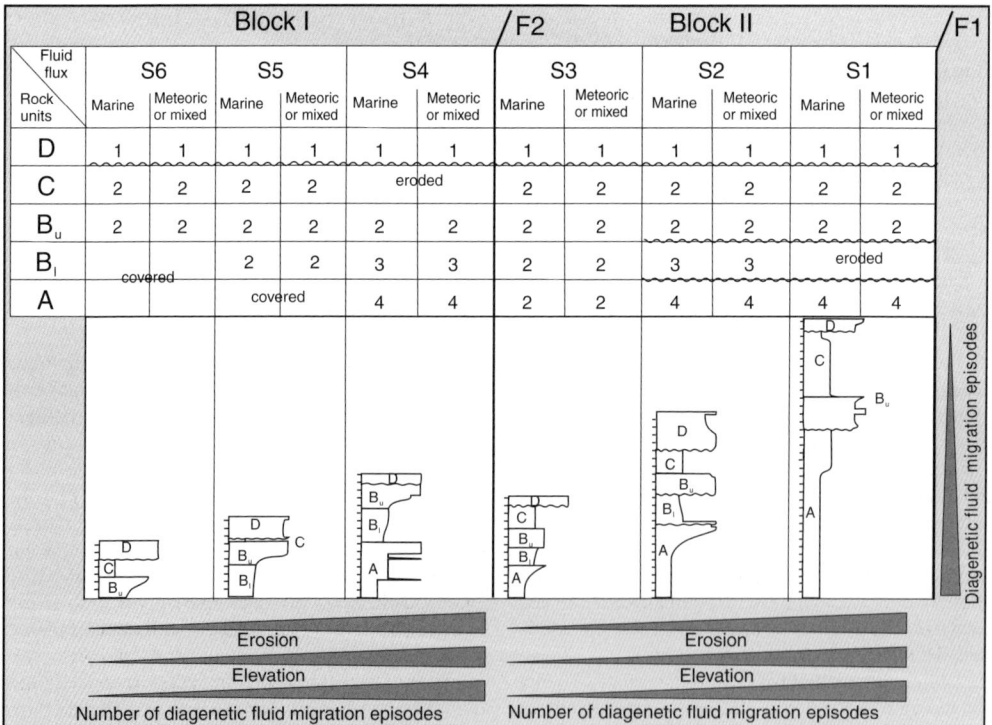

Block I						/F2	Block II					/F1
Fluid flux / Rock units	S6 Marine	S6 Meteoric or mixed	S5 Marine	S5 Meteoric or mixed	S4 Marine	S4 Meteoric or mixed	S3 Marine	S3 Meteoric or mixed	S2 Marine	S2 Meteoric or mixed	S1 Marine	S1 Meteoric or mixed
D	1	1	1	1	1	1	1	1	1	1	1	1
C	2	2	2	2	eroded		2	2	2	2	2	2
B$_u$	2	2	2	2	2	2	2	2	2	2	2	2
B$_l$	covered		2	2	3	3	2	2	3	3	eroded	
A			covered		4	4	2	2	4	4	4	4

Erosion — Elevation — Number of diagenetic fluid migration episodes (Block I)

Erosion — Elevation — Number of diagenetic fluid migration episodes (Block II)

Diagenetic fluid migration episodes

Fig. 9. Repetitions of the various diagenetic environments based on the proposed tectono-sedimentary model in each sedimentary column (S1–S6) for each rock unit.

number of diagenetic phases along the coastline, they increase in number toward the eastern part of each fault-limited block (Fig. 9), where the rock has been exposed to more erosion (e.g. there are no erosion surfaces in S3, whereas rock unit B_1 and the erosion surface overlying rock unit A have been eroded in S1; Figs 2, 7 & 9) and a higher local uplift rate (higher elevation in the eastern part of each block) at least in the two earliest stages of reverse faulting (Hosseini-Barzi & Talbot 2003). Therefore, the regional factors for relative sea-level changes are stronger in the latest episodes and produced a similar pattern of pumping along the coastline. Consequently, I relate the eight episodes of diagenetic fluid changes (Fig. 7) to eight episodes of relative sea-level changes occurring in the study area (see Hosseini-Barzi & Talbot 2003, fig. 3). Thus not only have the patterns of deposition (facies changes and rock unit thicknesses) and erosion of these sediments been controlled by relative sea-level changes (Fig. 9) (Hosseini-Barzi & Talbot 2003, fig. 3), but also the pattern of diagenetic features shows the same evolution in the coastal escarpments.

Origin of dolomite and methane seepage

The petrography of thin sections implies that although dolomite is not present in all samples, it has precipitated in all facies as two distinct phases, which indicates that it formed in the final stages of cementation (Fig. 7). Accordingly, the dolomite has undergone the minimum diagenetic alteration. Therefore, recognition of its origin and its relation to relative sea-level changes is decisive for the reconstruction of diagenetic fluid migration and can be used as a base-line for the correlation of diagenetic features (as described above).

The oxygen isotopic signature of dolomite cements is comparable with that reported in previous studies of Pleistocene–Holocene dolomite of various types (e.g. Humphrey & Quin 1989; Gregg et al. 1992; Warren 2003). Their high Mn contents and bright CL images confirm a mixing and/or seepage origin in a suboxic environment (Morad 1998; Warren 2003). Such mixing processes and migration of fluids could have been enhanced by fluctuations of sea level (e.g. Allan & Matthews 1977, 1982; Tucker 1993, Morad 1998).

The light carbon isotopic signature of dolomites (e.g. Baker & Kastner 1981; Friedman 1991; Morad 1998) results from methane seepage in the study area. This is consistent with the suboxic conditions, high Mn contents and bright luminescence of these dolomites. In spite of the small number of available isotopic data from these dolomites, they can be compared with dolomite with similar carbon isotopic composition that has been studied where active

fluid venting occurs over other accretionary prism settings in subduction zones (e.g. Russel et al. 1967; Kulm & Suess 1990; Deyle & Kopp 2001). In subduction zones, methane is expelled from sediments subducting below the wedge along the detachment zone (Oliver 1992). In Makran, accreted sediments also release fluids upwards, and this dewatering is accompanied by seepage of methane from a gas hydrate-bearing horizon (Harms et al. 1984; Fowler et al. 1985; Platt et al. 1985; Minshull & White 1989; Fruehn et al. 1997; von Rad et al. 2000). These fluids can be channelled through the prism along the fault systems and may drive sedimentary volcanoes on land and on the abyssal plain (e.g. von Rad et al. 2000; Wiedicke et al. 2001). The more common existence of dolomite in S1 (near F1) (Table 2) may support a heterogeneous distribution related to channelized methane seepage.

The dolomite phase, which is marked by a considerable negative carbon isotope signature, precipitates in all types of lithofacies with a variety of porosity and permeability related to different redox conditions and different organic matter preservation, but with the same isotopic signature. Therefore, the more negative isotopic signature is probably not due to organic matter oxidation. However, the light carbon isotopic signature of calcite in dolomite-bearing samples could be ascribed to methane input and diagenetic overprinting of meteoric fluids (Fig. 5).

Therefore, the carbon isotopic signatures of the studied samples may suggest that during a phase of upward seepage of methane, dolomite cements precipitated from fluids with a more negative carbon isotopic signature within a bacterial sulphate reduction zone and/or, more likely, in a suboxic zone (Morad 1998). Because the electron microprobe data and CL images indicate Mn enrichment of these non-stoichiometric dolomites (Fig. 4), formation of dolomite in an oxic zone is ruled out. The high accumulation rate of the sediment (Harms et al. 1984) implies a suboxic (to anoxic) condition (which is seen in the light grey to light olive colour of most samples) in a mixed marine–meteoric environment (Fig. 10). The upward seepage of methane, probably channelized along the faults while they were active, enhanced precipitation of dolomites during the latest stages of diagenesis (Fig. 7) and suggests diagenetic phases that correlate with the last stages of shortening (stages 6 and 8 of Hosseini-Barzi & Talbot 2003). This shortening began with the emergence of sediment drowned earlier in a marine environment and continued with its upward movement to a mixed marine–meteoric and meteoric environment until the sediment rose high enough to be exposed along the escarpments (Fig. 7). Sedimentary records (Hosseini-Barzi & Talbot 2003) and diagenetic features (Fig. 9)

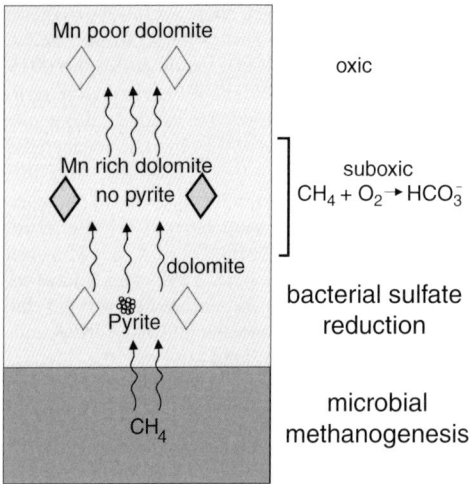

Fig. 10. Relationship between methane seepage and occurrence of dolomite in different geochemical environments. The studied dolomites are related to a suboxic environment.

suggest that regional controlling factors (eustatic or regional uplift), as well as local ones (local reverse slip along F1 and F2), elevated the sediment during the two latest episodes of uplift and exposure. Therefore, these strong driving mechanisms for fluid migration could have triggered the upward methane seepage in this period and produced a lower carbon isotopic signature in the syntectonic dolomites.

The methane seepage enhanced the alkalinity (Friedman 1991). Moreover, in this semi-arid area, with evaporation in a marine or (more likely) mixed environment, an increase in Mg/Ca ratio enhanced dolomite formation. The absence of dolomite in the uppermost part of the sections (rock unit D and to a lesser extent the upper part of unit C) could indicate the cumulative release of methane (Friedman 1991). Accordingly, there is a possibility that at the beginning of the last stage of shortening (Hosseini-Barzi & Talbot 2003) methane release was insufficient to enhance the dolomite precipitation. Later, when the methane release rate increased, unit D already cropped out. Therefore, I suggest that relative sea-level changes, and more likely episodic pulses of tectonic movement, acted as the driving mechanism for diagenetic fluid flow and methane seepage. This may have occurred during variations in fluid flow and/or channelized fluids along the faults. Although the dolomite is more common in S1 (near F1) than at other locations, the evidence is insufficient to determine the details of the methane seepage mechanism.

The suggested tectonic pulses above the subcretion zone are likely to pump internal fluids around

the Makran accretionary prism (Harms et al. 1984; Fowler et al. 1985; Platt et al. 1985; Minshull & White 1989; Fruehn et al. 1997; von Rad et al. 2000). Such a pump is unlikely to be unique, and may provide the mechanism to expel sufficient methane from this and other accretionary prisms to affect the climate (Hudson & Magoon 2002).

Concluding remarks

In diagenetic processes, each dissolution phase has been considered to have occurred in meteoric conditions and each micrite rim or fringe cement is considered to have formed in a marine environment, related to eight episodes of change in fluid flow.

Depositional facies controlled all the diagenetic processes and their paragenetic sequences, and changes in diagenetic fluid chemistry.

Facies of high porosity (with large inter-particle porosity) have open diagenetic systems, such as facies 1 and 2. In facies with small inter-particle porosity, despite of the dissolution of fossils, precipitation of early cements changed these sediments to semi-closed systems, and the development of the diagenetic process would then be limited to the mouldic porosity (such as facies 3 and 4).

Electron microprobe geochemical data in conjunction with the open diagenetic system suggest that zoning of cements observed in CL images is probably not related to rock–water interaction, as might otherwise be considered. Instead, it may imply a change in fluid chemistry (redox condition) as a result of fluid flux during sea-level fluctuation.

The two dolomitic phases recognized in the diagenetic history of the sediment have been utilized as base-lines for correlating paragenetic sequences between sedimentary columns. Such correlations indicate that the fluctuations of sea level recorded by the diagenetic fluid migration are mostly related to the fault movements. In other words, the eastern part of each fault-bounded block, which has relatively higher elevation, shows more contrasting layers of fluid chemistry.

Temporal study of paragenetic sequences in each column indicates that the younger rock units overlying unconformities show fewer contrasting layers of fluid chemistry than older units underlying the unconformities.

This study provides evidence for the occurrence of up to four episodes of meteoric fluid flux to marine diagenetic environment. These episodes can be correlated with the four phases of reverse faulting (Hosseini-Barzi & Talbot 2003), associated with the four escarpments that expose the sediment.

The last two episodes of relative sea-level fall, which were more important than the first two (Hosseini-Barzi & Talbot 2003), triggered methane seepage and are recorded in dolomites with lower

$\delta^{13}C$ values. These dolomites, however, resemble those previously found near active fluid vents on other accretionary prisms during relative sea-level fall.

The author gratefully acknowledges P. H. Molnar and C. Garzione for their constructive comments and suggested improvements to an early draft of the manuscript. Long discussions with S. Morad, M. Rezaee and H. Rahimpour are gratefully acknowledged. I am especially grateful to S. Morad for help with SEM and EMP laboratory work at Uppsala University, Sweden, and to I. S. Al-Aasm for help with CL and oxygen and carbon stable isotope analysis at the University of Windsor, Ontario, Canada. M. Esfehani-Nejad is thanked for help during fieldwork. I also thank R. Swennen and M. Pagel for careful reviews. The manuscript has greatly benefited from constructive comments, corrections and help from the GSL editors C. Robin and P. Leturmy.

References

AL-AASM, I. S., TAYLOR, B. & SOUTH, B. 1990. Stable isotope analysis of multiple carbonate samples using selective acid extraction. *Chemical Geology*, **80**, 119–125.

ALLAN, J. R. & MATTHEWS, R. K. 1977. Carbon and oxygen isotopes as diagenetic and stratigraphic tools: data from surface and subsurface of Barbados, West Indies. *Geology*, **5**, 16–20.

ALLAN, J. R. & MATTHEWS, R. K. 1982. Isotope signatures associated with early meteoric diagenesis. *Sedimentology*, **29**, 797–817.

BAKER, P. A. & KASTNER, M. 1981. Constraints on the formation of sedimentary dolomite. *Science*, **213**, 214–216.

BATHURST, R. G. C. 1975. *Carbonate Sediments and their Diagenesis*. Developments in Sedimentology, **12**.

BRAITHWAITE, C. J. R. 1993. Cement sequence stratigraphy in carbonate. *Journal of Sedimentary Petrology*, **63**, 295–303.

BRAND, U. & VEIZER, J. 1980. Chemical diagenesis of a multicomponent carbonate system—1: trace elements. *Journal of Sedimentary Petrology*, **50**, 1219–1236.

BRAND, U. & VEIZER, J. 1981. Chemical diagenesis of a multicomponent carbonate system—2: stable isotopes. *Journal of Sedimentary Petrology*, **51**, 987–997.

BUDD, D. A. & HARRIS, P. M. 1990. *Carbonate–Siliciclastic Mixtures*. SEPM (Society for Sedimentary Geology) Reprint Series, **14**.

BYRNE, D. E., SYKES, L. & DAVIS, D. M. 1992. Great thrust earthquakes and aseismic slip along the plate boundary of the Makran subduction zone. *Journal of Geophysical Research*, **98**, 449–478.

CANDER, H. C., KAUFMAN, J., DANIELS, L. D. & MEYERS, W. J. 1988. Regional dolomitization of shelf carbonates in the Burlington–Keokuk Formation (Mississippian), Illinois and Missouri: constraints from cathodoluminescent zonal stratigraphy. *In:* SHUKLA, V. & BAKER, P. A. (eds) *Sedimentology and Geochemistry of Dolomites*. SEPM, Special Publications, **43**, 129–144.

CARON, V. & NELSON, C. S. 2003. Developing concept of high resolution diagenetic stratigraphy for Pliocene cool-water limestones in New Zealand, and their sequence stratigraphy. *Carbonate and Evaporites*, **18**, 63–85.

CRITELLI, S., DE ROSA, R. & PLATT, J. P. 1990. Sandstone detrital modes in the Makran accretionary wedge, southwest Pakistan: implication for tectonic setting and long-distance turbidite transportation. *Sedimentary Geology*, **68**, 214–260.

DEYLE, A. & KOPP, A. 2001. Deep fluid and ancient pore water at the backstop: stable isotope systematic (B, C, O) of mud-volcano deposits on the Mediterranean Ridge accretionary wedge. *Geology*, **29**, 1031–1034.

DICKSON, J. A. D. 1965. Carbonate identification and genesis as revealed by staining. *Journal of Sedimentary Petrology*, **36**, 491–505.

ESTEBAN, M. & KLAPPA, C. F. 1983. Subaerial exposure environment. *In:* SCHOLLE, P. A., BEBOUT, D. J. & MOORE, C. H. (eds) *Carbonate Depositional Environments*. American Association of Petroleum Geologists, Memoirs, **33**, 1–95.

FARHOUDI, G. & KARIG, D. E. 1977. Makran of Iran and Pakistan as an active arc system. *Geology*, **5**, 664–668.

FOWLER, S. R., WHITE, R. S. & LOUDEN, K. E. 1985. Sediment dewatering in the Makran accretionary prism. *Earth and Planetary Science Letters*, **75**, 427–438.

FRIEDMAN, G. M. 1991. Methane-generated lithified dolostone of Holocene age: eastern Mediterranean. *Journal of Sedimentary Petrology*, **61**, 188–194.

FRUEHN, J., WHITE, R. S. & MINSHULL, T. A. 1997. Internal deformation and compaction of the Makran accretionary wedge. *Terra Nova*, **9**, 101–104.

GEOLOGICAL SURVEY OF IRAN 1996. *Geological map of Chabahar Quadrangle, 1:100 000 scale*. Geological Survey of Iran, Tehran.

GREGG, J. M., HOWARD, S. A. & MAZZULLO, S. J. 1992. Early diagenetic recrystallization of Holocene peritidal dolomites, Ambergris, Belize. *Sedimentology*, **39**, 143–160.

HANOR, J. S. 1978. Precipitation of beach rock cements: mixing of marine and meteoric waters vs. CO_2-degassing. *Journal of Sedimentary Petrology*, **48**, 489–501.

HARMS, J. C., CAPPEL, H. N. & FRANCIS, D. C. 1984. The Makran coast of Pakistan: its stratigraphy and hydrocarbon potential. *In:* HAQ, B. U. & MILLIMAN, J. D. (eds) *Marine Geology and Oceanography of the Arabian Sea and Coastal Pakistan*. Van Nostrand Reinhold, New York, 3–26.

HARRIS, P. M., KENDALL, C. G. St. C. & LERCHE, I. 1985. Carbonate cementation—a brief review. *In:* SCHNEIDERMAN, N. & HARRIS, P. M. (eds) *Carbonate Cements*. SEPM, Special Publications, **36**, 79–95.

HARWOOD, G. M. & SULLIVAN, M. 1991. Sedimentary history of the Moyvoughly area, County Westmeath, Ireland: evidence for syn-sedimentary fault movements in a mixed carbonate–siliciclastic system of the Courceyan age. *In:* LOMAND, A. J. & HARRIS, P. M. (eds) *Mixed Carbonate–Siliciclastic Sequences*. SEPM (Society for Sedimentary Geology), Special Publications, Core Workshop No. 15, 353–384.

HOSSEINI-BARZI, M. & TALBOT, C. 2003. A tectonic pulse in the Makran accretionary prism recorded in Iranian coastal sediments. *Journal of the Geological Society, London*, **160**, 903–910.

HUDSON, T. L. & MAGOON, L. B. 2002. Tectonic controls on greenhouse gas flux to the Paleogene atmosphere from the Gulf of Alaska accretionary prism. *Geology*, **30**, 547–550.

HUMPHREY, J. D. & QUIN, T. M. 1989. Coastal mixing zone dolomite, forward modeling, and massive dolomitization of platform-margin carbonate. *Journal of Sedimentary Petrology*, **59**, 438–453.

JORGENSEN, O. O. 1976. Recent high magnesian calcite/ aragonite cementation of beach and submarine sediments from Denmark. *Journal of Sedimentary Research*, **46**, 940–951.

KAUFMAN, J., CANDER, H. S., DANIELS, L. D. & MEYERS, W. J. 1988. Calcite cement stratigraphy and cementation history of Burlington–Keokuk Formation (Mississippian), Illinois and Missouri. *Journal of Sedimentary Petrology*, **58**, 312–326.

KIDD, R. G. W. & MCCALL, G. J. H. 1985. Plate tectonics and the evolution of Makran. *In*: MCCALL, G. J. H. (ed.) *East Iran Project, Area No. 1*. Geological Survey of Iran, Report, **1**, 564–618.

KUKOWSKI, N., SCHILLHORN, T., HUHN, K., VON RAD, U., HUSEN, S. & FLUEH, E. R. 2001. Morphotectonics and mechanics of the central Makran accretionary wedge off Pakistan. *Marine Geology*, **173**, 1–19.

KULM, L. D. & SUESS, E. 1990. Relationship between carbonate deposits and fluid venting: oregon accretionary prism. *Journal of Geophysical Research*, **95**, 8899–8915.

LINCOLN, J. M. & SCHLANGER, S. O. 1991. Atoll stratigraphy as a record of sea level changes: problems and prospects. *Journal of Geophysical Research*, **96B**, 6727–6752.

MACHEL, H. G. & BURTON, E. A. 1991. Factors governing cathodoluminescence in calcite and dolomite, and their implications for studies of carbonate diagenesis. *In*: BARKER, C. E. & KOPP, O. C. (eds) *Luminescence Microscopy and Spectroscopy: Qualitative and Quantitative Applications*. SEPM, Short Course Series, **25**, 37–57.

MCCALL, G. J. H. 1993. New insight into the plate tectonics of southern Iran and Persian Gulf region. *Earth Science Journal*, **9**, 59–62.

MEYERS, W. J. 1978. Carbonate cements: their regional distribution and interpretion in Mississippian limestones of south western New Mexico. *Sedimentology*, **25**, 371–400.

MEYERS, W. J. 1989. Trace element and isotope geochemistry of zoned calcite cement, Lake Valley Formation (Mississippian, New Mexico). *Sedimentary Geology*, **65**, 355–370.

MEYERS, W. J. 1991. Calcite cement stratigraphy: an overview. *In*: BARKER, C. E. & KOPP, O. C. (eds) *Luminescence Microscopy and Spectroscopy: Qualitative and Quantitative Applications*. SEPM, Short Course Series, **25**, 133–148.

MINSHULL, T. & WHITE, R. 1989. Sediment compaction and fluid migration in the Makran accretionary prism. *Journal of Geophysical Research*, **94**, 7387–7402.

MORAD, S. 1998. Carbonate cementation in sandstones distribution patterns and geochemical evolution.

In: MORAD, S. (ed.) *Carbonate Cementation in Sandstone*. International Association of Sedimentologists, Special Publications, **26**, 1–26.

MOSS, S. & TUCKER, M. E. 1995. Diagenesis of Barremian–Aptian platform carbonates (the Urgonian Limestone Formation of SE France): near-surface and shallow-burial diagenesis. *Sedimentology*, **42**, 853–874.

MUCCI, A. 1988. Manganese uptake during calcite precipitation from sea water: conditions leading to the formation of a pseudokutnahorite. *Geochimica et Cosmochimica Acta*, **52**, 1859–1868.

OLIVER, J. 1992. The spots and stains of plate tectonics. *Earth-Science Reviews*, **32**, 77–106.

PAGE, W. D., ALT, J. N., CLUFF, L. S. & PLAFKER, G. 1979. Evidence for the recurrence of large-magnitude earthquakes along the Makran coast of Iran and Pakistan. *Tectonophysics*, **52**, 533–547.

PINGITORRE, N. E., EASTMAN, M. P., SANDIDGE, M., ODEN, K. & FREIHA, B. 1988. The coprecipitation of manganese(II) with calcite, an experimental study. *Marine Geochemistry*, **25**, 107–120.

PLATT, J. P., LEGGETT, J. K. & ALAM, H. R. S. 1985. Large-scale sediment underplating in the Makran accretionary prism, southwest Pakistan. *Geology*, **13**, 507–511.

RAO, C. P. & ADABI, M. H. 1992. Carbonate minerals, major and minor elements and oxygen and carbon isotopes and their variation with water depth in cool and temperate carbonates, eastern Tasmania, Australia. *Marine Geology*, **103**, 249–272.

REYSS, J. L., PIRAZZOLI, P. A., HAGHIPOUR, A., HATTE, C. & FONTUGNE, M. 1998. Quaternary marine terraces and tectonic uplift rates on the south coast of Iran. *In*: STEWART, I. S. & VITA-FINZI, C. (eds) *Coastal Tectonics*. Geological Society, London, Special Publications, **146**, 225–237.

RUSSEL, K. L., DEFFEYES, K. S. & FOWLER, G. A. 1967. Marine dolomite of unusual isotopic composition. *Science*, **155**, 189–191.

SANDBERG, P. 1985. Aragonite cements and their occurrence in ancient limestones. *In*: SCHNEIDERMAN, N. & HARRIS, P. M. (eds) *Carbonate Cements*. SEPM, Special Publications, **36**, 33–58.

SIBSON, R. H., MOORE, J. M. & RANKIN, A. H. 1975. Seismic pumping—a hydrothermal fluid transport mechanism. *Journal of the Geological Society, London*, **131**, 653–659.

STONELEY, R. 1974. Evolution of the continental margin bounding a former southern Tethys. *In*: BURK, C. & DRAKE, C. L. (eds) *The Geology of Continental Margins*. Springer, Berlin, 889–903.

STRASSER, A. E. & DAVAUD, E. 1989. Formation of Holocene limestone sequences by progradation, cementation, and erosion: two examples from the Bahamas. *Journal of sedimentary petrology*, **56**, 422–428.

STRASSER, A. E., DAVAUD, E. & JEDOUI, Y. 1989. Carbonate cements in Holocene beachrock: example from Bahiret el Biban, southeastern Tunisia. *Sedimentary Geology*, **62**, 89–100.

TUCKER, M. E. 1993. Carbonate diagenesis and sequence stratigraphy. *In*: WRIGHT, V. P. (ed.) *Sedimentology Review 1*. Blackwell Scientific, Oxford, 51–72.

VECSEI, A. & HOPPIE, B. W. 1996. Sequence stratigraphy and diagenesis of the Miocene–Oligocene below the New Jersey continental slope: Implications of physical properties and mineralogical variations. *In*: MOUNTAIN, G. S., MILLER, K. G., BLUM, P., POAG, C. W. & TWICHELL, D. C. (eds) *Proceedings of the Ocean Drilling Program, Scientific Results, 150*. Ocean Drilling Program, College Station, TX, **150**, 361–376.

VEIZER, J. 1983. Trace elements and the stable isotopes in sedimentarycarbonates. *In*: REEDER, R. J. (ed.) *Carbonates: Mineralogy and Chemistry*. Reviews in Mineralogy, **11**, 265–299.

VERNANT, P., NILFOROUSHAN, F. *ET AL*. 2004. Present-day crustal deformation and plate kinematics in Middle East constrained by GPS measurements in Iran and northern Oman. *Geophysical Journal International*, **157**, 381–398.

VITA-FINZI, C. 1987. [14]C deformation chronologies in coastal Iran, Greece and Jordan. *Journal of the Geological Society, London*, **144**, 553–560.

VON RAD, U., BERNER, U. *ET AL*. 2000. Gas and fluid venting at the Makran accretionary wedge of Pakistan. *Geo-Marine Letters*, **20**, 10–19.

WARREN, J. 2003. Dolomite: occurrence, evolution and economically important associations. *Earth-Science Reviews*, **52**, 1–81.

WHITE, R. S. 1982. Deformation of the Makran accretionary sediment prism in the Gulf of Oman (north–west Indian Ocean). *In*: LEGGETT, J. K. (ed.) *Trench and Forearc Geology: Sedimentation and Tectonics on Modern and Ancient Active Plate Margins*. Blackwell Science, Oxford, 357–372.

WHITE, R. S. & KLITGORD, K. 1976. Sediment deformation and plate tectonics in the Gulf of Oman. *Earth and Planetary Science Letters*, **32**, 199–209.

WHITE, R. S. & ROSS, D. A. 1979. Tectonics of the western Gulf of Oman. *Journal of Geophysical Research*, **84**, 3479–3489.

WIEDICKE, M., NEBEN, S. & SPIESS, V. 2001. Mud volcanoes at the front of the Makran accretionary complex, Pakistan. *Marine Geology*, **172**, 57–73.

Current distribution of oil and gas fields in the Zagros Fold Belt of Iran and contiguous offshore as the result of the petroleum systems

M. L. BORDENAVE[1]* & J. A. HEGRE[2]

[1]*MouvOil SA, 24 Avenue du Président Kennedy, 75016—Paris, France*

[2]*GdF Suez, Exploration Production Division, 361 Avenue du Président Wilson, 93211—Saint Denis La Plaine, France*

Corresponding author (e-mail: Max.Bordenave@wanadoo.fr)

Abstract: In the current Zagros Fold Belt of Iran and in its contiguous offshore, five petroleum systems caused an impressive gathering of oil and gas fields that represent some 8% and 15% of global oil and gas reserves, respectively. Almost all the oil fields are located in the relatively small Dezful Embayment, which extends over 60 000 km^2, whereas most of the gas fields are concentrated in Central and Coastal Fars and in the contiguous offshore area. This paper describes the functioning of the various petroleum systems through time, each petroleum system having its own specificity, and reconstructs the succession of events that explains the current location of the oil and gas fields and the reservoirs in which oil and/or gas accumulated. In addition to the classical description of the petroleum systems (distribution and organic composition of the source rocks, evolution of their maturity through time, geometry of drains and reservoirs, and trap availability at the time of migration), the influence of tectonic phases (Acadian, Hercynian, Late Cenomanian to pre-Maastrichtian, and Late Miocene to Pliocene Zagros phases) on the various systems are discussed. As the time of oil and/or gas expulsion from the source rocks is necessary to reconstruct migration paths and to locate the traps available at the time of migration, extensive modelling was used. The timing of oil or gas expulsion was compared with the timing of tectonic events. For the older systems, namely the Palaeozoic (Llandovery source rocks), Middle Jurassic (Sargelu), Late Jurassic (Hanifa–Tuwaiq Mountain–Diyab) and Early Cretaceous (Garau), oil and/or gas expulsion occurred before the Zagros folding. Oil migrated over long distances, according to low-angle geometry, towards large-scale low-relief regional highs and salt-related structures. In the current Zagros Fold Belt, oil and gas remigrated later to the closest Zagros anticlines. In contrast, for the prolific Middle Cretaceous to Early Miocene System (Kazhdumi, Pabdeh), oil expulsion occurred almost everywhere in the Dezful Embayment after the onset of the Zagros folding. Oil migrated vertically towards the closest anticlines through a system of fractures. A comparison was made between the oil expelled from the source rocks, as calculated by the model, and the initial oil in place discovered in the fields. Oils were grouped into families based upon isotopic composition (carbon and sulphur), and biomarkers. Correlation between pyrolysates and oils verifies the origin of the oils that was proposed to explain the current location of the oil (and gas) fields.

Most of the Iranian oil fields are situated in the Dezful Embayment, a depressed area limited to the NW by the mountains of Lurestan, to the SE by the Province of Fars, and to the NE by the Khuzestan 'Mountain Front' (Figs 1 and 2). Ninety-five per cent of the oil accumulated in the Asmari (Early Miocene) and Bangestan (which includes the Cenomanian–Turonian Sarvak Formation and the Santonian Ilam Formation) limestone reservoirs (Fig. 3). In addition, a few middle-sized oil fields are scattered in SE Lurestan, SW of the High Zagros Thrust, and SE of Shiraz. They produce from the Khami (Early Cretaceous), Sarvak and/or Ilam reservoirs.

Fifteen gas fields were discovered to date in Permo-Triassic carbonate reservoirs; 14 are located in Central and Coastal Fars and the contiguous offshore area, including South Pars, the Iranian part of the giant North Dome–South Pars Field, and one in SW Lurestan. In addition, four gas fields located in the Bandar Abbas–Qeshm Island area contain gas in younger reservoirs (Cretaceous to Middle Miocene). In the Dezful Embayment, Pazanan is a gas field with a small oil leg, and Kuh-e Bangestan on the Khuzestan Mountain Front is gas-bearing in the Early Jurassic.

Five petroleum systems were defined, based on the association of source rocks, reservoirs and seals (Fig. 3), as follows.

(1) The Palaeozoic Petroleum System, for which Permo-Triassic carbonate reservoirs capped by the thick anhydrite of the Dashtak Formation of

From: LETURMY, P. & ROBIN, C. (eds) *Tectonic and Stratigraphic Evolution of Zagros and Makran during the Mesozoic–Cenozoic.* Geological Society, London, Special Publications, **330**, 291–353.
DOI: 10.1144/SP330.14   0305-8719/10/$15.00 © The Geological Society of London 2010.

Fig. 1. General location map of the Middle East region; the boxed area shows the location of Figure 2.

Middle Triassic age were charged by Early Silurian (Llandovery) source rocks.

(2) The Middle Jurassic Petroleum System, for which Sargelu source rocks developed in the Mesopotamian Depression are limited above and below by thick evaporitic series and therefore are not connected with reservoirs, except at the edge of the depression.

(3) The Late Jurassic Petroleum System, which is charged by the Hanifa–Tuwaiq Mountain–Diyab source rocks that extend on both side of the Qatar Arch. It caused the formation of most of the giant oil fields in Abu Dhabi and Saudi Arabia, both offshore and onshore. The Late Jurassic Petroleum System has an impact in the Iranian

waters, where oil, often heavy, and bitumen are found in Early Cretaceous and younger reservoirs.

(4) The Early Cretaceous Petroleum System, in which excellent source rocks were deposited in the lower part of the Garau Formation, during the Berriasian–Valanginian period. These source rocks occur in Lurestan, the Dezful Embayment, and the northern part of the Gulf. The Garau source rocks are limited below by the thick Gotnia evaporites, and above by thick basinal marls that continued to be deposited until Santonian time in most of Lurestan. In the Dezful Embayment, the porosity in the Early Cretaceous interval is generally limited to the thin limestone of the Khalij Member of the Gadvan Formation (Barremian).

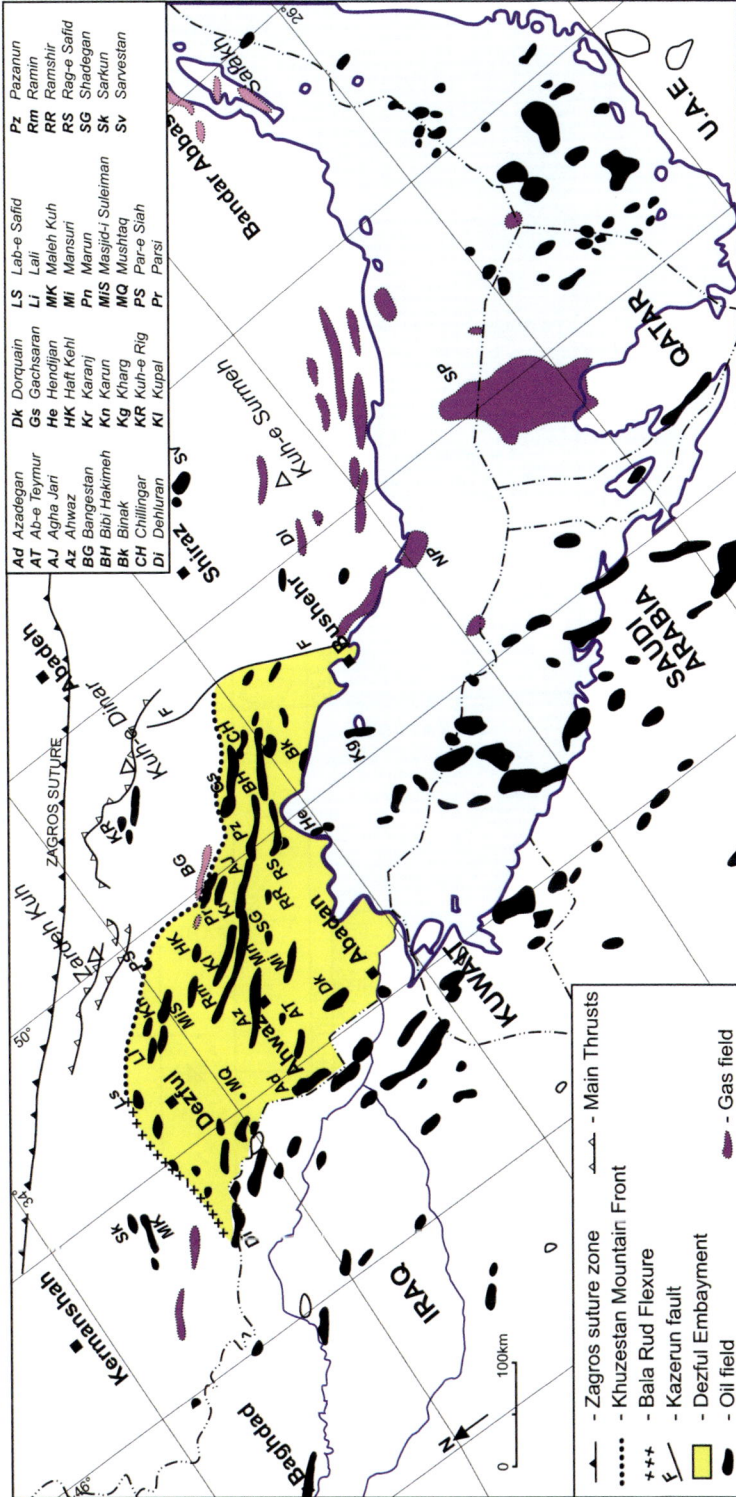

Fig. 2. Location map of the oil and gas fields in the Zagros foothills and contiguous offshore (after Bordenave & Hegre 2005).

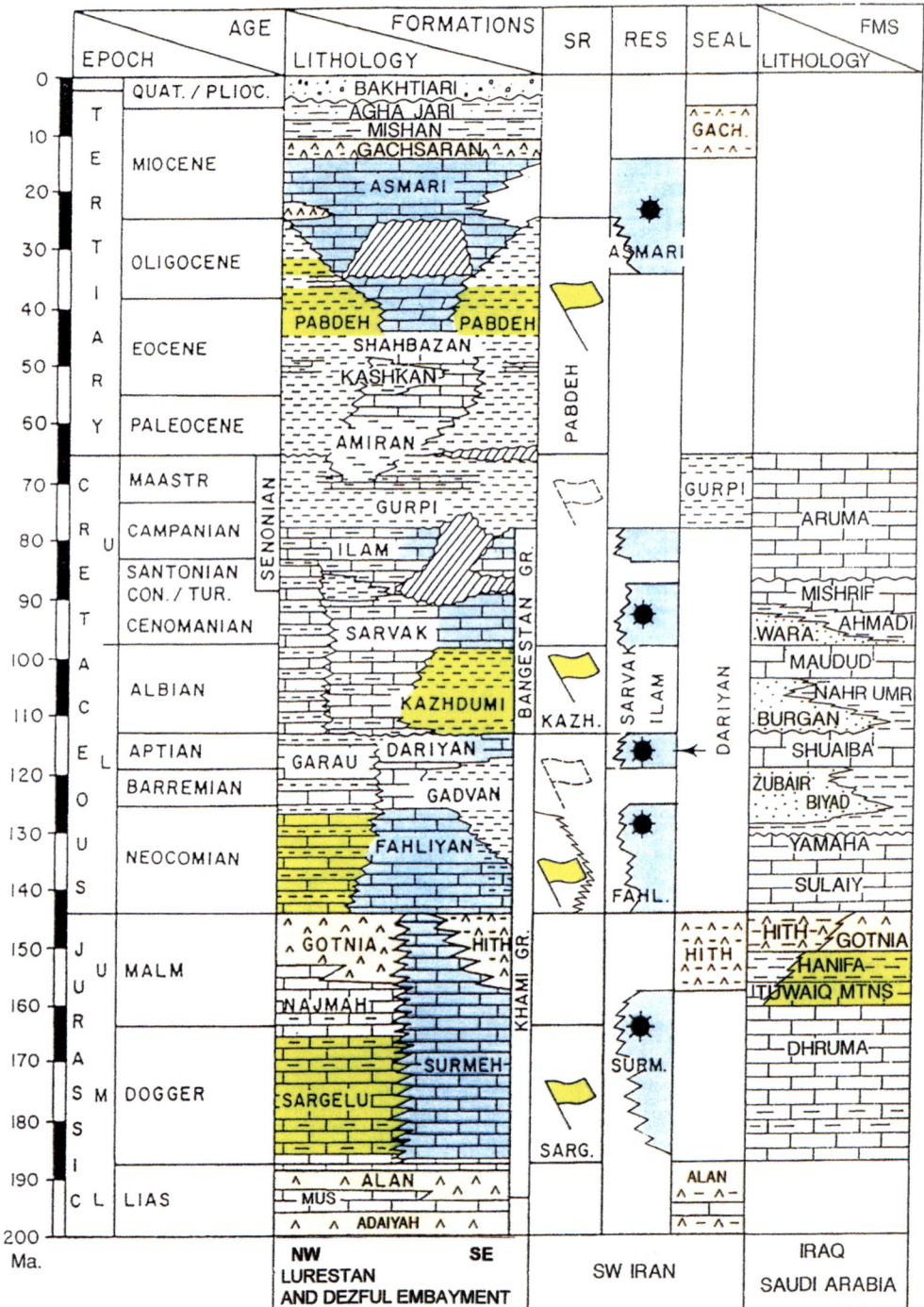

Fig. 3. Schematic stratigraphy and source rocks–reservoir–seal relationship for the Dezful Embayment and neighbouring areas. The main source rocks are indicated by green flags and marginal ones by white flags; reservoirs (Res), and seals are also indicated (after Bordenave & Burwood 1995).

(5) The Middle Cretaceous to Early Miocene System, which includes two main source rocks, the Kazhdumi Formation of Albian age and the upper part of the Pabdeh Formation (Middle Eocene to Oligocene), and a third one developed locally, known as the Ahmadi Member of the Sarvak Formation (Cenomanian) in Iran, and as the Shilaif Formation in Abu Dhabi. These source rocks charged two main reservoirs, the Asmari and Bangestan carbonates. This system is capped by the evaporites of the Gachsaran Formation (Early Miocene).

For each system, the sequence of events that caused the current distribution to oil and gas fields is described. Three of these systems were, at least partly, described in previous publications: Bordenave (2002a, 2008) for the Palaeozoic Petroleum System; Bordenave & Huc (1995) for the Early Cretaceous System; Bordenave & Burwood (1990, 1995), Bordenave (2002b) and Bordenave & Hegre (2005) for the Middle Cretaceous to Early Miocene System. These publications were used in this study and are supplemented by the present paper.

Material and methods

An inventory of the source rocks that charged the various reservoirs of the Zagros Fold Belt prolific fields was based upon extensive field work, and on the study of wells drilled in the Zagros Fold Belt and in the Gulf contiguous areas (Bordenave & Sahabi 1971; Bordenave et al. 1971; Bordenave & Nili 1973). A database in excess of 4500 field samples, together with core and ditch cuttings was used to study the distribution, thickness, and organic characteristics of the source rocks occurring in the Zagros Fold Belt of Iran (Bordenave & Burwood 1990, 1995). This database was complemented by the study of thin sections from high-density ditch cuttings from 55 key wells, together with a review of their gamma-ray logs, using the excellent correlation observed between uranium and carbon content in the Jurassic, Cretaceous and Eocene source rocks (Brosse & Bordenave 1993). The isotopic composition of 40 oil samples and of source rock pyrolysates was also measured. Vitrinite profiles were established for selected wells (Burwood 1978).

The thermal maturation of the source rocks was estimated by using the GENEX software (Ungerer 1993) developed by IFP (Institut Français du Pétrole). Several assumptions of variable and constant heat flow values were tested. Eventually, a constant heat flow was used because of the relatively constant rate of subsidence observed in the current Zagros Fold Belt (Bordenave & Hegre 2005). A $36 \, \text{mW m}^{-2}$ mantle heatflow ($64 \, \text{mW m}^{-2}$ total heatflow) corresponding to the best fit with the current values of thermal markers (vitrinite reflectance, Rock-Eval T_{max} and hydrogen index profiles) in Dezful Embayment wells was selected. Thermal markers reflecting the cumulative thermal history of the sediments were preferred to current geothermal gradients that depend upon Late Miocene–Pliocene subsidence. Because of the limited availability of current thermal maturation parameters, the heat flow calculated for the Dezful Embayment was also used for Fars. The thickness of the eroded section in these wells was estimated from a set of regional isopach maps. The onset of oil expulsion was assumed to occur for a 30% oil saturation of the pore space of the source rocks.

The relative timing of the onset of the oil expulsion window and of the availability of traps was compared (and particularly with the onset of the formation of the Zagros large-scale anticlines) as it is essential to establish the type of functioning of the various petroleum systems. According to the timing, oil migrated either over long distance, according to a low-angle geometry, towards large regional highs where it accumulated, or subvertically to the nearby anticlines. In the first case, part of the oil (and gas), accumulated in pre-Zagros orogeny regional highs, was later re-accumulated in the Zagros anticlines. In the second case, when migration and entrapment occurred after the onset of the Zagros orogeny, the concept of drainage area was used and the transformation ratio (TR) of the source rocks was calculated at the crestal part of the anticlines as well as in the synclines. The calculated amount of oil expelled from the source rocks was then compared with the initial oil in place (IOIP) of the discovered fields. Modelling was applied to 21 fields that contain 86% of the IOIP evidenced in the Iranian Zagros Fold Belt. Altogether, more than 200 profiles were used. Oils were grouped into families using isotopic composition (carbon and sulphur), and biomarkers. Correlations between pyrolysates and oils were established to verify the scenario proposed to explain the current location of the oil (and gas) fields.

The Palaeozoic Petroleum System

The Palaeozoic Petroleum System provides an example of a system based upon long-range migrations, formation of large accumulations on regional highs long before the formation of the Zagros folds, and the late re-accommodation of part of the gas in anticlines that resulted from the Late Miocene to Pliocene orogeny (Bordenave 2002a, 2008).

Since the discovery of Kangan in 1973, an increasing number of gas fields were discovered in the Permo-Triassic carbonates of Iran. Current gas reserves of these fields, in excess of 600 trillion cubic feet (Tcf), represent some 10% of the global gas reserves. The Iranian part of the huge North Dome–South Pars field that extends over the Iranian and Qatari waters contains 436 Tcf, and the reserves of other fields located either in the SW part of the Fars Province of Iran and or in its contiguous offshore amount to an additional 150–200 Tcf (Fig. 4). The Permo-Triassic Iranian gas fields are grouped into two areas; (1) the southeastern part of the Province of Fars and its contiguous offshore; (2) SW Lurestan. In addition, a group of gas fields was discovered in younger reservoirs in the Bandar Abbas–Qeshm Island area. Unlike what could have been expected, no source rocks were evidenced in those wells that reached pre-Palaeozoic layers. In contrast, excellent source rocks were described at the base of the Silurian (Llandovery) at Kuh-e Gahkum and Kuh-e Faraghan, NNW of Bandar Abbas (Fig. 5).

Stratigraphy, source rocks, reservoirs, and caprocks

The role of the layers critical for the functioning of the Palaeozoic Petroleum System was considered either for their source rock potential, their reservoir characteristics, their seal capacity or their ability to form early traps (Fig. 6).

The Proterozoic and Early Palaeozoic sediments. The Late Proterozoic to Early Cambrian Hormuz Salt formed deep-seated salt-pillows that ranges, sometimes as early as during the Triassic. These structures were available for trapping oil and gas long before the Zagros folding.

In SW Iran, Cambro-Ordovician layers crop out at the base of the High Zagros thrusts (Setudehnia 1976; Huber 1978; Sharland et al. 2001), whereas Ordovician sediments crop out at Kuh-e Surmeh. These layers include fluviatile reddish cross-bedded sandstone, micaceous mudstone, and shallow marine silty shale and dolomite (Fig. 6). Samples we analysed are organic-lean with total organic carbon (TOC) values being lower than 0.4%. However, some dark lacustrine to deltaic shales layers of the Mila Formation recently sampled in the north of the Izeh zone showed TOC values up to 11% for T_{max} in the 450 °C range (Rudkiewicz et al. 2007).

The Silurian source rocks. At the end of the Ordovician, the northern margin of Gondwana, which included North Africa and the Arabian Platform, was situated close to the South Pole, at a latitude of c. 60°S (Scotese & Golonka 1993). Major glaciations, which peaked during the Late Ordovician (Ashgill), caused the formation of a huge ice cap that extended widely over Gondwana (Al Husseini

1990; Brenchley et al. 1994). The melting of the ice cap during the Early Llandovery induced a series of sea-level rises (Loydell 1998) and a progressive inundation of the lowlands of the already peneplaned Gondwana surface. The rapid transgression caused the deposition of low-energy laminated graptolitic shales at the edge of the Gondwana continent extending from Morocco to Saudi Arabia and Oman. These shales, deposited in an oxygen-deprived environment at the early stage of the transgression, are organic-rich and highly pyritic. They were deposited only on low-lying areas, whereas the Ordovician palaeoreliefs were not affected by the early phase of the transgression (Aoudeh & Al-Hajri 1995; Lüning et al. 2000; Bordenave 2008). The anoxia was probably caused by water layering resulting from the flood of fresh water originating from the melting of the ice cap. Such an assumption could explain the relative brevity of the first anoxic event, 1–2 Ma at most. When the melting ceased, the anoxia also ceased (Lüning et al. 2000).

The organic-rich graptolitic shales contain abundant algal material, mostly *Tasmanacea* (Bordenave et al. 1970; Combaz 1986). They are highly radioactive (up to 400 API units) as a result of their uranium content, and are easily identified in wells from gamma-ray logs. They were called 'hot shales' because of their radioactivity. In Saudi Arabia, a 150 API units cut-off corresponding to 2% TOC was used (Jones & Stump 1999). The contrast between the acoustic impedance of the hot shales as compared with those of older layers results in a well-marked seismic reflection at the base of the hot shales. The geographical extension of this seismic marker permits the evaluation of the extension of the hot shales beyond wells.

In Saudi Arabia, light oil–condensate and sweet gas discoveries made in Early Permian sandstone of the Unayzah Formation and in basal Khuff limestone originated from Llandovery Qusaiba hot shales, as demonstrated by the similar isotopic carbon (−29.0‰ to −30.5‰) and biomarker compositions of the Qusaiba hot shale extracts and the oil samples (Abu-Ali et al. 1991; Mahmoud et al. 1992; Cole et al. 1994). In western Oman, oil accumulations in the Haushi clastic deposits of Carboniferous age are also related to Silurian hot shales, as shown by oil-to-source rock correlation (Grantham et al. 1987).

The Silurian sediments of Saudi Arabia are described as the Qalibah Formation (Vaslet 1987; Mahmoud et al. 1992). The coarsening-upward Qalibah Formation, deposited unconformably over the Ashgill glacial and periglacial clastic deposits, is divided into a lower Qusaiba Member, which is dominantly shaly, and a sandy upper Sharawra Member, which was part of a prograding

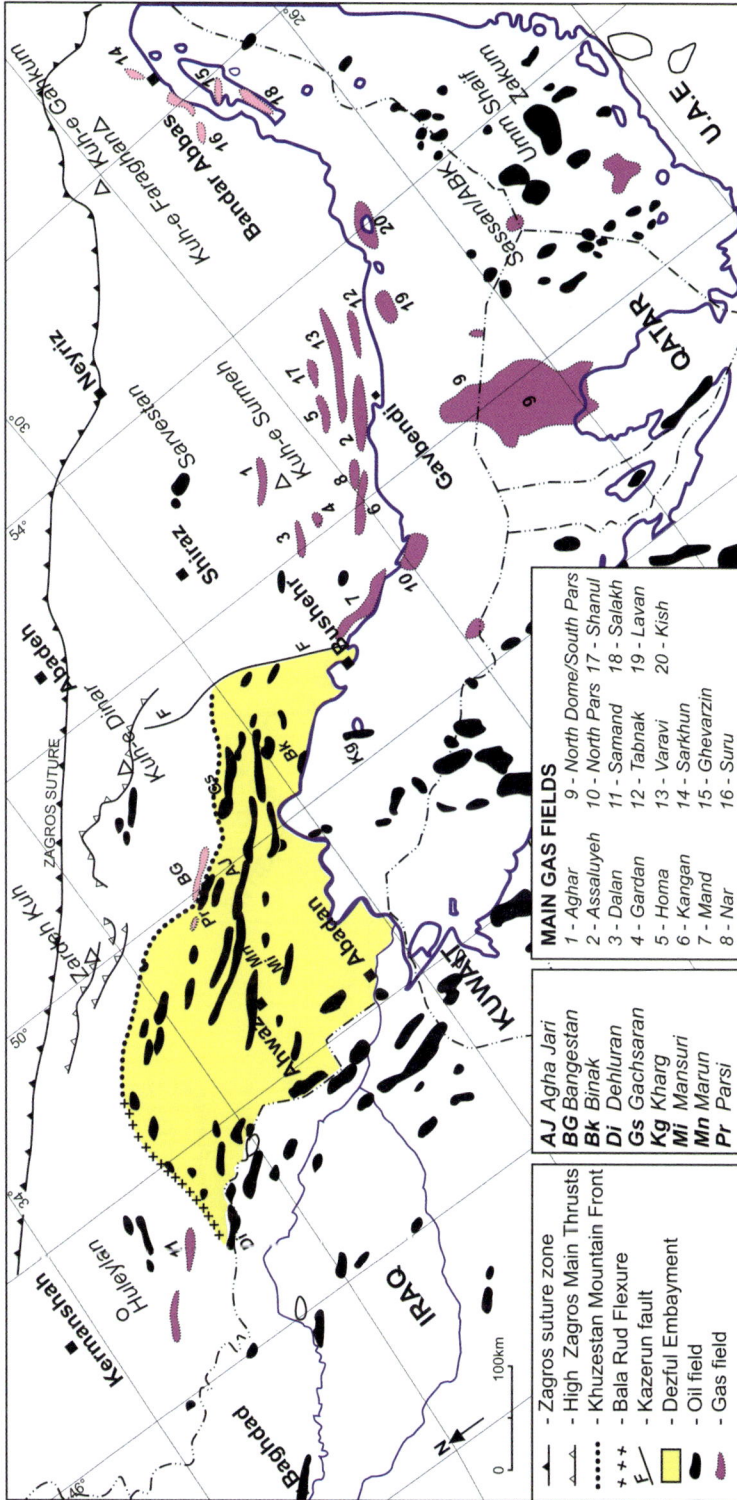

Fig. 4. The Zagros Fold Belt and contiguous offshore areas: location of the gas fields assumed to be related to the Palaeozoic Petroleum System (after Bordenave 2008).

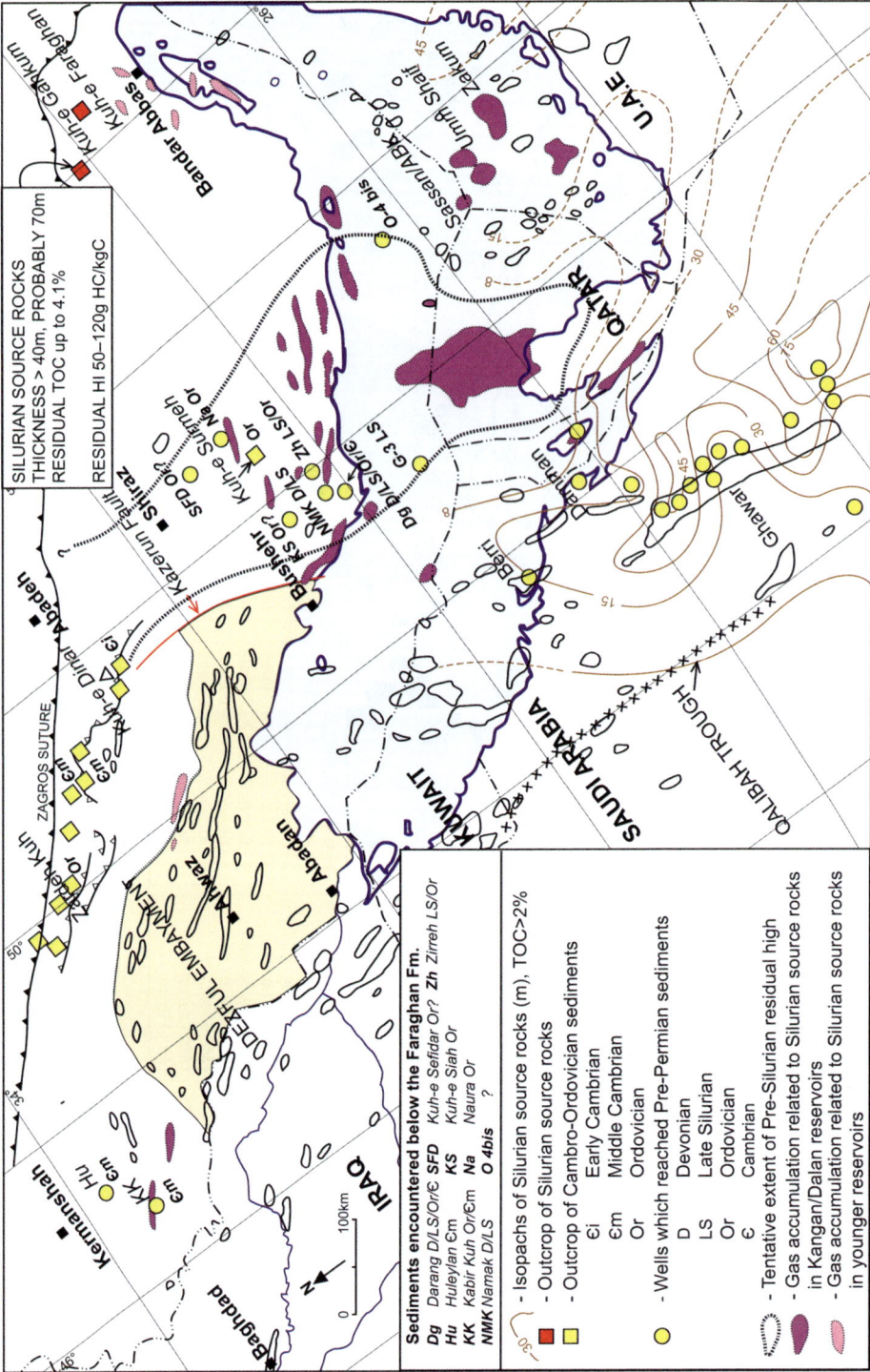

Fig. 5. Extension and thickness of the Early Llandovery source rocks (cut-off TOC = 2% or 150° API gamma ray), before the erosion that followed the Hercynian phase (after Bordenave 2008).

(1) PALAEOZOIC ROCKS: CENTRAL ARABIA

Age (Ma)	stage	Formation (member)	Lithology	(m)
PERMIAN 250	Tatarian / Kazanian / Kungurian	KHUFF		500
270	Artinskian / Sakmarian / Asselian	UNAYZAH (A) (B) (C)	alluvial fluvial / fluvial	180
290				
CARBONIFEROUS 315	Stephanian / Westphalian / Namurian / Visean / Tournaisian	"HERCYNIAN" EVENT		
360				
DEVONIAN 395	Famennian / Frasnian / Givetian-Eifelian / Emsian	JUBAH		340
	Siegenian / Gedinnian	JAUF		300
410	Pridolian	TAWIL	fluvial	180
SILURIAN 430	Ludlovian / Wenlockian	SILURIAN Hiatus		
	Llandoverian	QALIBAH SHARAWRA QUSAIBA / SARAH	marine	350 / 600
440			S	
ORDOVICIAN 450	Ashgillian	ZARQA	glacial	265
460	Caradocian	QUWARAH RA'AN KAHFAH HANADIR (QASIM)	marine	200+
470	Llandeilian			
480	Llanvirnian / Arenigian / Tremadocian	SAJIR (SAQ)	littoral	300
500				
CAMBRIAN 525 UPPER / 545 MIDDLE / LOWER		RISHA	fluvial-deltaic	300
		CAMBRIAN Hiatus		
>600		PRECAMBRIAN BASEMENT		

(2) IRAN - HIGH ZAGROS AND FARS

Formation (member)		(m)
DALAN Upper / Nar Mbr / Lower	restricted shallow marine	750
FARAGHAN	shallow marine	100/ 510
PRE-PERMIAN HERCYNIAN HIATUS		
ZAKEEN	fluvial/ shallow marine	285
PRE-DEVONIAN ACADIAN HIATUS		
SARCHAHAN	marine	70 / 50
SEYAHOO/ ZARD-e KUH	marine / marine	? / 325
ILEBEYK	shallow marine	270
BAZUFT	marine	420
LALUN	fluvial	1500+
HORMUZ	brackish	1000+

S: Source rocks ☀ Oil and gas-bearing reservoirs ☼ Gas-bearing reservoirs

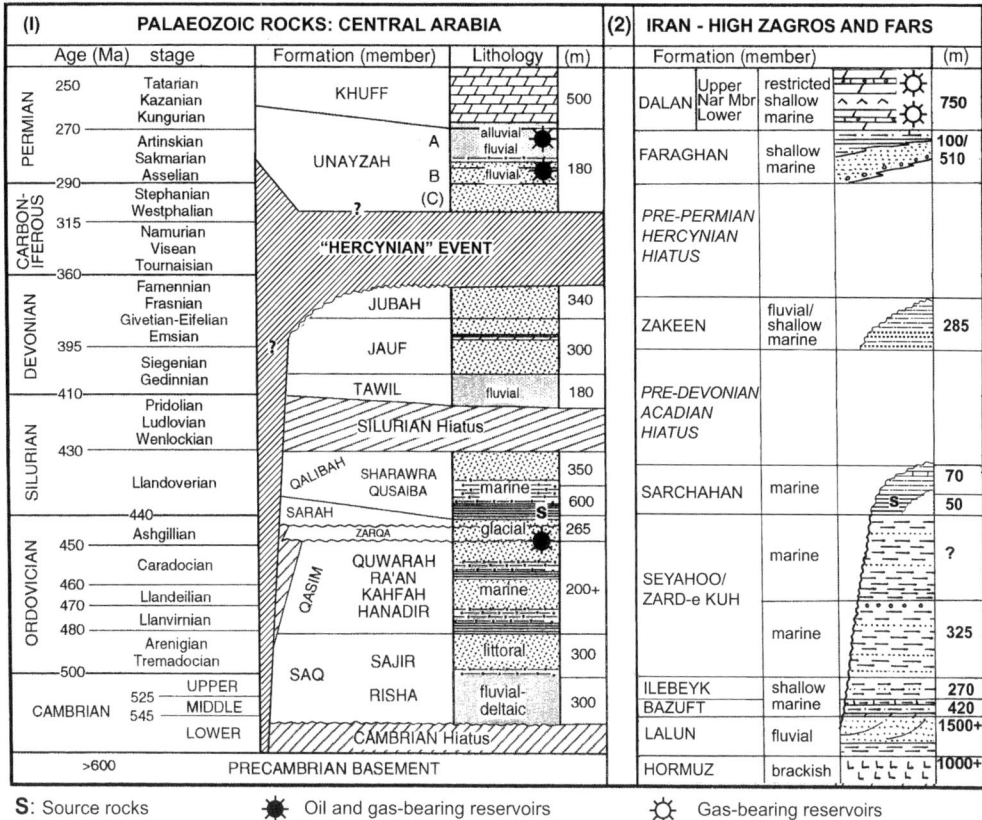

Fig. 6. Stratigraphy of the Palaeozoic sediments (1) in Central Arabia (modified after McGillivray & Husseini 1992) and (2) in the High Zagros, Iran, from Zardeh Kuh to Kuh-e Faraghan (data from Setudehnia 1976; Huber 1978 Ghavidel-Syooki 2003)

fluvio-deltaic system. Variable thickness of hot shales was deposited near the base of the Qusaiba Member. When immature, the basal Qusaiba hot shales generally contain between 4 and 7% TOC (average 4.7%), with local values that may reach 20%. The highest values of both organic content and radioactivity of the hot shales are observed near the base of the transgression; later these values decrease rapidly. The average hydrocarbon potential of immature hot shales (S_2) is 22 g HC kg^{-1} of rock, and their hydrogen index (HI) values vary, for most of the samples, from 400 to 600 g HC kg^{-1} C, classifying them as excellent type 2 oil-prone kerogen (Cole *et al.* 1994).

In the Iranian Zagros foothills, Silurian source rocks crop out at Kuh-e Gahkum and Kuh-e Fara-ghan, north of Bandar Abbas (Bordenave *et al.* 1971; Ala *et al.* 1980). At Kuh-e Faraghan, 600 m of barren shales are overlain by 70 m of organic-rich black shales that contain a Llandovery fauna. Well-preserved chitinozoans recently collected at

Kuh-e Faraghan confirm the Llandovery age of the Silurian layers. The Early Palaeozoic sediments of Kuh-e Faraghan were divided into an Ordovician Seyahou Formation and a Silurian Sarchahan Formation that includes the basal black shales (Ghavidel-Syooki 2000). At Kuh-e Gahkum, at least 40 m of black platy-bedded shales with a 'coaly' aspect are radioactive and contain residual values of TOC that are currently between 1.5 and 4.2% even though the organic matter is over-mature (Rock-Eval T_{max} is around 457 °C and HI is between 40 and 110 g HC kg^{-1} C). Cumulative isopachs revealed that the Silurian source rocks of Kuh-e Gahkum were buried deeper than 6000 m prior to Zagros folding. The δ^{13}C value of the Kuh-e Gahkum Silurian shales ($-30.8‰$) is comparable with that of the Safiq Silurian source rocks of Oman (Bordenave & Burwood 1990).

Silurian sediments were identified in two wells, Darang-1 and Zirreh-1 drilled in Coastal Fars (Fig. 5), but no hot shales were described in these wells. These sediments are probably equivalent to the organic-lean shales of the Sharawra Formation that probably cover the Ordovician shales in the North Dome field of Qatar (Konert *et al.* 2001; Abu Ali & Littke 2005). The absence of basal Qusaiba shales would indicate that the Qatar Arch and the NW part of Coastal Fars were still in emergence during the Early Llandovery. With the exception of Darang and Zirreh, the other wells drilled in the Zagros Fold Belt and offshore encountered organic-lean Cambrian and Ordovician sediments. In depressions, such as the Dezful Embayment, Palaeozoic sediments are too deep to be reached.

Data currently available on the extension and the thickness of the Silurian source rocks are assembled in Figure 5. Thickness does not take into account pre-Permian erosion. In Saudi Arabia, a distribution map of the Qusaiba hot shales was established from geochemical analyses, core sedimentological studies, interpretation of gamma-ray logs, and seismic ties between wells (Jones & Stump 1999). West of Qatar, hot shales (cut-off 150° API or 2% TOC) are well developed, with thicknesses varying from 30 to 75 m. Their extension towards the Dezful Embayment was postulated in a context of water stratification related to the melting of the ice cap and a condensed section. Moreover, the existence of Sarchahan Llandovery hot shales, north of Bandar Abbas, in a more distal situation, and the discovery of gas fields in Lurestan, such as Samand, and Kabir Kuh in the Dalan Formation give more credibility to this assumption. East of Qatar, in the Abu Dhabi onshore and offshore, no well was drilled deep enough to reach pre-Devonian sediments. In the absence of drilling information, the existence of large gas accumulations in the Khuff Formation of salt-related structures, such as Zakum or Umm Shaif, is considered as indirect evidence of the existence of Silurian source rocks.

Two Silurian basins are assumed, being separated by the Qatar Arch, one basin extending from Oman, where Silurian source rocks are well developed, with thickness varying from 15 to 80 m (Hughes Clark, 1988), to Abu Dhabi and to SE Fars, and the second extending from Iraq to Saudi Arabia and probably to the Dezful Embayment.

Devonian sediments. Early Devonian to Frasnian sediments were identified from pollen found in shallow-marine sandstones and shales resting unconformably on Llandovery silty shales at Kuh-e Faraghan (Ghavidel-Syooki 2003). The 285 m thick Kuh-e Faraghan Devonian sediments were named the Zakeen Formation (Ghavidel-Syooki 2003). Devonian sediments were also found in wells drilled in Coastal Fars, SE of Bushehr, and in Abu Dhabi (Ali & Silwadi 1989). The existence

of Early Silurian, Devonian and Carboniferous sediments in the Dezful Embayment is likely. Based on available data, this area was subsiding persistently more than Fars at least from the Early Jurassic, as shown by the basinal facies of both Jurassic and Early Cretaceous sediments, and by the isopachs of the Early Cretaceous formations.

The Acadian and Hercynian phases. The existence of Devonian sediments permits the identification of two tectonic phases that corresponded to periods of sedimentation hiatus; that is, a first hiatus between the top of Silurian Sarchahan Formation and the base of the Devonian Zakeen Formation, and a second between the top of the Zakeen Formation and the base of the Faraghan Formation. The first hiatus, which covered the Middle and Late Silurian period, resulted from a gentle Acadian phase. The second hiatus, which covered the Famennian and the entire Carboniferous, is related to an intense Hercynian phase. These two tectonic phases observed in the Zagros Fold Belt are similar to those described in Saudi Arabia (McGillivray & Husseini 1992; Jones & Stump 1999), and in the Western Desert of Iraq (Aqrawi 1998).

During the Namurian and the Westphalian, regional uplifts, basement-core uplifts, evidence of folding, and reverse faulting suggest that the Arabian plate underwent compressive phases as the result of the Hercynian orogeny. These phases resulted in the uplift of the Aleppo and Mardin Highs in Syria, and of the Arabian Shield (Konert *et al.* 2001). They caused the formation of a system of north–south horsts, grabens, and east-tilted fault blocks in Central and Western Arabia, probably along pre-existing Late Proterozoic faults. These phases were followed by a period of intense erosion prior to the deposition of the Late Carboniferous to Early Permian Unayzah Formation (Fig. 7). In Saudi Arabia, more than 1000 m of sediments were removed from the highest blocks, where the erosion reached the basement (Al Husseini 1989; McGillivray & Husseini 1992; Abu Ali & Littke 2005).

In the High Zagros, a marked high was centred on Kuh-e Dinar, where 1000 m of sediments were removed, as compared with Zardeh Kuh, located 220 km to the NNW (Szabo & Kheradpir 1978). The erosion took place prior to the Early Permian, as the eroded Cambro-Ordovician sediments are overlain with an angular unconformity by basal Early Permian clastic deposits dated by pollen at Chal-i Sheh, NE of Zardeh Kuh (Ghavidel-Syooki 1993). The progressive decrease of the erosion amplitude in a NNW direction suggests that Silurian shales may have been deposited and later eroded.

The Kuh-e Dinar High was the highest part of the north–south high that extended over Central

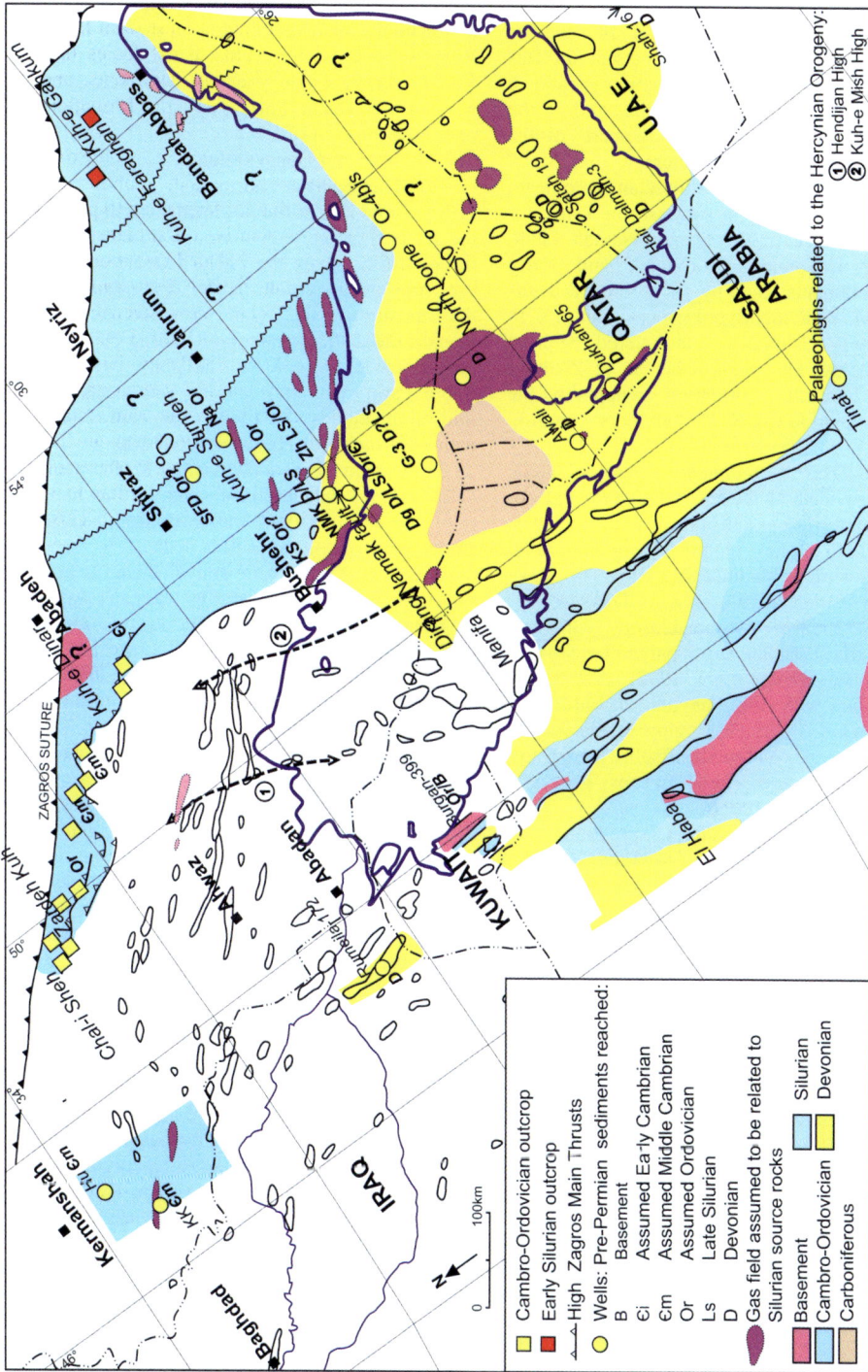

Fig. 7. Pre-Permian subcrop map showing the post-Hercynian intense erosion on north–south-oriented horsts. In contrast, north–south directions hidden in the Zagros Fold Belt are indicated by Jurassic and Cretaceous isopach maps (Hendijan High, Kuh-e Mish High). The major north–south-oriented Kazerun Fault was also related to the Hercynian orogeny and perhaps to an older tectonic phase. A well-identified regional uplift occurred as the result of the Hercynian orogeny, extending from Kuh-e Dinar to Kuh-e Surmeh and towards Lavan Island. Silurian source rocks, when deposited, were eroded on the main horsts (after Bordenave 2008).

and Western Fars and over the Qatar Arch before the deposition of the Permian sediments. This high was clearly demonstrated at Kuh-e Surmeh, where truncated Ordovician shales form a 10° angular unconformity with the overlying conglomerates of the base of the Faraghan Formation (Szabo & Kheradpir 1978). The extension to the south of the Kuh-e Surmeh High is confirmed by nearby wells, Kuh-e Siah-1, Dalan-2 and Naura-1, that found sediments attributed to the Ordovician underneath the Faraghan Formation. In contrast, the Darang-1 well, located close to the Darang-Namak Fault, encountered Devonian, Ordovician and Early Cambrian sediments dated by palynological determination (M. Ghavidel-Syooki, pers. comm.). In the same way, Namak-1 encountered Silurian sediments below Devonian, whereas at Zirreh-1 the Silurian Sarchahan Formation is capped by the Faraghan Formation (M. R. Kamali, pers. comm.).

The Permo-Triassic sediments. Permo-Triassic sediments provide both the reservoirs of the Palaeozoic Petroleum System and its cap rock (Szabo & Kheradpir 1978). They include, from bottom to top, the following formations (Fig. 8).

(1) The Faraghan Formation. Transgressive fluviatile to shallow marine clastic sediments, dated as Early Permian from palynological evidence (Ghavidel-Syooki 1993, 1997) were deposited, sometimes with strong angular unconformity, over a peneplaned surface ranging in the Zagros Fold Belt and in the contiguous offshore from Cambrian to Devonian age. Their thickness generally varies between 50 and 100 m.

(2) The Dalan Formation, predominantly carbonates, dated as Middle to Late Permian (Kungurian–Kazanian to Early Djulfian), according to Szabo & Kheradpir (1978). During the Permian, a reefoid belt was developed along the newly formed edge of the Tethys Ocean. This belt, well observed in the High Zagros, caused the formation of a restricted lagoon extending along the whole Arabian Platform (Fig. 9). As a result of warm and dry climatic conditions, the lagoon turned into a sabkha-type environment during periods of Sealevel lowstand, with the deposition of thick evaporites, whereas fossiliferous and peloidal–oolitic limestones were deposited in oxic clear water during periods of sea-level rise. Most of the time, massive low-energy dolomitic limestone accumulated in the lagoon. In the context of cyclic sedimentation, sedimentological models were used for each cycle to tentatively untangle the extremely complex and very variable distribution of the various facies (Insalaco et al. 2006). Massive dolomitic layers (dolomudstone) are generally tight, whereas grainstone layers, oolitic and peloidal, have fair to good reservoir characteristics.

In most of Fars, the Dalan Formation includes two carbonate members, made of thick massive dolomitic limestone and thin grainstone layers, separated by the massive anhydrite of the Nar Member. When porous and permeable, the Upper Dalan and Kangan form a single reservoir. The average thickness of the Dalan Formation is 750 m, roughly one-third for each of the three units (Lower Dalan, Nar Member and Upper Dalan).

(3) The Kangan Formation, which consists of limestone and dolomite, locally argillaceous, of the Early Triassic (Scythian). A downwarping of the platform during the Early Triassic caused the drowning of the reefoid barrier that existed during the deposition of the Dalan Formation. As a consequence, the shallow-water high-energy grainy facies developed in the SW of the platform (Kangan, South Pars) pass progressively into deeper-water mid-ramp muddy facies in the Kuh-e Dinar–Kuh-e Surmeh area (Insalaco et al. 2006).

Three main facies of the Kangan Formation were identified. (a) 'Clean facies' made of bioclastic to oolitic grainstone, and partly dolomitized mudstone, with minor evaporitic beds. This facies is described at Kuh-e Siah and Kangan. It is also present at South Pars, where the Kangan Formation is grain-dominated. (b) Argillaceous facies, with laminated dark grey marls, argillaceous and dolomitic limestone with abundant *Claraia*. This facies is observed at Kuh-e Surmeh (Szabo & Kheradpir 1978), and at Kuh-e Dinar (Insalaco et al. 2006). (c) Carbonate–evaporites facies developed in Lurestan. The thickness of the Kangan Formation varies from 150 to 250 m.

Prolific gas fields found to date on the Gavbendi High–Qatar Arch, offshore (North Dome–South Pars, North Pars and Balal) as well as onshore (Kangan, Nar, Aghar, Dalan, Varavi, Assaluyeh, Gardan, Tabnak, Homa, Lavan and Kish), accumulated either in the Upper Dalan and Kangan reservoir only, or in both the Dalan and Kangan and the Lower Dalan reservoirs.

(4) The Dashtak Formation, attributed to the later part of the Early to Middle Triassic age, in the absence of diagnostic fossils. The rusty-brown or varicoloured Aghar Shales, which form the base of the Dashtak Formation, are followed by a succession of massive anhydritic and calcareos–dolomitic intervals. Evaporitic zones, named A, B, C and D, consist of massive-bedded anhydrite with subordinate tight dolomitic and limestone beds. Evaporite A varies from 220 to 380 m. Evaporite B is almost isopach in the study area (40 m). Evaporites C and D vary considerably in thickness, C from 0 to 250 m, and D from 0 to 170 m. Both of them were eroded on the Qatar–Gavbendi High, prior to the deposition of the Neyriz Formation of Liassic age, demonstrating the existence of this

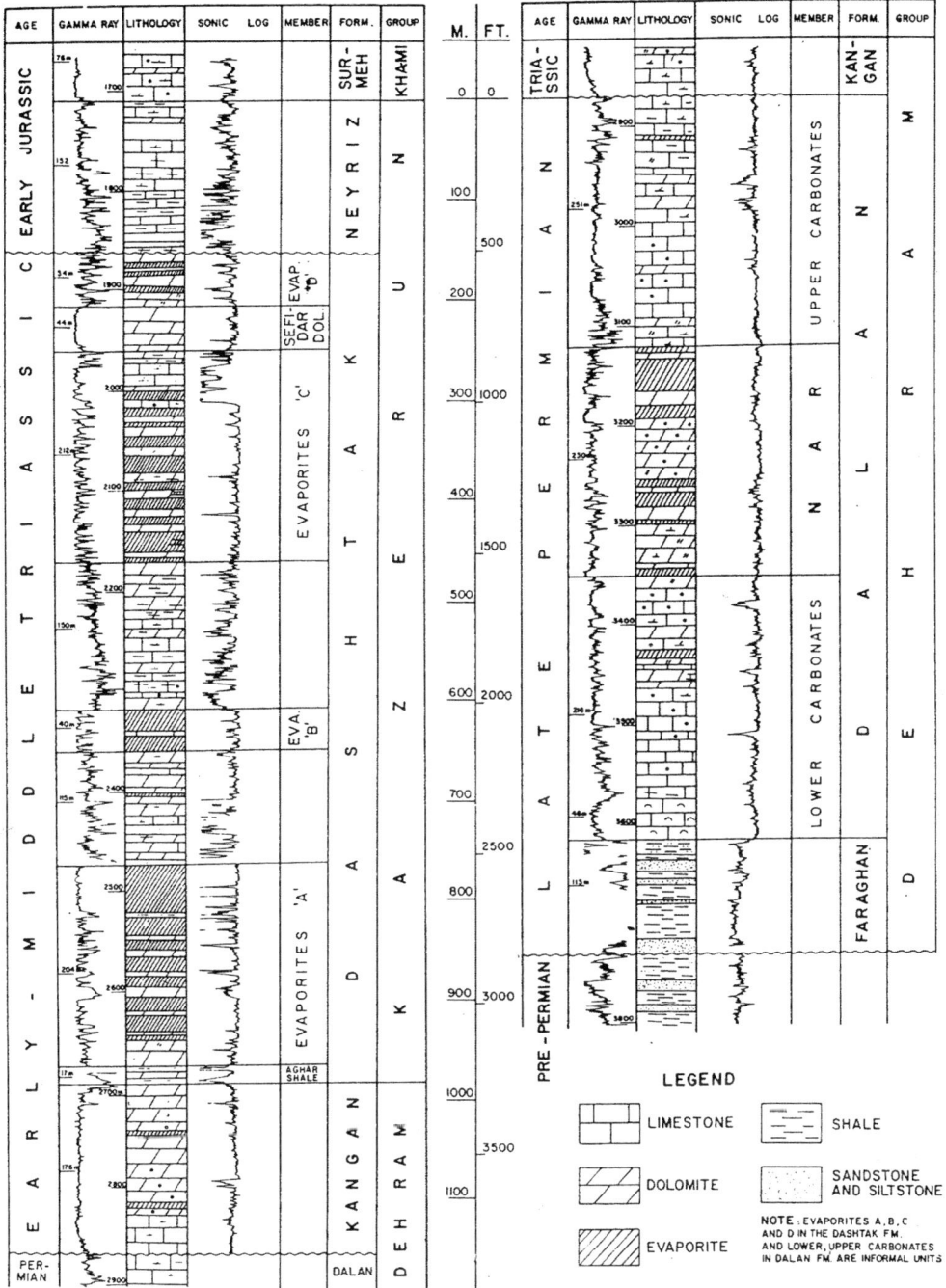

Fig. 8. Kuh-e Siah-1 well type-section showing the lithostratigraphic subdivision of the Permian and Triassic (after Szabo & Kheradpir 1978).

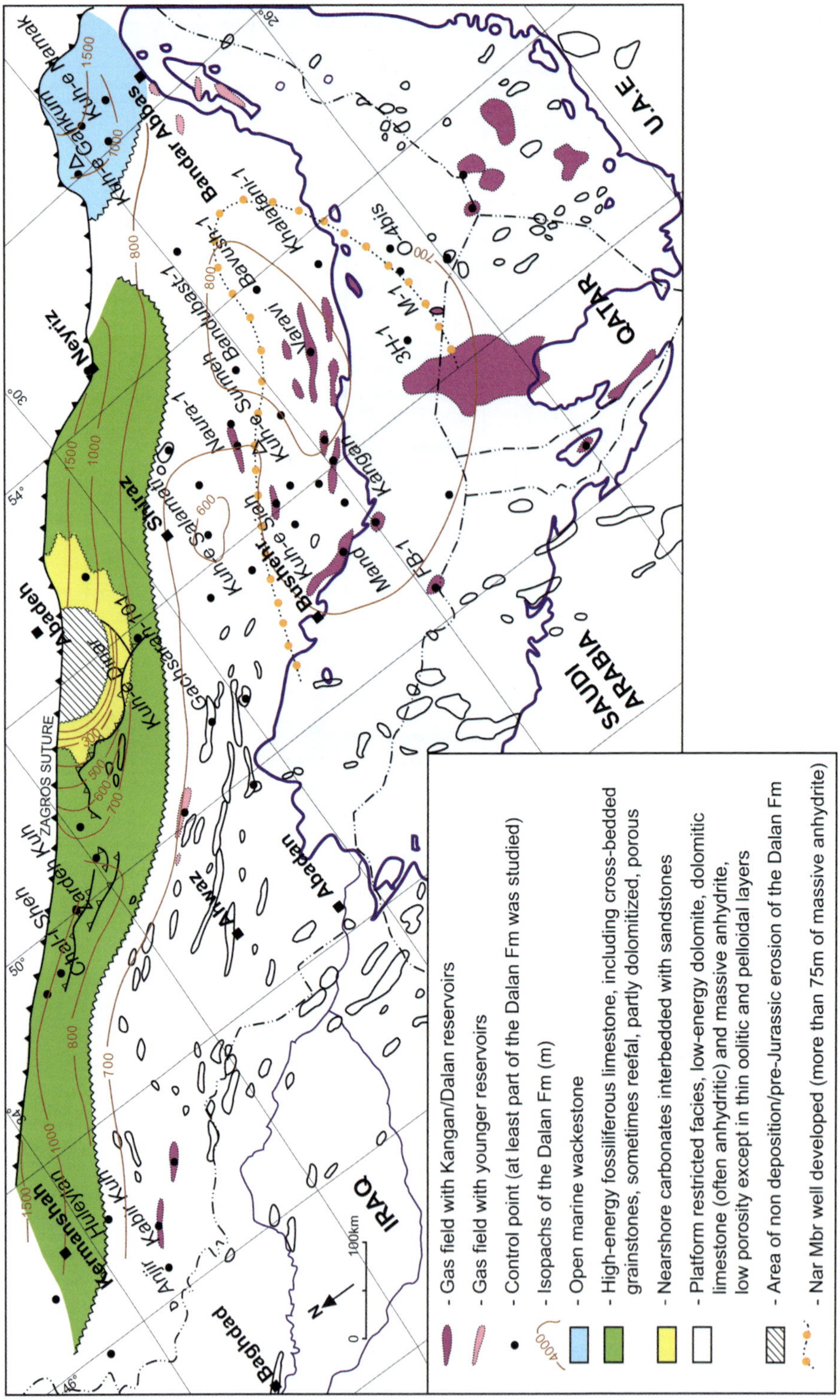

Fig. 9. Isopachs and facies distribution of the Permian Dalan Formation (modified after Szabo & Kheradpir 1978).

high in pre-Jurassic time (Szabo & Kheradpir 1978). Evaporite zones are separated by 40–150 m thick dolomitic layers, often argillaceous, containing subordinate anhydritic and marly intervals. C and D are separated by the fairly constant Sefidar dolomite, which was also eroded on the Qatar–Gavbendi High.

Similarly to what happened during the Permian, a reefoid barrier, described as the Khaneh Kat Formation, located on the NE edge of the Arabian Platform, caused the formation of a huge sabkha. The Dashtak forms an excellent seal where it is well developed, in contrast to the Khaneh Kat Formation, which does not contain any anhydrite (Fig. 10).

Analyses of samples of the Faraghan, Dalan and Dashtak Formations from Kuh-e Surmeh, Kuh-e Gahkum and Mand-2 have generally shown low organic content (TOC lower than 0.4%), according to Bordenave et al. (1971). However, thin organic-rich layers were observed in the Upper Dalan Member and in the Kangan Formation, at Kuh-e Dinar, Kuh-e Surmeh and at South Pars. They are made of mudstone deposited in local restricted poorly oxygenated hypersaline lagoons, and millimetre-scale laminae of algal mats, deposited in a subtidal to tidal flat environment. The TOC values of these layers may reach locally 10% (Insalaco et al. 2006). The cumulative thickness of these organic-rich layers is very limited. Small amounts of oil and gas could have been generated from them, supplementing the oil and gas generated by the main source rock, the Llandovery shales.

Chronology of migration, entrapment and late displacement of hydrocarbons originating from llandovery source rocks

The thermal evolution of the Llandovery source rocks, and more precisely the time of their oil expulsion and the onset of their gas window, is one of the critical factors in reconstructing the sequence of events that led to the formation of current hydrocarbon accumulations originating from these source rocks.

The Zagros Fold Belt and the Persian Gulf maintained an almost constant subsidence during the period extending from the partition between Central Iran and the Arabian Platform, during the Permian, to the onset of the Zagros folding at the end of the Middle Miocene. Only a short tectonic phase, observed at the end of the Cenomanian, temporarily downwarped the Zagros domain (Fig. 11). Because of the stability of the NE part of the Arabian Platform, the use of a constant heat flow was considered to be reasonable. A value of 36 mW m^{-2} constant mantle heat flow, which resulted in the best fit between the calculated and the observed values of the thermal indicators in the Dezful Embayment (Bordenave & Hegre

2005), was applied to Silurian source rocks in the Zagros Fold Belt and offshore. This heat flow is comparable to heatflow values calculated in Saudi Arabia. It is in line with the generally accepted range of 41–61 Wm^{-2} for continents (Sass et al. 1971).

In the Zagros Fold Belt, current geothermal gradients are probably influenced by the downwarping of the NE edge of the Arabian Platform, which caused the higher subsidence observed during the deposition of the syn-orogenic Agha Jari Formation (Late Miocene and Pliocene). Accordingly, it has been preferred to estimate heat flows from thermal markers (vitrinite reflectance, Rock-Eval T_{max}) rather than from current geothermal gradients.

Three deep wells, located in Fars (i.e. Well A (Varavi-1), which was drilled to the base of the Dalan Formation, Well B (Dalan-2) and Well C (Kuh-e Salamati-1), both drilled to the Faraghan Formation), were selected to construct a model of the thermal evolution for a layer located below the Permian transgressive Faraghan sands. The Varavi-1 well is located on the Gavbendi High, whereas Dalan-2 and Kuh-e Salamati-1 are situated 140 and 200 km to the NNE, respectively (Bordenave 2008). The onset of the oil window was reached in all three wells when the base of the source rocks was buried near 3000 m, and some oil began to be expelled at about 3200 m burial. The gas window was reached at 5600 m burial. According to the model, oil began to be expelled when the transformation ratio (TR) attained 27–30%; that is, 137 Ma ago at Well A, towards the limit of the Jurassic–Cretaceous, 150 Ma at Well B, during the Oxfordian, and 160 Ma at Well C, during the Bathonian. The gas window has not yet been reached at Well A. It was reached 20 Ma ago in the Early Miocene at Well B and 50 Ma ago in the Early Eocene at Well C. When the compaction of the sediments is taken into account, the 3200 m burial necessary for oil expulsion to commence would correspond roughly to a current sediment thickness of 2700 m, whereas for the gas window, the 5600 m burial would correspond to 5000 m of sediment thickness (Table 1).

Once the timing of the oil and the gas expulsion was established for the three wells, the results were extrapolated to the Silurian basins, by using a suite of cumulative isopachs, between the base of the Faraghan Formation and the tops of the Jurassic, the Early Cretaceous, the Cretaceous, the Asmari Formation (Early Miocene) and the Mishan Formation (Middle Miocene) successively. These isopachs illustrate the thermal evolution of Silurian source rocks through time in the various areas, according to subsidence variations.

Moreover, because Permian and Early Triassic sediments maintain a relatively constant thickness

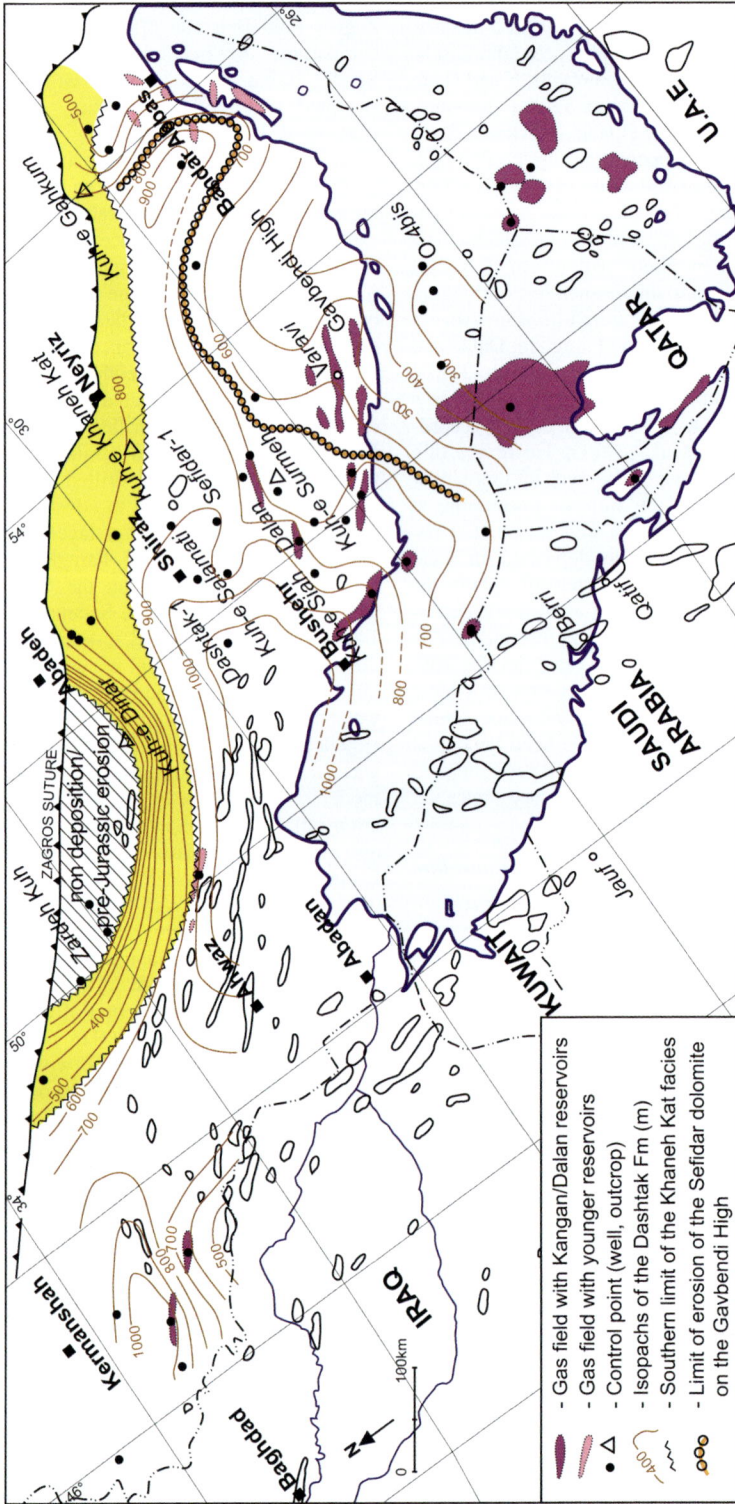

Fig. 10. Isopachs and facies distribution of the Dashtak Formation (Triassic) (modified after Szabo & Kheradpir 1978).

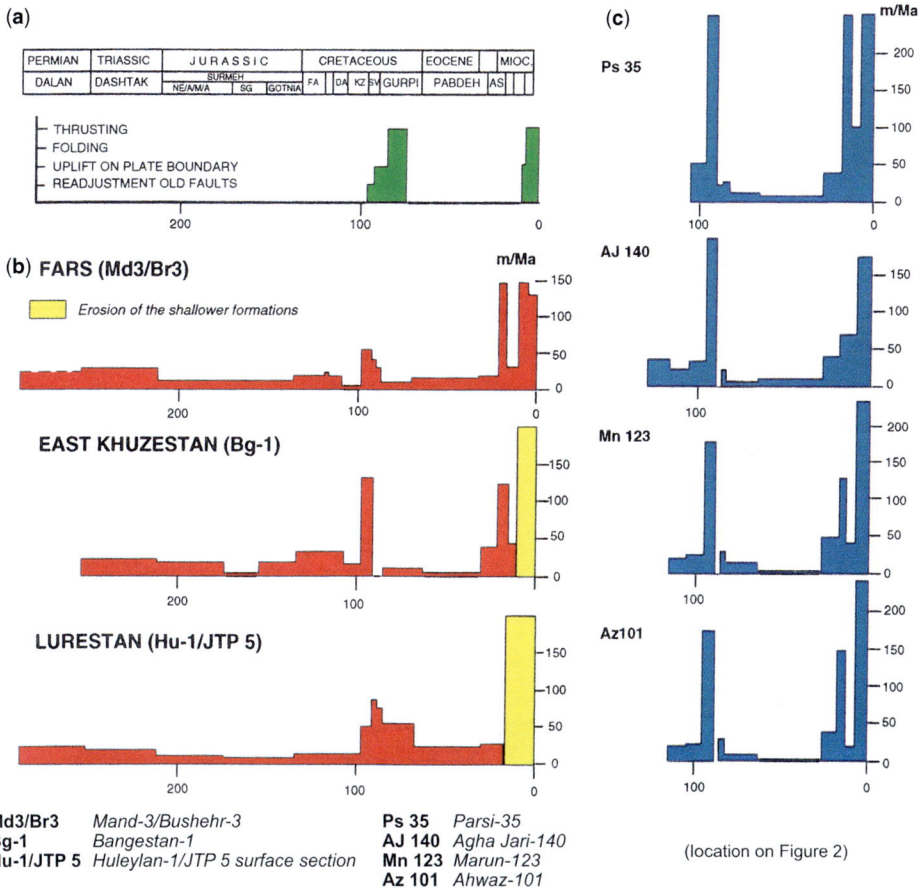

Fig. 11. Relationship between subsidence variations in the Iranian Zagros Domain and orogenic events observed on its NE limit (after Bordenave 2005). (**a**) Orogenic events observed close to the platform edge; (**b**) apparent subsidence (formation thickness/time of deposition ratio) observed from deep wells and surface sections; (**c**) apparent subsidence (m/Ma) in Dezful Embayment wells, from West to East.

throughout most of the Zagros foothills and of the Iranian offshore (900–1200 m), the isopachs between the base of the Faraghan Formation and the top of a given formation reflect the geometry of drains and reservoirs at the end of deposition of this formation, and permit the definition of migration paths of hydrocarbons at that time.

Two of these maps are provided (Bordenave 2008), for the cumulative isopachs at the top of the Jurassic sediments and at the top of the Asmari Formation (Figs 12 & 13). The cumulative isopach maps show the permanence of well-marked regional highs. The largest high covers the northern part of the Qatar Peninsula, the offshore area

Table 1. *Thermal evolution of the Silurian source rocks (after Bordenave 2008)*

	Onset of oil window		Expulsion window		Gas window	
	Time (Ma)	Depth of burial (m)	Time (Ma)	Depth (m)	Time (Ma)	Depth (m)
Well A	160	2900	137	3100	N.R.	N.R.
Well B	170	3000	150	3200	20	5600
Well C	180	3000	160	3200	50	5600

N.R., not reached.

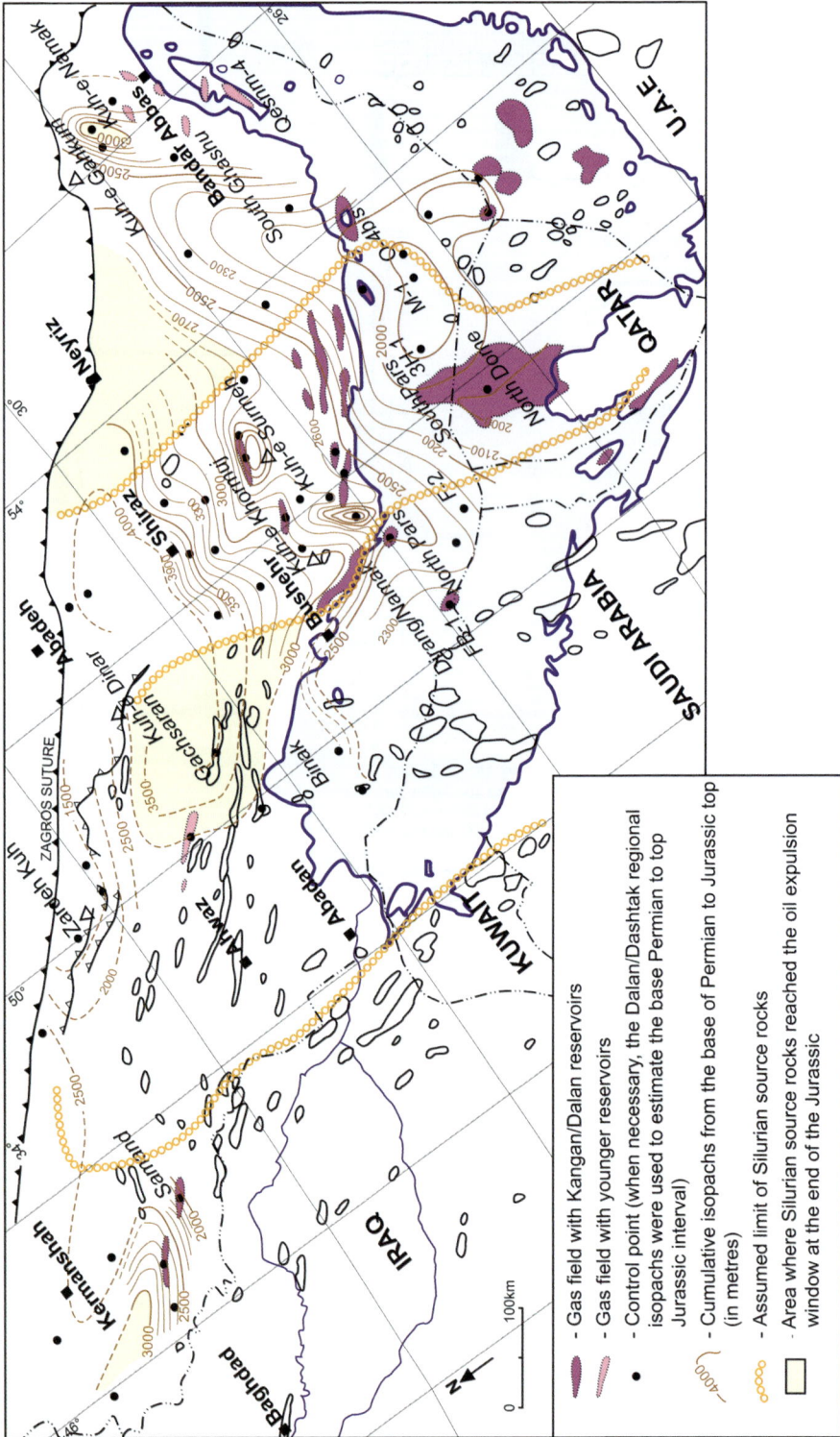

Fig. 12. Cumulative isopachs from the base Permian to the top Jurassic and the maturation of the pre-Permian layers (after Bordenave 2008).

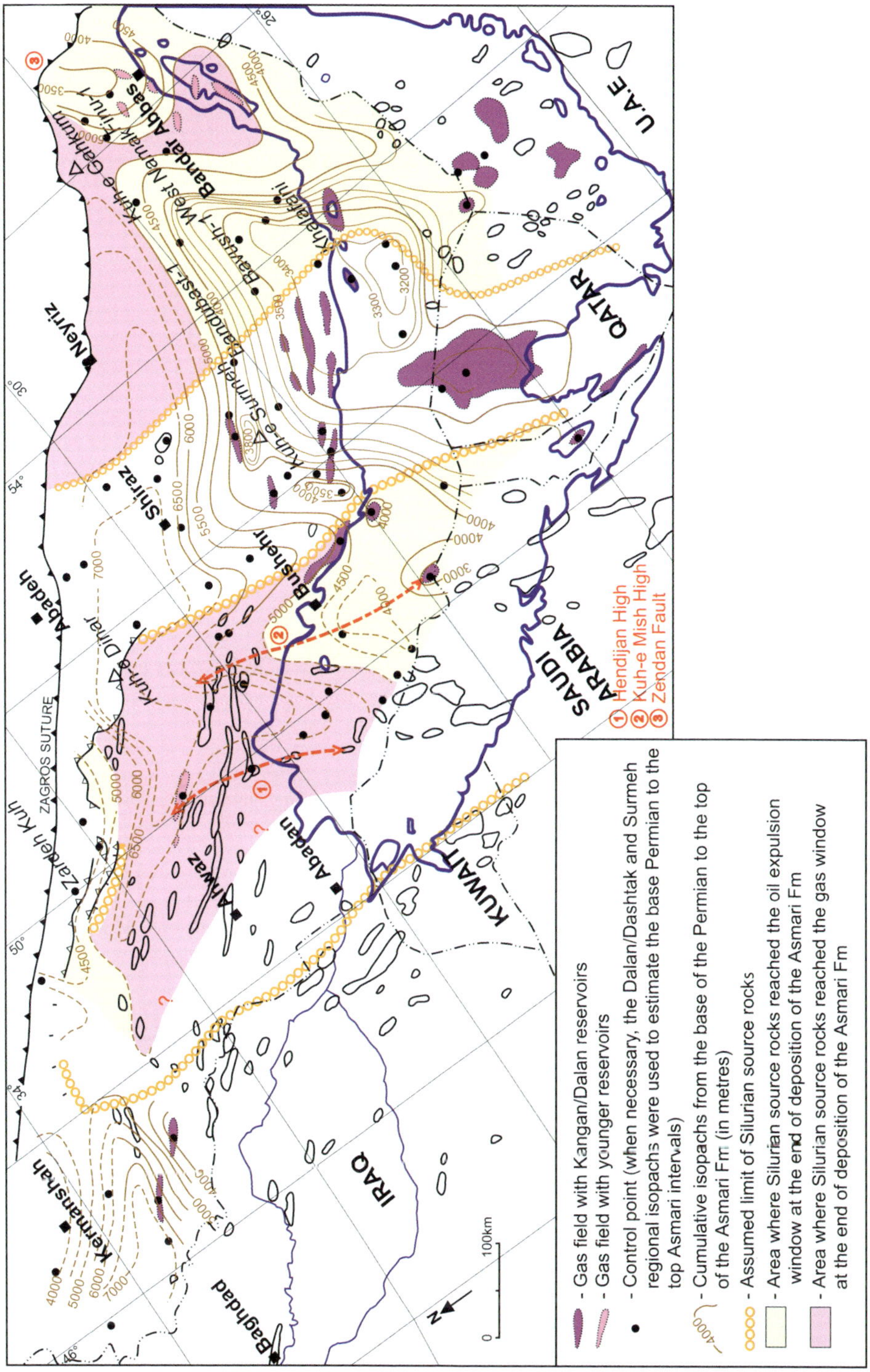

Fig. 13. Cumulative isopachs from the base Permian to the top Asmari Formation (Early Miocene) (after Bordenave 2008).

extending south of Lavan Island (O-4bis area), and part of Coastal Fars (Gavbendi High–Qatar Arch). The Gavbendi High was named after the village of Gavbendi in Coastal Fars; the subsidence on the Gavbendi High was very low during the Jurassic and the Cretaceous (Bordenave et al. 1971). A low-relief high is observed offshore, between Bushehr and the FB-1 well, at the limit of the Iranian waters (South Bushehr High). In SE Fars, a high is located north of Bandar Abbas. In southern Lurestan, a high is suggested SW of the Samand gas field (South Samand High).

The isopachs from the base of the Permian sediments to the top of the Jurassic show that by the end of the Jurassic, some oil was expelled from the depression located in NE Fars, south of Neyriz, and probably in the Dezful Embayment (Fig. 12). In Lurestan, some oil was expelled from the centre of the depression located NE of the Iraq border. Nowhere in the study area was the gas window reached. By the end of the Jurassic, some oil migrated upwards to the top of the Gavbendi–Qatar High, to the South Bushehr High, to the North Bandar Abbas High, to salt-related domes, such as North Pars, and to the North Samand High.

The cumulative isopach map at the end of the Middle Cretaceous shows a structural pattern similar to that established at the end of the Jurassic. The oil expulsion window had not yet been reached on the Gavbendi–Qatar High. In contrast, large amounts of oil were generated from the Eastern Silurian basin, from the northern part of the Gulf and from the SW part of the Dezful Embayment, where the gas window was reached. The size of the oil accumulation increased on the four highs and on salt-related structures, such as the north–south Darang–Namak ridge, and the North Pars, FB, F and 3H domes.

An increasing instability was observed during the Cenomanian. This instability was marked by the downwarping of the Platform, which induced a significant increase of subsidence (Fig. 12), and by the deposition of brecciated and conglomeratic facies at the top of the Sarvak Formation. This instability culminated during the Campanian, with the uplift and obduction of the former Radiolarite Trough, and of the ophiolites that were formerly located in the SW part of South Tethys (Ricou 1974; Stoneley 1981; Braud 1987; Bordenave & Hegre 2005). The readjustment of Hercynian (or earlier) north–south faults, such as the Kazerun and the Darang–Namak Faults, during the Late Cenomanian was accompanied by widespread movement of the Hormuz Salt. These movements induced the breaching of additional salt plugs (Kent 1979), and the enhancement of existing salt-related structures.

The existence of large oil accumulations at the end of the Middle Cretaceous was demonstrated by the impressive palaeo-seepage of Kuh-e Khormuj, NNE of the Mand field, in Western Fars (R. Player, unpubl. data, cited by Kent 1979). In the cliffs bordering the Kuh-e Khurmuj salt plug, two layers of bitumen-impregnated limestone, respectively 10 m thick in the Cenomanian Sarvak Formation and 5 m thick in the Early Miocene Asmari limestone, were observed. In both cases, the bituminous limestone has its porosity fully saturated with bitumen, and is limited both above and below by porous limestone devoid of bitumen. The limit between the bituminous limestone and the clean one is abrupt, linear, and parallel to the stratification. Observed in thin sections, the bituminous limestone looks like an accumulation of extremely shallow fossil tests slimed all over with bitumen. The Khormuj salt plug was already breached during the Cenomanian, as demonstrated by the extremely shallow beach facies that accumulated around it, and formed an island. Similar islands are observed today around the current salt plugs located in the Persian Gulf. The bitumen is assumed to be detrital, resulting from a huge natural pollution caused by the breaching of existing field(s). This pollution probably continued for some ten thousands of years. A rough estimate of the volume of oil spilled around Kuh-e Khormuj proves that the field was in the supergiant class. The bitumen could have originated from a field that was located on the current salt-related Darang–Namak Ridge, which is only 45 km from Kuh-e Khormuj. The fact that the Jurassic and the Early Cretaceous sediments cropping out on the two sides of the Ridge are much thinner that in the wells drilled in the vicinity proved the existence of a high-relief NNE–SSW elongated structure at the end of the Middle Cretaceous. Moreover, the Darang–Namak Ridge is cut at present by an axial north–south fault.

Cumulative isopachs at the end of the Cretaceous show a structural setting similar to that described at the end of the Middle Cretaceous. However, the considerable increase of subsidence in NW–SE narrow troughs located close to the NE edge of the Arabian Platform was marked by the accumulation of thick flysch deposits derived from the erosion of obducted radiolarites and ophiolites. An increase of subsidence is also observed in the north–south Binak Depression. On the crestal part of the Gavbendi–Qatar High, where the subsidence was extremely low during the Late Cretaceous, pre-Permian sediments were still immature. In the northern part of the Gulf and in the SE Fars–Qeshm area, pre-Permian sediments were still in the oil window. In contrast, they probably reached the gas window in the Dezful Embayment and in the Lurestan Depression. Oil continued to accumulate in the highs, accompanied by an increasing volume of gas.

During the Palaeocene and the Eocene, thick flysch deposits continued to accumulate on the NE edge of the current Zagros Fold Belt. By the end of the deposition of the Asmari Formation (Early Miocene), Silurian source rocks reached the gas stage in the Eastern Basin, in the Binak Trough, and in the Dezful Embayment (Fig. 13). As a result of the generation of huge quantities of gas, a large gas cap was formed on the Gavbendi–Qatar High, and oil was pushed downward to form a large oil leg. Gas caps were also formed in both the North Bandar-Abbas and the South Bushehr Highs. Another breach caused a second occurrence of sedimentary bitumen in the Asmari cliff near the Khormuj Salt dome. This observation suggests that the Darang–Namak High was again oil-bearing at the beginning of the Early Miocene, and that this oil was lost, probably together with large amounts of gas.

The Middle Miocene was marked by the existence of an extremely subsiding trough in Eastern Fars, extending from Kuh-e Faraghan towards Farur Island. In this trough, the deposition of 1000–2700 m of Mishan marls was caused by a local downwarping of the SE edge of the Arabian Platform, on the western side of the Zendan Fault. With the exception of this trough, there was little change in the structural setting, although the Gavbendi–Qatar High became progressively steeper. The map of the cumulative isopachs from the base of the Permian to the top of the Mishan provides a picture of the geometry of the Kangan–Dalan Formations, before the onset of the paroxysmal phase of the Zagros orogeny and the formation of the current large anticlines.

Most of the current gas fields discovered in Fars (Aghar, Assaluyeh, Dalan, Gardan, Homa, Kangan, Mand, Nar, Shanul, Tabnak, and Varavi) and in the Persian Gulf (Balal, Kish, Lavan, North Dome–South Pars, and North Pars) are located near the top of the Gavbendi–Qatar High as it was defined at the end of the deposition of the Mishan Formation. The small Namak gas accumulation, discovered in the Dalan–Kangan Formations, is located on the North Bandar Abbas High, east of the Mishan Trough. The origin of the gas accumulated in reservoirs younger than the Early Triassic (i.e. the Khami Group (Salakh anticline of Qeshm Island, South Gashu, and Suru), Asmari–Jahrum Formations (Gevarzin, Sarkhun, and West Namak), and even the Guri Member of the Mishan Formation (Sarkhun)) is not yet determined, in the absence of $\delta^{13}C$ isotopic measurements. However, the maturity of younger source rocks that possibly exist in the area and the results of the wells drilled

in the surrounding structures (Hulur-1, HD-1 and HA-1) suggest a probable Silurian origin.

The existence, prior to the Zagros folding, of huge gas fields and associated oil legs over four regional highs was considered as a reasonable assumption (Bordenave 2002a). The extension of the pre-Zagros Gavbendi–Qatar gas field could be deduced from the distribution of the current gas fields, assuming that only the Zagros anticlines formed within the limit of the pre-existing field could be charged with gas. If this assumption is correct, Mand, Aghar and Dalan, which are the fields located in the flank of the Gavbendi–Qatar High, would provide the limit of the pre-Zagros field. Down-dip of this limit, the wells drilled are dry (Kuh-e Siah-1, Naura-1, Dashtak-1, Kuh-e Salamati-1, Kuh-e Sefidar-1 and Ahmadi-1). The pre-Zagros Gavbendi–Qatar field could have extended over more than $100\,000\;km^2$, almost 10 times the surface of the present-day 800 Tcf North Dome–South Pars gas field. Pre-Zagros gas fields, although smaller, existed on the other regional highs, as well as in salt-related structures. The North Bandar Abbas pre-Zagros accumulation would have been developed updip of Salakh, Gevarzin and South Gashu. In Lurestan, a pre-Zagros gas field probably extended updip of Samand.

This assumed extension of pre-Zagros fields could be used for guiding future exploration, as only those of the Zagros folds formed within the perimeter of the extension of the pre-Zagros gas accumulations, or adjacent to these accumulations, would have trapped large amounts of gas. It is not possible to evaluate the proportion of the pre-Zagros gas accumulation that was lost as the result of the Late Cenomanian breaching and Zagros folding and thrusting.

According to geothermal modelling, as well as from calculations using current thermal gradients measured in wells located SW of the Zagros Front, the oil leg that surrounded the pre-Zagros Gavbendi–Qatar High gas field was subject to temperatures ranging from 165 to 195°C (Bordenave 2008). These temperatures caused, before the Zagros folding, the cracking of the oil into pyrobitumen and either very light oil–condensate or gas. The estimate of the temperatures of the reservoirs before and after the Zagros folding could also be used to evaluate the intensity of the thermal reduction of sulphates reaction (TSR), assumed to commence at 140 °C (Worden et al. 1995).

The Middle Jurassic Petroleum System

Sargelu source rocks were deposited from the end of the Lias to the end of the Bathonian in the

Mesopotamian Depression. The Sargelu Petroleum System provides an example of source rocks not connected with reservoirs, except for those close to the edge of the depression. A large part of the oil generated has not been expelled and was cracked *in situ* into pyrobitumen and gas. Some oil migrated laterally towards porous platform facies of the Surmeh–Khami Formations, accumulated on regional highs before the Zagros folding, and eventually moved into Zagros folds.

Distribution and characteristics of the Sargelu source rocks

Similarly to what was observed during the Permian and the Middle–Late Triassic, an elongated NW–SE shallow-water carbonate shoal was located on the NE edge of the Arabian Platform during the Middle and Late Jurassic. This shoal hampered the water circulation between the epicontinental sea that extended over the Arabian Platform and South Tethys. Low-energy oxygen-deprived facies were deposited in the subsiding Mesopotamian Depression (James & Wynd 1965). Thick evaporites accumulated during periods of sea-level falls at the bottom of hypersaline waters, whereas anoxic episodes corresponded to highstand periods. During the Bajocian–Bathonian, anoxic conditions that prevailed in the depression led to the deposition of the Sargelu Formation.

To the east of the depression, a wide platform covered Fars and the SE part of the Dezful Embayment (Fig. 14). The depositional environment remained oxic, and no source rocks were deposited. In the NE part of Saudi Arabia, the limit between the shallow-water Dhruma Formation of Bajocian and Bathonian age and the Sargelu basinal facies passes to the north of Abu Adriyah, whereas at Samawa-1, in southern Iraq, a thick basinal Sargelu (447 m) has been described (Ibrahim 1978, 1983).

The Sargelu Formation, defined in NE Iraq, is composed of thin-bedded black bituminous limestone, dolomitic limestone and black papery organic-rich shales (Dunnington *et al.* 1959). The Sargelu Formation is conformably overlain by a condensed section of Late Jurassic age represented by only 14 m of shaly laminated beds of which the lowermost 7 m of layers are highly bituminous (Naokelekan Formation). The facies of the Naokelekan Formation proves that anoxic conditions continued in the Mesopotamian Depression during at least part of the Kimmeridgian. In the central part of the depression, TOC values of Sargelu samples reach 20%. The kerogen, dominantly algal and/or sapropelic, is of type 2. On the edge of the depression, the kerogen is mixed with terrestrial plant debris. The hydrogen index varies widely from 100 to 750 g HC kg^{-1} C for immature kerogens.

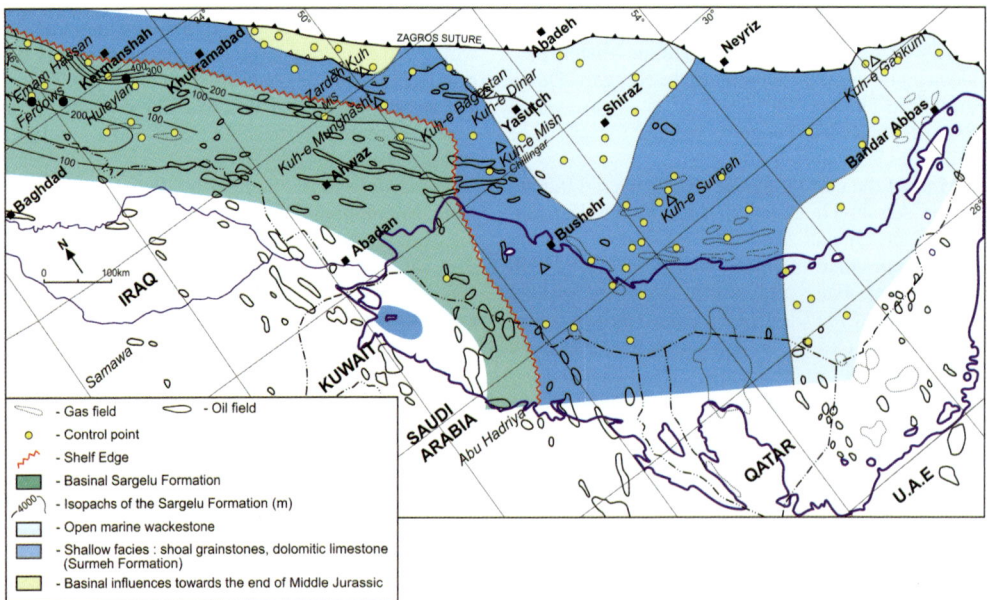

Fig. 14. Palaeogeography reconstruction during the Middle Jurassic and distribution of the Sargelu source rocks in the Mesopotamian Depression. Field sections and wells where the Sargelu Formation was available are indicated (after Bordenave & Huc 1995).

In Lurestan, the Sargelu Formation is 100–300 m thick. It is thinner in the central part of the depression, indicating a starved basin; that is, the supply of fine-grained sediments, coming from the distant Arabian Shield, was not sufficient to match the subsidence. At the Emam-Hassan-1, Kabir Kuh-1, and Huleylan-1 wells, the uppermost 60 m of the Sargelu Formation consists of dark brownish to black organic-rich shales, whereas the lowermost 150 m of shales are still interbedded with fine-grained pyritic limestone and thin anhydritic layers. Samples from Kuh-e Bangestan-1 and Samand-1 were analysed. At Samand, where the Sargelu is overmature (T_{max} between 462 and 476 °C, HI <10 g HC kg^{-1} C), the residual TOC varies from 1.2 to 3.0%. At Kuh-e Bangestan, the oil window was reached (T_{max} 443 °C), current TOC varies between 3.1 and 4.4%, and current HI varies from 190 to 260 HC kg^{-1} C.

Expulsion of oil and gas, timing and entrapment

GENEX thermal modelling was applied to the Sargelu Formation of the Kuh-e Bangestan-1 well, using a heat flow value similar to that used in the Dezful Embayment. The Sargelu Formation reached the onset of the oil window 65 Ma ago, by the end of the Cretaceous, for a burial of 2900 m, the onset of the expulsion window was reached 50 Ma ago, during the Eocene for a burial of 3050 m, and the gas window 25 Ma ago, during the Early Miocene for a burial of 5200 m (Fig. 15). Taking into account the compaction of the sediments, the expulsion window and the gas window of Sargelu source rocks were reached when the current thickness of sediments that capped them attained 2500 and 4700 m, respectively.

Keeping in mind that the folding commenced more or less after the end of deposition of the Mishan Formation, towards the end of the Middle Miocene (Hessami *et al.* 2001; Bordenave & Hegre 2005), the cumulative isopach map between the top Sargelu and the top of the Mishan Formation is indicative of the thermal evolution of the Sargelu source rock at the onset of the Zagros folding. In central Lurestan (Emam Hassan-1, Ferdows-1, Huleylan-1 and Kabir Kuh-1 wells) the cumulative current thickness between the top Sargelu and the top Gachsaran Formation varies between 4000 and 4900 m. Moreover, as the Mishan is not generally identified in Lurestan wells, because of the

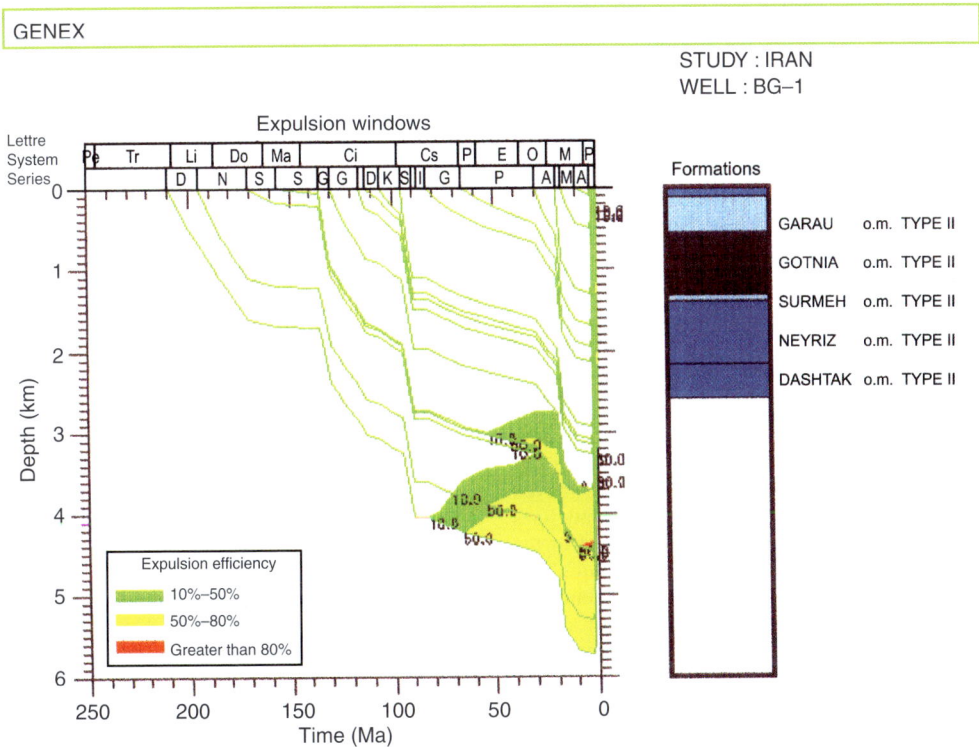

Fig. 15. Results of the Genex modelling at the Kuh-e Bangestan-1 well.

absence of deposition of the marine Guri Limestone Member, part of the Agha Jari Formation should instead be included in the calculation of the Sargelu depth of burial before onset of Zagros folding. This thickness could be estimated to be 300 m. Cumulative isopachs suggest that by the onset of the Zagros folding the Sargelu Formation had expelled most of its oil and had already reached the gas stage in the central part of the Lurestan Depression. This is in agreement with the overmature character of the Sargelu samples analysed at Samand.

In the central part of the Lurestan Depression, the Sargelu Formation is limited below and above by layers of massive anhydrite, thicker than 100 m, the Alan Formation at its base and the Gotnia Formation at its top (Fig. 16). Most of the oil generated in the central part of the Mesopotamian Depression was probably unable to escape from the Sargelu source rocks because of the absence of a reservoir, and was cracked into pyrobitumen and gas. The existence of a 2 m thick layer of pure asphaltite at the top of the Sargelu Formation observed at Qaleh Kuh in SE Lurestan (120 km ESE of Khorramabad) is a relic of oil that could not migrate (Bordenave & Sahabi 1971).

In contrast, at the edge of the depression, the 20–30 m thick shallow-water carbonate of the Najmah Formation is intercalated between the Sargelu Formation and the Gotnia evaporites (Samand, MIS and Kuh-e Bangestan). It formed a drain that permitted some oil and gas expelled from the Sargelu and the Naokelekan equivalent source rocks to migrate laterally towards the edge of the depression, and towards porous platform facies, on the other side of the shelf edge. Oil accumulated on low-relief regional highs adjacent to the edge of the depression (Fig. 17).

Current location of fields charged from Sargelu source rocks

As no Gotnia evaporites were deposited NE of the Kermanshah–Khurramabad area, the first seal encountered by migrating oil and gas was the thick Gurpi–Amiran argillaceous complex. Oil and gas accumulated in the Sarvak and Ilam reservoirs prior to the Zagros folding, and later moved into the Zagros folds. However, the anticlines located in this area are currently breached to the Sarvak and even deeper, and the oil accumulated in these anticlines has been lost. In the same way, to the ENE of the depression, on the other side of an abrupt shelf edge, the excellent reservoirs formed by the thick and porous shoal grainstone of the Surmeh and Khami Formations at the Kuh-e Munghasht anticline are breached.

To the east, the Sargelu oil migrated towards the north–south high that extended from Kuh-e Mish,

north of Gachsaran, towards the offshore south of Bushehr (Fig. 17). On this high, the Surmeh and Khami reservoirs are sealed either by the Hith Anhydrite to the south or by the organic-rich Kazhdumi to the north. Large amounts of oil accumulated on this high. Part of this oil was later reaccumulated in the Zagros high-relief anticlines located on this high (Garangan–Chillingar and Sulabedar anticlines). These fields produce from a single Khami–Surmeh reservoir. Significant oil columns with gross thickness ranging between 400 and 1500 m were found in these fields. Test results showed that several layers have excellent reservoir characteristics, with flows in the 5000–18 000 barrels per day range. Between Sulabedar and Kharg, no wells drilled deep enough to reach the Hith Formation.

The north–south low-relief Kharg anticline, formed before the Zagros folding, was located on the migration paths of oil originating from the Sargelu Formation as well as from Late Jurassic source rocks. The Kharg structure trapped oil in the excellent Khami–Surmeh single reservoir that delivered between 5000 and 27 000 barrels per day of oil in production tests (Kharg-1 and Darius-1 wells). The gross thickness of the oil column is close to 400 m.

Another north–south high extended from the Hendijan field to Kuh-e Bangestan. Some oil may have migrated north of Kuh-e Bangestan. The Bangestan-1 well found mediocre gas-bearing reservoirs below the Sargelu Formation, in the Neyriz Formation.

The Late Jurassic Petroleum System in Iran

The kitchens of the Hanifa–Tuwaiq Mountain and Diyab source rocks that charged the Late Jurassic Petroleum System are mostly located outside Iran (Fig. 18). Oil migrated over long distances towards the southern part of the Gavbendi and South Bushehr regional highs. Heavy oil and bitumen accumulated in various reservoirs of salt-related domes, because of the absence of efficient seals at the time of migration.

As the result of local subsidence, anoxic conditions prevailed in intrashelf depressions during the Late Callovian, Oxfordian and Early Kimmeridgian. Two depressions extended on both sides of the Qatar Arch; one into the west of Abu Dhabi (offshore and onshore), and the other extending into the eastern part of Saudi Arabia and its offshore (Murris 1980; de Matos & Hulstrand 1994). This latter depression communicated with the Mesopotamian basin (Fig. 18). Excellent source rocks were deposited in the Hanifa–Tuwaiq Mountain

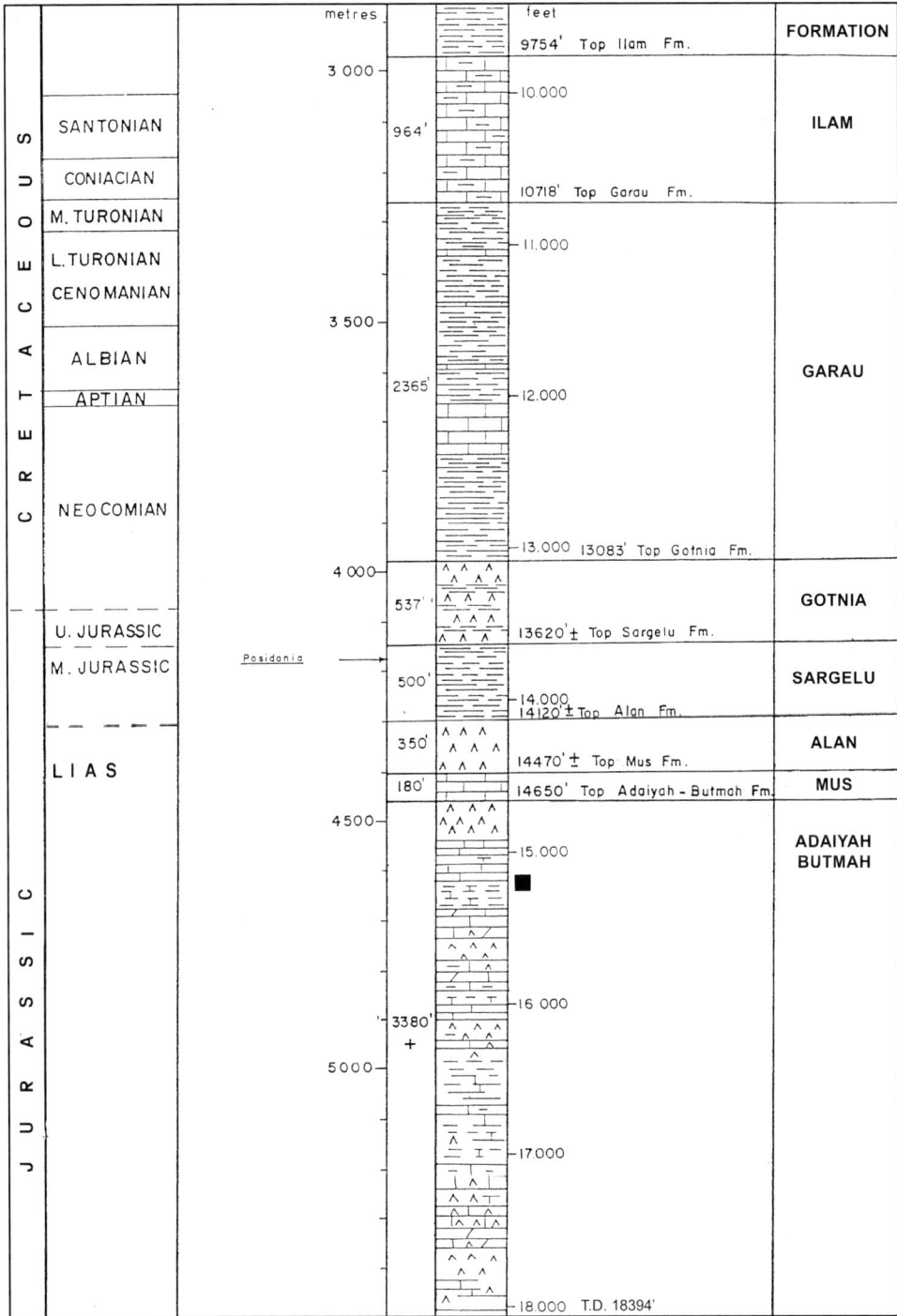

Fig. 16. Sketch log of the Emam Assan-1 well.

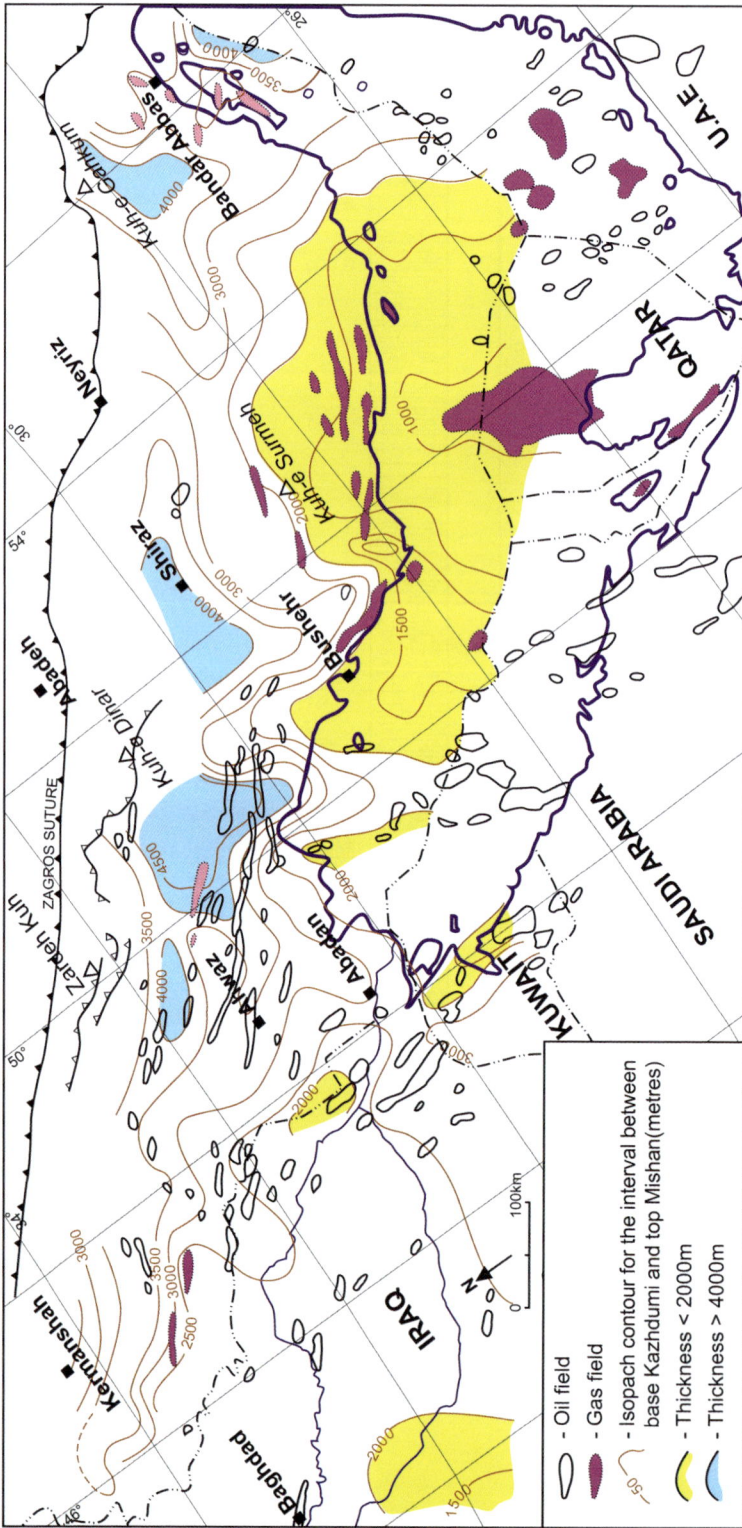

Fig. 17. Cumulative isopach map from the top of the Dariyan Formation (Aptian) to the top of the Mishan Formation (Middle Miocene) showing the regional highs before the Zagros folding.

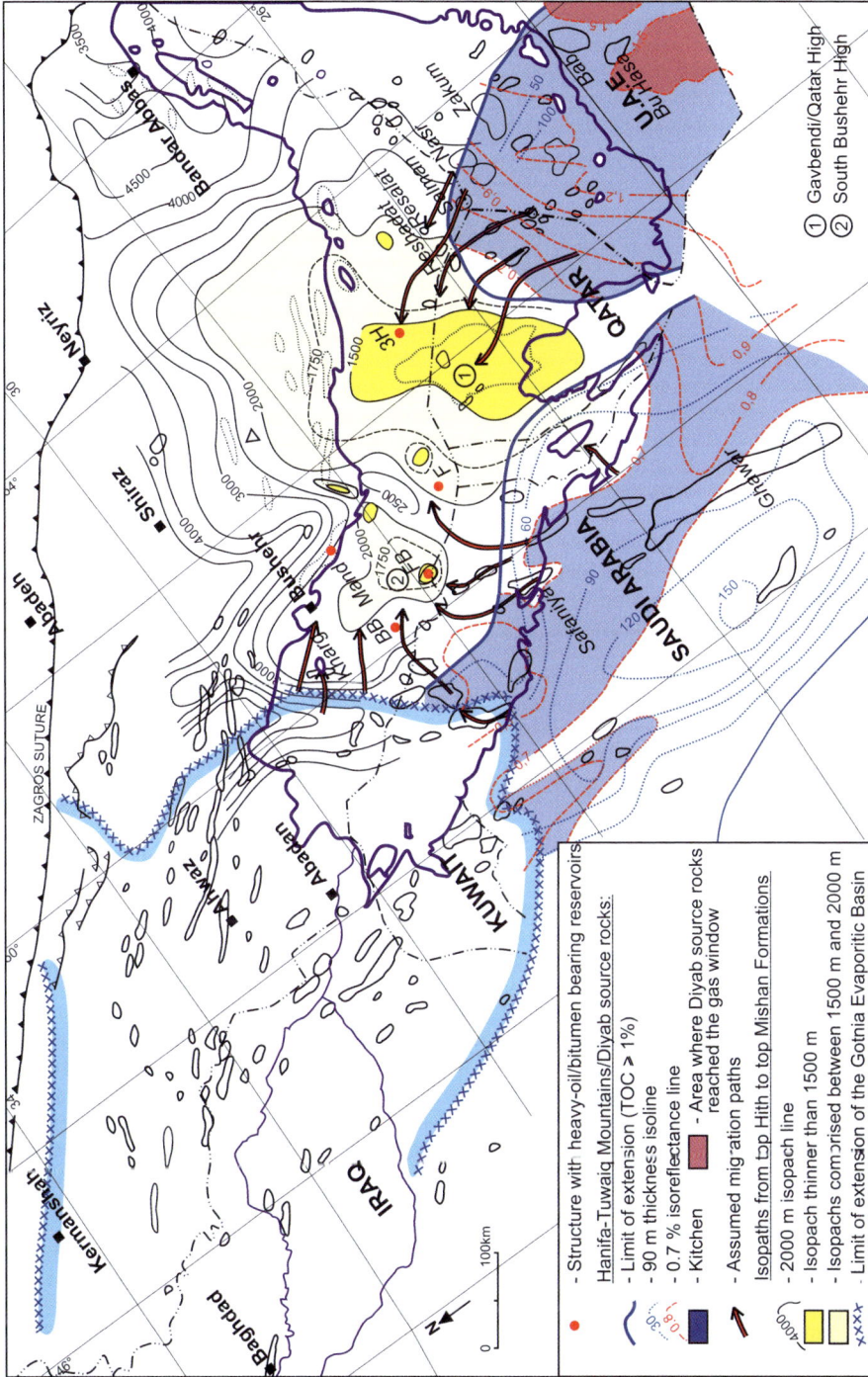

Fig. 18. Cumulative isopach map from the top of the Hith Anhydrite to the top of the Mishan Formation showing the geometry of the Arab reservoir before the Zagros folding and the isopach map of the Hanifa–Tuwaiq Mountain. The Diyab source rock layers (TOC in excess of 1%), and the current thermal maturity of the Hanifa source rocks, expressed as vitrinite reflectance equivalent (after Ayres *et al.* 1982; Carrigan *et al.* 1995, for Saudi Arabia; and Loutfi & El Bishlawy 1986, for Abu Dhabi) are indicated, together with the migration paths from the Hanifa–Tuwaiq Mountain and the Diyab kitchens towards the Gavbendi High.

Formations (Fig. 19) of Saudi Arabia, west of Qatar and in the northern part of the Persian Gulf, and in the time-equivalent Diyab Formation in Abu Dhabi. Several excellent and prolific reservoir layers in the Arab Formation are sealed by the Hith Anhydrite. In the Mesopotamian Depression, in Iraq and in Iran, the Late Callovian to Early Kimmeridgian period corresponded to the deposition of the condensed organic-rich Naokelekan Formation, as mentioned above.

Distribution and thermal evolution of the Hanifa–Tuwaiq Mountain–Diyab source rocks

In the Abu Dhabi waters, west of Sassan, the transgressive Diyab Formation consists of 200–400 m of finely laminated, dark grey, organic-rich lime mudstone and calcareous shales. The basal part, 45–85 m thick, is radioactive and rich in organic matter (Hassan & Azer 1985). To the south, the Diyab Formation is even thicker, 500–600 m at the Bab and Shah fields (Gumati 1993). The Diyab Formation is overlain by the regressive Arab Formation composed of alternating anhydritic and shallow-water dolomitic zones. East of Salman (ex Sassan), the Diyab Formation passes into clean sucrosic dolomite and dolomitized packstone and wackestone. Almost no source rocks exist in this formation at the Zakum and Nasr fields. The kerogen of the Diyab Formation is composed of 60–80% of oil-prone sapropelic type 2 in the centre of the depression. Actual TOC values vary between 0.3 and 5.5%, with most of the values being lower than 2%. However, as the Diyab Formation is already mature (R_o between 0.8 and 1.7%) the initial TOC was probably 1.5 times to twice the present values. The residual S_2 varies from 1.9 to 16.4 g HC kg^{-1} rock (Hassan & Azer 1985; Alsharhan 1989).

The maturity of the Diyab Formation increases from north to south. Computed values of vitrinite reflectance vary between 0.7 and 0.8% in the northern and the northwestern part of the offshore where the Diyab has currently reached the peak of oil generation. Vitrinite values increase to the south to 1.2–1.3%, and even to 1.7% west of Bu Hasa (Fig. 18). They reach 1.5% offshore Dubai (Alsharhan 1989). At the offshore ADNOC-1b well, the top Diyab was reached at 3250 m, and oil expulsion from the Diyab source rocks probably began towards the end of the Eocene (Gumati 1993). This statement is in agreement with palaeomaturity maps proposed by Loutfi & El Bishlawy (1986).

West of Qatar, up to 150 m of organic-rich and radioactive layers were deposited in the Hanifa–Tuwaiq Mountain Formations. The intrashelf basin, where source rocks were deposited, was flanked by a shoal grainstone facies capped by anhydrites. The source rock layers are located in the upper part of the transgressive system tracts, whereas evaporitic units could be considered as lowstands (Droste 1990). Petrographic examination showed that the kerogen is mostly made up of lamalginite (Cook & Sherwood 1991). TOC values are generally lower than 7.5%, having an average value of 3%. The Rock-Eval S_1 and S_2 peaks would categorize the kerogen as a rich type II, with S_2 generally ranging from 5 to 40 g HC kg^{-1} rock, averaging 25 g HC kg^{-1}, and HI values varying between 600 and 800 g HC kg^{-1} C, averaging 640 g HC kg^{-1}. The carbon isotopic composition of the source rock samples varies from $-27.5‰$ to $-26‰$ PDB (average $-26.4‰$) for the kerogen, and from $-28‰$ to $-26‰$ (average $-26.6‰$) for the extracted bitumen. These values are compatible with the carbon isotopic composition of oils found in the Jurassic reservoirs of Saudi Arabia (Carrigan et al. 1995).

Modelling applied to the Hanifa source rocks, using a constant heat flow calibrated with vitrinite reflectance, Rock-Eval T_{max} and thermal alteration index (TAI), shows that oil expulsion began in the Ghawar and Safaniya areas 75 Ma ago, at the beginning of the Maastrichtian (Carrigan et al. 1995). The peak of the oil generation, assumed to occur for a vitrinite reflectance of 0.75%, was reached by the Middle Eocene (50 Ma). However, oil generation continued until the beginning of the Miocene (25 Ma), and probably later (Fig. 18). These results are comparable with those obtained in Bahrain, where oil expulsion began between the early Eocene and the early Miocene (Chaube & Al Samahiji 1994).

Fields charged by the Hanifa–Tuwaiq Mountain and Diyab source rocks

In Eastern Saudi Arabia, its contiguous offshore, and Bahrain, oil charged by the Hanifa–Tuwaiq Mountain Petroleum System accumulated in Late Jurassic reservoirs of north–south structures that existed before the end of the Early Cretaceous (Carrigan et al. 1995). The huge Ghawar field is the most impressive example of these fields. The Ghawar is the largest global known oil accumulation, 225 km long, extending over 2275 km^2, and with an oil column as thick as 433 m. Similarly, most of the Abu Dhabi offshore oil fields that produce from the Arab Formation are interpreted as having been sourced from the Diyab Formation. The gas discovered at Bab, with 30% H_2S, was also assumed to have originated from Diyab source rocks (Loutfi & El Bishlawy 1986).

In the SE part of the Iranian waters, the Resalat, Reshadat and Salman fields were probably sourced from the Diyab Formation. They are oil-bearing, in both the Arab and the Shuaiba Formation. The presence of oil in the Shuaiba reservoir is due to the poor efficiency of the Hith potential caprock. Several structures drilled in the western part of the Iranian offshore are heavy oil- or bitumen-bearing, in all porous zones ranging from the Arab to the Asmari Formations (Bashari 1989), as follows.

(1) The 3H domal structure contains 10 m of producible $18°$ API oil in the Asmari (top at 452 m) and 10 m of $20°$ API oil in the Katiyah Formations (top 1077 m). Residual oil-staining was observed in the Arab Formation, and 145 m of the Upper Thamama Formation showed continuous heavy oil-staining. The Buwaib, Zubair, Ilam, and Upper to Middle Dammam (formation names are given in Fig. 19) were also heavy oil-bearing. In this structure, the 68 m thick Hith Formation is probably not a valid seal.

(2) The north–south elongated FB structure is heavy oil-bearing in almost all reservoirs, from the Shuaiba (top at 1190 m) up to a level located only 56 m from the sea bottom. Specifically, 230 m are impregnated in the Asmari–Jahrum Formations and 155 m in the Gurpi to Kazhdumi interval. The structure is faulted, and the Hith Formation is only 22 m thick.

(3) The F domal structure, which showed a period of growth during the Cretaceous, contains 170 m of $6°$ API oil, with 3.65% sulphur in the Arab Formation (top at 1783 m). The reservoir is excellent (20%). The Ratawi (Fahliyan, top at 1475 m) contains 600 m of oil column. The oil is $8–16°$ API, its sulphur content is 5%, and the reservoir porosity is in the 20% range. The Kharaib (Gadvan) contains 170 m of $16°$ API oil with a 3.5% sulphur content. All these accumulations would extend over an area of 266 km^2.

(4) In coastal Fars, SE of Bushehr, the large whaleback anticline of Kuh-e Mand, oriented NNW–SSE, 90 km long and 16 km wide, with a vertical closure of nearly 2500 m, was formed as the result of the Zagros orogeny. It contains heavy oil in the Jahrum (top at 304 m) and Sarvak (top at 1174 m) Formations. The Jahrum oil is immobile ($8°$ API), with an oil column probably in excess of 350 m. The Sarvak oil is mobile ($14.5°$ API) and flowed in tests. Its sulphur content is 4.9%. The oil in place was estimated by NIOC at 6 billion barrels, evenly divided between the Jahrum and the Sarvak accumulations (Mostaghian et al. 1989).

(5) The relatively small Bushehr dome (16 km long and 9 km wide, with a 70 m vertical closure) is oil-bearing in the Surmeh Formation. The oil column is 40 m thick. The gravity of the oil is $37°$ to $39°$ API, and its sulphur content is 0.85%. Its

carbon and sulfur isotopic composition, $-27‰$ and $-1.5‰$ respectively, are compatible with oil derived either from the Hanifa–Tuwaiq Mountain, the Garau, or the Kazhdumi source rocks, but not with Palaeozoic source rocks (Bordenave & Burwood 1990).

The large heavy oil or bitumen accumulations observed in the 3H, FB, F, BB, and Kuh-e Mand structures are assumed to derive from low-maturity source rocks. Moreover, the lighter fractions of migrating oil were lost to the surface prior to the deposition of the Gachsaran Formation (Early Miocene) that formed the first efficient caprock. No isotopic measurement or biomarker determination of these heavy oil and bitumen is available to us. However, it may be assumed that these heavy products, rich in sulphur, were generated from calcareous sulphur-rich source rocks deposited in euxinic environments, where the concentration of H_2S dissolved in the seawater was high (Bordenave 1993). Therefore, they could not have originated from the Silurian shales, which generated low-sulphur products. Furthermore, the oil window for the Silurian shales was reached before the end of the Jurassic, and the heavy oil and bitumen accumulated in the Late Jurassic and younger reservoirs.

In Saudi Arabia and Abu Dhabi, both onshore and offshore, north–south trends of early salt-related structures collected part of the oil that originated from the Hanifa–Tuwaiq Mountain–Diyab source rocks until their spill-point was reached. The excess of oil moved towards the Gavbendi High–Qatar Arch, according to the geometry of the Arab Formation, which acted as reservoir as well as drain for migrating oil. Migration paths can be reconstructed from the geometry of the Arab Formation. The cumulative isopach map from the base of the Hith Formation to the top Mishan gives a good picture of this geometry, prior to the Zagros folding (Fig. 18). As shown on this map, migrations occurred on a low-angle ramp towards the regional highs, such as the Qatar Arch–Gavbendi High, and the South Bushehr High, over distances of 100 km or more. Structures already existing in the Middle Cretaceous, such as Resalat, Salman, Reshadat, 3H, F, FB and Bushehr, were located on the migration paths. The case of the Kuh-e Mand Zagros Fold is of particular interest. The area located west of Mand is extremely flat, and it is likely that at least part of the oil that accumulated in the South Bushehr High prior to the Zagros folding moved at a later time towards the high-relief Mand anticline. At Kharg, in the northern part of the Gulf, oil generated by the Sargelu and Garau source rocks was complemented by oil that originated from the Hanifa–Tuwaiq Mountain source rocks.

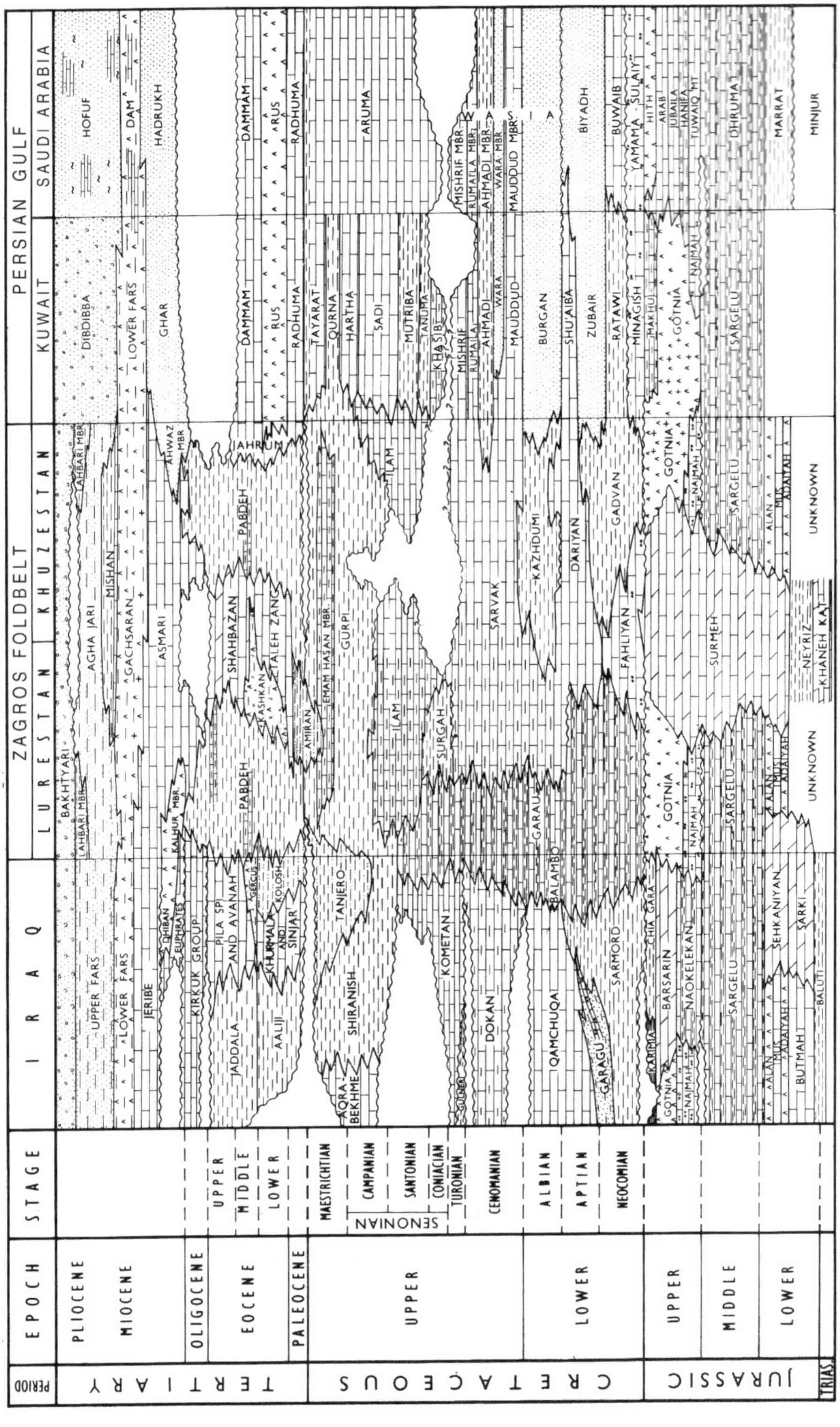

Fig. 19. Formation names used in the Persian Gulf and surrounding areas (after James and Wynd 1965).

The Early Cretaceous Petroleum System

Excellent source rocks, deposited during the Neocomian, in the lowermost part of the Garau Formation charged the Early Cretaceous Petroleum System. These source rocks are widely distributed in the Lurestan Depression and in the NW part of the Dezful Embayment (Fig. 20). The Early Cretaceous Petroleum System provides another example of source rocks not or poorly connected with reservoirs, for which a large part of the oil generated was cracked *in situ* into pyrobitumen and gas. As for the Middle Jurassic System, some oil migrated laterally towards porous platform facies of the Khami reservoirs, accumulated on regional highs before the Zagros folding, and eventually moved into Zagros folds.

Distribution of the Garau source rocks

The Garau Formation was defined at Tang-e Garau in Kabir Kuh, Central Lurestan (James & Wynd 1965; Bordenave & Sahabi 1971; Bordenave & Burwood 1995). It consists of a 700 m thick interval, predominantly argillaceous, capped by the Sarvak limestone. At the type section, the base of the formation does not crop out. However, the nearby Kabir Kuh-1 well drilled an additional 100 m of Garau organic-rich marls, before reaching the Gotnia Anhydrite. The Garau Formation of the type section comprises, from bottom to top, the following units.

(1) A 230 m thickness of monotonous dark grey to black, paper-like to platy-bedded bituminous marls, and micritic often pyritic argillaceous limestone. This interval, dated as Neocomian, is extremely rich in organic matter and uranium, especially towards the base. Residual TOC values vary from 4.5 to 9% in marls, whereas argillaceous limestone contains only 1–2% organic carbon. Palynological investigation indicates a type 2 algal-rich kerogen, already at the gas–condensate limit, according to its Thermal Alteration Index (TAI). Assuming a TR close to 0.80, the initial TOC could have been between 7.5 and 15% in marls and 1.7 and 3.4% in limestone (Fig. 21).

(2) A 265 m thickness of marls and pyritic argillaceous limestone, rich in radiolarians, that is no longer black in colour. Although the energy of the depositional environment remained very low, this interval contains little organic matter (TOC $\leq$ 0.5%). From correlations with other sections, this interval is assumed to belong to the Barremian.

(3) A 55 m thickness of cliff-forming low-energy limestone containing black cherts, which are the time-equivalent of the Dariyan Formation of Aptian age (TOC $\leq$ 0.3%).

(4) A 140 m thickness of medium grey slightly sandy marls, dated as (latest) Aptian to Albian, which are time-equivalent to the Kazhdumi Formation (TOC $\leq$ 0.6%). The Kazhdumi equivalent layers are overlain by the massive limestone of the Sarvak Formation.

The thickness of the Garau Formation, between the top of the Gotnia Anhydrite and the base of the Sarvak Formation, varies from 300 to 750 m. In the northeasternmost part of Lurestan, the Garau facies extended up to the Coniacian, as the Sarvak Formation consists of a basinal facies ('*Oligostegina* facies' containing *Rotalipora* and *Hedbergella*). On the NE edge of the Lurestan Depression, surface sections show an abrupt change in facies from low-energy basinal marls to shallow oolitic and peloidal limestone. As for the Sargelu, the isopach map of the Garau Formation shows a marked thinning in the central part of the Lurestan Depression, indicative of a starved basin (Fig. 20).

In the Dezful Embayment, several wells drilled thick Garau sediments. However, the occurrence of reverse faults, with repeated units makes estimation of the thickness of the formation difficult. A maximum thickness of 500 m seems likely. No Garau Formation was found in wells drilled SW of the Dezful Embayment. These wells penetrated instead thick layers of Fahliyan limestone. A tentative isopach map of the organic-rich Garau facies is presented in Figure 21. In addition to the Tang-e Garau analyses previously discussed, two cores cut in the Naft-1 well, in Central Lurestan, found TOC values between 1 and 5.5%. Thin sections prepared from cuttings and the gamma-ray logs from the MIS-306 well indicated the existence of thick Garau source rocks (500 m total thickness, of which 9 m of cored interval contains 1.7–4.3% residual TOC for source rocks already mature to the gas–condensate window). At the Khaviz-1 well, more than 1000 m of Garau facies were drilled. However, this thickness is not representative, because of tectonic complexity in the lower part of the drilled interval. From examination of both thin sections and gamma-ray logs, the Garau Formation drilled at Khaviz is assumed to have a relatively low organic content. In the northern part of the Gulf, more than 100 m of bituminous marls were observed at the Bahregansar-5 well, both from thin sections and gamma ray log examination. In the eastern part of the Dezful Embayment and in Fars, platform shallow-water carbonates of the Fahliyan Formation contain less than 0.5% TOC.

Thermal evolution of the Garau source rocks

At Huleylan-1, in Central Lurestan, the base of the Garau source rocks was buried deeper than 4700 m before the beginning of the Zagros folding, and reached the onset of the gas stage. At the

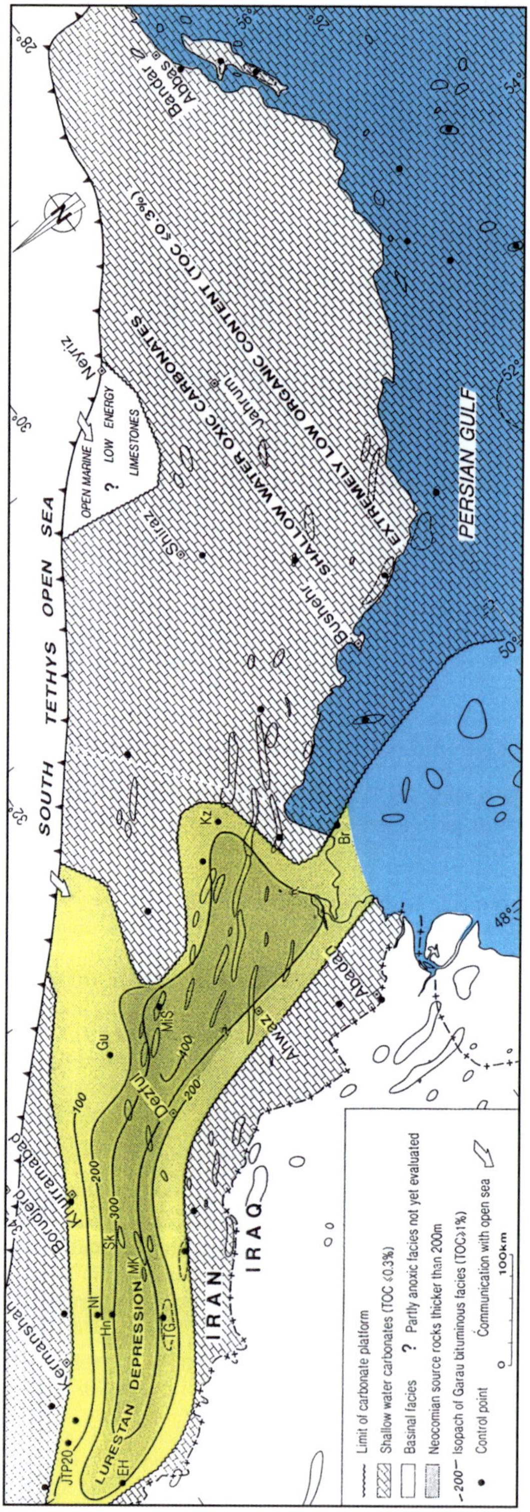

Fig. 20. Isopachs of the organic-rich lower part of the Garau Formation and source rock characteristics (after Bordenave & Huc 1995).

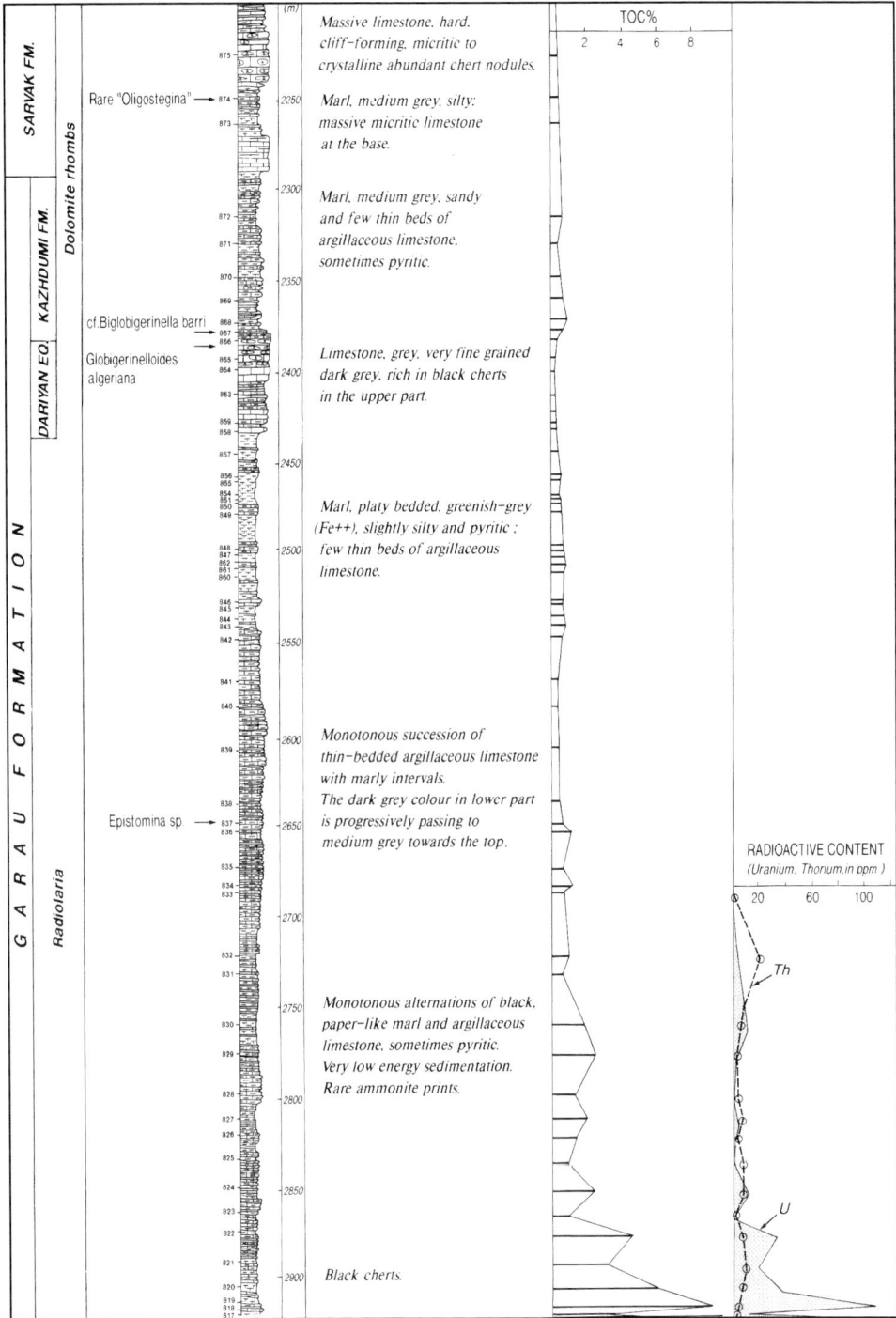

Fig. 21. Garau section of Tang-e Garau (Lurestan): organic carbon, uranium and thorium content (after Bordenave & Huc 1995).

nearby Naft-1 well, the Garau source rocks are in the dry gas zone (Rock-Eval T_{max} between 538 and 551 °C). This is confirmed by a residual HI of the order of 40 g HC kg^{-1} C. The oil expulsion window was probably reached during the Palaeocene (3400 m burial), as the result of the deposition of the thick Amiran Flysch. On the SW side of the depression, the thickness from the base of the Garau Formation to the top of the Gachsaran Formation is 3100 m. The source rocks, if present, would have reached the oil window during the deposition of the Gachsaran Formation, or slightly later. In the Dezful Embayment, a few wells (Agha Jari-140, Ahwaz-101, Pazanan-17) showed that the Garau facies, developed at the lowermost part of the Fahliyan Formation, was buried between 4500 and 5000 m at the end of the deposition of the Mishan Formation, prior to the Zagros folding. The oil window was reached in these wells during the Late Cretaceous. Almost everywhere in the Lurestan–Dezful Embayment Depression, the Garau source rocks reached the oil expulsion window between the Late Cretaceous and the Early Miocene, prior to the Zagros folding.

Source rocks, reservoirs and seals

In the Lurestan Depression, the Garau source rocks are overlain by thick marls and argillaceous limestone deposited during the Barremian, Aptian and Albian, and even up to the Coniacian in the central part of the depression. This 450–700 m thick argillaceous interval isolated the Garau source rocks from potential reservoirs. The first reservoir that could have drained the source rocks was either the high-energy limestone of the Sarvak Formation in the southern part of the depression or the Ilam limestone in its central part. In the absence of a reservoir closely associated with the Garau source rocks, a good part of the oil could not be expelled and was probably cracked *in situ*, causing the formation of pyrobitumen and gas. However, some oil was expelled through a network of linear microfaults and fractures, and formed impressive dykes, sometimes up to 20 cm thick, of gillsonite-type material as observed in several Lurestan breached anticlines (Bordenave & Sahabi 1971).

In the Dezful Embayment, the thin shallow-water limestone in the Khalij Member of the Gadvan Formation facilitated the migration of some oil or gas towards the platform facies of the Khami Group, and its accumulation on local or regional highs that existed before the Zagros folding (Fig. 17), according to a scenario similar to those of the Middle Jurassic Petroleum System. On the highs located in the Dezful Embayment, Khami reservoirs are efficiently sealed by the organic-rich Kazhdumi Formation, which generated a high pore pressure barrier.

The petroleum problems related to the Early Cretaceous System

The exploration of prospects related to the Garau source rocks has been so far disappointing. A tenth of wells, drilled in the NW part of the Lurestan Depression, did not found any reservoir, except in the Ilam Limestone, which crops out in the cores of most of the anticlines. In the southern part of the Lurestan Depression, two middle-sized fields, Sarkan and Maleh Kuh, produce light oil (42–46° API with 0.45–0.65% sulphur) from the Ilam and Sarvak connected reservoirs. Their carbon isotopic composition ($-26.9‰$ to $-26.5‰$) is compatible with the Garau source rocks. In the area located NE of the edge of the Lurestan Depression (Kermanshah–Khurramabad area) anticlines are deeply breached, at least to the Sarvak Formation. The same conclusion is applicable for most of the anticlines located north of the Mountain Front, which bounds the Dezful Embayment to the north.

In most of the Dezful Embayment, porous intervals of the Khami Formation are oil- and/or gas-bearing. However, good reservoirs are generally limited to the 14–30 m thick shallow-water limestone of the Khalij Member, and the reserves evidenced in the Khami reservoirs of the largest anticlines, either oil or gas, are limited.

Oil expelled from Garau source rocks complemented the oil that originated from the Sargelu source rocks in charging the pre-Zagros Kuh-e Mish and South Bushehr Highs. Later, part of the oil re-accumulated in Zagros folds, such as Gachsaran, Garangan–Chillingar, Sulabedar, Nargesi, Dara, and Milatun. The light oil (38–42° API) of Garangan–Chillingar and Sulabedar, which accumulated in the excellent Surmeh and Khami reservoirs, probably has a dual origin, from Sargelu, as explained above, and from Garau source rocks. In the same way, the north–south Kharg anticline was strategically located to trap oil originating from the Garau, the Hanifa–Tuwaiq Mountain and Sargelu source rocks. The carbon isotopic composition of the Kharg oil ($-26.8‰$) is compatible with both the Hanifa–Tuwaiq Mountain and the Garau source rocks.

The areal extent of the Garau facies was closely similar to those of the Sargelu Formation. However, the basinal area became narrower during the Neocomian. The NE shelf edge prograded slightly over the depression and passed to the west of MIS, east of Kuh-e Bangestan and Khaviz, east of Binak and west of Kharg Island (A. J. Wells, unpubl. data). Moreover, to the SW of the depression, a SW shelf edge was formed that extended from the Naft Khaneh area, close to the Iraq border, to Samand, Omid, Abadan, and to the east of Burgan. Beyond the shelf edge, the platform facies of the Fahliyan, Gadvan and Dariyan Formations overlies basinal facies (Sargelu and the Gotnia Formations). There, Garau-related oil could have accumulated in the porous layers of the Khami Formation. Oil found at Darkhovin and Jufeyr in the Khami Formation was probably charged laterally from Garau source rocks. In the same way, the Zubair sands, which have a total thickness between 50 and 100 m, could have trapped some Garau oil in low-relief north–south-trending pre-Zagros structures.

The Gadvan marginal source rocks

During the Barremian, oxic shallow-water Gadvan limestone covered most of Fars, NE Khuzestan, the Kermanshah–Khurramabad area, and SW Lurestan. In contrast, in the SW Dezful Embayment, between the prograding Zubair sands and the limestone platform, a slightly deeper water trough existed, where prodelta dark brown laminated low-energy argillaceous limestone was deposited under anoxic conditions. Part of this argillaceous limestone, which is sometimes pyritic, was described as 'slightly bituminous' to 'bituminous' based upon the examination of a large number of cuttings thin sections and of the gamma-ray logs from a few wells (Bordenave & Huc 1995). The thickness of the 'bituminous' layers varies from 20 m to a maximum of 100 m at Bushgan-1.

Few samples were analysed at Gachsaran GS-83 and Bushgan-1. The organic content is low, with only a few values slightly higher than 1%. Little oil could have been generated by the Gadvan source rocks. Perhaps the small 32° API oil accumulation of Bushgan in the Asmari originated from the Gadvan source rocks, in the absence of other source rocks. Moreover, the carbon and sulphur isotopic composition of the Bushgan oil ($-24‰$ and $1.2‰$, respectively) is different from that of the other oils analysed in the Zagros Domain. At Bushgan, the Gadvan to Mishan interval is 3350 m thick, and the oil expulsion window was probably reached during the deposition of the Mishan. However, because of their low organic content, thinness, and local extension, the Gadvan source rocks are marginal.

The Middle Cretaceous to Early Miocene Petroleum System

The Middle Cretaceous to Early Miocene Petroleum System provides an example of almost vertical oil migration, enhanced by fracturing (Bordenave &

Hegre 2005). As a consequence, a theoretical amount of oil expelled by various source rocks in the drainage area of one anticline, estimated by thermal modelling, can be compared with the initial oil in place (IOIP) discovered in the corresponding field. Moreover oil to source rock correlation, using isotopic composition ($\delta^{13}C$, $\delta^{34}S$) and biomarker composition, makes possible the verification of the oil origin, and determination of the validity of the proposed geological assumptions.

About 8% of the global oil reserves are concentrated in a small geographical area, the Dezful Embayment, which extends over 60 000 km^2 (Fig. 3). Oil accumulated in two carbonate reservoirs: the Asmari Formation, which contains 75% of the reserves, and the Bangestan Group (Sarvak and Ilam Formations), which contains another 23%. Seven fields, Marun, Ahwaz, Gachsaran, Mansuri, Rag-e Safid, Bibi Hakimeh and Parsi, grouped in an area of only 12 000 km^2, contain 72% of the IOIP. The reserves of each of these fields are between 60 and 10 billion barrels IOIP.

In most of the Dezful Embayment fields, accumulated in high-relief asymmetric often thrust anticlines, the Asmari and Sarvak reservoirs are interconnected and have the same oil water level (OWL), as the result of the intense fracturing of the Pabdeh–Gurpi marls in the crestal part of the anticlines. In contrast, the Asmari and Bangestan Formations form distinct reservoirs in low-relief anticlines, such as the Ab-e Teymur and Mansuri. In the Ahwaz and Kupal fields, the OWL of the Asmari and Bangestan reservoirs are different, but oils contained in these two reservoirs have the same origin, as shown by their carbon and sulphur isotopic composition. The Asmari reservoir is covered by the thick evaporites (halite and anhydrite) of the Gachsaran Formation.

The oil currently trapped in the Asmari and Bangestan reservoirs in the Dezful Embayment originated from two main source rocks, the Kazhdumi Formation of Albian age and the Pabdeh Formation (Middle Eocene and Early Oligocene). Marginal amounts of oil came from the Ahmadi Member of the Sarvak Formation (Middle Cenomanian), and perhaps from the Late Cretaceous Gurpi Formation.

The high pressure developed in the pore space of the Kazhdumi organic-rich marls as the result of oil generation formed a barrier that prevented oil generated by deeper source rocks from reaching the Sarvak and Asmari reservoirs. In contrast, limited amounts of oil generated by the Kazhdumi may have reached by downward migration the few porous zones included in the Khami Formation. Therefore, the Middle Cretaceous to Early Miocene Petroleum System can be considered to be independent from earlier petroleum systems, with a few probable exceptions (Pazanan, Chesh-meh Khush) that are discussed below.

Distribution of the source-rock facies

Kazhdumi source rocks

The Kazhdumi Formation provides one of the world's best examples of prolific source rocks deposited in an intrashelf silled depression (Bordenave & Burwood 1995). The Kazhdumi Formation was defined in the Khuzestan Mountain Front as a low-lying more argillaceous interval situated between the top of the Aptian Dariyan Limestone and the base of the cliff formed by the limestone in the Sarvak Formation of Cenomanian–Early Turonian age (James & Wynd 1965). The Kazhdumi Formation is dated as Albian. However, in the deeper part of the basin, where no breaks in sedimentation are recorded between the Dariyan Limestone and the Kazhdumi Marls, the

argillaceous deposition may have commenced in the Late Aptian. The uppermost part of the Kazhdumi Formation, which passes gradationally to the Sarvak Limestone, may belong to the Early Cenomanian (James & Wynd 1965).

In the central part of the Dezful Embayment, cuttings thin sections from deep wells have shown that the entire Kazhdumi interval was deposited in low-energy anoxic conditions (Bordenave & Nili 1973), with only pelagic fauna recorded (globigerina and radiolarian). More than 300 m of paper-like, thinly laminated, dark grey shales and fine-grained argillaceous limestone were deposited in a depression coincident with the area of the main field (Figs 22 and 23). These marls are radioactive and have a low density. As such, they are well characterized by gamma-ray and density logs. Locally, the basinal anoxic facies continued into the Early Cenomanian with the deposition of Sarvak units in 'Oligostegina facies'. However, this Sarvak basinal facies generally shows only moderate organic content.

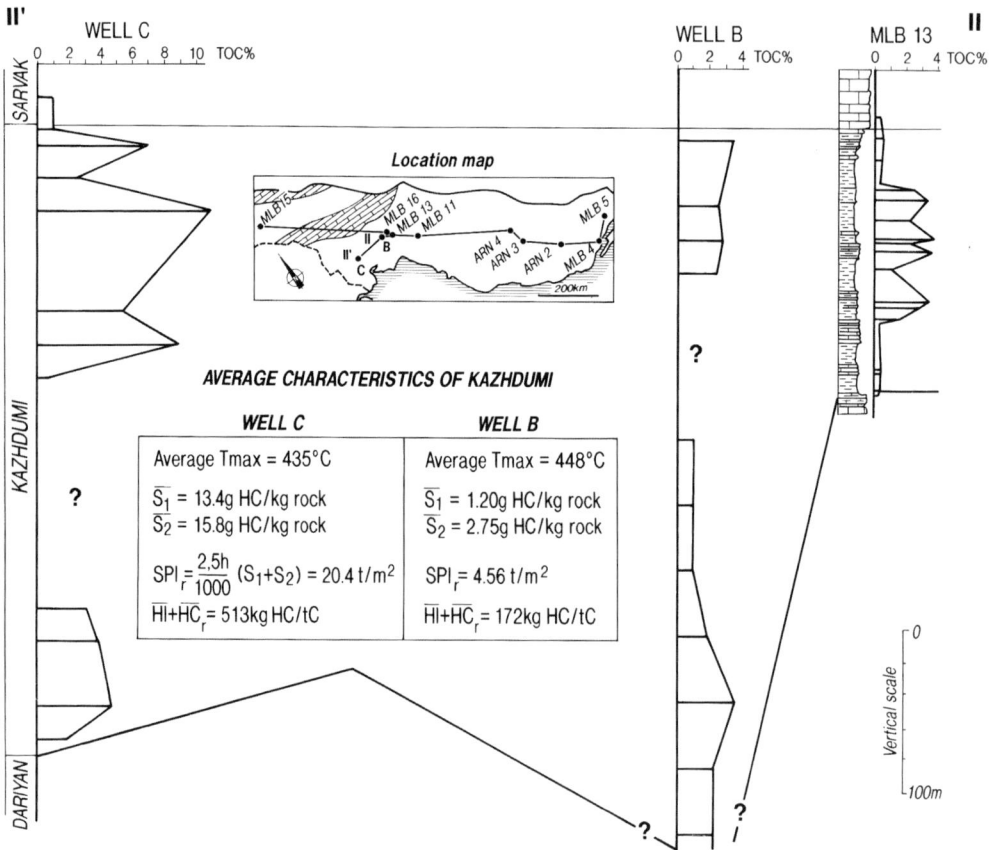

Fig. 22. Cross-section of the Kazhdumi Formation from Kuh-e Bangestan on the Mountain Front to the Dezful Embayment (after Bordenave & Burwood 1995).

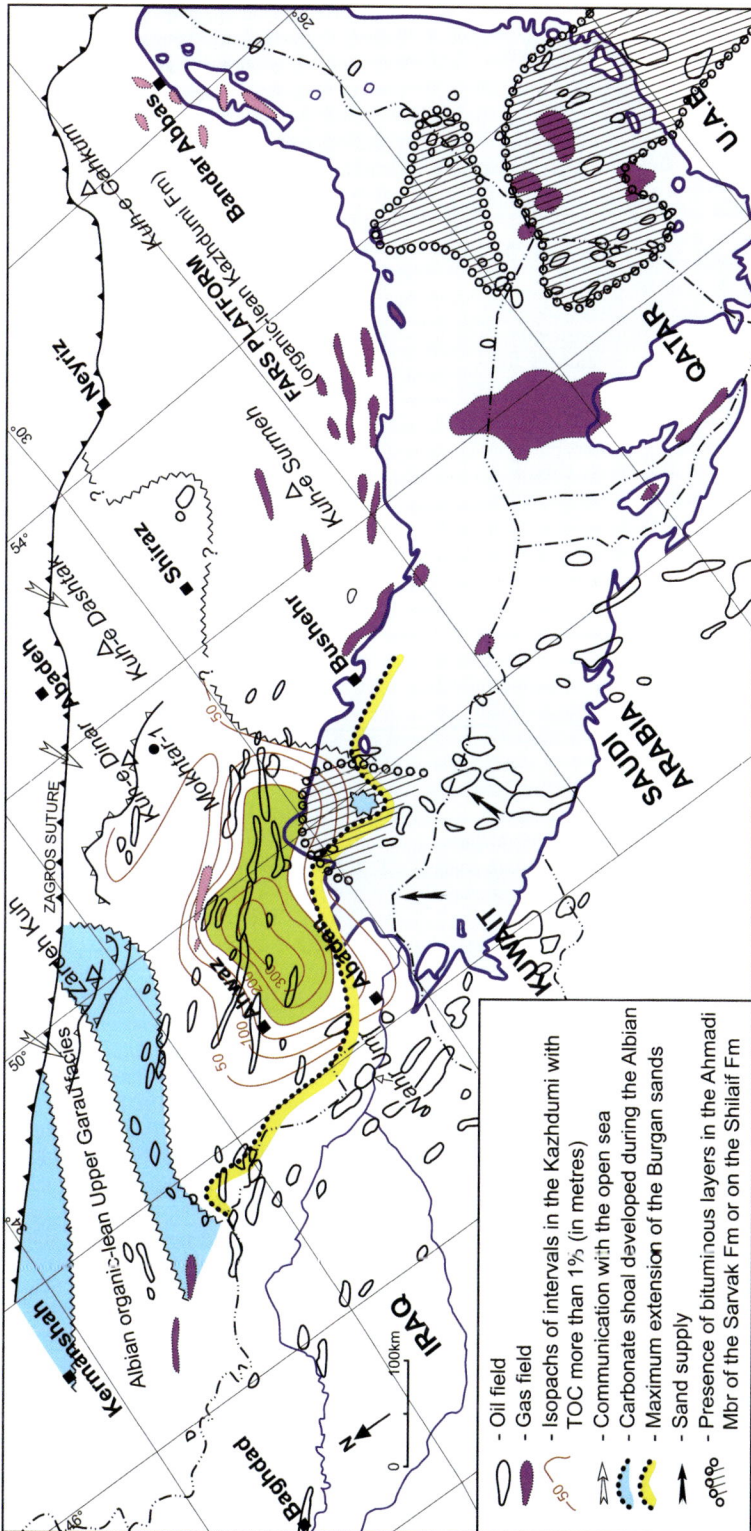

Fig. 23. Isopach map of organic–rich layers (TOC higher than 1%) in the Kazhdumi Formation (modified after Bordenave & Hegre 2005).

The Kazhdumi Dezful Depression was bordered to the north by the Bala-Rud Shoal, where shallow-water limestone accumulated during the Aptian, Albian and Cenomanian period, without any marked argillaceous breaks. This shoal separated the Dezful Depression from the pre-existing Lurestan Depression, where Garau facies continued to accumulate during the Albian, and even later. As mentioned above, the 'Albian Garau', time-equivalent of the Kazhdumi Formation, does not contain any source rocks (Fig. 21).

On its SW side, the Dezful Depression was bordered by the Safaniya–Burgan Delta, which extended over the northernmost part of the Gulf. In Kuwait, the Nahr Umr Formation, time-equivalent to the Kazhdumi Formation, consists mostly of sandstones averaging a thickness of 350 m. In contrast, in the Nahr Umr field (SE Iraq), pyritic black shales are interbedded with sandstones containing pieces of lignite and amber, with a sand/shale ratio averaging 50% (Dunnington et al. 1959; Ibrahim 1978).

Towards the SE, the Dezful Depression was bordered by the Fars Platform, where the deposition of organic-lean marls in shallow-water oxic conditions was often interrupted by temporary emergent episodes. On the Gavbendi High, the Kazhdumi Formation is extremely thin, only 15–50 m (Fig. 24). Although the environment was oxic, the fauna was only pelagic. North of Bandar Abbas, thicker oxic brownish red marls are organic-lean (150 m). In North Fars (Sabzpushan–Buzpar area), 250–300 m of organic-lean oxic marls were deposited in the Kazhdumi (Bordenave et al. 1971).

On the Mountain Front that borders the Dezful Embayment to the NE, at both Kuh-e Mish and Kuh-e Bangestan, a 'hard ground' located at the top of the Dariyan limestone reveals a temporary emergence, followed by an abrupt change of sedimentation conditions, with strong currents and still oxic conditions. The progressive deepening of the water column, as shown by the replacement of benthic by pelagic species, caused anoxic conditions to prevail. This deepening resulted in the deposition of organic-rich marls. The upper part of the Kazhdumi Formation contains a mixture of benthic and pelagic fauna showing the return to oxic conditions (Fig. 25). Comparison between the Mountain Front outcrops and wells located near the centre of the Dezful Depression shows that both the Kuh-e Mish and the Kuh-e Bangestan areas were located on highs, where the thickness of the Kazhdumi was thinner (160–260 m) as compared with 350–400 m in the Depression.

Isopachs of the organic-rich layers (TOC ≥ 1%) were prepared using geochemical analyses, gamma-ray logs, and examination of cuttings thin sections

(Fig. 23; See also Table 2). On the edge of the Dezful Embayment, at Khaviz-1, Gachsaran-36, Dorquain-1, Khorramshahr-1, and Barhregansar-5, anoxic and oxic facies alternate. There are no Kazhdumi source rocks at Naft-e Safid, MIS-306, Lab-e Safid-1, Danan-1, and Kabud-1 to the north of the Dezful Embayment, nor at Sulabedar-1, on its southeastern edge. No source rocks were indicated at Susangerd-1 to the west. Limited information is available to the east of Gachsaran. In the Izeh zone, north of the Mountain Front, Mokhtar-1 drilled an unusually thick sequence of Kazhdumi marls (1100 m). These marls include some high-organic layers. They correspond there to a complex detachment zone that formed the caprock of the gas-bearing Fahliyan reservoir (Sherkati & Letouzey 2004; Bosold et al. 2005). In addition, high-organic layers (TOC up to 5%, for T_{max} of 440 °C) have been described in the Kazhdumi outcrops at Kuh-e Dashtak, 100 km north of Shiraz (Rudkiewicz et al. 2007). The existence of Kazhdumi source rocks in the Shurom-Kuh-e Rig area and north of Shiraz is likely. The Dudrou field produced from the Fahliyan, whereas the Sarvak is breached. At Kuh-e Rig, both the Dariyan and the Sarvak produced, and at Shurom the Sarvak is oil-bearing, whereas the Dariyan is dry. Commercial oil was found in the Sarvak reservoir at Sarvestan near Shiraz. Significant viscous oil seepages observed east and NE of Shiraz may come from the Kazhdumi Formation.

Almost no source rocks were observed at Kharg. SE of Kharg Island, numerous control points showed the absence of Kazhdumi source rocks (Mand-2, Bushehr-3, FR-B-1, IMINOCO D-1, Lavan-1, IMINOCO T-2, PEGUPCO M-1, and Sassan-2), according to thin-section examination (Bordenave et al. 1971; Bordenave & Nili 1973). Offshore of Abu Dhabi, grey to greenish grey marls and mudstone with lenses of Orbitolina limestone graded into red–brown, splintery shales, deposited in a subtidal setting (Alsharhan 1989). However, the existence of low-energy, perhaps anoxic or dysoxic facies, should be noted in the Sirri A-1, Rostam-32, and Fateh-1 wells, in which the organic content is locally slightly higher.

At Mansuri, where the Kazhdumi source rock has already reached the onset of oil window (T_{max} 435 °C, R_o 0.71%), S_1 varies from 1.18 to 28.8 g HC kg^{-1} rock (average value 13.4 g HC kg^{-1} rock) and S_2 varies from 0.34 to 38.7 g HC kg^{-1} rock (average value 15.8 g HC kg^{-1} rock).

Using the Mansuri values, the residual source rock potential index (SPI) (Demaison & Huizinga 1991) of the Kazhdumi is higher than 20 t m^{-2}, and may reach 25–30 t m^{-2}, as only part of the section was sampled there, one of the highest values observed in the world. The average sum of the

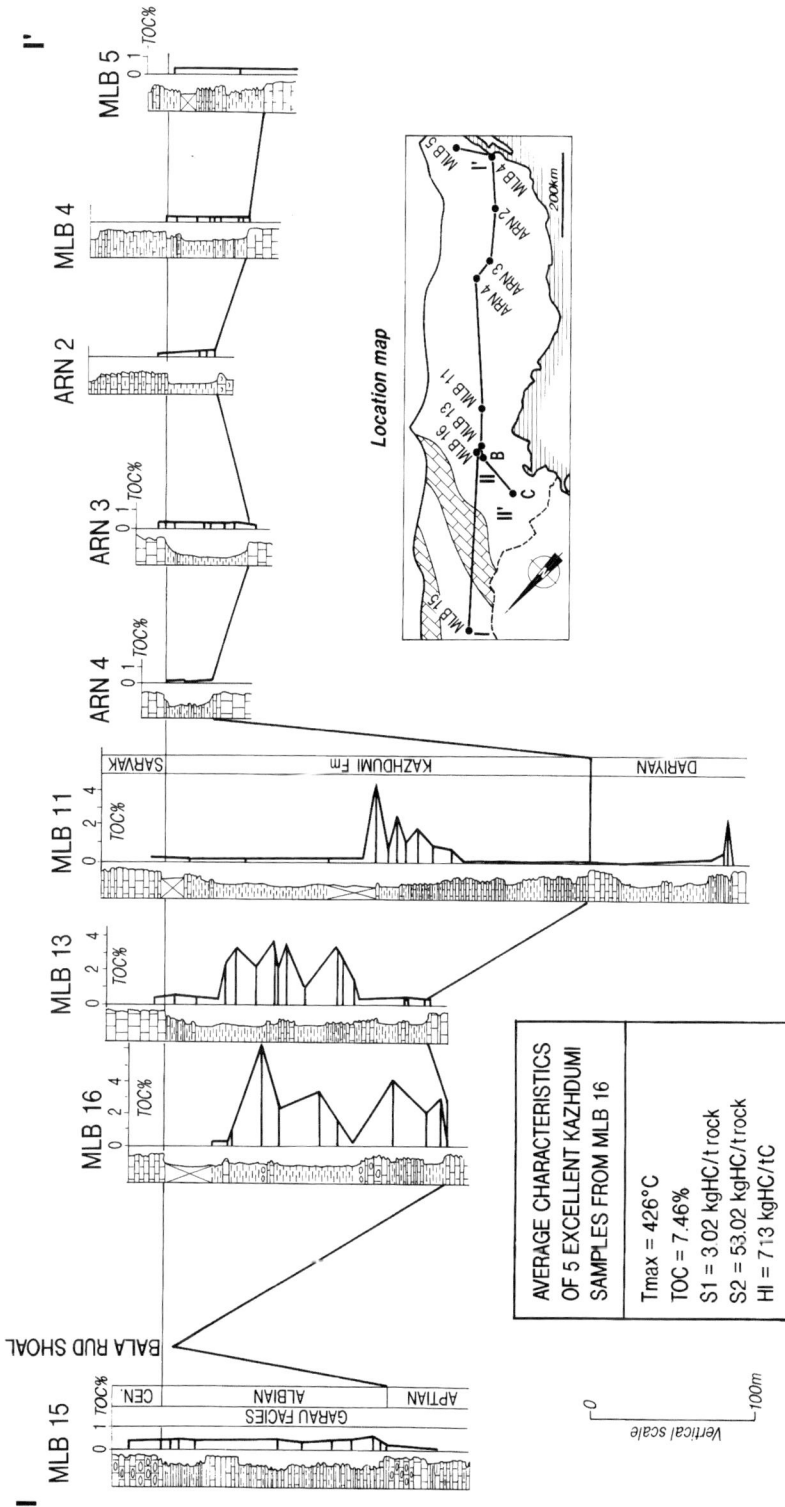

Fig. 24. Cross-section of the Kazhdumi Formation from Lurestan to Fars, indicating the evolution of the organic content from NW to SE (modified after Bordenave & Huc 1995).

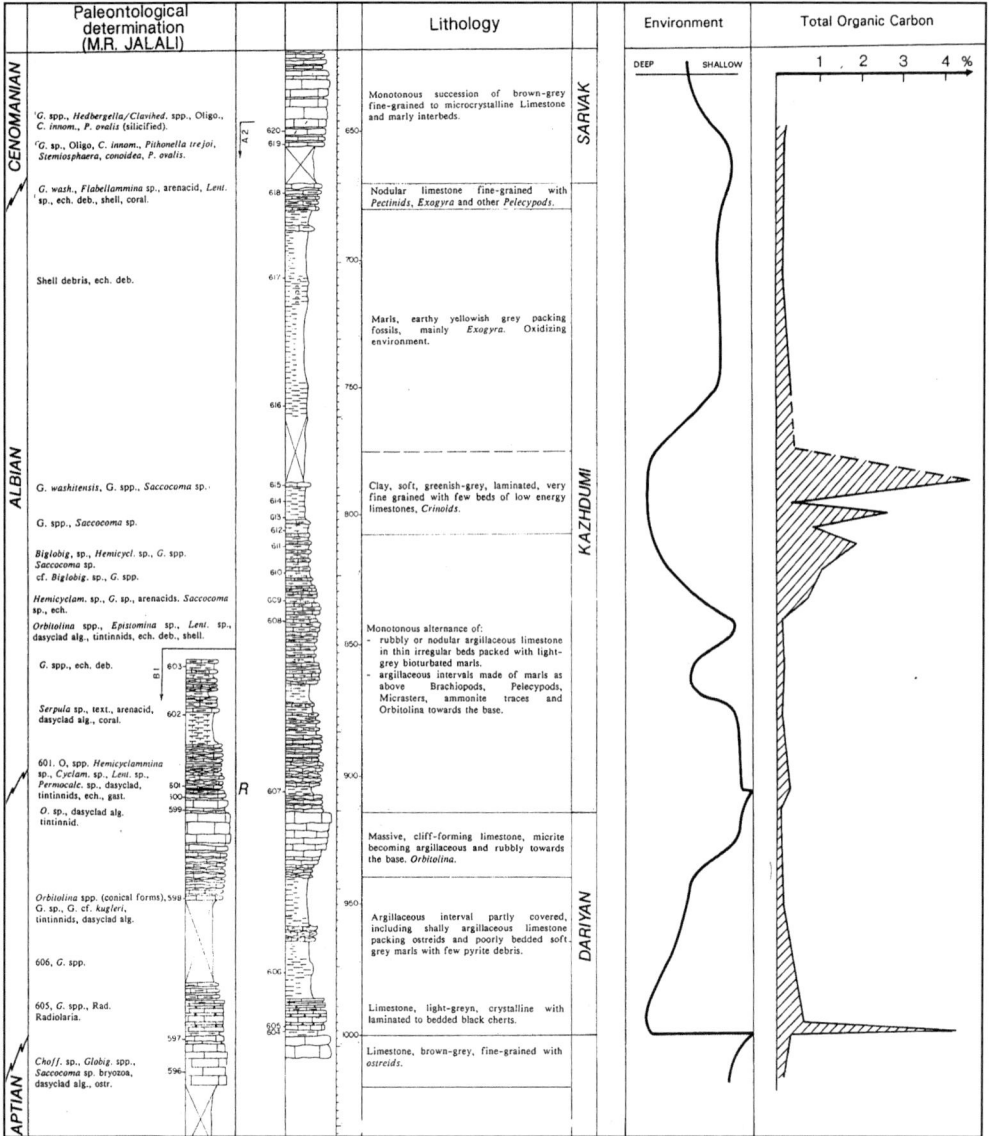

Fig. 25. Lithological description and stratigraphy of the Albian Kazhdumi Formation at Kuh-e Mish (after Bordenave & Burwood 1995).

hydrogen index (HI = S_2/TOC) and of the hydrocarbon index (HC = S_1/TOC) is high at 513 g HC kg^{-1} C (Fig. 22). At Parsi, close to the Mountain Front, the source rock characteristics seem rather poor at first glance. The average value of $S_1 + S_2$ is only 4.0 g HC kg^{-1} rock, and the average HC + HI is 172 g HC kg^{-1} C, with an SPI of 4.56. However, these source rocks already have attained the maximum oil generation zone (T_{max} = 448 °C) and have expelled most of their oil. Therefore,

initial TOC values were only slightly lower than those measured at Mansuri.

The Kazhdumi organic matter is mostly algal, without traces of terrestrial plants (Ala *et al.* 1980). The high organic sulphur content of the kerogen measured in surface samples collected at Kuh-e Bangestan varies from 4.95 to 5.5% (Bordenave & Hegre 2005) and its high HI values (500–700 g HC kg^{-1} C) classify the Kazhdumi as a 2S type source rock as defined by Orr (1986).

Table 2. *Organic content of the Kazhdumi Formation (modified from Bordenave & Burwood 1995)*

Location	Thickness of organic layers, h (m)	Residual TOC %			h TOC/100 if TOC $\geq$ 1%
		Minimum	Average	Maximum	
MLB-11	50	1	2	4.5	1.0
MLB-13	80	1	3.6	10.6(1)	2.9
MLB-16	120	1	3	3.5	3.6
Gachsaran	300	1	1.2	2	3.6
Parsi	240+	1	2.2	4	5.3+
Mansuri	280+	1	5.1	11	14.3+

(1) including the results of analysis of new samples from Kuh-e Bangestan, which showed the Kazhdumi to be immature (T_{max} 426 °C) and having an excellent potential (S_2 and HI respectively 50 and 700 g HC kg^{-1} C).
h is the thickness of the Kazhdumi source rocks, and toc is the average value of its total carbon content.

This is also shown by the high sulphur content of the derived oils. These oils contain 3–4% sulphur when immature and a minimum of 0.6% sulphur when derived from deeply buried source rocks (Bordenave & Burwood 1995).

During the Albian, the Dezful Depression was located near the Equator, according to plate-tectonic reconstruction (Scotese & Golonka, 1993). An equatorial humid climate is demonstrated by the large amount of freshwater and clastic sediments that reached the Dezful Depression through a network of rivers. The low-density oxygenated fluviatile water that originating from the wide Arabo-Nubian landmass was nutrient-rich (nitrates and phosphates), creating ideal conditions for high phytoplankton productivity. As the result of the global sea-level rise at 107.5 Ma, the sea floor of the central part of the Dezful Depression became deeper than the euphotic zone. A countercurrent poured heavier oceanic waters into the basin, creating density stratification between the freshwater input from the Arabo-Nubian Shield and the salty oceanic water. Sulphate-reducing bacteria were extremely active, as demonstrated by the sulphur incorporated into the Kazhdumi source rocks. The hydrogen sulphide produced by these bacteria poisoned the bottom waters; this explains the disappearance of benthic fauna and the occurrence of planktonic forms only (*Globigerina* and Radiolaria). While a delta prograded from the SW, fine-grained clastic deposits, silts and clays accumulated in the depression in a low-energy environment, enhancing the incorporation of organic matter into sediments. In contrast, regional highs or north–south salt-related structures remained slightly shallower, and accommodated, at least part of the time, shallow-water benthic fauna (*Orbitolina* and Arenacids).

In the Lurestan Depression, although deep-water conditions persisted, the organic content of the sediments remained lower than 0.6%. No water stratification was established, because of climatic factors and/or absence of rivers feeding the depression.

The Ahmadi–Shilaif source rocks

The Ahmadi Formation, defined in Kuwait, corresponds to Early Cenomanian marls that formed the caprock for the oil accumulated in the Burgan Sands (Owen & Nasr 1958). In Iran, the Ahmadi Member of the Sarvak Formation is an argillaceous interval identified in Fars and in the northern part of the Persian Gulf (James & Wynd 1965). Thin-section study showed that organic-rich facies were developed in the subsiding Binak Trough (Bahregansar and Binak, 105 m and 75 m of black bituminous marls, respectively). Two samples analysed at Binak-5 showed TOC values of 1.8 and 5.4%, S_2 up to 8 g HC kg^{-1} rock, and relatively low HI, 60 and 160 g HC kg^{-1} C, for a maturity that corresponded to the onset of the oil expulsion. In the central part of the trough, where the Ahmadi Formation is thicker (up to 300 m thick at H-1), the source rocks are probably richer (Fig. 23). In the Binak Trough, the Ahmadi source rocks reached the onset of oil generation towards the deposition of the end of the Agha Jari Formation (Late Miocene to Pliocene). The base of the Ahmadi Member is currently buried between 3700 and 3800 m at the Binak-4, Tanb-1, and H-1 wells.

At Kharg, the Ahmadi Member degenerated into grey–green marls, with only a few thin layers of dark grey, slightly bituminous marls. At Mand-2, the middle part of the Sarvak Formation is more argillaceous and of '*Oligostegina* facies'. The TOC of the marls remains lower than 1%. The Binak Trough was probably connected to the Parsi–Karanj–Khaviz subsiding area, where the Sarvak Formation was deposited in a dysoxic '*Oligostegina* facies'. The Ahmadi Member exists in the southern-most part of offshore Fars, but is organic-lean. However, the study of thin sections shows the existence of low-energy anoxic/dysoxic facies in the Ahmadi Member at Kish-1, Sirri A-1, Rostam-32 and Fateh-1, in which the organic content could be significant (Fig. 23). In contrast, in Abu Dhabi, the

Shilaif Formation (Late Albian and Early Cenomanian), between the Mauddud and the Mishrif Formations, is composed of organic-rich, laminated mudstone (Alsharhan 1989). It was deposited in a depression that extended in the western part of offshore Abu Dhabi, where anoxic conditions prevailed (Patton & O'Connor 1988). In the depression, TOC values generally vary from 1 to 6%, with a few maximum values of 15%. In the Umm Shaif–Zakum area, the petroleum potential of S2 is excellent (average 25 g HC kg^{-1} rock). Optical examination showed that the Shilaif kerogen belongs to sapropelic oil-prone type 2 (Hassan & Azer 1985; Loutfi & El Bishlawy 1986).

The excellent source rocks deposited in the western part of offshore Abu Dhabi have not yet reached the oil expulsion window (vitrinite reflectance is 0.3–0.6%). The oil expulsion window was reached in the eastern Abu Dhabi and Dubai offshore areas, where the Shilaif Formation is thinner than 15 m (Patton & O'Connor 1988). According to a fluid inclusion study carried out on the Mishrif reservoir of the Fateh field, oil migration would have taken place during the Oligocene and the Miocene (Videtich et al. 1988).

The marginal source rocks of the Gurpi Formation

Basinal, oxygen-deprived conditions are observed in a narrow elongated WNW–ESE trough extending from the south of Shiraz to the NW Dezful Embayment and to Lurestan. Gurpi marls, of Santonian and Campanian age, contain between 1 and 2% TOC. The HI varies from 150 to 400 g HC kg^{-1} C for immature or marginally mature marls. Their Rock-Eval S$_2$ values are low (2–3 g HC kg^{-1} rock). Pyrolysis of kerogen shows that the Gurpi marls are mostly gas-prone and have an extremely low potential (Bordenave & Huc 1995). The thickness of 'bituminous' marls (TOC ≥ 1%) increases from SE to NW: 60 m at Gachsaran and Binak, 120 m at Marun-21, 160 m at Qaleh Nar, and greater than 200 m in Lurestan. Optical examination showed the Gurpi organic matter is mainly of terrestrial origin, especially in the northern part of the dysoxic trough, with the presence of angiosperm pollen. However, in the central part of the trough, the kerogen is more algal (A. Combaz, unpubl. data).

The Pabdeh source rocks

As a consequence of the uplift and obduction of the former radiolarite trough on the NE edge of the Arabian Platform during the Campanian, the direct access of the Arabian Platform to Southern Tethys ceased. A thick sequence of flysch, in large part composed of radiolarite debris, accumulated in front of the newly emergent area, which was being heavily eroded. A NW–SE-trending depression, parallel to the Zagros suture, extended from Fars to Lurestan during the Palaeocene and Eocene, and continued during the Oligocene in Lurestan. This depression was asymmetrical, with its higher subsidence axis being located near its NE limit. Up to 1000 m of monotonous grey marls containing rich planktonic (Globorotalia and Globigerina) fauna accumulated there. The depression was bordered to the SW by a platform that covered most of the current Persian Gulf. On this platform, the evaporites of the Rus Formation (Palaeocene–Eocene boundary) were limited below and above by the shallow-water limestone of the Dammam and Radhuma Formations. To the ENE, the depression was bordered by the Fars Platform, where the shallow-water dolomitic limestone of the Jahrum Formation surrounded a subsiding sabkha where Sachun evaporites locally reached a thickness of 1400 m.

Euxinic conditions prevailed during the Late Eocene, NE of a threshold located in the Nar-Kangan area, where the two carbonate platforms of the Jahrum Formation to the east and the Umm al Radhuma–Dammam Formations to the west are almost contiguous. Euxinic conditions continued during the Early Oligocene in Lurestan. Up to 200 m of organic-rich marls accumulated in a trough extending from the Yasutch area (NW Lurestan) to Iraq (Fig. 26). Surface sections in the Yasutch area and in Lurestan show that the base of the Pabdeh Formation (namely, the basal dark purple marls dated as Palaeocene and the argillaceous limestone with brown cherts dated as Middle Eocene) were still being deposited in an oxic environment and are organic-lean, although they contain pelagic fauna.

Analyses performed on samples collected in 13 field locations, eight in the Yasutch–Kuh-e Bangestan area (Bordenave et al. 1971) and five in Lurestan (Bordenave & Sahabi 1971), showed that the average TOC of the Pabdeh source rocks varied from 2 to 5%, with maximum values attaining 8.6%. At Kuh-e Bangestan, where the Pabdeh is immature (T_{max} 425 °C), eight samples provided the following results: TOC has a minimum of 2.71%, average of 4.90% and maximum of 8.60%; S$_2$ has a minimum of 11.6 g HC kg^{-1} rock, an average of 25.7 HC kg^{-1} rock, and a maximum of 45.4 HC kg^{-1} rock; HI has a minimum of 434 HC kg^{-1} C, an average of 521 HC kg^{-1} C, and a maximum of 605 HC kg^{-1} C; organic sulphur content is 6.46–7.0 wt%.

Analyses made at Dehluran-4, Ab-e Teymur-1, Mansuri-6, Parsi-35, Gachsaran-83 and Binak-4

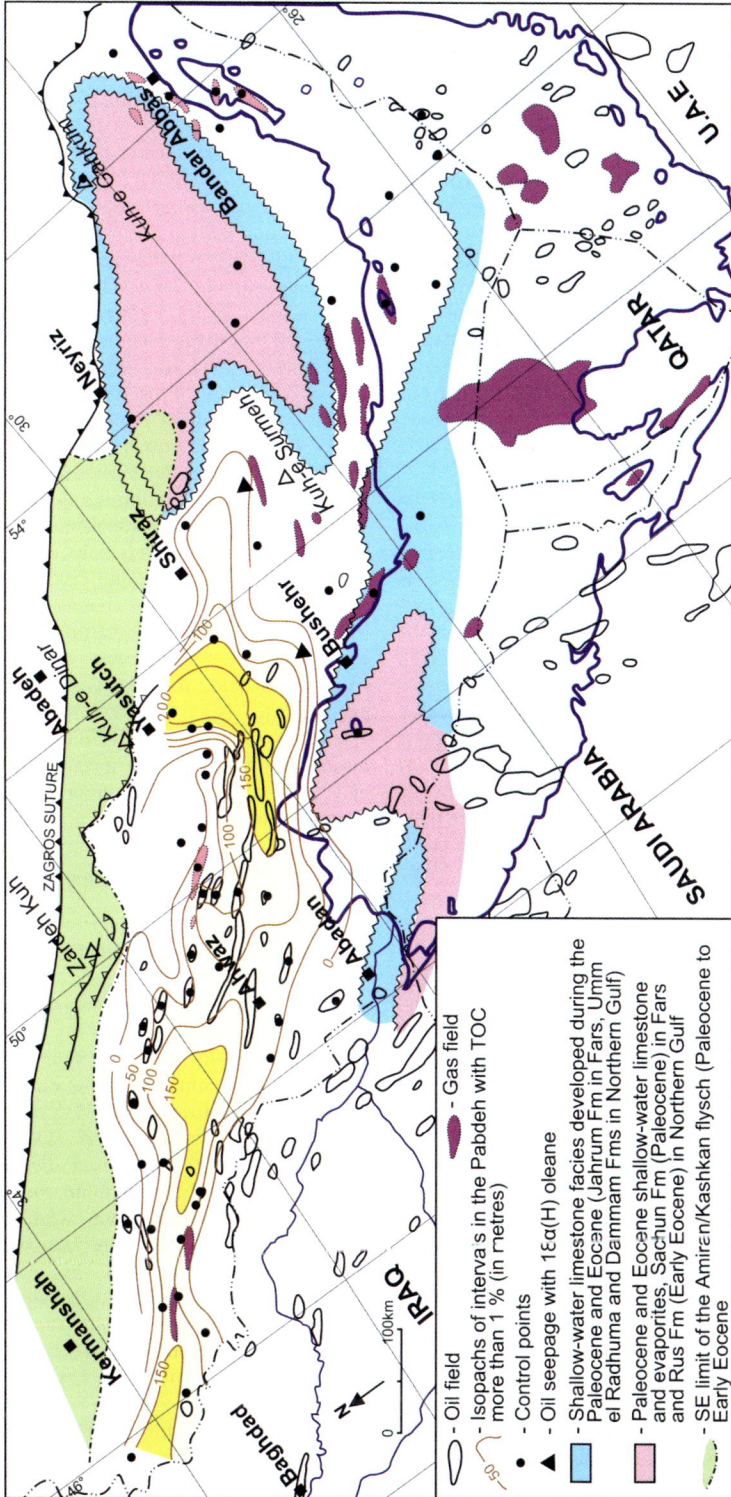

Fig. 26. Isopach map of organic-rich layers (TOC higher than 1%) in the Middle Eocene to Oligocene Pabdeh Formation (after Bordenave & Hegre 2005).

for immature or marginally mature Pabdeh found that average TOC values vary from 2 to 3.1% (maximum 6.2%), the average S_2 is between 5.2 and 15.9 HC kg^{-1} rock (maximum 36.2 HC kg^{-1} rock), and the average HI is 355 and 570 HC kg^{-1} C (maximum 646 HC kg^{-1} C). As indicated in Figure 26, the thickness of the Pabdeh source rocks (TOC $\geq$ 1%) reached a maximum of 230 m at Kuh-e Zard, 195 m at Bibi Hakimeh-1 and 100–150 m in Lurestan.

Excellent Pabdeh mature source rocks exist throughout Lurestan, but most of the Lurestan anticlines are breached to the Asmari Formation or deeper, and are not prospective for oil originating from the Pabdeh source rocks. The only known exception is the small Naft-e Shah–Naft Khaneh oil-bearing anticline, located at the Iran–Iraq border, where the Asmari Formation is well capped by a thick evaporites of the Gachsaran Formation.

In contrast, the Pabdeh marls are organic-lean (TOC < 0.8%) in the surface sections of coastal and SE Fars, from Kuh-e Khormuj to Kuh-e Gahkum. North of Bandar Abbas, the Palaeocene and Eocene interval is under shallow-water carbonates of the Jahrum facies. The transition from the all-limestone facies at Kuh-e Genow to a thick accumulation of Pabdeh marls in the Suru-Qeshm area is abrupt (in less than 20 km). At Kuh-e Khamir, WSW of Bandar Abbas, 80 m of marls contain up to 1% TOC (Bordenave *et al.* 1971). At Suru-1 and Ghevarzin-1, thin bituminous layers having a cumulative thickness of 15–30 m were observed from thin sections. The Pabdeh Formation thickens south of Qeshm, to 1800 m at HA-1 and HE-1. This increase in subsidence may correspond to stronger euxinic conditions in the depocentre and to the deposition of source rocks. At the Suru and Gevarzin wells, the base of the Padbeh Formation was buried to only 2000–2700 m. However, in the nearby synclines and south of the Qeshm Island, it was buried to at least 3500–4000 m.

Modelling and time of onset of oil expulsion

Methods and modelling parameters

The thermal evolution of the Kazhdumi and the Pabdeh source rocks, and the time of the onset of their oil expulsion were calculated with the 36 m W m^{-2} constant mantle heat flow that corresponds to the best fit with the current thermal markers. Kinetic parameters (i.e. the activation energy and the Arrhenius constant) were measured on rock pyrolysates and on kerogens isolated from the rock matrix by using the IFP OPTKIN program of pyrolysis, with a range of heating time varying from 10 to 25 °C min^{-1} and temperatures from 300 to 650 °C. Histograms of the Kazhdumi activation energy are similar, for both pyrolysates and kerogens. They are typical of algal type-2 kerogen, with a narrow spectrum (50–56 kcal mol^{-1}) and a pronounced 52 kcal mol^{-1} peak (Fig. 27). For the Pabdeh, the spread of the spectrum is wider (44–70 kcal mol^{-1}). The maximum at 52 kcal mol^{-1} is the same, although not as pronounced (Fig. 28). This configuration indicates type-2 kerogen mixed with terrestrial organic matter, which is in agreement with palynological observations showing the presence of plant debris, *Tasmanacea* algae and angiosperm pollen (A. Combaz, unpubl. data).

The estimate of the source rock burial before the end of deposition of the Asmari Formation is relatively easy, as many control points either from wells or from outcrops are available. However, it is difficult to estimate the burial at the end of the Mishan Formation, as the Gachsaran and its overlying beds are unconformably folded and thrust over the large Asmari anticlines, and the evaluation of the true thickness of the Gachsaran Formation before folding is difficult. For modelling purposes, we have used an average value at the wells, in the absence of a more precise evaluation. (At the Gachsaran field, the Gachsaran Formation was reconstructed as having a total thickness of 2000 m. (James & Wynd 1965). However, the vitrinite reflectance profile does not confirm this thickness, but rather a thickness of 1000 m, with erosion of some 200 m at most.)

Timing of the onset of oil expulsion

As explained above, the onset of oil expulsion was assumed in the model to occur for a 30% oil saturation of the source rock pore space. This means that the onset of expulsion was reached earlier for rich source rocks. The transformation ratio (TR) of the organic matter necessary to reach the onset of expulsion was estimated by the model to be in the 16–20% range for a rich 6% TOC source rock, 30–35% for an average 3–4% TOC, and 45–55% for a lean 2% TOC source rock. It also means that the burial necessary to reach the onset of oil expulsion is deeper for lean source rocks.

The results of modelling may be summarized as follows.

(1) On the external trend, at Ab-e Teymur and Mansuri, the onset of oil expulsion for Kazhdumi source rocks was reached between 1 Ma and 6.5 Ma when the base of the source rocks was buried between 4100 and 4150 m at crestal wells, and between 4500 and 4600 m in synclines. At Susangerd, where Kazhdumi source rocks are assumed to be lean, the expulsion was attained only in the deeper part of the syncline at 4 Ma, for a burial

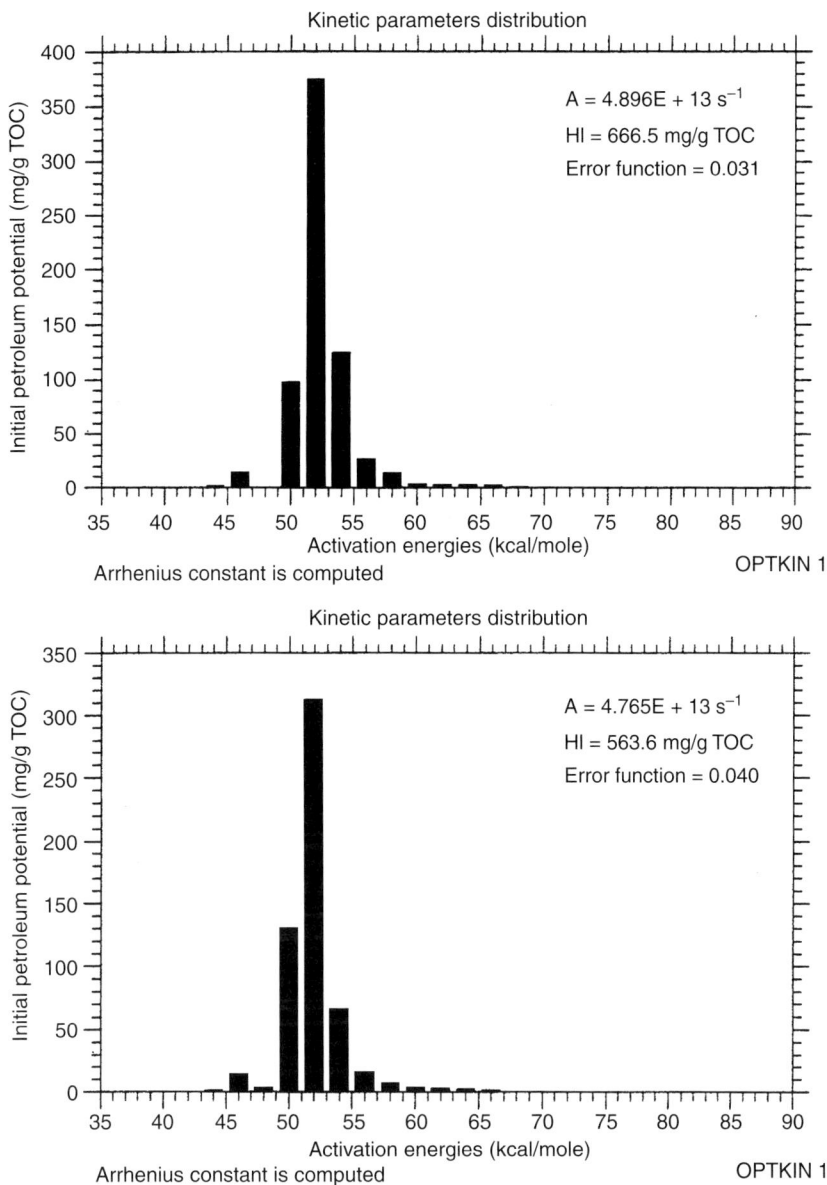

Fig. 27. Activation energy histograms for the Kazhdumi source rocks (top, rock sample; bottom, total organic matter).

depth of 5300 m. No oil was generated from Pabdeh source rocks on the external trend.

(2) On the Ahwaz trend, which includes Mushtaq, Shadegan, and Ramshir, the expulsion window was reached for the Kazhdumi between 3 and 7 Ma ago (4250–4500 m burial) at crestal wells, and between 5 and 8 Ma ago (4350–4800 m burial) in the deeper part of the synclines. At Mushtaq, poor-quality source rocks required deeper burial (5050 and 5500 m). Pabdeh source rocks expelled oil only in the deeper part of some of the synclines. Pabdeh oil represented less than 1% of the total oil expelled.

(3) At the Rag-e Sefid and Bibi Hakimeh fields, the Kazhdumi has not yet reached the onset of expulsion at crestal wells. In synclines expulsion commenced 6 Ma ago, at a depth varying from 4000 to 5000 m. The relative contribution of the Pabdeh was 6.5% for the Bibi Hakimeh field and 10% for the Rag-e Sefid field.

Kinetic parameters distribution

A = 9.483E + 13 s^{-1}

HI = 472.0 mg/g TOC

Error function = 0.063

Activation energies (kcal/mole)

Arrhenius constant is computed   OPTKIN 1

Kinetic parameters distribution

A = 5.072E + 13 s^{-1}

HI = 483.0 mg/g TOC

Error function = 0.117

Activation energies (kcal/mole)

Arrhenius constant is computed   OPTKIN 1

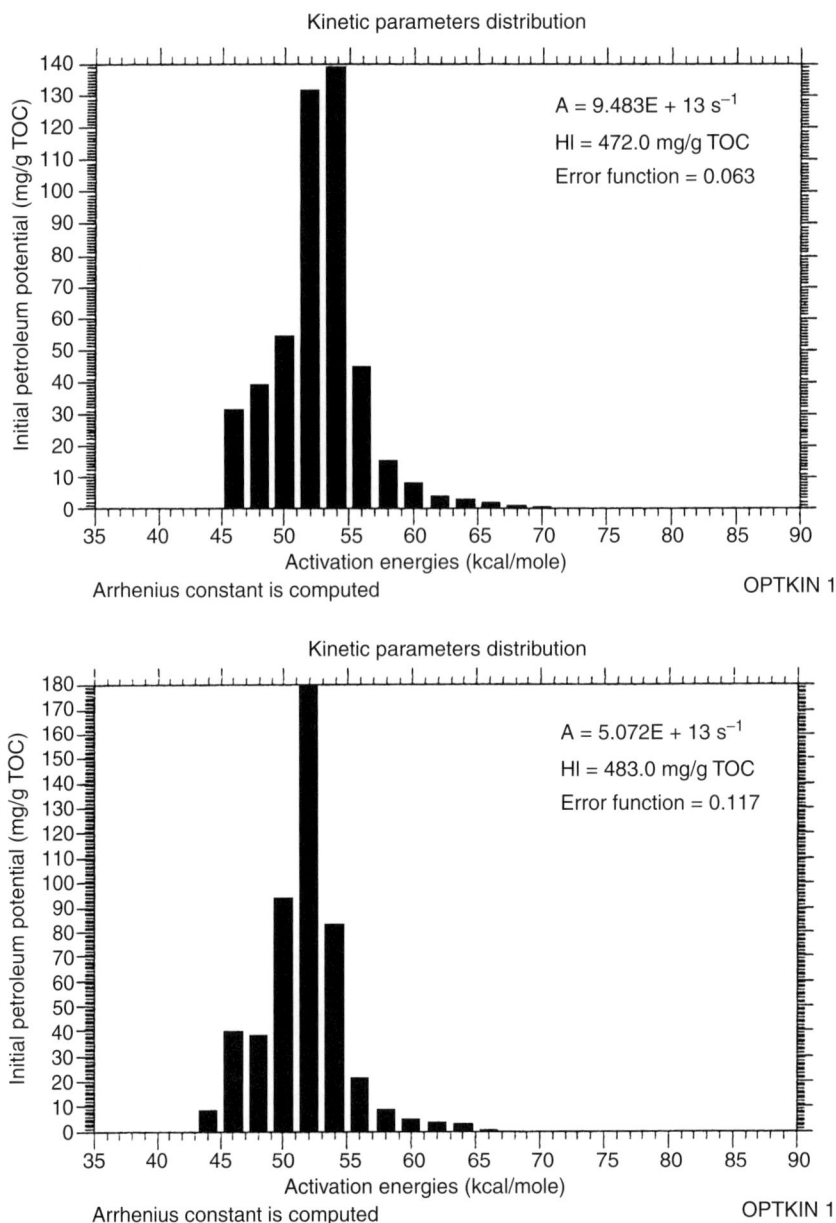

Fig. 28. Activation energy histograms for the Pabdeh source rocks (top, rock sample; bottom, total organic matter).

(4) At the Marun, Agha Jari, Pazanan and Kupal fields, the onset of the Kazhdumi oil expulsion began 5–7 Ma ago on crestal wells (3900–4250 m burial) and 6–9 Ma ago in synclines (4200–5100 m). At Ramin, where the Kazhdumi is assumed to be lean, the burial necessary to reach the oil expulsion was of the order of 5000 m. The contribution of Pabdeh source rocks was noticeable, from 2.6% at Agha Jari to 14% at Ramin and Pazanan.

(5) At Kilur Karim and Binak, the Asmari and Sarvak Formations form separate reservoirs. Oil expulsion from the Kazhdumi began 2 Ma ago at Binak wells, whereas the Kazhdumi is still immature at Kilur Karim wells. In the syncline, 0.7 expulsion commenced between 6.5 and 8 Ma ago at a depth of 5000 ± 200 m.

(6) At Karanj-12 and Parsi AF-1, expulsion commenced between 14 and 16 Ma ago (4000 ± 100 m

burial), before the Zagros folding. According to the model, about 30% of the oil was expelled before the folding (Fig. 29).

(7) In the NE Dezful area, the Lab-e Safid field was charged from Pabdeh source rocks only, as the Kazhdumi Formation time-equivalent is of a shallow-water limestone facies. At the LS-1 well, the Pabdeh is still immature, but in the deeper part of the syncline, expulsion began 7 Ma ago.

The Zagros folding commenced near the Mountain Front, which limits the Dezful Embayment to the NE, immediately after the end of deposition of the Mishan Formation 12.5 Ma ago. SW of Gachsaran, it commenced before the deposition of the Labhari Member of the Agha Jari Formation, perhaps 7 Ma ago, whereas in SW Lurestan it is dated between 8.1 and 7.2 Ma (Hessami et al.

2001; Homke et al. 2004; Bordenave & Hegre 2005).

For the Dezful Embayment fields (Parsi and the Karanj fields excepted), oil expulsion, either at the wells or in deeper part of the synclines, commenced later than 8 Ma ago, after the beginning of the Zagros folding. Fracturing of the Sarvak to Asmari interval observed in the crestal part of high-relief structures (McQuillan 1973, 1974) facilitated the expulsion of oil and its upward migration until the first caprock (Gachsaran Formation) was encountered. Source rocks were more easily drained in the crestal part of the area of drainage than in its synclinal area, which remained in a strongly compressive environment (Colman-Sadd 1978). In low-relief anticlines, such as Ab-e Teymur, Mansuri, Binak, Kilur Karim, and probably

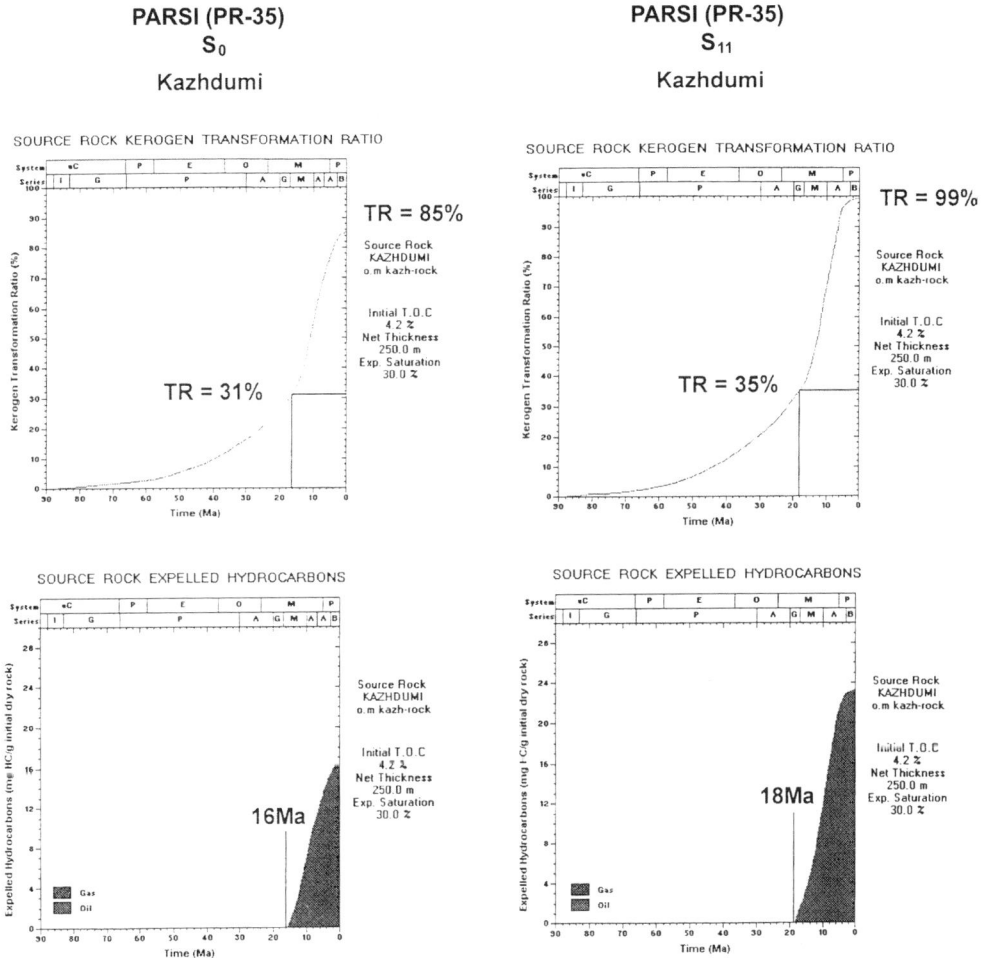

Fig. 29. Evolution through time of the transformation ratio (TR) and of the oil expulsion from the Kazhdumi source rocks, modelled at Parsi PR-35 (S0 corresponds to a well located at the crest of the anticline, and S11 corresponds to the deeper part of the syncline).

Ramin, fracturing is probably limited, and Asmari and Bangestan reservoirs are independent.

In the central part of the Dezful Embayment, the onset of oil expulsion was attained for a burial of 4200 ± 300 m at crestal wells, and 4500 ± 500 m in the deeper synclinal areas. In contrast, for areas located at the edge of the Dezful Albian Depression, where the Kazhdumi Formation is relatively lean (TOC averages of the order of 2–3.5%), the onset of oil expulsion was attained for a burial of 5000 ± 100 m at the wells and 5300 ± 300 m in the deeper part of the synclines. When a correction is applied to account for the compaction, areas where oil was expelled prior to the formation of the folds corresponded to areas where the current thickness of the interval base Kazhdumi–top Mishan is thicker than roughly 3900 m. The isopachs of this interval show that no oil was expelled in the Binak Trough before folding occurred (Fig. 17). In contrast, some oil could have been expelled in the Shiraz–Kazerun area before the Zagros folding, provided the Kazhdumi Formation includes layers of organic-rich source rocks.

Estimate of the amount of oil generated in the drainage area of each field

In the context of vertical migration, enhanced by fracturing, the oil that accumulated in an anticline originated from the source rocks located in its 'drainage area'. The drainage area of each anticline was limited by the axes of the neighbouring synclines, and by a line perpendicular to the anticlinal axis passing through the saddles limiting the anticline (Fig. 30). The concept of 'drainage area' makes possible the calculation of the estimated amount of oil expelled from source rocks existing in each anticline, based upon their maturity and their transformation ratio (Bordenave & Hegre 2005). This

The calculated amount of oil and gas generated by a source rock for the whole area of drainage of a field, Qc would be:

$Q_c = h \ (TOC \times HI) \ \rho_R \ \Sigma_i S_i \ TR_i$

- where h is the source rock thickness in the drainage area.
- TOC, the initial Total Organic Carbon content of the source rock per weight of rock (an average value was used for the whole drainage area)
- ρ_R the rock specific gravity
- S_i the surface of the slice i
- TR_i the transformation ratio calculated by the model for the slice i

If Q_c is expressed in barrels, h in metres, TOC and TR as proportions, HI in kg HC/ton C, ρ_R = 2.5 t/m^3 and S_i in sq km, Q_c would be:

$Q_c = 19.65 . 10^3 \ h \ (TOC \times HI) \ \Sigma_i S_i \ TR_i$

The amount expelled Q_e expressed in barrels would be:

$Q_e = 19.65 . 10^3 \ h \ (TOC \times HI) \ \Sigma_i S_i \ Exp_i$

- $Q_e \ Exp_i$ is the coefficient of expulsion for slice i

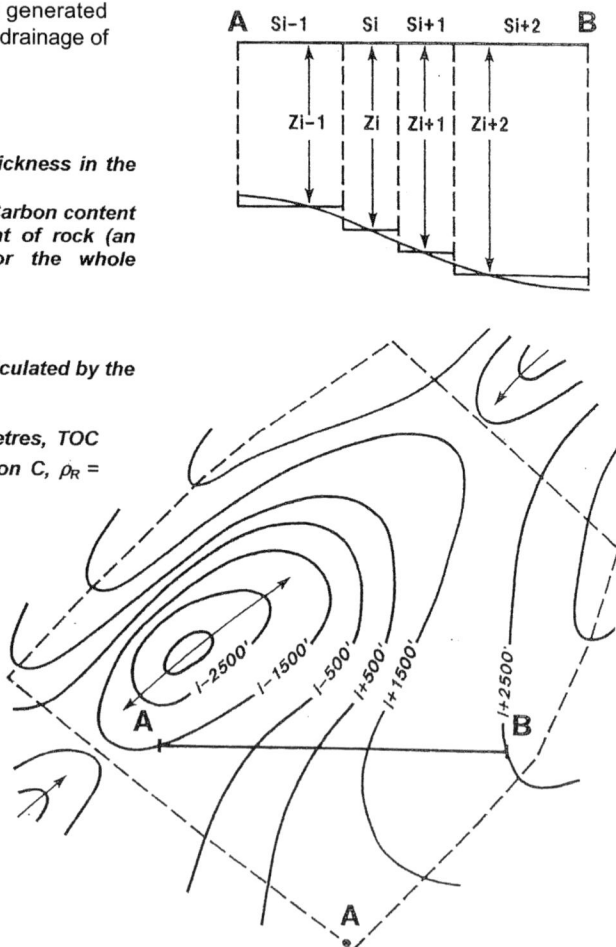

Fig. 30. Theoretical example of drainage area and calculation of the oil expelled from source rocks.

calculated amount of oil expelled can be compared with the IOIP of the corresponding field.

Assuming that oil expulsion occurred after the onset of the Zagros folding, the present-day geometry of the Kazhdumi and Pabdeh source rocks was used to model their TR as a function of their burial. The erosion for each anticline was estimated, keeping in mind that the initial thickness of the Agha Jari Formation was much thinner in crestal areas than in synclines. The geometry of the area of drainage was deduced from the top of Asmari seismic marker, which was well defined on pre-1973 seismic maps, in the absence of better data for the deeper layers. Each area of drainage was cut into 1000 foot (304.8 m) thick slices. The surface of each slice was measured. A burial profile was set to calculate the evolution of the TR through time for both the Kazhdumi and the Pabdeh source rocks, and the amount of oil generated and expelled (Fig. 30). The number of slices varied from three to five for the low-relief anticlines of the Mansuri–Ab-e Teymur trend, to 19 in high-relief anticlines, such as Naft-e Safid. Modelling was applied to 21 fields that contain 86.5% of the currently discovered IOIP for the Zagros Fold Belt. Altogether, 195 burial profiles were used. An example of the calculation for one of the major fields, Marun, was provided by Bordenave & Hegre (2005).

The total IOIP of the currently discovered fields is only 17.3% of the calculated amount of oil assumed to have been expelled from the Kazhdumi and Pabdeh source rocks in the drainage areas of the 21 fields that were modelled. At first glance, the percentage between the oil-in-place actually accumulated in single fields (Q_R) and the calculated value of the expelled oil in their area of drainage (Q_{EC}) varies widely from one field to the other. However, before drawing any conclusion, fields should be examined case by case. Their relationship to neighbours and the possibility of secondary migration from one field to another are discussed below.

Ab-e Teymur and Mansuri. The Q_R/Q_{EC} ratio of Ab-e Teymur is one of the highest at 48.5%. However, the saddle between Ab-e Teymur and Mansuri is not pronounced. The OWL occurring in the Ilam reservoir is identical for these two fields. The same observation is valid for the two separate Sarvak reservoirs. When considered as a single accumulation, the Q_R/Q_{EC} of Mansuri–Ab-e Teymur would be 11.8%, still slightly lower than the average.

Ahwaz. The Q_R/Q_{EC} ratio for this field is also one of the highest, at 38%. However, as indicated in Figure 31, the axis along the Ahwaz to Ramshir trend shows that the Asmari reservoir at Shadegan is full to the spill point and that oil moved from the Shadegan area of drainage towards Ahwaz. It

is also likely that the same OWL as in the Sarvak reservoir exists between Ramshir and Ahwaz and that the oil accumulation is continuous between the three fields, although the reservoir characteristics at Shadegan and Ramshir are mediocre. When the Q_R/Q_{EC} ratio is calculated for Ahwaz, Shadegan and Ramshir together, a value of 17.6% is obtained, which is close to the above-mentioned average.

Bibi Hakimeh and Rag-e Sefid. These fields have Q_R/Q_{EC} ratio of 18.2% and 20.9%, respectively.

Karanj. The Q_R/Q_{EC} ratios of Karanj (4.1%) and Parsi (21.2%) are very different. First, the Bangestan Formation is oil-bearing in these two fields, but as the reservoir characteristics are poor from a production standpoint (*Oligostegina* facies), no IOIP was accounted for this reservoir. Second, the Parsi anticline is thrust over Karanj, and some of the oil probably moved from the Karanj area of drainage to Parsi. As a consequence, the two fields should be considered jointly. Thus, the Q_R/Q_{EC} ratio is 7.96%, well below the average. Third, some oil was lost laterally prior to the Zagros folding. As mentioned above, perhaps 30% of the oil generated in the Karanj and Parsi drainage area has been lost laterally prior to the onset of folding. The Q_R/Q_{EC} ratio would become 11.4% using the assumption of folding at 12.5 Ma.

Kupal–Agha Jari–Pazanan trend. In the Kupal area of drainage, some of the oil was expelled from Pabdeh source rocks (13%), and the rest from the Kazhdumi. However, both the Asmari and the Bangestan oils are identical from an isotopic point of view ($\delta^{13}C$ and $\delta^{34}S$), although they form two separate accumulations (the Asmari oil is undersaturated, whereas a small gas cap exists in the Bangestan). When calculated together, the Asmari and Bangestan accumulations of Kupal corresponded to a Q_R/Q_{EC} ratio of 12.07%. As the Kupal anticline is full to the spill point, some oil may have moved toward Agha Jari, which is 2000 m higher. However, the Q_R/Q_{EC} of Agha Jari is only 11.5%. Moreover, the Pazanan field does not fit with the assumption of a dual charge from the Kazhdumi and the Pabdeh. The huge gas cap (44 Tcf) cannot be explained by the oil-prone Kazhdumi source rocks, even if the TR of the Kazhdumi may have reached 99% in the deeper part of the Pazanan area of drainage. The Pazanan gas probably has a deeper origin, either from the Garau, the Sargelu, or even from the Silurian source rocks. The crest of Pazanan is lower than that of Agha Jari. The two fields have a similar OWL. The Q_R/Q_{EC} ratio of Pazanan is fairly low, only 2.8%. The gas with a deeper origin could have pushed down the oil accumulated in the Pazanan anticline until the spill point was reached, toward the NE,

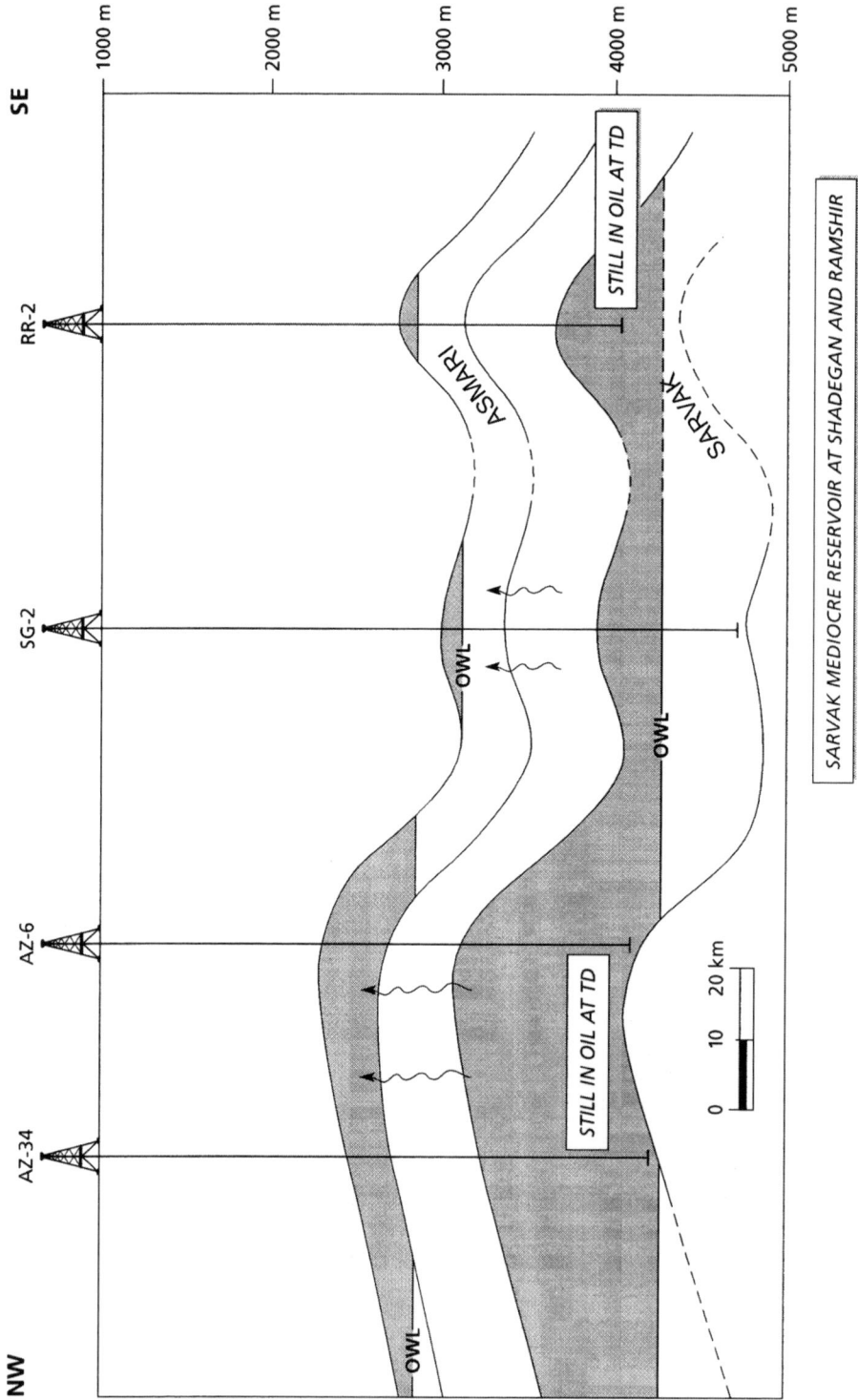

Fig. 31. Schematic axial section through the Ahwaz to Ramshir trend.

in the direction of the Garangan–Chillingar trend. Some oil may have been lost in that area, as the Asmari and Bangestan are heavily fractured, water-bearing, and covered by a relatively thin Gachsaran Formation. Oil may also have moved towards the Gachsaran anticline where the crestal part is more than 1200 m higher than at the Pazanan top. This assumption would explain the anomalous case of Gachsaran, where the Q_R/Q_{EC} ratio is greater than unity. However, the poor quality of the seismic data available north of Pazanan makes it difficult to confirm the validity of one or the other of these assumptions.

Marun. This is the second best field in the Dezful Embayment, and shows a very high Q_R/Q_{EC} ratio, about 50%. Some of the oil probably came from the Ramin area of drainage. Ramin is 1500 m lower than Marun. The calculation of Marun and Ramin together is not sufficient to lower the Q_R/Q_{EC} ratio, which would still be 46.5%. Another possibility would be to extend the area of drainage in Ramin to the NW or to extend the Marun area of drainage underneath the Agha Jari field, where it is thrust over Marun, as observed on the SE flank of Marun.

Lab-e Safid. Of the modelled fields, only Lab-e Safid was charged only by Pabdeh source rocks. The Q_R/Q_{EC} ratio is more than 50%, but the seismic quality in this part of the Dezful Embayment is extremely poor, and the syncline may be deeper.

Binak and Kilur Karim. These fields have separate Asmari and Bangestan reservoirs. The Asmari oil assumed to come from the Pabdeh showed a Q_R/Q_{EC} ratio of 28.3% and 12% respectively, whereas the Bangestan oil assumed to come from Kazhdumi and Ahmadi source rocks has a surprisingly low Q_R/Q_{EC} ratio (4.3% and 2.7%). At Binak, the Sarvak oil column is at least 750 m thick. However, it may be much thicker, up to an additional 450 m (pressure extrapolation), as the OWL was not established. No OWL was found in Kilur Karim either, where the Bangestan seems to have poor reservoir characteristics.

A review of the complex relationship between single fields shows that between 11.5% and 21% of the oil calculated to have been expelled from the Kazhdumi and the Pabdeh source rocks in the drainage area of a field (or group of fields) is currently seen as IOIP in the corresponding field or group of fields.

The apparent deficit of IOIP as compared with the calculated amount of oil expelled from the source rocks could be due to many reasons, such as the loss of oil during primary and secondary migrations, and leakage through cap rocks. More

important, the absence of fracturing in synclines, which remained under compressional stress from the beginning of the folding, made the expulsion from thick and homogeneous source rocks more difficult (Durmish'yan 1973). Once expelled, oil has to find its way through the Sarvak Formation, which is often argillaceous and tight. As a consequence, some oil remained trapped in the Sarvak Formation that was not accounted in reserve estimates. Even near the crest of the anticlines, calcite-coated fractures contain some bitumen, as observed by one of the authors at the Kuh-e Khaviz outcrop. This means that even in fractured areas, some oil never reached the reservoirs.

Other reasons could be inaccuracies related to both the model and data entered in the model. Among possible inaccuracy we can cite the validity of the algorithm on which the model is based. This algorithm requires additional confirmation, especially for the conditions that determine the onset of oil expulsion. There are not yet enough laboratory experiments available in this field of research to justify the 30% saturation of the source rock pore space expulsion threshold.

However, in spite of these uncertainties, the model permits an evaluation of a rough order of magnitude of the oil-in-place expected in an anticline, using the characteristics and the burial profile of the source rock as well as the geometry of its area of drainage, provided that the trap is large enough (geometry and reservoir characteristics) to contain it. More accuracy would require a systematic sampling and analyses of the source rocks in a few critical wells drilled in the Dezful Embayment as well as in the Shiraz–Kuh-e Rig area. In addition, a review of the E-logs, gamma-ray logs and an examination of thin sections on at least one deep well for each field is required. Better resolution seismic surveys and well-established velocity surveys would be essential to draw balanced sections, to better understand the geometry of the drainage areas and their extension below the thrust units, and to decipher the disharmonic folding of the Gachsaran Formation and of its Mishan–Agha Jari cover.

Characteristics of the iranian oils and petroleum systems

According to modelling carried out for the Middle Cretaceous to Early Miocene Petroleum System in the Dezful Embayment, oil expulsion from the Kazhdumi and the Pabdeh source rocks occurred after the onset of the Zagros folding (except in the Karanj–Parsi area). Oil expelled from the Kazhdumi source rocks, and locally from the Pabdeh, migrated almost vertically through a network of fractures to the Asmari and Bangestan reservoirs. It was assumed that the Middle Cretaceous to Early Miocene Petroleum System is independent from previous systems, and no deeper source rocks, either Garau or Sargelu, have charged the Asmari and Bangestan reservoirs in the Dezful Embayment. The Sargelu Formation was overlain in Lurestan and in the Dezful Embayment by the thick Gotnia evaporites, which isolated it from younger reservoirs. Similarly, in the Dezful Embayment, the pressure barrier formed by maturing Kazhdumi source rocks did not permit the oil expelled from Garau source rocks to reach the Sarvak and Asmari reservoirs. If this scenario is confirmed, the oils discovered in the Dezful Embayment should have characteristics fairly close to those of the pyrolysates of the source rocks from which they were assumed to originate. In contrast, older sources expelled their oil before the folding and migration took place over long distances on low-angle ramps towards potential traps, either salt-related early structures or towards large-scale regional highs, and these oils have to be compared with distant source rocks.

Iranian oils accumulated in reservoirs ranging from 500 to 4100 m depth, where temperatures vary from 40 to 120 °C. No significant biodegradation was observed in the shallower fields, even at MIS, Haft Kehl and Maleh Kuh (the heavy oil of Mand was not studied in this paper). In the deepest fields, the Asmari reservoir, which is in contact with the Gachsaran Anhydrite, is not buried deeply enough for thermal sulphate reduction (TSR) to commence. TSR requires a temperature of the order of 140 °C (Worden et al. 1995; Worden & Smalley 1996). Consequently, the isotopic composition, $\delta^{13}C$ as well as $\delta^{34}S$, of an oil sample should be close to the isotopic composition of the source rock from which it is originated. No in situ thermal cracking of the oil was observed in the reservoirs. The shallower reservoirs contain lighter oils, and not the contrary, as would be the case for in situ thermal cracking (Fig. 32). The API gravity depends upon the geometry of the source rocks after the Zagros folding and especially on the depths of the synclines.

Oil characteristics (carbon and sulphur isotopic fingerprint, vanadium-to-nickel ratio and biomarker composition) permit the grouping of the oils into families, and their relationship with specific source rocks.

Isotopic composition

The carbon isotopic composition ($\delta^{13}C$, PDB) was measured on both kerogen and kerogen pyrolysates in organic-rich Sargelu, Kazhdumi and Pabdeh source rocks. It varied from −29.2 to −28.2‰ for Sargelu pyrolysates, and kerogens gave similar results (−29.2 to −28.0‰). Kazhdumi pyrolysates varied from −27.9 to −26.5‰, and kerogen were slightly lighter (−27.4 to −25.6‰). Pabdeh pyrolysates varied from −25.9 to −24.6‰, and kerogens from −25.3 to −23.9‰. Kerogens of the Garau and the Gadvan Formations were also analysed, but a large range of variation was found, from −29.2 to −25.6‰ for the Garau, and from −26.5 to −23.7‰ for the Gadvan, which is not considered as representative (Fig. 33). The isotopic composition of the Hanifa–Tuwaiq Mountain source rocks (−27.5 to −26‰ for kerogen and −28 to −26‰ for extracted bitumen) is close to that of the Kazhdumi (Carrigan et al. 1995).

The carbon isotopic composition of the oils originating from the oil fields as well as from surface oil seeps, sampled in Lurestan, in the Dezful Embayment and in Iraq close to the Iran border, varied widely from −28.5 to −23.6‰ (Bordenave & Burwood 1990; Connan & Deschesne 1996). The histogram of the carbon isotopic values shows that most of the values are situated within a narrow interval (i.e. −27.4 to −26.2‰) and two poorly marked peaks are observed, one around −28.0‰ and the other between −26.2 and −23.6‰. Oil samples from oil fields located SW of the city of Dezful have isotopic values corresponding to the histogram's main peak (−27.4 to −26.2‰). These values are compatible with the isotopic composition of Kazhdumi pyrolysates. In the same way, the $\delta^{13}C$ of samples from the NE part of the Dezful Embayment (Lab-e Safid, Par-e Siah, Karun, Lali, and Qaleh Nar) corresponds to the isotopic composition of Pabdeh pyrolysates (−26.2 to −23.6‰). Values measured at Buzurgan and Jebel Fauqi, close to the Iran–Iraq border (−28.2 to −27.7‰) may correspond to Sargelu pyrolysates, although being slightly lighter (Connan & Deschesne 1996).

Sulphur isotopic composition ($\delta^{34}S$ CDT) was measured for oil samples only (W. L. Orr, unpubl. data). It varies from −10 to +12‰. Most of the samples have values between −4 and +2‰. Oil samples that have $\delta^{13}C$ between −27.4 and −26.2‰ (i.e. assumed to be related to the Kazhdumi source rocks) have $\delta^{34}S$ varying between −5 and +3‰. In contrast, samples assumed to have been generated from the Pabdeh Formation (−26.2 to −23.6‰) are isotopically heavier as far as sulphur is concerned (+1 to 13‰). No sulphur isotopic measurement was available for the Buzurgan and Jebel Fauqi fields.

Isotopic composition does not permit the differentiation of the oils originating from Garau source rocks from those originating from the Kazhdumi or from the Hanifa–Tuwaiq Mountain source rocks. Oil sampled at the Sarkan and Maleh Kuh fields, in SE Lurestan, assumed for geological

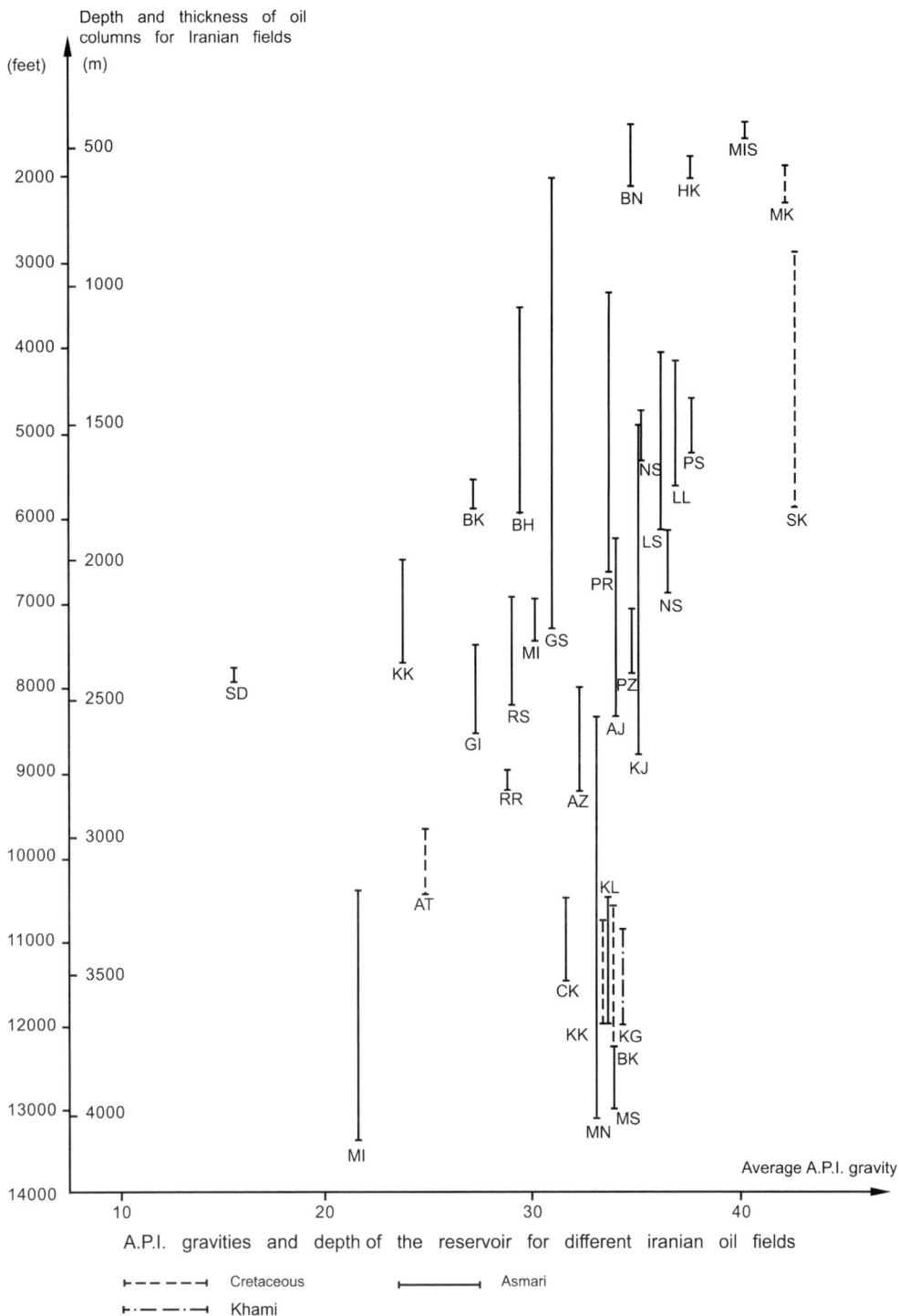

Fig. 32. Depth of the reservoirs of the Iranian onshore fields, as compared with the API gravity of the oil they contain (field symbols in Figs 2 and 34).

AGE	FORMATION	LITHOLOGY	MAXIMUM EFFECTIVE SOURCES (m)	SOURCE ROCK CHARACTERISTICS			ISOTOPIC SIGNATURE KEROGEN ◇ / KEROGEN PYROLYSATE ◆ δ C‰ (PDB)
				KEROGEN TYPE	TOC (%WT)	S2 (g/kg)	
EOC/OLIGO	PABDEH [3]	MARL/ARG.LST	230+	IIs	1 - 12	7 - 40	
LATE CRET.	GURPI [3]	MARL/ARG.LST	150 (N. FARS)	II/III	0,5 - 2	2 - 10	
	SURGAH/ [3] LAFFAN	MARL	150 (S. LURESTAN)	III	1,5 - 3	3 - 8	
MID CRET.	SHILAIF [2]	MARL/ARG.LST	60 (ABU DHABI)	IIs	1 - 6	3 - 47	
	KAZHDUMI [3]	MARL/ARG.LST	300+ (DEZFUL EMB.)	IIs	3 - 12	17 - 40	
EARLY MED.	GADVAN [3]	MARL/ARG.LST	100 (NW FARS)	IIs	1	5	
	GARAU [3]	MARL/ARG.LST	300+ (LURESTAN.)	IIs	1,5 - 10	ND	
LATE JUR.	HANIFA/ [1] TUWAIQ MTS	MARL/ARG.LST	150 (SAUDI ARABIA)	IIs	2 - 7,5	5 - 40	
	DIYAB [2]	MARL/ARG.LST	85 (ABU DHABI)	IIs	1 - 5,5	3,7 - 16,4	
MID JUR.	SARGELU [3]	MARL	150+ (LURESTAN)	IIs	3 - 8	18	
SIL.	SARCHAHAN [3]	SHALES	70+ (SE FARS)	II	2,5 - 4,3	ND	

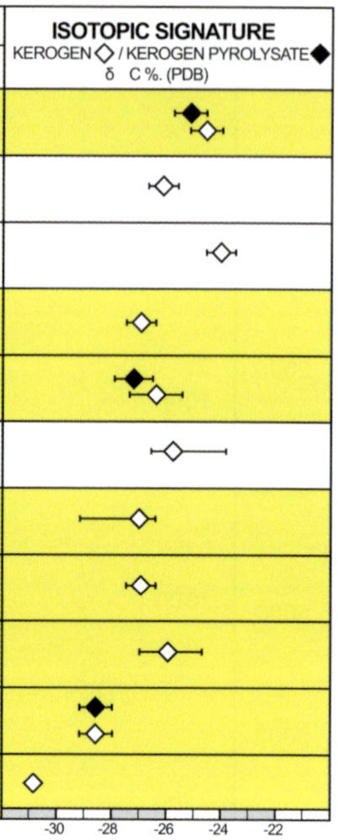

-30　-28　-26　-24　-22

(1) Data after Carrigan et al (1998)
(2) Data after Hassan and Azer (1985)
(3) Data after Bordenave and Burwood (1990)

Fig. 33. Characteristics and isotopic signatures of the source rocks that occurred in the Zagros Fold Belt.

reasons to come from the Garau source rocks (the Kazhdumi is replaced by oxic shallow limestone of the Bala Rud shoal), has an isotopic composition similar to those of the Kazhdumi ($\delta^{13}C$ −26.7 ±0.2‰, $\delta^{34}S$ + 0.2 ± 0.05‰). In the same way, the isotopic composition does not allow us to categorize the oils from the Kharg–Dorood and Bushehr fields. In these two fields, production comes from the Surmeh and Fahliyan reservoirs, and oil may derive from Sargelu, as well from Garau and/or from Hanifa–Tuwaiq Mountain source rocks. Similarly, we cannot categorize the oils found in the northern part of the Gulf. Based upon biomarkers, two groups of oils were found: (1) the Abouzar, Hendijan and Bahregansar oils produced from the Ghar–Asmari reservoirs, which are likely to come from the Kazhdumi, the Ahmadi, and perhaps the Garau source rocks; (2) Nowrouz and Foroozan, which produce from Burgan sands and Fahliyan limestone, originated from Sargelu, Hanifa–Tuwaiq Mountain and perhaps Garau source rocks (Rabbani & Kamali 2005).

Two fields are anomalous, Chesmeh Khush, which has an unusually heavy $\delta^{34}S$, −7.3‰, and Bushgan, which has a carbon composition close to that of the Pabdeh Formation (−24.0‰), and a sulphur isotopic composition that resembles that of the Kazhdumi Formation (+1.2‰). However, no Kazhdumi or Padbeh source rocks exist at Bushgan. The only rock that may have sourced this marginal field of Central Fars is the Gadvan Formation.

Oil families

According to their carbon and sulphur isotopic fingerprints (Figs 34 and 35), their chemical composition, and biomarkers they contain, Iranian onshore oils are grouped into seven families, as follows.

Family A. This corresponds to oils originated from the Asmari or Bangestan reservoirs in the NE Dezful Embayment fields (Lab-e Sefid, Karun, Par-e Siah, Lali and Qaleh Nahr). These oils have similar isotopic signature, both for carbon and sulphur: $\delta^{13}C = -24.8 \pm 0.4$‰ PDB; $\delta^{34}S = -7.6 \pm 4.4$‰ CDT. The $\delta^{13}C$ of both the oils of Family A and the Pabdeh pyrolysates are similar (Fig. 35). Moreover, the occurrence of 18α(H) oleanane in the m/z 191 terpane fragmentogram shows that the age of the progenitor source rocks is probably Cenozoic, when angiosperm plants became widespread (Ekweozor et al. 1979; Bordenave & Burwood, 1990). It also shows the relative vicinity of a coastline. Among other biomarkers, the abundance of diasteranes (C_{27} and C_{28} dia)

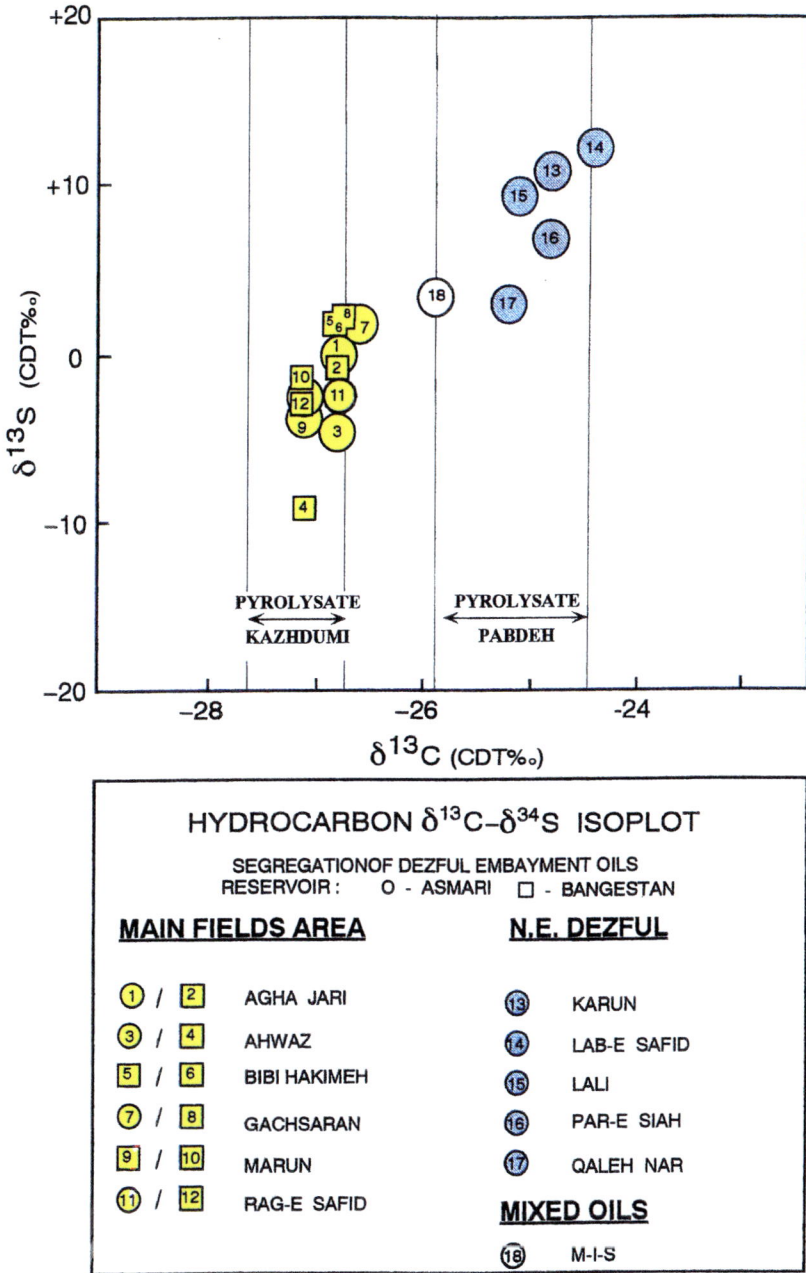

Fig. 34. Relationship between carbon and sulphur isotopic composition of the Dezful embayment oils (after Bordenave & Burwood 1990).

would indicate relatively argillaceous source rocks deposited in slightly anoxic conditions. Short chain steranes are also abundant. A preference of hopane over norhopane, and of C_{35} hopane over C_{34} hopane is also noted. The oil seepage

in Mamatain also belongs to this type, with $T_m/T_s < 2$ (Connan & Deschesne 1996). The vanadium/nickel ratio is low (less than 1.70).

In the NE Dezful Embayment, the Kazhdumi Formation is replaced by shoal limestone devoid

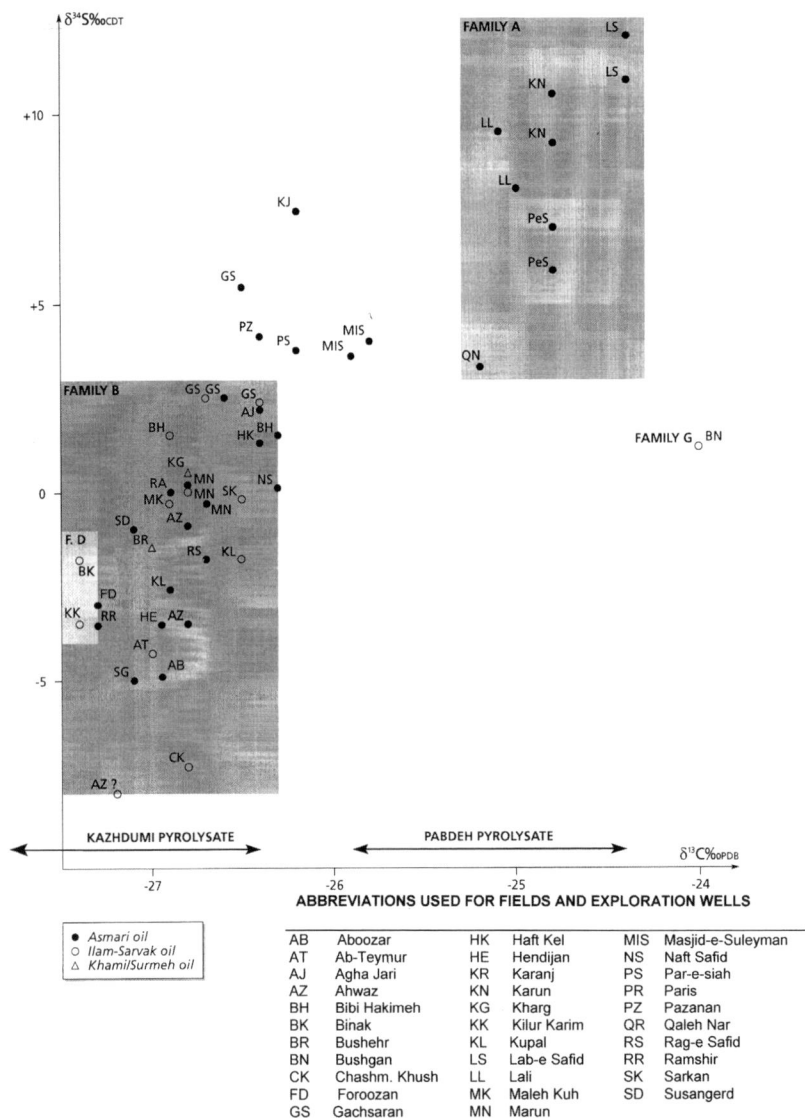

Fig. 35. Oil families as the result of the carbon and sulphur isotopic composition.

of organic matter, whereas the Pabdeh source rocks are excellent and mature. Deeper source rocks (Garau and Sargelu) did not participate in the charge of the Asmari and Bangestan reservoirs, because of thick evaporites of the Gotnia Formation and thick Garau marls.

Family B. This includes most of the fields located in the central and SW part of the Dezful Embayment, which produce from the Asmari and Bangestan reservoirs. The fields grouped into Family B contain nearly 90% of the oil discovered in onshore

Iran (excepting Azadegan and Yadavaran). The isotopic fingerprint of Family B is well marked: $\delta^{13}C = -26.8 \pm 0.5\%$ PDB; $\delta^{34}S = -2.5 \pm 5.5\%$ CDT. The $\delta^{13}C$ of Family B is compatible with that of the Kazhdumi pyrolysates (Figs 34 and 35).

No oleanane was found in oil samples from fields such as Ahwaz, Marun, Haft Kehl, Ramshir and Ab-e Teymur, whereas traces of oleanane were found at Bibi Hakimeh, Gachsaran and Agha Jari. The $C_{27}-C_{29}$ diasteranes are present. An enhancement of $35\alpha\beta$ hopane is observed, in contrast to

Family A. The T_m/T_s ratio is less than two. The vanadium/nickel ratio, 3.3 ± 1.0, is higher than for Family A.

Whereas the two main source rocks, Kazhdumi and Pabdeh, are well represented in the Family B domain, vitrinite reflectance and Rock-Eval T_{max} values show that the Kazhdumi is mature and the Pabdeh is immature at crestal wells. Modelling showed that in the deeper part of some synclines, the Pabdeh source rock has reached the oil expulsion window. Therefore, in some fields, the oil generated by the thick and mature Kazhdumi Formation was complemented by minor amounts of Pabdeh oil. The geological setting (i.e. the geometry of the two source rocks) explains the homogeneous isotopic fingerprint of Family B, and the biomarker study confirms the minor charge coming from the Pabdeh.

The isotopic characteristics of MIS are halfway between those for Family A and Family B, which makes sense because only the southeastern part of its area of drainage contains thin Kazhdumi source rocks. Parsi and Karanj, still very close to Family B, received a noticeable amount of oil from the Pabdeh.

Family C. This is exemplified in Iraq by the Buzurgan and Jebel Fauqi oils and in Iran by solid veins of bitumen from the Lalah Bisheh mine, 27 km from Pol-e Dokhtar in Lurestan. Family C oils have a $\delta^{13}C = 28.15 \pm 0.35‰$ compatible with $\delta^{13}C$ of the Sargelu source rock pyrolysates. It would also be compatible with the Garau source rocks, but the Garau facies did not extend into the Buzurgan–Djebel Fauqi area.

Biomarkers show the absence of oleanane. The existence of gammacerane suggests slightly more evaporitic depositional conditions. The molecular configuration of Family C is different: terpanes are led by $\alpha\beta$ hopanes, methyl $\alpha\beta$ hopanes and hexahydrobenzohopanes, absence of diasteranes (dominantly calcareous source rocks deposited in strict anoxic conditions), and a strong distribution of short-chain steranes (C_{21}–C_{22} St, C_{22}–C_{23} 4 Me St). The T_m/T_s ratio is very high (Connan & Deschesne 1996).

The biomarker of the Susangerd and Chesmeh Khush oils resembles those of the Pol-e Dokhtar sample and of the Buzurgan and Abu Ghirab oils. The anomalous biomarker composition of the Chesmeh Khush and Susangerd Sarvak oils may suggest at least a partial Sargelu origin, although the oils have an isotopic composition similar to that of Family B.

Family D. This is represented by only two samples; the Binak and Kilur Karim Bangestan oils. It is characterized by $\delta^{13}C = -27.4‰$ PDB and $\delta^{34}S = -2.65 \pm 0.85‰$ CDT. The vanadium/nickel ratio is also quite low at 1.1–1.8.

From a geological standpoint, the Family D oil may come from a mixture of the two source rocks existing in the Binak Trough, the Ahmadi Member of the Sarvak Formation and the Kazhdumi Formation. No sample from the Ahmadi source rocks was analysed and no oil-to-source rock correlation is available.

Family E. The isotopic composition of oil samples originated from the Khami–Surmeh reservoirs of Kharg and Bushehr is similar to that of the oils of Family B. However, the isotopic composition of the Hanifa–Tuwaiq Mountain source rocks (-27.5 to $-26‰$ for kerogen and -28 to $-26‰$ for extracted bitumen) and those of the Kazhdumi are similar, and oils that originated from these two source rocks cannot be separated by using only $\delta^{13}C$. A biomarker study of the Kharg oil (Rabbani 2008) shows that the diasteranes/steranes ratio is quite low, indicating carbonate source rocks deposited in strict anoxic conditions. For geological reasons, this oil is assumed to come from a mixture of Sargelu, Hanifa–Tuwaiq Mountain and Garau source rocks.

Family F. This includes oils from the Sarvak and Ilam reservoirs of the Sarkan and Maleh Kuh fields. Their isotopic composition, $\delta^{13}C = -26.7 \pm 0.2‰$ and $\delta^{34}S = -0.2‰$, is similar to that of Family B. In contrast, the sulphur content for a given API Gravity is higher in Family F than in Family B. The vanadium/nickel ratio is similar to that of Family B. From a geological standpoint, the oil is assumed to come from the Garau source rocks. The Kazhdumi marls of the Lurestan Depression are organic-lean and the deeper Sargelu Formation, separated from the Garau by thick Gotnia anhydrites, is not isotopically compatible with the Sarkan and Maleh Kuh oils.

Family G. This includes only the marginal Bushgan field. Its $\delta^{13}C$ value of $-24‰$ puts it close to Family A, whereas its $\delta^{34}S$ value of $-1.85‰$ would place it in Family B. The carbon isotopic composition would be compatible with the Pabdeh and the Gadvan marginal source rock. The Pabdeh is not mature at Bushgan-1 well and likely in the adjacent syncline. In consequence, this oil is assumed, with caution, to come from marginal Gadvan source rocks.

The Sarvestan and Saadat Abad oils, and the oil seeps in the Shiraz area were not sampled; nor were the Kuh-e Rig, Shurom and Dudrou oils in the High Zagros analysed.

The origin of the onshore Iranian oils is summarized in Figure 36. The isotopic and chemical

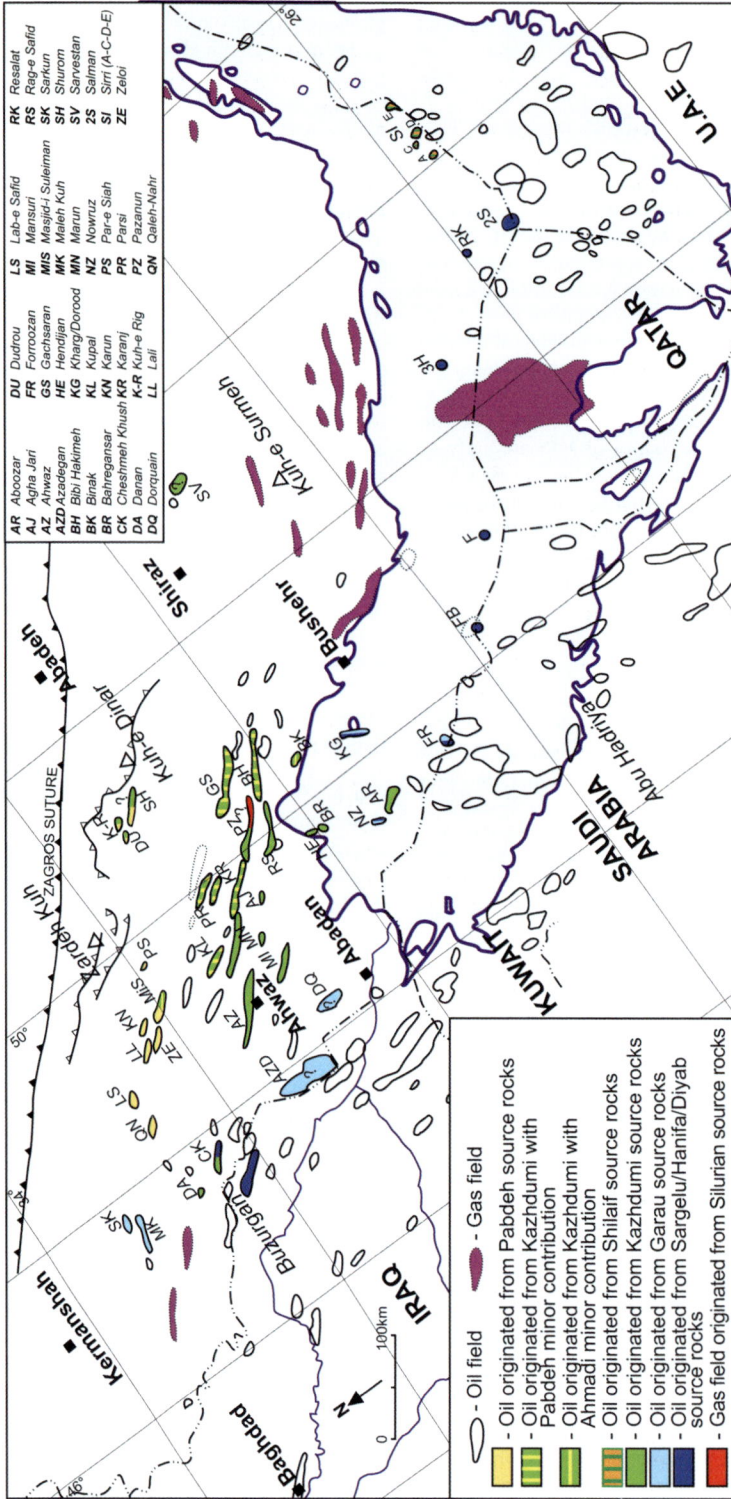

Fig. 36. Origin of the oils accumulated in the Zagros Fold Belt and in contiguous offshore areas.

composition of the oils and the oil to source rock correlations are in agreement with the geological and geochemical information and with the results of the modelling applied to source rocks. A few questions remain regarding the Pazanan gas, which may have a deep origin, the oil found in the High Zagros (Kuh-e Rig, Dudrou, and Shurom) and south of Shiraz (Sarvestan and Saadat Abbad) fields, which may have been, at least partly, charged from an eastern extension of the Kazhdumi anoxic facies.

Conclusions

Two types of petroleum systems charged the Iranian Zagros and contiguous offshore: (1) the systems using long-distance migrations along gently dipping ramps towards regional highs, and salt-related structures where oil and/or gas accumulated; in the Zagros Fold Belt, oil and/or gas was later re-accommodated in Zagros anticlines; (2) the systems using vertical migration to the large-scale Zagros anticlines according to their drainage. The four older petroleum systems (Palaeozoic, Middle Jurassic, Late Jurassic and Early Cretaceous) belong to the first type, whereas the Middle Cretaceous to Early Miocene System belongs to the second. For this reason the age dating of the oil expulsion is extremely important for comparison with the onset of the Zagros folding.

Among the petroleum systems, the Middle Jurassic and the Early Cretaceous Systems were less important for the charge of the Iranian oil reserves because their source rocks, even excellent ones, were not or poorly connected with reservoirs. In contrast, almost the entirety of the oil reserve comes from the Middle Cretaceous to Early Miocene System. In the same way, the gas reserves originated almost entirely from the Palaeozoic System.

The Palaeozoic Petroleum System formed by Llandovery source rocks and Dalan and Kangan carbonate reservoirs is sealed by the thick massive anhydrite of the Dashtak Formation. As the result of the north–south residual high that extended between Central Fars and the Qatar Arch before the Llandovery transgression, and the Hercynian uplift followed by intense erosion, Llandovery source rocks are confined in two basins, one extending from Oman to Abu Dhabi and to the north of Bandar Abbas, and another including Saudi Arabia, west of Qatar, Iraq, Jordan, and most probably the Dezful Embayment. The Palaeozoic System does not generally interfere with younger petroleum systems, at least in areas where the Dashtak evaporites are well developed. In contrast, in the Bandar Abbas coastal area, gas probably of Llandovery origin found its way upward to younger reservoirs.

Oil, and later gas, expelled from the Llandovery kitchens migrated over long-distance paths towards gentle regional highs, located either in Fars and contiguous offshore (Gavbendi–Qatar High, South Bushehr High, North Bandar Abbas High) or in Lurestan (South Samand High), where they accumulated. Early salt-related structures situated on migration paths also trapped large amounts of gas. Oil accumulation began during the Middle Jurassic. The gas window was reached in increasing areas from the Early Cretaceous onwards. Progressively, larger amounts of gas invaded the upper part of the highs, pushing the oil down-dip in the form of peripheral oil legs. The existence of early oil accumulations was confirmed by major detrital bitumen deposited during the Cenomanian, as the result of the onset of the pre-Maastrichtian orogenic phase, and during the Early Miocene. Part of the gas moved at later date into the large-scale Zagros folds formed during the Late Miocene to Pliocene. The reconstruction of the gas pools that were formed prior to the Zagros orogeny remains one of the main keys to be used to evaluate the potential of the current Zagros anticlines, as only anticlines connected with pre-Zagros accumulations have trapped gas.

In the Middle Jurassic Petroleum System, Sargelu source rocks, deposited in the Mesopotamian Depression, are limited above and below by thick evaporitic layers. As the result of the absence of reservoirs, a large part of the oil generated has not been expelled and was cracked *in situ* into pyrobitumen and gas. However, close to the edge of the depression, some oil migrated laterally towards the porous platform facies of the Surmeh–Khami Formations, and accumulated on regional highs such as the Kuh-e Mish-Kharg High before the Zagros folding. Eventually, some oil moved into Zagros anticlines.

Oils of Buzurgan and Djebel Fauqi on the Iraqi side of the Iran–Iraq border were probably charged by the Sargelu source rocks, as they are compatible with the Sargelu pyrolysate. Some Sargelu oil could have supplemented Kazhdumi oil at Cheshmeh Khush and Susangerd, and Garau oil at Chillingar–Garangan, Sulabedar, and Kharg.

In the Late Jurassic Petroleum System, the Hanifa–Tuwaiq Mountain–Diyab source rocks accumulated in the two sides of the Qatar Arch. In the Mesopotamian Depression they corresponded to the condensed Naokelekan Formation. In Iran, oil migrated from the Hanifa–Tuwaiq Mountain–Diyab kitchens towards both the Gavbendi and South Bushehr Highs, and accumulated in salt-related structures located on the southern edge of the highs, such as F, FB and 3H. At the time of migration (Eocene), no efficient caprock was available and the oil lost most of its lighter fractions.

Therefore, heavy oil and bitumen remained trapped in porous zones from the Arab zone to sometimes up to almost the sea floor. The large Mand Zagros anticline probably collected the oil accumulated prior to the Zagros folding in the South Bushehr High.

In the Early Cretaceous Petroleum System, similarly to the Sargelu, the Garau source rocks are not connected with reservoirs in the centre of the Lurestan Depression, and a good part of the Garau oil could not migrate and was cracked *in situ*. However, on the SW edge of Lurestan, two middle-sized fields, Sarkan and Maleh Kuh, were charged by the Garau source rocks. In the Dezful Embayment, Garau oil was drained by the thin, but porous, limestone of the Khalij Member of the Gadvan Formation. Oil and gas were found in this layer in all the deep wells drilled there. However, the reserves are relatively small. Some oil migrated laterally through the shelf edge, and accumulated in north–south regional highs such as in the Kuh-e Mish–Kharg High. The oil of the Chillingar–Garangan, Sulabedar, and Kharg fields probably originated from Garau source rocks, where they supplemented the oil that originated from Naokelekan–Sargelu source rocks.

In the Middle Cretaceous to Early Miocene Petroleum System, two main source rocks, Kazhdumi and Pabdeh, charged the Asmari and Bangestan reservoirs, capped by the Gachsaran evaporites. Up to 300 m of outstanding Kazhdumi source rocks accumulated in an intrashelf silled depression that corresponded almost exactly to the current Dezful Embayment. The Kazhdumi source rocks probably extended towards the High Zagros and the Shiraz area. Pabdeh sources were deposited in a NW–SE trough extending from Lurestan to the Shiraz area, where a maximum of 200 m of excellent source rocks accumulated. In addition, in the NE of the Gulf (Binak Trough), the argillaceous Ahmadi Member of the Sarvak Formation represents anoxic conditions and includes up to 100 m of source rocks that generated an additional charge to the Asmari and Bangestan reservoirs.

The onset of the Zagros folding occurred in the NE part of the Dezful Embayment after the end of the deposition of the Mishan Formation, 12.5 Ma ago, and in South Lurestan between 8.1 and 7.2 Ma. Modelling showed that oil began to be expelled from Kazhdumi source rocks between 8 and 3 Ma, for the various producing trends of the Dezful Embayment, after the onset of the Zagros folding. The only exception corresponded to the Karanj and Parsi area, where expulsion commenced between 18 and 12 Ma.

The Middle Cretaceous to Early Miocene Petroleum System provides an example of a system for which the onset of oil expulsion from source rocks

occurred after the beginning of the formation of the Zagros large-scale anticlines. Oil migrated almost vertically to the closest anticlines, enhanced by fracturing developed in reservoirs, as well as in more argillaceous intervals, in the crestal part of the high-relief anticlines. Oil that accumulates in an anticline originates from the source rocks located in its 'drainage area'. The concept of 'drainage area' permits a comparison between the calculated amount of oil expelled from source rocks existing in one anticline, based upon their areal extent and maturity, and its current oil in place (IOIP).

According to modelling results, the oil of the main fields was charged from the Kazhdumi source rocks (90% of the total) with only minor amounts of oil originating from the Pabdeh in some deep synclines. This result was confirmed by the oil to source rock correlation: the oil of the main fields is compatible with the carbon isotopic composition of the Kazhdumi pyrolysate, whereas the oil from the fields located in the NE Dezful Embayment is compatible with the Pabdeh pyrolysate and contains $18\alpha(H)$oleanane, a biomarker characteristic of the angiosperm plants that flourished from the Early Tertiary onwards.

Modelling was applied to 21 fields that contain 86.5% of the oil in place for the Zagros foldbelt. The total IOIP of the currently discovered fields is only 17.3% of the calculated amount of oil that was supposed to have been expelled from the Kazhdumi and the Pabdeh source rocks in the drainage areas of the 21 fields that were modelled. At first glance, the ratio between the oil in place actually accumulated in single fields and the calculated value of the expelled oil in their area of drainage varies widely from one field to the other. However, a case-by-case field examination showed that some fields are full to the spill point. Secondary migration occurred from one field to another, and the ratio of oil expelled to IOIP remained fairly constant when fields of the same trend are considered together.

The apparent deficit between the IOIP of the fields when compared with the calculated amount of oil expelled from the source rocks could be due to loss of oil during migrations, leakage through cap rocks, but mostly to the difficulty of oil expulsion from thick source rocks in synclines, where no fracturing is expected. Other reasons could be related either to the model and/or to the inaccuracy of the data entered in the model. The validity of the algorithm on which the model is based could be questioned, especially for those conditions that determine the onset of oil expulsion.

However, in spite of these uncertainties, the model permits an evaluation of the order of magnitude of the oil in place expected in undrilled anticlines, using the characteristics and the burial profile of the source rocks as well as the geometry of its area of drainage, provided that the trap is large enough (geometry and reservoir characteristics) to contain it. More accuracy would require systematic analyses of the source rocks in critical wells. Better seismic resolution would be essential to draw balanced sections, to better reconstruct the geometry of the drainage areas and their extension below the thrust units.

The authors gratefully acknowledge referees A. W. Bally (Rice University) and J. L. Rudkiewicz (IFP) for their constructive suggestions and advice, which helped to improve the manuscript significantly. They would also like to thank C. Tiratsoo and the *Journal of Petroleum Geology* for granting permission to reuse in this paper some figures that were previously published in papers written for *JPG* in 2005 and 2008.

References

ABU-ALI, M. A. & LITTKE, R. 2005. Paleozoic petroleum systems of Saudi Arabia: a basin modeling approach. *GeoArabia*, **10**, 131–168.

ABU-ALI, M. A., FRANZ, U. A., SHEN, J., MONNIER, F., MAHMOUD, M. D. & CHAMBERS, T. M. 1991. Hydrocarbon Generation and Migration in the Palaeozoic Sequence of Saudi Arabia. *Society of Petroleum Engineers*.

ALA, M. A., KINGHORN, R. R. F. & RAHMAN, M. 1980. Organic geochemistry and source rock characteristics of the Zagros petroleum province, Southwest Iran. *Journal of Petroleum Geology*, **3**, 61–89.

AL HUSSEINI, M. I. 1989. Tectonic and depositional model for the Late Precambrian–Cambrian Arabian and adjoining plates. *AAPG Bulletin*, **73**, 1117–1131.

AL HUSSEINI, M. I. 1990. The Cambro-Ordovician Arabian and adjoining plates: a glacio-eustatic model. *Journal of Petroleum Geology*, **13**, 267–288.

ALI, A. R. & SILWADI, S. J. 1989. Hydrocarbon potential of Palaeozoic pre-khuff clastics in Abu Dhabi, UAE. SPE Middle East Technical Conference and Exhibition, Manama, Bahrain, 11–14 March 1989, 819–832.

ALSHARHAN, A. S. 1989. Petroleum geology of the United Arab Emirates. *Journal of Petroleum Geology* **12**, 253–288.

AOUDEH, S. M. & AL-HAJRI, S. A. 1995. Regional Distribution and Chronostratigraphy of the Qusaiba Member of the Qalibah Formation in the Nafud Basin, NW Saudi Arabia. *Middle East Petroleum Geosciences*, **1**, 143–154.

AQRAWI, A. A. M. 1998. Paleozoic Stratigraphy and Petroleum Systems of the Western and Southwestern Deserts of Iraq. *GeoArabia*, **3**, 229–248.

AYRES, M. G., BILAL, M., JONES, R. W., SLENZ, L. W., TARTIR, M. & WILSON, A. O. 1982. Hydrocarbon habitat in main producing areas, Saudi Arabia. *AAPG Bulletin*, **66**, 1–9.

BASHARI, A. 1989. Occurrence of heavy crude oil in the Persian Gulf. *In*: MEYER, R. F. & WIGGINS, E. J. (eds) *Fourth UNITAR/UNDP International Conference on Heavy Crude and Tar Sands; Volume 2,*

Alberta Oil Sando Technology and Research Authority, Edmonton, Alberta, 203–214.

BORDENAVE, M. L. 1993. Screening techniques for sourcerock evaluation—sampling and validity of results of analyses. *In*: BORDENAVE, M. L. (ed.) *Applied Petroleum Geochemistry*. Technip, Paris, 217–233.

BORDENAVE, M. L. 2002a. Gas prospective areas in the Zagros Domain of Iran and in the Gulf Iranian waters. Presented at the AAPG Convention, Houston, 10–13 March 2002 (extended abstract). World Wide Web address: www.aapg.org/datasystems/abstract/13annual_/extended/42471.pdf.

BORDENAVE, M. L. 2002b. The Middle Cretaceous to Early Miocene Petroleum System in the Zagros Domain of Iran, and its prospect evaluation. Presented at the AAPG Convention, Houston, 10–13 March 2002 (extended abstract). World Wide Web Address: www.aapg.org/datasystems/abstract/13annual_/extended/42471.pdf

BORDENAVE, M. L. & HEGRE, J. A. 2005. The influence of tectonics on the entrapment of oil in the Dezful Embayment, Zagros Foldbelt, Iran. *Journal of Petroleum Geology*, **28**, 339–368.

BORDENAVE, M. L. 2008. The Paleozoic Petroleum System in the Zagros Foldbelt of Iran and Contiguous Offshore. *Journal of Petroleum Geology*, **33**, 3–42.

BORDENAVE, M. L. & BURWOOD, R. 1990. Source rock distribution and maturation in the Zagros belt; provenance of the Asmari and Bangestan reservoir oil accumulations. *Organic Geochemistry*, **16**, 369–387.

BORDENAVE, M. L. & BURWOOD, R. 1995. The Albian Kazhdumi Formation of the Dezful Embayment, Iran: one of the most efficient petroleum generating systems. *In*: KATZ, B. J. (ed.) *Petroleum Source Rocks Series, Case Book in Earth Sciences*. Springer, Berlin, 183–207.

BORDENAVE, M. L. & HUC, A. Y. 1995. The Cretaceous source rocks in the Zagros Foothills of Iran: an example of a large size intracratonic basin. *Revue de l'Institut Français du Pétrole*, **50**, 727–753.

BORDENAVE, M. L. & NILI, A. R. 1973. *Geochemical Project, Review and Appraisal of the Khuzestan Province*. Geology and Exploration Division, Iranian Oil Operating Companies, Report, **1194**.

BORDENAVE, M. L. & SAHABI, F. 1971. *Geochemical project, Appraisal of Lurestan*. Geology and Exploration Division, Iranian Oil Operating Companies, Report, **1182**.

BORDENAVE, M. L., COMBAZ, A. & GIRAUD, A. 1970. Influence de l'origine des matières organiques et de leur degré d'évolution sur les produits de pyrolyse du kérogène. *In*: HOBSON, G. D. & SPEARS, G. C. (eds) *Advances in Organic Geochemistry*, 389–405.

BORDENAVE, M. L., NILI, A. & FOZOONMAYEH, C. 1971. *Geochemical Project, Appraisal of Fars Province*. Geology and Exploration Division, Iranian Oil Operating Companies, Report, **1181**.

BOSOLD, A., SCHWARZHANS, W., JULAPOUR, A., ASHRAFZADEH, A. R. & EHSANI, S. M. 2005. The Structural Geology of the High Central Zagros revisited (Iran). *Petroleum Geosciences*, **11**, 225–238.

BRAUD, J. 1987. *La suture du Zagros au niveau de Kermanshah (Kurdistan iranien), reconstitution paléogéographique, évolution.* PhD Thesis, Université de Parsi Sud, Orsay.

BRENCHLEY, P. J., MARSHALL, J. D. ET AL. 1994. Bathymetric and isotopic evidence for a short-lived Late Ordovician glaciations in a greenhouse period. *Geology*, **22**, 295–298.

BROSSE, E. & BORDENAVE, M. L. 1993. Isotopic composition of organic constituents. *In*: BORDENAVE, M. L. (ed.) *Applied Petroleum Geochemistry.* Technip, Paris, 315–334.

BURWOOD, R. 1978. *Source rock potential evaluation and characterization.* Technical notes 3 (Parsi-35 well), 6 (Agha Jari-140 well), 17 (Gachsaran-83 well), 19 (Mand-5 well), 22 (Binak-5 well) and 25 (Samand-1 well). Iranian Oil Operating Company Notes.

CARRIGAN, W. J., COLE, G. A., COLLING, E. L. & JONES, P. J. 1995. Geochemistry of the Upper Jurassic Tuwaiq Mountain and Hanifa Formation Petroleum Source Rocks of Eastern Saudi Arabia. *In*: KATZ, B. J. (ed.) *Petroleum Source Rock Series, Case Book in Earth Sciences*, Springer, Berlin, 67–87.

CHAUBE, A. N. & AL SAMAHIJI, J. 1994. Jurassic and Cretaceous of Bahrain. *In: Geology and Petroleum Habitat, Geo 94.* Gulf PetroLink, Manama, Bahrain, 292–305.

COLE, G. A., ALPERN, H. I., AOUDEH, S. M., AL-HAJJI, A. A., CARRIGNAN, W. J. & GWATHNEY, W. J. 1994. The use of Biomarkers and Gas Chromatography to imply new Source Rocks in the Paleozoic Sequence of Saudi Arabia. *In*: Proceedings, 2nd International Conference on Chemistry in Industry, Vol. 1, 24–26 October 1994. Manama Bahrain, 523–536.

COLMAN-SADD, S. P. 1978. Fold development in Zagros simply folded belt, Southwest Iran. *AAPG Bulletin*, **62**, 984–1003.

COMBAZ, A. 1986. Les 'zones Gamma' du Silurien des régions sahariennes. Contenu organique et conditions de dépôt. *In*: BRÉHÈRET, J. G. (ed.) *Les couches riches en matières organiques et leur conditions de dépôt* Documents Bdu RGM, Paris, 239–258.

CONNAN, J. & DESCHESNE, O. 1996. *Le bitume à Suse, collection du musée du Louvre.* Réunion des Musées Nationaux, Paris.

COOK, A. C. & SHERWOOD, N. R. 1991. Classification of oil shales, coals and other organic-rich rocks. *Organic Geochemistry*, **17**, 211–222.

DE MATOS, J. E. & HULSTRAND, R. H. 1994. Regional Characteristics and Depositional Sequences of the Oxfordian and Kimmeridgian, Abu Dhabi. *In*: AL-HUSSEINI, M. I. (ed.) *Middle East Petroleum Geosciences.* Gulf Petrolink, Manama, 346–356.

DEMAISON, G. T. & HUIZINGA, G. T. 1991. Genetic classification of petroleum systems. *AAPG Bulletin*, **75**, 1626–1643.

DROSTE, H. 1990. Depositional cycles and source rock development in an epeiric intra-platform basin, The Hanifa Formation of the Arabian Peninsula. *Sedimentary Geology*, **69**, 281–296.

DUNNINGTON, H. V., VAN BELLEN, R. C., WETZEL, R. & MORTON, D. M. 1959. Iraq. *In*: DUBERTRET, L. (ed.) *Léxique Stratigraphique International: Vol. III, Asie, Iraq.* CNRS, Paris, fascicule 10a.

DURMISH'YAN, A. G. 1973. The compaction of the clay rocks. *Izvestiya Akademii Nauk SSSR, Seriya Geologicheskaya*, **8**, p. 85–89 [in Russian]. (Translated in *International Geological Review*, **16**, 650–653.)

EKWEOZOR, C. M., OKOGUN, J. I., EKONG, D. E. U. & MAXWELL, J. R. 1979. Preliminary organic geochemical studies of samples from the Niger Delta (Nigeria) 1—Analyses of crude oils for triterpanes. *Chemical Geology*, **27**, 11–28.

GHAVIDEL-SYOOKI, M. 1993. Palynological study of Paleozoic sediments of the Chal-I Sheh area, Southwestern Iran. *Journal of Sciences Islamic Republic of Iran*, **4**, 32–46.

GHAVIDEL-SYOOKI, M. 1997. Palyno-stratigraphy and paleogeography of Early Permian strata in the Zagros Basin, Southeast–Southwest Iran. *Journal of Sciences Islamic Republic of Iran*, **8**, 243–262.

GHAVIDEL-SYOOKI, M. 2000. Biostratigraphy and paleogeography of Late Ordovician and Early Silurian chitinozoans from the Zagros Basin, Southern Iran. *Historical Biology*, **15**, 29–39.

GHAVIDEL-SYOOKI, M. 2003. Palyno-stratigraphy of Devonian sediments in the Zagros Basin, Southern Iran. *Review of Paleobotany and Palynology*, **127**, 241–268.

GRANTHAM, P. J., LIJMBACH, G. W. M., POSTHUMA, J., HUGUES CLARKE, M. W. & WILLINK, R. J. 1987. Origin of crude oils in Oman. *Journal of Petroleum Geology*, **11**, 61–80.

GUMATI, Y. D. 1993. Kinetic Modeling, Thermal Maturation and Hydrocarbon Generation in the United Arab Emirates. *Marine and Petroleum Geology*, **10**, 153–161.

HASSAN, T. H. & AZER, S. 1985. The occurrence and origin of oil in Offshore Abu Dhabi. *In: Paper SPE 13696, Middle East Technical Conference and Exhibition, Bahrain*, Society of Petroleum Engineers, 143–160.

HESSAMI, K., KOYI, H. A., TALBOT, C. J., TABASI, H. & SHABANIAN, E. 2001. Progressive unconformities within an evolving foreland fold–thrust belt, Zagros Mountains. *Journal of the Geological Society, London*, **158**, 969–981.

HOMKE, S., VERGES, J., GARCES, M. & KARPUS, R. 2004. Magnetostratigraphy of Miocene–Pliocene Zagros foreland deposits in the front of the Pusht-e Kuh Arc (Lurestan Province, Iran). *Earth and Planetary Science Letters*, **225**, 397–410.

HUBER, H. 1978. *Geological and tectonic map of Southwest Iran: Explanatory Notes and Cross sections, with the participation of the Geological Survey of Iran, and of Oil Companies.* NIOC, Tehran.

HUGHES CLARKE, M. W. 1988. Stratigraphy and rock unit nomenclature in the oil-producing area of Interior Oman. *Journal of Petroleum Geology*, **11**, 5–60.

IBRAHIM, M. W. 1978. *Petroleum Geology of South Iraq.* PhD thesis, Imperial College, London.

IBRAHIM, M. W. 1983. Petroleum geology of southern Iraq. *AAPG Bulletin*, **67**, 97–130.

INSALACO, E., VIRGONE, A. ET AL. 2006. Upper Dalan Member and Kangan Formation between the Zagros Mountains and Offshore Fars, Iran: depositional system, biostratigraphy and stratigraphic architecture. *GeoArabia*, **11**, 75–175.

JAMES, G. A. & WYND, J. G. 1965. Stratigraphic nomenclature of Iranian Oil Consortium Agreement Area. *AAPG Bulletin*, **49**, 2182–2245.

JONES, P. J. & STUMP, T. E. 1999. Depositional and tectonic setting of the Lower Silurian source rock facies, Central Saudi Arabia. *AAPG Bulletin*, **83**, 314–332.

KENT, P. E. 1979. The emergent Hormuz salt plugs of Southern Iran. *Journal of Petroleum Geology*, **2**, 117–144.

KONERT, G., AFIFI, A. M., AL-HAJRI, S. A. & DROSTE, H. 2001. Paleozoic Stratigraphy and Hydrocarbon Habitat of the Arabian Plate. *GeoArabia*, **6**, 407–442.

LOUTFI, G. & EL BISHLAWY, S. 1986. Habitat of Hydrocarbon in Abu Dhabi, UAE. *In*: *Symposium on Hydrocarbon Potential of Intense Thrust Zones, Abu Dhabi, December 1986*, 63–124.

LOYDELL, D. K. 1998. Early Silurian sea-level changes. *Geological Magazine*, **135**, 447–471.

LÜNING, S., CRAIG, J., LOYDELL, D. K., STORCH, P. & FITCHES, B. 2000. Lower Silurian 'hot shales' in North Africa and Arabia: regional distribution and depositional model. *Earth-Science Reviews*, **49**, 21–200.

MAHMOUD, M. D., VASLET, D. & HUSSEINI, M. I. 1992. The Lower Silurian Qalibah Formation of Saudi Arabia. *AAPG Bulletin*, **76**, 1491–1506.

MCGILLIVRAY, J. G. & HUSSEINI, M. I. 1992. The Paleozoic petroleum geology of Central Arabia. *AAPG Bulletin*, **76**, 1473–1490.

MCQUILLAN, H. 1973. Small-Scale Fracture Density in Asmari Formation of Southwest Iran and its Relation to Bed Thickness and Structural Setting. *AAPG Bulletin*, **57**, 2367–2385.

MCQUILLAN, H. 1974. Fracture Patterns on Kuh-e Asmari Anticline, Southwest Iran. *AAPG Bulletin*, **58**, 236–246.

MOSHRIF, M. A. 1987. Sedimentary history and paleogeography of Lower and Middle Jurassic Rocks. *Journal of Petroleum Geology*, **10**, 335–350.

MOSHTAGHIAN, A., MALEKZADEH, R. & AZARPANAH, A. 1989. Heavy oil discovery in Islamic Republic of Iran. *In*: MEYER, R. F. & WIGGINS, E. J. (eds) *Fourth UNITAR/UNDP International Conference on Heavy Crude and Tar Sands; Vol. 2*, Alberta oil sando technology and research authority, Edmonton, Alberta, 235–243.

MURRIS, R. J. 1980. Middle-East, Stratigraphic evolution and oil habitat. *AAPG Bulletin*, **64**, 587–618.

ORR, W. L. 1986. Kerogen, asphaltenes, sulfur relationship in the Monterrey Shales. In: *Advances in Geochemistry 1985, Organic Geochemistry 10*, 499–516.

OWEN, R. M. S. & NASR, S. N. 1958. Stratigraphy of the Kuwait–Basra area. *In*: WEEKS, L. G. (ed.) *Habitat of Oil*, AAPG, Special Publication, 1252–1278.

PATTON, T. L. & O'CONNOR, S. J. 1988. Cretaceous flexural history of Northern Oman Mountains foredeep, United Arab Emirates. *AAPG Bulletin*, **7**, 797–809.

RABBANI, A. R. 2008. Geochemistry of crude oils samples from the Iranian Sector of the Persian Gulf, *Journal of Petroleum Geology*, **31**, 303–316.

RABBANI, A. R. & KAMALI, M. R. 2005. Source rock Evaluation and Petroleum Geochemistry, Offshore SW Iran. *Journal of Petroleum Geology*, **28**, 413–428.

RICOU, L. E. 1974. *L'étude géologique de la région de Neyriz (Zagros iranien) et l'évolution structurale des Zagrides*. PhD Thesis, Université de Paris Sud, Orsay.

RUDKIEWICZ, J. L., SHERKATI, S. & LETOUZEY, J. 2007. Evolution of Maturity in Northern Fars and in the Izeh Zone (Iranian Zagros) and Link with Hydrocarbon Prospectivity. *In*: LACOMBE, O., LAVÉ, J., ROURE, F. & VERGES, J. (eds) *Thrust Belt and Foreland Basins*. Springer, Berlin, 1–17.

SASS, J. H., LARENBRUCH, A. H., MUNROE, R. J., GREENE, G. W. & MOSES, T. H. 1971. Heat flow in the western United States. *Journal of Geophysical Research*, **76**, 6376–6413.

SCOTESE, C. R. & GOLONKA, J. 1993. Paleogeographic Atlas on the evolution of Southern Tethys, Paleomap project, Department of Geology, University of Texas, Arlington. *Journal of the Geological Society, London*, **138**, 509–526.

SETUDEHNIA, A. 1976. The Paleozoic Sequence at Zard-e Kuh and Kuh-e Dinar. *Bulletin of the Iran Petroleum Institute*, **60**, 16–33.

SHARLAND, P. R., ARCHER, R. *ET AL*. 2001. *Arabian Plate Sequence Stratigaphy*. GeoArabia, Special Publication, **2**.

SHERKATI, S. & LETOUZEY, J. 2004. Variation of structural style and basin evolution in the Central Zagros (Izeh zone and Dezful Embayment), Iran. *Marine and Petroleum Geology*, **21**, 517–640.

STONELEY, R. 1981. The geology of the Kuh-e Dalneshin area of Southern Iran and its bearing on the evolution of Southern Tethys. *Journal of the Geological Society, London*, **138**, 509–526.

SZABO, F. & KHERADPIR, A. 1978. Permian and Triassic Stratigraphy, Zagros Basin, South-West Iran. *Journal of Petroleum Geology*, **1**, 57–82.

UNGERER, Ph. 1993. Modeling of petroleum generation and migration. *In*: BORDENAVE, M. L. (ed.) *Applied Petroleum Geochemistry*. Technip, Paris, 395–442.

VASLET, D. 1987. *The Paleozoic (Pre-Late Permian) of Central Arabia, and correlations with neighbouring regions*. Saudi Arabian Directorate General for Mineral Resources Technical Record, **BRGM-TR-07-2**.

VIDETICH, P. E., MCLIMANS, R. K., WATSON, H. K. S. & NAGY, R. M. 1988. Depositional, diagenetic, thermal, and maturation history of Cretaceous Mishrif Fm, Fateh field, Dubai. *AAPG Bulletin*, **72**, 1143–1159.

WORDEN, R. H. & SMALLEY, P. C. 1996. H_2S producing reactions in deep carbonate gas reservoirs: Khuff Fm, Abu Dhabi. *Chemical Geology*, **133**, 157–171.

WORDEN, R. H., SMALLEY, P. C. & OXTOBY, N. H. 1995. Gas souring by thermochemical sulfate reduction at 140 °C. *AAPG Bulletin*, **79**, 854–863.

Index

Note: Page numbers denoted in *italics* indicate figures and those in **bold** indicate tables.